P9-DEQ-987

PRINCIPLES OF ANIMAL BEHAVIOR

Principles of Animal Behavior

LEE ALAN DUGATKIN
University of Louisville

 W. W. Norton & Company | New York | London

Editor: Jack Repcheck
Developmental Editor: Sandy Lifland
Production Manager: JoAnn Simony
Photo Researchers: Penni Zivian, Nathan Odell
Book Designer: Leelo Marjamaa-Reintal/Blue Shoe Studio
Artists: J. B. Woolsey and Associates; John McAusland
Editorial Assistant: Julia Hines
Composition: UG / GGS Information Services, Inc.
Manufacturing: RR Donnelley

Printed in the United States of America
First Edition

Cover photo: Male chameleon parenting its three-week-old offspring. AFP/Corbis.

Library of Congress Cataloging-in-Publication Data

Dugatkin, Lee Alan, 1962–
Principles of animal behavior / by Lee Alan Dugatkin.
p. cm.
Includes bibliographical references and index.
ISBN 0-393-97659-9
1. Animal behavior. I. Title.

QL751.D748 2003
591.5—dc21

2003048797

W. W. Norton & Company has been independent since its founding in 1923, when William Warder Norton and Mary D. Herter Norton first published lectures delivered at the People's Institute, the adult education division of New York City's Cooper Union. The Nortons soon expanded their program beyond the Institute, publishing books by celebrated academics from America and abroad. By mid-century, the two major pillars of Norton's publishing program—trade books and college texts—were firmly established. In the 1950s, the Norton family transferred control of the company to its employees, and today—with a staff of four hundred and a comparable number of trade, college, and professional titles published each year—W. W. Norton & Company stands as the largest and oldest publishing house owned wholly by its employees.

W. W. Norton & Company, Inc., 500 Fifth Avenue, New York, N.Y. 10110
www.wwnorton.com

W. W. Norton & Company Ltd., Castle House, 75/76 Wells Street,
London W1T 3QT

1 2 3 4 5 6 7 8 9 0

For Dana, My True Helpmate

Contents in Brief

1 Principles of Animal Behavior 2

2 Natural Selection 30

3 Proximate Factors 72

4 Learning 108

5 Social Learning and Cultural Transmission 144

6 Sexual Selection 176

7 Mating Systems 218

8 Kinship 252

9 Cooperation 294

10 Foraging 338

11 Antipredator Behavior 378

12 Communication 410

13 Habitat Selection, Territoriality, and Migration 446

14 Aggression 476

15 Play 508

16 Aging and Disease 540

17 Animal Personalities 566

Contents

Preface xvii

PRINCIPLES OF ANIMAL BEHAVIOR 2

Three Foundations 6
Foundation 1—Natural Selection 6
Foundation 2—Learning 12
Foundation 3—Cultural Transmission 15
Conceptual, Theoretical, and Empirical Approaches 18
Conceptual Approaches 18
Theoretical Approaches 20
Empirical Approaches 22
Overview of What Is to Follow 25
INTERVIEW WITH DR. E. O. WILSON 26

NATURAL SELECTION 30

Artificial Selection 33
The Process of Natural Selection 35
Selective Advantage of a Trait 35
How Natural Selection Operates 37
Behavioral Genetics 44
Mendel's Rules 45
Locating Genes for Polygenic Traits 45
Dissecting Behavioral Variation 46
Animal Behavior and Natural Selection 49
Sociobiology and Selfish Genes 50
Natural Selection and Antipredator Behavior in Guppies 51
Adaptation 57
Genetic Techniques as Tools 61
Kinship and Naked Mole Rat Behavior 62
Coalition Formation 65
INTERVIEW WITH DR. RICHARD ALEXANDER 66

PROXIMATE FACTORS 72

Ultimate and Proximate Perspectives 75
Hormones and Proximate Causation 80
Hormones and Helping-at-the-Nest 82
Testosterone and Play Fighting in Rats 85
Neurological Underpinnings of Behavior and Proximate Causation 87
The Nervous Impulse 88
Vocalizations in Plainfin Midshipman 90
Mushroom Bodies, Insects, and Learning 92
Sleep and Predation in Mallard Ducks 94
Biochemical Factors 95
Ultraviolet Vision in Birds 96
Environmental Sex Determination and Sex Ratios 97
Reptiles, Sex Determination, and Temperature 97
Red Deer, Dominance Status, and Sex Ratios 99
Genes and Proximate Explanations 101
Learning as a Proximate Factor 104
INTERVIEW WITH DR. TIMOTHY CLUTTON-BROCK 102

LEARNING 108

What Is Individual Learning? 111
How Animals Learn 113
Learning from a Single-Stimulus Experience 114
Pavlovian (Classical) Conditioning 115
Instrumental (Operant) Conditioning 119
Why Animals Learn 121
Within-Species Studies and the Evolution of Learning 122
Population Comparisons and the Evolution of Learning 127
A Model of the Evolution of Learning 130
What Animals Learn 132
Learning Where Home Is Located 132
Learning about Your Mate 133
Learning about Familial Relationships 135
Learning about Aggression 136
INTERVIEW WITH DR. SARA SHETTLEWORTH 138

SOCIAL LEARNING AND CULTURAL TRANSMISSION 144

What Is Cultural Transmission? 149
Animal Culture 150
Why All the Fuss about Culture? 151
Types of Cultural Transmission 152
Social Learning 152
Teaching in Animals 158
Modes of Cultural Transmission 161
Vertical Cultural Transmission 162
Horizontal Cultural Transmission 162
Oblique Cultural Transmission 163
The Interaction of Genetic and Cultural Transmission 164
The Grants' Finches 164
Whitehead's Whales 165
Genes for Cultural Transmission? 167
Cultural Transmission and Brain Size 169
INTERVIEW WITH DR. BENNET "JEFF" GALEF 170

SEXUAL SELECTION 176

Intersexual and Intrasexual Selection 178
Genetics and Mate Choice 181
Direct Benefits and Mate Choice 182
Good Genes and Mate Choice 186
Runaway Sexual Selection 193
Learning and Mate Choice 194
Sexual Imprinting 195
Learning and Mate Choice in Japanese Quail 196
Learning and Mate Choice in Blue Gouramis 197
Cultural Transmission and Mate Choice 198
Defining Mate-Choice Copying 198
Mate-Choice Copying in Grouse 200
Mate-Choice Copying in Guppies 202
Song Learning and Mate Choice in Cowbirds 203
Male-Male Competition and Sexual Selection 205
Underground Mating and Male-Male Competition in Fiddler Crabs 206

Red Deer Roars and Male-Male Competition 206
Male-Male Competition by Interference 208
Male-Male Competition via Cuckoldry 209
Neuroethology and Mate Choice 211
Swordtails, Platyfish, and Sensory Bias 211
Frogs and Sensory Bias 214
Zebra Finches and Sensory Bias 215
INTERVIEW WITH DR. MALTE ANDERSSON 212

MATING SYSTEMS 218

Different Mating Systems 220
Monogamous Mating Systems 221
Polygamous Mating Systems 223
Promiscuous Mating Systems 226
Choosing Polygyny 228
Polygyny and Resources 228
The Polygyny Threshold Model 229
Surreptitious Promiscuity 236
Extrapair Copulations 236
Sperm Competition 238
Multiple Mating Systems in One Population? 245
Dunnocks 245
INTERVIEW WITH DR. NICK DAVIES 248

8

KINSHIP 252

Kinship and Animal Behavior 255
Kinship Theory 258
Relatedness and Inclusive Fitness 259
Family Dynamics 261
Conflict within Families 276
Parent-Offspring Conflict 276
Sibling Rivalry 282
Kin Recognition 286
Matching Models 286
Rule-of-Thumb Models 290
INTERVIEW WITH DR. STEPHEN EMLEN 288

COOPERATION 294

The Range of Cooperative Behaviors 297
Helping in the Birthing Process 297
Social Grooming 298
Group Hunting 299
Nest Raiding 300
Three Paths to Cooperation 300
Path 1: Reciprocity 301
Path 2: Byproduct Mutualism 309
Path 3: Group Selection 315
Phylogeny and Cooperative Breeding in Birds 321
Hormones, Reproductive Suppression, and Cooperative Breeding 323
Coalitions 326
Coalitions in Baboons 328
Alliances and "Herding" Behavior in Cetaceans 330
Interspecific Mutualisms 331
Ants and Butterflies—Mutualism with Communication? 331
INTERVIEW WITH DR. HUDSON KERN REEVE 332

FORAGING 338

Optimal Foraging Theory 342
Basic OFT: What to Eat and Where to Eat It 343
Specific Nutrient Constraints 351
Risk-Sensitive Foraging 353
Learning and Foraging 356
Foraging, Learning, and Brain Size in Birds 357
Learning and "Work Ethics" in Pigeons 359
Foraging and Group Life 360
Group Size 361
Social Learning and Foraging 364
Public Information and Foraging 369
Molecular, Neurobiological, and Hormonal Aspects of Honeybee Foraging 371
The Period Gene, mRNA, and Foraging 371
Juvenile Hormone, "Mushroom Bodies," and Foraging 373
INTERVIEW WITH DR. JOHN KREBS 374

ANTIPREDATOR BEHAVIOR 378

Behavioral Tradeoffs Associated with Predation 384
Predation and Foraging 385
Predation and Hatching Time in Wasps 387
Alarm Signals 389
Vervet Alarm Calls 389
Tail Flagging 391
Prey Approaching Their Predators 392
Costs and Benefits of Thomson's Gazelles Approaching a Predator 393
Interpopulational Differences 395
Interpopulational Differences in Antipredator Behavior in Minnows 396
Predator vs. Prey Arms Races 400
Learning and Antipredator Behavior 402
The Direct Fitness Consequences of Learning about Predators 402
Social Learning and Antipredator Behavior 406
INTERVIEW WITH DR. MANFRED MILINSKI 404

COMMUNICATION 410

Communication and Honesty 415
Communication Venues 418
Foraging 418
Play 423
Mating 424
Aggression 427
Predation 434
Songs 436
INTERVIEW WITH DR. AMOTZ ZAHAVI 440

HABITAT SELECTION, TERRITORIALITY, AND MIGRATION 446

Models of Habitat Choice 451

The Ideal Free Distribution Model and Habitat Choice 451

The IFD Model and Foraging Success 452

Territoriality 454

Territoriality and Learning 455

Territory Owners, Satellites, and Sneakers 458

How to Keep a Territory in the Family 462

Conflict in Family Territories 463

Migration 464

The Challenges of Migration 465

The Heritability of Migratory Behavior 469

Learning and Migration in Fish 470

INTERVIEW WITH DR. JUDY STAMPS 472

AGGRESSION 476

Game Theory Models of Aggression 479

The Hawk-Dove Game 482

The War of Attrition Model 487

The Sequential Assessment Model 489

Winners, Losers, Bystanders, and Aggression 491

Winner and Loser Effects 491

Bystander Effects 498

Endocrinology, Neurotransmitters, and Aggression 499

Corticosterone and Aggression 500

Testosterone and Aggression 501

Neurotransmitters and Aggression 504

INTERVIEW WITH DR. JOHN MAYNARD SMITH 502

PLAY 508

Defining Play 511
Types and Functions of Play 513
Object Play 513
Locomotor Play 518
Social Play 522
A General Theory for the Function of Play 527
Some Proximate Aspects of Play 529
Hormones, Energy, and Play in Young Belding's Ground Squirrels 529
The Neurobiology of Play in Young Rats 530
A Phylogenetic Approach to Play 533
INTERVIEW WITH DR. BERND HEINRICH 534

AGING AND DISEASE 540

Senescence in the Wild? 543
Theoretical and Empirical Perspectives on Senescence 544
The Antagonistic Pleiotropy Model of Senescence 544
Disposable Soma Theory and Longevity 546
Longevity and Extending Life Spans 552
Hormones, Heat-Shock Proteins, and Aging 552
Glucocorticoids and Aging 553
Heat-Shock Proteins and Aging 554
Disease and Animal Behavior 555
Avoidance of Disease-filled Habitats 555
Avoidance of Diseased Individuals 557
Self-Medication 558
Why Some Like It Hot 559
INTERVIEW WITH DR. RICHARD WRANGHAM 560

ANIMAL PERSONALITIES 566

Boldness and Shyness 570
Bold and Inhibited Pumpkinseeds 572
Guppies, Boldness, and Predator Inspection 574
Some Case Studies 578
Hyena Personalities 578
Octopus Personalities 581
Ruff Satellites 582
Learning and Personality in Great Tits 585
Chimpanzee Personalities and Cultural Transmission 587
Coping Styles 588
Some Practical Applications of Animal Personality Research 590
Predators and Domesticated Prey 590
Guide Dog Personalities 591
INTERVIEW WITH DR. JEROME KAGAN 592

Glossary A1
References A5
Credits A53
Index A55

Preface

This is a book about the empirical, theoretical, and conceptual foundations upon which the field of animal behavior rests. In it, we shall analyze the extent and manner in which natural selection, learning theory, and cultural transmission shape animal behavior in fascinating, interactive ways. Animal behavior books written by biologists tend to focus on genetics and behavior, while those authored outside of biology, usually in psychology, tend to focus on classic learning theory as well as cultural transmission (that is, social or observational learning). Biologists tend to concentrate on how natural selection has shaped animal behavior and often relegate learning to a separate chapter specifically on that topic, rarely even touching on cultural transmission. Psychologists generally devote the vast majority of their space to learning per se, with only a chapter or two on "evolution and genetics." Yet, such compartmentalization of the different forces that shape animal behavior simply will not do. Rather, animal behavior can best be understood as a tapestry composed of the interwoven threads of natural selection, learning theory, and cultural transmission. These forces all act to shape everything from neurobiology in animals to the social milieu in which they act. The tapestry of animal behavior cannot be separated into components without losing its beauty.

To accomplish the task of integrating natural selection, learning, and cultural transmission into a text on animal behavior, I have adopted a new approach. After an initial introductory chapter, I present the reader with four "primer" chapters that provide an overview of natural selection and animal behavior (Chapter 2), proximate issues (Chapter 3), learning (Chapter 4), and cultural transmission (Chapter 5). That much is not new. Subsequent to the primer chapters, however, these forces are not set off in chapters of their own, and that is novel. Rather, whenever possible, natural selection, learning, and cultural transmission are woven into each and every chapter of the text itself. In some chapters, this was easier than others, and, indeed, in some it was impossible, for lack of evidence. But the effort was made to be as integrative as possible on this front.

Subsequent to the primer chapters, I have not attempted to wall off some chapters for proximate issues and other chapters for ultimate issues. Rather, for every important class of behaviors, I try to weave together proximate factors and ultimate factors. Often proximate and ultimate perspectives are finely intermeshed, which is an even stronger reason to discuss them in unison. That said, the different sort of explanatory power associated with proximate and ultimate forces is always made crystal clear to the reader in chapters on sexual selection, mating systems, kinship, cooperation, foraging, antipredator behavior, communication, habitat selection, aggression, and play. Moreover, these factors are also both considered in chapters on aging and disease and on animal personalities.

Another important new approach I will adopt here centers on the relationship between nonhuman animal behavior and human behavior. Textbooks on animal behavior tend to have a closing chapter on what the rest of the book (which has focused on nonhumans) means for understanding human behavior. Rather than adopting this segmented view of behavior, I integrate nonhuman and human behavior throughout the book. The vast majority of cases focus on nonhumans, but only when such studies are juxtaposed with work on humans can students receive a truly integrative view of animal (nonhuman and human) behavior.

One of the parts of this book that I am most excited about are the two-page interviews at the end of each of the seventeen chapters. I had hoped to convince leaders in the field to submit to interviews designed to excite students, and my hopes were not only met, but exceeded. The following individuals took time out of their very busy schedules to be interviewed: E. O. Wilson, Richard Alexander, Tim Clutton-Brock, Sara Shettleworth, Jeff Galef, Malte Andersson, Nick Davies, Steve Emlen, Kern Reeve, John Krebs, Manfred Milinski, Amotz Zahavi, Judy Stamps, John Maynard Smith, Bernd Heinrich, Richard Wrangham, and Jerome Kagan.

There is little question that few subjects in the sciences lend themselves to wonderful storytelling as does the study of animal behavior, and this no doubt accounts for the strong interest students show in this topic. In that light, one overarching goal of this textbook has been to present animal behavior in a narrative manner that is accessible and entertaining to the student—a book that students in a course would recommend to their friends as interesting bedtime reading. That said, there is much more to the study of animal behavior than weaving together a good story. To that end, each chapter in this book has a sound theoretical and conceptual basis upon which the empirical studies rest. The presentation of theory, often in the form of mathematical models is not meant to intimidate students, but rather it is designed to mesh with the glorious examples of intricate animal behavior that will surround it. Indeed, we have commissioned a noted art studio to illustrate the chapters in this book with more than 100 original works of art designed to convey various theoretical, empirical, and conceptual aspects of behavior.

In an undertaking of this size, the number of people that need to be thanked is staggering. First, and foremost, though, I need to thank my editor, Jack Repcheck, for all the time and effort that he has invested in this project. Jack has been a proponent of this book since he first read the proposal I sent to Norton, and since that time, he has read and edited all seventeen chapters of this book, *many times*. The result is a product that is infinitely better than what would have been produced without Jack's help. The developmental editor on this project, Sandy Lifland, has been just fantastic. Sandy's eye for detail, and her sense of how a good book should read, have been very valuable. Likewise, the artistic skills of John Woolsey, whose studio was responsible for the beautiful drawings that adorn this book, and John McAusland, who created the graphs for the book, and the keen eye of my photo editors, Penni Zivian and Nathan Odell, and of my production manager, JoAnn Simony, have taken the text and brought it to life. And all of this—the whole book—might have turned out differently had it not been for my remarkable agent, Susan Rabiner, who knew immediately that I needed to do this book with Norton.

I don't have any objective statistics to back it up but I am convinced that my colleagues in the field of animal behavior are more generous than a random sample of the population. My convictions were only bolstered when I saw the extent that my friends and colleagues were willing to help put this project together This book has been reviewed by a suite of colleagues on numerous occasions. My sincere thanks to the following individuals for reading all or parts of this book: Jerry Wilkinson, Geoff Hill, Naomi Pierce, Ann Hedrick, Anne Magurran, Geoff Parker, Anne Houde, Curt Lively, Allen Moore, Marc Bekoff, Dan Papaj, Anne Clark, Michael Mesterton-Gibbons, Fred Dyer, David Pfennig, Manfred Milinski, Locke Rowe, Michelle Scott, Rudolph Jandler, Samuel Beshers, Max Terman, Deborah Gordon, Nick Fuzessery, and Susan Foster. Please credit these folks with all that is good about this book, and assign any problems you have to my hand.

In addition to reviewing the text, my colleagues have been generous to a fault providing me with photos for use in this book. Special thanks are due to: Michael Alfieri, Alex Basolo, Andrew Bass, Marc Bekoff, Chris Boesch, Charles Brown, John Byers, Nancy Burley, Roy Caldwell, Tim Caro, Richard Connor, Richard Coss, Scott Creel, Phil Crowley, Cameron Currie, Nick Davies, Florentino de Lope, Frans de Waal, Hugh Drummond, Reuven Dukas, Magnus Enquist,

Stephen Emlen, Chris Evans, Kevin Foster, Jeff Galef, Jim Gilliam, Luc-Alain Giraldeau, Samuel Gosling, Bernd Heinrich, Geoff Hill, Bert Holldobler, Karen Hollis, John Hoogland, Yu-Ying Hsu, Michael Huffmann, Fred Janzen, Joseph Kiesecker, Steve Lima, Karen McComb, Marco Masseti, Andrew Meltzoff, Frank Moore, Ronald Mumme, Scott Nunes, Craig Packer, Geoff Parker, Serge Pellis, David Pfennig, Naomi Pierce, Daniel Promislow, Francis Ratnieks, Jan Rees, Stephen Rothstein, Michael Ryan, Gordon Schuett, Michael Seres, Paul Sherman, Steve Siviy, Judy Stamps, Jeff Stevens, Alan Templeton, Jennifer Templeton, Joseph Terkel, Kaci Thompson, Eckart Voland, Fred vom Saal, Karen Warkentin, David Westneat, Hal Whitehead, Stim Wilcox, Jerry Wilkinson, and Christian Ziegler.

Lastly, special thanks go to my wife Dana, who helped with almost every aspect of this project, and to my son Aaron for his smile and the sparkle in his eyes.

L.A.D.
May 2003

PRINCIPLES OF ANIMAL BEHAVIOR

1

Three Foundations

- Foundation 1—Natural Selection
- Foundation 2—Learning
- Foundation 3—Cultural Transmission

Conceptual, Theoretical, and Empirical Approaches

- Conceptual Approaches
- Theoretical Approaches
- Empirical Approaches

An Overview of What Is to Follow

Interview with Dr. E. O. Wilson

Principles of Animal Behavior

FIGURE 1.1. American cockroach. Almost everyone is familiar with the roach, often a pest in households around the world. *(Photo credit: E. R. Degginger/Color-Pic, Inc.)*

I grew up in the heart of New York City. Metropolitan New York may be a melting pot of human diversity, but it is anything but a hotbed of animal diversity. One animal, however, that my family and I did encounter on a fairly regular basis was the American cockroach (Figure 1.1). Much to my mother's chagrin, we seemed locked in a never-ending battle with the roaches, a battle I might add, that we usually lost.

Even as a young child I was able to draw some inferences and formulate some hypotheses about cockroach *behavior* by watching my mother put out the bug traps. First, it seemed to me that roaches liked to spend their time in dark places, and second, it appeared that most roaches agreed on what was a good place for roaches to be, as Mom kept putting the traps out in the same place. These two thoughts on cockroach behavior could readily be developed into the following hypotheses: (1) cockroaches will choose dark places over light places, and (2) roaches tend to return to the same places, rather than move randomly through their environment.

Of course, as a young boy, I didn't formally sit down and generate these hypotheses, and I surely didn't run the controlled experiments that a scientist studying behavior would run to test these hypotheses, but I was nonetheless dabbling with scientific hypotheses about animal behavior, and that made me an animal behaviorist, or more technically, an ethologist.

There are many kinds of ethologists: from my mother, who understood roach behavior; to the farmer, who has detailed knowledge about farm animal behavior; to the boy who works training his dog; to the outdoorsman, who on his camping vacation searches for some animals and tries to avoid others. What's more, humans have always been ethologists. If our hunter-gatherer ancestors had not been ethologists, and had not, for example, understood the behavior of the prey they were trying to catch, as well as the behavior of the predators that were trying to catch them, we wouldn't be here today.

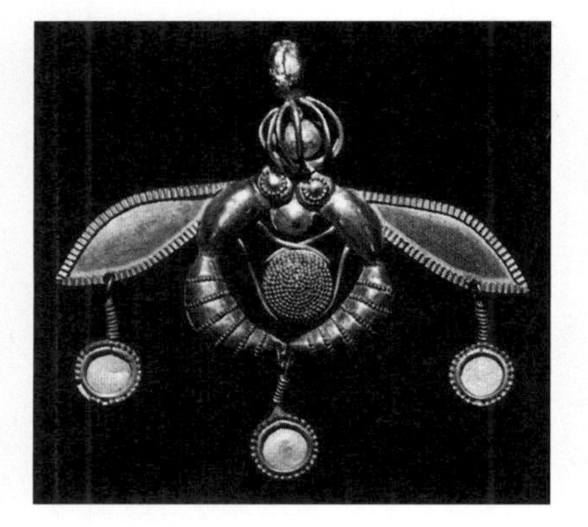

FIGURE 1.2. Art captures animal behavior. This pendant from the Chrysolakkos funeral complex in Crete suggests that some members of the ancient culture had a detailed knowledge of bees and animal behavior in general. *(From Masseti, 2000, p. 90)*

The study of animal behavior appears to have been so fundamental to human existence that the earliest cave paintings tend to be stock full of animal drawings. This is equally true of early human civilization. For example, using artifacts from 4,000-year-old Minoan cultures, Marco Masseti presents the case that the Minoans had an advanced understanding of animal behavior (Masseti, 2000). One fascinating example supporting this claim is a golden pendant from a Cretan cemetery that depicts two wasps transferring food to one another (Figure 1.2). Masseti hypothesizes that this kind of knowledge of insect behavior could only have come from people who observed and studied the details of wasp life. A similar sort of argument is offered regarding a beautiful wall painting of "white antelopes" (Figure 1.3). This painting most likely depicts gazelles in the early stages of an aggressive interaction, and again it is the sort of art that is associated with an in-depth knowledge of the subject in question.

Spanning the millennia between ancient Cretan civilization and the present, there have been literally thousands of amateur and professional naturalists who have made some contribution to the study of animal behavior. These contributions have enabled ethologists to draw on a rich trove of information that has greatly expanded our understanding of animal behavior (Figure 1.4). Aristotle's work on animals, for example, though 2,500 years old, is a veritable treasure chest of ethological tidbits.

FIGURE 1.3. Wall paintings of "white antelopes." The drawing may depict a "lateral intimidation" during an aggressive encounter between antelopes. *(From Masseti, 2000)*

This is a textbook about animal behavior, and in many ways, a course in animal behavior is where all the other biology and psychology classes that students have sat through up to this point in their academic career come together. Genetics, development, anatomy, physiology, endocrinology, neurobiology, evolution, learning, and social theory congeal into one grand subject—animal behavior. The field of **ethology** is integrative in the true sense of the word, in that it combines the insights that biologists, psychologists, anthropologists, and mathematicians have made to the discipline. To cover such a dizzying array of topics is a daunting task, but in the case of ethology, case studies, field studies, and experiments, which drive the narrative in this book, should make the learning enjoyable.

FIGURE 1.4. Fantastic images from a cave. A beautiful drawing of a herd of antelope found on the walls of a cave at Dunhuang, China. *(Photo credit: Pierre Colombel/Corbis)*

Three Foundations

Incredible tales and fascinating natural history make the study of animal behavior different from the study of organic chemistry or molecular genetics. What links animal behavior to all scientific endeavors, however, is a structured system for developing and testing hypotheses and a bedrock set of foundations on which such hypotheses can be built. Throughout this book, we shall see that the force of natural selection, the ability of animals to learn, and the power of transmitting learned information to others serve as the foundations upon which we shall build our approach to ethology.

Charles Robert Darwin accomplished many things in his classic *On the Origin of Species,* a book widely regarded as the most influential biology book ever written, but perhaps the most important was to put forth his argument on evolution by natural selection (Darwin, 1859). In a nutshell, Darwin argued that any trait that caused its possessor to have some sort of reproductive advantage would be favored by the process he dubbed natural selection. **Natural selection** is thus the process whereby traits that confer the highest relative reproductive success (that is, the greatest relative fitness) on their bearers and that can be passed down across generations increase in frequency over many generations (evolutionary time). Again and again, we shall turn to natural selection to understand animal behavior.

In addition to natural selection changing the frequency of different behaviors over the course of many generations, **individual learning** can alter the frequency of behaviors within the lifetime of an organism. Animals learn about everything from food and shelter to predators and familial relationships. Such individual learning represents a second major force we shall examine in depth throughout this book.

Cultural transmission is a third major force affecting animal behavior. While definitions of cultural transmission vary across disciplines, when we speak of this phenomenon, we shall be referring to situations in which animals learn something by copying the behavior of others through what is typically referred to as **social learning.** Cultural transmission can allow newly acquired traits to spread through populations at a very quick rate, as well as permit the transmission of information across generations rapidly.

FOUNDATION 1—NATURAL SELECTION

Charles Darwin recognized all too well that his theory of natural selection applied to behavioral traits as well as morphological, anatomical, and developmental traits. Darwin's ideas on evolution, natural selection, and behavior were revolutionary, and ethology today would look very different were it not for the ideas that Darwin set forth in *On the Origin of Species.* To see how natural selection operates in the wild, let's focus on beak size in Galápagos finches, also known as Darwin's finches.

Evolutionary biologists, inspired by Darwin and those who followed him, have continued to use the Galápagos Islands as a sort of evolutionary proving grounds. At least fourteen species of finches live on the Galápagos, and they are still studied extensively by scientists. Since 1973, Peter and Rosemary Grant have led a team of investigators who have focused on numerous aspects of evolution, natural selection, and finch life on the Galápagos (Grant, 1986; Weiner, 1995). They have found, for example, that there is a nice fit between a particular species' beak dimensions and the type of food it consumes (Figure 1.5). Here we shall focus on two finch species, *Geospiza magnirostris* and *Geospiza fortis,* and the role of natural selection in their beak size.

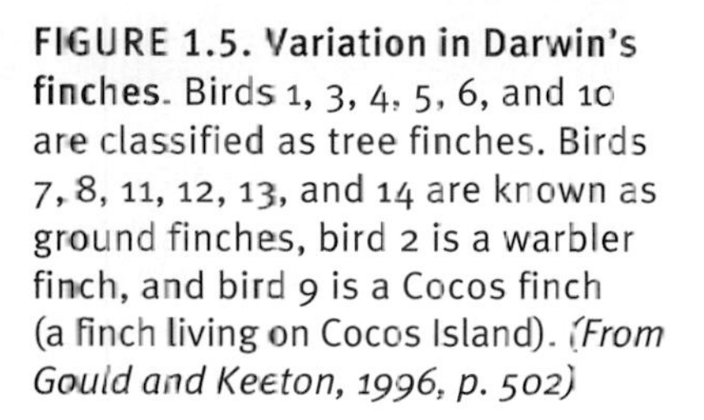
FIGURE 1.5. Variation in Darwin's finches. Birds 1, 3, 4, 5, 6, and 10 are classified as tree finches. Birds 7, 8, 11, 12, 13, and 14 are known as ground finches, bird 2 is a warbler finch, and bird 9 is a Cocos finch (a finch living on Cocos Island). *(From Gould and Keeton, 1996, p. 502)*

G. magnirostris is the bigger of the two species and, because of its relatively large beak, it can crack open the large, but very tough fruit of the caltrop (*Tribulus cistoides*) much more quickly and efficiently than its smaller counterpart *G. fortis*. Conversely, while the evidence is not as strong, it appears that *G. fortis* is more efficient when it comes to small seeds. Given that beak length is passed down across generations (Grant, 1986; Grant and Grant, 1995), we expect that natural selection should favor larger beaked birds in times when caltrops and other large seeded plants are abundant, as it is then that large beaked birds should outcompete others, and that smaller beaked birds should be favored when smaller seeded plants are more plentiful (Figure 1.6). This prediction should hold true with respect to both a comparison *between* these two species, as well as *within* each species. That is, not only should natural selection favor *G. magnirostris* over *G. fortis* when larger seeds are the norm, but larger *G. magnirostris* should do better than smaller *G. magnirostris,* and larger *G. fortis* should outcompete smaller *G. fortis*.

One of the many incredible aspects of working on evolution in the Galápagos Islands is that the intense variation in weather from year to year allows one to test the effect of natural selection directly over fairly short periods of time. This variation in weather allowed the Grants to examine natural selection when small seeds were abundant and when large seeds were abundant. For instance, in 1977–1978, a drought hit the Galápagos, and an entire wet season was missed. The finches were hard hit, and their overall population shrank by 80 percent. During this crunch, larger birds with larger beaks were favored, since in a short time all that was left for

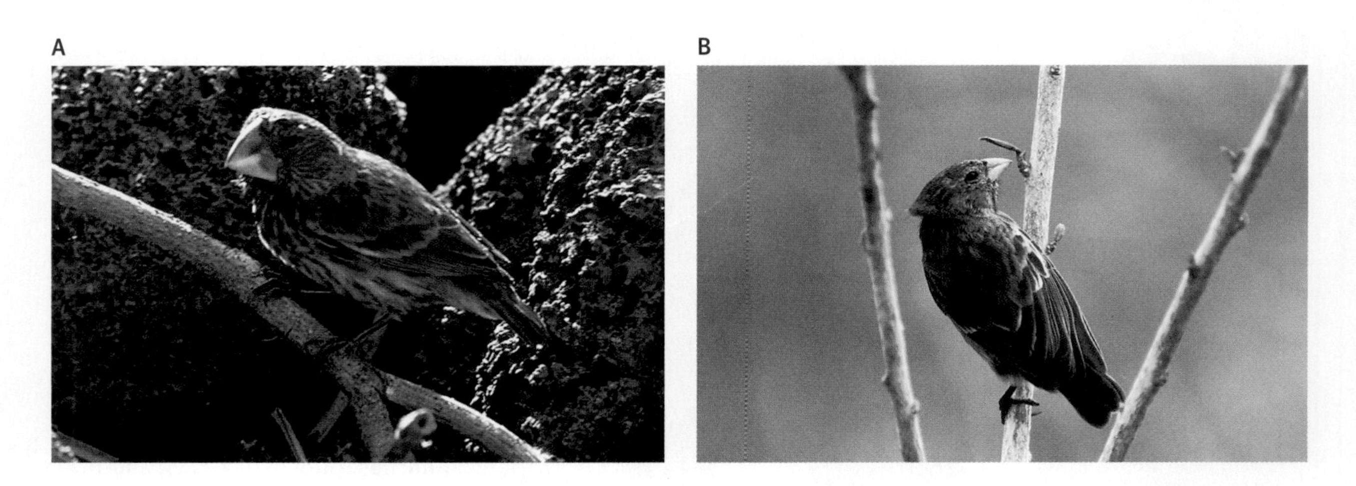

FIGURE 1.6. Finches and natural selection. (A) Larger finches *(G. magnirostris)*, which have large beaks, are more likely to survive and to reproduce when conditions favor large seeds. **(B)** Smaller finches *(G. fortis)*, which have smaller beaks, are more likely to survive and reproduce when conditions lead to an excess of small seeded plants. *(Photo credits: Ann and Rob Simpson)*

finches to eat were large seeds. As a consequence, the frequency of larger birds in general rose, as natural selection thinking would predict. In 1982, the torrential rainstorms of El Niño hit the Galápagos hard, and the weather eventually led to a huge excess of small seeded plants. Sure enough, over the next few generations, the frequency of smaller individuals then increased, just as natural selection thinking would lead us to expect.

As a second example of how natural selection might operate on animal behavior, let's examine the evolution of xenophobia—the fear of strangers—in the common mole rat (*Cryptomys hottentotus*). For good or bad, a single motto that unites many social living species is "beware of strangers." Whether they bear gifts or otherwise, strangers—unknown individuals from outside one's group—are often threatening: they may compete for scarce resources one's group needs, disrupt group dynamics, and so on. As such, natural selection should favor the expression of extreme xenophobia when group resources are limited and groups are tightly knit. It is in such cases that fearing strangers should have the largest positive effect on reproductive success, and hence it is where natural selection should be strongest.

To test this idea, Spinks and his colleagues (1998) examined xenophobia in the common mole rat. Common mole rats, who live in underground colonies of two to fourteen individuals in South Africa (Figure 1.7), are an ideal species in which to examine xenophobia for two reasons. To begin with, all populations are "tightly knit" in the sense that each group has only a single pair of reproductive individuals, and hence group members tend to be very related. Furthermore, populations of common mole rats differ in terms of limited group resources. Some populations inhabit mesic (moderately moist) environments that present only moderate resource limitations, while other populations live in arid environments and face intense limitations on their resources. Mesic environments have about four times as much rainfall as arid environments. It is this variation that allowed Spinks and his colleagues to test a specific hypothesis regarding xenophobia: populations from arid areas should be more xenophobic than those from mesic environments because resources are more scarce in arid regions.

FIGURE 1.7. Common mole rats. This xenophobic common mole rat *(Cryptomys hottentotus)* is showing an aggressive stance in response to a stranger. *(Photo credit: Graham Hickman)*

Spinks and his collaborators conducted 206 "aggression" trials. The protocol for these experiments was quite simple. Two individuals, one from the arid and one from the mesic environment, were placed together and aggression was recorded. Results were clear-cut: fear of strangers, and rejection as shown by aggression toward such individuals, was much more pronounced in the common mole rats from the arid environment where resources were more limiting. This result was not a function of individuals from arid populations just being more aggressive in general. Control experiments demonstrated

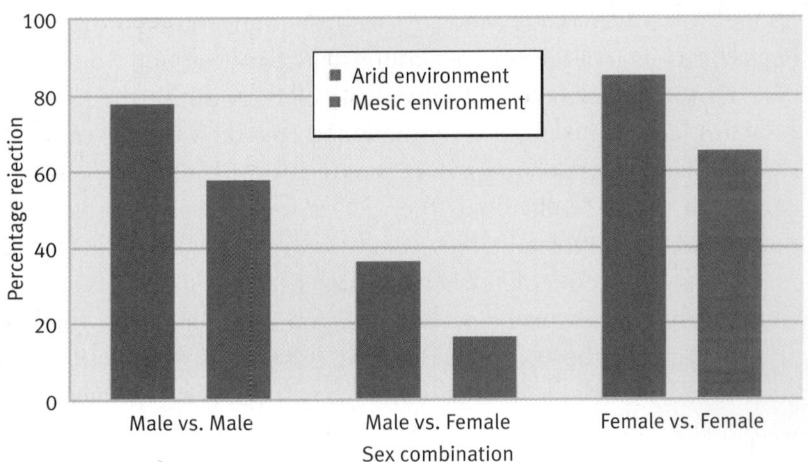

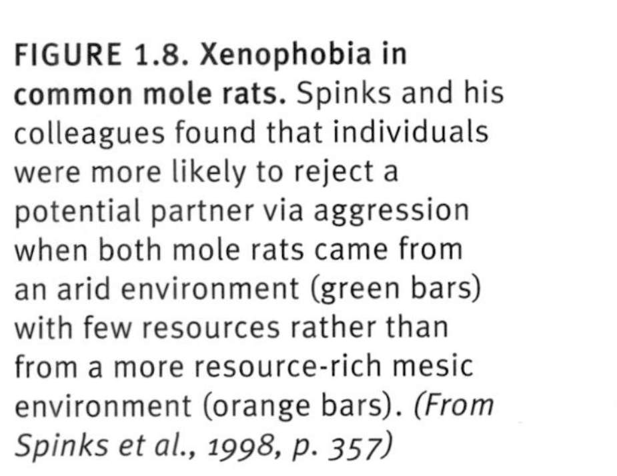

FIGURE 1.8. Xenophobia in common mole rats. Spinks and his colleagues found that individuals were more likely to reject a potential partner via aggression when both mole rats came from an arid environment (green bars) with few resources rather than from a more resource-rich mesic environment (orange bars). *(From Spinks et al., 1998, p. 357)*

that when two individuals who knew each other from the arid population were tested together, aggression disappeared—that is, identifying a stranger initiated aggression. This was precisely the sort of behavior that natural selection should favor.

A second important finding from this study can only be understood as the likely product of the selection process. Those common mole rats who are lucky enough to end up reproducing always move from their home colony to find a mate. This means that some strangers are potential mates and hence perhaps worth tolerating. Natural selection then should not simply favor all xenophobia, but a xenophobia that is sensitive to the sex of the stranger. In support of this, Spinks and his colleagues found that while aggression was still uncovered in the arid population when the two individuals tested were a male and a female, the level of aggression decreased dramatically when compared to same-sex interactions (Figure 1.8). Common mole rats then temper their fear of strangers as a function of both where they live and the sex of the strangers.

Our understanding of how natural selection operates on animal behavior was greatly advanced with the appearance of sophisticated, usually mathematical, models for the evolution of social behavior in animals and humans in the late 1960s and early 1970s. This work is most closely associated with George C. Williams, William D. Hamilton, John Maynard Smith, Robert Trivers, E. O. Wilson, and Richard Alexander. As we shall see throughout the course of the book, the models developed by this group in the 1960s and 1970s revolutionized the way that ethologists looked at almost every type of behavior they study.

For a brief sortie into the utility of models, let us consider the evolution of altruistic and cooperative behavior. While W. D. Hamilton tackled the question of how altruistic behavior could evolve among blood relatives (see Chapter 8), animal cooperation and altruism among unrelated individuals had long been a paradox of sorts for evolutionarily oriented ethologists (Figure 1.9). How, the argument goes, could any behavior that was detrimental to the actor, but beneficial to the recipient, be favored by natural selection? Shouldn't natural selection weed out such actions, favoring those who receive but fail to give? Robert Trivers's (1971) "reciprocal altruism" model provided an initial answer to this riddle by suggesting that altruism can be favored when altruistic acts are exchanged by individuals, a sort of "you scratch my back, and I'll scratch yours" system. To resolve this seeming paradox, Trivers (1971) and then Robert Axelrod and W. D. Hamilton (1981) relied on a mathematical technique called game theory (see Chapter 9). Using what is known as the **prisoner's dilemma game,** in which two individuals are given conditions in which they have to decide whether to cooperate or not, these researchers were able to predict that animals should exchange acts of altruism and cooperation *under a limited set of conditions* (we shall examine this case in detail in Chapter 9).

In retrospect, the reciprocity that emerges from the prisoner's dilemma seems obvious. Brilliant ideas, however, often seem obvious with the benefit of time and hindsight. What is most important is that while experiments played a fundamental role in understanding a critical issue in ethology—the evolution of altruism and cooperation—mathematical theory played an equally important role.

FIGURE 1.9. Cooperation among sticklebacks. Three sticklebacks cooperatively approach a potential predator in what is called predator inspection. This is a dangerous behavior, but it provides an array of benefits to cooperating inspectors. *(Photo credit: Manfred Milinski)*

FOUNDATION 2—LEARNING

Individual learning can take many forms (see Chapter 4), but let us begin by considering a hypothetical case of learning in the context of mating. Let's imagine, as is the case for most animals, that females mate with numerous males throughout the course of a lifetime. Suppose that female birds were somehow able to keep track of how many chicks fledged their nest when they mated with male 1, 2, 3, and so forth. Further suppose that older females preferred to mate with the males that fathered the most successful fledglings. That is, what if the females changed their mating behavior as a result of direct personal experience? In this case, we would see that learning had changed the behavior of an animal within the course of a lifetime (Figure 1.10). Over and over, we shall see how important a factor learning is in the study of ethology.

The learning example above was chosen to emphasize an important relationship between learning and natural selection. In our example, females changed their preference for mates as a result of prior experience, and so learning affected behavior frequencies within a generation. But just because the frequency of a behavior is changing within the course of an individual's lifetime does not mean that natural selection is removed from the picture, for it is certainly possible for natural selection to operate on the ability to learn. To see this, we must realize that the behavior "change your preference based on personal experience" not only can shift mate choices within a generation, but it also can change the ability to learn, which if genetically coded, can be subject to natural selection. That is, natural selection might very well favor the ability to

FIGURE 1.10. A role for learning. Imagine a female who mates with different males over the course of time. Such a female might learn which male is a good mate by keeping track of the number of eggs she laid when associated with each male. During future mating opportunities, female bird 1 (pictured here) would be likely to choose male bird 2 because she had learned that she lays the most eggs after mating with him.

learn about mates over, say, the lack of such an ability If this were the case in the example above, learning would change behaviors within a generation, but natural selection might be changing the frequency of different learning rules, such as "change your preference based on . . ." across generations.

As a good example of how learning and natural selection can be intimately tied together, we turn to an ingenious protocol designed by Dukas and Bernays. While learning in insects is well-documented (Quinn et al., 1974; Papaj and Prokopy, 1989; Papaj and Lewis, 1993; Lee and Bernays, 1990; Lewis and Takasu, 1990; Turlings et al., 1993; Raubenheimer and Tucker, 1997; Dukas, 1998c, 1999), pinning down the potential fitness-related benefits of learning has proved more difficult. To tackle this question directly, Dukas and Bernays (2000) examined the potential fitness-related benefits of learning in foraging grasshoppers (*Schistocerca americana*; Figure 1.11).

FIGURE 1.11. Some aspects of foraging in grasshoppers are learned. *Schistocerca americana* grasshoppers learned to associate various cues with food sources. *(Photo credit: L. Bernays)*

In their experiment, Dukas and Bernays placed two food dishes in each grasshopper's cage. In one dish was a balanced diet (b) containing proteins and carbohydrates—a diet that promotes maximal growth rates in *S. americana*. The balanced diet also contained one of two non-nutritive flavorings (coumarin or citral; these flavorings provided grasshoppers with a specific odor). In the other dish was a "deficient" diet (d) that contained flavoring and protein, but no carbohydrates. In addition, next to each dish a colored card (either brown or green) was placed. That is, each diet type (balanced or deficient) could be paired up with one flavor (coumarin or citral) and one colored card (brown or green).

Dukas and Bernays designed two treatments, which they denoted as the "learning" and "random" treatments (Figure 1.12). In the learning treatment, grasshoppers would have the balanced diet dish always associated with one specific flavor and one specific colored card. Twice a day, grasshoppers would be presented with the two diet dishes and allowed to choose from which to forage. In the random treatment, the cues (flavor and card color) associated with each diet were assigned randomly, such that the grasshoppers were prevented from associating either cue with a particular diet, and as a result, no associative learning would be possible.

Major differences between grasshoppers in the learning and random treatments were uncovered. Grasshoppers in the learning treatment almost always went straight for the balanced diet dish, and they did so to a much greater extent than did the individuals in the random treatment (Figure 1.13). Over the course of the experiment, individuals in both treatments increased the proportion of time they spent feeding on the balanced diet, but grasshoppers in the learning treatment did so more quickly than those in the

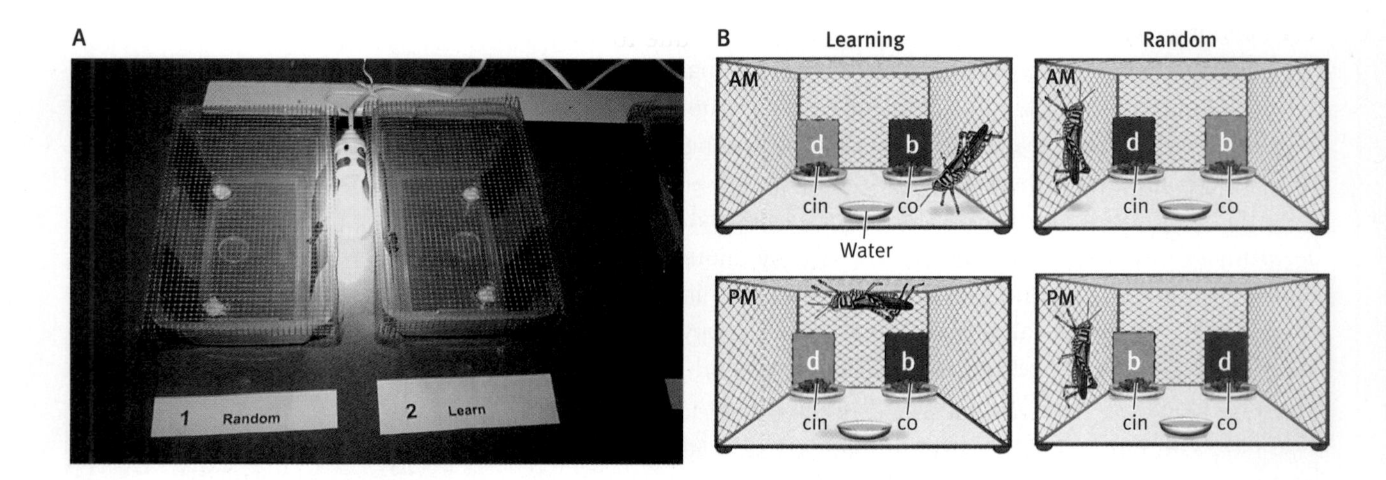

FIGURE 1.12. Learning, foraging, and fitness in grasshoppers. (A) A photograph of the experimental set-up. *(Photo credit: R. Dukas)* **(B)** A schematic of the set-up showing the learning and random conditions. In the learning condition, the set-up consisted of a water dish in the center of the cage and a nutritionally balanced dish (b) on one side of the cage and a nutritionally deficient dish (d) on the other side of the cage. Each dish was paired with one flavor and one colored card (brown or green). In the learning condition, when the dishes were removed and replaced by new dishes, the new balanced and deficient dishes were flavored in the same way as the old balanced and deficient dishes and were placed on the same sides of the cage and next to the same colored cards as the old dishes. In the random condition, location, colored card, and flavor were all changed when the new dishes were put into the cage later in the day. *(Adapted from Dukas and Bernays, 2000)*

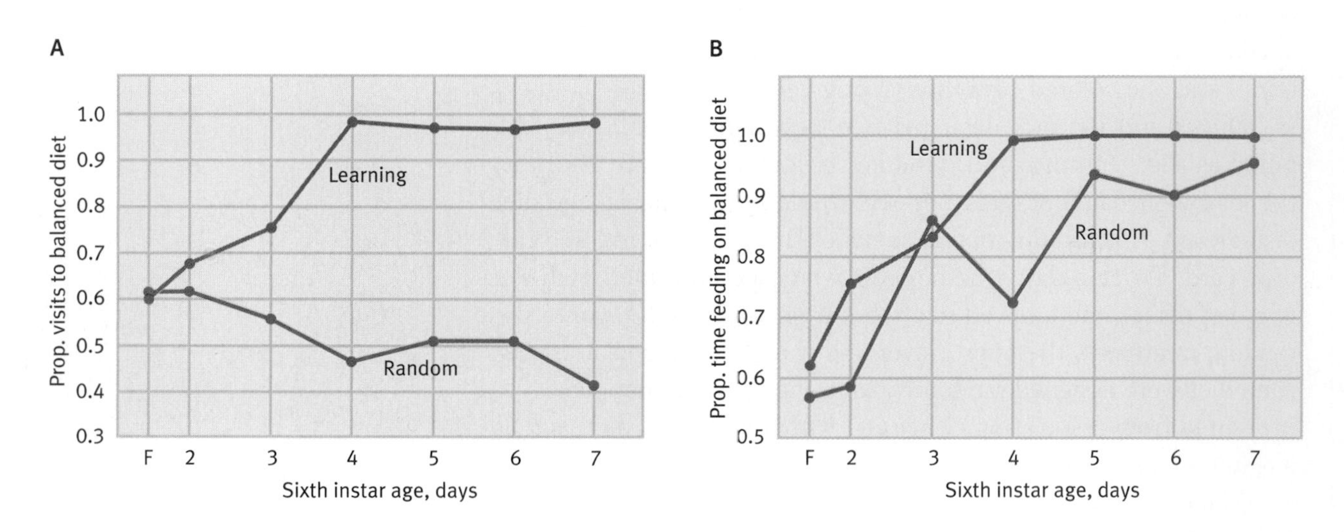

FIGURE 1.13. A balanced diet in grasshoppers. Grasshoppers in the instar stage of insect development were given a choice between a balanced diet or a deficient diet, and researchers recorded the proportion of visits and feeding times of those in a learning treatment and those in a random treatment. In the learning condition, the food was presented in a way in which grasshoppers could learn to associate colored background cards and flavorings with balanced and unbalanced diets, while this was not possible in the random condition. **(A)** Grasshoppers in the learning condition visited the balanced diet dish more often than those in the random condition. **(B)** Grasshoppers in the random condition eventually also ended up feeding more often on the balanced diet dish, but it took more trial and error for them to recognize and feed on the balanced diet. *(From Dukas and Bernays, 2000)*

random treatment. This difference was most likely due to the fact that grasshoppers in the learning treatment went to the balanced diet dish almost immediately when feeding, while those in the random treatment ended up at the balanced diet dish, but only after much sampling of the deficient diet dish. Perhaps most important of all, the individuals in the learning treatment had a growth rate that was 20 percent higher than that of the grasshoppers in the random treatment (Figure 1.14).

The ability to learn about food in *S. americana* translated into significant gains, with a significant increase in growth rate observed in those individuals in the learning treatment (see Pravosudov and Clayton, 2001, for a related experiment on learning and fitness in mountain chickadees, *Poecile gambeli*). This difference in growth rate likely translates into greater fitness later in life, as growth rate is positively correlated with the number and size of eggs laid over the course of an individual's life (Slansky and Scriber, 1985; Atkinson and Begon, 1987).

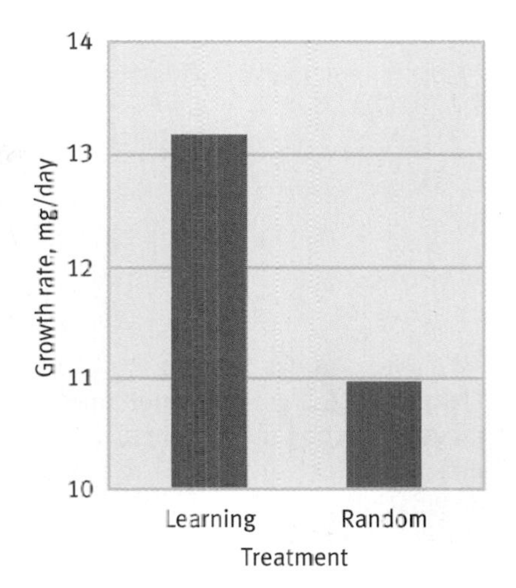

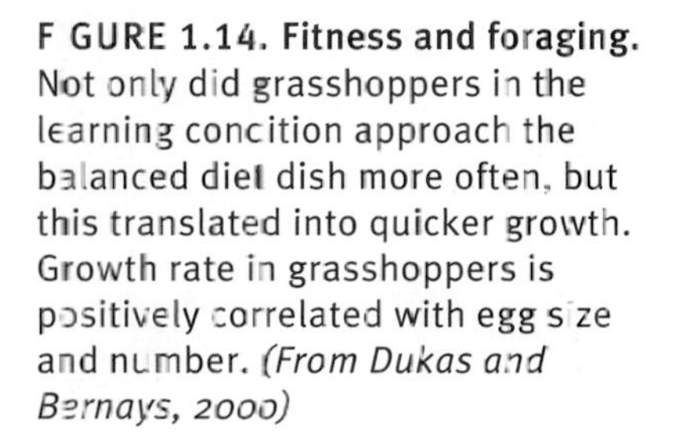

FIGURE 1.14. Fitness and foraging. Not only did grasshoppers in the learning condition approach the balanced diet dish more often, but this translated into quicker growth. Growth rate in grasshoppers is positively correlated with egg size and number. *(From Dukas and Bernays, 2000)*

FOUNDATION 3—CULTURAL TRANSMISSION

Recall that when we speak of cultural transmission, we are referring to situations in which animals learn something by copying the behavior of others through what is known as social learning.

For an interesting case study illustrating the importance of cultural transmission and social learning in animals, let's consider Jeff Galef's work on foraging in rats. Being scavengers, rats are often presented with opportunities to sample new foods. This probably has been true for most of the rat's long evolutionary history, but it has been particularly the case over the last few thousand years, during which time humans and rats have had a close relationship (Figure 1.15). Scavenging presents a foraging dilemma. On the one hand, a new food source may be an unexpected rat bounty. On the other hand, new foods may be dangerous, either because they contain elements inherently bad for rats, or because the rats don't know how a new food should smell, and so it is difficult to tell if something they come upon is fresh or spoiled. This is the ideal environment for cultural transmission and social learning.

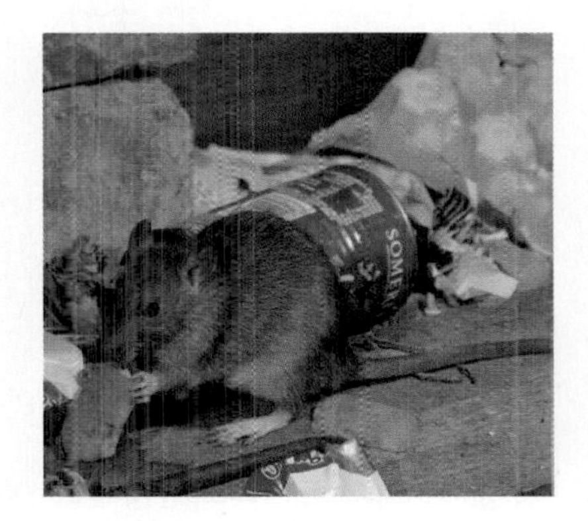

FIGURE 1.15. Scavenging rat. A black rat scavenges at a food dump. New food items are often encountered during such foraging bouts. *(Photo credit: Paul Hobson/ Nature Picture Library)*

The study of social learning and food preferences in rats began with a test of what is known as the information-center hypothesis (Ward and Zahavi, 1973). The idea here is that in species where environmental cues about food are constantly shifting, foragers may learn critical tidbits about the location and identity of food by interacting with others who have recently returned from a foraging bout. Galef and his colleagues tested this hypothesis in the Norway

FIGURE 1.16. Two rats smell one another. Among other things, olfactory cues provide information on what others are eating. *(Photo credit: Jeff Galef)*

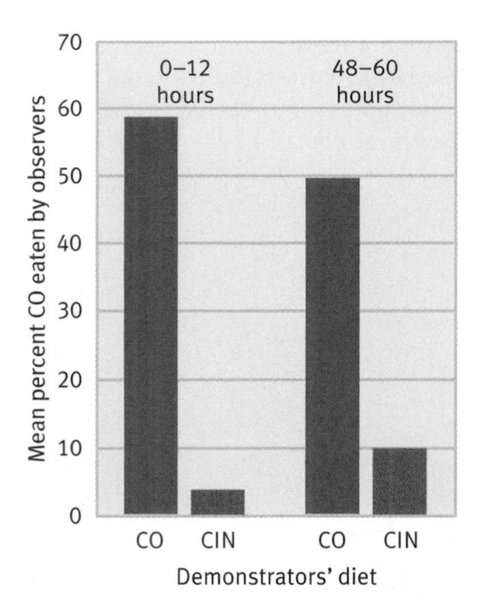

FIGURE 1.17. Social learning and foraging in the Norway rat. Observer rats had a "tutor" (demonstrator) who was trained to eat rat chow containing either cocoa (co) or cinnamon (cin) flavoring. Once the observer rats had time to interact with a demonstrator rat, the observer rats were much more likely to add their tutor's food preferences to their own. Rats with "cinnamon tutors" preferred cinnamon-flavored food, and rats with "cocoa tutors" preferred cocoa-flavored food. *(From Galef and Wigmore, 1983)*

rat, a species that finds itself in this precise scenario often (Galef and Wigmore, 1983). To test whether cultural transmission via social learning played a role in rat foraging, rats were divided into two groups—observers and demonstrators (tutors). The critical question was whether observers could learn about a new, distant food source simply by interacting with a demonstrator who had experienced such a new addition to his diet.

After living together in the same cage for a few days, a demonstrator rat was removed and taken to another experimental room, where he was given one of two new diets—either rat chow flavored with Hershey's cocoa (eight demonstrators) or rat chow mixed with ground cinnamon (eight demonstrators). The demonstrator was then taken back to his home cage and allowed to interact with the observer for fifteen minutes (Figure 1.16). Then the demonstrator was removed from the cage. For the next two days, the observer rat was given two food bowls, one with rat chow and cocoa, the other with rat chow and cinnamon. Keep in mind that observers had no personal experience with either of the novel food mixes they were experiencing, nor had they ever seen their tutor eat these new food items. Galef's results were clear. Observer rats, via olfactory cues, were influenced by the food their tutors had eaten, as shown by the fact that they were now more likely to eat the food themselves (Figure 1.17).

As was the case for individual learning, it is certainly possible that if the behavior "copy the diet choice of others" is genetically coded, this rule might increase in frequency through natural selection. But the case of cultural transmission is more complicated than that of individual learning. The reason is that what an animal learns via individual learning is lost when that animal dies. The actual information that one learns via individual learning never makes it across generations. This is not the case with cultural transmission. What a single animal does, if copied, can affect individuals many generations down the road (see Chapter 5). Suppose adult rat A (in generation 1) adopts a new, formerly uneaten, type of food into his diet because he smelled this food on a nestmate. Now suppose young individuals (generation 2) in the same colony as A, add this new food to their diet because they smell it on A. When individual A eventually dies, the cultural transmission chain he began may still be in force, as the young individuals who copied A are still around. In other words, a culturally learned preference in generation 1 may make it to generation 2 (Figure 1.18). If generation 3 individuals learn from generation 2 individuals, then our culturally derived preference has been transmitted to two generations. And potentially so on down the generations. Cultural transmission itself, in other words, has both within- and between- generation effects (see Chapter 5).

It is important to note that we shall constantly be turning to natural selection, learning, and cultural transmission in our analysis of animal behavior, and all are critical to any broad-based overview of ethology. Yet, by necessity, each chapter of this book does not give equal weight to each of these three forces; in fact, in

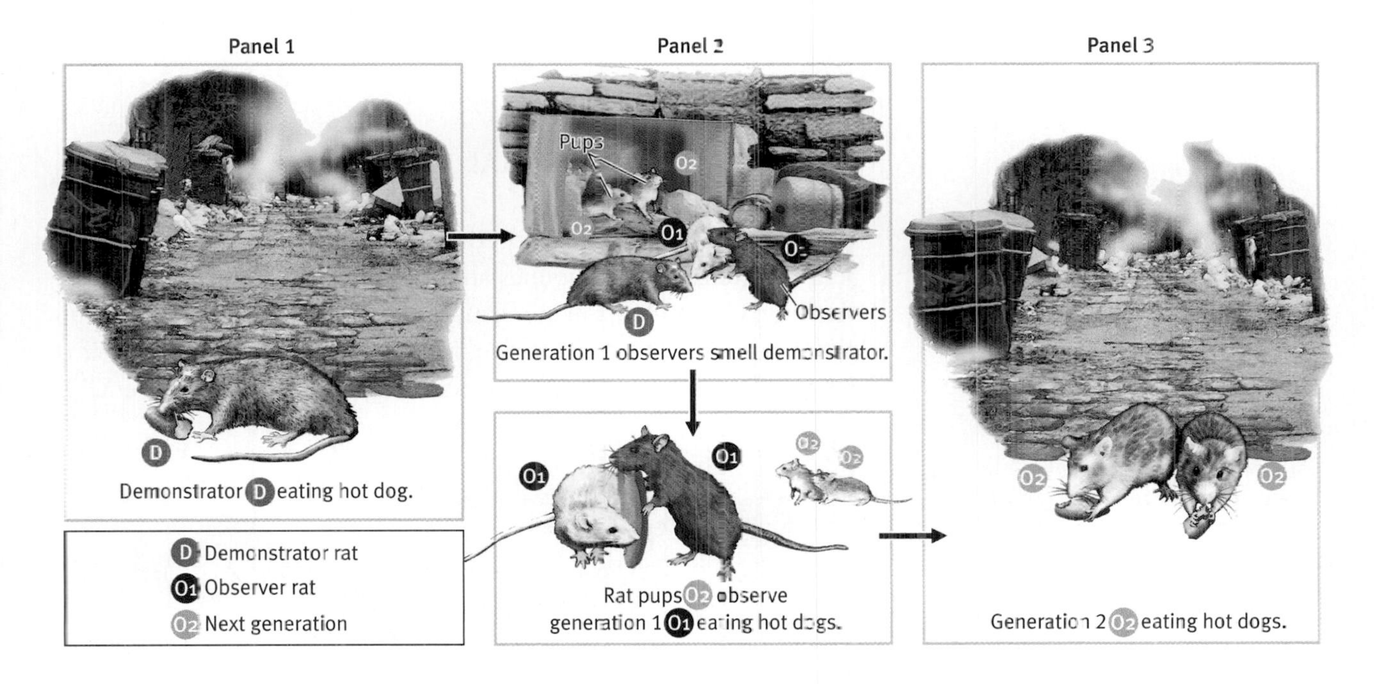

FIGURE 1.18. A role for cultural transmission. In panel 1, a rat eats a new food type (hot dog). When this rat (D for demonstrator rat) returns to his nest (panel 2), observer rats (O1) smell him, and then are more likely to add hot dogs to their diet when they encounter such an item. Multigenerational cultural transmission occurs when rats from the next generation (O_2) smell generation O1 rats after they have eaten hot dogs and subsequently add hot dogs to their own diet (panel 3).

some chapters we will not even touch on all three. Historically, there have been many more studies on the role of natural selection on nonlearning-related behaviors (see Chapter 2), and hence this approach will often make up the largest part of most chapters. Conversely, while social psychologists have been studying social learning for more than a century (see Chapter 5), it is only over the last few decades that animal behaviorists have seriously investigated social learning, or more generally cultural transmission, and this is reflected in the text.

Conceptual, Theoretical, and Empirical Approaches

Any field of scientific endeavor can be studied using conceptual, theoretical, and empirical approaches (Figure 1.19). This is true in ethology as well, and in fact, the best studies in animal behavior tend to use all three of these approaches to one degree or another (Dugatkin, 2001b). In addition to focusing on natural selection, learning, and cultural transmission, the conceptual-theoretical-empirical axis will form another bedrock on which much of what follows is based.

CONCEPTUAL APPROACHES

Conceptual approaches to ethology tend to import ideas generated in different subdisciplines and to combine them in a new, cohesive way. Generally speaking, natural history and experimentation do play a role in concept generation, but a broad-based concept itself is not usually directly tied to any specific observation or experiment.

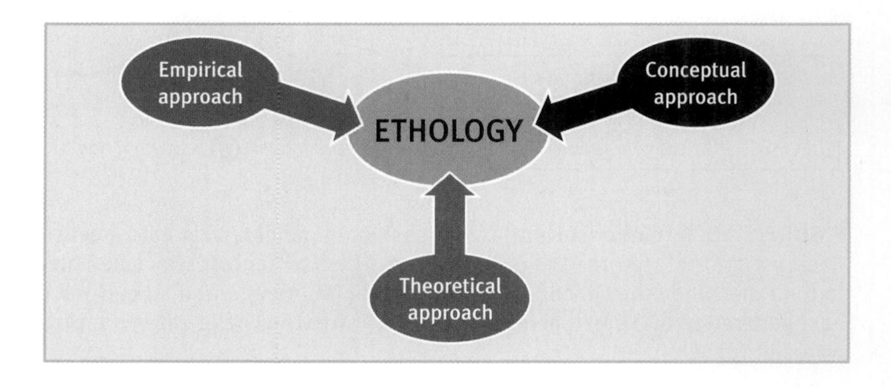

FIGURE 1.19. Different approaches to ethology. Ethology can be studied from a conceptual, theoretical, or empirical approach.

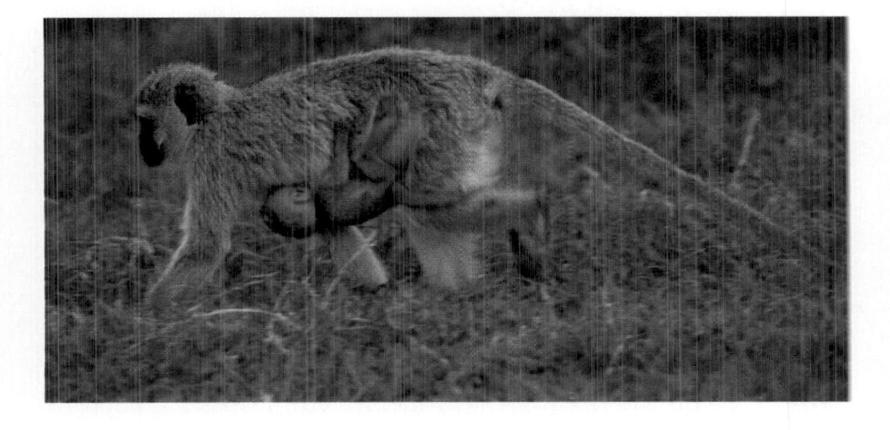

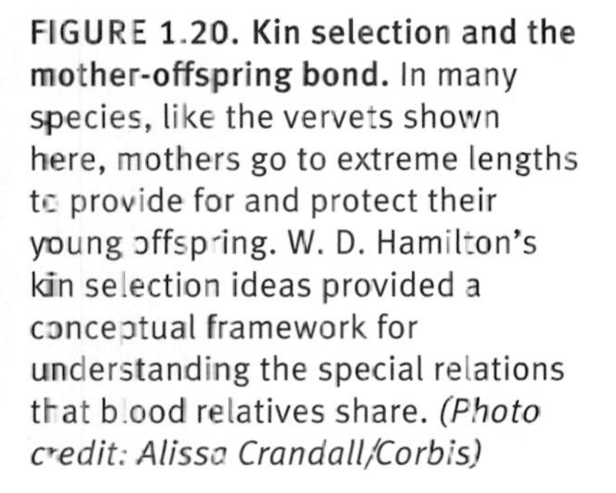

FIGURE 1.20. Kin selection and the mother-offspring bond. In many species, like the vervets shown here, mothers go to extreme lengths to provide for and protect their young offspring. W. D. Hamilton's kin selection ideas provided a conceptual framework for understanding the special relations that blood relatives share. *(Photo credit: Alissa Crandall/Corbis)*

Not only do major conceptual advances tend to generate new experimental work, but they also reshape the way that a discipline looks at itself (Figure 1.20). The clearest case for a concept that has made behavioral ecologists rethink the basic way they approach their science is the late William D. Hamilton's ideas on kin selection (Hamilton, 1964, see Chapter 8). Hamilton hypothesized that an individual's total fitness, measured by its genetic contribution to the next generation, is not simply a function of the number of viable offspring that it produces (Figure 1.21). Rather,

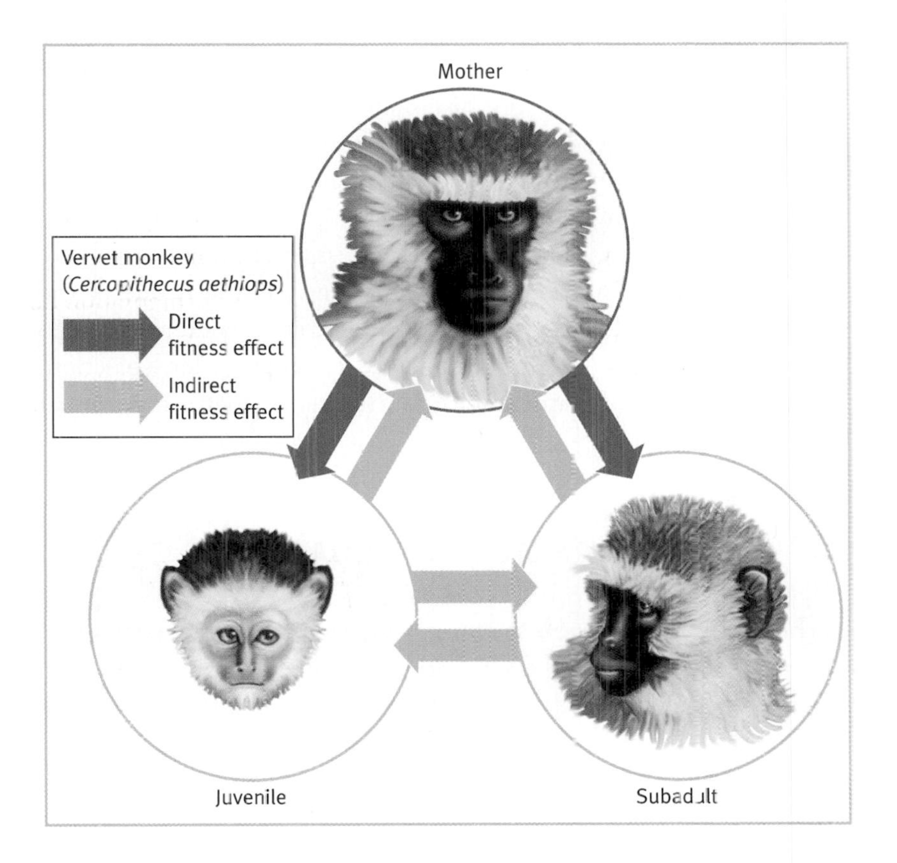

FIGURE 1.21. Two components to fitness. Three vervet monkeys are pictured—a mother, her juvenile offspring, and her subadult female offspring. Direct fitness effects are a consequence of mother-offspring interactions, while the more mature offspring can increase her indirect fitness by helping her younger sibling. Direct effects are assigned to individuals that physically transmit their genes or their own sperm or eggs to the affected individual. Indirect effects do not require the transmission of gametes between affected individuals (Brown, 1987).

Hamilton's inclusive fitness hypothesis argues that an individual's total fitness is a combination of the number of young it raises, plus some benefit assigned for any help it provided in raising the offspring of blood relatives.

We shall explore the logic of inclusive fitness in detail in Chapter 8, but the kernel of this powerful idea is that, evolutionarily speaking, blood relatives are important because of shared genes—genes held in common by relatives and inherited from some common ancestor. So, imagine for a moment a Mexican jay bird, a species that has been the subject of much inclusive fitness work (J. L. Brown, 1987). A jay's inclusive fitness is a composite of the number of offspring it has, plus some fraction of the number of offspring it helped a relative raise. Let's say that a jay helps its parents raise an additional brood of two birds, above and beyond what mom and dad could have raised on their own. Our helper is related to itself by a value of 1, and it is related to its siblings by a value of 0.5 (see Chapter 8 for more on this calculation). By helping Mom (and perhaps Dad) raise two additional offspring, it has contributed 2×0.5 or 1 offspring "equivalent" to its inclusive fitness. If this is the only help that it undertakes, our jay's inclusive fitness is calculated by adding its indirect fitness (from helping raise its siblings) to its direct fitness (measured by success at raising its own offspring).

Inclusive fitness thinking affects almost every aspect of animal behavior, and this concept has made ethologists and behavioral ecologists rethink their entire discipline. The mathematical theory underlying kin selection is critical for specific predictions in any particular animal system, but kin selection's strength lies not only in its predictive power in any given system, but also in helping rethink the way that we approach relatedness in ethology in general. "What role does kin selection play in your system?" is one of the most common questions asked at any ethology conference. The ability of ethologists to understand the concept of kin selection, even outside of the mathematics, is a fine sign that a conceptual idea has made its mark.

THEORETICAL APPROACHES

A **theoretical approach** to animal behavior entails the generation of some sort of model of the world. This usually means constructing a mathematical model. For example, during the formative years of behavioral ecology, much theoretical work focused on animal foraging behaviors (Stephens and Krebs, 1986; Kamil, Krebs, and

Pulliam, 1987). One question of particular interest was: Given a list of potential edible items, which ones should a foraging animal add to its diet, and under what conditions? To tackle this, mathematical optimality theory was used (see Chapter 10). For example, one such model could use the energy (e) provided, the "handling time" (h), and the "encounter rate" (λ) associated with various items to predict which such items should be added to an animal's diet to optimize some quantity, such as energy intake per unit time. These variables would then be placed into a fairly complex algebraic inequality (Figure 1.22). Solving this inequality would produce numerous general, and often counterintuitive, predictions. For example, the model predicted that whether or not a low-ranking food item (one that provides little energy per unit time) would be added to an animal's diet depended on the availability of high-ranking items, not on the availability of the low-ranking item itself. In other words, if rabbits provided more energy per unit time for wolf predators than did chickens, then the availability of rabbits,

FIGURE 1.22. Mathematical optimality theory and foraging. Cheetahs can feed on many different prey items, including gazelle fawn. Ethologists have constructed mathematical models of foraging that determine which potential prey items should be taken. The value assigned to each prey is a composite of energy value (e), handling time (h), and encounter rate (λ). We shall delve into the mathematical theory underlying foraging behavior in Chapter 10.

not chickens, would determine whether chickens would be added to the wolf's diet.

Whether an animal encounters a chicken a minute or a chicken a year has no effect on its decision of whether or not to add chickens into its diet. Without a mathematical model of foraging behavior, it is unlikely that anyone could have used intuition to come up with that hypothesis. Tests of this "prey selection" model demonstrated its general applicability, as well as how new models of foraging could improve upon it (Stephens and Krebs, 1986).

One final comment for now on how theoreticians, including those who work on ethological questions, operate. Theoreticians are not interested in mimicking the natural world in their models, but rather in condensing a difficult, complex topic to its barest ingredients in an attempt to make specific predictions. A good theory will whittle away the details of specific systems, but just enough to allow for general predictions that can apply to many systems.

EMPIRICAL APPROACHES

The lion's share of this book will be devoted to empirical work in the area of ethology. Over and over, we shall examine **empirical studies** that are often, but not always, designed to test the theories and concepts that have been proposed as explanations for behavior. Empirical work in ethology can take many forms, but essentially it can be boiled down to one of two types—either observational or experimental. Both have been, and continue to be, extraordinarily important to the field of ethology.

While empirical studies in ethology preceded the work of Karl von Frisch, Niko Tinbergen, and Konrad Lorenz, modern ethological experimentation is often associated with these three Nobel Prize winners (particularly Tinbergen and von Frisch). They were extraordinary naturalists who had a fundamental understanding of the creatures they worked with and the world these creatures lived in. Above and beyond this, they were able to translate their talents as naturalists into asking fundamentally important questions about animal behavior—questions that could be addressed by a combination of observation and experimentation.

Observational work entails watching and recording what animals do, but making no attempt to manipulate or control any ethological or environmental variable. For example, I might go out into a marsh and record every action that I see redwinged blackbirds do from 9 A.M. to 5 P.M. In so doing, I might note foraging behavior, encounters with predators, the feeding of nestlings, and so forth. From this sort of work, I might be able to piece together the time budget of the redwinged blackbirds in my study population. While this sort of work can be an end in and of itself, it is usually the first step in a long process. Next, from my observations, I might hypothesize that redwinged blackbird males, for example, seem to make few foraging bouts when predators are present in their general area. To empirically examine the relationship between foraging and predation pressure, I could make detailed observations on how much food males eat and how many predators I could spot. I could then look for a relationship between these two variables, and test the notion that they rise concurrently (Figure 1.23).

Let's say that when I graph male foraging behavior against predation pressure, I do in fact find that they are correlated. Males do increase and decrease their foraging behavior in a manner correlated with changes in predation pressure. At times when lots of predators are around, males forage infrequently, but when preda-

FIGURE 1.23. Observation and experimentation. Imagine your observations led you to believe that redwinged blackbirds decrease foraging when under predation pressure. To experimentally examine causality, you could use a trained falcon to fly over a redwinged blackbird area and observe how this affected the amount of foraging.

tors are few and far between, males forage significantly more often. What then can I conclude? Is it fair to say that increased predation pressure causes decreased foraging? No, the data we have so far cannot be used to infer causation. We can say that predation and foraging are correlated, but from the data we have, we can't speak to the subject of what caused what. It might be that some other variable is causing predation pressure to increase and foraging to decrease. For example, it might be that when the temperature rises, redwing predators become more active, but redwings become less active. Increased predation pressure and foraging would still be correlated, but now the former wouldn't be seen as causing the latter; rather, they would both be seen as caused by changes in the weather.

In order to examine causality, we must experimentally manipulate our system. We might, for example, experimentally increase the number of redwing predators in area 1, but not in area 2, and see how redwing foraging is affected in these populations (Figure 1.23). This could be done by using trained predators or by simulating increased predation pressure by flying realistic predator models in area 1, but not in area 2. In either case, if we see redwing foraging behavior decrease in area 1, but not in area 2, we can feel much more confident that increased predation pressure causes decreased foraging in redwinged blackbirds.

Before completing this section on conceptual, theoretical, and empirical perspectives in ethology, let us touch on one more topic—the question of whether there is any natural ordering when it comes to the latter two approaches. Does theory come before or after empirical work? The answer is, "It depends." Good theory can precede or postdate data collecting and hypothesis testing. On some occasions, an observation or experiment will suggest to a researcher that the results obtained beg for a model of behavior to be developed. Models of reciprocity and cooperation, for example, originally emerged from anecdotal evidence that many animals appeared to sacrifice something in order to help others. Given that natural selection should typically eliminate such unselfish actions, the anecdotes cried out for a mathematical model to explain their existence. The prisoner's dilemma, as well as many other mathematical models, provided some very useful insights on this question, as well as creating the impetus for future empirical work.

On the other hand, the foraging models we discussed above preceded the large number of empirical studies on foraging that ethologists and behavioral ecologists continue to undertake. While it is true that ethologists have long studied what and when animals eat, controlled experimental work designed to test specific predictions about foraging were initially spurred on by the theoretical work in this area. Regardless of whether theoretical work predates or postdates empirical work, a very powerful feedback loop typically emerges wherein advances in one realm (theoretical or empirical) lead to advances in the other realm.

An Overview of What Is to Follow

We shall begin the remainder of this book with four "primer" chapters that provide an overview of natural selection and animal behavior (Chapter 2), proximate issues in animal behavior (Chapter 3), learning and animal behavior (Chapter 4), and cultural transmission from an ethological perspective (Chapter 5). Subsequent to the primer chapters, however, these forces will not be set off in chapters of their own. Post-primer chapters will cover sexual selection, mating systems, kinship, cooperation and coalitions, foraging, antipredator behavior, communication, habitat selection and territoriality, aggression, play, aging and disease, and animal personalities. Textbooks on animal behavior tend to have a closing chapter on what the rest of the book (which has focused on nonhumans) means for understanding human behavior. Rather than adopting this segmented view of behavior, studies of human behavior (though in the minority in this book) will be woven into the fabric of each chapter. Only then can the reader gain a true integrative view of animal (nonhuman and human) behavior.

In examining the many kinds of animal and human behavior we shall touch on, ethologists need to be as clear as possible about whether a particular study of behavior is examining ultimate questions, proximate questions, or both. Generally speaking, ultimate questions focus on how evolutionary forces like the force of natural selection have shaped a trait, while proximate causes deal with more immediate causative factors. For example, in answering the question "How is it that cheetahs run so fast?," we can frame our answer in terms of the selective forces that have produced fast cheetahs (an ultimate perspective), or by describing how the muscles, bones, and tendons of a cheetah interact to create a speedy animal (a proximate perspective). It is to the issue of ultimate and proximate approaches to ethology that we now turn in the first two primer chapters.

Interview with Dr. E. O. WILSON

The 25th Anniversary edition of your classic book *Sociobiology*, a landmark book in the field of animal behavior, was recently published. What prompted you to write *Sociobiology*?

In the 1960s, as a young researcher working in the new field of population biology, which covers the genetics and ecology of populations of organisms, I saw the logic of applying the principles of making that discipline the foundation of the study of social behavior in animals. At that time a great deal was known about societies of bees, ants, fish, chimpanzees, and so forth, but the subject largely comprised descriptions of each kind of society in turn, and with few connections. There had been little effort to tie all that information together. I had the idea of analyzing animal societies as special kinds of populations, with their characteristics determined by the heredity of behavior of the individual members, the birth rates of the members, together with their death rates, tendency to emigrate or cluster, and so forth—in other words, all the properties we study and put together in analyzing ordinary, nonsocial populations.

Sociobiology as a discipline grew from this idea and was born, not in my 1975 book with that name (*Sociobiology: The New Synthesis*), but in my 1971 *The Insect Societies*. In this earlier work I synthesized available knowledge of the social insects (ants, termites, the social bees, and the social wasps) on the base of population biology. I defined the term sociobiology that way, and predicted that

if made a full unified discipline it would organize knowledge of all animal societies, from termites to chimpanzees. In *Sociobiology: The New Synthesis* I added the vertebrates to the social insects (and other invertebrates) to substantiate this view, then in the opening and closing chapters, the human species. In the latter chapters, I suggested that sociobiology could (and eventually would) serve as a true scientific foundation for the social sciences. This was a very controversial notion then, but it is mainstream today.

What do you see as *Sociobiology*'s legacy to date?

The legacy of *Sociobiology*, which took hold and generated interest and discussion as *The Insect Societies* never could, is indeed the discipline of sociobiology, with journals and many new lines of research devoted to it. This advance was greatly enhanced by the rapid growth of studies on animal communication, behavioral ecology, and, in population genetics, kin selection. Of ultimately equal and probably even greater importance, it showed how to create a link of cause-and-effect explanation between the natural sciences, including especially the study of animal social behavior, on the one side and the social sciences on the other.

What sort of debt do ethologists owe Charles Robert Darwin?

Ethologists owe an enormous debt to Darwin, by encouraging the deep and now well-established concept that instincts are biological traits that evolved by natural selection. A word on terminology is worth introducing here. Ethology is the systematic (i.e., scientific) study of the behavior of animals (including, by extension, humans) under natural conditions. Sociobiology is the study of the biological basis of all forms of social behavior and social organization in all kinds of organisms, including humans, and organized on a base of ethology and population biology. Evolutionary psychology is a spin-off of both ethology and sociobiology, including both social and nonsocial behavior with special links to traditional studies of psychology.

After Darwin, whose work has had the most profound impact on the scientific study of animal behavior?
In 1989 the Fellows of the International Animal Behavior Society voted *Sociobiology: The New Synthesis* the most influential book on animal behavior of all time. The most important individual discoveries of all time would have to include sign stimuli, ritualization, the multiple modalities of nonhuman communication, the neurological and endocrinological basis of many forms of behavior, and the amazingly diverse and precise manifestations of kin selection.

Why should a talented undergraduate studying biology care about animal behavior?
Animal behavior is of course a fundamental and extraordinarily interesting subject in its own right. But it is also basic to other disciplines of biology, all the way from neuroscience and behavioral genetics to ecology and conservation biology.

Why should social scientists pay attention to what is happening in the field of animal behavior? What can they gain by doing so?
The social sciences desperately need biology as their foundational discipline, in the same way and to the same degree as chemistry needed physics and biology needed chemistry. Without biology, and in particular genetics, the neurosciences, and sociobiology, the social sciences can never penetrate the deep wells of human behavior; they can never acquire the same solidity and explanatory power as biology and the other natural sciences.

You and Bert Hölldobler won a Pulitzer prize for *The Ants*. Why have you devoted so much time and effort to studying this taxa?
There are two kinds of biologists, those who select a scientific problem and then search for the ideal organism to solve it (such as bacteria for the problems of molecular genetics), and those who select a group of organisms for personal aesthetic reasons and then search for those scientific problems which their organisms are ideally suited to solve. Bert Hölldobler and I independently acquired a lifelong interest in ants as children, and added science to that fascination later.

SOCIOBIOLOGY IS THE STUDY OF THE BIOLOGICAL BASIS OF ALL FORMS OF SOCIAL BEHAVIOR AND SOCIAL ORGANIZATION IN ALL KINDS OF ORGANISMS, INCLUDING HUMANS, AND ORGANIZED ON A BASE OF ETHOLOGY AND POPULATION BIOLOGY.

You have recently written much on the subject of conservation biology. How does work in animal behavior affect conservation biology studies, and vice versa?
The understanding of animal behavior is crucial to conservation biology and its applications. Consider how important to ecosystems and species survival are the behaviors of mating, territorial defense, dispersal, pollination, resource searching, and predation. To be successfully grasped, these phenomena have to be studied in an organized, scientific manner, not just added haphazardly to conservation strategies.

What do you believe will be the most important advance in animal behavior in the next twenty-five years?
My prediction: the complete linkage of a number of complex behavior patterns from genes to proteonones to sensors and neuron circuits to whole patterns of behavior. Biologists will learn how to scan the whole range of levels of organization to account for each animal behavior in turn.

Will animal behavior be a discipline fifty years from now, or will it be subsumed by other disciplines?
Today the study of animal behavior is the broad gateway to a wide array of different modes of study. But in fifty years—who knows? It may well be subsumed by other disciplines, some as yet undefined.

DR. E. O. WILSON is an emeritus professor at Harvard University and a member of the National Academy of Sciences. He is the recipient of two Pulitzer prizes, and his book Sociobiology *(Harvard University Press) is regarded as one of the most important books on evolution and behavior ever written.*

SUMMARY

1. The scientific study of animal behavior, which dates back hundreds, if not thousands of years, is called ethology.
2. The force of natural selection, the ability of animals to learn, and cultural transmission serve as the foundations upon which we shall build our approach to ethology.
3. Work in ethology, like in all scientific fields can be conceptual, theoretical, or empirical. Empirical work can be further subdivided into observational and experimental studies.

DISCUSSION QUESTIONS

1. Take a few hours one weekend day and focus on writing down all the behavioral observations you've made recently, as well as any, even indirect, behavioral hypotheses you have constructed. Think about your interaction with both people and with nonhumans. How has your very brief introduction into ethology reshaped the way you observe behavior?
2. Why do we need a science of ethology? What insights does this discipline provide both the scientist and the layperson?
3. Imagine you are out in a forest, and you observe that squirrels there appear to cache their food only in the vicinity of certain species of plants. Construct a hypothesis of how this behavior may have been the result of (a) natural selection, (b) individual learning, and (c) social learning.
4. Why do you suppose that mathematical theories play such a large part in ethology? Couldn't hypotheses be derived in their absence? Why does mathematics force an investigator to be very explicit about his or her ethological hypotheses?

SUGGESTED READING

Alexander, R. D. (1974). The evolution of social behavior. *Annual Review of Ecology and Systematics,5,* 325–383. This paper, published just before E. O. Wilson's *Sociobiology,* provides the reader with a good overview of how one approaches behavior using "natural selection thinking."

Dewsbury, D. (Ed.) (1985). *Studying animal behavior: Autobiographies of the founders.* Chicago: University of Chicago Press. A fascinating introduction to the lives of early ethologists.

Galef, B. G. (1996). Social enhancement of food preferences in Norway rats: A brief review. In C. M. Heyes & B. G. Galef (Eds.), *Social learning in animals: The roots of culture* (pp. 49–64). London, Academic Press. A short, but comprehensive account of one of the best long-term studies of social learning.

Heinrich, B. (1999). *Mind of the raven.* New York: HarperCollins. Heinrich is a top-notch writer, who gives the reader a sense of both the science of animal behavior and the beauty of nature.

Tinbergen, N. (1963). On aims and methods of ethology. *Zeitschrift fur Tierpsychologie 20,* 410–440. A classic paper that outlines Niko Tinbergen's approach to animal behavior.

2

Artificial Selection

The Process of Natural Selection

- Selective Advantage of a Trait
- How Natural Selection Operates

Behavioral Genetics

- Mendel's Rules
- Locating Genes for Polygenic Traits
- Dissecting Behavioral Variation

Animal Behavior and Natural Selection

- Sociobiology and Selfish Genes
- Natural Selection and Antipredator Behavior in Guppies

Adaptation

Genetic Techniques as Tools

- Kinship and Naked Mole Rat Behavior
- Coalition Formation

Interview with Dr. Richard Alexander

Natural Selection

As an undergraduate with a burgeoning interest in animal behavior, I naturally asked my professors what books I should read to get a better feel for the subject. Every single one mentioned Darwin's *On the Origin of Species,* and so I bought a copy—but not without a bit of trepidation. Even as a novice biologist, I was aware that *On the Origin of Species* was considered one of the greatest, if not *the* greatest science book ever written, and so I was worried that it would simply be above me. This book was not only readable, however, it was one of the best reads I have ever had. But what surprised me most about Darwin's book was not the ease with which it read, but rather the subject of the first chapter. The opening chapter of the most significant book ever written in biology talks at length about pigeon breeding in Victorian England (Figure 2.1).

The reason for this seemingly odd subject matter for the opening chapter of *On the Origin of Species* was that Darwin was bracing the reader for what was to come. Darwin knew that his readers would feel at home with a discussion of pigeon breeding, a popular pastime in Victorian days. If he could convince them that the process leading to the extraordinary *variants* produced by breeding pigeons was similar to the process leading to *variation* in nature, his task would be a little simpler. The process leading to pigeon variation is artificial selection, while the process leading to the wide variety of traits we see in nature is natural selection.

This chapter will serve as an introduction or "primer" to the manner in which ethologists think about natural selection, genes, and animal behavior. Few concepts have had a more profound impact on the way that ethologists operate than natural selection. Once Darwin's ideas were widely disseminated and integrated into the heart of biology (Huxley, 1942), animal behaviorists possessed

FIGURE 2.1. Natural and artificial selection. Both natural and artificial selection have produced many variants of the pigeon. *(Photo credits: Ann and Rob Simpson)*

A

B

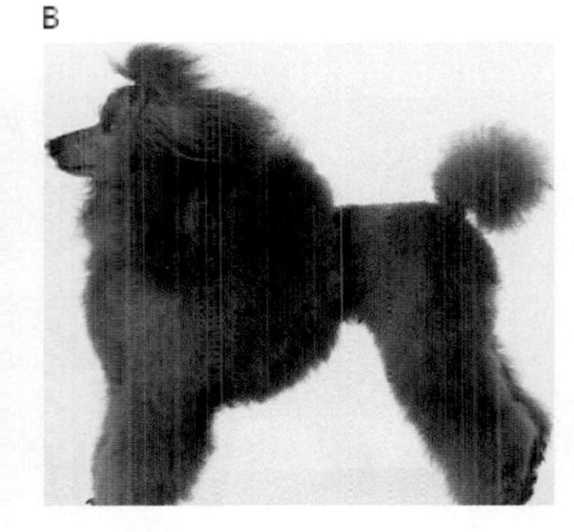

FIGURE 2.2. The power of artificial selection. **(A)** Artificial selection has produced dogs of all shapes and sizes. **(B)** Many generations of dog breeding have produced "miniature" varieties of poodles, such as this miniature gray poodle. *(Photo credits: Getty Images; Yann Arthus-Bertrand/Corbis)*

a theory that helped explain not only *what* animals do, but *why* they do it. In fact, questions surrounding natural selection, called **ultimate questions,** are often also labeled "why" questions. The term "ultimate" does not infer a greater importance attached to such questions than to any other questions in animal behavior, but rather a focus on evolutionary forces per se. Throughout this chapter, we shall be outlining, step-by-step, how one tackles "why" questions. We will begin this process by discussing how selection operates when humans, rather than "nature," are the selective force, and then we will move directly to the case of natural selection.

Artificial Selection

Artificial selection, like the pigeon case Darwin described, is simply the process of humans choosing certain varieties of an organism over others—for example, large versus small, hairy versus hairless—by implementing breeding programs that cause one or more varieties to increase in frequency. For the better part of 10,000 years, humans have been shaping animals and plants by this process of artificial selection. Ever since we selected some varieties of wheat, corn, and rice over others, and systematically planted their seeds, we have been involved in artificial selection. The question "Why do we see particular forms of grain today?" is answered by referring to the process of artificial selection. The same can be said of our systematic preference to breed certain varieties of dogs over others, as well as hundreds of other examples.

Consider the case of dogs (Figure 2.2). Suppose we begin our artificial selection process with a standard, large breed of poodle,

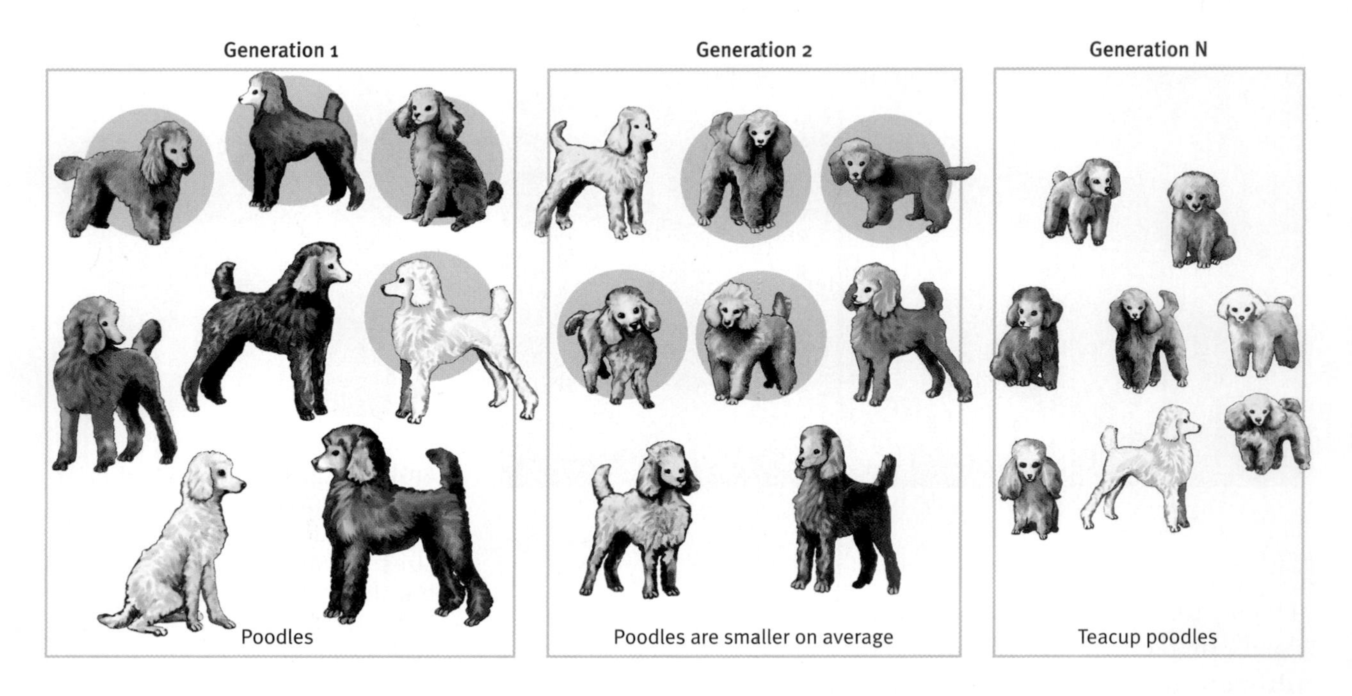

FIGURE 2.3. Artificial selection and poodles. Each generation, our dog breeder allows only the smallest half of his poodles to breed (those in circles). If this continues over many generations, artificial selection may produce very small (teacup) poodles.

and we wish to create a variety of poodle that is much smaller. How would we go about this through artificial selection? First, we'd begin our breeding program by choosing the individuals that were already closest to our ideal tiny poodle. In our case, this would amount to choosing the smallest males and females from our initial population of poodles. Then, assuming that size was under some sort of genetic control, we would preferentially breed small dogs. That is, in every generation we would sort the dogs we had, choosing those that met our breeding criteria (small individuals) and allowing them the chance to mate, while at the same time we would deny the breeding option to those dogs that failed to meet our criteria. With each new generation of dogs, we would repeat this process. Along the way, we would be producing individuals that were coming closer and closer to our ideal dog variety. Eventually, we would recognize that we were as close as we were ever going to get to our idealized tiny poodle, something akin to the extraordinarily small "teacup" variety of poodle (Figure 2.3). As with the case of grains, we can answer our "why" question—why do we see the breeds of dog we see today—by referring to a selection process—in this case, artificial selection.

With a basic understanding of the nuts and bolts of artificial selection in hand, we, like Darwin's original readers, are ready to move on to the main show—natural selection.

The Process of Natural Selection

We shall begin our analysis of natural selection at the broadest level possible, and only then examine this process and its role in understanding ethology in detail.

Given the monumental impact that Darwin's two greatest works, *On the Origin of Species* (1859) and *The Descent of Man and Selection in Relation to Sex* (1871), have had, it is often surprising to many that Darwin's ideas with respect to natural selection are very straightforward, particularly once you've understood artificial selection.

SELECTIVE ADVANTAGE OF A TRAIT

Consider any trait—for example, height, weight, visual acuity, coat color, speed, seed size, and so forth—and instead of imagining humans as the selective agent, allow the selective agent to be nature itself. If one variety of a trait helps individuals survive and reproduce better in their environment than another variety of the same trait, and if the trait in question can be passed across generations, then natural selection is occurring. For example, imagine there are *differences* in tooth sharpness among individuals in a population of predators and that sharper canine teeth translate into killing more prey (Figure 2.4). When individuals with the trait in question

FIGURE 2.4. Small differences matter. In carnivores like wild dogs, the sharpness of an individual's teeth can be critical to survival. If this trait is under genetic control, we expect that even small differences in sharpness will be selected. Over time, wild dogs would start to resemble the dog on the left, who has slightly sharper canine teeth.

(tooth sharpness) produce *offspring that resemble themselves,* then any variant (for example, sharper canine teeth) that somehow enables an individual to *outreproduce* others will spread through the population over time through natural selection. This holds true even if having sharper teeth produces only a very slight edge in terms of the number of offspring one raises.

Even a selective advantage of 1 percent per generation is more than sufficient for sharper teeth to evolve in our population. For example, for the sake of simplicity, let's assume that tooth sharpness is controlled by a single gene. In reality, of course, there are probably dozens of genes that work together to control this trait. The logic we are invoking would work equally well for traits controlled by dozens of genes; it's just that the math would be more difficult, and so for our purposes, let's imagine a single gene at work. If an *allele* (a gene variant) that codes for slightly sharper teeth provides its possessors with just an average of 1 percent more offspring per generation, then all else being equal, this allele will eventually increase in frequency to the point where virtually everyone in the population has it. That is, we will end up with a population of individuals with slightly sharper teeth than their ancestors as a result of natural selection. This is because the selective advantage conferred by the sharper teeth makes those with this allele more likely to survive and reproduce offspring. These offspring, who in turn have the allele coding for sharper teeth, are more likely to survive and produce more offspring with that allele, and so on down through the generations.

The above scenario assumes that once a new, superior genetic variant—in this case, for slightly sharper canine teeth—arises, no new alleles enter our population. We simply have two alleles for our tooth sharpness trait (one of which codes for slightly sharper teeth), and over time, natural selection replaces one with the other. That is, our above example assumes that a genetic variant might get culled out along the way but that no new variants are added in the process once it begins. We start with two alleles, none are added, and we end up with one allele. To make our example a bit more realistic, however, imagine now that new genetic variants affecting tooth sharpness are always being added to our population, and that, by chance, every hundred generations or so, a new allele comes about that produces slightly sharper canine teeth than the other variants produce. Now each time a new superior allele is added to the pool, it will begin to increase in frequency, as in our original example, and over time, our predator will be getting sharper and sharper teeth. Over evolutionary time, small differences can accumulate into large changes.

HOW NATURAL SELECTION OPERATES

The above example was meant to give you a flavor for how natural selection operates. At this point, however, we need to delve into the details a bit. We will examine what it takes for the process of natural selection to operate and what the end product of this process is (Williams, 1966; Endler, 1986; Ridley, 1996; Brodie et al., 1995; Bell, 1997; Mousseau et al., 1999).

To understand how natural selection operates, the first thing any ethologist must do is to specify which trait is of interest. That is, we don't so much speak of "natural selection" as we do of "natural selection on trait X," where the researcher fills in the X, depending upon his or her particular interest. Once we have picked a trait, the process of natural selection requires three prerequisites in order to operate: variation, fitness consequences, and a mode of inheritance. Technically speaking, a fourth requirement exists, and that is that resources must be limited with respect to the trait one is examining. So, if one is studying natural selection and foraging, food in some sense must be limited, while if one is studying natural selection and mating, there must be a finite set of individuals available to mate with, and so on. In practice, our limited resources requirement is almost always met.

To make the process of natural selection easier to follow, let's examine a specific trait: ovipositor length in insects (Figure 2.5).

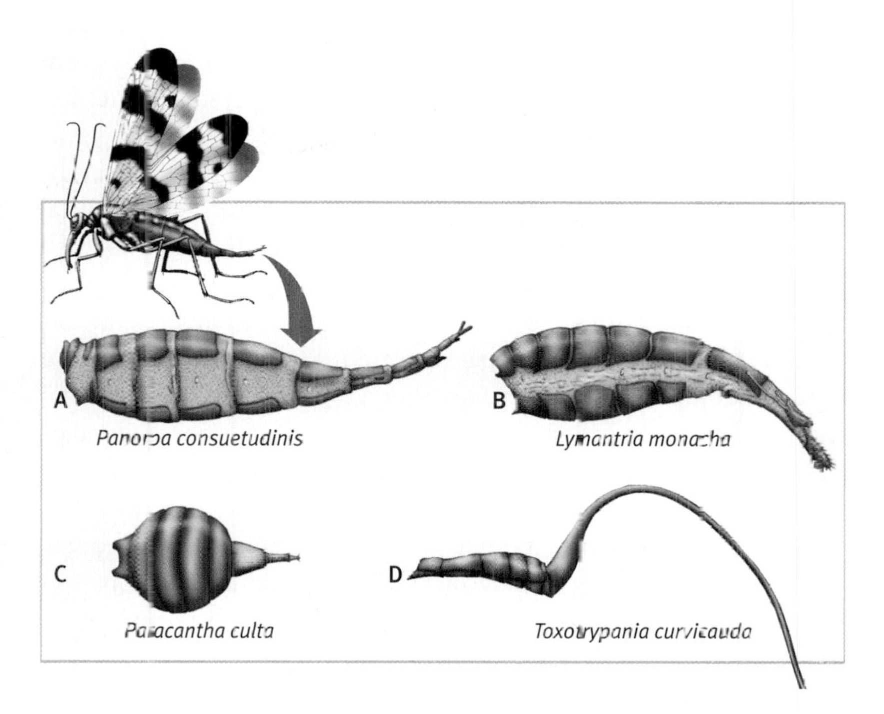

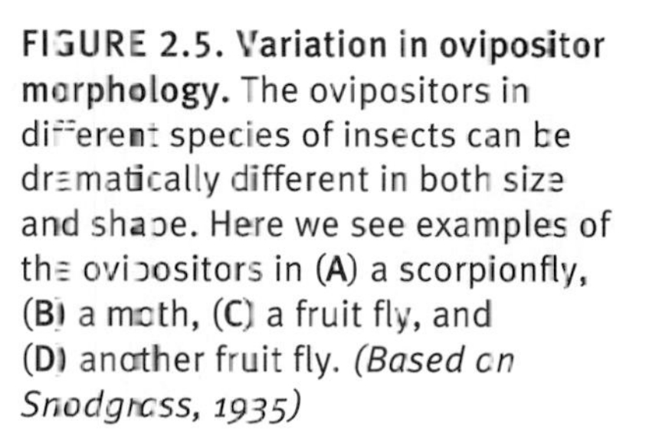

FIGURE 2.5. Variation in ovipositor morphology. The ovipositors in different species of insects can be dramatically different in both size and shape. Here we see examples of the ovipositors in (A) a scorpionfly, (B) a moth, (C) a fruit fly, and (D) another fruit fly. *(Based on Snodgrass, 1935)*

In many insect species, females have a modified organ called an ovipositor, which is a sometimes elaborate extended tube through which eggs are deposited (Snodgrass, 1935). In some species eggs are placed on the surface of an object, while in others the ovipositor probes inside an object—for example, dead wood or fruit—and eggs are deposited directly within the wood or fruit. Ovipositor length is easy to measure and has a direct link to reproductive success. Depending on the species and environment, long, short, or medium ovipositor lengths may be directly favored by natural selection (Figure 2.6).

FIGURE 2.6. Ovipositing. Many female insects such as this slender meadow grasshopper deposit their eggs via an ovipositor. *(Photo credit: Dwight Kuhn)*

We shall now examine how natural selection operates to change ovipositor length in a hypothetical population of insects, walking through the three prerequisites—variation, fitness consequences, and modes of inheritance—as we do.

VARIATION. For natural selection to act, there must be variation in the trait under investigation (Mousseau et al., 1999). Without such variation, there is nothing for natural selection to *select* between. If every female insect in our hypothetical population had an ovipositor length of 10 millimeters, it would be ridiculous to study natural selection and ovipositor length in that population because there would be nothing for natural selection to select between with respect to ovipositor length.

Variation in a particular trait can be either environmental or genetic. We shall return to these two different forms of variation a bit later, but here we shall focus on **genetic variation,** which is variation that is caused by genetic differences in the trait in question. Genetic variation is generated in a number of ways. For example, **mutation,** which is any change in genetic structure, automatically creates new variation in a population.

Mutation can take many different forms and can be caused by excessive exposure to X-rays, cosmic rays, radioactive decay, ultraviolet light, or various mutagenic chemicals. Addition and deletion mutations occur when a single nucleotide is either added or deleted from a stretch of DNA. This type of mutation typically causes the production of an inactive enzyme. Base mutations occur when one nucleotide replaces another. About half of the base mutations that occur will strongly affect protein function. This number would be higher still if it were not the case that some base mutations cause no change in which amino acid is produced (so-called silent mutation). Another form of mutation is labeled transposition, where a long section of DNA is moved from one part of the genome to the other. Rates of mutation, though fairly low per gene, are sufficiently high to create significant amounts of new variation over evolutionary time.

In addition to mutation, another factor that affects variation is **genetic recombination.** In sexually reproducing organisms, when pairs of chromosomes line up during cell division, sections of one chromosome may "cross over" and swap positions with sections of its pairmate. This swapping potentially creates new genetic variation. Crossing-over points are essentially random (Gould and Keeton, 1996), and so virtually every possible crossover between a pair of chromosomes is possible in principle. As such, crossing over creates huge amounts of new genetic variation.

New genetic variants of a trait can also enter a population via nongenetic pathways the most common of these being **migration.** That is, when we study how natural selection is operating on a trait, we are studying selection in a given population. One way for the variation in this population to increase is for individuals to migrate in from other populations. For example, suppose that in our population individuals had ovipositor length anywhere from x to y millimeters, but in a neighboring population some individuals had ovipositor length that measured $y + 1$ millimeter. If such a female with an ovipositor $y - 1$ millimeter migrated to the population we are studying, variation would increase.

Generally speaking, the prerequisite for genetic variation is almost always met (Darwin, 1868; Endler, 1986, 1999; Clark and Ehlinger, 1987; Mousseau et al., 1999), as it is rare to find a trait that doesn't display at least some variation. Pick up any journal in ethology, evolution, or behavioral ecology, and virtually every study speaks of variation in the trait being studied. In fact, the lack of variation is so uncommon that demonstrating it is often thought to be a significant finding in and of itself (O'Brien et al., 1983; Figure 2.7).

FIGURE 2.7. A cheetah on the Serengeti Plain of Africa. Cheetahs have markedly low levels of genetic variance. *(Photo credit: Tim Caro)*

FITNESS CONSEQUENCES. For natural selection to operate, it must also be the case that the genetic variation we see in a trait has **fitness** consequences. Broadly speaking, when we talk of fitness we are referring to the expected contribution of an individual to the next generation, relative to other individuals (Fisher, 1958; Williams, 1966; Clutton-Brock, 1988; Grafen, 1988; Reeve and Sherman, 1993; Dejong, 1994). In later chapters, we shall broaden this definition, but for now, "contribution" will refer to the mean number of reproductively viable offspring that an individual produces.

It is important to note that fitness is a relative term. We know nothing of fitness when we say that an individual produced 10 viable eggs in its lifetime. Its fitness will depend on whether others are producing 5 viable eggs or 100 viable eggs; compared to the former, our producer of 10 eggs has high fitness; compared to the latter, it has a lower fitness (Figure 2.8).

Returning to ovipositor length in female insects, there must not only be differences in ovipositor length, but these differences must translate into fitness differences. That is, the length of the ovipositor must affect its lifetime reproduction, even if only in a very slight way. Without this translation from variation to fitness differential, natural selection cannot act on variation in a trait because such variation would have no evolutionary significance. For example, suppose we have 100 female fruit flies, and to make things simple, let's imagine that 50 female flies have ovipositors that are 10 mm long and 50 flies have ovipositors that measure

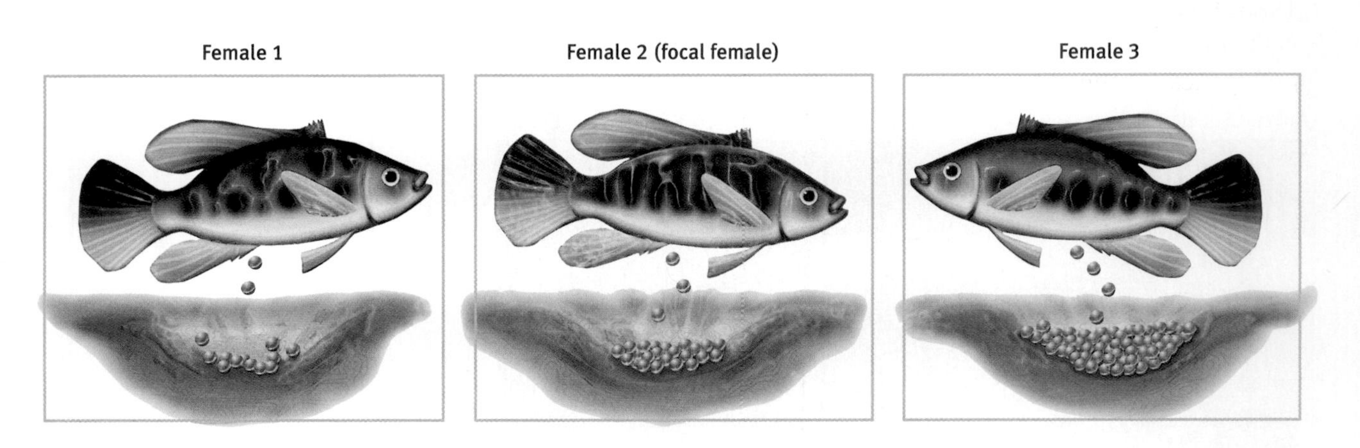

FIGURE 2.8. Fitness is a relative concept. Here are three female fish laying eggs. Assuming every egg has an equal shot at being fertilized, the absolute number of eggs our focal female lays is not a useful fitness parameter per se. The focal female's fitness is high compared to female 1, but low compared to female 3, as fitness is a relative measure.

15 mm. Suppose individuals with 10 and 15 mm long ovipositors all produced an average of 8 offspring that were identical except for ovipositor differences. Since the two kinds of fruit flies produced the same number of offspring, ovipositor length did not affect reproductive success, and hence there were no fitness consequences of varying ovipositor length. In this case, natural selection couldn't operate on ovipositor length because there was nothing of evolutionary significance to *select* between.

Conventional wisdom in ethology is that fitness differences are almost always present if the investigator searches hard enough (see Gould and Lewontin, 1979, for a dissenting view, and Mayr, 1982, 1983, for a defense of this position). Recent work suggests that this is in fact often, but certainly not always, the case (Endler, 1999; Mousseau et al., 1999).

MODES OF INHERITANCE. The last prerequisite that needs to be in place before natural selection can operate is a mode of inheritance. Without a mode of inheritance, fitness differences are washed away, and natural selection cannot move our hypothetical population toward a longer or shorter ovipositor. To see this, imagine that fruit flies with ovipositors that are 10 mm long have 12 offspring on average and fruit flies with 15 mm long ovipositors have on average 10 offspring. If female offspring don't resemble their mothers with respect to ovipositor length, our 10 mm mothers are no more likely to produce daughters with 10 mm long ovipositors than are our 15 mm mothers, and vice versa. Any fitness difference that exists between females with different ovipositor lengths is then lost (Figure 2.9).

Genes are passed down from generation to generation, and many traits are known to be under some type of genetic control. As such, genes are the most obvious candidate for a mode of transmission. We shall focus on genetic transmission here, holding off our discussion of other modes of transmission until our sortie into cultural transmission in Chapter 5. One way to measure genes as a mode of transmission is through **heritability** analyses, which measure the proportion of variance in a trait attributable to genetic variance (Hartl and Clark, 1989).

In the literature on behavioral and population genetics, there are two different types of heritability—broad sense and narrow sense. We shall primarily be dealing with narrow-sense heritabilities, but let's examine each of these concepts in turn. **Broad-sense heritability** measures the total proportion of variance in a trait—that is, genetic variance as opposed to environmental variance. Suppose, for example, that we raise a group of mice in

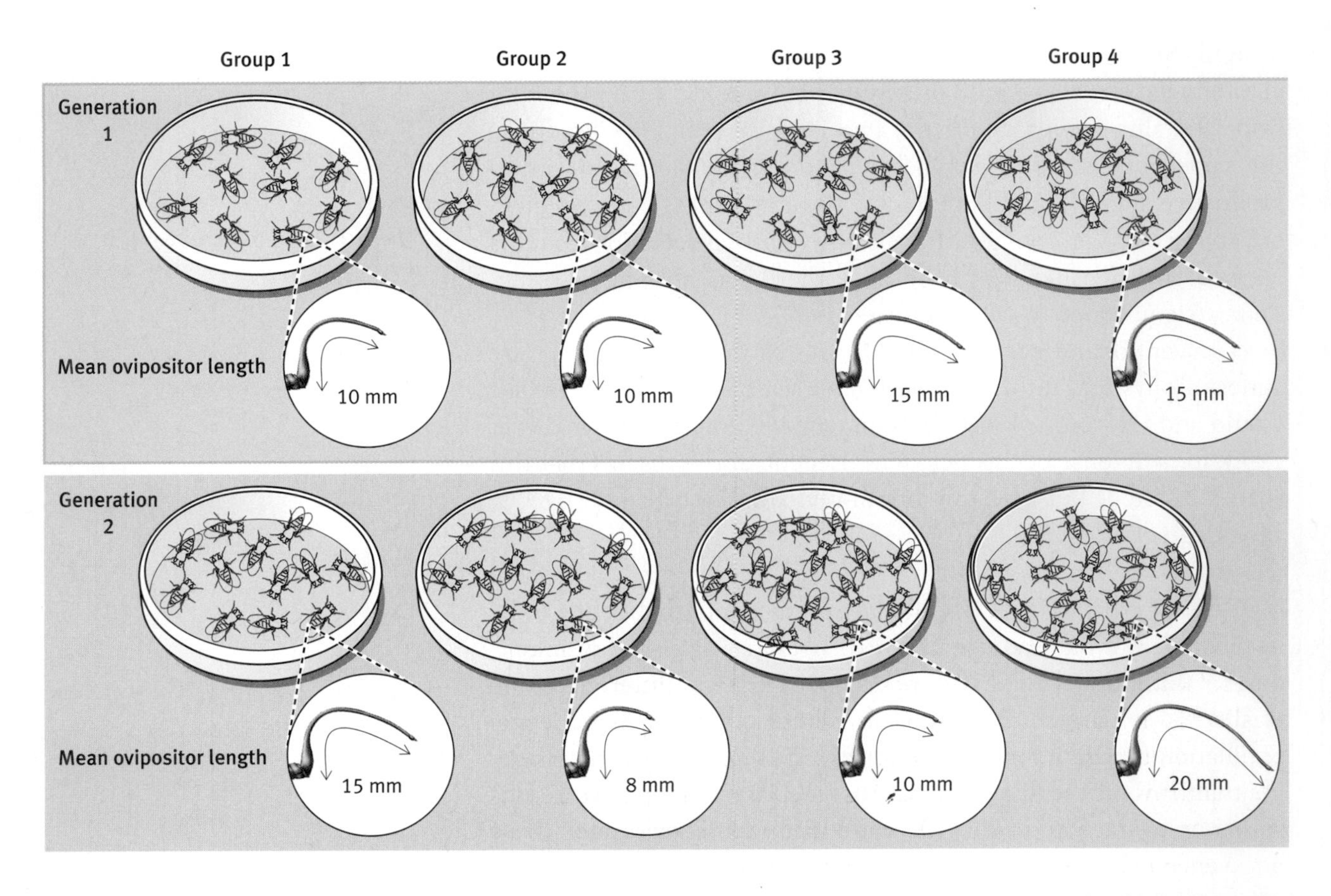

FIGURE 2.9. Parent-offspring similarity. One prerequisite for natural selection to operate is a mode of inheritance. In the above example, ovipositor lengths in parents and offspring are uncorrelated, suggesting that ovipositor length is not a trait transmitted across generations.

identical environments and then measure their foraging behavior. If our environments were truly identical, then differences across our mice in their foraging behavior would most likely be due to genetic differences between individuals—that is, due to genetic variance.

After measuring genetic variance in foraging behavior, let's say I raise another lot of mice, but now I raise individual mice in dramatically different environments. Differences in mouse foraging behavior would now be due to both genetic variation (which we identified above) and environmental variation. Having obtained a measure of genetic variation from my first experiment, I could then estimate environmental variance (total variance minus genetic variance).

Broad-sense heritability is scored by taking the genetic variance and dividing by the sum of genetic and environmental variance (this sum is referred to as phenotypic variance). This measure then is akin to what we read about when we hear in the media that

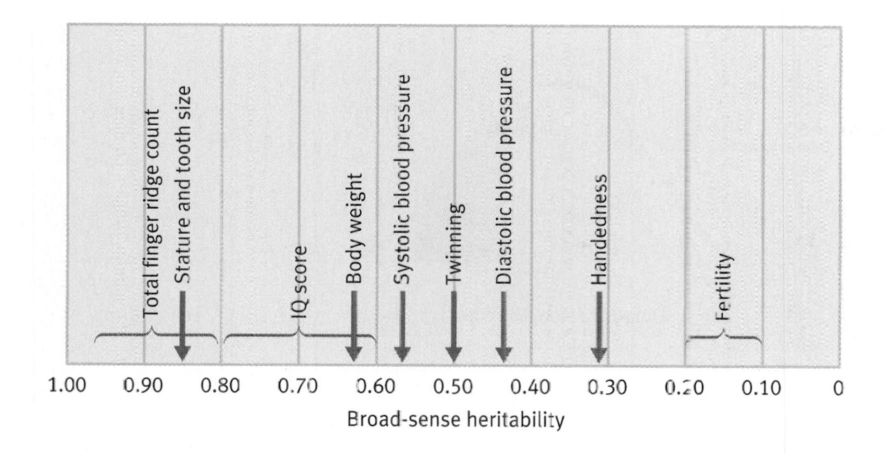

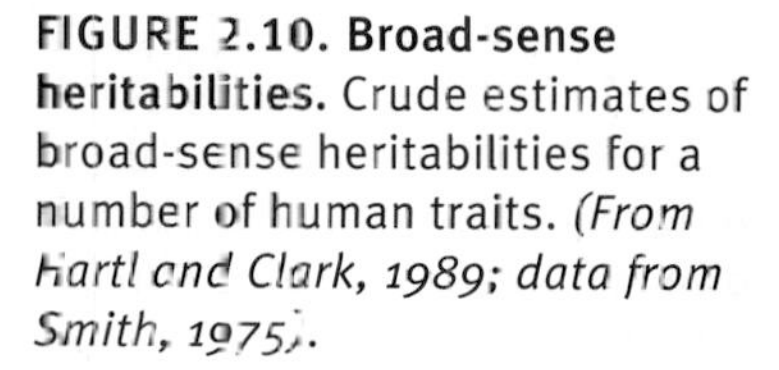
FIGURE 2.10. Broad-sense heritabilities. Crude estimates of broad-sense heritabilities for a number of human traits. *(From Hartl and Clark, 1989; data from Smith, 1975).*

a trait "has a strong genetic component." Figure 2.10 illustrates a few measures of broad-sense heritability in humans.

Narrow-sense heritability can be envisioned in numerous ways. Unfortunately, defining narrow-sense heritability typically requires one to define the concept of additive genetic variance, which is admittedly difficult to get across without going into genetic models that are beyond the scope of this book. In essence, additive genetic variance is that portion of genetic variance that is accessible to natural selection. Measuring narrow-sense heritability is much more straightforward than defining it. One means for measuring narrow-sense heritability is by designing a "truncation selection" experiment.

Imagine that we start with generation 1 adult females, and in step 1 we measure ovipositor length in every individual in our population. This gives us a mean ovipositor length for generation 1. Let's say that the mean length equals 12 mm, and let's label this mean value x_0. Step 2 is to *truncate,* or cut off, the population variation in ovipositor length by allowing only those individuals with ovipositors greater than 14 mm to breed. We then calculate the mean ovipositor length of those females that we allow to breed. Let us label that mean x_1, and imagine it equals 17 mm in our population. The difference between x_1 and x_0 is referred to as the selection differential, or S. In our case, then S = 5 mm.

In step 3, we measure the ovipositor length of all the adult females in generation 2, recalling that the generation 2 females are all descended from generation 1 mothers that were above our truncation value. Let's label the mean ovipositor length of adult females in generation 2 as x_2, and imagine this value to be 12.5. The difference between this mean (x_2) and the mean in the first generation (x_1) is referred to as the response to selection, or R. For our hypothetical population of insects, R = 0.5 mm.

Narrow-sense heritability is simply R/S, or how much our pop-

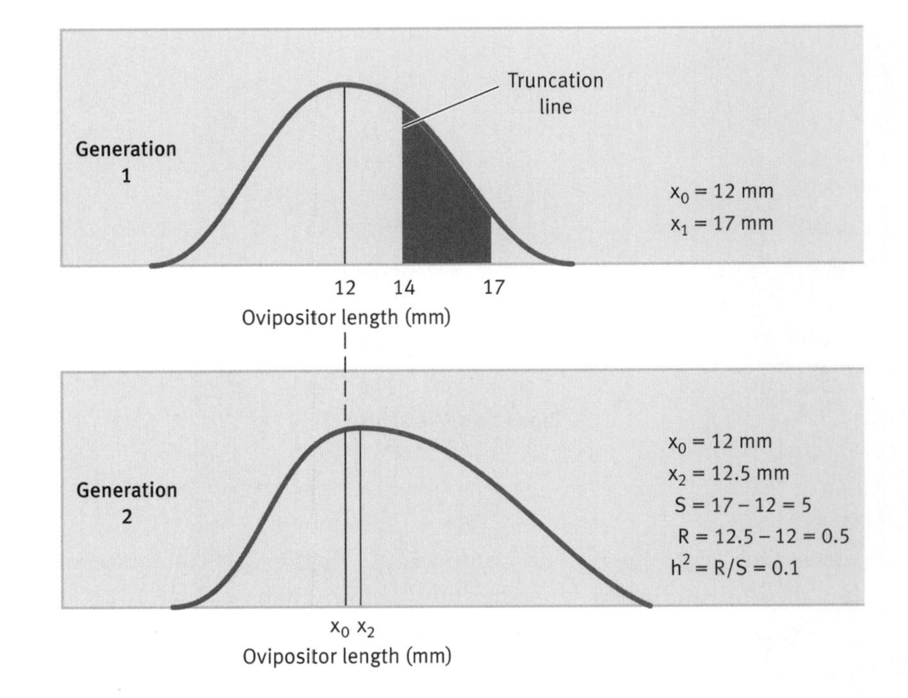

FIGURE 2.11. Narrow-sense heritability (h^2). One method for calculating h^2 is through truncation selection. In generation 1, only females with ovipositors greater than 14 mm (orange area) were allowed to breed. The selection differential (S) is equal to the difference between the mean ovipositor length of those that breed and the entire population. The response to selection (R) is equal to the difference between the mean of generation 1 and generation 2. Narrow-sense heritability (h^2) equals R/S.

ulation has moved in the direction we selected. For our hypothetical case, the narrow-sense heritability is 0.1 (Figure 2.11).

Work in both evolutionary biology, population genetics, and animal behavior all suggest that many traits show low to moderate degrees of heritability (Mousseau and Roff, 1987; Price and Schulter, 1991; Hoffmann, 1999; Weigensberg and Roff, 1996).

Behavioral Genetics

Heritability analyses are often undertaken by behavioral geneticists. As such, before we move on to specifically examining the details of how natural selection works with respect to animal behavior, we need to take a slight detour and devote a little time to understanding what behavioral geneticists do, and how their work interfaces with that done in ethology (Ehrman and Parsons, 1981; Plomin et al., 2000). Behavioral geneticists do many things, but for our purposes three of the most salient include:

- Using Mendel's laws of genetics to predict the distribution of behavioral phenotypes.
- Examining the effect of sex-limited trait expression (traits that are expressed only in males or only in females).
- Determining what percentage of the variation in a behavioral trait is genetic and what percentage is environmental.

MENDEL'S RULES

Behavioral geneticists often use Mendel's rules of genetics to predict the distribution of behavioral phenotypes—the manifestation of genotypes in a given environment—based on the knowledge of whether an allele is dominant, requiring just a single copy of an allele to express itself, or recessive, requiring two copies of an allele for expression. For example, Mendel's first law, *the principle of segregation,* essentially states that individuals have two copies of every gene, that such genes remain distinct entities, and that they segregate fairly during formation of eggs or sperm. As a simple example, consider the case of Huntington's disease in humans. Huntington's disease is controlled by a single gene with two alleles, H and h, where H is dominant. HH and Hh genotypes have Huntington's disease, while hh genotypes do not. With just this single bit of information, if behavioral geneticists know the genotype of a given set of parents, they can predict what proportion of their children will have Huntington's disease.

Now let's consider a trait encoded by a recessive allele. Here, in order for the trait to be expressed, an individual must have two copies of the allele in question (unless the gene lies on a sex chromosome). For example, in mute swans (*Cygnus olor*), about 75 percent of young individuals molt and take on what is referred to as "subadult" gray feather coloration (known as the subadult plumage or SAP phenotype), while about 25 percent molt directly into the white plumage associated with adults (adult plumage or AP phenotype); (Figure 2.12). A number of hypotheses have been put forth to try to delineate how natural selection could favor both these plumage types (Conover et al., 2000), but for our purposes, the critical piece of the story is that behavioral genetic analysis demonstrates that in mute swans the AP phenotype results from a recessive allele on the X chromosome (Munro et al., 1968; Kear, 1972). With such information in hand, if we know the genotype of the parents, we can again predict the frequency of SAP and AP phenotypes in the next generation.

FIGURE 2.12. Mute swan young in a nest. Plumage coloration (gray or white) in this species depends on a sex-linked recessive allele. *(Photo credit: Jan Rees)*

LOCATING GENES FOR POLYGENIC TRAITS

With respect to traits as complex as behavioral ones, it isn't surprising that oftentimes more than one gene is responsible for the expression of a behavior. Using a number of experimental and quantitative techniques, behavioral geneticists are often able to delineate a range for the number of genes underlying some trait of interest to ethologists, and they can even pinpoint the location of such genes.

When many genes are involved in the expression of a behavioral trait—that is, when the trait is **polygenic**—behavioral geneticists often search for a set of genes that each contribute a small amount to the expression of the trait of interest. Such a search is said to be looking for quantitative trait loci (QTL). Sometimes behavioral geneticists actually locate QTL, in the sense of delineating the exact region on a chromosome where QTL lie; at other times, resolution is such that behavioral geneticists can place the QTL underlying a trait on a specific chromosome, but can be no more specific than that. To see how this works, let's briefly examine Flint's work on QTL and fear and fearlessness in mice (Flint et al., 1995).

Flint's group used "open field behavior"—a measure of fear expression when animals are placed in new environments—in two lines of mice, one selected for high open field activity and one selected for low open field activity. After a long controlled regimen of breeding, the researchers were able to identify 84 DNA markers for open field behavior. QTL for fear and fearlessness were initially found to be located on mouse chromosomes 1, 12, and 15. More specifically, QTL for open field behavior were located on chromosome 15, while those on chromosomes 1 and 12 were found to be associated with other fear-related traits. Subsequent follow-up work employing even more sophisticated molecular genetic techniques confirmed the QTL on chromosomes 1 and 12, but not on 15 (Talbot et al., 1999).

With the breakneck speed at which molecular genetics is progressing, it is only a matter of time before our understanding of the genetics of complex behavioral traits becomes much more sophisticated and comprehensive.

DISSECTING BEHAVIORAL VARIATION

Behavioral geneticists are particularly interested in dissecting the variation underlying ethological traits. We have already touched on this point, but let's look at it in a bit more detail. Many studies in behavioral genetics focus on what proportion of the variance in a behavioral trait is due to genetic variation that natural selection can act upon—what we labeled earlier as additive genetic variance. As such, these sorts of studies measure the narrow-sense heritability of some behavioral trait.

In addition to the truncation method we outlined earlier, another classic way to measure narrow-sense heritability is through **parent-offspring regression.** The idea here is fundamentally simple. Parents pass genes down to their offspring, and so if one

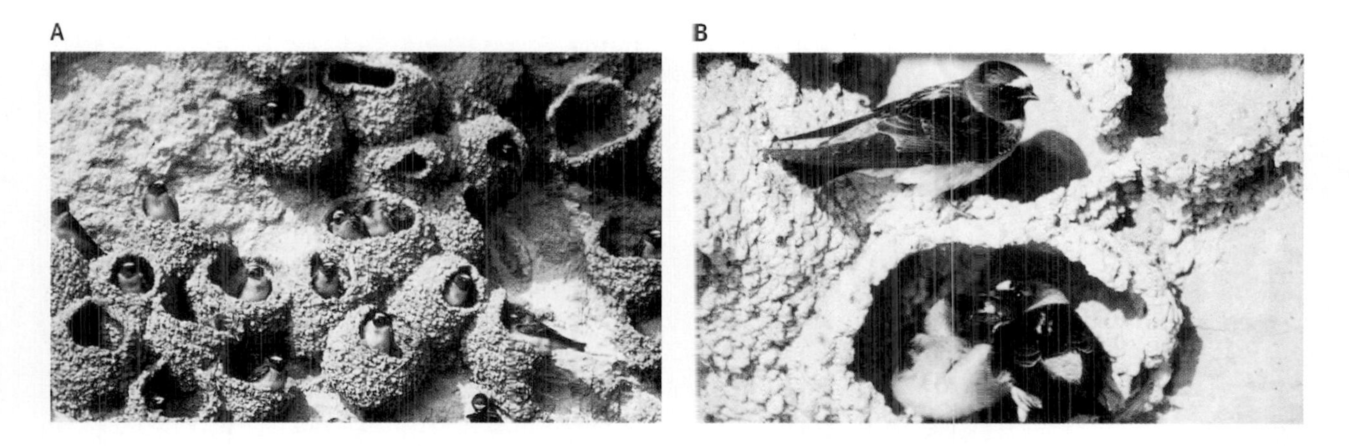

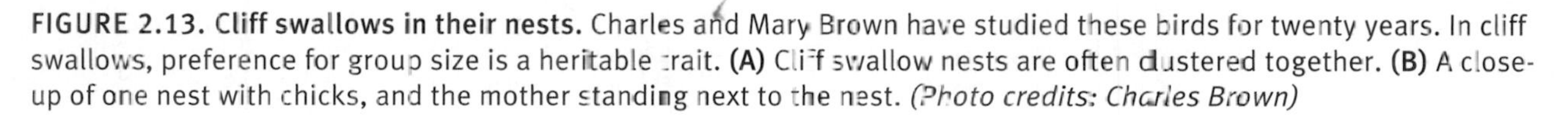

FIGURE 2.13. Cliff swallows in their nests. Charles and Mary Brown have studied these birds for twenty years. In cliff swallows, preference for group size is a heritable trait. **(A)** Cliff swallow nests are often clustered together. **(B)** A close-up of one nest with chicks, and the mother standing next to the nest. *(Photo credits: Charles Brown)*

can account for environmental variation, the genetic variation one sees in offspring should map onto the genetic variation observed in parents when narrow-sense heritability is high. Let's consider a case in which narrow-sense heritability has been examined in a behavioral trait: group size preference in the colonial breeding cliff swallow (*Petrochelidon pyrrhonota*).

Ethologists and behavioral ecologists have had something of a love/hate relationship with animal social groups. On the one hand, group living, and particularly group size, seem so basic to an animal's well-being—in terms of predation, disease transmission, mating opportunities, and so forth—that it seems eminently reasonable to suggest that natural selection should operate in ways to modulate group size, according to ecological, genetic, and social circumstances. On the other hand, putting together all the pieces necessary to demonstrate that animals modulate their group size and that such a modulation is heritable (in terms of narrow-sense heritability) is a tall measure, but one that appears to have finally been met.

Charles and Mary Brown (2000) have been studying cliff swallow birds in the field for more than twenty years (Figure 2.13). Their recent work tackled the question of group size preferences and heritability head-on in two related experiments. In experiment 1, using five clusters of cliff swallow colonies and an incredible 2,581 birds, Brown and Brown found that the group size in which individual swallows lived was statistically similar to the group size in which their parents lived. This was true for birds that bred at the same site as their parents as well as for birds that emigrated, suggesting that the correlation between parent and

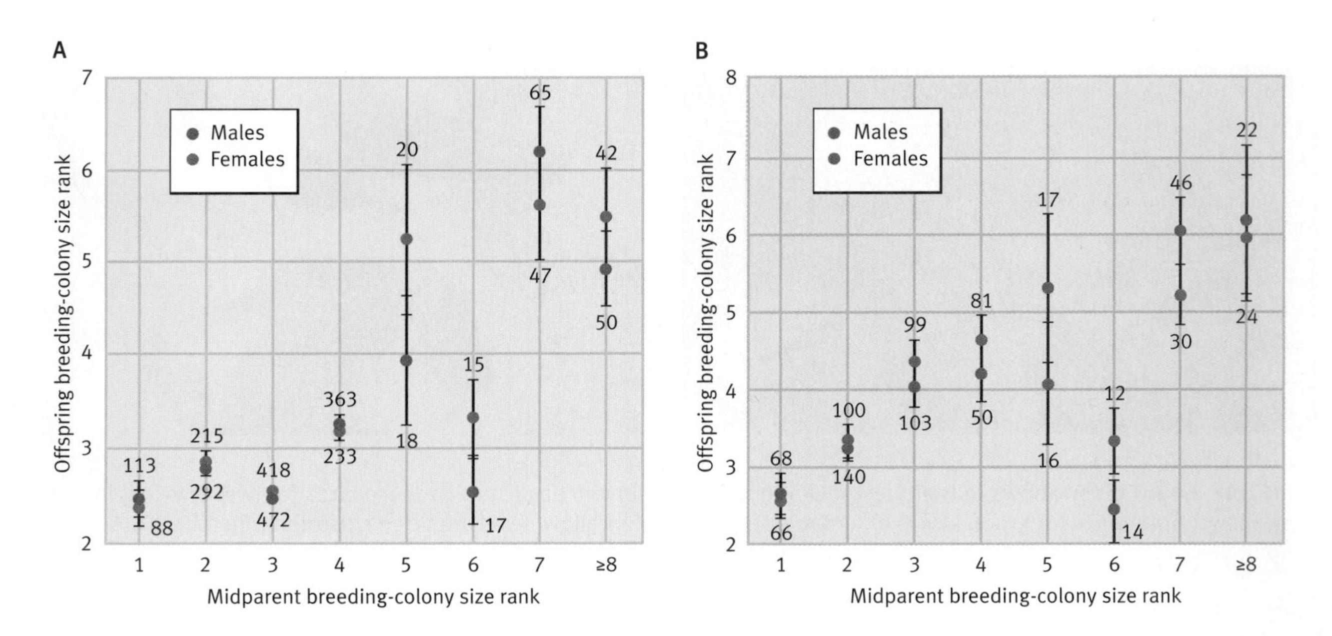

FIGURE 2.14. Breeding colony size in parents and offspring. There is a correlation between breeding colony size in parents and offspring in cliff swallows. Charles and Mary Brown sampled thousands of birds in their native habitat and found a strong correlation between parent and offspring colony size. This held true for all offspring **(A)**, as well as for offspring who bred away from their natal site **(B)**. *(From Brown and Brown, 2000)*

offspring was not a function of living in the same place per se (Figure 2.14).

In their second experiment, Brown and Brown used nestlings from two large and five small colonies. Half the young from the many nests in the large colonies were removed, and the young from small colonies were placed in their stead. Likewise, half the young in small colonies were replaced by young from large colonies. As such, in all their manipulated nests, Brown and Brown had offspring from both large and small colonies. This is referred to as a cross-fostering experiment, and of the almost 2,000 birds in this experiment, 721 were recaptured.

When examining the preference for group size in cross-fostered birds, a positive correlation with the group size preferences in biological parents was found; a negative correlation was uncovered when the group size preference of nestlings and foster parents were compared. In other words, even after the cross-fostering, the young displayed the same group size preference as their true parents (not their foster parents), suggesting a significant heritability for this complex behavioral trait.

One thing to keep in mind when we are discussing behavioral genetics and heritability is that, in addition to parsing variance into that which can be acted on by selection and that which can't, environmental variance in a behavioral trait can also be subdivided.

This subdivision involves splitting environmental variance into two categories: (1) that due to "shared environments," and (2) that due to "nonshared environments." Shared environments are made up of environmental effects that are due to family members developing in very similar environments. For example, some of the variance (or lack of variance) in a behavioral trait may be due to the fact that a clutch of offspring were all raised in the same nest, and hence they experienced a very similar environment. The extreme case of this would be twins, who share similar environments, not only after birth, but in utero! Nonshared environmental effects are made up of what is left after genetic variation and shared environment are taken into account.

Even when shared environments exist, they don't necessarily account for significant amounts of behavioral variance. For example, research in chimpanzee personality suggests that the chimp persona can be decomposed into six components that include agreeableness, dominance, and dependability (Figure 2.15). Let's say we were trying to determine how much of the variance in "agreeableness" was due to genetic versus environmental variation. We might hypothesize that if one group of chimps were raised in environment 1 and another in environment 2, environment might play a critical role in decomposing the variance underlying chimp agreeableness. Yet, a detailed behavioral genetic analysis in chimps within and across zoo environments shows that a shared environment—living in the same zoo—had only a negligible effect on differences and similarities in chimp personality (Weiss et al., 2000).

FIGURE 2.15. Dominant chimp. The dominant chimp has priority access to food and mates. *(Photo credit: A.J. Copley/Visuals Unlimited)*

While we shall be returning to some of the above behavioral genetic topics—heritability, for example—more often than others throughout the remainder of the text, a short overview of the field is something all animal behavior students should experience (at a minimum). We now turn back to specifically looking at how natural selection operates on animal behavior.

Animal Behavior and Natural Selection

In addition to operating on morphological, physiological, anatomical, and developmental traits, natural selection also operates on behavioral traits. Any gene(s) coding for a behavior that causes its bearer to increase its relative reproductive success—its fitness—even slightly is favored by natural selection. Instead of imagining individuals that differ in tooth sharpness, imagine that we have two different types of lions—some hunt prey by hiding in the brush, while others stand out in the open while hunting. Now, suppose genetic variation exists for this behavioral trait, and that each of our types produces offspring that use the same hunting strategy as their parents. If ambushers do better than their nonambushing counterparts, ambush-hunting behavior will be favored by natural selection and will, over time, increase in frequency.

Despite the fact that he didn't know about genes per se, the idea that natural selection operated on animal behavior goes back at least as far as Darwin. *On the Origin of Species* is peppered with many examples of how selection might operate on behavior (Darwin, 1859). Darwin didn't need to know about modern-day genetics for his theory to work; all he needed to realize was that somehow behavioral traits that affected reproductive success were passed from parents to offspring. Any Victorian naturalist worth his salt would know that offspring resemble their parents, and Darwin was more than a good naturalist, he was a great naturalist (Darwin, 1845).

The modern theoretical framework for how natural selection applies to genes coding for animal behavior can be traced to George C. Williams's book, *Adaptation and Natural Selection* (Williams, 1966), William D. Hamilton's papers on kinship and social behavior (Hamilton, 1964; see Chapter 8), and Richard Dawkins's conceptually based book, *The Selfish Gene* (Dawkins, 1976; see also Dawkins, 1982, 1987). But perhaps the most important contribution to both the theoretical and empirical framework underlying natural selection and animal behavior was laid out in E. O. Wilson's masterful synthesis, *Sociobiology* (1975). In *Sociobiology,* Wilson demonstrated the power of natural selection to shape behavior from the simplest creatures to humans (Alcock, 2001). While there was an early social science backlash against Wilson's claim that sociobiological thinking can shed light on humans, ethologists and behavioral ecologists grabbed on to Wilson's compelling arguments on the power of natural selection to shape behavior, particularly social behavior (hence the term "sociobiology").

SOCIOBIOLOGY AND SELFISH GENES

The sociobiological notion that natural selection acts on genes that code for animal behavior is often referred to as the selfish gene approach to ethology. The phrase "selfish genes" was first coined by Richard Dawkins in 1976. As Dawkins himself makes crystal clear, genes aren't "selfish" in any emotional or moral sense. In fact, genes aren't anything but a series of tiny bits of DNA put together in a particular sequence and orientation, and somehow distinct from other such tiny tidbits of DNA. Yet, genes can sometimes be treated as though they were selfish in that the process of natural selection favors those genes that increase the expected relative reproductive success of their bearers. From the gene's-eye perspective, that is all that matters.

Any gene that codes for a trait that increases the fitness of its bearer above and beyond that of others in the population will increase in frequency. So natural selection often, but not always, produces genes that *appear* to be selfish. Apply this approach to animal behavior, particularly animal social behavior (Alexander, 1974; J. L. Brown, 1975; E. O. Wilson, 1975; Alexander and Tinkle, 1981; Krebs and Davies, 1979, 1984, 1991, 1997; Alcock, 1998), and you have one of the main ways in which ethologists think about genes and animal behavior. But it is worth emphasizing again that genes have no inherent qualities such as selfishness. Thinking of a gene as "selfish" just provides a convenient means of conceptualizing some problems in animal behavior.

Since the sociobiological approach underlies many, but by no means all, of the examples found throughout this book, it is useful to examine here a case study of how natural selection works on animal behavior in order to understand the application of the approach in future chapters. Before that, however, it is important to note that finding *concrete experimental evidence* that natural selection operates on animal behavior *in the wild* is no easy task (Endler, 1986; Mousseau et al., 1999). This is not to say that there are not many, many cases in which we can *infer* how natural selection has operated on behavior in the wild, but rather that experimental studies that do so directly in the field are harder to come by. Many times, such studies focus on smaller animals, who by way of their shorter generation life spans, are more amenable to multigenerational studies.

NATURAL SELECTION AND ANTIPREDATOR BEHAVIOR IN GUPPIES

The guppy has often been used by ethologists and behavioral ecologists to study aspects of natural selection, with dozens of papers a year published on some aspect of guppy ethology. One reason for this interest in guppies is not only that they breed quickly, facilitating multigenerational studies, but also that their population structure is ideal for studies of natural selection. Most guppy studies use fish from the Northern Mountains of Trinidad and Tobago. In

FIGURE 2.16. Different guppy environments. (A) An upstream, low-predation stream in Trinidad, and **(B)** a downstream, high-predation stream. Guppies in these streams have been subject to different selection pressures. *(Photo credits: Lee Dugatkin)*

many of these streams, guppies can be found both upstream and downstream of a series of waterfalls (Seghers, 1973; Houde, 1997; Figure 2.16). These waterfalls act as a barrier to many of the guppy's predators. Upstream of such waterfalls, guppies are typically under only slight predation pressure from larger species of fish, while downstream populations of guppies are often under severe predation pressure from numerous piscine (that is, fish) predators.

High-predation and low-predation sites in the same streams are often kilometers apart, and as such, we might expect natural selection to operate differently in these populations. If the type of predators differ dramatically based on site, then natural selection should favor different traits in upstream and downstream guppy populations. Indeed, this turns out to be the case, and such between-population comparisons in guppies have found differences with respect to many traits, including color (Figure 2.17), number of offspring in each clutch, size of offspring, and age at reproduction (Seghers, 1973; Endler, 1980, 1995; Reznick, 1996; Reznick et al., 1997).

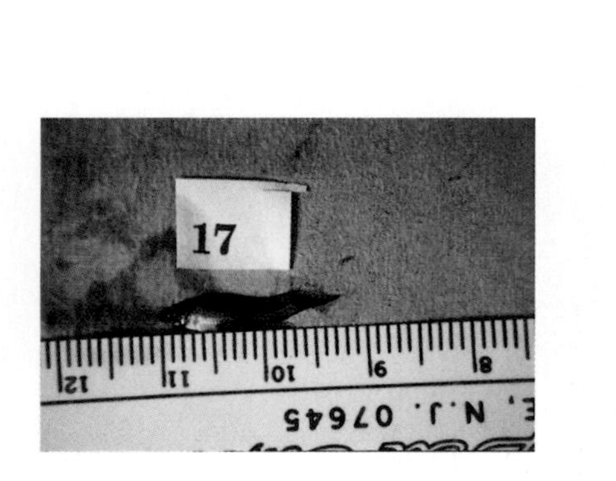

FIGURE 2.17. Colorful guppies. Guppies can differ on traits like color when between-population comparisons of guppies are made. Here is a colorful male guppy from the Paria River in Trinidad. *(Photo credit: Lee Dugatkin)*

David Reznick and his colleagues have found that guppies from high-predation sites reach sexual maturity at a younger age and produce more broods with more but smaller offspring, and tend to channel their resources to early reproduction when compared to guppies from low-predation sites (Figure 2.18). Why? At high-predation sites, guppy predators tend to be much larger (Figure 2.19A), and can eat a guppy no matter how large it gets. At such sites, producing many smaller fish would seem to be favored, as this is akin to buying lots of lottery tickets and hoping that one is a winner. At low-predation sites, only a single small fish predator

FIGURE 2.18. Natural selection and predation. Natural selection acts differently on guppy populations from high-predation sites (with *Crenicichla alta*) and low-predation sites (with *Rivulus hartii*). At high-predation sites, selection favors guppies producing many small young, but at low-predation sites, selection pressure is flip-flopped, favoring fewer, but larger, offspring.

(*Rivulus hartii*) of guppies exists (Figure 2.19B). If guppies can get past a certain size threshold, they are safe from *R. hartii*. As such, natural selection favors females producing fewer, but larger offspring, that can quickly grow large enough to be out of the zone of the danger associated with *R. hartii*, and this is precisely what we see (Reznick, 1996).

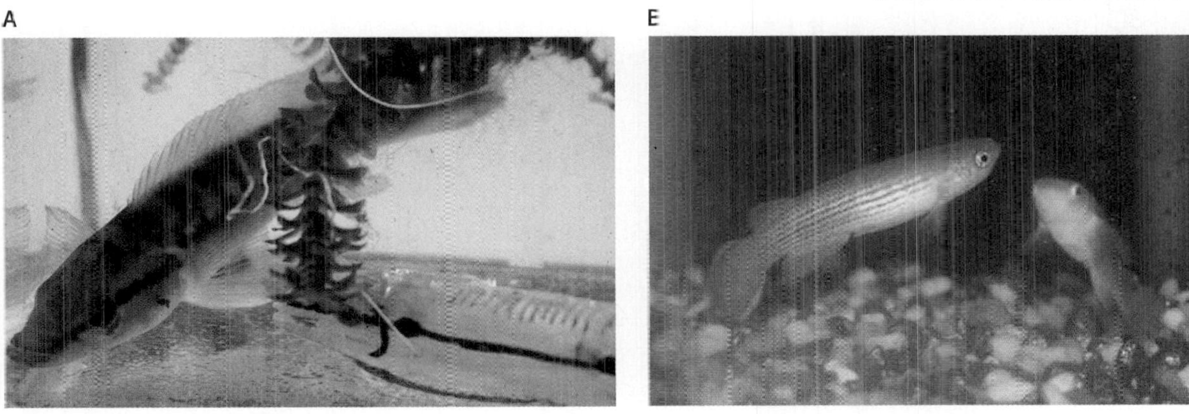

FIGURE 2.19. Guppy predators. **(A)** A pike cichlid, *Crenicichla alta*. These predators are common in downstream sites native to guppies. **(B)** A killifish, *Rivulus hartii*. These small, fairly innocuous predators can eat only tiny guppies and are found in upstream (low-predation) sites in the rivers of Trinidad. *(Photo credits: James Gilliam)*

Natural selection has operated on various aspects of guppy *behavior* as well (Seghers, 1973; Endler, 1995; Endler and Houde, 1995; Houde and Endler, 1990; Magurran et al., 1995; Reznick 1996; Reznick et al., 1997). One suite of behaviors that has been studied extensively in natural populations of guppies is their antipredator activities (Seghers, 1973; Liley and Seghers, 1975; Magurran and Seghers, 1991; Magurran et al., 1995; Houde, 1997). Guppies, as well as sticklebacks and minnows, modify their behavior as a function of predation risk (Table 2.1). In particular, two components of antipredator behavior have come under close scrutiny in guppies—shoaling and predator inspection. Shoaling—sometimes referred to as schooling—is a measure of group cohesiveness (Keenleyside, 1955; Pitcher, 1986), and predator inspection behavior refers to the tendency for individuals to move toward a predator to gain various types of information (George, 1960; Pitcher et al., 1986; Pitcher, 1992; Dugatkin and Godin, 1992).

TABLE 2.1 The effects of predation. An abbreviated list of some behaviors that are affected by predation pressure in guppies (*P. reticulata*), sticklebacks (*G. aculeatus*), and minnows (*P. phoxinus*). *(Adapted from Magurran et al., 1993, p. 30)*

BEHAVIOR	EFFECT OF INCREASED PREDATION PRESSURE	SPECIES
Schooling	Larger and more cohesive schools	*P. reticulata* *P. phoxinus*
Evasion tactics	More effectively integrated in high-risk populations	*P. reticulata* *G. aculeatus* *P. phoxinus*
Inspection and predator assessment	Increase in inspection frequency	*G. aculeatus* *P. phoxinus*
	Increase in inspection group size	*P. phoxinus*
Habitat selection	Remain near surface and seek cover at edge of river	*P. reticulata*
Foraging	Increased feeding tenacity	*P. reticulata*
Female mating choice	Preference for less brightly colored males	*P. reticulata*
	Avoidance of sneaky mating attempts	*P. reticulata*
	Compromised by predator avoidance	*P. reticulata*
Male mating tactics	Increased use of sneaky mating tactics in high-risk populations	*P. reticulata*

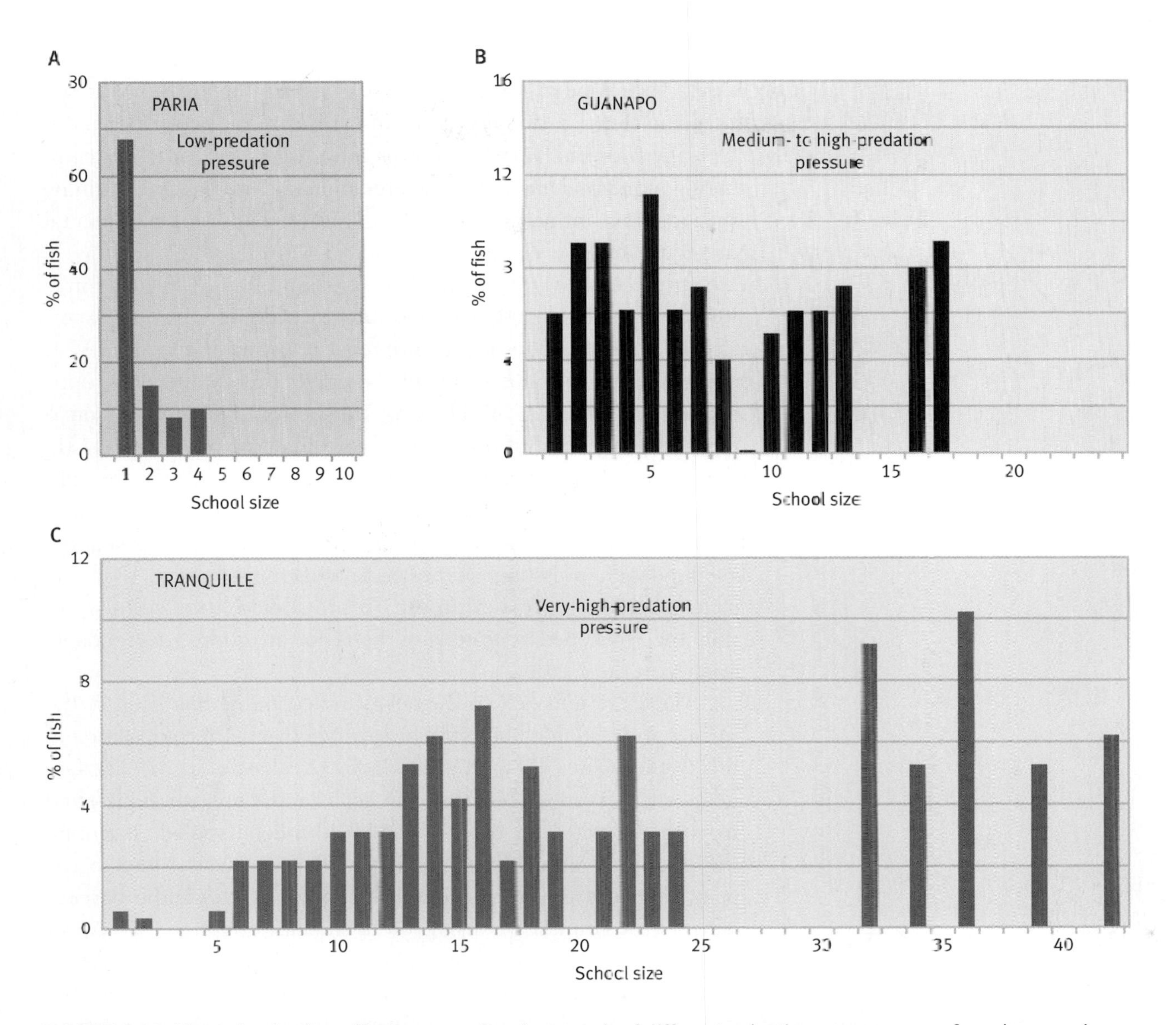

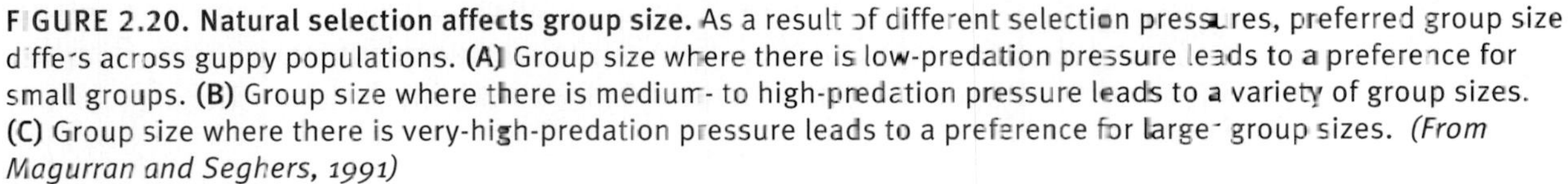

FIGURE 2.20. Natural selection affects group size. As a result of different selection pressures, preferred group size differs across guppy populations. **(A)** Group size where there is low-predation pressure leads to a preference for small groups. **(B)** Group size where there is medium- to high-predation pressure leads to a variety of group sizes. **(C)** Group size where there is very-high-predation pressure leads to a preference for larger group sizes. *(From Magurran and Seghers, 1991)*

Generally speaking, guppies from high-predation sites school more tightly and in greater numbers than guppies from low-predation sites (Magurran et al., 1995; Houde, 1997; Figure 2.20). Furthermore, fish from high-predation sites inspect a predator less closely, but more often than their low-predation counterparts. Again, both of these results are in line with what sociobiological thinking would lead us to expect. At high-predation sites, swimming in large groups, which tends to provide more protection than swimming in small groups, and seriously examining any potential threat should be at a higher premium than in areas of low predation, and that is just what we see in natural guppy populations. But typically these behavioral patterns have been uncovered by comparing numerous high- and low-predation sites and then examining the guppies' antipredator behavior, rather than via experimental manipulation.

In the early 1990s, Anne Magurran and her colleagues fortuitously learned of a unique opportunity to examine a "natural experiment" on the evolution of antipredator behavior in guppies. It seems that back in 1957, one of the original researchers on guppy population biology, C. P. Haskins, transferred 200 guppies from a high-predation site (in the Arima River) to a low-predation site (in the Turure River) that had been unoccupied to that point by guppies. Magurran realized that this was a golden opportunity to examine natural selection on antipredator behavior. If natural selection, via predation pressure, shapes antipredator responses, then the lack of predation pressure in the Turure should have led to selection for weakened antipredator behavior in guppy descendants studied in the 1990s.

Magurran and her colleagues sampled numerous sites in the Turure and examined both the behavior and genetic composition of fish at these sites (Magurran et al., 1992; Shaw et al., 1992). Genetic analysis suggested that the high-predation fish transferred from the Arima River back in 1957 had indeed spread all around the previously guppy-free site in the Turure River. More to the point, because of strong natural-selection pressures, the descendants of the Arima River fish evolved shoaling and predator inspection behaviors that were more similar to guppies at low-predation sites than they were to the behaviors of their ancestors from the dangerous sites in the Arima River. This is precisely the sort of thing that had been found in other transplant experiments that had examined shifts in morphology and life history (Reznick et al., 1990; Reznick, 1996; Figure 2.21).

Magurran's group uncovered another curious finding. In addition to colonizing the low-predation areas of the Turure River (located upstream, where there are few predators), over the course of time the descendants of the Arima River fish moved downstream in the Turure River, back into areas of greater predation pressure. When tested, these fish showed antipredator behavior similar to that of their ancestors from the original high-predation site in the Arima River. One possible explanation is that the original colonizers spread quickly, and since their antipredator behavior was beneficial when they reached high-predation sites in the Turure River, selection simply maintained such behavior. A more tantalizing, but

FIGURE 2.21. Transplants and natural selection. One way to examine the force of natural selection is to do reciprocal transplant experiments, which in effect switch selection pressures. Here, some guppies from high-predation sites (with *C. alta*) are transplanted to low-predation sites, and vice versa. Over the course of years, transplanted populations converge on the characteristics of the fish in the populations into which they were transplanted.

to date untested hypothesis, is that the Arima River fish and their descendants colonized their new habitat at a much slower rate. If this were the case, natural selection may have shifted the colonizers and early descendants one way—toward the norm for low-predation sites—and then later on, shifted the late descendants back in the opposite direction—toward the norm for fish from high-predation sites.

Adaptation

Now that we have a feel for how natural selection operates in shaping behavior, what shall we call the traits, behavioral or otherwise, that are almost always the result of this process? The term typically reserved for these traits is **adaptation.** While there are many definitions of adaptation in the literature (Leigh, 1971; Krimbas, 1984; Sober, 1984; Wallace, 1984; Fisher, 1985; Endler, 1986; Brandon, 1990; Symons, 1990; Sober, 1987; Mitchell and Valone, 1990; Baum and Larson, 1991; Reeve and Sherman, 1993; Sherman and Reeve, 1997), for our purposes, we shall simply note that natural selection generally produces adaptations, where adaptations are traits that provide their bearers with the highest relative fitness in an environment. For the guppies mentioned above, tight schooling and vigilant predator inspection are adaptations to life in high-predation streams.

Let's look at one particular definition of adaptation, and focus our attention on behavioral adaptations. Kern Reeve and Paul Sherman (1993) argue that an adaptation is a trait—in our case, a behavior—that results in the highest fitness among a specified set of behaviors in a particular environment. One critical aspect of this definition is that it shows that, while natural selection produces adaptations, it is not necessarily true that adaptations are always the product of natural selection. If adaptations are traits that provide their bearers with the highest relative fitness in an environment, then any number of forces, including chance, could produce an adaptation. Mind you, as long as the environment we are studying is the same environment in which natural selection operated, it's likely that most adaptations *are* the product of natural selection. Nonetheless, adaptations *don't have to be* the product of selection.

Reeve and Sherman's definition allows for the possibility that the most common behavioral variant is not adaptive, and this definition is thus immune to many of the criticisms raised about other definitions of adaptation (Gould and Lewontin, 1979). To see this, let us examine the question of behavioral adaptation and the egg-dumping behavior found in one population of wood ducks (*Aix sponsa*).

In 122 species of birds, females undertake a behavior called **brood parasitism,** wherein they "dump" their eggs into the nests of others in their population (Payne, 1977; Yom-Tov, 1980; Lyon and Eadie, 1991; Eadie et al., 1998; Rohwer and Freeman, 1989; Figure 2.22). If these dumped eggs are then raised by foster parents, the biological parents receive a huge benefit (new chicks), but they avoid the normal costs of parenting. Such parasitism is particularly common in waterfowl (Beauchamp, 1997). While waterfowl make up only about 2 percent of bird species, they account for more than 25 percent of the cases of parasitic egg dumping. In particular, brood parasitism is most prevalent in waterfowl that nest in tree cavities and man-made nest boxes. In fact, Brad Semel and Paul Sherman studied one population of wood ducks in Missouri that nest in man-made boxes, and they found that a remarkable 95 percent of active nests were parasitized, despite many empty nest boxes that could be used to raise young. This produced the odd situation of some nests having thirty to forty eggs, as opposed to the normal clutch of ten to twelve (Semel and Sherman, 1986; Figure 2.23). Many heavily parasitized nests were simply abandoned by the wood ducks. A given female was often both parasitized and an egg dumper herself. For females that returned to this site over a one- to four-year period, 75 percent had served as a host once and 54 percent had deposited at least one egg in another

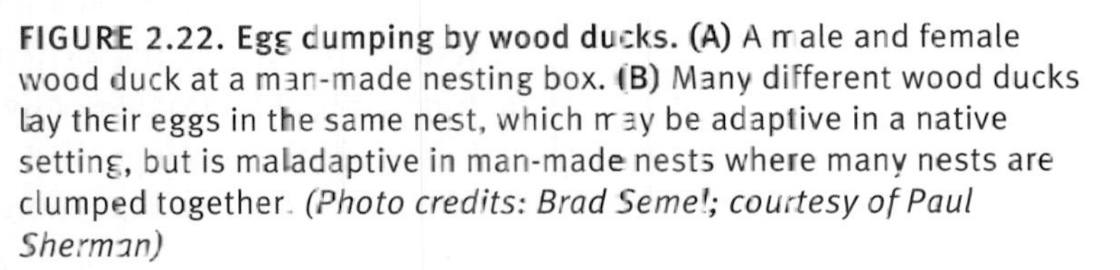

FIGURE 2.22. Egg dumping by wood ducks. (A) A male and female wood duck at a man-made nesting box. **(B)** Many different wood ducks lay their eggs in the same nest, which may be adaptive in a native setting, but is maladaptive in man-made nests where many nests are clumped together. *(Photo credits: Brad Semel; courtesy of Paul Sherman)*

FIGURE 2.23. Maladaptive behavior in wood ducks. Wood ducks typically nest in tree cavities (left), In such cavities, females generally lay ten eggs. When females nest in densely packed man-made boxes (right), a great deal of “egg dumping” occurs, where many females lay eggs in one box. This causes a decrease in reproductive success for all females and is not adaptive.

female's nest. Twenty-three percent acted as both host and parasite in the very same season!

Wood ducks typically nest in tree cavities created by woodpeckers or by limb breakage (Sherman, 2001). During the breeding season, females lay one egg per day and spend the remainder of their waking hours foraging. Males stay by their mates during the egg-laying period and *then abandon the nest when females begin to incubate.* A full clutch of ten to twelve eggs makes up more than 75 percent of a female's body mass (Bellrose and Holm, 1994).

Given the ecology and breeding biology of the wood duck, Semel and Sherman were interested in how such a system could lead to 95 percent of the active nests being parasitized when empty nest boxes were also present. Were there some benefits accrued by females for egg dumping in an environment with nest boxes? Being good "adaptationists," Semel and Sherman laid out five potentially adaptive reasons that females might dump their eggs. They hypothesized that females

- egg dump to reduce the risk of predation on all their eggs at once, thereby engaging in what is called "bet hedging,"
- are forced to parasitize the nests of others for lack of available nests to call their own,
- parasitize when their own nest gets destroyed (by predation, storms, and so on),
- may gain some benefit by placing eggs in the nests of relatives,
- may enhance their reproductive success in general by simply laying eggs in any available nest they come across.

Using a brilliant combination of observation, experiment, and inference, Semel and Sherman were able to systematically discount each of the above five hypotheses. In fact, what they found was that the nests that were parasitized had *dramatically reduced reproductive success* (Semel et al., 1988, 1990; Figure 2.24). With such information in hand, we can ask whether the egg-dumping behavior that Semel and Sherman saw is adaptive. According to Reeve and Sherman (1993): "In applying our definition, the behavioral set includes parasitizing versus nonparasitizing conspecifics, the measure of fitness is hatchability of eggs, and the environmental context being examined is grouped, visible nest boxes. Under these conditions extreme brood parasitism lowers individual success. . . . Thus dump nesting is nonadaptive." In other words, the prevalence of brood parasitism in the population Semel and Sherman studied must be due to processes other than natural selection operating on parasitism and egg dumping in that population.

Sherman (2001) hypothesizes that when nest boxes were not present (as was the case for most of the wood duck's history) and

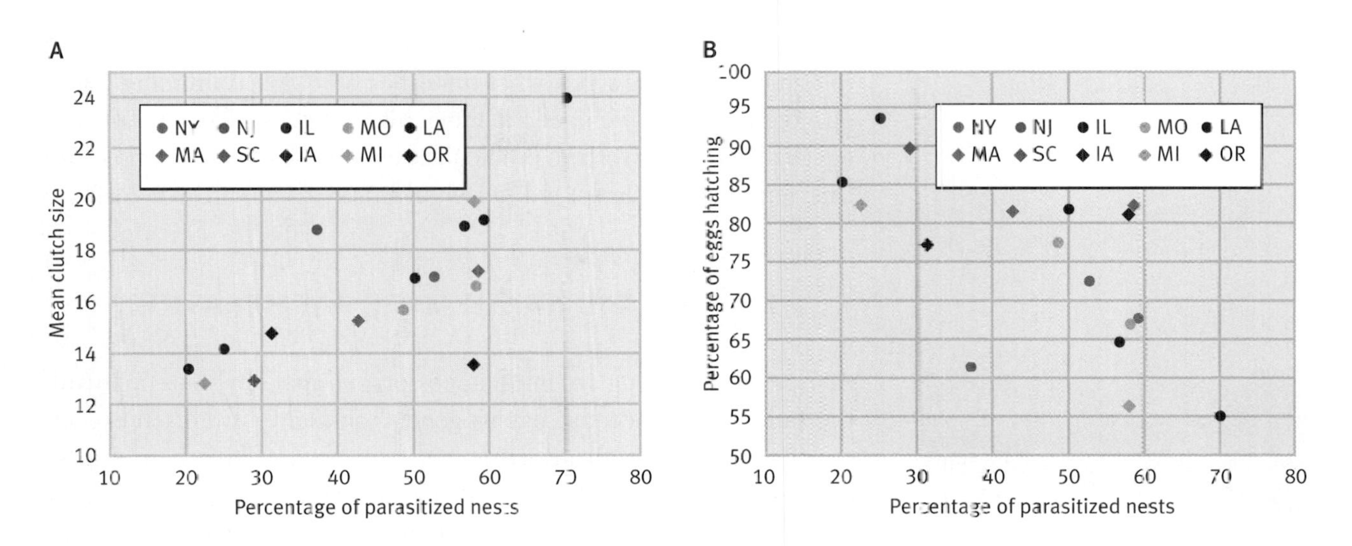

FIGURE 2.24. Egg dumping and rearing young. In wood ducks, the level of nest parasitism affects both the number of eggs laid **(A)** and the proportion of eggs that hatch **(B)**. *(From Sherman, 2001, p. 318)*

good territories were not clumped, there was likely a very high premium placed on good cavities in which to nest. Thus, females may have egg dumped in such cavities, and potentially benefited from such parasitism (Figure 2.25). But when nest boxes are present and they are clumped, as in Semel and Sherman's population, such egg dumping is no longer adaptive under Reeve and Sherman's definition. Natural selection may very well have favored brood parasitism in the wood duck's past, but the nest boxes radically changed the population ecology of this species, such that egg dumping was no longer adaptive in Semel and Sherman's population.

While most behaviors that we see commonly adopted in animals will likely be adaptations and the product of natural selection, the wood duck example shows this need not be the case. This is particularly likely to occur when man-made changes cause an unnatural scenario to suddenly be common, as in the case of the wood ducks.

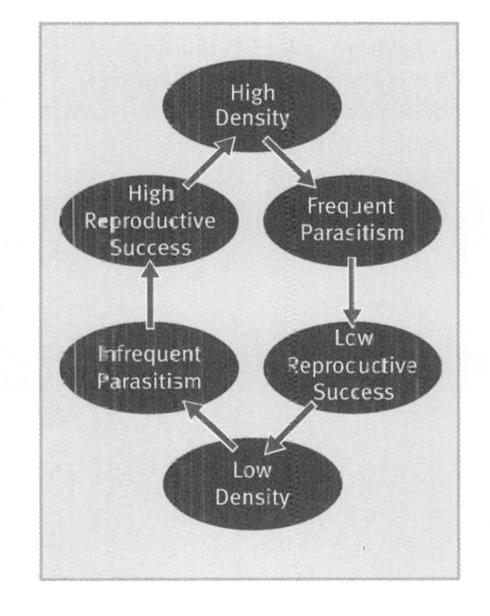

FIGURE 2.25. Population dynamics and nest parasitism. Sherman and Semel argue that both nest parasitism and population density cycle together. *(From Sherman, 2001)*

Genetic Techniques as Tools

In addition to genes playing a fundamental role in the process of natural selection per se, animal behaviorists also employ genetic techniques as *tools* to test specific hypotheses that often center on the notion of genetic relatedness. As we shall see in much more detail in Chapter 8, genetic relatedness plays a large role in much of the theory that underlies ethological thinking. Various techniques developed in molecular genetics allow us to use genes as a tool for testing ethological hypotheses. Two examples of this approach are presented below.

KINSHIP AND NAKED MOLE RAT BEHAVIOR

FIGURE 2.26. A naked mole rat worker. Naked mole rats show very high within-colony relatedness. This maps nicely onto numerous cooperative and altruistic behaviors common to this species. *(Photo credit: Stanley Braude)*

As a result of having common ancestors, animal kin share many of the same genes. Because of this genetic similarity, we often expect more cooperative and altruistic behavior among kin than among unrelated individuals. In fact, as we shall see in Chapter 8, according to basic kinship theory, the more related individuals are, the more we expect to see cooperation and altruism (Hamilton, 1964). To see how we can use genes as a tool to test this idea, we shall take a brief foray into the fascinating, but bizarre life of naked mole rats (*Heterocephalus glaber*; Figure 2.26).

Few mammals have captured the fancy of both scientists and laymen to the extent of the naked mole rat (Jarvis, 1981; Sherman et al., 1991). These small, hairless rodents of tropical Africa display **eusociality,** an extreme form of sociality that is present in many social insect groups (Table 2.2). Naked mole rats were the first vertebrates to display the three characteristics associated with this extreme form of sociality:

- a reproductive division of labor, wherein some individuals reproduce and some don't,
- overlapping generations, such that individuals of different generations are alive at the same time,
- communal care of young.

Naked mole rats live in large groups, wherein a *single queen* and one to three males have exclusive breeding rights for the entire colony. Although intra-colony aggression certainly exists in this species (Reeve and Sherman, 1991; Reeve, 1992), cooperation is especially notable among naked mole rats. Nonreproductive male and female naked mole rats, who live much shorter lives than reproductive naked mole rats, undertake a wide variety of cooperative behaviors, such as digging new tunnels for the colony, sweeping debris, grooming, and predator defense (Lacey and Sherman, 1991; Pepper et al., 1991). But why? Why do workers yield exclusive reproduction to a single queen and a few males? The answer in part centers on blood relatedness, and to see why, molecular genetic techniques are needed.

TABLE 2.2 The relationship between social behavior in naked mole rats and social insects. Comparisons between naked mole rats and eusocial insects indicate basic similarities in such characteristics as reproductive division of labor, overlapping generations, and communal care of the young, as well as other similarities and differences. *(From Lacey and Sherman, 1991)*

SPECIES	SIMILARITIES TO NAKED MOLE RATS	DIFFERENCES FROM NAKED MOLE RATS
Paper wasp (*Polistes fuscatus*)	Single breeding female per colony Aggressive domination of other colony members by reproductive female No permanent sterility in subordinate foundresses Size-based subdivision of nonbreeding caste Similar colony sizes (20–100 workers) Slightly larger size in queens than in workers	All female workers Haplodiploid genetics New nests each spring Outcrossing promoted by dispersal of reproductives from nest site Carnivore
Honeybee (*Apis mellifera*)	Single breeding female per colony Colony reproduction by fissioning (swarming) Age polyethism among nonreproductives Slightly larger size in queens than in workers	All female workers Haplodiploid genetics Primarily chemical, not behavioral, reproductive suppression of workers Permanently "sterile" nonreproductive females Vastly larger colonies in honeybees Outcrossing promoted by aerial mating aggregations Nectarivore
Wood termite (*Kalotermes flavicollis*)	Single breeding female per colony Diploid genetics Male and female workers Delayed caste fixation Division of labor among nonbreeders Extremely long life of reproductives Diet of plant material; breakdown of cellulose by gut endosymbionts Allo- and autocoprophagy Opportunity for workers to become reproductives if they escape the breeder's influence Working behavior begins before adulthood	Chemical, not behavioral, reproductive suppression Vastly larger colonies in termites Queens many times larger than workers Outcrossing promoted by aerial mating aggregations

Basic kinship theory suggests that the more genetically related individuals are, the more cooperation they will show with each other. Using this as their theoretical framework, Kern Reeve and his colleagues predicted that since naked mole rats are very cooperative with others in their colony, they might also be close blood kin (Reeve et al., 1990). To test this idea, Reeve's group used a technique called "DNA fingerprinting" (Jeffreys et al., 1985; DeSalle and Schierwater, 1998). The DNA of fifty naked mole rats was sampled. Nuclear DNA for liver, muscle, or brain samples was used, and after a series of intermediate steps, three DNA "probes" were

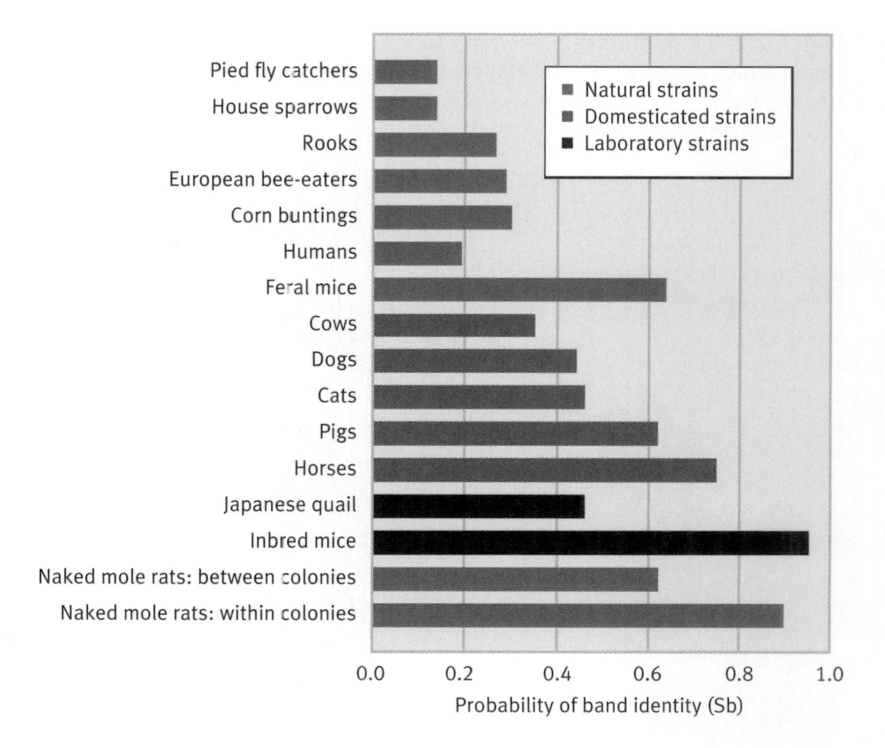

FIGURE 2.27. Naked mole rats show high levels of genetic relatedness. Reeve and his colleagues used what is known as Jeffrey's molecular genetic probe to ascertain relatedness among individuals. They found that naked mole rats in a colony were more genetically related to one another than any non-inbred strain of animal known. *(From Reeve et al., 1990)*

used to separate out three distinct DNA fragments ("minisatellites") for each individual. These probes produced a series of prominent black bands, which represented between 1.6 and 1.8 kilobytes of DNA. Individuals had a mean of about twenty-nine such distinct bands, and these bands together were their DNA fingerprint. The DNA fingerprints of different individuals could then be compared—the better the bands matched up, the more closely related were the individuals (Reeve et al., 1992).

Reeve and his colleagues found that because of very high levels of inbreeding—as a result of mating between relatives—average relatedness in these colonies of cooperative naked mole rats was extremely high, as predicted by kinship theory. The exact value they came up with was an average relatedness of 0.81 within naked mole rat colonies. To put this in some perspective, unrelated individuals have a value of 0.0 for this indicator, brothers score (on average) 0.5, and the most related of all individuals, identical twins, score 1.0 (Figure 2.27). Thus, naked mole rat individuals, on average, fall between normal siblings and identical twins on a relatedness scale, and they even lean toward the identical twins' side of the equation. In the case of naked mole rats, then, genetic techniques were successfully used to test a theory derived from sociobiological and selfish gene thinking.

COALITION FORMATION

Another context in which relatedness and genetic techniques to assess it often emerge is coalition formation in animals. In coalitions, the result of a two-party interaction is affected by a third individual that aids one of the other interactants. The recipient of such aid is then more likely to obtain the resource for which it is vying. Coalitions have been examined most extensively in primate species (Harcourt and de Waal, 1992; Chapais, 1992), but they have also been the subject of study in hyenas (*Crocuta crocuta*; Zabel et al., 1992), wolves (*Canis lupus*; Fentress and Ryon, 1986), lions (*Panthera leo*; Packer and Pusey, 1982), cheetahs (*Acinonyx jubatus*; Caro, 1994a; Figure 2.28); coatis (*Nasua narica*; Russell, 1983), and dolphins (*Tursiops truncatus*; Connor, 1992; Connor et al., 1992, 1999).

Aiding another individual against a third party is often a dangerous activity, and injuries to coalition members are not uncommon. As such, behavioral ecologists and ethologists have hypothesized that coalitions should often be formed among kin (Harcourt and de Waal, 1992), who stand to gain the most by aiding each other (Chapter 8). That being said, determining which animals are blood kin can be very tedious, if not impossible, in natural populations. Unless a thorough genealogy is available, it is often difficult for ethologists to know exact relatedness among coalition partners, and hence they rely on molecular genetic techniques to assay relatedness.

Relatedness is often quite important in structuring lion communities (Packer and Pusey, 1982; Packer et al., 1988), but it is

FIGURE 2.28. Coalitions. Males, often brothers, form long-lasting coalitions that have significant fitness effects on all members. **(A)** Two cheetahs in a coalition survey the area, looking in different directions for the appearance of prey. **(B)** The two cheetahs in the coalition both flush out and capture prey. *(Photo credits: Tim Caro)*

Interview with Dr. RICHARD ALEXANDER

Your work on natural selection and behavior came on center stage in the late 1960s and early 1970s. What was it about that time period that spurred so much interest in behavior and evolution?
There are two questions here. For myself, I decided consciously as a graduate student in zoology, about 1955, that evolution has to be the central theory in biology. Therefore, I reasoned, I needed to decide what the most interesting topic in biology is and proceed to try and understand it in evolutionary terms. I chose behavior. I also deliberately set for myself the goal of tackling increasingly challenging problems, using evolution as the vehicle. I reasoned that if evolution really is biology's central theory, then there should be no biological topic it cannot explain.

I started with behavior in insects, later wrote about learning, then culture (that is, I recognized that humans become the most difficult biological topic). More recently I tried to understand morality, then realized that music and all the arts, as well as religion, still lay before me. I think I've written on all those topics, but only rudimentarily and imperfectly.

Those topics came to center stage in biology when they did, however, because of George Williams's 1966 book and William Hamilton's 1964 papers. Charles Darwin and Sir Ronald Fisher presaged these later authors on both levels of selection and kin (nepotistic) selection. But people didn't understand what Darwin meant, and scarcely anyone understood Fisher until the mid-1960s.

Williams convinced us that we had to regard it as parsimonious to assume that selection is most potent at lower levels in the hierarchical organization of life. Until then everyone—including me—invoked selection at any seemingly convenient level and failed to recognize the conflicts among different levels. Our explanations were all over the map, and mostly seriously flawed.

Should every university student, regardless of discipline, take some time to read parts of *On the Origin of Species?* If so, why, and what parts?
It would be nice if people simply thought it valuable to examine the greatest books of all time. Darwin's book is one of the greatest. Anyone who doesn't at least glance at Darwin's writings now and then misses the chance to recognize what marvelous things can come about as a result of superior intellect combined with superior determination and perseverance.

Some have argued that the process of natural selection is a tautology—something that is true by definition. Where have they gone wrong?
I explained why natural selection is not a tautology in *Darwinism and Human Affairs* (1979), and I gave a simple example: Knowing about how selection works when there are visual predators on animals like mice, we can predict that white mice will do better on white sand than other shades or colors. The more we understand the details of how selection works, the more we can predict (and understand) about the nature of life.

In the 1950s, I predicted the existence, appearance, general distribution, and song of a species of 13-year cicada that had not yet been discovered. That prediction didn't get published before it was tested, but my co-worker, Tom Moore, can verify that I made the prediction, and when we drove into 13-year cicada range in 1959 we heard the undescribed species within five minutes. Many taxonomists have done the same. In the 1970s I predicted about 14 attributes of the first known eusocial mammal (the naked mole rat) before anyone knew that any such animal existed.

No single kind of knowledge gives as much predictive power about life as an understanding of natural selection. The Austrian mathematical ecologist, Karl Sigmund, said this case was comparable to astronomers predicting the planet Neptune before they had direct evidence of its existence. I relayed to him the reply, only half-jokingly, that predicting 14 or so of the basic traits of a hypothetical species (among millions) that turned out to be the naked mole rat is surely more difficult than adding another planet. I was assured that he would laugh with me.

If natural selection is so powerful a force, why should anyone bother studying questions at the proximate level?

Proximate mechanisms evolve too; they occur in hierarchies, some proximate to others. Understanding their nature, at every level, is essential not just for full understanding of life, but as well for knowing how to change things most efficiently. A wonderful example is kin helping. I used a full chapter of *Darwinism and Human Affairs* to discuss its proximate mechanisms. It still looks as though we learn who our kin are, usually by direct association with them. Knowing that, we can start to understand why in modern society so many kin relations go awry. Because sociality is so central to human happiness, not many things are more important to us.

Can it really be true that the same process—natural selection—operates on everything from amoebas to humans?

Yes. How can it be otherwise, when all life continues to survive only by reproducing? No form of life lacks any of the attributes that make differential reproduction (selection) inevitable.

NO SINGLE KIND OF KNOWLEDGE GIVES AS MUCH PREDICTIVE POWER ABOUT LIFE AS AN UNDERSTANDING OF NATURAL SELECTION.

In what way has our fundamental understanding of the process of natural selection been changed by the revolution in molecular genetics?

Molecular genetics is close to the most reductionist level of meaningful analysis of living creatures. It is the base from which to start a process of understanding by induction, by understanding progressively the hierarchical interactions of units we only found out about recently. The wonderful barrier that lies in front of initial molecular approaches is the one that Theodore Dobzhansky identified 40 years ago: He said that heredity is particulate, but development is unitary. This unity of development is certainly one of the most amazing and difficult of all human realizations. Molecular geneticists start with perhaps 40,000 genes. Unity of development means that in a multitrillion celled organism such as ourselves—with every cell containing the full complement of those 40,000 genes, in an organism that can continue to develop (or change) across 100 years—there may be vastly more than 40 000 interactions taking place even in fractions of seconds, and across all our lives. This is the daunting barrier faced by current efforts to invent medicines that can improve on natural selection. As one of my students said, taking some of these medicines is like lubricating a fine watch with water: it works for a few seconds but then the machinery disintegrates dramatically because of it. All people who watch drug advertisements, are prescribed man-made drugs by a physician, or wonder whether medicine is going to double or triple human lifetimes should understand that almost any humanly imaginable thing that is easy to generate would have been generated by natural selection long ago. The concept of selective trade-offs, in which there is always an expense for every presumed benefit, is essential to everyone who wishes to live the longest possible happy life. Hence, the importance of knowing what is a benefit and what is a cost—in what senses, and under different circumstances. Hence, the central importance of understanding ourselves by understanding the history of natural selection.

DR. RICHARD ALEXANDER is a professor at the University of Michigan and a member of the National Academy of Sciences. He is well known for his seminal work on natural selection and behavior in animals and humans.

FIGURE 2.29 A pride of lions. In lions, males form coalitions and patrol and guard their territories. *(Photo credit: Craig Packer)*

often difficult to assess relatedness in large groups of lions without the aid of some sort of molecular genetic tool. Male lions form cooperative coalitions in patrolling their group's territorial borders (Figure 2.29). Although territorial defense is dangerous, larger coalitions are particularly adept at defending their territories against intrusions from other lion groups (Bygott et al., 1979; Packer et al., 1988). One of the questions that Craig Packer and Anne Pusey and their colleagues addressed as part of their long-term study of lion behavior was whether large male coalitions were typically made up of relatives (Packer et al., 1991).

To address this question, they turned to DNA fingerprinting and examined the number of DNA bands shared by coalition members. Sure enough, while small coalitions were sometimes made up of relatives and sometimes not, larger coalitions were always composed of relatives (Packer et al., 1991). It seems that when coalitions are small, each male in the coalition has about equal reproductive success. On the other hand, when coalitions are large, some individuals have very low reproductive success, while others have very high reproductive success. The only way then to promote the formation of large groups, given that individuals risk low reproductive success by joining such groups, is to have some added benefit (Figure 2.30). Kinship provides that benefit, as even when individuals have low reproductive success in large coalitions, they are in fact helping raise the fitness of their relatives. Since their relatives share many of the genes they themselves have, those on the low end of the coalition totem pole still get some indirect fitness benefits by joining such coalitions. This adaptationist per-

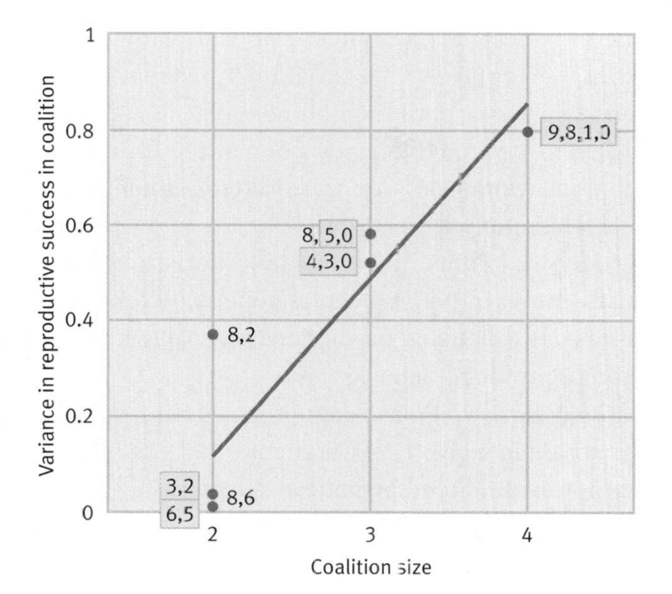

FIGURE 2.30. Coalition size and kinship. The variance in reproductive success increases as coalition size goes up in lions. In large groups, one or two individuals produce most of the young, with other coalition members fathering far fewer offspring. Larger coalitions are more likely to be made up of kin, providing an indirect benefit to those males that do not father many cubs. The numbers by each point indicate the number of cubs fathered by each male in that coalition, and a box around the number indicates that the males are close relatives. *(From Packer et al., 1991)*

spective on coalition formation is possible only as a result of the now readily available molecular genetics techniques that ethologists have at their disposal.

SUMMARY

1. Artificial selection is the process of humans choosing certain trait varieties over others through implementing breeding programs that cause one or more selected varieties to increase in frequency. We have been shaping animals and plants by this process of artificial selection for the better part of 10,000 years.
2. Much (but as we shall soon see, not all) of animal behavior work revolves around the notion that behavior, just as any trait, has evolved by the process of natural selection. The process of natural selection requires limited resources, variation, fitness differences, and heritability. For example, if we have a number of different foraging variants displayed by individuals in a population, and these variants translate into fitness differences and are heritable, then natural selection will operate to change the gene frequencies of the genes underlying our foraging variants.
3. Even a small selective advantage—on the order of 1 percent per generation—is sufficient for natural selection to dramatically change gene frequencies over evolutionary time.
4. Among other things, behavioral geneticists use Mendel's laws of genetics to predict the distribution of behavioral phenotypes, to

examine the effect of sex-limited trait expression, and to determine what percentage of the variation in a behavioral trait is genetic and what fraction is environmental.

5. In addition to viewing the gene as one of the fundamental units by which behavioral traits are transmitted, ethologists also employ molecular genetics as a sort of tool to allow them to test basic models developed by animal behaviorists and behavioral ecologists. For example, molecular genetic techniques are often employed to test detailed predictions regarding kinship and the evolution of prosocial behavior.
6. Natural selection produces adaptations—traits with the highest relative fitness in a given environment—but not all adaptations are necessarily produced by natural selection.

DISCUSSION QUESTIONS

1. How was it possible for Darwin to come up with his theory of natural selection in the complete absence of a science of genetics? Many modern studies in ethology rely on genetics, particularly molecular genetics, as a critical tool. How did Darwin manage without this tool?
2. Jacques Monod once referred to natural selection as a tinkerer. Why is this a particularly appropriate analogy for how the process of natural selection operates?
3. Read Reeve and Sherman's (1993) paper, "Adaptation and the goals of evolutionary research," in volume 68 (pp. 1–32) of the journal *Quarterly Review of Biology*. After reviewing all the definitions of "adaptation," do you agree with Reeve and Sherman's approach to this subject? If you do not agree, explain why not by listing your reasons.
4. When considering how much variation in a behavioral trait is genetic and how much is environmental, how does the "uterine environment" complicate matters?
5. Secure a copy of Gould and Lewontin's (1979) paper, "The spandrels of San Marcos and the Paglossian paradigm: A critique of the adaptationist programme" in volume 205 (pp. 581–598) of the *Proceedings of the Royal Society of London, Series B*. List both the merits and flaws in Gould and Lewontin's approach to "adaptationist thinking." Overall, do you think their critique is a fair one?

SUGGESTED READING

Darwin, C. (1859). *On the origin of species*. London: J. Murray. The starting point for modern ethology and a joy to read.

Dawkins, R. (1976). *The selfish gene*. Oxford: Oxford University Press. The book that introduced the term "selfish gene" and all its implications.

Reeve, H. K., Westneat, D. F., Noon, W. A., Sherman, P. W., & Aquadro, C. F. (1990). DNA "fingerprinting" reveals high levels of inbreeding in colonies of the eusocial naked mole rat. *Proceedings of the National Academy of Sciences, U.S.A.*, 87, 2496–2500. A nice overview of the manner in which molecular genetics can shed light on behavioral adaptations.

Weiner, J. (1995). *The beak of the finch: A story of evolution in our time*. New York: Vintage Books. A Pulitzer prize–winning book on Peter and Rosemary Grant's work on evolution in the Galápagos finches.

Williams, G. (1966). *Adaptation and natural selection*. Princeton: Princeton University Press. A critical approach to the adaptationist perspective. Regarded by many as a must-read for those studying animal behavior.

3

Ultimate and Proximate Perspectives

Hormones and Proximate Causation

- ▸ Hormones and Helping-at-the-Nest
- ▸ Testosterone and Play Fighting in Rats

Neurological Underpinnings of Behavior and Proximate Causation

- ▸ The Nervous Impulse
- ▸ Vocalizations in Plainfin Midshipman
- ▸ Mushroom Bodies, Insects, and Learning
- ▸ Sleep and Predation in Mallard Ducks

Biochemical Factors

- ▸ Ultraviolet Vision in Birds

Environmental Sex Determination and Sex Ratios

- ▸ Reptiles, Sex Determination, and Temperature
- ▸ Red Deer, Dominance Status, and Sex Ratios

Genes and Proximate Explanations

Learning as a Proximate Factor

Interview with Dr. Timothy Clutton-Brock

Proximate Factors

Suppose that I was sitting in a room with an ethologist, an ophthalmologist, a neurobiologist, and a molecular biologist, and I asked them to explain an animal's—let's make it a robin's—visual acuity (Figure 3.1). Further, let's imagine that we have a lecturer who has just provided information on the anatomy of a robin's eye, the neuronal underpinnings of robin vision, and even information about the molecular biology underlying robin vision.

Going around the room, we first come to our ethologist, who, without any reference to what the lecturer has said, proceeds to explain how natural selection produced an increase in visual acuity over time. She outlines the process of natural selection on vision, and describes how better vision produces a robin that is a superior forager and more adept at avoiding predators. Next, our ophthalmologist stands up and explains how the curvature of the eye of our robin has changed over the course of time, and how that change helps explain why modern-day robins see better than their ancestors. The neurobiologist then describes how the neural circuitry underlying vision has changed in robins, and how that change will

FIGURE 3.1. Different perspectives on a trait. Four different scientists studying a robin's eye could come up with different, but mutually compatible, explanations for the bird's visual acuity.

shed light on increased visual acuity in these birds. Lastly, our molecular biologist proceeds to describe changes to the molecular pathways that are associated with robin vision, and how changes in these pathways are critical to understanding changes in robins' vision.

Who's right? They all are. The ethologist has provided an "ultimate," natural selection-based answer, while the neurobiologist, ophthalmologist, and molecular biologist have provided answers, all of which may be correct, at the "proximate" level.

Ultimate and Proximate Perspectives

Undoubtedly the most common error made by those not trained in the field of animal behavior is to confuse the level at which ethologists analyze a question. For example, we can answer the question "why do songbirds sing?" in terms of physiology, neurobiology, survival value, relationship to other species of birds, and so on (Sherman, 1988). Confusion arises when I answer a question at a level that you are not expecting. I might give you an answer to why songbirds sing based on the bird vocal box, its musculature, and its connections to the brain, when in fact you were looking for a reply that explains birdsong in terms of why natural selection seems to have favored this trait in many different species. In an attempt to be clear about the precise nature of the questions being posed in ethology, there have been many attempts to solve what Sherman (1988) refers to as the "levels of analysis" problem in behavior (Huxley, 1942; Mayr, 1961; Orians, 1962; Hailman, 1982; Dewsbury, 1992, 1994; Reeve and Sherman, 1993; Alcock and Sherman, 1994; Hogan, 1994).

Here, following Ernst Mayr's (1961) general discussion of cause and effect in biology, we shall focus on two levels of analysis—proximate and ultimate. While this dichotomy has its detractors, it is generally a well-accepted one (Alcock, 1998). When we speak of ultimate causation, we are asking how *evolutionary forces* have shaped a behavior over time. The primary evolutionary force we shall be focusing on is natural selection, but other such forces can, and do, shape animal behavior as well (Huxley, 1932; Wright, 1969; Gould, 1977; Harvey and Pagel, 1991).

In this chapter, we shall focus on proximate causation, as ultimate causation runs as an underlying current in much of Chapter 2

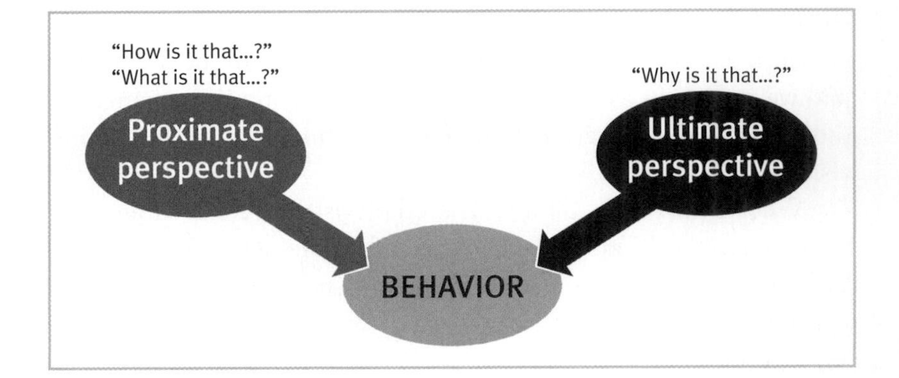

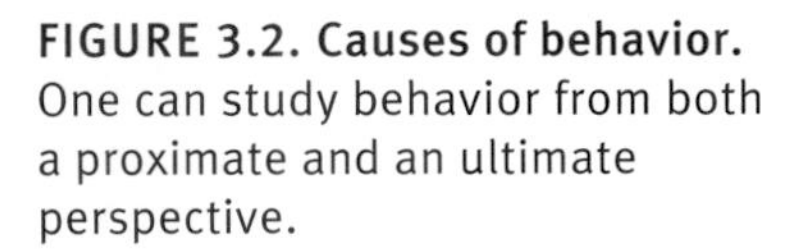
FIGURE 3.2. Causes of behavior. One can study behavior from both a proximate and an ultimate perspective.

and parts of Chapters 4 and 5. In the most general sense, **proximate** causes include all those factors that are not evolutionary in nature. Another way to think of the distinction between proximate and ultimate perspectives is to recognize that proximate questions include "How is it that . . . ?" and "What is it that . . . ?," while ultimate questions include "Why is it that . . . ?" (Alcock, 2001; Figure 3.2).

Before examining proximate causation per se, an example of employing both proximate and ultimate perspectives to address a single trait may help elucidate the relationship between these two approaches, and will serve as a nice starting point for an in-depth look at the proximate level of analysis. To do this, we shall consider Geoff Hill's (1990, 1991) thorough and ingenious work on plumage coloration in the house finch (*Carpodacus mexicanus*). As is the case for many bird species, male house finches are more brightly colored than female finches. The general problem Hill addressed was the nature of differential plumage coloration in male and female house finches. This problem can be examined from both a proximate and an ultimate perspective. That is, we can ask *what* causes males and females to differ in plumage coloration and also examine *why* such color differences persist over evolutionary time (Figure 3.3). In addressing these questions, we shall venture into a discussion that involves foraging behavior, mate choice, and parental care, all of which play a part in understanding the proximate and ultimate underpinnings of plumage coloration in male and female house finches.

FIGURE 3.3. Natural variation. Significant natural variation exists in house finch coloration. This set the stage for Hill's work on the proximate and ultimate questions on plumage coloration. *(Photo credit: Geoff Hill)*

To examine the *proximate basis* for female plumage, Hill employed intra- and interpopulation comparisons and controlled feeding experiments. To begin our proximate analysis of coloration, we must recognize that, from the start, Hill knew that plumage coloration in the male house finch was a result of carotenoid color pigments that were ingested (Brush and Power, 1976; Hill, 1992, 1993a). What is particularly interesting about this is that birds are unable to synthesize their own carotenoid pigments, and they rely completely on diet for this substance (Goodwin, 1950; Brush and Power, 1976; Brush, 1990). Hill's work, and that of others, demonstrated that, at the proximate level, differences in *male* plumage within and between populations of male finches were correlated with the amount of carotenoids in their diet, although precisely which foods provide the finches with the carotenoids is still not clear (Hill, personal communication). Differences in plumage had yet to be addressed in females, and it is to the cause of such differences that we now turn (Hill, 1993b, 1993c; Brush and Power, 1976; Butcher and Rohwer, 1989).

Hill (1993c) found between-population differences in plumage coloration in females from different localities in Michigan, New York, and Hawaii. Here, we shall focus on Hill's controlled feeding experiments on two groups of females who were fed a fixed diet while housed in aviaries at the University of Michigan. Both groups were fed a commercially made finch food, and both groups had their diets supplemented with water and apples. One of these groups, however, had their water and apples treated with canthaxanthin—a red carotenoid pigment. Females in the canthaxanthin treatment developed much brighter plumage after their diet was supplemented, while females who were just fed normal apples and water maintained a drab plumage pattern.

From a proximate perspective, we can now address two issues: (1) between-population differences in female coloration, and (2) differences in plumage coloration between males and females. With respect to differences between females across populations, it appears that the differential availability of carotenoid pigments in food across populations is critical. Differences in color between males and females appear to be due to differences in the dietary intake of carotenoid pigments. More specifically, Hill suggested that the proximate explanation for the differences in plumage coloration between males and females is due to males' active search for carotenoid-based foods. While females don't avoid carotenoid foods, they are much more passive in terms of searching for such items. This proximate explanation for differences in male and female coloration leads us nicely to our discussion of ultimate questions. Hill suggested that the reason that females don't actively search for carotenoid-based foods is that whereas colorful males receive significant benefits for being colorful (see below), females do not. As such, females focus on maximizing the calories they get per unit time, not on the amount of carotenoids they ingest.

Hill conducted further research to examine the ultimate causes of male and female differences in coloration in detail (Hill, 1990, 1991, 1993b). That is, once he knew that diet intake was correlated with plumage color and that males actively searched for

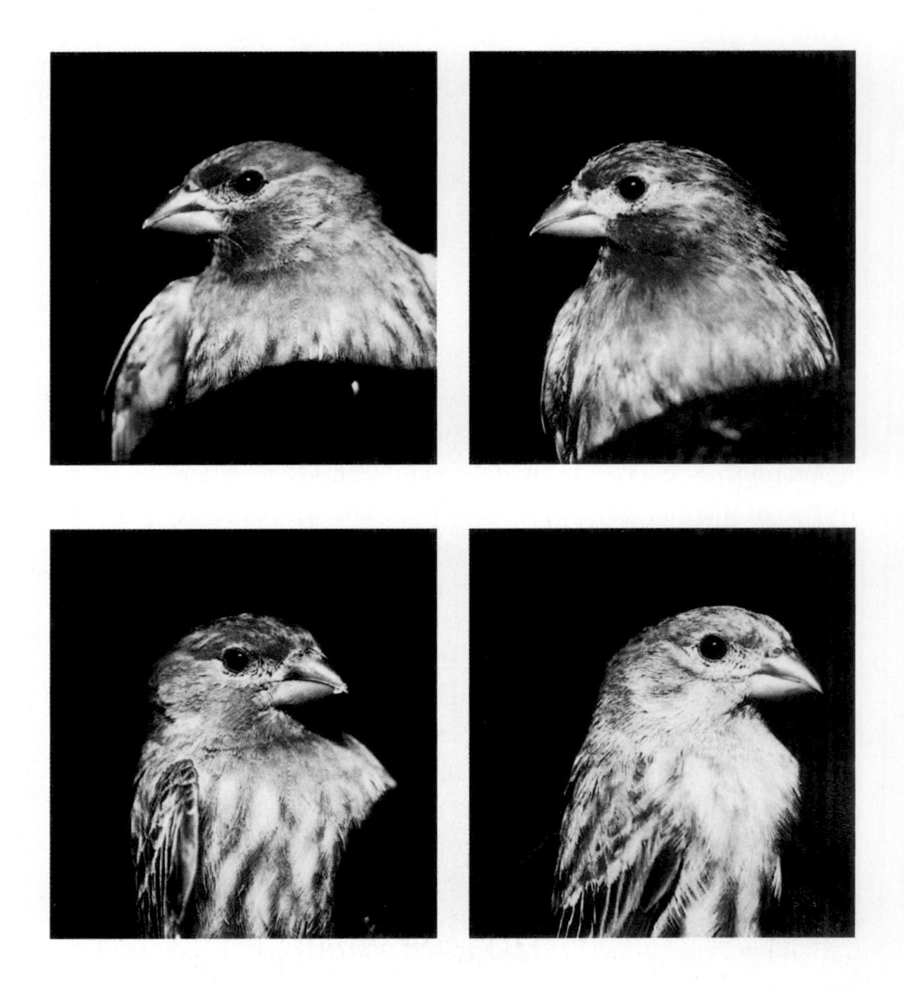

FIGURE 3.4. Plumage manipulation. As part of Hill's study on plumage coloration, he artificially brightened or lightened the plumage coloration of house finches. *(Photo credits: Geoff Hill)*

carotenoid-rich foods, he could ask why. Why do males actively search for carotenoids that increase their plumage coloration? What's in it, fitness-wise, for males to do this?

To uncover the *ultimate* reason that males are more colorful than females, Hill used hair dyes to either brighten or lighten the plumage coloration of a group of wild male finches and then examined how such manipulations affected female mate choice (Figure 3.4). Brightly colored males were much more likely to get a mate than were lightened (or control) males (Table 3.1). This result alone provided powerful evidence that one of the ultimate reasons that males search for carotenoid-rich food is that females are attracted to males with bright plumage. Hill went on to examine precisely *why* females prefer males with bright plumage. This is an ultimate question regarding mate choice in females. What's in it for females who make good choices with respect to male plumage coloration? When examining the relationship between male color and male parental care, Hill (1991) found that the mean number of times a male feeds a chick at his nest, as well as his mate, was positively correlated with the intensity of plumage coloration, with

TABLE 3.1 Plumage manipulations. Hill examined how brightening and lightening plumage coloration affected a suite of variables in male house finches. There were 40 males in the brightened condition, 20 males in the sham control, and 40 males in the lightened condition. *(From Hill, 1991)*

MALE CHARACTERISTICS	BRIGHTENED	SHAM CONTROL	LIGHTENED	STATISTICAL SIGNIFICANCE (P)
Original plumage score	140.7	139.9	141.00	0.95
Manipulated plumage score	161.6	139.9	129.40	0.0001
Proportion paired	1.0	0.6	0.27	0.0001
Time to pair (days)	12.1	20.2	27.80	0.07

brighter males feeding their mates and chicks more than twice as often as drab males (Figure 3.5A). Females may then prefer red males because such males make good mates and fathers, at least with respect to feeding. Females may also be obtaining more indirect benefits by mating with brightly colored males, as such males appear to have increased probabilities of survival. If the traits responsible for this increased survival are passed on to offspring, then mating with colorful males may produce very fit offspring. Hill found that males with lots of red plumage produced sons with colorful plumage (Figure 3.5B). Given that red plumage coloration can't be inherited (remember, such coloration is diet dependent),

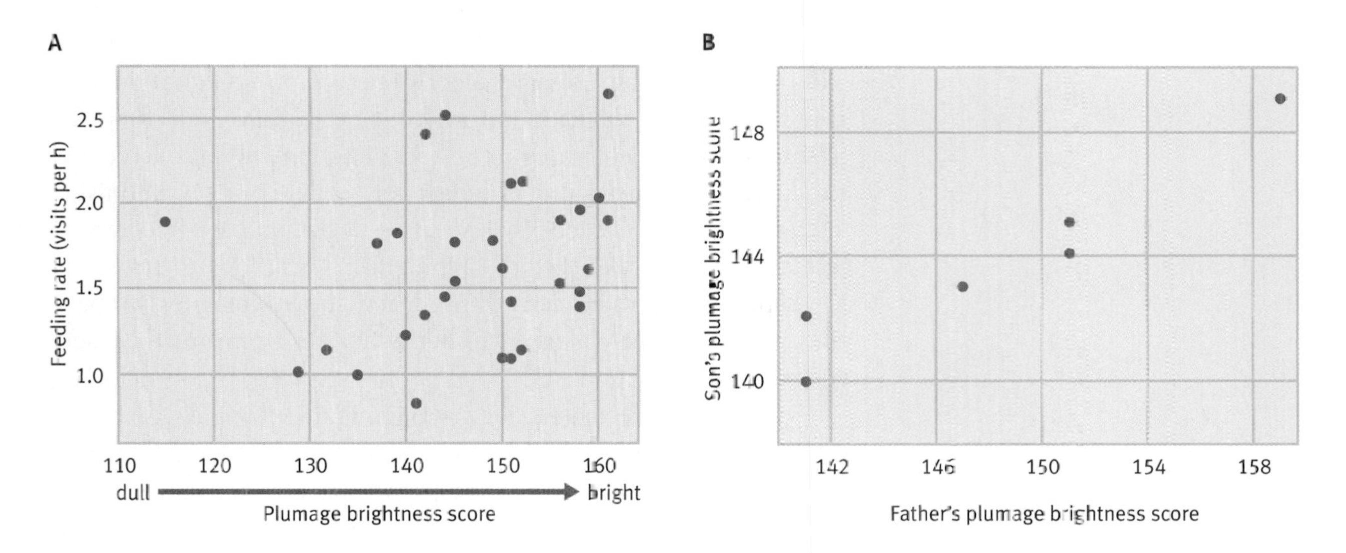

FIGURE 3.5. Plumage, feeding, and heritability. (A) Brighter male house finches feed their young at higher rates than drabber male finches. **(B)** A significant positive relationship exists between father and son plumage brightness scores in house finches. *(From Hill, 1991)*

this suggests that males who are good foragers, and thus have the carotenoids needed to produce colorful plumage, produce sons who are also good foragers.

Hill's work is an excellent example of how ultimate and proximate perspectives can provide an in-depth picture of animal behavior. Unfortunately, it is a rare case for an investigator to do in-depth work on both proximate and ultimate causation. Rather, ethologists are generally trained to focus on either proximate *or* ultimate questions.

In the remainder of this chapter, we shall examine the following proximate factors affecting behavior:

- Hormones,
- Neuronal underpinnings,
- Biochemical factors,
- Environmental correlates,
- Genes,
- Learning.

Hormones and Proximate Causation

The **endocrine system** is composed of a group of ductless glands that secrete hormones—molecular messengers—directly into an animal's bloodstream. Correct functioning of this system is of fundamental importance to behavioral functions and to the modification of rates and directions of various cellular functions. Malfunctioning of hormones, either through diminished secretion (hyposecretion), or excessive secretion (hypersecretion), affects functions like growth, metabolism, reactions to stress, aggression, and reproduction. Thus, when something goes wrong in these areas, we can ask whether hormones are a proximate cause of the malfunction.

Hormones affect behavior in two fundamental ways. In adults, and to a lesser extent in juveniles, hormonal state can affect both the strength and form of a behavioral response. For example, when androgens (male sex hormones) like testosterone, which initiates and maintains male sex characteristics, are at high circulating levels prior to an aggressive interaction—that is, when baseline levels of testosterone are high—the aggressive behavior that may follow is likely to be more intense than when baseline levels of testosterone are lower. With testosterone, as with many hormones, there is a feedback loop in play. Winning a fight, partly as a result of the behaviors triggered by high baseline levels of testosterone, can lead to even higher levels of this hormone in the bloodstream, at least temporarily (Figure 3.6).

In addition to the feedback loop described above, during early developmental stages, hormones affect the organization of behavior systems. For example, in a case of multiple fetuses, a male fetus may be surrounded by female fetuses, and consequently exposed to lower levels of circulating testosterone during this critical stage. Such a male may then be less likely to be aggressive throughout his life (see below). The entire behavioral repertoire associated with aggressive interactions is fundamentally altered by hormonal effects early on in development. Similarly, if certain endocrine glands malfunction early in development or are experimentally removed, long-lasting behavioral effects may result.

It is important to keep in mind that the endocrine system is not a disconnected amalgam of ductless glands. Oftentimes, hormones secreted by one endocrine gland stimulate the production and secretion of hormones from another endocrine gland. As a case in point, the anterior pituitary gland in humans secretes a series of "tropic" hormones (for example, human growth hormone, thyroid stimulating hormone, and adrenocorticotropic hormone) that result in the secretion of a number of other hormones throughout the body, including thyroid hormones, estrogen, testosterone, and stress hormones (glucocorticoids). Conversely, the secretion of some hormones inhibits the production of hormones in other ductless glands, rather than promoting such production.

While the endocrine system will vary from species to species, as well as across larger taxa (invertebrate versus vertebrate), in humans

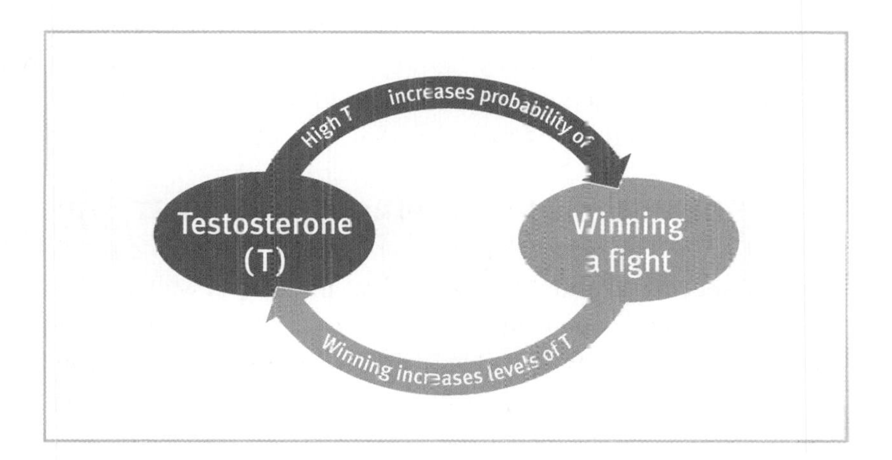

FIGURE 3.6. Testosterone (T) and aggression. A positive feedback loop exists between levels of T and probability of winning a fight. High levels of T increase the probability of winning, while winning further increases circulating levels of T.

the endocrine system and hormones play a role in the growth of body cells, tissue repair, ovulation and milk secretion in females, sperm production in males, as well as in the stress response. To see the importance of hormones, we can examine the "fight or flight" stress response exhibited by humans and other animals in reaction to a perceived threat.

When a stressor—a predator, for example—is sensed by an individual, it stimulates the hypothalamus to initiate what is called the general adaptation syndrome (GAS), which works along two pathways. First, adrenocorticotropic hormone (ACTH) stimulates the adrenal glands to secrete epinephrine and norepinephrine, leading to a quick and very large increase in blood sugar, which along with oxygen is also delivered quickly to vital organs. In particular, the brain, skeletal muscles, and heart are recipients of both increased blood sugar and oxygen. Circulation increases and "nonessential" systems—for example, the digestive and reproductive systems—are shut down. All of this allows for an appropriate behavioral response to whatever the stressor may be.

A second "responsive reaction" chain is also put in motion when a stressor is sensed. Here corticotropin-releasing hormone (CRH), growth hormone releasing hormone (GHRH), and thyrotropin-releasing hormone (TRH) are secreted by the hypothalamus. CRH in turn stimulates the anterior pituitary gland to increase the production of ACTH, which then stimulates the adrenal gland to secrete cortisol, which converts noncarbohydrates into sugars—energy that can be used to handle the stressor in question. The adrenal gland also increases its production of aldosterone, which increases water retention and hence reduces bleeding if the stressor causes injury. The body's endocrine system then is well designed to handle any number of different stressors. It is only when stress is chronic that the endocrine system fails to respond in the appropriate manner.

Understanding how hormones play such a crucial role in the fight or flight response, we can now move on to examine the manner in which ethologists study hormones and animal behavior, examining two cases: helping-at-the-nest and play fighting.

HORMONES AND HELPING-AT-THE-NEST

In many species of birds, mammals, and insects, as well as a few species of fish, young individuals act as helpers-at-the-nest (Skutch, 1935; J. L. Brown, 1987; Taborsky, 1994; Solomon and French, 1996). Helpers are often young, but sexually mature, individuals that remain at their natal nest and assist in the raising of younger siblings. The aid that such helpers produce may take many forms, including feeding and nest defense (J. L. Brown, 1987). As we shall see in Chapter 8, many evolutionary (ultimate) hypotheses have been put forth to explain helping, with most surrounding the notion that helpers are often blood relatives. Helpers are hypothesized to make cost-benefit decisions about the quality of available open territories for their own breeding versus what they get from helping their parents (Emlen, 1995b). Similarly, there have been numerous proximate-level theories put forth to explain helping-at-the-nest.

Here we shall examine how developmental factors affect circulating levels of testosterone, which in turn affect whether male Mongolian gerbils (*Meriones unguiculatus*) decide to help or not to help (Figure 3.7). But first, one quick, but important point to make in terms of the generality of work on testosterone and other hormones. And that is that testosterone is identical across vertebrate species, as are the other steroid hormones like estrogen, progesterone, cortisol, and corticosterone. The case is somewhat different for peptide hormones such as lutenizing hormone, follicle stimulating hormone, oxytocin, and vasopressin (Norris, 1996; Bentley, 1998; Knapp, personal communication), but testosterone in one species is like testosterone in another. This finding suggests that testosterone is a potent factor in shaping animal behavior, particularly the behavior of males, in many different species.

FIGURE 3.7. Helping in male Mongolian gerbils. Male helping behavior as a function of prior intrauterine position has been examined in Mongolian gerbils. (Photo credit: G&H Denzau/ naturepl.com)

The story of male Mongolian gerbils, development, testosterone, and helping behavior starts early—in fact, it all begins with a male's position in the uterus of its mother. In a number of rodents, including gerbils, the sex of the two siblings surrounding an individual in utero can have dramatic effects on its behavior after birth (vom Saal, 1989; Figure 3.8). In gerbils, adult males that were surrounded by two females (2F males) in utero have half the level of circulating testosterone of males that were surrounded by two other males (2M males) (vom Saal, 1989; Clark et al., 1991, 1992; Even et al., 1992). With this in mind, Clark and Galef (2000) examined whether testosterone differences mediated by intrauterine position (IUP) affect a male's decision to help at the nest. Male Mongolian gerbils are known to huddle over their young and groom them. Clark and Galef tested whether such actions were affected by IUP-mediated testosterone effects.

Clark and Galef undertook two experiments using 2F males. In the first of these, eighteen 2F males were tested in terms of both mating and parental care. The researchers discovered that, of the ten males that failed to impregnate a female, four of these exhibited sexual behavior that was so compromised that they were

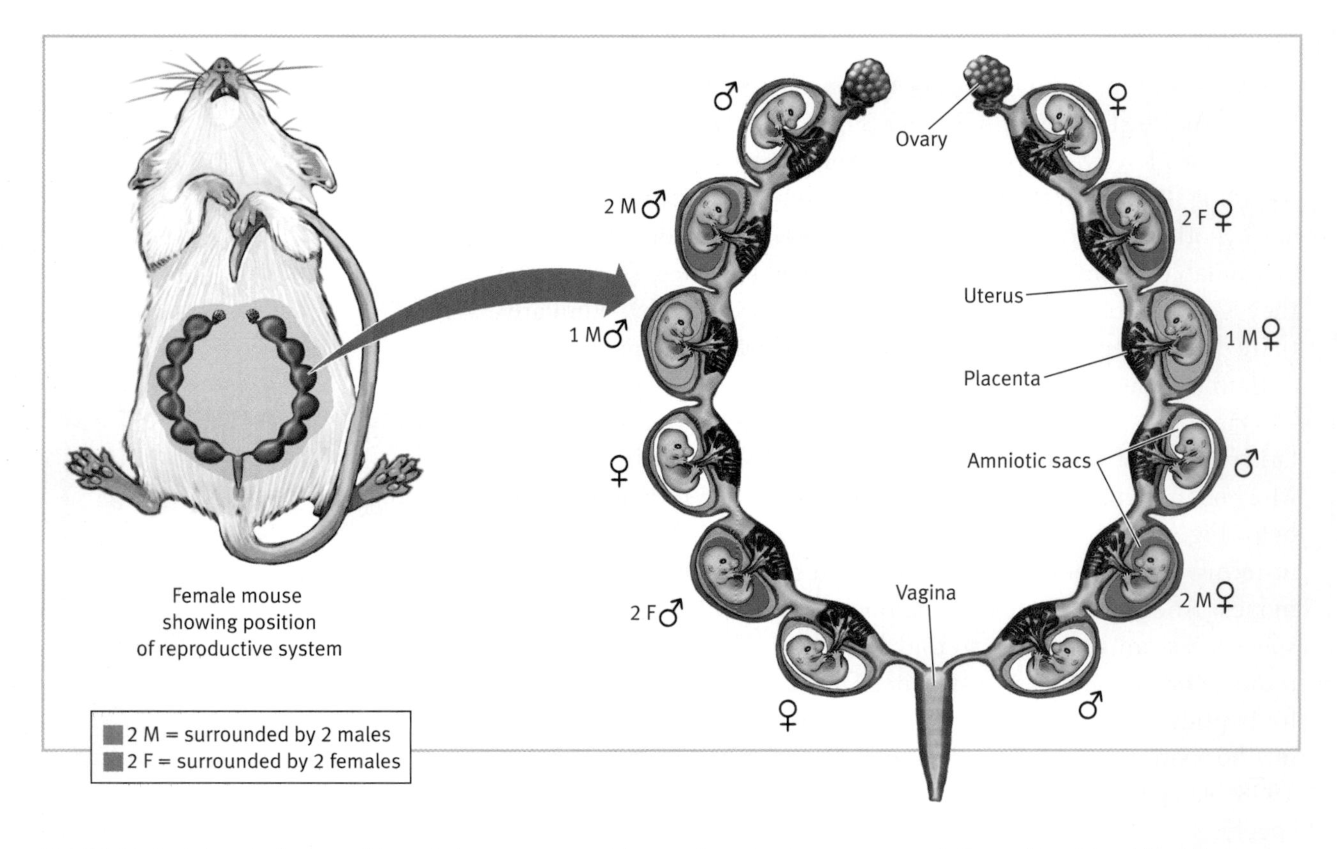

FIGURE 3.8. Intrauterine position. Males surrounded by two females in utero act relatively "feminized," while females surrounded by two males act relatively "masculinized." These behavioral differences are typically a result of differential exposure to hormones while in utero. 1M = surrounded by one male; 2M = surrounded by two males; 2F = surrounded by two females. *(From vom Saal, 1989)*

completely uninterested in mating with opposite-sex individuals. These males did not just fail to mate with females, they didn't even attempt to mate, and they were labeled "asexual."

To examine the parental behavior of all males, individuals were placed with a female that they had not encountered before on the day she delivered a litter of pups. This procedure assured that all males, even "asexual" ones, could have their parental care abilities measured. What Clark and Galef found was that the degree of parental care was negatively correlated with the strength of a male's interest in mating in the earlier part of the trial. In other words, the most parental care was displayed by asexual males, and the least by those who mounted and inseminated females. Furthermore, parental care displayed by asexual males was significantly greater than that seen in other experiments with 2M males (Clark, et al., Vonk, and Galef, 1998). These differences appear to be related to circulating levels of testosterone, as this hormone is known to increase sexual, but decrease parental behavior. The second of Clark and Galef's experiments directly examined circulating testosterone in males. As expected, asexual males had the lowest circulating testosterone levels of any group of males tested. It should be noted that examining circulating levels of testosterone in *adult* males makes it difficult to distinguish between cause and effect here. That is, differences in circulating levels of testosterone may cause differences in male parenting behavior or they may be the result of such behavioral differences. In support of the causal role of testosterone in Mongolian gerbil parental behavior, Clark and Galef (1999) have found that when adult or adolescent males are castrated, their parental behavior increases. When such castrated males have their testosterone levels brought back to normal, through silicon implants, their parental behavior decreases.

TESTOSTERONE AND PLAY FIGHTING IN RATS

Male rodents fight with one another from early on in development. While such fights among adult males can be very dangerous, aggression among younger rodents often takes the form of a much less dangerous "play fighting" (see Chapter 15; Figure 3.9). Fighting ability can have significant effects on male reproductive success in many rodent species (an ultimate issue), but here we are interested in what the proximate factors affecting fighting behavior might be. Testosterone has long been associated with many stereotypic male behaviors, particularly aggression (Moore, 2000).

To best examine if and how testosterone affects aggressive behavior, experimental *manipulations* are ideal, as examining the relationship between testosterone and aggression otherwise makes it difficult to assign cause and effect. For example, suppose you hypothesize that males with high levels of testosterone are more likely to win fights than are males with lower levels of testosterone. You then measure testosterone in animals after a fight is complete and find that indeed males that won aggressive encounters had higher levels of testosterone than their losing counterparts. In this scenario, you could not distinguish cause and effect with respect to testosterone's effect on fighting—winning might be a result of high levels of testosterone, or high testosterone levels might be a result of winning. Furthermore, even if you were able to measure testosterone levels in males before they fought, cause and effect would still be difficult to ascertain whether or not victors had high levels of pre-fight testosterone. In such a case, it might be that some other factor increased both testosterone and the probability of

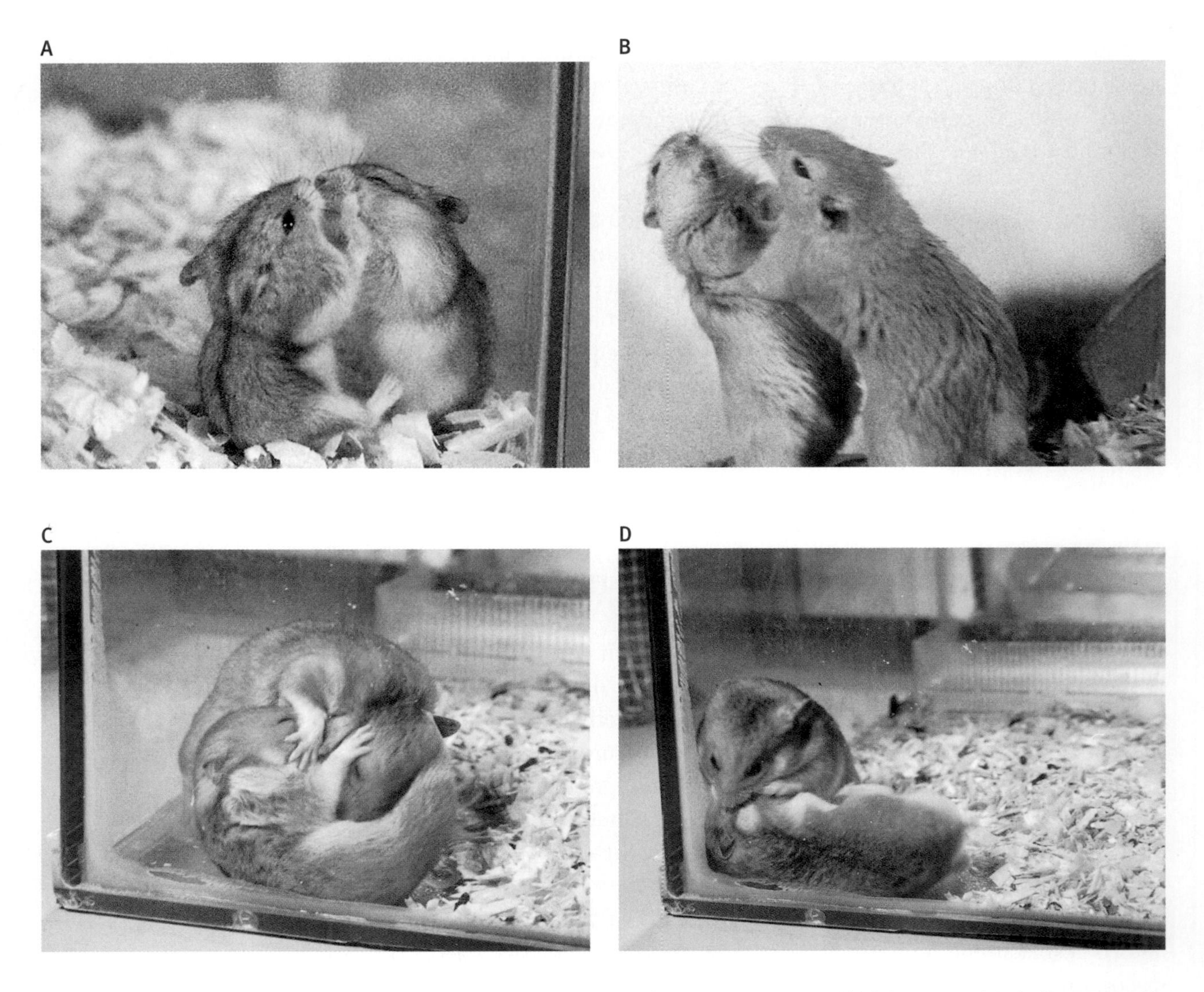

FIGURE 3.9. Play fighting in rodents. (A) Djungarian hamsters (*Phodopus campbelli*); **(B)** Fat sand jirds (*Psammomys obesus*); **(C** and **D)** Syrian golden hamsters (*Mesocricetus auratus*). *(Photo credits: Serge Pellis)*

victory, but that testosterone level per se didn't impact fighting abilities (Figure 3.10).

To see how experimental work can distinguish between cause and effect with respect to hormones and aggression, let us examine the work of Serge Pellis and his colleagues on fighting, play fighting, and androgens in rats (Pellis et al., 1992a). Pellis and his colleagues injected neonatal male rats with either testosterone propionate (TP) or an oil substance (as a control). That is, they manipulated the system in order to experimentally examine the effect of testosterone on aggression. During weaning, each juvenile male was paired up with an untreated one of his brothers. Play fighting

FIGURE 3.10. Causation and correlation. Just because animals with naturally high levels of testosterone (T) are more aggressive does not mean that high T causes aggression. Some other factor may cause both high T and increased aggression. Controlled experiments are often needed to distinguish between correlation and causation.

of TP- and oil-treated rats was compared between days thirty and thirty-six, and true fighting was observed when the males reached adulthood. From a proximate perspective, it appears that testosterone levels affect play fighting, but they do not affect dominance rank in adulthood. Pellis and his colleagues found that the rate of initiating playful attacks was significantly greater for TP-treated rats, but the eventual dominance rank in adulthood was independent of which treatment individuals were placed in early in life, suggesting that different (but still unknown) proximate factors control fighting as individual male rats mature.

Neurological Underpinnings of Behavior and Proximate Causation

In addition to the endocrine system, the nervous system also clearly plays an important role in understanding proximate aspects of animal behavior. A strictly chemical communication system, such as that of the endocrine glands, has its limitations, particularly with respect to speed of response when an instantaneous behavioral response is required. The nervous system provides an electrical impulse system ideally designed for instantaneous responses (Gould and Keeton, 1996; Gazzaniga et al., 1998; Table 3.2).

Animals possess specialized nerve cells called neurons that share certain similarities, regardless of what message they conduct. Each neuron has a cell body that contains a nucleus and one or more nerve fibers. Fibers that receive electrical information from other cells are called **dendrites,** while those fibers that transmit information to the next cell in the electrochemical cascade that makes up the nervous system are referred to as **axons.** A neuron may have thousands of dendrites, thereby receiving information from a correspondingly large number of other cells. Each neuron, however, typically has only a single axon, which is generally much thicker and longer than the average dendrite. At the tip of the axon are synaptic terminals, which convey electrical impulses to the next cell in the sequence (Figure 3.11).

THE NERVOUS IMPULSE

Let's trace what happens when the nervous system responds to something in the environment in which an animal lives. In response to an external stimuli—let's say a touch—a wave of electri-

TABLE 3.2 Comparisons of nervous system and endocrine system. The nervous and endocrine systems differ with respect to such variables as the mechanism of control, affected cells, type of action that results, time to onset of the action, and the duration of the action. *(From Tortora and Grabowski, 1996)*

CHARACTERISTIC	NERVOUS SYSTEM	ENDOCRINE SYSTEM
Mechanism of control	Neurotransmitters released in response to nerve impulses.	Hormones delivered to tissues by blood.
Affected cells	Muscles, glands, other neurons.	Almost all body cells.
Resulting action	Muscles contract or glands secrete.	Changes in metabolic activities.
Time to onset	Generally milliseconds.	Seconds to hours or days.
Duration	Brief.	Long.

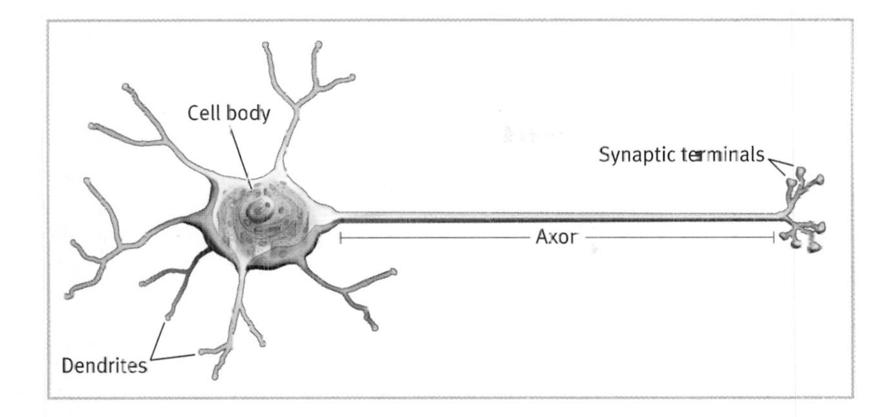

FIGURE 3.11. Nerve cell. Information is collected by dendrites, conducted along an axon, and transmitted to neighboring cells from the synaptic terminals.

cal activity sweeps down along the axon of a sensory nerve cell. For this process to begin, however, the stimulus must exceed the nerve cell's "threshold." Stimuli that don't meet this threshold fail to cause the nerve cell to fire, and stimuli above the threshold always cause the nerve cell to fire.

Nerve cell thresholds are the equivalent of on-off switches, rather than a dimmer switch. Any stimulus greater than the threshold, regardless of how much greater, causes the nerve cell to fire. In other words, the nerve cell fires in exactly the same manner, whether the stimulus is 1 percent greater than the threshold or 1,000 percent greater. Organisms can, however, use their nervous system to gauge the strength of a stimulus—how hard they are being touched in our case—in at least two ways: (1) the number of times that a neuron fires increases with the strength of the stimulus, and (2) the number of neurons that fire also increases as a function of stimulus strength.

Once an impulse has reached the end of an axon, it must be transmitted to the next nerve cell. This can be accomplished by an electrical impulse jumping across a synaptic gap between neurons, or more commonly by the release of a neurotransmitter chemical—acetylcholine, for example—that is released by thousands of synaptic vesicles located at the end of an axon. Neurotransmitters migrate across the synaptic gap and are absorbed by the membrane of the next neuron, which in turn causes that neuron to fire. Once the neurotransmitter has served its function, it is destroyed by the nerve cell. This process continues over and over as the impulse migrates along its neural path.

The neural pathway that was initiated by our animal being touched may end when the terminal neuron in the pathway innervates (stimulates) some effector—in our case, it might be the jaw muscle (if the animal was searching for food) or the muscles associated with fear avoidance (if the touch was that of a potential predator). In other instances, depending upon the stimulus, nerve pathways might end with the secretion of a neurotransmitter that causes an endocrine gland to secrete a hormone.

At a very general level, nervous systems across many animals show a number of consistent evolutionary trends (Gould and Keeton, 1996):

- Nervous systems become centralized, with longitudinal nerve cords becoming the major highway across which nervous impulses travel.
- Electrical currents begin to travel only in one direction, as opposed to nerve nets that have bidirectional travel of electrical impulses.
- Nerve cells that serve different functions become segregated, so that functional groupings of neurons become clear.
- The front end of the longitudinal nerve cord becomes dominant, leading to the creation of a brain.

With this brief review of the animal nervous system in hand, we can now move on and examine how ethologists might address proximate questions using a neurobiological, or neuroethological, approach.

VOCALIZATIONS IN PLAINFIN MIDSHIPMAN

Many questions of the form "how do animals undertake behavior X" can be analyzed at the neurobiological level (see Bass, 1998), including at the level of neurotransmitters (Edwards and Kravitz, 1997). Bass's work on the vocalizations of the plainfin midshipman (*Porichthys notatus*) demonstrates just how nicely neurobiology can be employed to address both proximate and ultimate questions (Bass, 1996; Figure 3.12). In this species of fish, there are what Bass and his colleagues refer to as type I and type II males (Ibara et al., 1983; DeMartini, 1990; Brantley and Bass, 1994). Type I males build nests and produce sounds to court females (Table 3.3). They generate either short duration grunts when engaged in aggressive contests with other males, or longer duration "hums" when courting a female (Brantley and Bass, 1994; Lee and Bass, 1994; McKibben et al., 1995). If females choose a type I male, they remain on the nest and lay eggs there. Type II males are "sneakers" (see Chapter 6 for more on this behavioral strategy) that do not build nests or sing, but rather lurk around type I nests, where they shed sperm in an attempt to fertilize the nesting female.

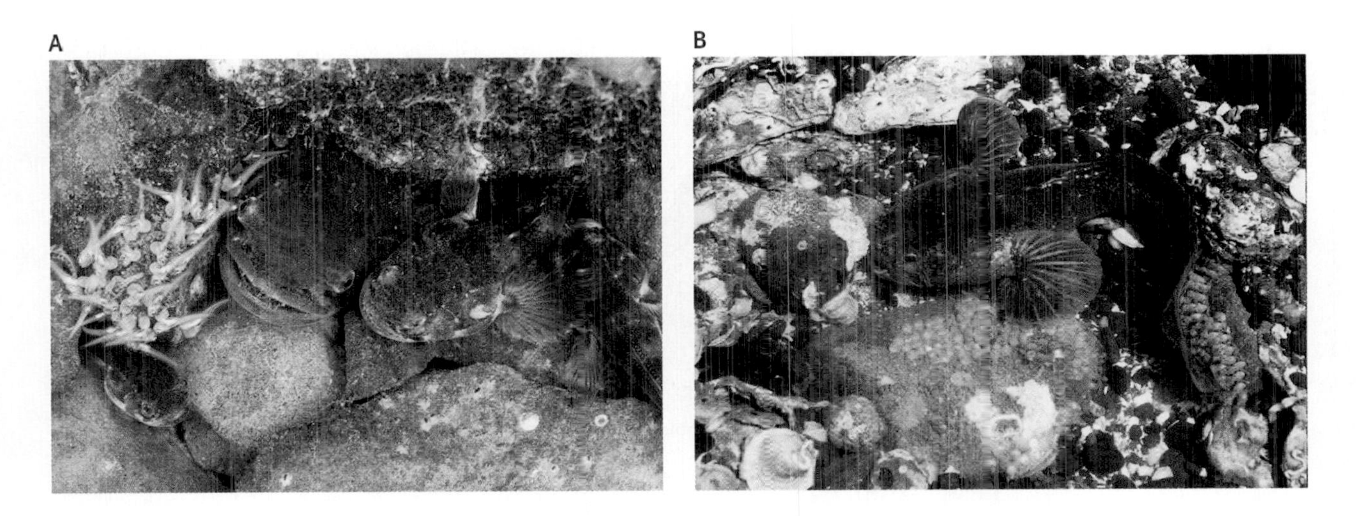

FIGURE 3.12. Vocal fish. In plainfin midshipman, some male types produce vocalizations while others do not. **(A)** The two smaller fish on the ends are type II sneaker males (who do not sing), while the fish that is second from the left is a "singing" type I parental male. **(B)** A type I male in his nest with his brood attached to the rocks. *(Photo credits: Andrew Bass)*

TABLE 3.3 Traits of type I and type II males. A summary of the differences between type I and type II plainfin midshipman males, and a comparison with plainfin midshipman females. Type II males are more similar to females than to type I males.

SEXUALLY POLYMORPHIC TRAITS	TYPE I MALE	TYPE II MALE	FEMALE
nest building	yes	no	no
egg guarding	yes	no	no
body size	large	small	intermediate
gonad-size/body-size ratio	small	large	large (gravid), small (spent)*
ventral coloration	olive-gray	mottled yellow	bronze (gravid), mottled (spent)*
circulating steroids	testosterone, 11-ketotestosterone	testosterone	testosterone, estradiol
vocal behavior	hums, grunt trains	isolated grunts	isolated grunts
vocal muscle	large	small	small
vocal neurons	large	small	small
vocal discharge frequency	high	low	low

**Gravid connotes pregnant; spent connotes post-pregnant.*

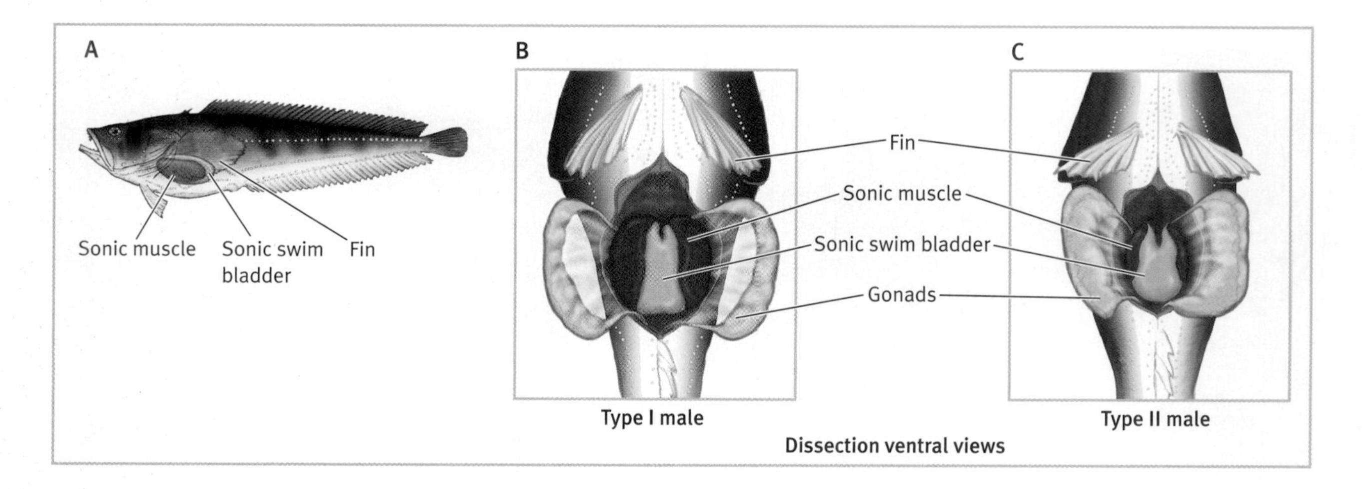

FIGURE 3.13. Sonic muscles and vocalization. (A) The vocal organ of the plainfin midshipman is made up of a pair of sonic muscles attached to the walls of the swim bladder. **(B)** Sonic muscles of type I males are well developed in comparison with muscles from type II males **(C)**. *(From Bass, 1996, p. 357)*

Type I and II males differ in many regards. For example, type I males are two to four times larger than type II males, but the gonad-to-body ratio of type II males is *nine times greater* than that of their type I counterparts. In males, the vocal organ of the midshipman is a set of paired sonic muscles attached to its swim bladders (Figure 3.13). But what causes the differences in sound production between the two types of males? From a proximate perspective, at least part of the answer lies in the fact that type I males have larger sonic muscles, more muscle fibers, different neuronal rhythmic firing properties, differences in a "vocal pacemaker-motor-neuron circuit," and myofibrils that have subtle, but important differences from those of type II males, (Bass and Marchaterre, 1989; Bass and Baker, 1990, 1991; Bass, 1992, 1996).

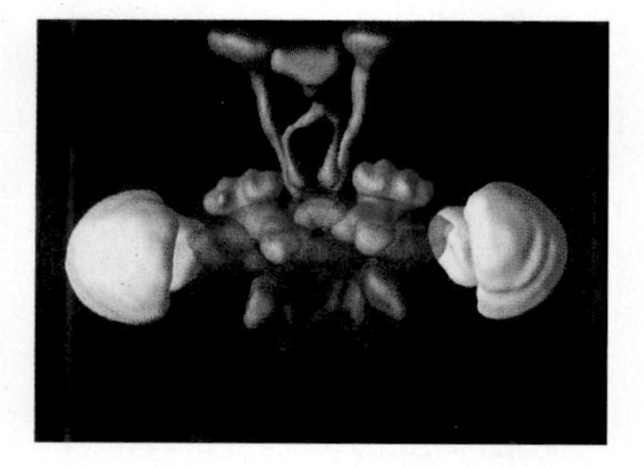

FIGURE 3.14. Mushroom bodies. This small cluster of neurons located at the front of the bee brain is involved in spatial learning. *(Photo credit: Birgit)*

MUSHROOM BODIES, INSECTS, AND LEARNING

The ability to remember and retrieve information from the environment is obviously critical for many aspects of animal behavior. In vertebrates, this spatial learning is often associated with the hippocampus. In invertebrates, however, spatial learning is most often linked with a cluster of small neurons located at the front of the brain. This cluster, known technically as the corpora pedunculata, is often referred to as the **mushroom bodies** (Capaldi et al., 1999; Figure 3.14). Both the mushroom bodies and the hippocampus are characterized by elevated levels of the products of genes that are often associated with learning-related activities (Davis, 1993; Kandel and Abel, 1995).

One species in which the mushroom bodies play a role in spatial learning is the honeybee (*Apis mellifera*). These tiny animals often travel many kilometers in search of food (Visscher and Seeley, 1982). Honeybees clearly learn various aspects about their foraging environment, and hence they are ideal for examining the neuroethology of spatial learning. Whereas younger bees generally reside within their nest, older bees primarily give up "in hive" tasks and take to foraging for the colony. Such foragers then use both visual and olfactory cues in their search for food.

One very prominent type of honeybee learning is the "orientation" flight that foragers make when first leaving a hive to forage. When they first leave a hive, rather than making a typical "beeline" for food sources, would-be foragers often turn back toward the hive and hover up and down for several minutes. This behavior apparently orients the foragers to the position of their nest. Bees often begin undertaking orientation flights at about one week old and commence foraging when they are about three weeks old. When examining the size of mushroom bodies in bees of different ages and bees that undergo different tasks within a colony (foraging versus "nursing"), Withers and his colleagues (1993) found that the mushroom bodies of foragers were 14.8 percent larger than those of the other groups they measured, suggesting a strong proximate role for mushroom bodies in honeybee spatial learning and foraging. This is a particularly striking finding because the relative volume of other nerve clusters in the bee brains remained relatively unchanged in foragers, with only the mushroom bodies increasing in relative size.

Having found that mushroom body size increases with age (Farris et al., 2001) and with increases in juvenile hormone, the researchers examined whether foraging-related experience also affected growth in mushroom body volume (Withers et al., 1993). To do this, they induced early foraging behavior in a colony of bees such that seven-day-old bees began foraging, whereas normal bees begin this task at about twenty days of age. The configuration of the mushroom bodies in these precocious foragers resembled that of older foragers, suggesting that activities surrounding foraging, including both foraging per se and an increase in juvenile hormone associated with foraging, may also directly trigger a series of neural-based changes in mushroom body volume (Withers et al., 1993; Sigg et al., 1997). Further support for this hypothesis comes from the finding that similar changes in mushroom body growth occur in drones (Coss et al., 1980; Brandon and Coss, 1982), as well as in other insects such as ants (Gronenberg et al., 1996). Moreover, corresponding experience-based changes in brain neuronal structure have been uncovered in the hippocampus of birds and rodents (Greenough and Chang, 1988; Black et al., 1990; Krebs et al., 1996; Basil et al., 1996).

SLEEP AND PREDATION IN MALLARD DUCKS

Having read thousands of papers on animal behavior, I am seldom "wowed" by a study. One exception is John Rattenborg and his colleagues' studies of sleep behavior and antipredator strategies in mallard ducks (*Anas platyrhynchos*). Birds, it seems, have the incredible ability to sleep with one eye open and one hemisphere of the brain awake (Ball et al., 1988). In essence, these birds sleep with half their brain awake and half their brain asleep (Rattenborg et al., 1999a, 1999b). The researchers examined unihemispheric sleep in mallards and found that not only could they sleep with one eye open, but that mallards that were in dangerous positions in a group used such a technique more than birds in the center of a group, and that birds in danger slept with the appropriate eye—the eye facing the area of potential danger—open (Figures 3.15 and 3.16).

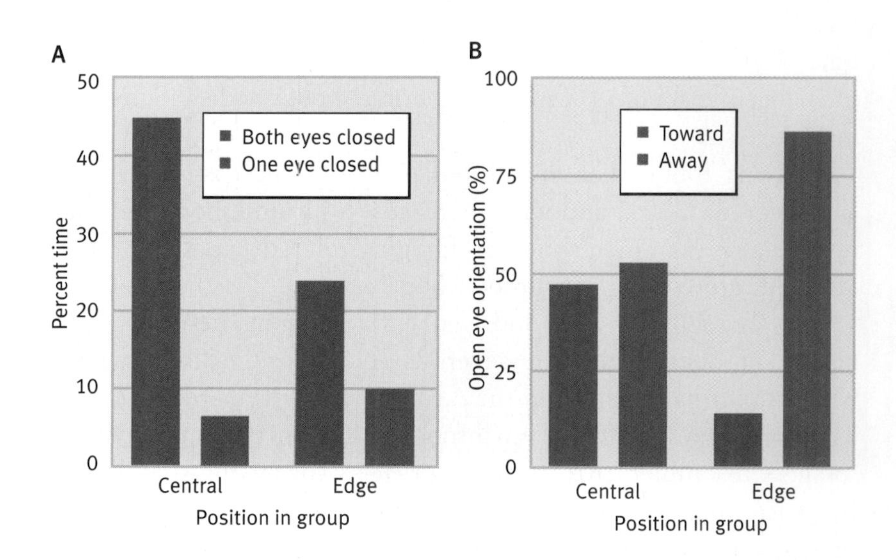

FIGURE 3.15. Unihemispheric sleep. (A) Percent of time ducks spent with one eye closed or both eyes closed as a function of position in the group (at the group's center or on its edge). **(B)** When ducks were at the edge of a group and had one eye open, they spent much more time looking away from the group's center than when they had one eye open and were at the center of the group. *(From Rattenborg et al., 1999a)*

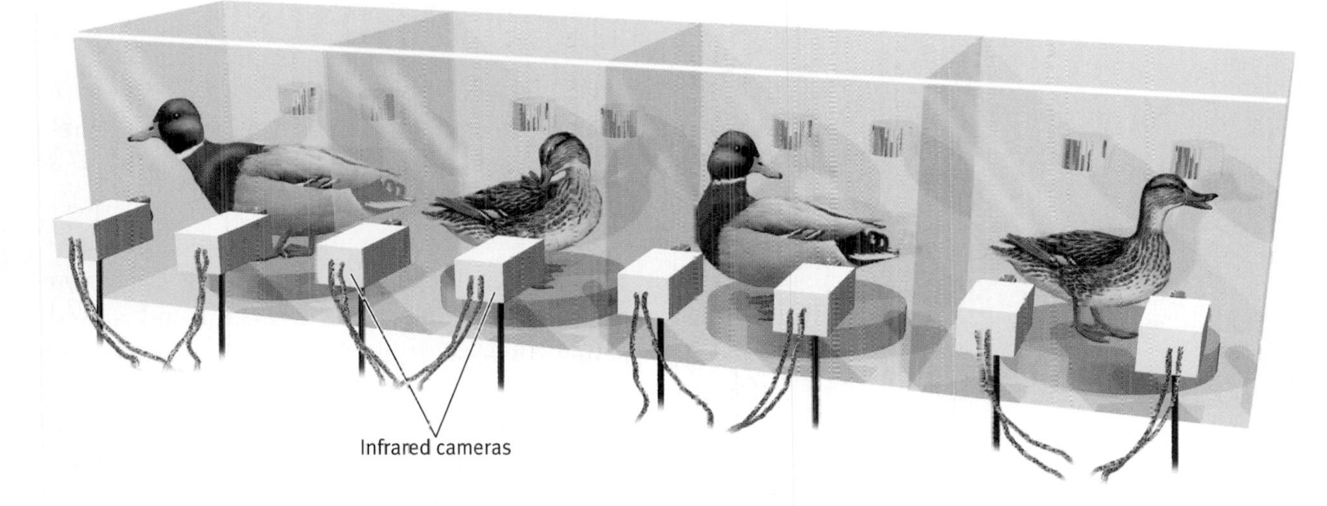

FIGURE 3.16. Sleeping apparatus. This experimental housing unit was employed to record eye state and electrophysiology of four mallard ducks. Eight infrared cameras were used to allow the movement of each eye of each mallard to be recorded. Birds on the extreme left and right were considered to be on the edges of the group. *(From Rattenborg et al., 1999a)*

How do mallards manage this split-brained sleep? It seems that these birds are capable of putting one hemisphere of the brain into what is called slow-wave sleep. This state allows quick responses to predators, but it does not interfere with the sleeping half of the bird's brain until danger is present. Rattenborg and his colleagues demonstrated this by measuring electroencephalograms (EEG) and electromyographic (EMG) activity. Slow-wave sleep in birds has a characteristic high power in the low frequency range of 1–6 Hz, and this frequency is quite different from other states of sleep or wakefulness. EEG recordings show that the part of the brain controlling the open eye during unihemispheric sleep showed the low frequency range characteristic of slow-wave sleep.

Biochemical Factors

Many proximate explanations of behavior can be analyzed at the biochemical level. Thus, hormonal and neurological explanations may be analyzed at the biochemical level, as well as at the endocrinological and neuronal levels. Other phenomena, however, by their nature *need* to be addressed at the biochemical level, as they are difficult to interpret at any higher level of complexity.

ULTRAVIOLET VISION IN BIRDS

The effect that a single change at the biochemical level can have on traits that are important to animals (as well as ethologists) is remarkable. Consider the case of ultraviolet vision in birds. To the surprise of many animal behaviorists, it turns out that ultraviolet (UV) vision is found in a wide array of animals, including fish, amphibians, reptiles, birds, and mammals (Jacobs, 1992). UV vision in vertebrates is determined by retinal visual pigments and is used in the context of mating (Bennett et al., 1996, 1997; Hunt et al., 1997), foraging (Church et al., 1998), hunting (Viitala et al., 1995), and social signaling (Fleishman et al., 1993). As such, UV vision has clear ultimate consequences, but until recently, little was known about the proximate side of the coin. What allows some organisms, but not others, to see in this wavelength?

Using molecular genetic cloning techniques, Yokoyama and his colleagues (2000) examined UV vision in zebra finches (*Taeniopygia guttata*; Figure 3.17). Zebra finches are able to see in UV, and via *a single amino acid change* (C84S), the researchers were able to transform the ultraviolet pigment—the substance that allows the birds to see in UV—in finches into a violet pigment that doesn't allow for UV. Thus, with the change of a single amino acid—the swoop of a genetic magic wand of sorts—ultraviolet light perception was gone and violet perception was present.

Even more remarkably, Yokoyama and his colleagues performed the reverse experiment. To accomplish this, they worked with chick-

FIGURE 3.17. A small change goes a long way. A single amino acid change is all that separates the coding for ultraviolet versus violet perception in the zebra finch. *(Photo credit: Ann and Rob Simpson)*

ens and pigeons who possess violet pigments, but not true ultraviolet pigments (pigeons can actually see in UV, but they don't possess true UV pigments; Kawamura et al., 1999). Using sections of DNA that were very similar to those found in finches, they were able to generate true ultraviolet pigments from violet pigments, again by changing a single amino acid. One change in amino acids, and pigeons and chickens were able to see in the UV wavelength via true ultraviolet pigments. This sort of proximate level analysis has clear implications for our understanding of the evolution of UV vision in birds, for if a single change in amino acids causes such a dramatic shift in vision, this undermines arguments that UV vision is "too complex" a trait to evolve via natural selection.

Environmental Sex Determination and Sex Ratios

There is a somewhat indirect way in which the environment itself can be cast in the role of a proximate agent of behavior—sex determination. To be fair, we are interested in animal behavior, and the sex of an individual is not a behavior per se. But the sex of an individual is so fundamental to its behavioral processes, and hence to ethology, that it's worth examining how the environment itself can affect sex determination.

REPTILES, SEX DETERMINATION, AND TEMPERATURE

For most animals, sex determination is a function of which sex chromosomes are inherited from the mother and the father, and the determination of sex is not directly influenced by the environment into which the young are born. This, however, is not the case for some reptile species.

In at least forty families of reptiles—most particularly turtles (Figure 3.18) and crocodilians—sex determination is strongly influenced by temperature and nest placement (Janzen and Paukstis, 1991; Ciofi and Swingland, 1997). These groups typically lack distinct sex chromosomes. Ambient nest temperature during the middle third of incubation influences the development of gonads in developing fetuses, and hence determines sex. Jim Bull (1980,

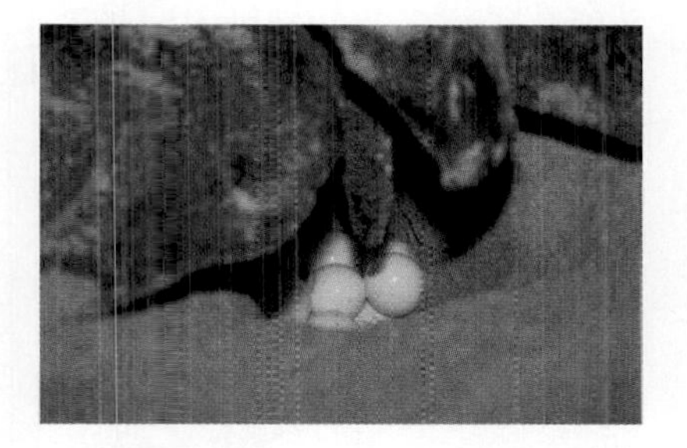

FIGURE 3.18. Atlantic leatherback sea turtle laying eggs. In many species of turtles, the sex of the young is dependent on temperature. *(Photo credit: Ann and Rob Simpson)*

1983) has identified three general patterns of reptilian temperature-dependent sex determination. In Type A groups, composed primarily of lizards and crocodiles, high nest temperatures produce males and low temperatures increase the frequency of females. Type B groups, which include many turtle species, are the converse, with females resulting from high nest temperatures, and males the product of low nest temperatures. Finally, Type C groups, which are a mix of crocodile, turtle, and lizard species, produce females at high and low temperatures, but males at intermediate temperatures (Figure 3.19).

The evolutionary history and ultimate reason for sex determination remain hotly debated, but a proximate understanding is

A

Type A:
Hot sun (high nest temperatures)—male-biased sex ratio.

Type A:
Weak sun (lower nest temperatures)—female-biased sex ratio.

B

Type B:
Hot sun (high nest temperatures)—female-biased sex ratio.

Type B:
Weak sun (lower nest temperatures)—male-biased sex ratio.

FIGURE 3.19. Temperature-dependent sex ratios in reptiles. The sex ratio in clutches of different types of reptiles is a function of temperature. **(A)** Crocodiles are Type A, and **(B)** turtles are Type B.

coming into clear focus with more and more work being undertaken on precisely how temperature mediates sex determination. Since numerous reptilian (especially turtle) species are on the endangered species list, an understanding of proximate causation and sex determination is critical. Conservationists incubate perhaps millions of turtle eggs in artificial hatcheries and have only recently started paying close attention to how temperature may dramatically skew the sex ratio of turtles that emerge from such incubations, and how such skews may even be employed to promote conservation causes (Janzen and Paukstis, 1991; Girondot et al., 1998).

RED DEER, DOMINANCE STATUS, AND SEX RATIOS

Although temperature has no effect that we know of on mammalian *sex determination*, environmental factors do influence *sex ratio* as well as mating (Crews and Moore, 1986). To understand how, we must understand basic sex ratio theory, first elucidated by Sir Ronald Fisher, a mathematician and evolutionary biologist. From an evolutionary perspective, Fisher's sex ratio theory predicts that natural selection should generally favor a sex ratio of unity (Fisher, 1930). Why? Imagine a population in which the sex ratio deviated from 1:1. Now, any genes that coded for overproducing the rarer sex would be favored, and this would quickly bring the population back to 1:1, as Fisher's general theory predicts. The catch is that the above theory hinges on the phrase "all else being equal." When all other things aren't equal, Fisher's theory no longer predicts a 1:1 sex ratio. In particular, we are interested in how sex ratio is affected if one sex can differentially benefit from more **parental investment**.

If, in fact, one sex does differentially benefit from each unit of additional parental investment, natural selection will favor such parental investment in the sex that benefits more (Trivers and Willard, 1973; Charnov, 1982; G. R. Brown 2001). For example, imagine a case in which male offspring benefit more from being large than do female offspring. In this situation, every unit of parental effort put into raising large offspring pays more when that effort is directed at male offspring. Now a mother in good condition—that is, one with sufficient resources to produce large males—may invest more in male offspring, leading to a male-biased sex ratio in her offspring (Clutton-Brock and Iason, 1986; Clutton-Brock and Godfrey, 1991). Mothers in poor condition, on the other hand, may be favored to invest more in daughters, since they don't have the resources to produce large, and hence competitively superior, male offspring. In other words, if larger male offspring obtain the lion's share of critical resources (mates, food, and so on), then a mother in poor condition should divert resources away from males—males that wouldn't do particularly well anyway—and toward daughters.

From a proximate perspective, we can ask what social/environmental cue prompts females to invest more in males or females, and to do so we shall examine Tim Clutton-Brock and his colleagues' (1984) work on red deer (*Cervus elaphus*; Figure 3.20). In red deer, there is much more variation in male than female reproductive success. Large male red deer have much greater reproductive success than their smaller counterparts. The variance that exists in female reproductive success does not appear to be tied to the size of females. Among red deer females, rank in a hierarchy does, however, affect the reproductive success of offspring, and it does so in a sex-biased way. In terms of lifetime reproductive success, sons of dominant females benefit more than do daughters. Adopting the logic described earlier, from an evolutionary perspective, we expect dominant females to invest more in male offspring, thereby biasing the sex ratio of their offspring toward males, and subordinate females to invest more in female offspring, thereby biasing the sex ratio of their offspring toward females. And indeed, this is precisely what Clutton-Brock and his colleagues found when analyzing twenty years of sex ratio data in red deer. It appears that a female's rank in the dominance hierarchy, and the rank's

FIGURE 3.20. Red deer. Tim Clutton-Brock's long-term study of red deer has, among other things, examined the role of female condition and position in a hierarchy on the sex ratio of her offspring. *(Photo credit: Ann and Rob Simpson)*

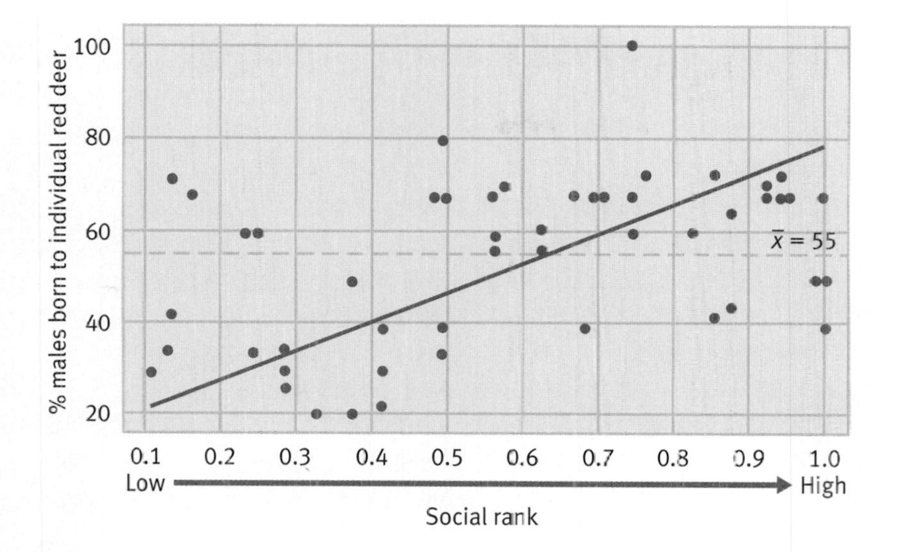

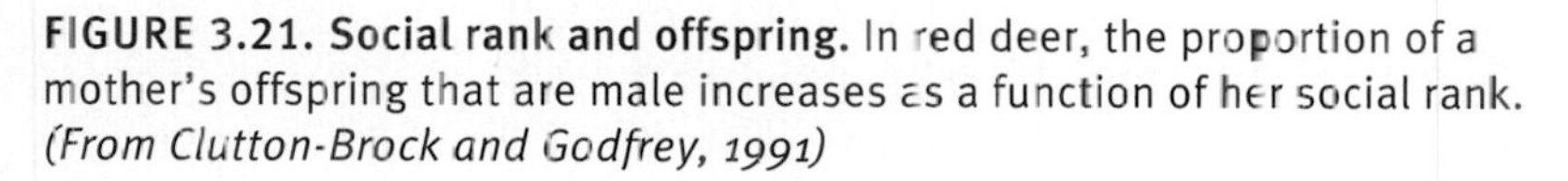

FIGURE 3.21. Social rank and offspring. In red deer, the proportion of a mother's offspring that are male increases as a function of her social rank. *(From Clutton-Brock and Godfrey, 1991)*

cascading effects on such things as hormone levels, provides the proximate cue that in the end produces skewed sex ratios. In other words, a female uses her low or high position in a hierarchy, as well as the physiological effects of her position in a hierarchy, to gauge how to allocate resources to produce more male or female offspring (Figure 3.21).

Genes and Proximate Explanations

It may seem strange to think of studies on genetics as somehow proximate in nature. After all, most of the primer chapter on natural selection (Chapter 2) was written in an evolutionary light, and typically focused on how selection favors one behavioral genetic variant over another. In Chapter 2, we also examined how molecular genetics can be used to test various ethological hypotheses. Genes, however, can also be used in the proximate explanation of a trait in a very straightforward matter. If, rather than expounding on what forces are involved in explaining gene frequency change, a study is instead designed to address which specific gene or set of genes code for a trait, then we are using genes as a proxi-

Interview with Dr. TIMOTHY CLUTTON-BROCK

While you have worked in a number of different systems, you are perhaps best known for your studies of red deer. What initially drew you toward working in this system?
I have always been drawn toward working on mammalian systems, perhaps partly because of their greater relevance to understanding human behavior. I had initially worked on primates but wanted to find a mammalian system which did not do everything 100 feet up above my head. In particular, I wanted to find one where visibility was good, animals could be caught and measured, and their food supplies could be monitored. Red deer populations living in the (virtually) treeless Highlands of Scotland filled this bill. They are also of economic importance, which probably made raising the necessary funding easier in the early days of the red deer project.

You've recently focused on working with cooperatively breeding meerkats, where a single pair of individuals dominates the reproductive landscape in a group. How has your work on the endocrinology of the species helped you better understand the complex behavioral dynamics that underlie meerkat society?
Our work on endocrinology in meerkats is still in its early stages, and our main discoveries have concerned the division and organization of labor. However, the work so far emphasizes the fundamental difference between the sexes: while subordinate males show levels of

steroid hormones similar to dominants, subordinate females show lower levels of circulating estrogen (though they are responsive to stimulation with GNRH) and may also show higher prolactin levels than dominants. Our results for females raise the issue of whether dominant females manipulate the endocrine status of subordinate females to maintain their contributions toward cooperative rearing, which we are investigating at the moment. This, in turn, raises the broader question of the extent to which the responses of dominant animals in social groups are adjusted to maximizing benefits for particular interactions or whether they are adjusted to maximizing longer-term benefits from the influence of their actions on the physiological status of other group members or even on the composition of the group itself.

What are your thoughts on presenting proximate problems as "how" questions, and ultimate problems as "why" questions? Generally speaking, how do you think proximate and ultimate perspectives interact in helping us to better understand animal behavior?
I avoid contrasting "proximate" and "ultimate" perspectives, for the distinction varies between scientists from different disciplines working at different levels. As a result, one person's proximate perspective is another's ultimate perspective. But I will make two comments that bear on this issue at different levels. First, it is obviously important to distinguish between causal and functional explanations of behavior. In the 1950s, this distinction was not always made and this led to considerable confusion. Today, few behavioral biologists would fail to differentiate between causal and func-

tional explanations, though it is important to remember that many other biologists work exclusively at the causal level and are less familiar with this distinction. Second, it is important to distinguish clearly between demographic trends (by which I mean patterns of reproductive success or survival) and the behavioral mechanisms underlying them. In most cases, several different behavioral mechanisms may generate similar demographic trends, and it is important to know what behavioral mechanisms are involved. For example, it was suggested that the widespread tendency in passerine birds for smaller young to show higher mortality was a result of parents preferentially provisioning larger chicks that had a better chance of survival. However, there was little evidence that parents preferentially fed larger young, and most studies that have been able to investigate this issue have found either that parents just feed the closest chick or that they favor small chicks or those that they have not just fed. There are quite a number of cases where suggestions about the behavioral mechanisms underlying demographic trends have been made on the basis of little or no direct behavioral evidence and have proved very resilient to counterevidence. My impression is that this is particularly likely to be the case where the suggested explanation is of particular theoretical interest.

IN MOST CASES, SEVERAL DIFFERENT BEHAVIORAL MECHANISMS MAY GENERATE SIMILAR DEMOGRAPHIC TRENDS, AND IT IS IMPORTANT TO KNOW WHAT BEHAVIORAL MECHANISMS ARE INVOLVED.

Could you give one example in which adopting both a proximate and an ultimate perspective in the red deer work proved to be particularly useful?

One case in the red deer project where knowing the behavioral mechanisms involved became important was the issue of sex difference in mortality among calves. Male deer calves are more likely to die in their first year of life than females. It had been suggested that where males required greater parental investment in order to breed successfully, mothers that could not afford to raise male calves would terminate care prematurely—and that this might explain why males showed higher mortality. However, a simpler explanation is that mothers do not discriminate between male and female calves but that male calves grow faster than females and are consequently more likely to die when resources are inadequate. In red deer, we found no evidence that poorer mothers discriminated against male calves, and there is virtually no evidence of parental discrimination against offspring of one sex in other vertebrates. In contrast, there is plenty of evidence that sexual selection can lead to increases in body size, growth rate, or reproductive strategies that are associated with reductions in survival when resources are in short supply.

DR. TIMOTHY CLUTTON-BROCK is a professor at Cambridge University, England. His long-term work on all aspects of red deer behavior and evolution is widely regarded as a classic case study.

mate causative factor. If a study finds that some variant of gene X is partly responsible for a variant of trait Y, we are casting genes in a proximate, rather than an ultimate light. Naturally, such studies abound in the genetic literature, and identifying the genes behind certain traits is a large-scale endeavor going on in hundreds of labs around the world today.

Another way in which we can think about genes as proximate causes of behavior is with respect to gene-based recognition. If animals can judge genetic similarity with conspecifics (others in the same species) based on cues, let's say the odor of urine, then the absence or presence of that cue can be used as a trigger for initiating or inhibiting some behavior. For example, suppose that an animal has a rule that it will act cooperatively with kin, but aggressively toward nonkin. Further suppose that individuals gauge relatedness by odor, and that differences in odor have a genetic basis. If we ask why individual A is cooperative with B, and answer, "because A sensed B had the same odor he did, and that indicates some level of genetic relatedness," we are speaking of genes in a proximate sense. If we then ask why individuals prefer to be altruistic or cooperative only with kin, we have switched levels and are now phrasing the question as an ultimate one.

Learning as a Proximate Factor

As we shall see in the next chapter (Chapter 4), thousands of experiments in psychology delve into how and what individuals learn by examining animals in scenarios in which they are rewarded for pushing down a bar (let's call it bar 1). In most instances, individuals quickly learn to push bar 1. In these cases, we can speak of learning being one proximate factor responsible for bar pushing. That is, we can answer the question "what causes individuals to push/peck bar 1?" with "they have learned that bar 1 provides a reward." Naturally, we can also ask why individuals have adopted such learning rules, but that would take us into the ultimate realm.

In addition to learning serving as a proximate factor at the level outlined above, researchers are also examining both the biochemistry (Fagnou and Tuchek, 1995) and neurobiology (Vanderwolf and Cain, 1994) of learning, thus expanding the way learning can be viewed as a proximate cause of behavior.

SUMMARY

1. Proximate causes include all those factors that are not evolutionary in nature. One way to distinguish between proximate and ultimate perspectives is to remember that the former tends to address "How is it that . . . ?" and "What is it that . . . ?" questions, while the latter addresses "Why is it that . . . ?" questions. If we were, for example, examining vision in birds, a proximate perspective might lead us to ask how the visual system is set up or what the neurobiological and molecular underpinnings of vision are. An ultimate, or evolutionary, approach, would focus on why the visual systems is designed as it is. One of the most frequently made errors by those who have not trained in the field of animal behavior is to confuse proximate and ultimate perspectives.
2. Proximate and ultimate approaches complement one another and together provide a comprehensive picture of the ethological trait under study. In order to understand why a trait is designed as it is (an ultimate question), an understanding of exactly how it is built (a proximate question) is often essential. Similarly, if we are interested in understanding how a trait is designed (a proximate question), our analysis can only be aided by knowing what the trait was designed to do (an ultimate question).
3. In this chapter we have seen how endocrinology, neurobiology, biochemistry, and environment can be construed as important factors in the proximate analysis of behavior.
4. The endocrine system is made up of ductless glands that secrete hormones into an animal's bloodstream. Correct functioning of the system is of critical importance to behavioral functions via the modification of rates and directions of various cellular functions. When the endocrine system malfunctions, through either diminished secretion (hyposecretion) or excessive secretion (hypersecretion), it may affect growth, metabolism, reactions to stress, aggression, and reproduction.
5. The nervous system provides an electrical impulse system ideally designed for instantaneous responses that are often a critical component to behavior. Animals possess specialized nerve cells called neurons that share certain similarities, regardless of what message they conduct.
6. One level of proximate analysis is the biochemical level. The effect that a single change at the biochemical level can have on traits that are important to animals (as well as ethologists) is remarkable, as was demonstrated by the study of ultraviolet vision in birds.

7. While genes are often thought of in terms of the role they play in natural selection, genes can also be employed in a proximate study of behavior. If I say "trait X is controlled by gene 1," I am analyzing genes through the lens of proximate causation.

DISCUSSION QUESTIONS

1. Make the case that with respect to any particular behavioral trait, one can't fully understand ultimate causation without understanding proximate causation and vice versa.
2. Find the 1998 special issue of *American Zoologist* (vol. 38), which was devoted to proximate and ultimate causation. Choose two different papers in this issue and compare and contrast how they try to integrate proximate and ultimate causation.
3. The lateral line in fish—a series of sensory hair cells running along the side of the body—appears to be used to detect various forms of motion. How might you experimentally examine whether one proximate explanation for the lateral line is predator detection? Could you then slightly modify your experiment to test the ultimate claim that predator detection via the lateral line has been favored by natural selection?
4. There has been some discussion, but much less experimental work, on the notion that animals shift the sex ratio of their offspring as a result of the sex ratio in their population. Construct at least three proximate explanations about how such a sex ratio shift might occur.
5. Elaborate on how learning itself may be a proximate explanation for certain animal behaviors but how the ability to learn is thought of in ultimate terms.

SUGGESTED READING

Adkins-Regan, E. (1998). Hormonal control of mate choice. *American Zoologist, 38,* 166–178. A review of the endocrinology of mate choice.

Alcock, J., & Sherman, P. (1994). The utility of the proximate-ultimate dichotomy in ethology. *Ethology, 96*, 58–62. A short defense of the proximate-ultimate classification.

Capaldi, E., Robinson G., & Fahrbach, S. (1999). Neuroethology of spatial learning: The birds and the bees. *Annual Review of Psychology, 50*, 651–682. A review of the neural underpinnings to spatial learning.

Finch, C., & Rose, M. (1995). Hormones and the physiological architecture of life history evolution. *Quarterly Review of Biology, 70*, 1–52. A thorough review of the effects of hormones on many aspects of development.

Gerhardt, H. C. (1991). Female mate choice in treefrogs: Static and dynamic acoustic criteria. *Animal Behaviour, 42*, 615–635. A nice illustration of how animal nervous systems play a critical role in behavioral decision making.

4

What Is Individual Learning?

How Animals Learn

- Learning from a Single-Stimulus Experience
- Pavlovian (Classical) Conditioning
- Instrument (Operant) Conditioning

Why Animals Learn

- Within-Species Studies and the Evolution of Learning
- Population Comparisons and the Evolution of Learning
- A Model of the Evolution of Learning

What Animals Learn

- Learning Where Home Is Located
- Learning about Your Mate
- Learning about Familial Relationships
- Learning about Aggression

Interview with Dr. Sara Shettleworth

Learning

Many putative prey species live in areas that contain both predatory and nonpredatory species, as well as predators that are in hunting mode at times, and nonhunting mode at other periods. If prey can distinguish between dangerous and benign encounters with potential predators, they may free up time for other activities. *Learning* about possible predation pressure may allow animals to more optimally handle the tradeoffs they face in everyday life.

FIGURE 4.1. Damselflies. Damselflies learn about predation threat through chemical cues. *(Photo credit: Ann and Rob Simpson)*

In aquatic systems, recognition of predators often involves processing chemical stimuli. If the food that a potential predator eats produces a chemical label that is recognizable to its prey, then this might provide an opportunity for such prey to learn what is dangerous and what isn't (Howe and Harris, 1978; Crowl and Covich, 1990; Keefe, 1992; Gelowitz et al., 1993; Mathis and Smith, 1993a, 1993b). Douglas Chivers and his colleagues (1996) examined this possibility by studying learning and antipredator behavior in the damselfly (*Enallagma spp*; Figure 4.1).

Damselfly larvae are found in ponds with minnows, and both species are often attacked and consumed by pike (*Esox lucius*). Chivers and his colleagues hypothesized that the damselfly larvae might learn about potential dangers posed by predators through some sort of chemical cues system. To test this hypothesis and gather background data on chemical cues and diet, the researchers fed pike either minnows, damselflies, or mealworms, where mealworms are a species not found in the ponds and served as a control. After four days on one of these diets, a pike was removed from its tank, and damselflies that had never before had any contact with a pike were exposed to the water from a pike's tank.

When exposed to the water containing chemical cues from a pike, damselflies were also given the opportunity to feed on brine shrimp. The researchers found that damselflies significantly curtailed their own foraging behavior when exposed to water with chemical cues from a pike that had eaten damselflies or a pike that had eaten minnows, but not when they were exposed to the water with chemical cues from a pike that had consumed mealworms (Figure 4.2). Since damselflies are found in the same ponds as minnows, but the damselflies here had never experienced a pike before, these results strongly suggest that damselflies innately associate the scent of pike plus damselfly or pike plus minnow with danger, but make no such association between pike, mealworm, and threat. That is, the damselflies here hadn't learned anything; they simply were predisposed to respond to the smell of pike and prey as dangerous.

Chivers and his colleagues followed up their initial experiment by examining the role of learning and antipredator behavior in damselflies. Here, they took the damselflies that had been exposed to

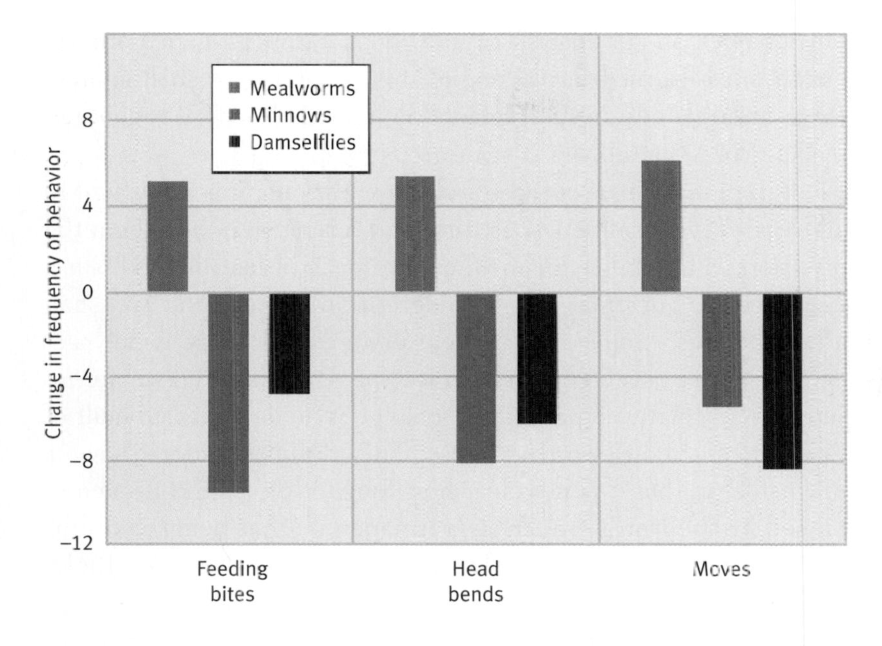

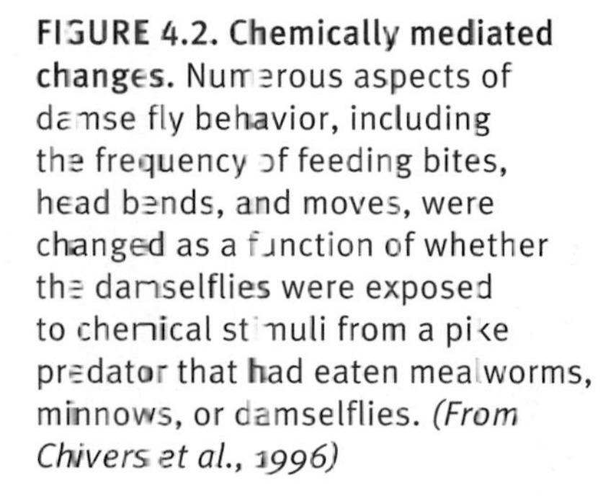
FIGURE 4.2. Chemically mediated changes. Numerous aspects of damse fly behavior, including the frequency of feeding bites, head bends, and moves, were changed as a function of whether the damselflies were exposed to chemical stimuli from a pike predator that had eaten mealworms, minnows, or damselflies. *(From Chivers et al., 1996)*

the three treatments above (water from pike plus damselfly, water from pike plus minnow, and water from pike plus mealworm) and isolated them for two days. Then each and every damselfly was exposed to water from a pike that had been fed mealworms. As such, we now have two particularly interesting groups of damselflies. One of these groups has individuals that were initially exposed to water from pike plus damselfly, but are now exposed to pike plus mealworm water, and one of the groups has individuals that were initially exposed to pike plus minnow, but are now exposed to pike plus mealworm. Damselflies in these two treatments responded to pike plus mealworm by decreasing their foraging activities. Recall that in the first experiment, damselflies did *not* curtail feeding when encountering the smell of pike plus mealworm. This suggests that based on their earlier experience, damselflies in the second experiment learned to associate pike plus any scent with danger, and this translated into a reduced foraging rate.

What Is Individual Learning?

In this chapter, we shall examine the role that individual learning plays in animal behavior. The role of social learning—that is, learning from other individuals—will be the subject of the next chapter. In our analysis of individual learning—henceforth referred to as learning in the rest of this chapter—we shall address three interrelated questions: How do animals learn? Why do animals learn? What do animals learn?

Before tackling these overarching questions, it is important to address a few baseline issues, first and foremost among them the question: What do we mean when we speak of learning? This is a complicated question, but the definition of learning we shall adopt here is simple and straightforward and fairly widely accepted within psychology. When we speak of learning, we are referring to a relatively permanent change in behavior as a result of experience (Shettleworth, 1998). This definition does have a downside, in that it is not clear how long a time period is encompassed by the words "relatively permanent." That being said, this is a working definition, already adopted implicitly by most ethologists, and it will serve our purposes here well.

It is interesting to note that the phrase "relatively permanent" was added to the older definition of learning, which was something akin to "a change in behavior as a result of experience." The insertion of the new phrase was meant to address a particularly dicey problem regarding what constitutes learning. Sara Shettleworth (1998) describes the problem as follows: A rat that experiences no food for twenty-four hours is more likely to eat than a rat that has just been fed. Most people would say that hunger per se, not any learning by the rat, explained its increased proclivity to forage, even though the experience of *not being fed* did affect the rat's behavior when it was presented with food. Insertion of the phrase "relatively permanent" into our description of learning eliminates this problem.

Adopting "a relatively permanent change in behavior as a result of experience" as our definition of learning, we see an interesting relationship between learning and what evolutionary ecologists refer to as "phenotypic plasticity" (Levins, 1968; Via and Lande, 1985; Stearns, 1989; West-Eberhard, 1989; Harvell, 1994; Tollrian and Harvell, 1998). A **phenotype** is typically defined as the observable characteristics of an organism (Walker, 1989), and **phenotypic plasticity** is broadly defined as the ability of an organism to *produce different phenotypes* depending on environmental conditions. For example, many invertebrates like the bryozoan *Membranipora membranacea* live in colonies. When living in such colonies, individuals typically lack the spines that are often used as an antipredator defense in related species. These spines are simply not grown when *Membranipora membranacea* develops in the absence of predators. Yet, individuals will grow spines relatively quickly when exposed to predatory cues (Figure 4.3; Harvell, 1994, 1998;

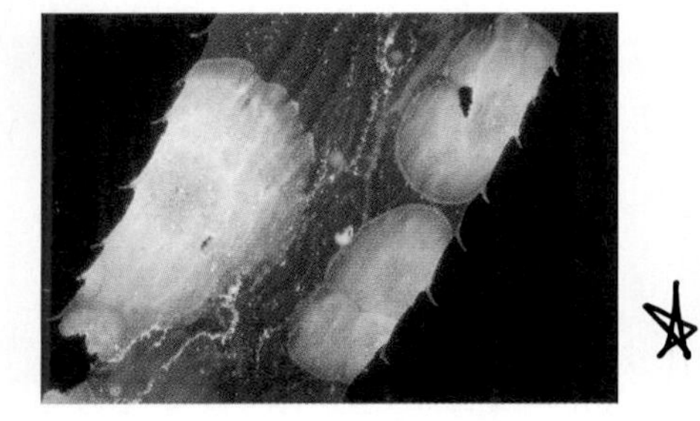

FIGURE 4.3. Inducible defenses. In some bryozoans, like *Membranipora membranacea*, colonies produce spines when predators are present. *(Photo credit: Ken Lucas/Visuals Unlimited)*

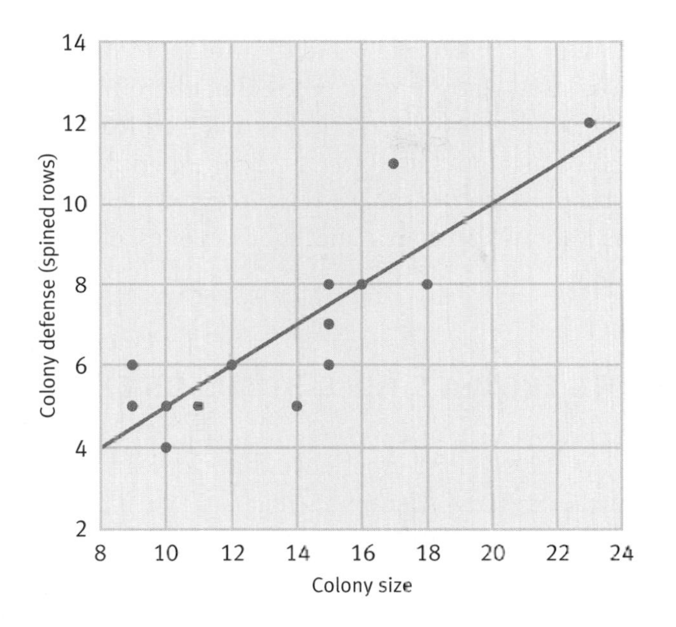

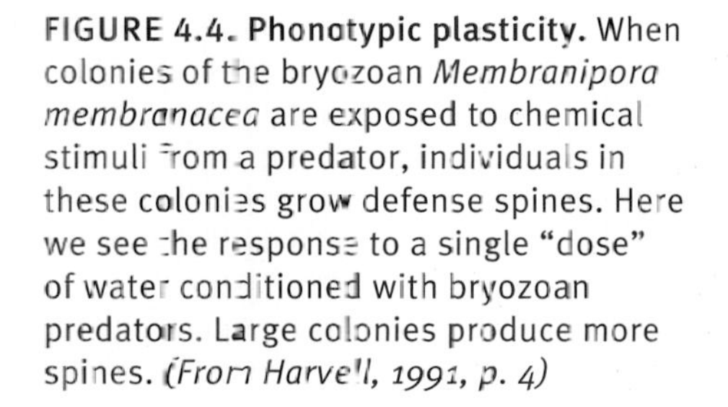
FIGURE 4.4. Phonotypic plasticity. When colonies of the bryozoan *Membranipora membranacea* are exposed to chemical stimuli from a predator, individuals in these colonies grow defense spines. Here we see the response to a single "dose" of water conditioned with bryozoan predators. Large colonies produce more spines. *(From Harvell, 1991, p. 4)*

Tollrian and Harvell, 1998). The resultant change, from spineless to spined, constitutes a case of phenotypic plasticity. The phenotype of this bryozoan shifts dramatically as a result of environmental changes—in our case the addition of a predator—and hence is thought of as "plastic" (Figure 4.4).

If learning is "a relatively permanent change in behavior as a result of experience," it then becomes one type of phenotypic plasticity *if* we think of behavior as part of a phenotype (Dukas, 1998a). Thus, if we replace "behavior" with "phenotype" in our definition of learning, phenotypic plasticity becomes the broader category under which learning is subsumed. Phenotypic plasticity then encompasses everything from relatively long-lasting morphological changes, as in the case of the bryozoans, to behavioral changes that come about as a result of experience.

How Animals Learn

In this section, we shall delve into *how* animals learn what they learn. There is a huge psychological literature on this, both theoretical and empirical, and I am making no pretensions of covering this issue in any detail. Rather, I hope to review some basic ideas on how animals learn, or what psychologists often refer to as the mechanisms for learning. Our discussion of how animals learn will follow an outline developed by Heyes (1994), both because of its conciseness and its attempt to tie *how* animals learn to *why* they learn. Heyes notes that there are three commonly recognized types of experience that can lead to learning. We shall now touch on these three types of experience—stimulus, stimulus-stimulus, and response-reinforcer—and in the process look at the various categories and subcategories of learning that they facilitate.

LEARNING FROM A SINGLE-STIMULUS EXPERIENCE

In many ways the simplest experience that can lead to learning involves a single stimulus—a stimulus that can take almost any form. For example, let's imagine that we are interested in learning in rats. Let's further imagine that numerous times throughout the day a researcher simply puts a blue colored stick into a rat's cage (Figure 4.5). Rats will often take note of such a disturbance and turn their heads in the direction of the blue stick. If *over time* the rats become more likely to turn their heads in the direction of the blue stick—that is, if they are more sensitive to the stimuli with time—**sensitization** has occurred. Conversely, if over time the animals become less likely to turn their heads, **habituation** is said to have taken place. Sensitization and habituation are two simple single-stimulus forms of learning.

The process of habituation is of great concern to experimental ethologists, particularly to those who work in the laboratory. It can become quite difficult to examine certain types of behaviors if animals habituate quickly to stimuli. For example, in many ethology experiments involving antipredator behavior, predators may be housed such that visual interactions between predator and prey are

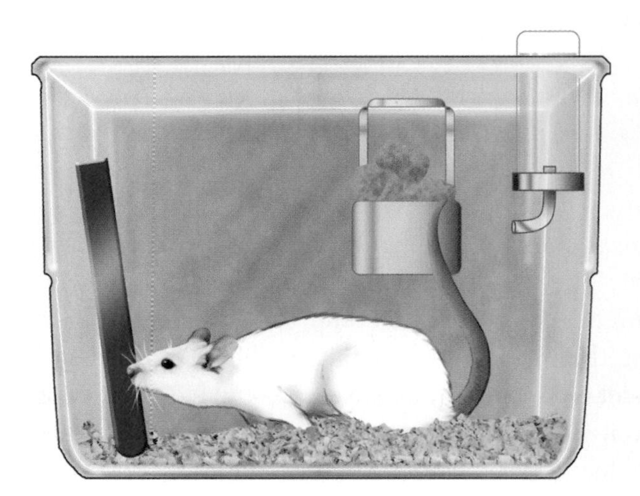

FIGURE 4.5. Habituation and sensitization. Numerous times each day a blue stick is placed in a rat's cage. If the rat takes less and less notice of the stick, habituation has occurred. If the rat pays more attention to the blue stick over time, sensitization has taken place.

FIGURE 4.6. Habituation as a problem. Suppose the antipredator behavior of the smaller fish on the right is the subject of investigation. If the experimental protocol involves keeping the study's prey species in one tank (right) and the putative predator in another aquaria (left), the prey may learn that the predator is not dangerous and habituate to its presence.

possible, but the predator can't actually harm the prey (Figure 4.6). This ethical compromise spares the life of the potential prey, but it creates a scenario in which the prey may now habituate to the predator, having learned that the predator cannot in fact move close enough to present any real danger (Huntingford, 1984). Because of these sorts of issues, ethologists often need to go to great lengths to be certain that habituation has *not* occurred in their study system (Rowland and Sevenster, 1985).

A single stimulus that results in habituation or sensitization may have other consequences for learning as well. If an animal habituates to a stimulus, this may interfere with *later* attempts to get the individual to associate that stimulus with another event. For example, if rats habituate to the blue stick, it might prove more difficult for them to subsequently learn that the blue stick signals the arrival of food. On the other hand, if sensitization to a single cue has occurred, it may facilitate the association of the sensitized stimulus and other events.

PAVLOVIAN (CLASSICAL) CONDITIONING

Suppose that rather than giving a rat a single stimulus like the blue stick, from the start we pair this stimulus with a second stimulus, let's say the odor of a cat—an odor that our rats have an inherent fear of.

FIGURE 4.7. Paired stimuli. Five seconds after a blue stick is placed in a rat's cage, the odor of a cat is sprayed in as well. The question then becomes: Will the rat pair the blue stick with danger (cat odor)?

Now, five seconds after the blue stick is in place, we spray the odor of a cat into one corner of the cage (Figure 4.7). If the rat subsequently learns to associate stimulus 1 (blue stick) with stimulus 2 (cat odor) and responds by climbing under the bedding of its nest as soon as the blue stick appears, but *before* the odor is sprayed in, we have designed an experiment in **Pavlovian** or **classical conditioning**.

Pavlovian conditioning, as is obvious from the name, was first developed by Ivan Pavlov in the late 1800s (Pavlov, 1927; Figure 4.8). At the most basic level, Pavlovian conditioning experiments involve two stimuli—the conditioned stimulus and unconditioned stimulus (Domjan, 1998). A **conditioned stimulus (CS)** is often defined as a stimulus that fails initially to elicit a particular response, but comes to do so when it becomes associated with a second (unconditioned) stimulus. In our rat case, the blue stick is the conditioned stimulus. The **unconditioned stimulus (US)** is a stimulus that elicits a vigorous response in the absence of training. In our rat case, the US would be the cat odor. Once the rat has learned to hide after the blue stick (CS) alone is in place, we can speak of a conditioned response (CR) (Figure 4.9).

Before examining Pavlovian conditioning any further, we must define a few more terms. In the learning literature, any stimulus that is considered positive, pleasant, or rewarding is referred to as

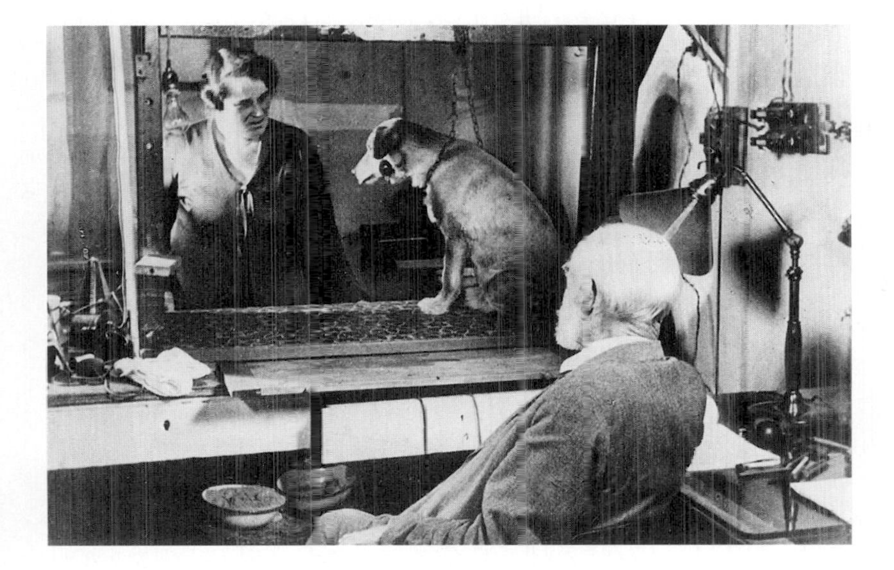

FIGURE 4.8. Ivan Pavlov watching a classical conditioning experiment as it was conducted in his laboratory. In the experiment, a device to measure salivation was attached to the dog's cheek, the unconditioned stimulus was a dish containing meat powder, and the conditioned stimulus was a light. *(Photo credit: Sovfoto)*

an **appetitive stimulus**. Appetitive stimuli include food, the presence of a potential mate, a safe haven, and so on. Conversely, any stimulus that is associated with some unpleasant event is labeled an **aversive stimulus**. A further distinction in the learning literature is made between positive and negative relationships. When the first event (placement of the blue stick) in a conditioning experiment predicts the occurrence of the second event (cat odor), we have a positive relation between events; if the first event predicts that the second event will *not* occur, we have a negative relationship. Positive relations produce excitatory conditioning, while negative relations produce inhibitory conditioning.

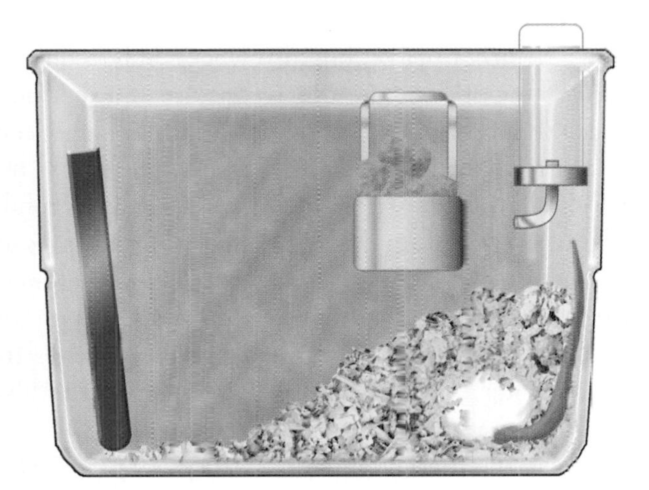

FIGURE 4.9. Conditioned response. If the rat pairs the blue stick (CS) and the cat odor (US), then it will hide under the chips when just the blue stick is presented. Such hiding represents a conditioned response (CR).

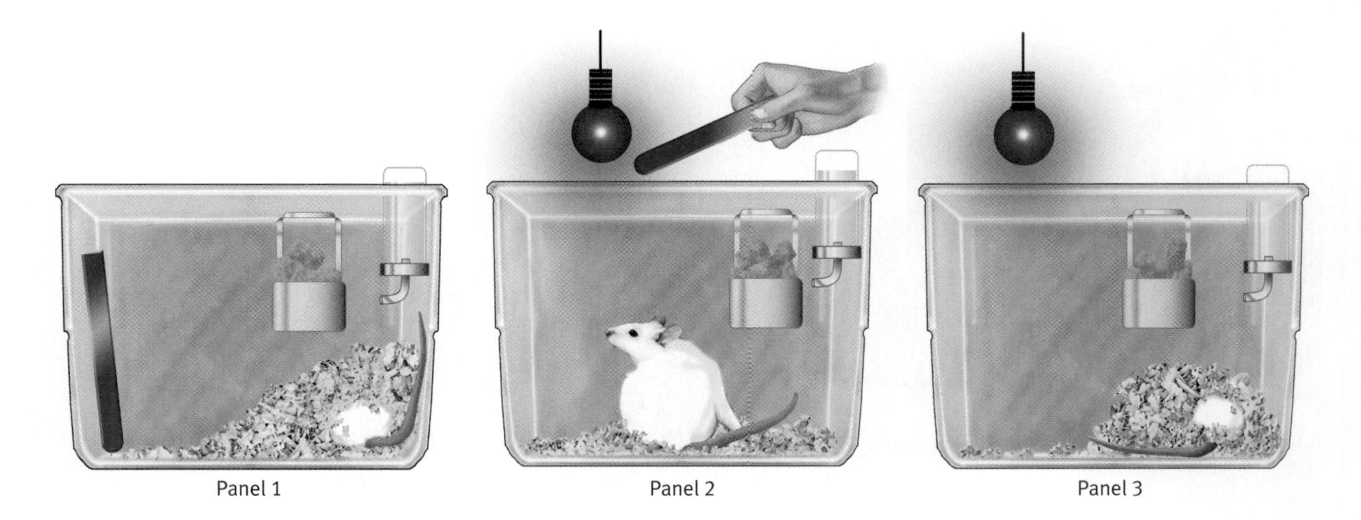

FIGURE 4.10. Second-order conditioning. In panel 1, a rat has already been conditioned to associate the blue stick with a cat odor. In panel 2, a red light is flashed on before the blue stick is placed in its cage. In panel 3, the rat displays a conditioned response when just the red light is turned on. In such a case, second-order conditioning to the red light has taken place.

Pavlovian conditioning experiments can become very complicated by adding second-order conditioning to the experimental equation. In second-order conditioning, once a conditioned response (CR) has been learned by pairing US and CS1, a new conditioned stimulus (CS2) is presented before CS1, and CS2 eventually also elicits the conditioned response. In our case, any rats that learned to hide when the blue stick was present might now see a red light (the second-order stimulus) that proceeds the blue stick. Once the rat has learned to pair the red light with the danger associated with the cat odor, we can speak of second-order Pavlovian conditioning (Figure 4.10).

BLOCKING AND OVERSHADOWING. Pavlovian conditioning affects not only behavior per se, but also what is referred to as *learnability*. To see how this operates, let's consider an experiment with three groups. Suppose that Group 1 individuals undergo a standard Pavlovian paradigm involving two stimuli, the placement of a blue stick (CS1), and a cat odor (US). In Group 2, a second conditioned stimulus (CS2)—let's make it a red light—is always presented simultaneously with the blue stick just prior to the presence of cat odor (Figure 4.11). Subjects from both groups are then tested in response to the blue stick. If the red light *overshadows* the blue stick, rats in Group 2 will respond less strongly to the blue stick than will rats in Group 1.

Now let us add a third group to our experiment. In Group 3, individuals are first trained to associate the blue stick with the cat odor,

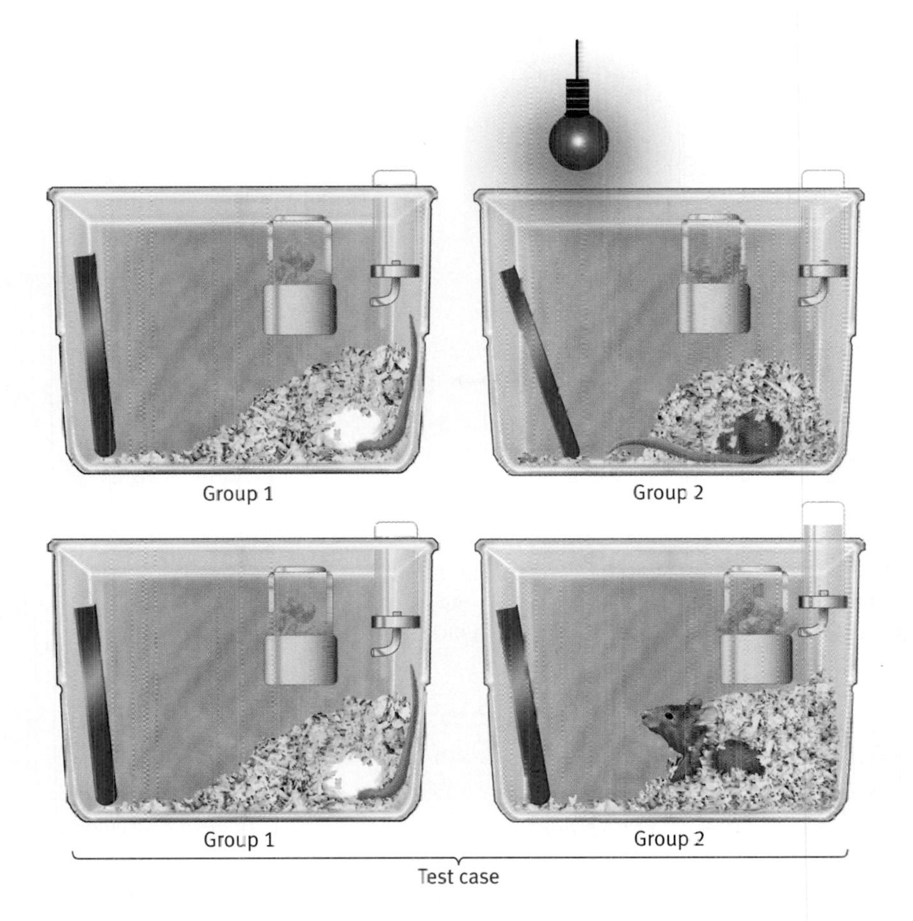

FIGURE 4.11. Overshadowing. Group 1 rats learn an association between a blue stick and a cat odor, whereas Group 2 rats learn to associate both a red light and a blue stick (together) with the cat odor. In the test arena, Group 1 and Group 2 rats are examined when just the blue stick is present. If Group 1 rats respond more strongly than Group 2 rats, overshadowing has taken place.

but subsequent to this, the red light is presented at the same time as the blue stick, and this compound stimulus is paired with the cat odor. If we compare individuals in Group 3 to those in Group 2 when they are presented with the red light, blocking occurs when those in Group 3 respond less strongly to the red light than individuals in Group 2 (Kamin, 1968, 1969). Thus, initially learning to associate the blue stick alone with the cat smell blocks the ability of individuals in Group 3 to pair the red light with the cat odor (Figure 4.12).

INSTRUMENTAL (OPERANT) CONDITIONING

Instrumental conditioning, also known as **operant** or **goal-directed learning,** occurs when a response made by an animal is somehow reinforced. One of the most fundamental differences between Pavlovian conditioning and instrumental learning centers on the fact that in the latter, the animal must undertake some action or response in order for the conditioning process to produce learning. The classic example of instrumental learning is a rat pressing

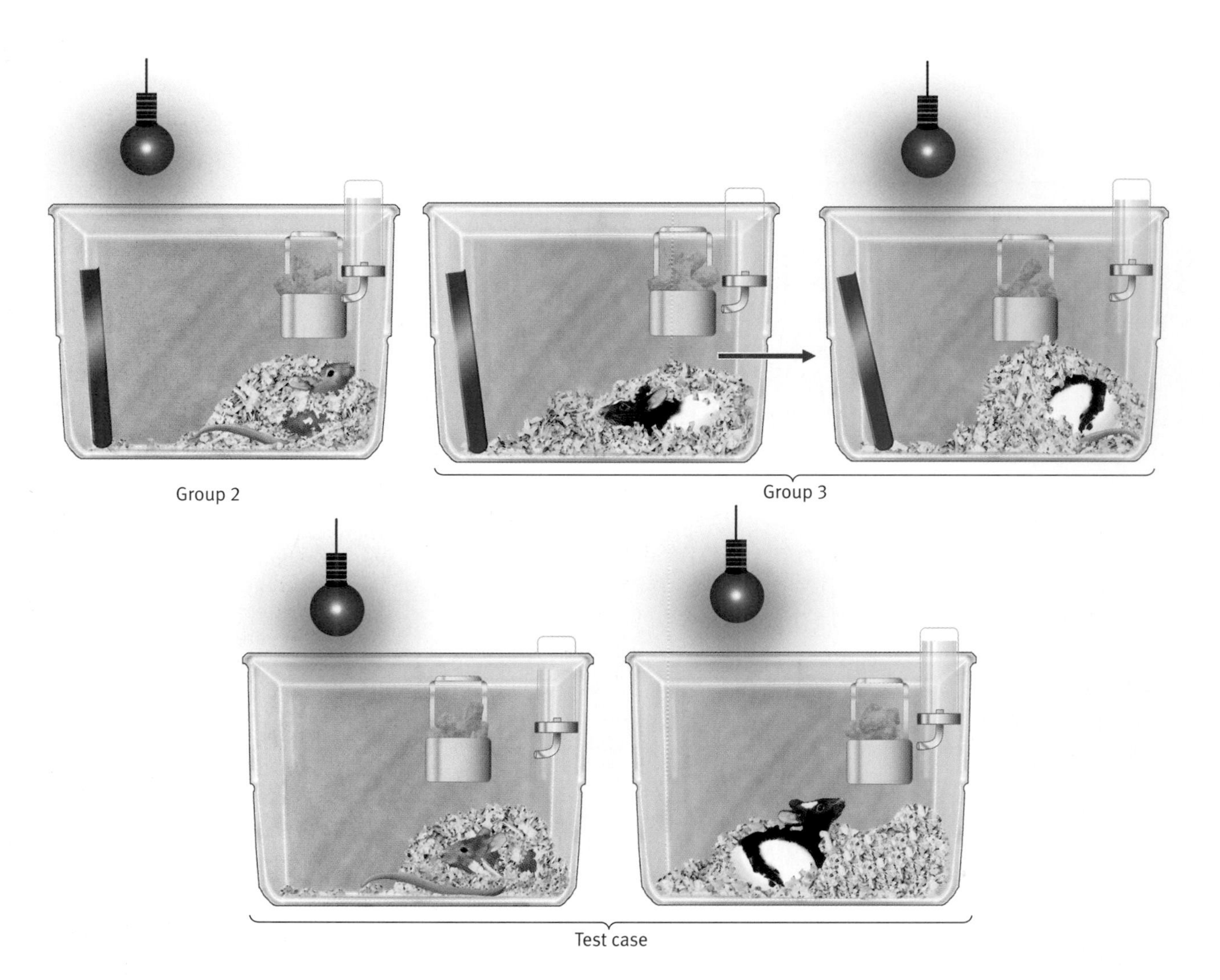

FIGURE 4.12. Blocking. Group 2 rats are trained to associate the red light and the blue stick (together) with the cat odor. Group 3 rats are initially trained to associate just the blue stick with the cat odor, but then they are trained to associate the joint presence of the stick and the light with the cat odor. Individuals from both groups are then tested in the presence of just the red light. If Group 2 individuals react more strongly to the light, blocking has occurred in the Group 3 individuals.

some sort of lever to get food to drop into his cage. Rats associate pressing on the lever (response) with food (outcome), and they quickly learn this task. But oftentimes instrumental learning is not as easy to quantify as in the rat case. Michael Domjan (1998) illustrates this nicely by considering a dog that barks at intruders until the intruders go away. Clearly, the dog has to bark (undertake an action) to make the intruder go away (get a response), but does this illustrate instrumental *learning* in nature? Did the dog learn that barking (the response) made the intruder leave (the outcome)? It is certainly possible; however, an equally plausible case could be made that a novel intruder induces barking and that this barking simply continues until the intruder disappears. In this case, the dog may have learned nothing. The difficulties inherent in these sorts of cases are why operant conditioning experiments are most often undertaken in the laboratory.

The earliest work on instrumental learning was that of Edward Thorndike (1898, 1911), and it involved testing how quickly cats could escape from "puzzle boxes" that Thorndike had constructed. Combining this with other results he obtained, Thorndike postulated the **law of effect**. This law states that if a response in the presence of a stimulus is followed by a satisfying event, the association between the stimulus and the response will be strengthened. Conversely, if the response is followed by an aversive event, the association will be weakened.

Thinking on the subject of instrumental learning was revolutionized by the work of B. F. Skinner (1938), who argued that much work in this area ignored the fact that behavior was a continuous, free-flowing variable. To remedy this problem, Skinner devised what is now known as a Skinner box, which allowed a *free-operant* procedure that he devised. Skinner's idea was to create a continuous *measure* of behavior that could somehow be divided into meaningful units. His solution was the operant response, defined by the effect that such a response has on the animal's surroundings. When a rat pushes down on a lever it is making an operant response because the action changes the rat's environment by adding food to it (Figure 4.13). The fact that "lever pushing" was a relatively unambiguous event that was easily measurable, and that it occurred in an environment over which the rat had control, has greatly facilitated the work of psychologists working within the instrumental learning paradigm.

FIGURE 4.13. Skinnerian rats. To test various theories of animal learning, rats are often placed in "Skinner boxes," where they have to take an action (here, pressing a bar) to get a response (here, the reward of food dropping into the cage). *(Photo credit: Eric S. Murphy)*

There is still some debate among psychologists over the relative merits of Pavlovian and instrumental learning techniques. From our perspective, however, much of this debate, though interesting in its details, is not relevant, in that over the years both instrumental and Pavlovian conditioning techniques have become very fine-tuned and have provided much in the way of understanding how animals learn what they learn.

Why Animals Learn

With an understanding of *how* animals learn, we can now take a definitively more evolutionary perspective and ask *why* animals learn. In so doing, we will need to recall how adaptationists think. If we imagine that the ability to learn is under some sort of complex genetic control, we can ask whether this ability is favored by natural selection, and if so, under what circumstances. In so doing, we will be addressing three related questions: (1) How can within-species studies, with a particular emphasis on the natural environment of an organism, help us understand learning and evolution? (2) How can population comparisons be used to shed light on the evolution of learning? (3) What theories are in place that examine how learning evolves in different environments?

WITHIN-SPECIES STUDIES AND THE EVOLUTION OF LEARNING

Psychologists have done thousands, perhaps tens of thousands, of studies of learning in animals, particularly in rodents and birds. For the most part, until recently, these experiments have not been designed to take into consideration the natural environment of the organism or the evolutionary forces acting on the organism. This is due, at least in part, to a history that includes Thorndike and Pavlov's work in which they argued that, aside from the details, the qualitative features of learning were the same in all animals, including man (Bitterman, 1975). This view became widely accepted, with such psychologists as Skinner (1959) and Harry Harlow (1959) promoting it. So much so that Bitterman, a leader in the field of comparative animal learning, notes that "work on learning has been dominated from the outset by a powerful theory which denies that learning has undergone any fundamental evolutionary change" (Bitterman, 1975, p. 699). If Thorndike and Pavlov are right that the particular environment an organism evolved in has no effect on learning, the same sort of learning should be seen in all creatures that learn, regardless of what sort of learning tasks they are presented with.

In adopting the adaptationist approach, the above claim seems hard to believe. Could it really be true that natural selection has not acted on the ability to learn? Rather, it seems that the ability to learn should be under strong selection pressure, with individuals who learn appropriate cues that are useful in their particular environment strongly favored by natural selection. Those who adopt this view are said to be in the "ecological learning" school (Johnston, 1985), and their influence is getting stronger as our knowledge about evolution and learning increases (Shettleworth, 1998; Dukas, 1998a).

GARCIA'S RATS. In the mid-1960s, John Garcia ran a series of experiments on learning in rats. Experiments on rat learning were hardly news in either the psychological or biological community, but Garcia's results made both groups rethink their approach to studies of learning (Garcia et al., 1972; Seligman and Hager, 1972). What is particularly amazing about Garcia's work is that in many ways the protocol he used was very similar to that already being used in psychology experiments centered on learning. Garcia and Koelling (1966) attempted to have rats form an association between two cues. One of those cues was water that was either "bright-noisy" water (water associated with a noise and an incandescent light) or "tasty" water (water with a particular taste). One of these water types was then paired up with one of the following negative stimuli: radiation, a toxin, immediate shock, or delayed shock. For example, in the bright-noisy water/radiation treatment, bright-noisy water would be presented and the rats who had been trained to this cue would drink the water and then be exposed to radiation that would make the rats ill.

Garcia and Koelling found a fascinating interaction between the type of punishment a rat received for drinking water and the type of water that the rat consumed. X-ray and toxin treatments, each of which made the rat *physically* ill, were easily associated with the tasty water (gustatory) cues, but not with the bright-noisy (audiovisual) cues. That is, the rats quickly learned that tasty water was to be avoided after this cue was paired with X-rays or toxins, but they did not learn to avoid bright-noisy water after it was paired with X-rays or toxins. In the latter case, rats had trouble pairing an audiovisual cue with X-rays or toxins that made them physically ill. In contrast, when Garcia and Koelling examined the rats who had shock treatments, they found that shock was easily paired with the audiovisual cue, but not with the gustatory cue.

Garcia and Koelling explained their views by adopting an adaptationist stance (see Chapter 2)—something quite unusual for psychologists of the 1960s. They argued that natural selection would favor pairing gustatory cues (tasty water) with internal discomfort (getting ill). After all, many instances of internal discomfort in nature are likely to be caused by what you have consumed, and rarely would a food cue be associated with audiovisual cues (as in the bright-noisy water treatment). On the other hand, peripheral pain like that caused by a shock might be more commonly associated with some audiovisual cue like a conspecific or a predator, and natural selection should favor pairing these cues together as well (see Wilcoxon et al., 1971, for similar results in rats and blue quail, and DeCola and Faneslow, 1995, for a dissenting view on what Garcia's work tells us about learning).

In another apparent blow to orthodoxy in psychological learning circles of the period, Garcia found that learning in rats will indeed occur without immediate reinforcement (Garcia et al., 1966).

While most psychologists believed that delays in reinforcement on the order of seconds can cripple animal learning, Garcia found learning even after delays of up to seventy-five minutes when injections of noxious substances were paired with drinking saccharin-flavored water. From an evolutionary perspective, the work of Garcia and his colleagues makes perfect sense. In nature, one would expect a delay between the time that a rat consumes a substance and any subsequent negative effect of such consumption. As such, natural selection would have favored rats that were able to associate what they ate with the effect of their diet, even if separated by significant time intervals.

OPTIMAL FORGETTING IN STOMATOPODS. The ability to remember events plays an obvious role in animal learning. In fact, depending on your precise definition of memory, learning is impossible without some form of memory. Psychologists have long studied memory and learning by, for example, looking at **extinction curves** in learning experiments. This typically consists of testing how long an animal will remember some paired association once the pairing itself has stopped. Such tests are critical to numerous aspects of psychological learning theory, but they fail to ask a question that jumps to the surface when an adaptationist approach to the subject of memory is invoked. Namely, given the ecology of the species, might natural selection shape an optimal memory span? That is, while psychologists tend to focus on how long associations last, an evolutionary animal behaviorist would ask why we see variation in the length of time that associations stay paired and how natural selection may have favored certain extinction curves over others.

It may be that certain events in an animal's life are timed such that it would be beneficial to remember them for X days, but after this time period, remembering and acting on such memories would be either of no benefit or perhaps even detrimental. For example, imagine a forager who is able to remember the location and amount of food in patches. If patches are renewable, then while it may pay to remember the location of a food patch, it doesn't necessarily pay to remember how much food was there last time you visited, as this might lead to erroneous decisions, based on outdated information (Figure 4.14).

Before proceeding to examine an example of how behavioral ecologists and ethologists might test the idea of optimal memory, it is important to realize that no natural selection model of any behavior predicts that an animal will reach some global optimum. Rather, natural selection models predict that, given the constraints that an animal faces, selection will favor moving as close to an opti-

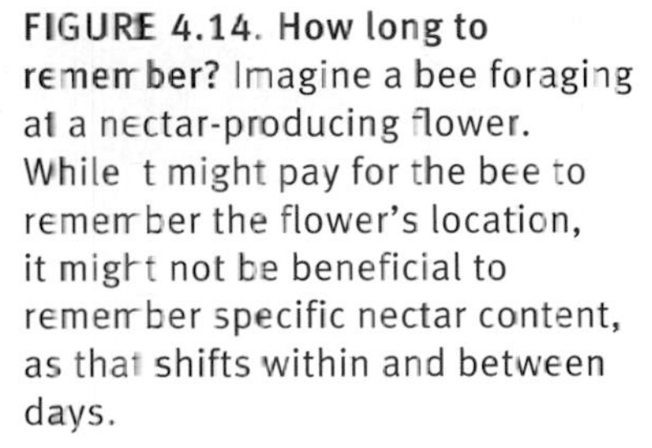
FIGURE 4.14. How long to remember? Imagine a bee foraging at a nectar-producing flower. While it might pay for the bee to remember the flower's location, it might not be beneficial to remember specific nectar content, as that shifts within and between days.

mum as possible (Mayr, 1982; Alcock, 1993). In other words, sociobiologists don't search for the best of all possible solutions to a problem and assume that selection will take us there, but rather they recognize that many constraints prevent selection from ever reaching such a global optimum. With respect to learning there are, in all likelihood, neurobiological constraints on memory and other cognitive functions, and these serve as guideposts for the limits of natural selection's ability to shape memory capacity (Shettleworth, 1998; Dukas, 1998b). That being said, it is interesting to note that many neurobiologists have found that a relatively small number of neurons seem needed to store a relatively large amount of memory-related pieces of information (Singer, 1994; Morton and Chiel, 1994).

To see how ethologists might tackle the question of natural selection and memory, consider Caldwell's (1992) study of aggressive behavior in stomatapod crustaceans (Healey, 1992). Male and female *Gonodactylus bredini* share a cavity for a few days before they breed (Dingle and Caldwell, 1972; Caldwell, 1986) and actively repel all intruders from this area (Shuster and

Caldwell, 1989). Very shortly after mating, however, the male leaves the breeding cavity and goes out in search of a new mate. Caldwell set up an experimental design that allowed him to examine whether males who left the breeding cavity and the females they left behind recognize one another after the male's desertion.

FIGURE 4.15. Stomatopod threat. Male and female stomatopod crustaceans share a cavity for a few days before they breed. Although the males leave the breeding cavity soon after mating, they tend to remember their former mates and to be less aggressive toward them during the four weeks that their brood remains in the cavity. This is a photo of a male *Gonodactylus smithii* in a threat position. *(Photo credit: R. Caldwell)*

It turns out that both male and female *Gonodactylus bredini* remember one another after the male has left. Males are less likely to be aggressive toward their former mates, while females put up a weaker resistance to such males. With respect to memory, males and females recognize one another for at least two weeks—the duration of the experiment. What makes this study fascinating is that, based on his thorough understanding of stomatopod biology, Caldwell hypothesizes that individuals might actually be able to recognize each other for four weeks. Why four weeks? It turns out that the females guard their brood for this period of time, after which the brood leaves its cavity for good. At the four-week point, then, males need not worry about being aggressive toward former mates, as such aggression will not harm their own brood (Figure 4.15). In other words, there is an optimal time to recognize former mates, and this optimum is set by the self-interest of the interacting parties. Caldwell argues that for *Gonodactylus bredini* this optimum is set at four weeks, and hence he predicts that male and female stomatopods will either forget each other after this time period, or at the very least act as though this they had forgotten each other.

Unfortunately, Caldwell's experiment did not run for four weeks. But suppose it had and suppose he found that at that point in time males and females no longer responded differently to former pairmates. It would be difficult, if not impossible, to determine whether the stomatopods actually forgot information or simply ignored information that they still retained. This is a general problem in animal learning studies, however, and is beyond the scope of what we are addressing here. That is, since we measure learning by what an animal does, in the absence of language it is hard to be specific about what is going on cognitively when an animal doesn't do something. Did it forget? Did it not act on the information it possessed? There are some ways around this, but they are often circuitous (Roper et al., 1995). For example, if an animal is placed in an obvious life-and-death situation and fails to perform a learned behavior that would save its life, we can, in all likelihood, infer that it simply forgot the appropriate life-saving behavior rather than that it recalled such behavior but ignored it. In any case, the stomatopod study raises some fascinating questions about how natural selection might be acting on some aspect of learning and memory in animals.

POPULATION COMPARISONS AND THE EVOLUTION OF LEARNING

In Chapter 2, we learned that one technique ethologists employ to understand the evolution of behavior is to compare behavior across different populations. Oftentimes such studies compare two or more populations to one another and make predictions about behavioral similarities and differences between these populations based on both a knowledge of their ecology and some hypotheses about how natural selection has operated in each (Bitterman, 1975; Marler and Terrace, 1984; Macphail, 1987; Rozin, 1988; Johnston, 1985; Dukas, 1998a, 1998c; Shettleworth, 1998; Balda et al., 1998). This is precisely the technique we shall use when examining the evolution of learning via population comparisons.

LEARNING, FORAGING, AND GROUP LIVING IN DOVES. Animals in groups often find food faster and have more time available for foraging (Krebs and Davies, 1993). With respect to learning, Carlier and Lefebvre (1996) predicted that individuals who live in groups should learn more quickly than territorial (and hence more isolated) individuals. Ideally, one would like to test this hypothesis in a single species, where natural selection has favored group living in some populations, but solitary living in others. Zenaida dove populations from Barbados (*Zenaida aurita*; Figure 4.16) fit the bill perfectly, as solitary living (via territoriality) is the norm in one dove population studied by Carlier and Lefebvre, while group living is the norm in the other. Yet, these two populations were separated by only 9 kilometers, and hence other environmental variables, above and beyond those associated with territoriality, were likely to be fairly minimal.

FIGURE 4.16. Zenaida doves. Zenaida doves from populations where individuals live in groups appear to be better at learning foraging tasks than individuals from populations where doves are territorial. *(Photo credit: James Mountjoy, courtesy of L. Lefebvre)*

Sixteen doves from each of the group-living and territorial populations were captured and brought into the laboratory. All subjects were then individually presented with the challenge of learning how to operate an experimental apparatus that required the birds to pull on a metal ring, which then opened a drawer containing food. Carlier and Lefebvre found clear evidence that group living facilitated quicker learning than did territorial life (Figure 4.17). Similar results have also been found in three other bird species (Sasvari, 1985). Furthermore, Carlier and Lefebvre found that the more difficult the learning task, the more pronounced the between-population differences.

These differences may be attributed to at least two possible factors (Carlier and Lefebvre, 1996). First, the animals in Carlier and Lefebvre's study may have already differed in foraging experience before their experiment, and hence some of the differences uncovered may have been due to what individuals had experienced, and

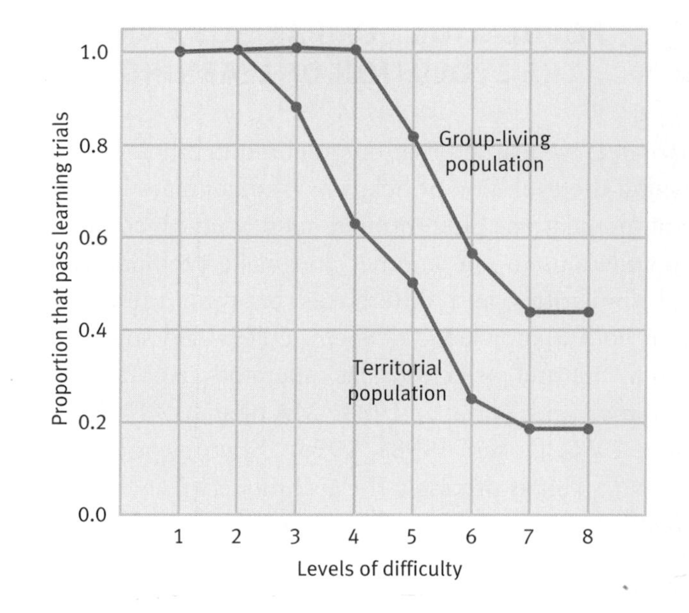

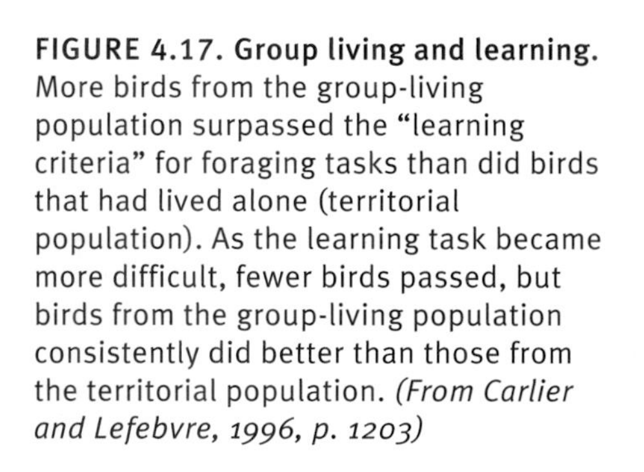
FIGURE 4.17. Group living and learning. More birds from the group-living population surpassed the "learning criteria" for foraging tasks than did birds that had lived alone (territorial population). As the learning task became more difficult, fewer birds passed, but birds from the group-living population consistently did better than those from the territorial population. *(From Carlier and Lefebvre, 1996, p. 1203)*

potentially learned, prior to being brought into the laboratory. Second, above and beyond what experiences any given set of birds took into Carlier and Lefebvre's experiment, natural selection may have operated on learning ability across these populations. The verdict is still out on which of these possibilities best explains differences between dove populations.

LEARNING AND ANTIPREDATOR BEHAVIOR IN STICKLEBACKS. In interpopulational studies of learning, one way to partially circumvent the confounding effects of personal experience per se and natural selection acting on learning abilities is through the use of laboratory experiments. In the lab it becomes possible to raise individuals from two very different populations in a similar environment, and hence minimize experience differences across populations. This is just what Huntingford and Wright (1992) did in their study of avoidance conditioning in two populations of three-spined stickleback fish (*Gasterosteus aculeatus*).

Along with guppies, sticklebacks have become a favorite species for testing ethologically based hypotheses regarding interpopulation differences (Wooton, 1976; Godin, 1997). The reason for the interest in using sticklebacks in such a context stems from the fact that many stickleback populations are very similar, with often only a single critical variable—predation pressure—differing across populations. Some sticklebacks live in locales that contain a variety of predators, and some populations live in lakes with virtually no predators.

Huntingford and Wright raised individuals derived from predator-rich and predator-absent streams in the laboratory, where neither had any interaction with predators. The evolutionary approach would suggest that selection would have acted more strongly on antipredator strategies—including *learning* about danger—in the sticklebacks from the predator-rich population.

Huntingford and Wright started by training eight sticklebacks in each group to associate one side of their home tank with food. They found no differences in learning across populations in the context of foraging alone. After a stickleback had learned that one side of its tank was associated with food, fish would be subject to a simulated attack from a heron. Huntingford and Wright then examined whether between-population differences existed in how long it took to learn to avoid the side of the tank associated with predation (and food). While all but one fish from both populations learned to avoid the dangerous end of their tank, fish from high-predation areas learned this passive avoidance task more quickly than did fish from predator-free populations (Figure 4.18; see Licht, 1989, and Magurran, 1990, for more on learning and antipredator behavior in guppies and minnows, respectively).

Two lines of reasoning support the hypothesis that natural selection has operated on learning and antipredator behavior in sticklebacks. First, Huntingford and Wright's laboratory protocol minimized the probability that individual experiences differed significantly across the populations they examined. Second, and

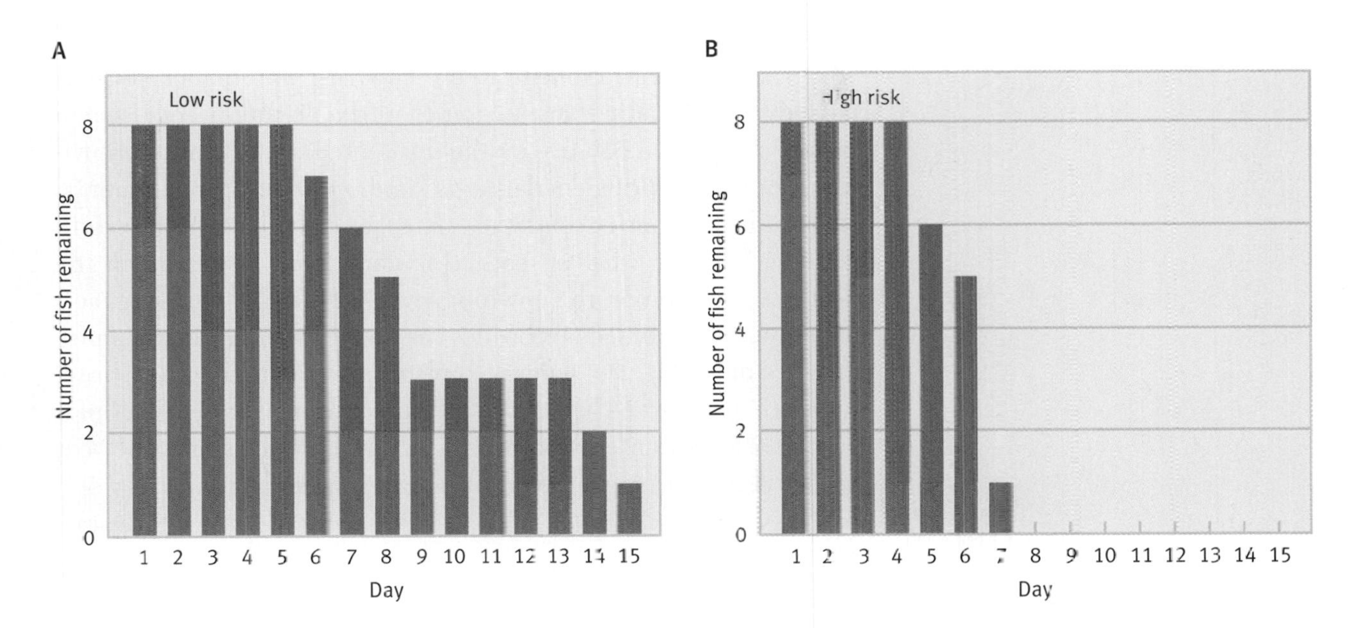

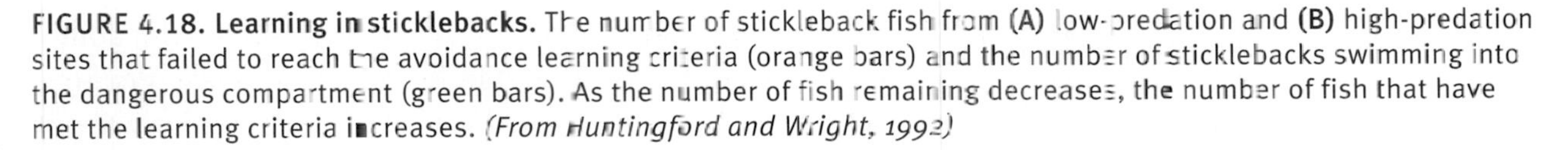
FIGURE 4.18. Learning in sticklebacks. The number of stickleback fish from **(A)** low-predation and **(B)** high-predation sites that failed to reach the avoidance learning criteria (orange bars) and the number of sticklebacks swimming into the dangerous compartment (green bars). As the number of fish remaining decreases, the number of fish that have met the learning criteria increases. *(From Huntingford and Wright, 1992)*

equally important, Huntingford and Wright did not find differences in all learning contexts (in this case, foraging and avoiding predators), but only in the category in which they hypothesized natural selection was operating.

A MODEL OF THE EVOLUTION OF LEARNING

In addition to empirical examinations that shed light on the evolution of individual learning, a number of mathematical models have been developed to examine the evolution of learning (Boyd and Richerson, 1985; Hinton and Nowlan, 1987; Stephens, 1991, 1993; Bergman and Feldman, 1995). One way theoreticians typically begin modeling the evolution of learning is to ask whether an organism would be better off possessing a behavior that was fixed by natural selection at some level, or whether learning the appropriate behavior is favored by natural selection.

Ethologists, behavioral ecologists, and psychologists have long argued that the ability to learn should be favored over the genetic transmission of a fixed trait when the environment an animal lives in changes often, but not too often. To see how such models work, we must recognize two assumptions that underlie many of them: (1) most models assume that there is some cost to learning, even if it is only a very small cost, and (2) when we speak of learning we are referring to a trait passed on genetically, but now the trait of interest is the *ability to learn.*

When the environment *rarely changes,* information is best passed on by genetic transmission of inflexible traits, since such a means of transmission avoids the costs of learning, and the environment the offspring encounter is similar to that of their parents. If the environment is *constantly changing,* there is nothing worth learning because what is learned is completely irrelevant in the next situation. When the environment is constantly changing, acting on past experience is worthless as past experience is of no predictive value. Hence, genetic transmission of a fixed response, rather than a learned response, is favored. Somewhere in the middle, in between an environment that never changes and one that always does, learning is favored over the genetic transmission of a fixed response. In this zone, it is worth paying the cost of learning. The environment is stable enough to favor learning, but not so stable as to favor genetic transmission.

David Stephens (1991, 1993) agrees that the starting point in modeling genetic transmission of fixed traits versus the ability to learn should indeed be the stability of an animal's environment. But Stephens challenges the model's assumptions about environmental stability as it is represented in these models, saying that the models conflate two types of stability that need to be separated.

Stephens's model breaks environmental predictability into two sections: (1) predictability within the lifetime of an individual, and (2) predictability between the environment of parents and offspring. These two types of predictability can be very different, and lumping them together may hinder our understanding of the evolution of learning (Table 4.1). For example, consider a case in which early in life the offspring of particular species move to environments that are far removed from those of their parents. Further suppose that during the adult lifetime of such individuals, their environment will be relatively stable and thus predictable, even though the environment is unpredictable across generations of this species. In such a case, between-generation environmental variability is quite high, while within-lifetime environmental variability is much lower.

Learning is favored in Stephens's model when predictability within the lifetime of an individual is high, but environmental predictability between generations is low. To see why, let's walk through each of the possible scenarios in Table 4.1. In boxes 1 and 2, predictability within generations is low, and so neither strategy does particularly well. But since learning has a cost associated with it, fixed genetic transmission is favored. Fixed genetic transmission is again favored in box 4 because with high predictability at all levels, the cost of learning is never worth the investment. Only in box 3, with high within-generation predictability but low between-generation

TABLE 4.1 Stephens's model for the evolution of learning. The key variables in this model are within-lifetime environmental predictability and between-generation environmental predictability.

		Within-lifetime predictability	
		Low	High
Between-generation predictability	Low	Ignore experience (1)	Learn (3)
	High	Ignore experience (2)	Ignore experience (4)

predictability, is learning favored. Learning is favored here because, once an organism learns what to do, it can repeat the appropriate behaviors during its lifetime. But isn't this the sort of predictability that usually favors a fixed genetic transmission? It is indeed, but now the environment changes so much between generations that genetic codes for fixed traits would fail at that level despite working within the lifetime of an individual.

What Animals Learn

As we have already seen, we don't necessarily expect to observe learning in every conceivable behavioral scenario, nor in every species. That being said, if the ability to change behavior based on prior experience is a significant force in shaping animal behavior, then we should generally expect to see learning in a wide variety of contexts. In order to provide a broad (but brief) overview of animal learning, in this section we shall examine evidence for learning with respect to habitat selection, mating, aggression, and familial relationships. One prominent omission from this list is "learning about food sources"—this topic was discussed earlier in the chapter, and is delved into in depth in a subsequent chapter.

LEARNING WHERE HOME IS LOCATED

Learning plays a fundamental role in habitat selection in many animal species. For example, while some aspects of bird migration are known to be instinctual, learning also plays a critical role in the often long journeys involved in migration (Dingle, 1996; Gauthreaux, 1997; Dyer, 1998; Able, 1999). More generally, learning likely plays an important role in habitat selection (Chapter 13), including learning territory boundaries, the identity (and possibly the personality) of neighboring individuals, and the nature and value of resources in a given area (Fisher, 1954; Getty, 1987; Godard, 1991; Eason et al., 1999; Beecher et al., 1998). Here we shall focus on the role of learning in the orientation and migratory behavior of fish (Dodson, 1988).

There are numerous theories on how some species of fish make the often very long migration back to their natal stream as adults. For example, there is some evidence that fish use celestial and magnetic compasses (Quinn, 1984), while other studies find that fish may use drifting currents (Miller et al., 1985) or some physiological change in response to their current position to guide

FIGURE 4.19. Pink salmon migration. In some species of fish, learning plays a role in orientation and migration. *(Photo credit: Tom Walker/Photri)*

them back home (Barkley et al., 1978). At the same time, there is more and more evidence that demonstrates that learning may play a role in fish migration (Dodson, 1988).

Juvenile salmon appear to learn the odors associated with their natal streams, and subsequently use such information to guide them home again (Dittman and Quinn, 1996; Figure 4.19). Dittman and colleagues examined the age at which young salmon learn about their home streams, and whether such learning is age-dependent in hatchery-reared coho salmon (*Oncorhynchus kisutch*). At different stages during development, young salmon were exposed to either natural odors or an artificial odor (beta-phenylethyl alcohol) and then tested in the presence of such odors when they matured into adults. Results from the beta-phenylethyl alcohol treatment show clearly that learning the odor of the natal stream was taking place when the young matured from the parr stage (a post-larval stage when fish have dark "parr bars" on their sides) to the smolt stage of salmon development (a stage immediately following the parr stage), and not earlier. Only young smolts exposed to this artificial odor were attracted to this smell when they matured. This age-dependent learning effect is particularly interesting, as learning occurs at the same time as a surge in plasma thyroxin levels, suggesting a possible hormonal underpinning to migratory learning.

LEARNING ABOUT YOUR MATE

Given the fundamental importance of finding a willing and able mate, one might expect that animals would use any possible information available to them to help in the process of searching for potential mates. If an individual could learn to associate certain cues with mating opportunities, this ability would surely reap long-term benefits (Zamble et al., 1985, 1986; Graham and Desjardins, 1980; West et al., 1992). Michael Domjan and his colleagues have been examining learning and mate choice in a number of distinctly different species, and here we shall touch on their work with the Mongolian gerbil (*Meriones unguiculatus*).

Mongolian gerbils are burrowing desert rodents who rely on chemical communication for many forms of social exchange, including the formation of pair bonds (Ågren, 1984; Thiessen and Yahr, 1977). Given this, Villarreal and Domjan (1998) ran an experiment that first allowed pair bonds to form between males and females. Males in one group were presented with an olfactory cue (mint or lemon), and then they were given access to their partners. Males in another (control) group were presented with the odor, but they were not provided with access to females following this presentation. Males that experienced the pairing of odor and subsequent access to a pairmate learned relatively quickly to approach an area where access to the female was signaled by odor, while males in the control group made no such association.

Villarreal and Domjan took their work a step further and went on to examine whether females learned to associate odor with the presence of their pairmates, and whether any male/female differences emerged. Females did learn to pair odor and access to their pairmates—conditioned females responded to odor cues by approaching the area associated with this cue, and any differences between males and females disappeared over time, suggesting that discrepancies in learning between males and females may not be great in nature.

Domjan and Hollis (1988) hypothesized that differences between males and females in their learning abilities should be positively correlated with differences in male and female parental investment. The more that parental investment is shared equally, the more the sexes should be similar in terms of learning ability, particularly with respect to learning the location of partners. One way to think about Domjan and Hollis's hypothesis is in terms of how much each sex is willing to invest in future offspring (Trivers, 1974). In the case where males and females both provide resources for offspring, selection pressure will be strong on both sexes. Assuming the appropriate cognitive infrastructure, in such a case we would expect males and females to have the ability to learn where their mate—the co-provider of resources for their offspring—is at any given time. This "learning the location of mates" ability may also be demonstrated during courtship.

FIGURE 4.20. Parental investment and learning ability. (A) Parental investment is shared in blue gourami, and differences in learning between the sexes is small. *(Photo credit: K. Hollis)* **(B)** In Japanese quail, there is no parental investment by males, and males show greater learning abilities than females. *(Photo credit: B. Galef)*

In many mating systems, only a single sex provides resources for offspring, and that sex is predominately the female sex. There are many reasons for this, reasons we shall explore more in Chapters 6 and 7, but for the purposes of Domjan and Hollis's hypothesis, what this translates into is that males in such systems should be better at learning about the location of mates than females. In terms of parental care for subsequent offspring, females are a valuable resource for males, as they alone provide food for offspring. From a parental care perspective, males are much less valuable a resource to females. Males then are under strong selection pressure to find receptive females, while females can almost always find males in such systems that are willing to mate. Hence, selection for learning locations associated with potential mates is stronger on males than females in such systems. In support of this hypothesis, in both gerbils (Villarreal and Domjan, 1998) and gourami fish (Hollis et al., 1989) parental investment is shared and learning differences between the sexes are small; while in Japanese quail, where there is no parental investment on the part of males, males show significantly greater learning abilities than females (Gutierrez and Domjan, 1996; Figure 4.20).

LEARNING ABOUT FAMILIAL RELATIONSHIPS

How animals recognize each other as kin or nonkin is still a subject of contention, but learning may play a role in such recognition in at

FIGURE 4.21. Java monkey mother and child. In Java monkeys, individuals appear to learn who is kin and who is not kin. *(Photo credit: E. R. Degginger/Color-Pic, Inc.)*

least some species (Fletcher and Michener, 1987; Hepper, 1991). If animals can learn how they themselves are related to others, as well as how different individuals in their group are related to one another, natural selection might favor altruistic and cooperative behavior being preferentially allocated to close blood kin.

Much work on primates demonstrates that they distinguish others based on kinship (Cheney and Seyfarth, 1990). For example, primates are able to recognize kin and to preferentially groom family members (Gouzoules, 1984; Gouzoules and Gouzoules, 1987). Armed with the information that kin-based individual recognition is important in primates, Dasser (1987, 1988) set out to examine whether Java monkeys (*Macaca fascicularis*) could *learn* the category "mother-offspring" when presented with slides of different individuals in their group (Figure 4.21). One monkey, Rini, was presented with "positive" slides (slides that were rewarded by honey water) depicting a mother and daughter from her own group, and "negative" slides (slides that had no reward) showing a pair of unrelated monkeys of a similar age to those in the positive slides. While this is an admittedly minuscule sample size (n=1 monkey), Rini was presented with fourteen sets of slides and asked to identify them based on whether or not they were mother and offspring by pushing a button located under the slides. Rini was correct in all fourteen trials and was rewarded (with honey water) in each case. A second Java monkey—Riche—went through a slightly different training procedure and was able to match mother-daughter pairs correctly in twenty of twenty-two trials.

Exactly *what* Rini and Riche learned that allowed them to differentiate mother-daughter pairs from others is hard to say. They may, for example, have learned from the photos that mother-daughter pairs share facial features, expressions, or any number of other traits. Or they may have learned from their own personal experience that mothers and daughters share some, still unknown, distinguishing features. One thing, however, is clear, and that is that Rini and Riche were indeed able to learn that a specific kind of relationship existed in some subset of slides, and they were both then able to correctly generalize what they had learned (the mother-daughter relationship) to other pairs of individuals.

LEARNING ABOUT AGGRESSION

Given the pervasiveness of aggressive behavior in even the most cooperative of animal societies (Dugatkin, 1997a; Gadagkar, 1997;

Huntingford and Turner, 1987; Archer, 1988), it should come as no surprise that ethologists and behavioral ecologists have had a long-standing interest in this phenomenon (Chapter 14). In the behavioral ecology literature, aggression is often partitioned into two components—intrinsic and extrinsic factors (Landau, 1951a, 1951b). Intrinsic factors usually refer to traits that correlate with an animal's fighting ability (Parker, 1974a), with the most common of these factors being some measure of size (Huntingford and Turner, 1987; Archer, 1988).

Extrinsic factors are encapsulated by what have come to be known as "winner" and "loser" effects (Chapter 14). Winner and loser effects are usually defined as an increased probability of winning, based on past victories, and an increased probability of losing based on past losses, respectively.

How much of the winner and loser effect is due to learning was addressed by Karen Hollis and her colleagues (Hollis et al., 1995). To begin with, prior work by these researchers had demonstrated that territorial male blue gourami fish (*Trichogaster trichopterus*) could be trained to associate a light with either the presence (+) or absence (−) of an intruder male. The researchers found that after the presentation of the light cue, two individuals that had both learned that a light meant the presence of another fish were much more aggressive than two fish that associated a light with the absence of another individual, presumably because the former were prepared for a fight (Figure 4.22).

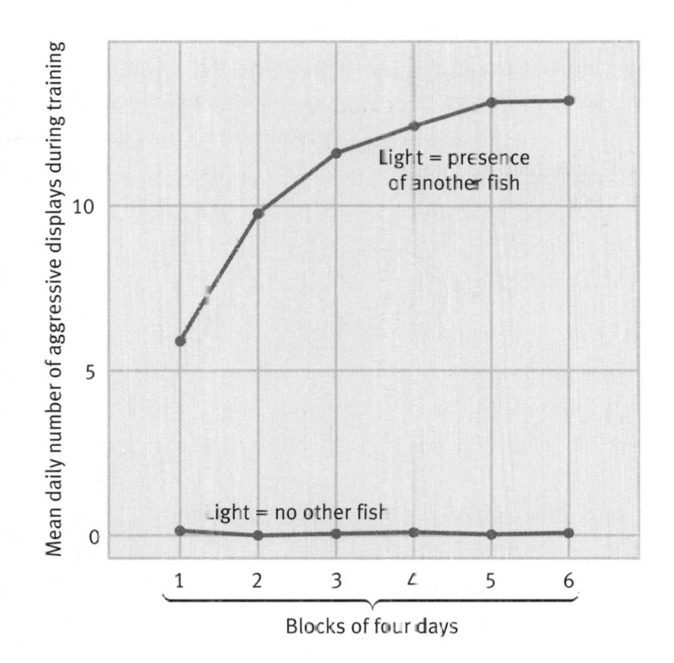

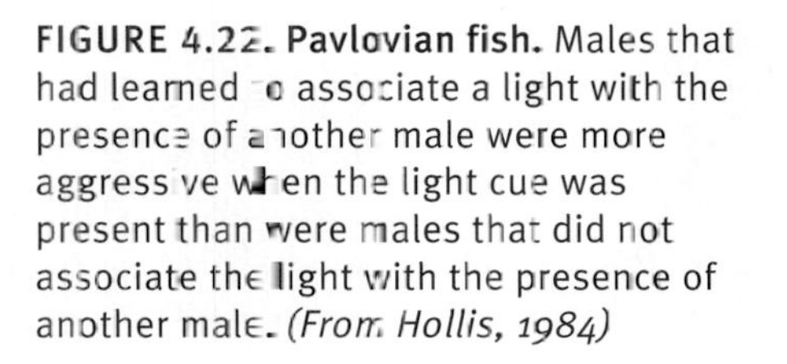
FIGURE 4.22. Pavlovian fish. Males that had learned to associate a light with the presence of another male were more aggressive when the light cue was present than were males that did not associate the light with the presence of another male. *(From Hollis, 1984)*

Interview with Dr. SARA SHETTLEWORTH

How did you become interested in studying learning? Were you trained as a psychologist or a biologist?
As an undergraduate, I was drawn to psychology by the introductory course taught by Henry Gleitman. He is a wonderful lecturer who went on to write a popular introductory textbook and win awards for his teaching. Much of the course dealt with the then-current controversies in learning theory, and I found the interplay of theory and experiments fascinating.

In graduate school I was exposed to biological approaches to behavior, which piqued my interest in what role learning played in the natural lives of animals. As it happened, it was at about that time that conditioned taste aversion was discovered by Garcia and his colleagues, and not long afterwards Brown and Jenkins first described autoshaping in pigeons. These discoveries were absolutely mind-blowing at the time because they seemed to show that learning was not a completely general process. Animals learned some things with very minimal input and others not at all. To understand these patterns, we might need to take into account the role that learning might play in the animal's natural life. Although I have worked on quite a number of different species and learning problems since then, it has always been with a commitment to this point of view. As for biological training, I have been fortunate to collaborate or otherwise be associated with biologists who are interested in psychological questions, and I think we have learned a lot from one another.

Do we know of any species of animal that can't learn something?
I don't know of any. Considering that even the single-celled organism *Stentor* can modify its behavior in simple ways, as shown by Jennings 100 years ago, it might be hard to find one that did not at least show habituation or sensitization.

Is there any reluctance in mainstream psychology to study learning from an evolutionary perspective? If so, why?
Historically, learning and memory have been studied pretty much in isolation from anything about the biology of the species doing the learning. People have suggested that the prominence of learning in American psychology and its abiological emphasis are attributable to the American faith in the importance of the environment as opposed to hereditary factors in individual development and adult achievement. In Henry Plotkin's excellent book about evolutionary psychology, *Evolution in Mind* (Penguin, 1997), there is a nice little history of the relationship between psychology and evolutionary thinking. This is definitely changing, partly due to discoveries within psychology like those I mentioned in answering question 1. The increasingly important role of neuroscience and genetic manipulations of the nervous system in psychology probably also play a role in making people think more and more of psychology as, in effect, part of biology. At the same time, behavioral ecologists have brought the evolutionarily based study of behavior full circle, back to an interest in causal mechanisms and development, which tended to be neglected in the early days of that field. Indeed, the term *cognitive ecology* has been invented to refer to the study of the role of cognitive mechanisms—perception, learning, memory, and the like—in solving ecological problems. This can only lead to more cross-disciplinary com-

munication, collaboration, and training, which in the long run may have an impact on thinking in psychology.

How much of a quagmire is the terminology used in the study of learning? To the outside reader it seems as if there are endless definitions and subdefinitions. Is that a fair statement?

No, I don't think it's fair. Fields of science all have their own specialized terminologies, which are necessary to convey specialized ideas and distinctions that can't be expressed concisely in ordinary language. When I ask psychology students to read literature from behavioral ecology, they find terms like ESS, MVT, conspecific, homology versus analogy, phylogeny, and the like, pretty baffling.

What do you think ethologists can glean from work on human learning?

In his famous paper "On Aims and Methods of Ethology," Tinbergen pointed out that a complete understanding of behavior includes answers to four questions. Two of them are "What is the current (or proximate) cause of this behavior?" and "How does it develop in the individual?" (The others are "How did it evolve?" and "How does it function?" or "What is its survival value?") Causation and development are essentially what psychologists study, even if the species and behaviors are somewhat limited. So it seems obvious that ethologists and psychologists (whether they study humans or other animals) should have a lot to learn from each other.

Could you weigh in on the "modular mind" debate? Do you think learning is better viewed as one all-purpose algorithm, or as a series of smaller programs, each designed by selection to allow animals to cope with particular sorts of problems (foraging, mating, etc.)?

I am definitely of the "modular mind" school because I think it makes more functional sense and because it also makes better sense of some data. It is also a more sensible way to approach a broad comparative psychology than the old idea that species differ in some single dimension like "intelligence." I have written about this in a couple of recent chapters (Shettleworth, 2000, 2002). However, along with other psychologists of this persuasion, I tend to think of modules more as cognitive subunits that perform distinct information-processing operations rather than as mechanisms for separate biological functions like mating and food-finding. For example, learning and using information from landmarks might be identified as a distinct (modular) part of spatial cognition, but it could be employed in finding food or mates or a nest. Of course, there have been rather few tests of whether this is really correct. At least in rats in laboratory tests, spatial learning seems to proceed similarly whether the animal is rewarded with food (i.e., foraging) or escaping from a water tank or some other aversive situation. There may also be some differences in the details, but it is not clear whether or not this is due to the different motivational systems in play or for some other reason like sheer strength of motivation.

What's the next breakthrough to look for in the study of animal learning?

Although there are still many unanswered questions about animal learning and cognition that need to be studied entirely at the level of behavior, at present there is probably more research on learning being done by behavioral neuroscientists and geneticists than by people like me who focus on behavior of normal intact animals. Thus, statistically it is most likely that the biggest breakthroughs will be made in studies of the neural and molecular basis of learning. However, while much work of this kind simply uses traditional behavioral tests like maze learning or Pavlovian conditioning, some behaviorally sophisticated researchers in this area are making novel observations about how animals learn and remember. For example, in a quest to develop simple tests for rats in which the animals would remember many items of information, Howard Eichenbaum and his colleagues discovered that—not surprisingly given their nocturnal way of life—rats are extraordinarily good at olfactory learning. A rat can learn and remember the significance of many different odors and at the same time. As well, when odors are used, rats can easily learn kinds of tasks that they would learn only with great difficulty, if at all, with visual stimuli, which have traditionally been used. This example illustrates how an integrated approach to learning can lead to new findings: by using ethologically relevant stimuli the researchers have revealed new facts about animal learning and memory, as well as making possible investigations of the neural basis of learning that would not be possible otherwise.

DR. SARA SHETTLEWORTH is a professor at the University of Toronto, Canada. Her work integrating biological and psychological approaches to the study of animal cognition and learning ranks her as a leader in that field.

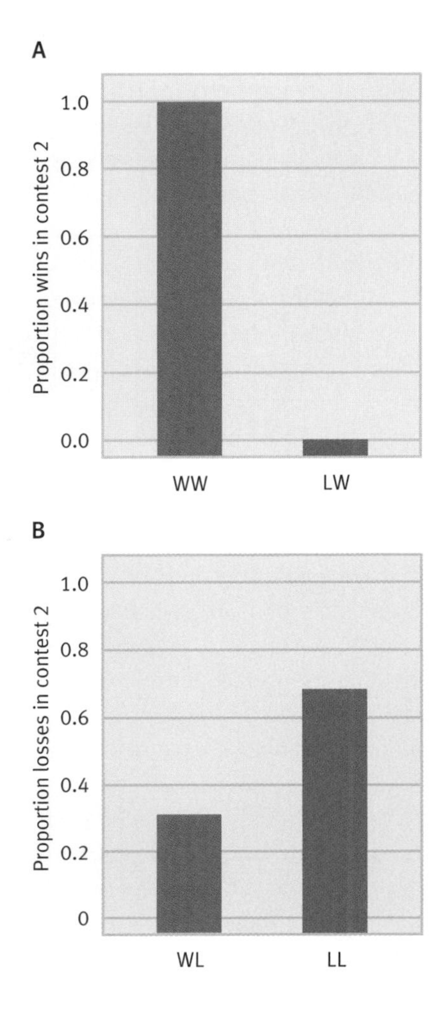

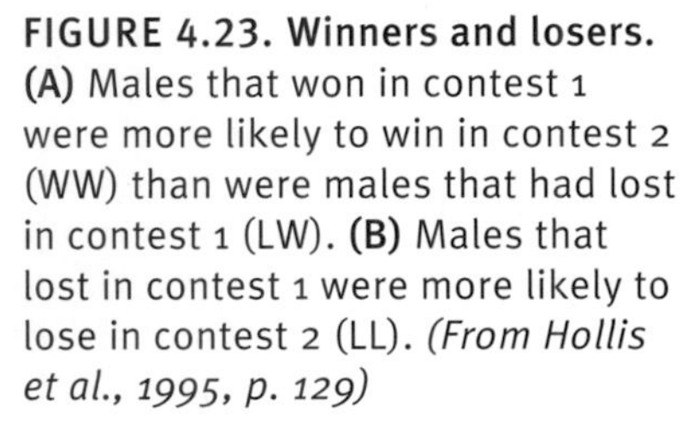

FIGURE 4.23. Winners and losers. **(A)** Males that won in contest 1 were more likely to win in contest 2 (WW) than were males that had lost in contest 1 (LW). **(B)** Males that lost in contest 1 were more likely to lose in contest 2 (LL). *(From Hollis et al., 1995, p. 129)*

Blue gourami males set up their territories well in advance of the point in time when females start visiting nests and, as such, males may have multiple encounters with numerous territorial intruders. Hollis and her colleagues (1995) used a similar protocol to the one described above to examine the effect of conditioned learning on winner and loser effects. The essential idea was to examine whether conditioning could have long-term effects on territorial males by creating the start of a series of encounters that would produce a winner or loser effect.

Half the gouramis in the study were trained to associate a red light with the presence of a territorial intruder. In a control group of males, the red light was presented six hours before seeing a territorial intruder, and males did not associate the red light with an intruder. In the first contest of this experiment, after a red light was turned on, trained males and controls were pitted against each other until a clear winner and loser emerged. As expected from earlier work, trained males were much more likely to win such encounters. Trained males presented with the red light before an encounter were also more likely to defeat trained males that were not shown the stimulus light before their encounter.

In the second contest staged by Hollis and her colleagues, winners and losers from the first contest were paired up with new intruders three days later. In these pairings, no red light was shown to either fish. All individuals that won their fights in the second contest turned out to be winners in the first contest as well, and these individuals were all trained males that had seen the signal light before their first encounter (Figure 4.23). That is, males trained through conditioning were not only more likely to win initial encounters, but they also won subsequent encounters in which the stimulus they had been trained with was now absent. A similar effect was found with respect to losing; losers were likely to lose to new intruders even in the absence of the stimulus. These results as a whole suggest that learning may be a powerful force in shaping aggressive interactions.

SUMMARY

1. Psychologists have defined learning in many ways. When we speak of learning here, we are referring to "a relatively permanent change in behavior as a result of experience." This definition does, however, have a downside, in that often it is difficult

to determine how long a time period constitutes "relatively permanent."
2. The ability of organisms to learn provides them with the opportunity to respond in a very flexible fashion to environmental change. Learning, in the most general sense, is considered a form of phenotypic plasticity.
3. The simplest form of learning involves a single stimulus. Sensitization and habituation are two simple, single-stimulus forms of learning. Habituation in particular is often of concern to experimental ethologists, whose interpretation of results can be confounded if animals habituate too quickly to stimuli.
4. Pavlovian, or classical, conditioning experiments involve two stimuli—the conditioned stimulus and the unconditioned stimulus. A conditioned stimulus (CS) is defined as a stimulus that fails initially to elicit a particular response, but comes to do so when it becomes associated with a second (unconditioned) stimulus. Second-order conditioning, excitatory conditioning, inhibitory conditioning, learnability, blocking, and overshadowing are issues often addressed in Pavlovian conditioning experiments.
5. Instrumental, also known as operant or goal-directed, learning occurs when a response made by an animal is somehow reinforced. One fundamental difference between Pavlovian conditioning and instrumental learning centers on the fact that in the latter the animal must undertake some "action" or "response" in order for the conditioning process to produce learning.
6. Interpopulation comparisons in learning, as in studies of sticklebacks and doves, are a powerful tool employed by ethologists interested in learning. By comparing across populations that differ in their abilities to learn, ethologists can address both proximate questions and ultimate questions (why do such differences occur?).
7. Ethologists, behavioral ecologists, and psychologists have long argued that learning is favored over the genetic transmission of a fixed trait when the environment an animal lives in changes often, but not too often. David Stephens has expanded on this idea by investigating within- and between-generation variability and finds that learning is favored when predictability within the lifetime of an individual is high, but predictability between generations is low.
8. Animals learn in many different contexts, including, but not limited to, foraging (what to eat?), habitat selection (where to live?), predators (what's dangerous?), mates (what constitutes a mate, and what constitutes a good mate?), and familial relationships (who is blood kin?).

DISCUSSION QUESTIONS

1. Following up on the "optimal forgetting" study in stomatopods that we discussed in this chapter, can you think of other situations in which it might pay for animals to forget, or at least not act on, information they have obtained? Try to come up with three cases and write a paragraph on each case justifying its selection.
2. Obtain a copy of parts I and II of Tooby and Cosmides's 1989 article "Evolutionary psychology and the generation of culture," in volume 10 (pp. 29–97) of the journal *Ethology and Sociobiology*. After reading this article, explain how "Darwinian algorithms" work and how they relate to our discussion of animal learning.
3. Read Domjan and Hollis's 1988 chapter "Reproductive behavior: A potential model system for adaptive specializations in learning," which appeared in Bolles and Beecher's book, *Evolution and learning* (pp. 213–237). Then outline how classic psychological models of learning can be productively merged with evolutionary approaches to learning.
4. Design an experiment that can distinguish between the two alternative explanations for interpopulational differences in dove foraging, as described in Carlier and Lefebvre's 1996 "Differences in individual learning between group-foraging and territorial Zenaida doves," which appeared in volume 133 (pp. 1197–1207) of the journal *Behavior*.

SUGGESTED READING

Balda, R., & Kamil, A. (1989). A comparative study of cache recovery by three corvid species. *Animal Behaviour, 38,* 486–495. A good example of integrating between-species comparisons with an evolutionary approach to learning.

Johnston, T., & Pietrewicz, A. (Eds.). (1985). *Issues in the ecological study of learning.* Hillsdale, NJ: Lawrence Erlbaum Associates. The "ecological" approach to learning is outlined in this edited volume.

Krebs, J. R., & Inman, A. J. (1992). Learning and foraging: Individuals, groups, and populations. *American Naturalist, 140,* S63–S84. A review of learning and its role in foraging studies.

Shettleworth, S., (1998). *Cognition, evolution and behavior*. New York: Oxford University Press. A very detailed treatment of behavior and evolution, with much on learning.

Stamps, J., & Krishnan, V. V. (1999). A learning-based model of territorial assessment. *Quarterly Review of Biology, 74,* 291–318. In this paper, a model of learning is constructed in the context of territoriality, particularly in lizards

5

What Is Cultural Transmission?

- Animal Culture
- Why All the Fuss about Culture?

Types of Cultural Transmission

- Social Learning
- Teaching in Animals

Modes of Cultural Transmission

- Vertical Cultural Transmission
- Horizontal Cultural Transmission
- Oblique Cultural Transmission

The Interaction of Genetic and Cultural Transmission

- The Grants' Finches
- Whitehead's Whales

Interview with Dr. Bennet "Jeff" Galef

Social Learning and Cultural Transmission

FIGURE 5.1. Imo the monkey. Imo, a Japanese macaque monkey, introduced a number of new behaviors (for example, potato washing) that spread through her population via cultural transmission. *(Photo credit: Umeyo Mori)*

In our discussion of genetics and heritability in Chapter 2, we noted that for natural selection to act on a behavior, a mechanism for transmitting that behavior across generations is required. When Mendel's work on genetics was rediscovered in the early 1900s, it became obvious that genes were one clear and easily interpretable means by which a trait could pass down across generations, enabling natural selection to act on genetically encoded traits. In fact, with respect to animal behavior, until recently evolutionary biologists and ethologists acted as if genes were not only *one* way to transmit information across generations, they were the *only* way. Slowly, this view is beginning to change, and there is a growing recognition that cultural transmission of behavior both within and between generations may be a potent force in shaping animal behavior (Bonner, 1980; Boyd and Richerson, 1985; Zentall and Galef, 1988; Heyes, 1994; Heyes and Galef, 1996; Dunbar et al., 1999; Avital and Jablonka, 2000; Dugatkin, 2000).

Consider the case of Imo the monkey (Kawamura, 1959; Kawai, 1965). Imo was a Japanese macaque that lived on Koshima Islet, Japan, in the 1950s (Figure 5.1). Researchers studying the macaques threw sweet potatoes on the sandy beach for the monkeys to gather and eat. When Imo was a year old, she began to wash the sweet potatoes before she ate them. This creative move allowed her to remove all the sand from the sweet potatoes before she ingested them. Soon enough, many of Imo's peers and relatives had learned the art of potato washing from watching little Imo. By 1959, most infant macaques in Imo's troop intently watched their moms, many of whom now had acquired Imo's habit, and they learned to wash their own sweet potatoes at early ages.

When Imo was four, she introduced an even more complicated new behavior into her group. In addition to the sweet potatoes that monkeys on Koshima Islet were given, they were also occasionally treated to wheat. The problem was that wheat was again provisioned to the monkeys on a sandy beach, and wheat and sand mixed together is not nearly as appealing as wheat alone. So, Imo came up with a novel solution—she tossed her wheat and sand mixture into the water, where the sand sank and the wheat floated. As with the sweet potatoes, it was only a matter of time before her groupmates learned this handy trick from Imo. It took a bit longer for this trait to spread through the population, though, as monkeys aren't used to letting go of food once it is in their possession. As such, it was hard for them to learn to throw the wheat and sand combination into the water. But eventually this new behavioral trait was spread to many group members via cultural transmission (Figure 5.2).

Imo's actions were not the only ones to garner attention about

FIGURE 5.2. Potato washing in monkeys. In Japanese macaques living on Koshima Islet, Japan, the skill of potato washing appears to be transmitted culturally. *(Photo credits: Frans de Waal)*

cultural transmission of behavioral traits, as many similar cases are now on record. Michael Huffman, for example, has found another case of cultural transmission in primates (Huffman, 1996). Huffman's study revolves around twenty years of work on the Japanese macaques of the Iwatayama National Park (Kyoto). Early on in his work, Huffman began to observe a behavior never before noted in macaques: individuals would play with stones, particularly right after eating (Figure 5.3). This bizarre behavior began in 1979 when Glance-6476, a three-year-old female macaque, brought rocks in

FIGURE 5.3. Stone play in monkeys. Michael Huffman has found that one population of Japanese macaques in the Iwatayama National Park in Kyoto has created a tradition of "stone play" in which individuals stack up stones and then knock them down. *(Photo credit: Michael Huffman)*

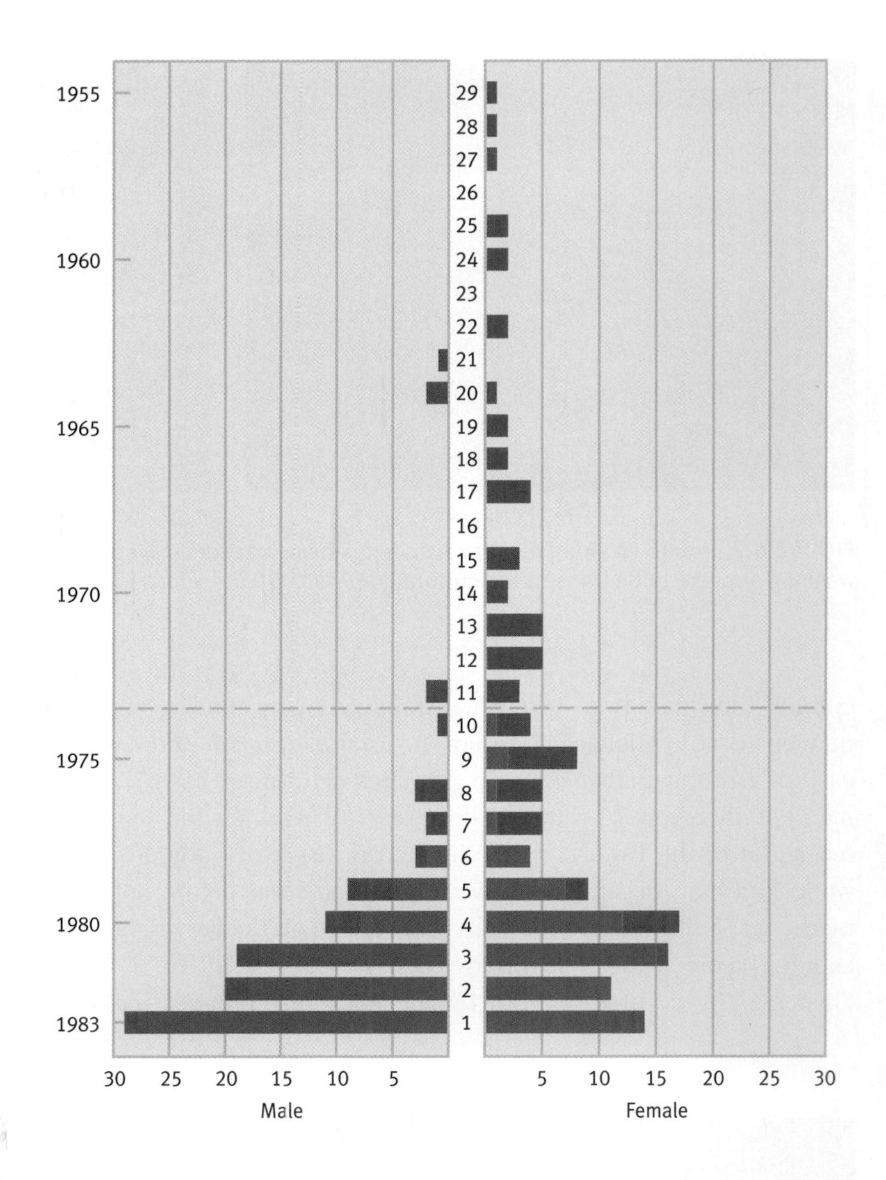

FIGURE 5.4. Stone play tradition spreads. Orange bars represent verified stone handlers, and green bars represent individuals not verified as stone handlers. Note the increase in the frequency of the tradition with time. *(From Huffman, 1996, p. 276)*

from the forest and started stacking them up and knocking them down. Not only that, Glance-6476 was very territorial about her stones and took them away when approached by other monkeys. When Huffman returned to Glance's troop four years later, stone play (also referred to as stone handling) "had already become a daily occurrence [and] was already being transmitted from older to younger individuals" (Figure 5.4). Interestingly, cultural transmission in this system seems to work down the age ladder, but not up. While many individuals younger than Glance-6476 acquired her stone play habit, no one over her age added this behavior to their repertoire.

The reason for opening a discussion of culture with the monkeys of Koshima Islet and the Iwatayama National Park is not because they are necessarily the most convincing cases of cultural transmission recorded in animals. As we shall soon see, these examples lack the controlled work that many other studies of cultural transmission contain. But food washing and stone handling do dramatically illustrate that culture can be a powerful force and that cultural transmission across generations occurs in nonhuman animals, as well as humans.

What Is Cultural Transmission?

George Romanes was one of the first to actually raise the possibility that cultural transmission plays an important role in animal life (Figure 5.5). In a series of books, Romanes (1884, 1889, 1898) argued that imitation, one form of cultural transmission, is integral to a wide variety of animal actions. It is perhaps worth noting that Romanes was a good friend and admirer of Charles Darwin, and Darwin's theory of natural selection had a huge impact on Romanes's view of the world (Romanes, 1892). The world of nineteenth-century science was a small one, and the impact of the theory of natural selection touched virtually every corner of it in one way or another.

FIGURE 5.5. What constitutes cultural transmission? Clearly the child is learning to use utensils by watching others (that is, through cultural transmission). The chimp on the right, however, also has learned his nut-cracking skills from watching others. Should this also count as a culturally transmitted behavior?

ANIMAL CULTURE

In order to understand the role of cultural transmission in animal societies, we need to begin with a definition. What exactly do we mean when we speak of animal **culture?** This turns out to be a more difficult question than it might appear to be at first glance. In fact, even back in 1952, anthropologists already had more than 150 definitions of culture (Kroeber and Kluckhohn, 1952). While there are no doubt even more definitions that exist today, evolutionary and behavioral biologists have tended to focus on one general definition of culture that is summed up nicely by Robert Boyd and Peter Richerson in their book *Culture and the Evolutionary Process* (1985).

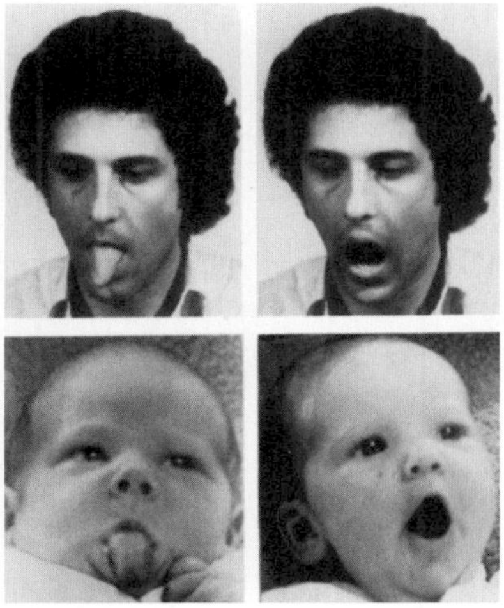

FIGURE 5.6. Imitation in newborns. The top two photos show Dr. Andrew Meltzoff demonstrating tongue protrusion and mouth opening to a twelve- to twenty-one-day-old baby. The bottom two photos show imitation on the part of the infant. *(Photo credit: Andrew Meltzoff)*

Slightly modifying Boyd and Richerson's definition, we shall speak of culture as a system of information transfer that affects an individual's phenotype, in the sense that part of the phenotype is acquired from others by teaching or social learning (Figures 5.6 and 5.7). For example, recall the Norway rats we discussed in Chapter 1. The upside of scavenging for food as these rats do is that they find all sorts of food. The downside is that a lot of what is found is novel, and in such circumstances, it can be difficult to determine what is safe to eat, and what isn't. Such a scenario is ideal for the cultural transmission of information, as new items are always being encountered, and information on their potential danger constantly needs to be updated. Obtaining some sort of information about food from others is one way to update knowledge about the suitability of novel foods. In fact, Norway rats learn what new foods to try by smelling their nest mates and subsequently trying the new food items that such nest mates have recently ingested. As

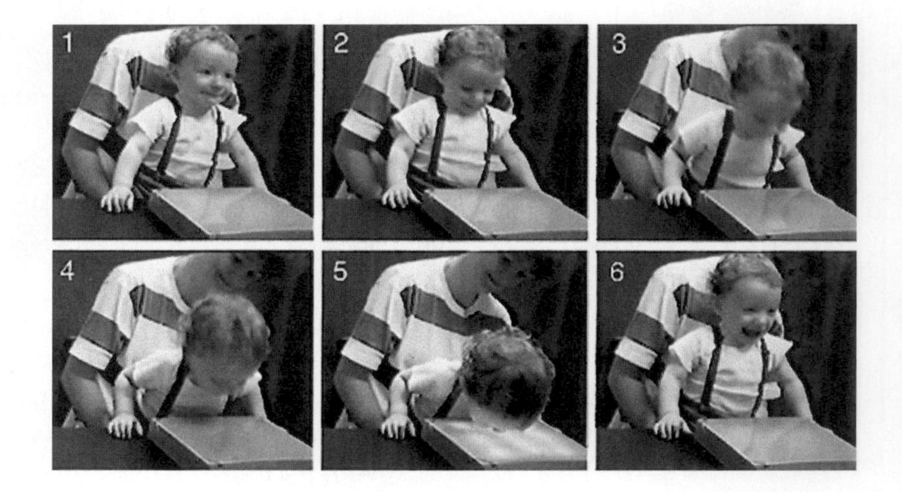

FIGURE 5.7. Imitation in toddlers. A fourteen-month-old baby imitates the novel behavior of "head touching." Such imitation often solicits a smile from the child subject. *(Photo credit: Andrew Meltzoff)*

such, part of the rat's phenotype—in this case what sorts of food it eats—is modified by information that it has learned from other Norway rats.

WHY ALL THE FUSS ABOUT CULTURE?

Before moving on to what exactly constitutes social learning and teaching in animals, let us briefly touch on an issue that biologists, psychologists, and anthropologists often raise when we speak of culture being tied to teaching and learning from others. "Why," the question goes, "all the special treatment for culture?" After all, we spent a great deal of time on the topic of learning in Chapter 4, and if cultural transmission is just one form of learning—that is, social learning—why not simply consider it a special kind of learning and move on? In other words, as Boyd and Richerson put it, "Why not simply treat culture as a . . . response to environmental variation in which the 'environment' is the behavior of conspecifics?" Why not think of culture as just another means by which organisms adapt to the environment, and leave it at that? Why all the hoopla?

The answer to these questions is that learning from other individuals differs dramatically from learning about other parts of the environment in two important ways. To begin with, cultural influences, unlike other environmental influences, *are passed on from individual to individual*. That is, if I learn from others and they learn from me, information can be spread in a very efficient manner. This potentially translates into the behavior of a single individual in a population dramatically shifting the behavior patterns seen in an entire group. This is simply not the case for other types of learning. What an individual learns via individual learning disappears when that individual dies (and perhaps before). When social learning is in play, however, what is learned by one individual may be passed on through endless generations. If Imo, the Japanese macaque monkey we discussed at the beginning of the chapter, had learned to wash her sweet potatoes by throwing them in water in an environment in which social learning did not occur, this amazing new trait would have vanished when Imo died. Social learning did play an important role in the population of monkeys in which Imo lived, however, and some fifty years later one can still go to Koshima Islet and see these monkeys washing their sweet potatoes.

What makes cultural evolution so important is the speed at which it can operate. When genetic evolution operates quickly, we

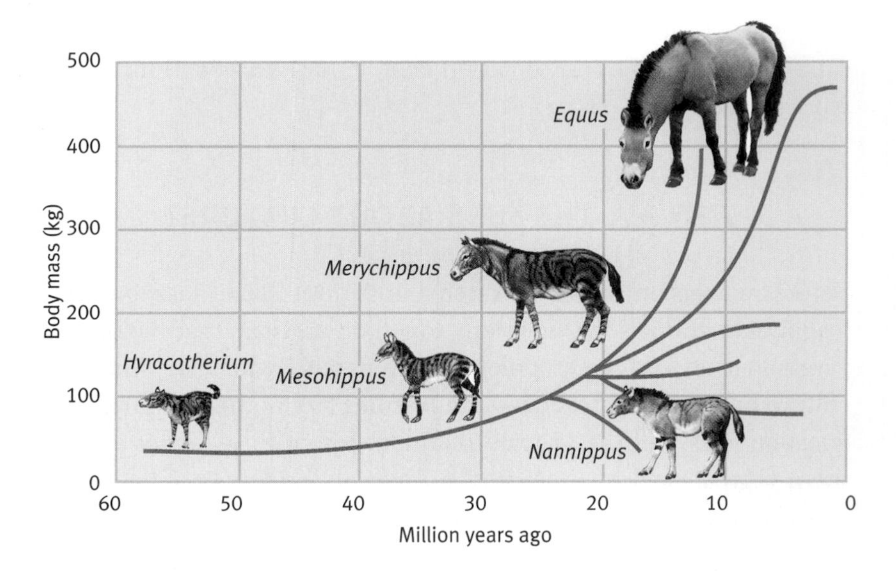

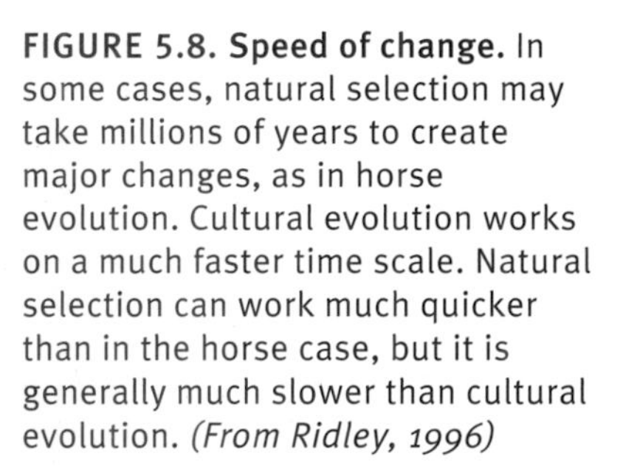

FIGURE 5.8. Speed of change. In some cases, natural selection may take millions of years to create major changes, as in horse evolution. Cultural evolution works on a much faster time scale. Natural selection can work much quicker than in the horse case, but it is generally much slower than cultural evolution. *(From Ridley, 1996)*

are usually talking about tens, hundreds, thousands, or perhaps tens of thousands of generations for natural selection to make a noticeable difference (Figure 5.8). Not so for cultural evolution, which can easily have a huge impact in just a handful of generations. In fact, cultural evolution can have a dramatic impact within a single lifetime.

Types of Cultural Transmission

We can now be a bit more specific about how different types of cultural transmission are defined, as well as how they operate. While in principle cultural transmission could be divided in many different ways, here we shall break cultural transmission into two parts, "social learning" and "teaching." Both social learning and teaching rely in one sense or another on a "model" individual (also called a demonstrator or tutor) and an "observer" (Figure 5.9). But teaching requires a much more active role on the part of the model.

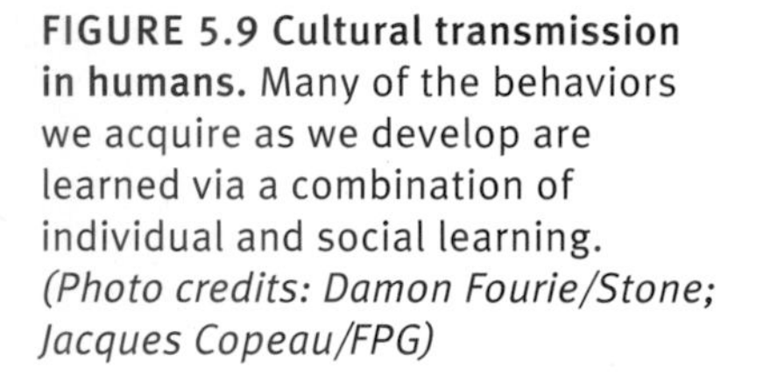

FIGURE 5.9 Cultural transmission in humans. Many of the behaviors we acquire as we develop are learned via a combination of individual and social learning. *(Photo credits: Damon Fourie/Stone; Jacques Copeau/FPG)*

SOCIAL LEARNING

Examples of social learning in humans abound in the psychological literature (Bandura, 1977, 1986), as do examples of social learning

FIGURE 5.10 Watch, learn, and decide. Chimps learn how to "fish" for termites by watching others. Individual chimps are clearly capable of judging how effective a foraging technique this is, and choosing whether to add it to their behavioral repertoire or not.

in animals (Zentall and Galef, 1988; Heyes and Galef, 1996; Dugatkin, 2000; Figure 5.10). To take a classic example, let's examine Albert Bandura's "Bobo" doll experiment (Bandura et al., 1961). In this experiment, children were exposed to one of two conditions. In one group, a child was involved in an art project when an adult in the room began punching and kicking a "Bobo" doll. This doll was inflatable and weighted at the bottom, so it snapped back up when knocked down. In the other group, children doing an art project saw a calm adult. In this treatment, the Bobo doll was still in the room; it just wasn't being assaulted by an adult. Children from both groups were then given the choice of playing with aggressive toys or nonaggressive toys. Not surprisingly, the children exposed to the "violent adult" treatment not only chose aggressive toys, but often beat up the doll and yelled the same things that the adult had yelled, while children in the other treatment played much more calmly (Figure 5.11).

No doubt, some readers will look at the Bobo doll experiment and think "Of course Bandura got the results he did; it's obvious to anyone that's what would happen." But two things should be kept in mind when looking at these results: First, Bandura didn't "have to" obtain the results he did. The kids could have ignored the adult punching the Bobo doll or could even have been particularly nice in response to the adult's aggressive behavior. In fact, some social psychology theories predicted that watching aggression would somehow reduce the observers' aggressive tendencies by providing an outlet. It could have worked that way in Bandura's experiment; it just didn't. Second, science is full of examples where researchers

FIGURE 5.11. The Bobo doll experiment. In Bandura's classic Bobo doll experiments, the power of social learning is frighteningly evident, as young children treated the Bobo doll as aggressively as the adult they observed. *(Photo credits: Albert Bandura)*

actually did the experiment to test "something that was obviously true" and found that it wasn't. The only way to know is to do the experiment. The same thing can be said of work on social learning in animals. In retrospect, it might seem obvious that rats learn what to eat and what not to eat by social learning. But it didn't have to turn out that way. Rats could have ignored any information they obtained from smelling their nest mates. The only way to really know is to do an experiment testing the hypothesis.

LOCAL ENHANCEMENT. Though interesting in their own right, here we shall not consider developmental psychology or cognitive processing approaches to social learning, but rather focus on the different means by which social learning, broadly defined, operates (Galef, 1988; Heyes and Galef, 1996). We begin with something called local enhancement.

The concept of **local enhancement,** first introduced by Thorpe (1956, 1963), suggests that individuals may learn from others, not so much by doing what they see, as by being drawn to a particular area because a model they were watching was in that location (Heyes, 1994; Figure 5.12). In other words, when local enhancement is in play, a model simply draws attention to some aspect of the environment by the action he undertakes there (for

FIGURE 5.12. Local enhancement. If fish 1 is drawn to where fish 2 is foraging (a stone), but then learns nothing else from fish 2, local enhancement is at work.

example, digging for worms). Once the observer is drawn to the area, he learns on his own.

As an example of local enhancement, consider foraging behavior in colonially nesting cliff swallows. These birds feed in groups that range in size from 2 to 1,000 individuals, and Charles Brown (1988) found that in addition to the actual transfer of information between groupmates (Brown, 1986), local enhancement facilitates foraging, as some individuals are drawn to good foraging areas because others are foraging there. Once this takes place, swallows may eventually learn to forage better in that area.

SOCIAL FACILITATION. As opposed to local enhancement, in which the action of a model draws attention to some aspect of the environment, under **social facilitation** the mere presence of a model, regardless of what he does, is thought to facilitate learning on the part of an observer (Zajonc, 1965). For example, there are many instances in the foraging literature in which increased group size causes increased foraging rates per individual (Stephens and Krebs, 1986; Giraldeau and Caraco, 2000). As we shall see in Chapter 10, there are many possible reasons why this may occur.

FIGURE 5.13. Social facilitation. In social facilitation, the mere presence of a model(s) draws in an observer. Here our lone starling is attracted to a group, not because of what group members are doing or where they are, but simply because it is drawn to the presence of others. "Safety in numbers" might be a benefit to such facilitation.

With respect to social facilitation, one possibility is that individuals have reduced fear levels when in groups, which allows them more time and energy to learn how to forage well (Galef, 1988). The mere presence of others causes a reduction in fear and an increase in foraging-related learning (Figure 5.13).

CONTAGION. First introduced by Thorpe (1963), **contagion**, also known as the "response facilitation effect" (Bandura, 1986), refers to behaviors in which "the performance of a more or less instinctive pattern of behavior by one will tend to act as a releaser for the same behavior in others . . ." (Thorpe, 1963, p. 163).

Jeff Galef (1988) notes that contagion itself does not qualify as social learning, but in conjunction with individual learning, contagion may be a form of social learning. If, for example, an animal runs away from a predator when he sees others do so, contagion is likely in play, as an instinctive "flight response" of others may serve to set off his own flight response. But suppose that individuals learn which conspecific's flight is the best indicator of true danger and flee primarily when such individuals flee. In this case, contagion and individual learning together may produce a form of social learning. As another example, if an animal just starts digging for food because he sees a conspecific do so, we have contagion; if he does so primarily when he sees others that he knows from experience are good foragers, contagion plus individual learning produce social learning.

IMITATION. The term **imitation** has been used many different ways in the psychological literature, but it is usually portrayed as the most difficult form of social learning. Following Galef, Celia Heyes (1994) argues that imitation "is now commonly understood to refer to the acquisition of a topographically novel response through observation of a demonstrator making that response, and to involve a 'goal-directed' psychological mechanism." As such, to demonstrate imitation (sometimes referred to as observational learning), there must be some new behavior learned from others, and that behavior must involve some sort of spatially new manipulation as well as lead to the achievement of some goal (Figure 5.14).

An interesting case of imitation in birds has been uncovered by Galef and his colleagues (1986). In this study, budgerigars observed others lifting a flat cover off a dish, using either their beak or their foot. Those who observed a model who used his beak to gain access to food were themselves likely to use their beaks when placed in the same situation, while those observing a model who used his foot to unlock hidden food also used their feet.

COPYING. When animals **copy** one another, an observer repeats what he has seen a model do. Oftentimes, the copier is then

FIGURE 5.14. Imitation. Here an observer bird watches a trained pigeon who must lift its foot and push on a lever to open a small circular entrance to a food source. Imitation occurs when the observer learns this new task by watching the model lift its leg and push the lever down.

rewarded for whatever behavior he has copied. For example, if an animal observes another individual forage in a certain way and then adopts such a foraging technique for himself after he has been rewarded for doing so (for example, he receives more food), copying has taken place. In the psychological literature, the rewards associated with copying can be extrinsic (such as the food items in the above example) or intrinsic (for example, related to animal emotions and feelings). Copying differs from imitation in that what is copied need not be novel. That is, an individual can copy the action of another, even if he already knows how to do what the model is doing. For example, animals generally know how to choose a mate, but they may still copy the mate choice of a model. In addition, imitation in its strictest form requires a "topographically novel response" (for example, some new combination of movements), while copying has no such requirement. Lastly, copying differs from observational conditioning in that what is being copied need not relate to a new stimulus.

TEACHING IN ANIMALS

At the most basic level, the power of teaching to shape behavior is clear. If individuals teach others, and then individuals who learn teach still others, you have a self-perpetuating system of cultural transmission. Not surprisingly, given the complexity that we associate with teaching, the notion that animals teach each other is one of the more contentious aspects of animal cultural transmission (Caro and Hauser, 1992; Hauser, 1997, 2000; Hauser and Konishi, 1999).

While there are many definitions of "teaching," the idea usually implies that one individual serves as an instructor and at least one other individual acts as a pupil who learns from the teacher (Caro and Hauser, 1992). Teachers then have a much more active, not to mention complicated, role than just being a model that someone mimics, as in social learning. In a review on teaching in animals, Tim Caro and Marc Hauser suggest the following definition of teaching:

> An individual actor **A** can be said to teach if it modifies its behavior only in the presence of a naive observer **B**, at some cost or at least without obtaining an immediate benefit for itself. **A**'s behavior thereby encourages or punishes **B**'s behavior, or provides **B** with experience, or sets an example for **B**. As a result, **B** acquires knowledge or learns a skill earlier in life or more rapidly or efficiently than it might other-

wise do, or that it would not learn at all. (Caro and Hauser, 1992, p. 153).

Essentially, then, to teach a naive individual, a teacher must provide an immediate benefit to the student but not to himself, he must teach only naive "students," and he must impart some new information to students faster than they would otherwise receive it. This definition is interesting, not only because of the emphasis on what must take place for teaching to occur, but also for what kinds of behaviors are excluded from the realm of teaching. For example, in the 1940s, bottles of milk in Britain were being tampered with by blue tits (*Parus caeruleus*). The foil caps on the top of milk bottles were being torn off before people could retrieve the freshly delivered bottles from their doorsteps (Figure 5.15). J. Fisher and Robert Hinde (1949) argued that this new behavior had been accidentally stumbled upon by a lucky blue tit and that others learned this nifty trick, at least in part, from watching the original milk thief. So, while blue tits learned how to open foil caps by observing others, those who opened the caps did so regardless of who watched them. Blue tits may have been imitating or perhaps copying one another, but according to Caro and Hauser's definition, they weren't teaching other birds anything.

FIGURE 5.15. Blue tit birds learn to open milk bottles. Blue tit birds learned to peck open the top of milk jugs decades ago. This behavior may have spread via cultural transmission. *(Photo credit: Sarah Colter)*

What sort of examples might fall under the Caro and Hauser definition of teaching? Consider a female cat who captures live prey and allows her young to play with this prey, making sure the prey doesn't escape along the way (Figure 15.16). If mother cats only engage in this behavior when in the presence of young cats,

FIGURE 5.16. Cheetah teaching? (A) A mother cheetah brings a Thomson's gazelle to her cubs, and allows them to "kill it," even though it was already dead. **(B)** Until the young cheetah is taught how to hunt, it can only kill small items like the hare shown here. *(Photo credits: Tim Caro)*

we may have a true case of teaching. Anecdotal examples of this genre of teaching have been documented to some extent in domestic cats (Ewer, 1969; Baerends-van Roon and Baerends, 1979; Caro, 1980), lions (Schenkel, 1966), tigers (Schaller, 1967), meerkats (Ewer, 1963), and otters (Liers, 1951). The most thorough study on this phenomenon has taken place in cheetahs. Consider Caro's description of three ways mothers used "maternal encouragement" to facilitate hunting skills in their offspring:

> Firstly, they pursued and knocked down quarry but instead of suffocating the victim allowed it to stand and run off. By the time the prey had risen, the cubs had normally arrived. Second, mothers carried live animals back to their cubs before releasing them, repeatedly calling (churring) to their cubs. Third, and less often, mothers ran slowly during their initial chase of a prey and allowed their cubs to overtake them and thus be the first to knock down the prey themselves. (Caro, 1994a, pp. 136–137)

Caro and Hauser listed other putative cases of animal teaching (Caro and Hauser, 1992). These include the following:

- In chimpanzees, gorillas, rhesus monkeys, yellow baboons, and spider monkeys, "mothers have been observed encouraging their young to walk and follow them, typically in the context of group movement or foraging." (p. 160)
- Adult chacma baboons and squirrel monkeys chase juveniles away from objects that the adults know are dangerous.
- Washoe the chimpanzee, an expert in sign language, may have taught Loulis (a less adept signer) the correct way to sign for "food."
- Vervets may teach their young the appropriate alarm call for different predators.
- Chimpanzees in the Tai National Forest may teach their young how to crack open nuts. For example, some females leave a cache of nuts on top of an anvil in the vicinity of young that are at approximately the age at which they should begin to learn nut-cracking skills.
- A number of species of raptor birds (for example, hawks, falcons, ospreys) appear to teach their young the difficult art of precise hunting while flying at high speeds.

Caro and Hauser also undertook a search for common themes that underlie these possible cases of teaching and found two candidates. First, almost all instances of animal teaching focus on the parent/offspring relationship. This may strike you as intuitive, but remember that in many animal societies, young can learn from others besides their parents. Furthermore, in principle, adult animals could presumably teach other adults. This certainly suggests something special about the costs and benefits of teaching in the parent/teacher, offspring/student relationship. The kinship that bonds teacher and student—that is, parent and offspring—may be the only benefit large enough to make up for the costs of teaching (Caro and Hauser, 1992).

Second, Caro and Hauser argue that cases of teaching tend to fall into one of two categories: "opportunity teaching" and "coaching." In opportunity teaching, teachers actively place others, the students, in a "situation conducive to learning a new skill or acquiring knowledge." Coaching, however, involves the teacher directly altering the behavior of students by encouragement or punishment. The majority of examples of animal teaching fall under opportunity teaching, presumably because this type of teaching is the simpler of the two.

Modes of Cultural Transmission

With a better understanding of what constitutes teaching and social learning, we are now ready to delve into cultural transmission in a bit more detail. To begin this process, we shall examine three different modes of cultural transmission: vertical, horizontal, and oblique (Cavalli-Sforza and Feldman, 1981; Figure 5.17).

FIGURE 5.17. Three types of cultural transmission. In vertical transmission systems (panel 1), information is passed from parent to child. When horizontal transmission (panel 2) is in operation, an individual learns from peers. In oblique transmission (panel 3), young individuals learn from adults who are not their parents.

FIGURE 5.18. Cultural transmission in finches. Birdsong in some species of finches is learned culturally. *(Photo credit: Gary Carter/Visuals Unlimited)*

VERTICAL CULTURAL TRANSMISSION

Vertical cultural transmission refers to the situation in which information is passed directly from parent(s) to offspring. This might take place in any number of different ways. For example, offspring may simply learn what their parents do and copy it. In some finch species, both males and females use this form of vertical transmission (Grant and Grant, 1996; Figure 5.18). Males learn the song that they will sing from their fathers, while females learn what songs male finches are supposed to sing also by listening to their dads.

Keep in mind that we have already touched on a more sophisticated form of vertical transmission—teaching. The finch example, however, does not qualify as teaching, as there is no evidence that adult birds sing differently when offspring are present, nor is there evidence of any of the prerequisites for teaching that we touched on earlier.

HORIZONTAL CULTURAL TRANSMISSION

Cultural transmission need not operate strictly in the context of parent/offspring interactions. In humans, for example, our everyday experiences demonstrate that most of the information we get in fact comes from peers—people who are in our own age group. This type of transmission of information, called **horizontal cultural transmission**, operates not only for adults, but for children as well. One powerful example of this is when young individuals learn how to behave by watching and imitating their peers. In fact, in humans, horizontal transmission of information is so powerful that adults spend much of their time trying to subdue its effects on their offspring. We want children to learn some things from their peers, but not everything.

Horizontal cultural transmission plays a role in animal behavior as well. Consider the case of horizontal transmission of information about foraging in guppies. Kevin Laland and Kerry Williams (1998) trained guppies of approximately the same age to learn different paths to a food source: a long path and a short path. Not surprisingly, it was more difficult to train fish to take the longer path when both paths were present, but Laland and Williams found a clever way (involving trapdoors) to do it.

Once the "long-path" and "short-path" groups were trained, over the course of time Laland and Williams slowly removed the original group members in each group and replaced them with new "naive" individuals who didn't know either of the paths to the food source. Initially, groups contained five trained guppies, then four trained individuals and one untrained, then three trained individuals and two untrained, and so on, until none of the original trained group members remained. The question, then, was whether the fish remaining at the end of the experiment—none of which were trained to a particular path—maintained the "traditional" path taken by the original fish in their group.

Laland and Williams found that in both the short-path and long-path groups, guppies at the end of the experiment still followed the path to which the original fish had been trained. Horizontal transmission of information was clearly in operation, as the only "models" from which to learn were same-aged individuals. What makes this experiment particularly interesting is that it demonstrates that cultural transmission can produce "maladaptive" (long-path use), as well as adaptive (short-path use) behavior. In fact, horizontal transmission in the long-path groups not only resulted in guppies acquiring the "wrong" information, it actually made it more difficult to learn subsequently what was the shorter path to take. Cultural information as a means of navigating around unknown spaces is not restricted to guppies; it has also been demonstrated in coral reef fish (Helfman and Schultz, 1984; Warner, 1988). In addition, while the evidence is not clear on this point, it may be that cultural transmission plays a role in the fidelity many birds and insects show to particular roosting sites.

OBLIQUE CULTURAL TRANSMISSION

Oblique cultural transmission refers to the transfer of information across generations, but not via parent/offspring interactions. Young simply get their information from adults that are not their parents. This sort of transmission might be particularly common in animal species where there is no parental care, and hence most young/old interactions will be between nonrelatives.

Oblique transmission has been uncovered in many scenarios, but let us just examine one for the moment—learned snake aversion in rhesus monkeys. Early work comparing wild- and laboratory-raised primates found that the lab-raised animals, who never saw a snake, did not respond to snakes in the same manner as the wild-raised individuals, who had the chance to experience snakes in nature (Hall and Devore, 1965; Struhsaker, 1967b; Seyfarth et al., 1980). Knowing this, Susan Mineka and her colleagues examined whether oblique cultural transmission played a role in the fear of snakes (Mineka et al., 1984; Cook et al., 1985; Mineka and Cook, 1988).

Mineka began her laboratory-based experiment with juvenile rhesus monkeys who were unafraid of snakes. Shortly after observing an adult model respond to snakes with typical fear gestures and actions, juveniles themselves adopted these same gestures (for at least three months). Importantly, it made no difference whether the individuals observed were the subjects' parents (vertical transmission) or unrelated adult monkeys (oblique transmission).

The Interaction of Genetic and Cultural Transmission

If, as seems to be the case, cultural and genetic transmission both operate on animal behavior, then we can search for cases in which both forms of transmission are in play in the same system. To better achieve a fundamental understanding of how such interactions work, let's consider two different case studies that examine the interaction of genes and culture in two dramatically different animal social systems: birds and whales.

THE GRANTS' FINCHES

Peter and Rosemary Grant's work probably best simulates what Charles Darwin might be doing if he were alive today (Weiner, 1995). Like Darwin, the Grants have been studying finches in the Galápagos, and like Darwin, their studies are already considered classics. Among the many problems the Grants have tackled is the role of cultural transmission in the evolution of finch songs. In Galápagos finches, it turns out, cultural transmission not only facilitates birdsong but interacts in a new way with the genetics of reproductive isolation in these birds (Grant and Grant, 1994).

The medium ground finch (*Geospiza fortis*) and the cactus finch (*G. scadens*) both inhabit the Galápagos Island of Daphne Major. Although these are considered to be different species, numerous cases of interbreeding between finch species have been uncovered. Yet, despite this, finch species manage to maintain their integrity. Without barriers blocking interbreeding, why don't the many species of finches just merge into one species?

What the Grants found was that the songs transmitted across generations via oblique and vertical transmission in these two species were quite different from one another. For example, the song of the cactus finch has shorter components that are repeated more often than in the song of the ground finch. These differences have a dramatic impact on gene flow across species. Of 482 females sampled, the vast majority (over 95 percent) mated with males who sang the song appropriate to their species. Cultural transmission of song allows females to recognize individuals of their own species and provides a barrier to forming hybrid offspring.

In their long-term study, the Grants uncovered eleven cases in which the male of one species sang the song of another species. In most of these cases, the result was cross-species breeding, resulting in a hybrid. In other words, remove the normal pattern of cultural transmission and the barrier to breeding across species disappears. In addition to their work on the interaction of genetic and cultural transmission in finch hybridization, the Grants have also found that this interaction plays a critical role in preventing inbreeding in finches (Grant and Grant, 1994).

WHITEHEAD'S WHALES

Despite the fact that much of the lore surrounding whales is based on anecdote and anthropomorphism (projecting human emotions onto nonhumans), there is now good evidence from Hal Whitehead that cultural transmission plays a large role in whale life—a role that interacts with whale genetics in a new and fascinating manner (Rendell and Whitehead, 2001). Whitehead (1998) began his work on cultural transmission by noting two important similarities that emerged from his work on pilot whales, sperm whales, and killer whales (Figure 5.19): (1) in these groups, females spend their lives with other females that are relatives, and (2) new groups in these species tend to form when an already existing group splits up. These sorts of societies are referred to as "matrilineal," and what is critical from our perspective is that matrilineal whale species have much less genetic diversity than other species of whales. Whitehead found that in matrilineal species, mitochondrial DNA (mtDNA) is ten times less diverse than the mtDNA in other whale species.

While nuclear DNA is inherited from both mothers and fathers, mtDNA is inherited only from mothers. Whitehead wanted to know whether there was any link between the fact that mtDNA was inherited from mothers, and the fact that mtDNA was less diverse in matrilineal species. Strangely enough, this search led him to a fascinating discovery of a new kind of interaction between genetic and cultural transmission.

To get an appreciation for how genes and culture interact in whales, we need to step back for a moment and review a concept

A

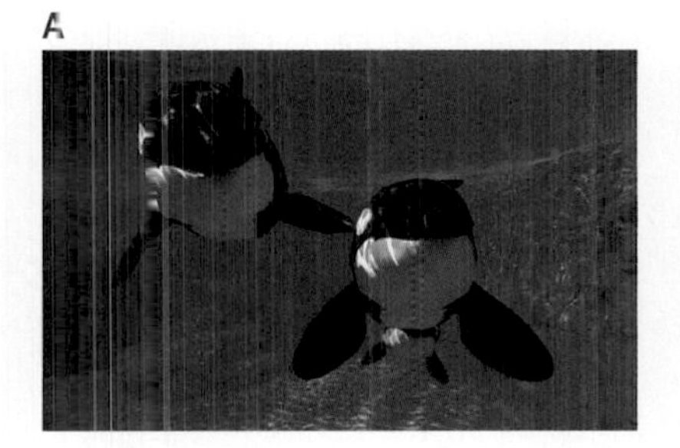

B

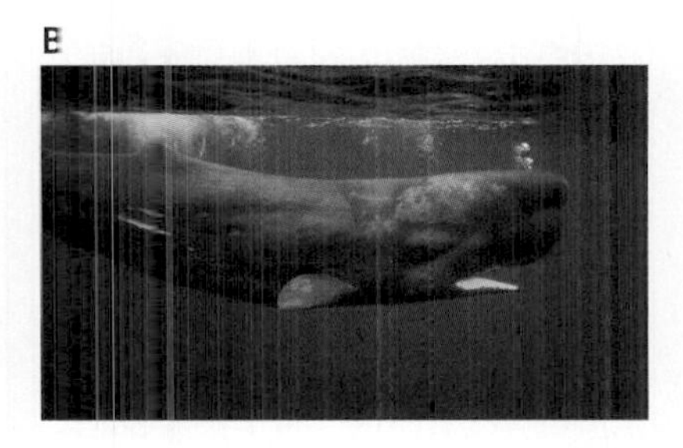

C

FIGURE 5.19. Matrilineal whales. In **(A)** killer (orca), **(B)** sperm, and **(C)** pilot whales, females live together in groups and tend to have less of certain types of genetic diversity than other whale species. *(Photo credits: Brandon Cole, François Gohier, Philip Rosenberg)*

called "hitchhiking" genes (Dugatkin, 2000). The easiest way to envision hitchhiking is to imagine two genes, let's call them gene 1 and gene 2, with each gene having two alleles, so that we have four total alleles—1a, 1b, 2a, and 2b. Now, suppose that our two genes (1 and 2) are genetically "linked" in the sense that when you find 1a, you find 2a, and when you find 1b, you find 2b (this linkage is technically referred to as "linkage disequilibrium"). We then have a situation in which the fate of genes 1 and 2 are not independent—what happens to gene 1 can have implications for gene 2, and vice versa.

Now imagine that all else being equal, allele 2a is much superior to 2b. Natural selection should cause an increase in 2a (over 2b). But what if 1a and 1b are "neutral," in that natural selection does not favor either type? That is, what if the fitness of any individual was unaffected by whether it had allele 1a or 1b? If gene 1 were the only gene we were considering, this would translate into chance factors determining whether we observe 1a or 1b. Sometimes by chance we would see 1a, and sometimes by chance we'd see 1b. When hitchhiking is occurring here, however, we expect something quite different. Even though 1a and 1b are neutral, we should still see 1a most of the time, because it is associated with 2a, and 2a is increasing in frequency due to natural selection. Gene 1a's increase in frequency here is said to have occurred because 1a hitchhiked a ride with 2a. This type of hitchhiking between genes is common, and it is described in most population genetic textbooks. Whitehead, however, suggests a new form of hitchhiking, in which we have an allele and a cultural variant being linked instead of the normal case of two alleles being linked and one hitchhiking a ride from the other.

Whitehead suggests that mtDNA varieties are linked to cultural traits in matrilineal species of whales. That is, rather than variants of gene 1 (1a and 1b) being linked to variants of gene 2 (2a and 2b), now variants of gene 1 (different mtDNA variants) are linked to cultural variants (for example, song type 1, song type 2). To see how this works, we need to begin by noting that young females, who will spend most of their lives with older female relatives, learn a lot from such relatives. For example, cultural transmission appears to play a significant role in foraging behavior, as well as in the distinct "dialect" used in vocal communication in whales. Now, given that young whales learn much via cultural transmission, suppose that one variety of a culturally acquired trait in whales is superior to another variety. For example, in some whales, females find certain varieties of male song much more attractive than others. Then we would expect the superior variety, even if culturally inherited, to increase in frequency. If we started out with many cultural variants, we would end up with the one whale song most attractive to females, thus producing a decrease in diversity.

Whitehead argues that some cultural variants have a greater effect on whale fitness than others, and as such, some cultural variation should be weeded out by "cultural selection"—selection among cultural variants that had different effects on fitness. We also know for certain that mtDNA is at the same time decreasing in variability. One last piece of information, and we can put the pieces of our "gene/culture" puzzle together. The final piece to our puzzle is Whitehead's assumption that different varieties of mtDNA are neutral—that is, that all mtDNA varieties have the same effect on fitness.

Whitehead postulates that somehow, and how this would work remains a mystery, mtDNA variants are "linked" to cultural variants. That is, one way or another, different genetic varieties are associated with different cultural variants. If Whitehead's model is correct in its assumptions, then as diversity in cultural variants decreases, so too would mtDNA diversity. That is, certain cultural variants are favored over others, and the mtDNA linked to those variants stick around in our population, while the mtDNA linked to those cultural variants that are not favored are eventually lost from the population. Cultural selection favoring certain cultural variants over others causes a decrease in cultural variation and, since cultural variants are hypothesized to be linked to neutral mtDNA varieties, we get a dramatic drop in mtDNA diversity, and we have our gene/culture interaction.

Genes for Cultural Transmission?

When discussing cases for cultural transmission, one question that inevitably rises to the surface is: Isn't it possible that the ability, or the tendency, to use cultural transmission is under some sort of genetic control? That is, could there be selection pressure for the ability to both send and receive culturally transmitted information? The first thing to note here is that there is no empirical evidence to date that a gene(s) directly or even indirectly codes for any of the cases of cultural transmission in the literature. That being said, we must still address the question in principle—that is, we must still tackle the possibility of a genetic variation underlying cultural transmission, and examine what would happen if this in fact proved to be true in the future.

Suppose that it turns out that for every study of social learning ever done we discover that there are genes coding for the tendency to use cultural transmission. This possibility raises two related questions. First, would this mean that cultural transmission was really just the long reach of the selfish gene at work? If so, why even speak of culture in animals? Second, is it even possible in principle to imagine how cultural transmission in animals would work if it weren't under some type of direct genetic control?

To answer these questions, consider a hypothetical situation. Imagine that approximately 80 percent of the hundreds of female guppies I have tested in my lab over the last ten years copy the mate choice of others when given the opportunity to do so. Now, imagine that I collaborate with a geneticist and a molecular biologist, and we actually locate, and even sequence, the hypothetical gene(s) in question—those coding for the tendency to copy the mate choice of others. If this were true, copying would clearly have a genetic basis in the guppy system. If we assume that copying is a good strategy, and that's why the alleles coding for it are still around, however, sequencing the genes involved in copying would do little in the way of allowing us to predict how female preference for male traits would change.

To see why the above genetic analysis tells us little with respect to predicted changes in female mate choice, remember that females can only copy the choice of others when they have someone to copy. In the wild, that opportunity will be present at certain times, and absent at others. So, we need to know the frequency with which females rely on copying to choose a mate in order to understand how male and female sexual traits will change over time (Chapter 6). This information has nothing to do with our hypothetical genes underlying copying. Furthermore, even if we did have this information, while we might know how the frequency of the genes underlying copying might change over time (although even this would be very hard to ascertain), we wouldn't be able to predict which mates females would prefer in any given population because this would depend on exactly which females they copied. And it's even more complicated than that, because even if we could know exactly which female everyone copied, we would still need to know whether the individual who was copied had copied someone else, and if so, who, and so on, backward ad infinitum. Again, this may have nothing to do with the genetic analysis we might undertake and everything to do with the dynamics of cultural transmission. Naturally, it would be great to know as much as possible about the genetics of copying, but that wouldn't obviate the need to speak of cultural transmission as a force in its own right.

When tackling the issue of a genetic component to cultural transmission, some strong advocates of the selfish gene school of thinking (see Chapter 2) will challenge the notion that we can even imagine how cultural transmission can work in the absence of some genetic underpinnings. Formalizing this as a question, we can then ask: Is it possible to have a system of cultural transmission that is completely divorced from any genetic underpinnings? The answer is "yes."

The first type of cultural transmission that can exist in the absence of any genes coding directly (or even indirectly) for it is one we have already discussed—teaching. If individuals teach others, and then those who have learned teach yet others, then we have a self-perpetuating system of cultural transmission completely divorced from genes. It is possible, of course, that the tendency to teach may have a genetic component (Ewer, 1969), but the point is that it is not necessary for this to be part of the system in order for it to work. If individuals start teaching each other for any reason, and if those who learn then teach others, and if this whole system is beneficial for those in its midst, then teaching will spread in a population.

Teaching is not a prerequisite for creating a system of cultural transmission that lacks a genetic underpinning. Imagine a case where some individuals start copying the actions of others. Now suppose that animals are also able to assess when copying the actions of others pays better than not copying. Once this is in place, cultural transmission operates. This system works particularly well when individuals copy the most successful strategy used by others.

Cultural Transmission and Brain Size

In this chapter, we have seen examples of cultural transmission in everything from fish to primates. Clearly, at least some forms of cultural transmission are possible even in animals with very tiny brains. That being said, it may very well be that some sort of relationship between brain size and cultural transmission exists, if we just search hard enough and ask precisely the right question.

E. O. Wilson has suggested that a population of large-brained animals might translate into new innovations popping up and spreading through the population more often than would happen in a population of small-brained animals. The data on this question

Interview with Dr. BENNET "JEFF" GALEF

Why study cultural transmission in animals? Isn't culture the sort of subject that sociologists examine?
To the extent that socially learned or traditional behaviors contribute to the ability of animals to survive and reproduce in their natural environments, if we ignore the role of social interactions in behavioral development, our understanding of the origins of animal behavior will be incomplete. Of course, sociologists, anthropologists, and other social scientists have an interest in culture and tradition. However, those trained in the natural sciences often have techniques at their disposal that are not part of the normal analytical repertoires of those with backgrounds in the social sciences. Conversely, those trained in the social sciences sometimes have ways of looking at the world that are foreign to many natural scientists. Culture and tradition are sufficiently complex topics that we will need all available tools to make progress in understanding them.

Why study this sort of thing in rats?
Systematic observation of rat colonies living in different parts of the world indicate that rats from different areas differ remarkably in their behavior. Some dive for fresh-water mussels in shallow streams. Others feed on fingerling trout, or sparrows, even ducks. Some of the most compelling evidence of a tradition in an animal is to be found in the work of Joseph Terkel. Terkel has studied extraordinary populations of wild rats that have learned how to take the seeds from pinecones in a way that permits a net energy gain when eating pine seeds. These rats live in forests where pine seeds are the only food available, and young rats learn how to open pine cones and eat the seeds found in them by interacting with adults of their colonies that have learned the trick from yet others.

The nice thing about rats, as Terkel has shown, is that you can study them in sufficient numbers under controlled conditions to answer all sorts of questions about how traditions are transmitted from generation to generation. These are questions about the development of behavior that are much harder to ask and answer in animals more difficult and expensive to keep in captivity than are rats.

Why is there such a strong bias toward studying culture and social learning in primates?
I'm somewhat uncomfortable with the current focus on primates in the study of culture. Of course, the apes are our closest living relatives, so if any animals share true culture with humans, it is likely to be chimpanzees, gorillas, or orangutans. That said, a great deal of recent work in primatology seems to be designed to show that apes have all the intellectual capacities of humans. This is probably a necessary antidote to a preceding period when some asserted that primates, even humans, had the same intellectual capacities as rats. The truth probably lies somewhere in the middle.

In the meanwhile, much progress can be made examining simpler systems, such as rats, to get as broad a picture as possible of the causes and functions of animal traditions. It will then be much easier to determine in what ways chimpanzee traditions (or cultures) resemble traditions in rats and in what ways chimpanzee traditions (or cultures) resemble culture in humans.

Do biologists and psychologists tackle the study of social learning and culture in animals differently? If so, why?
The distinction between psychologists and biologists studying animal behavior is becoming less meaningful over time as psychologists incorporate Darwinian thinking into their theories and research and biologists acquire the methodological sophisti-

cation traditionally associated with experimental psychology. Presently, psychologists tend to be more concerned with the causation and development of behavior than do biologists, and biologists tend to focus more on the functions of behavior than do psychologists. There's a difference in emphasis, rather than an absolute difference.

The problems we deal with are sufficiently complex to accommodate both approaches. As Niko Tinbergen made clear, we will have to understand development and causation, as well as function, if we are to have a complete science of animal behavior.

What's the most common misunderstanding in the study of social learning and culture in nonhumans?
The view that if primates have traditions, those traditions are evidence of human-like intellectual capacities. Rats have traditions. Reef fish have traditions. Songbirds have traditions. Yet, no one proposes that because reef fish, canaries, and rats have traditions, they have human-like intelligence. Clearly, the existence of tradition in animals tells us little about their intellectual capacities, though lots of people seem to think it does.

How would you describe the relationship between human and nonhuman culture? Is this a graded continuum, or is there a chasm here that separates these two types of cultures by orders of magnitude?
The question of the degree of similarity between human and nonhuman animals is one that has bedeviled biology since the time of Darwin, and I doubt that I will be able to resolve it here. Animals are both amazingly similar to humans (for example, in genetics or anatomy) and amazingly different from humans (for example, in use of language or production of artifacts).

As far as culture is concerned, I have yet to be convinced that, even in chimpanzees, there is much beyond a superficial similarity between human and animal "culture." Until evidence of some deeper similarity is available, I prefer to talk about animal tradition and human culture, so that it is evident in our language that we don't yet know the extent to which nonhuman animals and humans share a capacity for human-like culture.

It might sound like a strange question, given that we are talking about culture, but how do you think the various recent advances in molecular genetics like genome sequencing projects or development of knockouts will affect our understanding of cultural transmission in animals?
At this point, the field is not really ready to proceed to a molecular level of analysis. I surely don't expect to see any "culture genes" that one might insert into the drosophila genome to produce fruit flies that learn by observation or knockout mice that would lack animal traditions for any interesting reason.

It is, of course, easy to produce knockout mice that don't, for example, show social learning of food preferences of the sort my students and I have been studying in normal rats and mice for years. However, any number of uninteresting deficits can produce animals that can't learn food preferences socially. For example, the knockout mice might just have a poor sense of smell or a poor memory. Indeed, the latter is the case with one strain of knockout mice that fails to learn food preferences socially. That's why you have to have a rather deep understanding of the physiological and behavioral substrates of traditions before you can look at their molecular basis in a meaningful way.

If some animals have culture, does this have any implications for the "animal rights" community?
If, for example, it were to be shown that animal "culture" rests on teaching or observation and copying of the behavior of others, that would add some strength to arguments that we should treat animals humanely. However, there are already so many good reasons to be humane when interacting with animals—not causing unnecessary pain, weighing the possible benefits to our own species and others against the suffering we inflict—that I suspect any impact of the discovery of culture in animals would be marginal.

We already treat animals very differently, depending on how much we think they are like us. However, our treatment of different species is often based more on intuition and perceived self-interest than on science. Consequently, even if it were proven that guppies have "culture," I don't foresee humans changing their behavior toward fish in any very noticeable way. After all, we have known for a very long time that our fellow humans have culture, but that knowledge hasn't done a great deal to improve our treatment of one another.

DR. BENNET "JEFF" GALEF is a professor at McMaster University in Canada. He has worked on social learning in a variety of species (most notably, the rat) for twenty-five years and is regarded as a leader in a field that integrates biological and psychological approaches to animal learning.

are mixed (Sawaguchi and Kudo, 1990; Dunbar, 1992). Nonetheless, in the most comprehensive study to date, Reader and Laland (2002) found that across more than 100 species of nonhuman primates, there was a significant positive correlation between brain size and both innovation and tool use frequency, thus confirming the predicted trends. In examining the primate literature, they uncovered 533 recorded instances of innovation (discovery of "novel solutions to environmental or social problems"), 445 observations of social learning, and 607 episodes of tool use that covered 116 of the 203 known species of primates. They then mapped out these behaviors against something called "executive brain" volume, a measure that is a composite of the neocortex and striatum sections of the brain (Jolicoeur et al., 1984; Keverne et al., 1996).

Innovation, social learning, and tool use all had a positive correlation with absolute executive brain volume. That is, even when the large discrepancy in body size across the 166 species of primates included in the study is taken into account, it appears that it is the absolute size of the primate brain, not brain size in relation to body size, that correlates best with social learning, tool use, and innovation (Figures 5.20). Put simply, big brain size, per se, matters in primates. A similar trend between (fore)brain size and innovations was found for birds in North America, Britain, and Australia (Lefebvre et al., 1997a; 1997b).

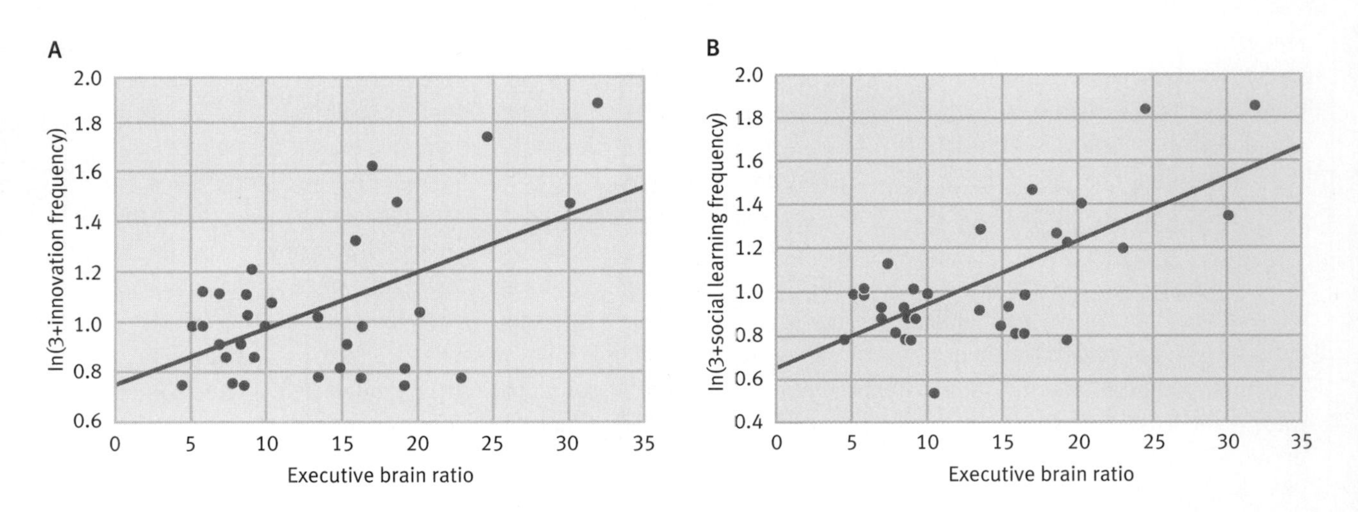

FIGURE 5.20. Effects of increasing brain size in primates. Reader and Laland used published measures of the size of the executive brain (the neocortex plus the striatum) and determined the executive brain ratio (the executive brain size divided by the brainstem size) to examine the relationship between innovations and social learning and brain size. They found that: **(A)** As the executive brain ratio of primates increases so too does the occurrence of new innovations (discovery of novel solutions to problems). This same relationship holds true when innovation is plotted against absolute executive brain size. **(B)** The occurrence of social learning also increases with an increase in primate brain size. This same relationship was found when social learning was plotted against absolute executive brain size. *(From Reader and Laland, 2002)*

Having worked our way through an introductory chapter and four primer chapters, we are now ready to explore specific animal behaviors in more depth, beginning with sexual selection.

SUMMARY

1. Cultural transmission involves the transfer and acquisition of information via social learning and teaching and may represent a powerful means of acquiring and spreading behaviors both within and between generations in animals.
2. Cultural transmission—learning from other individuals, either via social learning or teaching—differs from learning about other parts of the environment. Unlike other environmental influences, what is learned via cultural transmission is passed on from individual to individual. This can translate into the behavior of a single individual shifting the behavior patterns seen in an entire group. Furthermore, when cultural transmission is in play, what is learned by one individual may be passed on down through many generations.
3. Social learning can take many forms, including local enhancement and social facilitation, as well as some forms of contagion, imitation, and copying.
4. Teaching, when rigorously defined, implies that one individual serves as an instructor and at least one other individual acts as a pupil who learns from the teacher. Teaching then entails providing an immediate benefit to the student but not the teacher, instructing naive "students," and imparting some new information to students faster than they would otherwise receive it.
5. Cultural transmission may occur via vertical, horizontal, or oblique transmission. Vertical transmission involves the transfer of information from parent to child, horizontal transmission occurs when information moves across a cohort, and oblique transmission excludes the transfer of information from parent to offspring but encompasses all other information transferred from older to younger individuals.
6. Genetic and cultural transmission can operate independently on animals, but often they interact in interesting ways, as exhibited in Darwin's finches, and some species of pilot whales, sperm whales, and killer whales. As an example, Hal Whitehead's work suggests that mitochondrial DNA may be linked to cultural traits in matrilineal species of whales.

DISCUSSION QUESTIONS

1. Why do you suppose it took so long for ethologists to focus on the possibility that cultural transmission was an important force in animals? Can you imagine any biases—scientific, ideological, etc.—that could be responsible for this?
2. Suppose I run an experiment in which I take a bird (the observer) and let him view another bird (the demonstrator) opening a sealed cup by pecking at a circle on the cover of the cup. I then test the observer and see that he now opens the cup by pecking at the circle. What can I infer about social learning here? What other critical treatment is missing from this experiment?
3. Imagine that adults in some population of monkeys appeared to pick up new innovations (for example, potato washing, stone play) from observing others. How might you disentangle vertical, horizontal, and oblique cultural transmission as possible explanatory forces?
4. List the pros and cons of Hauser and Caro's definition of teaching. How might you modify this definition to address what you listed on the "con" side of your ledger?
5. Suppose that after extensive observations, you determine that certain animals in a population appear to rely on social learning much more often than other individuals and that such differences are due to genetic differences. How might you use the truncation selection technique described in Chapter 2 to examine the narrow-sense heritability of the tendency to employ social learning?

SUGGESTED READING

Bonner, J. T. (1980). *The evolution of culture in animals*. Princeton: Princeton University Press. One of the earliest books looking at culture from an evolutionary perspective. This is a fun read and informative as well.

Boyd, R., & Richerson, P. J. (1985). *Culture and the evolutionary process*. Chicago: University of Chicago Press. Probably the best book out there on the evolution of culture. This book can get very technical, but it is worth the effort to work through it.

Heyes, C. M. (1994). Social learning in animals: Categories and mechanisms. *Biological Reviews of the Cambridge Philosophical*

Society, 69, 207–231. Also Heyes C.M., & Ray, E.D. (2000). What is the significance of imitation in animals. *Advances in the Study of Behavior, 29,* 215–245. Two very nice overviews on social learning and cultural transmission.

Reader, S. M., & Laland, K. N. (2002). Social intelligence, innovation, and enhanced brain size in primates. *Proceedings of the National Academy of Sciences, U.S.A., 99,* 4436–4441. A well-done study that combines published data on primate brain size and primate behavior, including cultural transmission, to examine the relationship between the two.

6

Intersexual and Intrasexual Selection

Genetics and Mate Choice

- Direct Benefits and Mate Choice
- Good Genes and Mate Choice
- Runaway Sexual Selection

Learning and Mate Choice

- Sexual Imprinting
- Learning and Mate Choice in Japanese Quail
- Learning and Mate Choice in Blue Gouramis

Cultural Transmission and Mate Choice

- Defining Mate-Choice Copying
- Mate-Choice Copying in Grouse
- Mate-Choice Copying in Guppies
- Song Learning and Mate Choice in Cowbirds

Male-Male Competition and Sexual Selection

- Underground Mating and Male-Male Competition in Fiddler Crabs
- Red Deer Roars and Male-Male Competition
- Male-Male Competition by Interference
- Male-Male Competition via Cuckoldry

Neuroethology and Mate Choice

- Swordtails, Platyfish, and Sensory Bias
- Frogs and Sensory Bias
- Zebra Finches and Sensory Bias

Interview with Dr. Malte Andersson

Sexual Selection

In the first five primer chapters we have focused on basic issues in animal behavior (natural selection, learning, and so on) and proceeded to identify basic theories and then to examine the issues in question across a whole suite of behaviors. In the remaining twelve chapters, we once again shall return to theoretical, empirical, and conceptual questions, but each chapter will focus on a specific behavior, or more generally, a specific set of related behavioral issues (sexual selection, mating systems, kinship, cooperation, foraging, predation, and so on). We begin with the topic of **sexual selection**.

In his 1871 book, *The Descent of Man and Selection in Relation to Sex*, Darwin provided the first detailed evolutionary theory of mate selection. Darwin's work on sexual selection encompasses a theory that includes (1) **intrasexual selection**, whereby members of one sex compete with each other for access to the other sex, as well as (2) **intersexual selection**, whereby individuals of one sex choose which individuals of the other sex to take as mates. Recall that from the perspective of genetics, the number of copies of a gene passed down to the next generation is the currency of importance. Aspects of sexual selection are clearly important in measuring that currency, and by beginning here we have an excellent portal through which we can examine the processes that lead to the differential success of one gene over another.

In this chapter we shall examine sexual selection, with particular emphasis on: (1) the evolution of mating preferences, (2) learning and sexual selection, (3) cultural transmission and female mate choice, (4) male-male competition, and (5) the neuroethology of mate choice.

Intersexual and Intrasexual Selection

Darwin proposed that one factor that leads to differences in reproductive success is competition for mates—intrasexual selection. Competition for mates is ubiquitous in the animal kingdom, and anyone who walks through the woods, and knows where to look, can see it at every turn. In most species, it is the males that compete for the females as mates and not vice versa.

This fundamental difference between the sexes is due to a number of factors, but ethologists and evolutionary biologists believe that the primary one centers on the different type and number of gametes (sperm and eggs) produced by males and females.

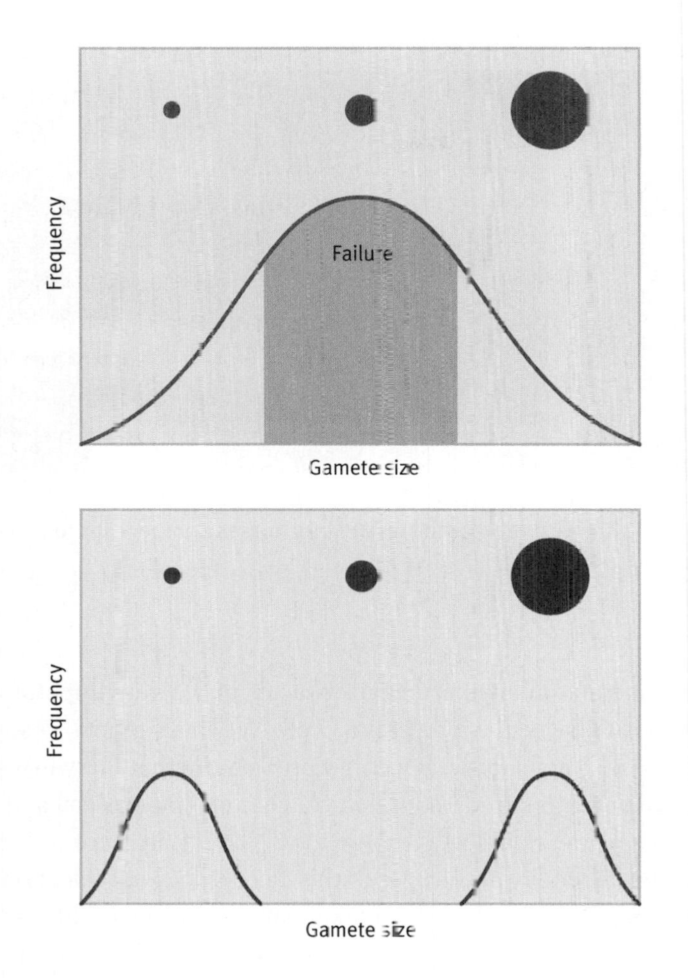

FIGURE 6.1. Natural selection and gamete size Natural selection has favored two types of gametes: small and quick (sperm), and large (egg). For gametes to be successful, they must be able to find another gamete and produce a healthy zygote. Small gametes (sperm) do the former, while large gametes (eggs) do the latter. Medium-sized gametes do neither and are selected against. *(From Low, 2000)*

By definition, females produce fewer, but larger gametes (Figure 6.1). In a sense, then, each egg is extremely valuable because of both its size and its relative rarity. Each sperm, on the other hand, requires much less energy to produce, and sperm are usually found in prolific quantities. This translates into females limiting male reproductive success through their much lower rate of gamete (egg) production compared with that of males. Thus, while males often produce millions of sperm, creating the possibility that some males will have extraordinary reproductive success, females' eggs are often, but not always, few and far between, causing severe competition for this scarce resource (Trivers, 1985)

According to Darwin, all else being equal, any heritable male trait that confers mating and fertilization advantages on its possessor will, over time, increase in frequency in a population, because males with such traits will simply produce more offspring than their competitors. Darwin's novel ideas about the struggle among males for mating opportunities forms one basic foundation of our current understanding of sexual selection. Naturally, the ways in which hormones, neurobiology, development, environment, and many other

FIGURE 6.2. Competition for mates. (A) Male deer battle with their antlers. **(B)** Male horses compete for females. *(Photo credits: Wallner; Rich Pomerantz)*

factors operate on any particular behavior play a very important role in the sexual selection process as well. We shall return to some of these factors very shortly, but it is worth noting that Darwin's theory of sexual selection predated much of what we know today about endocrinology, neurobiology, and so forth. This statement says much about the generality and power of the theory of sexual selection.

Competition for mates can take many forms depending on such factors as ecology, demography, and cognitive ability. For example, males may fight among themselves, occasionally in dramatic "battles to the death," but often in less dangerous bouts, to gain mating privileges with females (Figure 6.2). This less dangerous form of male-male sexual competition is illustrated by male stag beetles and deer, which use their "horns" (enlarged jaws) and antlers, respectively, in physical fights over females; the winners of such contests mate more often than the losers.

Until about thirty years ago, much of the work on mate selection focused almost exclusively on competition for mates, or intrasexual competition (Bradbury and Andersson, 1987; Andersson, 1994), and not on mate choice, or intersexual selection. Exactly why this was so is difficult to assess, but part of the explanation may be that male-male competition is very easy to observe in nature, and part of it may be due to the fact that some prominent evolutionary biologists of the 1930s dismissed mate choice as unimportant, thus focusing evolutionary biologists on the topic of male-male competition (Huxley, 1938).

Mate choice is a much more subtle form of sexual selection than is competition for mates. Mate choice, or intersexual selection, directly involves *both* males and females in the decision-making process. Just as intrasexual selection, in principle, could involve either males fighting for females or females fighting for males, so too does intersexual selection encompass female choice of a male and male choice of a female. Both clearly occur in nature, but female choice is much more prevalent. This is likely due to the fact that females stand to lose much more than males by making a bad choice of mates. In the first place, females invest much more energy in each gamete they produce, and hence should be choosier in terms of who has access to their gametes than should males. In addition, in species with internal gestation, females typically devote much more energy to offspring before they are born, and hence they should be under strong selection pressure to choose good mates, who will produce healthy offspring.

One important piece of information to keep in mind when thinking about female mate choice is that, above and beyond their sex organs, males often possess other traits that play an important role in mating. These traits are referred to as secondary (or epigametic) sexual characteristics. As we shall see, many conspicuous secondary sexual traits in males, such as ornamental plumage, bright colors, and courtship displays, evolved to influence female mate choice.

Intersexual and intrasexual selection both play a role in virtually all mating systems (see Chapter 7). Be the system monogamous, where a single male pairs up with a single female, or polygamous, where some subset of males mates with many females, or polyandrous, in which some subset of females mates with numerous males, sexual selection plays a role. Sexual selection will often be stronger in polygamous and polyandrous systems than in monogamous systems. In polygamous and polyandrous systems, some individuals obtain many mating opportunities and some obtain no mating opportunities, while there is generally less variation in reproductive success in monogamous systems. That being said, sexual selection acts in monogamous systems, just not as strongly.

Genetics and Mate Choice

Studying how females choose their mates has become one of the most active areas in the field of ethology. As mentioned earlier, male-male competition took center stage in the sexual selection theater until approximately thirty years ago. The tables have clearly turned of late, and female mate choice is now the most studied of the processes leading to sexual selection (Andersson, 1994). Until recently, one unifying theme running through

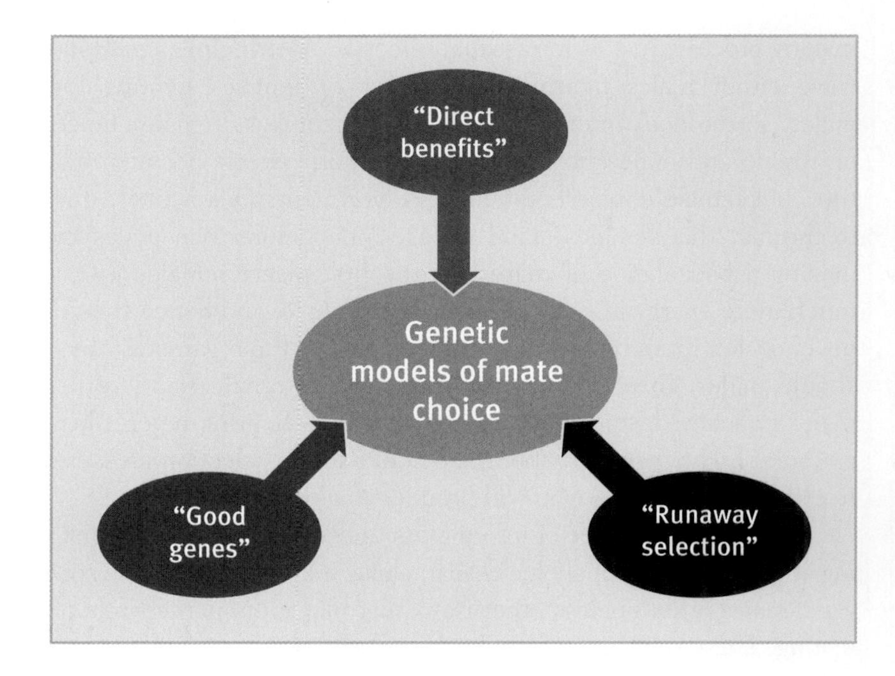

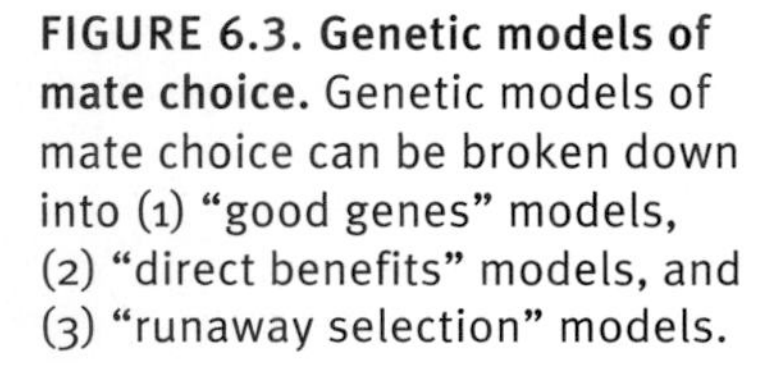
FIGURE 6.3. Genetic models of mate choice. Genetic models of mate choice can be broken down into (1) "good genes" models, (2) "direct benefits" models, and (3) "runaway selection" models.

virtually all such studies was that they assumed that a female's choice of mates was under some sort of genetic control (Hinde and Fisher, 1951). That is, females were assumed to be genetically programmed to prefer certain male traits over others. Naturally, the environment played a huge role in determining *which* genes were chosen, and proximate factors played a role in determining *how* genes affected mate choice, but at the bottom of it all there was assumed to be a set of genes that controlled the process of mate choice. Sometimes these genes were posited to be very flexible in the traits they coded for, but the process of sexual selection was nonetheless assumed to be under strict genetic control. We shall begin by examining the accepted doctrine regarding genes and sexual selection. Subsequent to this, we shall spend a significant amount of time questioning the universality of this doctrine.

Genetic models of female mate choice can be broken down into three classes: "direct benefits," "good genes," and "runaway selection" models (Figure 6.3). In many species, these three processes may be operating simultaneously, but we shall focus on the more clear-cut cases that fall under each of these models.

DIRECT BENEFITS AND MATE CHOICE

The direct benefits model of female mate choice is the most basic of the genetic models we shall examine (Kirkpatrick and Ryan, 1991; Price et al., 1993). In this model of the evolution of

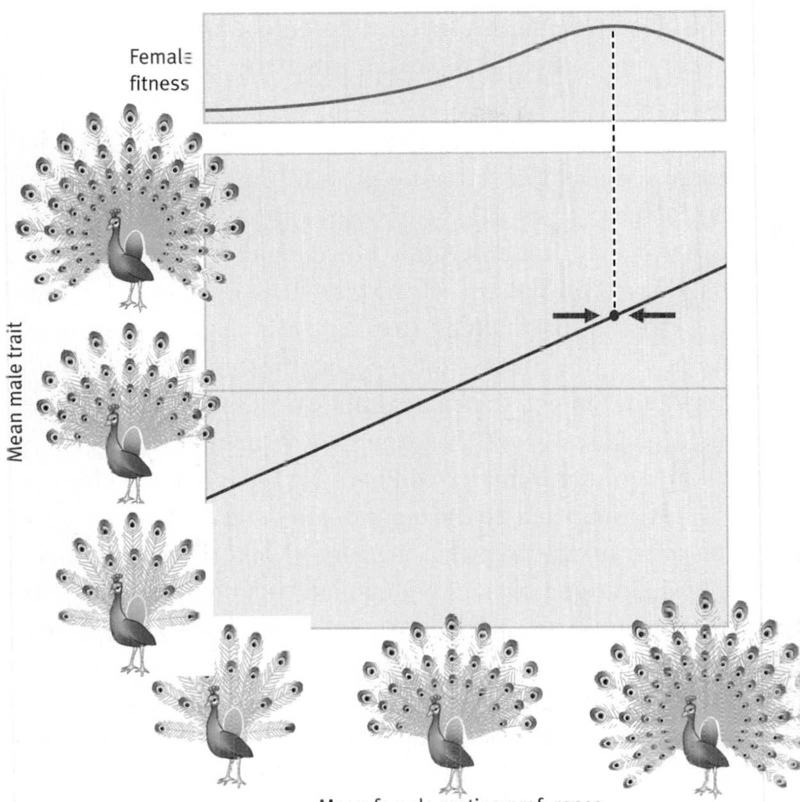

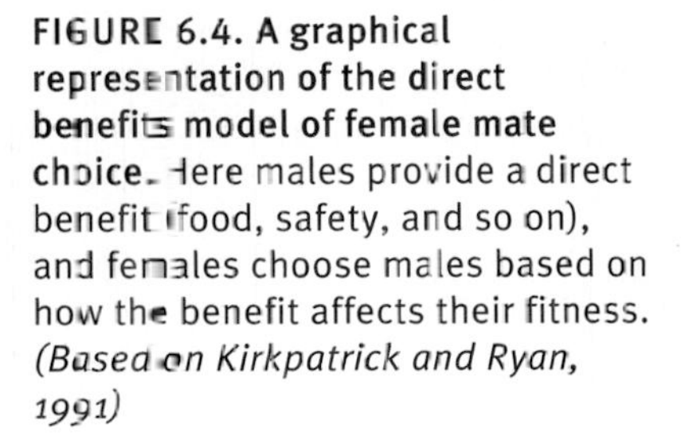

FIGURE 6.4. A graphical representation of the direct benefits model of female mate choice. Here males provide a direct benefit (food, safety, and so on), and females choose males based on how the benefit affects their fitness. *(Based on Kirkpatrick and Ryan, 1991)*

female mate choice, natural selection favors females who have a genetic predisposition to choose mates that provide them with a tangible resource (above and beyond sperm) that increases either their child-rearing potential, their survival, or both (Figure 6.4). That is, any female who has the tendency to choose males that provide her with some important resource will do better than her counterparts that are less choosy, and through time we should see more acutely picky females. Two fascinating cases of direct benefits center on nuptial gifts in scorpionflies and the dangers of parasite transmission in barn swallow birds.

DIRECT BENEFITS AND NUPTIAL GIFTS. Adult scorpionfly (*Hylobittacus apicalis*) populations number in the thousands per acre, and mate choice is an ongoing process in all such neighborhoods (Figure 6.5). Randy Thornhill and his colleagues have found that female scorpionflies choose their mates using a very basic rule: choose males that bring relatively large prey items during the courtship process (Thornhill, 1976, 1980a, 1980b; Thornhill and Alcock, 1983). This rule, which makes good sense from the female perspective, produces a dilemma from the male perspective, as hunting is quite dangerous. Given the distribution of

FIGURE 6.5. A scorpionfly. Male scorpionflies present females with nuptial gifts (not seen here). This system has been used to test the direct benefits model of sexual selection. *(Photo credit: Bill Beatty)*

prey items in nature, at any given time only about 10 percent of males are in possession of prey that meet the criteria laid down by females.

Nuptial gifts are the prey that males present to females during courtship; males that bring no prey are immediately rejected. Females do more than simply choose a mate based on nuptial gifts, however; they also determine how long they will mate with a given male based on the size of his gift. This is a critical component of courtship, as matings that last less than approximately seven minutes often involve no sperm transfer. The better the nuptial gift, the longer the mating time a female allocates to a given male (Figure 6.6A). For example, if a male brings a gift smaller than 16 mm^2, females will mate with such a male for considerably less time than the average matings that typically occur when the male brings prey that are larger than the critical mass. That is, if males don't bring a big enough prey item, they may receive an opportunity to mate, but the odds are low that such matings will produce any offspring (Figure 6.6B). From the gene's-eye perspective, this is bad news for males. In fact, strong selection pressure on males to provide large nuptial gifts to females may have favored prey-stealing on the part of males (Thornhill, 1979a). Under certain conditions, some males mimic female behavior by calling to other males and adopting female-like behavior. When

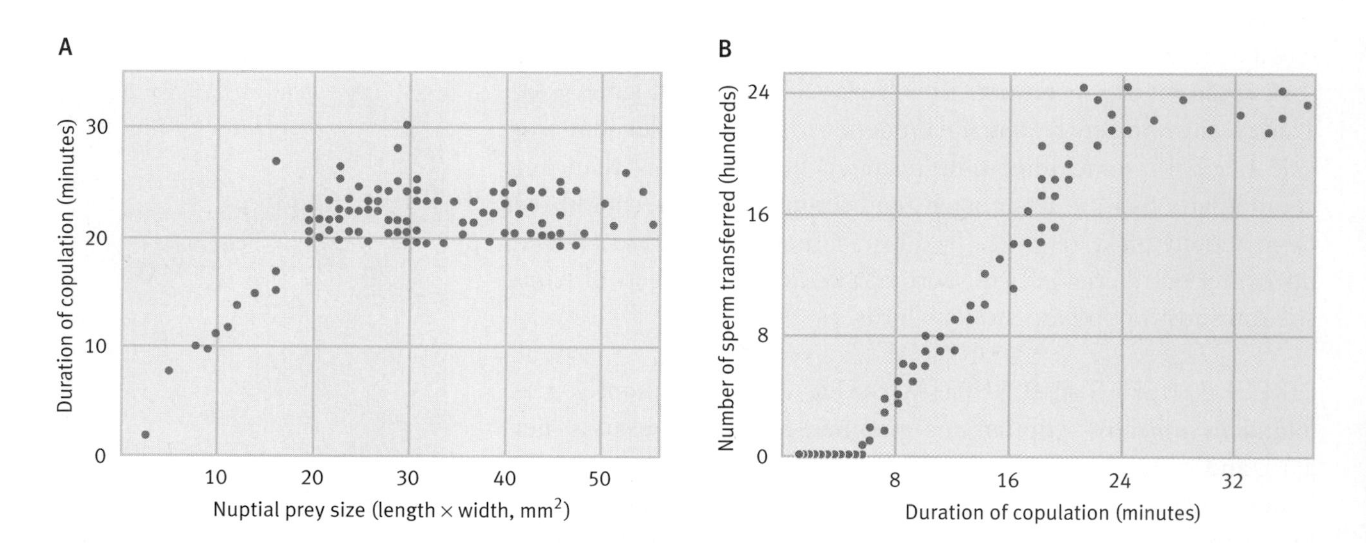

FIGURE 6.6. The direct benefits model and nuptial prey gifts. (A) Copulation time as a function of direct benefits. Male scorpionflies provide females with a nuptial prey gift. The greater the value of the gift, the longer a male is allowed to copulate with the recipient. **(B)** Longer copulation time leads to greater sperm transfer in scorpionflies. When males provide large nuptial gifts to females, the added copulation time they are granted translates into more sperm transferred to the female. *(From Thornhill, 1976, 1980b)*

successful at this ploy, such males simply steal the nuptial gift presented to them and use it when they themselves are courting females.

Given that females can clearly distinguish males with large gifts from other males, what benefits do they receive by doing so? It turns out that a female who actively chooses males that bring large nuptial gifts produces more eggs and, in all likelihood, has a longer life span, because of both the nutrition she receives directly and the decreased amount of time she must allocate to hunting (Thornhill, 1976, 1979b, 1980a, 1980b; Thornhill and Alcock, 1983).

BARN SWALLOWS AND PARASITE REMOVAL. Bird-watchers in Denmark and many other European countries have long known that in barn swallows, males have tails that are significantly longer than those in females. In every other way, however, the two sexes look remarkably similar. One unpleasant reality of barn swallow life in the wild is that these birds are engaged in a never-ending struggle with parasites. Barn swallows are infested with two different types of parasites: *ecto*parasites, which cling to the outside of a host and hence easily move from one host to another, and *endo*parasites, which reside inside a host. Here we shall focus on ectoparasites and the research of Anders Møller, who has done some very interesting work in this area (see Møller, 1994, for a review)

As in many species of birds, barn swallow females prefer to mate with males possessing elongated tail feathers. Such males are less infested with ectoparasitic mites than are their short-tailed counterparts in the same vicinity. Møller undertook a fascinating experiment to determine if females preferred longer-tailed males partly as a consequence of a direct benefit of mating with males with long tails and hence reduced ectoparasite load (Møller, 1990a). Barn swallow nests were lumped into two groups. In one group of selected nests, extra parasites were added in, while in the second group, nests were fumigated to destroy these pests. The results of this experimental manipulation were clear: fumigated nests had a much greater number of eggs reaching hatching stage than nests with experimentally increased mite load (Figure 6.7). Furthermore, the offspring raised in fumigated nests were generally much healthier, as measured by body weight, than those unfortunate young raised in superinfested areas. Natural selection then seems to have favored mating with males bearing a low parasite load, and this results in a clear benefit in terms of fitness—more and healthier chicks (Figure 6.8).

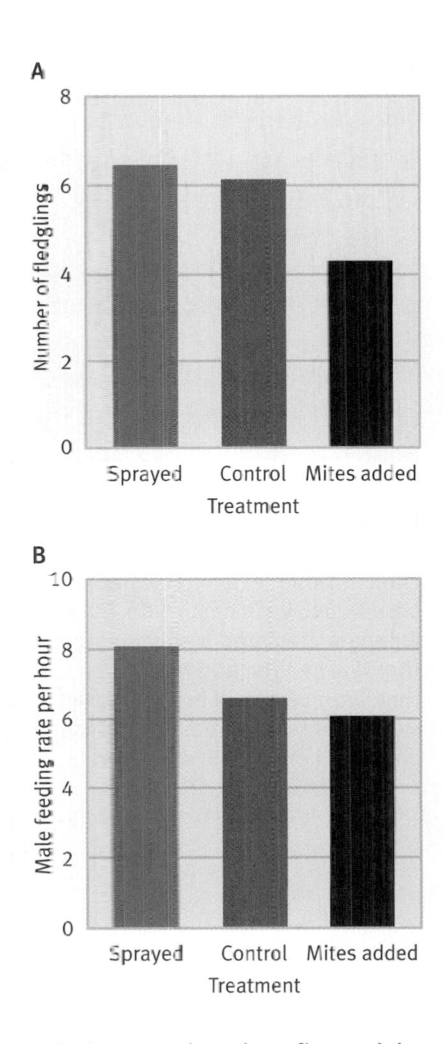

FIGURE 6.7. Direct benefits model and barn swallows. In a test of the direct benefits model, Møller found that females preferred longer-tailed barn swallows and that such males had lower ectoparasite loads. Møller experimentally removed parasites from some nests (by spraying the nests) and added parasites to other nests. **(A)** One direct benefit was increased reproductive success as a function of parasite removal in barn swallows. Females in sprayed nests had a higher mean number of offspring reach the age of fledglings. **(B)** Another direct benefit accrued by females having fewer parasites in their nest was an increased feeding rate on the part of the male. *(From Møller, 1994, pp. 126–127)*

FIGURE 6.8. Lousy males. In a test of the direct benefits model of mate choice, Møller examined whether barn swallows in nests with fewer ectoparasites produced more eggs that reached the hatching stage than barn swallows in nests with experimentally increased mite load, and whether birds in such nests produced healthier offspring. *(Photo credit: Maslowski/Visuals Unlimited)*

GOOD GENES AND MATE CHOICE

Females get more than direct resources such as food and safety from their mates; they also get sperm, and with it genes that are passed on to offspring. Much theory, and a little empirical work, suggests that females are favored to choose mates that possess "good genes" (Fisher, 1915; Kodric-Brown and Brown, 1984; Andersson, 1994). Good genes are those that code for some suite of favorable traits—traits that can be inherited by offspring of the appropriately choosy female. In **good genes models,** females who choose the males with genes best suited to their particular environment receive "indirect" benefits, in the sense that their offspring receive some of the good genes—genes, for example, for superior foraging skills or abilities to fend off predators—that led their mother to choose a particular male as a mate in the first place (Figure 6.9).

Good genes models of sexual selection apply to mating systems in which the only benefit received by females lies in the genes residing on male sperm. Good genes models are somewhat controversial, for all such models must overcome two potentially difficult hurdles. First, how can females determine which males possess "good" genes? This is clearly a more daunting task than distinguishing between males in some more direct fashion—for example, by seeing how much potential for providing food mates display during courtship. Second, given that females are attempting to choose males with the best genes, wouldn't natural selection favor males that cheat? That is, wouldn't males attempt to give the impression that they possess appropriately good genes, even if they don't? One theory developed to overcome these problems is the **handicap hypothesis**, also known as the handicap principle (Zahavi, 1975, 1997).

FIGURE 6.9. Good genes. If males differ in the symmetry of their tail feathers, and this is a reflection of underlying genetic quality, females may select more symmetric males.

The handicap hypothesis focuses on the notion of "honest advertising." With respect to mate choice, this hypothesis suggests that only traits that are true and honest indicators of male genetic quality should be used by females when choosing mates. If females focus on such honest indicators, they can overcome the male cheater problem—at least temporarily.

Let's explore the handicap hypothesis in a bit more depth.

PARASITES AND GOOD GENES. Even if the selection of honest traits in males is beneficial to females, females still face a difficult task. Which traits should females accept as honest indicators of a male's genetic quality? One theory is that honest indicator traits should be generally quite "costly" to produce—the costlier the trait, the more difficult it is to fake (Figure 6.10). One costly trait that has been studied in the context of the handicap hypothesis is endoparasite resistance. Recall that endoparasites reside within their hosts and generally don't move from host to host with the ease of ectoparasites (Hamilton and Zuk, 1982).

Females who choose males with strong resistance to endoparasites, a trait that would be difficult to fake, may receive indirect benefits in that they may be mating with individuals with "good genes"—in this case, genes that confer parasite resistance abilities. The predictions of the handicap model when applied to parasites go under the name of the "Hamilton-Zuk" hypothesis, after Bill Hamilton and Marlene Zuk, who first applied the idea of good genes to parasite resistance.

Given that endoparasites can't be seen, how do females know which males have good genes with respect to endoparasite resistance? After all, endoparasites are inside the male, making endoparasite

FIGURE 6.10. Peacock's tail. An example of an elaborate, costly trait that evolved in males as a result of sexual selection. *(Photo credit: Royalty-Free/Corbis)*

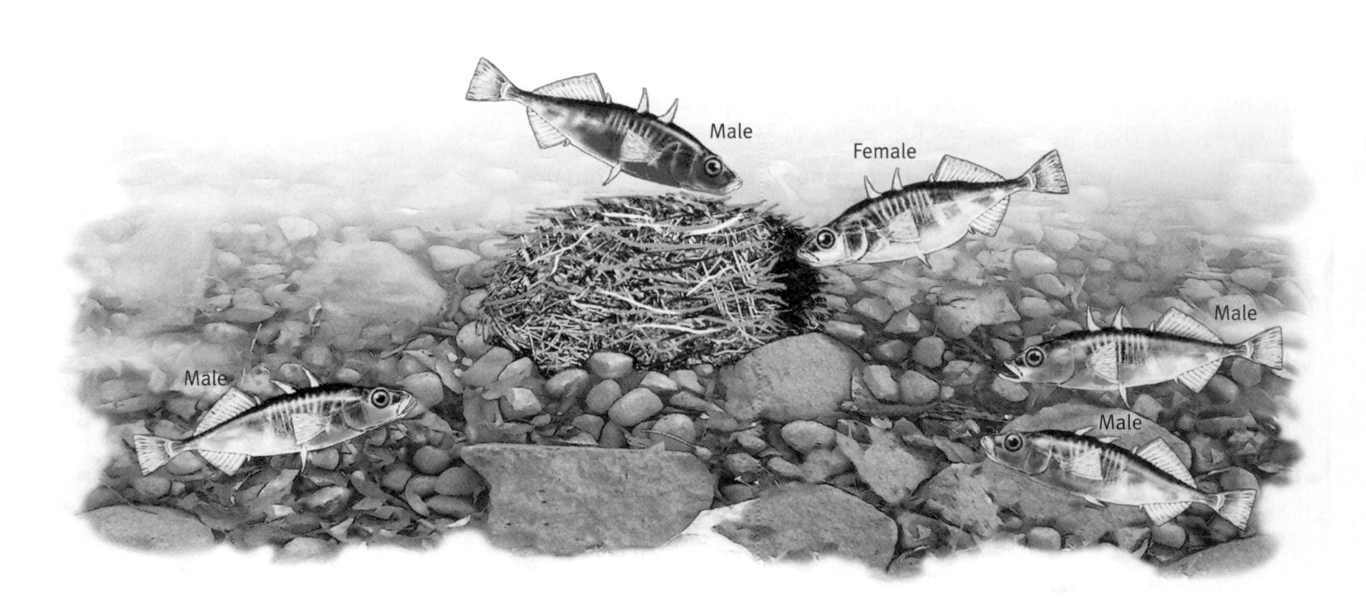

FIGURE 6.11. Color, parasites, and good genes. One reason stickleback females may prefer the most colorful (red) males is that color indicates resistance to parasites.

resistance a difficult thing for females to judge. The answer to this dilemma must in some way lie in the female's ability to use some *other* male trait that correlates with the ability to avoid internal parasitization (Figure 6.11). If possessing trait X also means that males are good at fighting endoparasites, females can use trait X as a proxy for judging what they really wish to learn. That is, if information on internal parasitization is unavailable, using some other trait (X) that correlates well with the one of interest should be favored by natural selection. One such proxy cue appears to be body coloration (Figure 6.12). Healthy

FIGURE 6.12. Elaborate coloration in males. In many birds, such as the bare-necked umbrellabird **(A)** and the cardinal **(B)**, males are much more colorful than females as a result of different sexual selection pressures. *(Photo credits: Michael and Patricia Fogden; Johann Schumacher)*

males tend to be very colorful, while infected males have much drabber colors (Milinski and Bakker, 1990). Although still the subject of some heated debate, numerous studies undertaken in birds and fish are consistent with the predictions of the Hamilton-Zuk hypothesis in that females often choose the most colorful (and least parasitized) males (Figure 6.13).

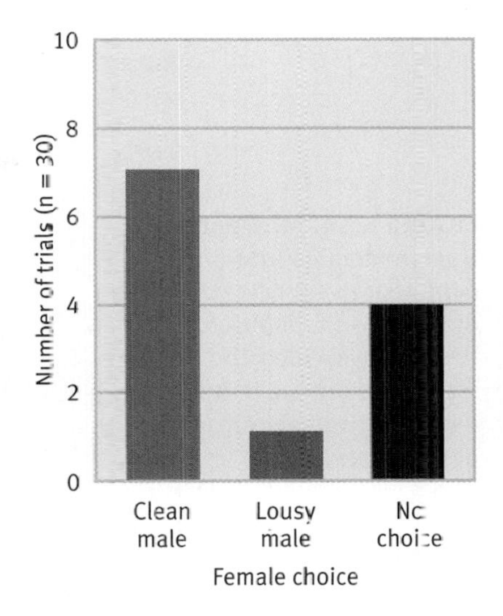

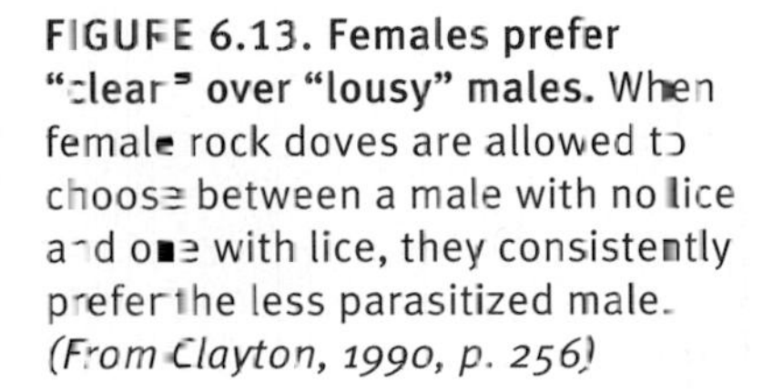
FIGURE 6.13. Females prefer "clean" over "lousy" males. When female rock doves are allowed to choose between a male with no lice and one with lice, they consistently prefer the less parasitized male. *(From Clayton, 1990, p. 256)*

MHC AND GOOD GENES. As we have seen above, if females are searching for honest indicators of good genes in males, honing in on traits associated with disease resistance makes perfect sense from a natural selection perspective. One set of genes that is involved in disease resistance in many animals is known as the Major Histocompatability Complex (MHC) (Brown and Eklund, 1994; J. L. Brown, 1998a; Penn and Potts, 1998, 1999). Proteins produced by MHC genes guide the body in identifying "self" versus "foreign" cells (Wills, 1991).

A unique aspect of the MHC is that it is the most variable set of genes ever uncovered. Very few (if any) individuals have exactly the same MHC. Given this incredible variability, biologists have predicted that individuals may prefer mating with others who have a dissimilar MHC. The reason such a preference is speculated to arise is that offspring resulting from a mating between individuals with very different MHCs will have a new combination of MHC genes. Such new MHC combinations might be particularly good at providing immune system protection, since diseases evolve so rapidly that it is as if animals are trying to hit a moving target. In this case, their weapon against such a moving target is a set of MHC genes constantly changing across generations.

Once again, females are faced with a problem—identifying which males are those that have MHCs that differ from their own. Although we don't know exactly how yet, mice and rats can use odors to determine if another individual is a good MHC match—that is, whether they have sufficiently different MHC genes (Penn and Potts, 1998). This work has spurred experiments on odor, MHC, and mate choice in humans. To test the hypothesis that humans use MHC when choosing mates, and that odor plays a role in the process, male and female undergraduate students in Switzerland were tested by Claus Wedekind (Wedekind and Furi, 1997). In Wedekind's study, men were instructed to wear a cotton T-shirt for two nights. Blood samples from each of these males were then analyzed to determine MHC. Females also had blood samples taken for MHC analysis. These women were then given T-shirts from males with MHCs similar or dissimilar to their own. Women not on oral contraceptives consistently found the odors of the T-shirts from males with dissimilar MHCs sexier, suggesting that

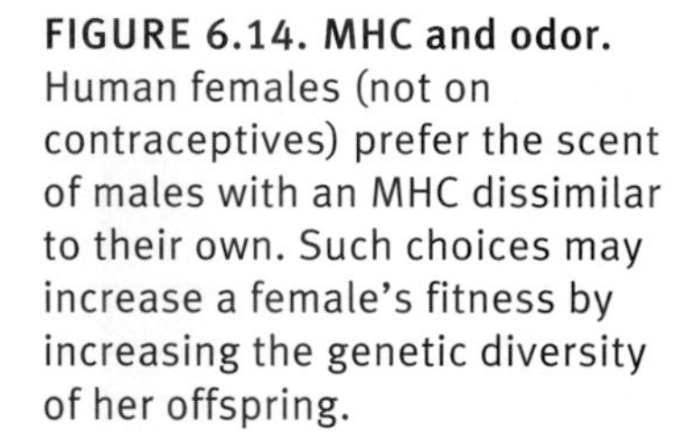

FIGURE 6.14. MHC and odor. Human females (not on contraceptives) prefer the scent of males with an MHC dissimilar to their own. Such choices may increase a female's fitness by increasing the genetic diversity of her offspring.

odor and MHC have a significant effect on female mate choice in humans (Figure 6.14). Wedekind found similar results when the men were subjects and women wore the T-shirts (Figure 6.15). Indeed, it seems that not only do human females choose males with dissimilar MHC alleles, but they themselves use perfumes that specifically magnify their own MHC-mediated odors (Milinski and Wedekind, 2001).

Reusch and colleagues argue that, although choosing a mate dissimilar in MHC related genes has fitness consequences (avoiding inbreeding), when it comes to choosing mates that help produce disease-resistant offspring, individuals should be preferring possible mates with a *high diversity* of MHC alleles (Reusch et al., 2001). They tested their "MHC allele counting" hypothesis using stickleback fish from a set of interconnected lakes. Female sticklebacks did not show a preference for males that had MHC alleles different from their own, but instead they consistently preferred males with the greatest diversity of MHC alleles.

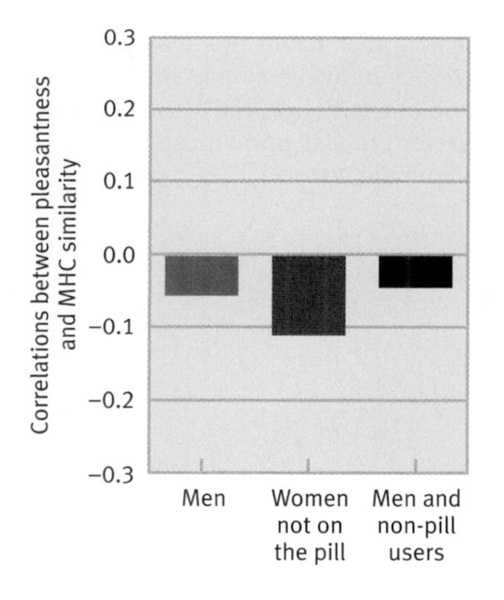

FIGURE 6.15. MHC and odor in humans. Wedekind and Furi examined whether humans used odor to try and select mates with a dissimilar MHC. Subjects were asked about the "pleasantness" of the odor associated with a T-shirt worn by someone else. With the exception of women on oral contraceptives, men and women found the odor associated with MHC dissimilar from their own to be the most pleasant. This manifests itself as a negative correlation when our y-axis is the correlation between pleasantness and MHC similarity (each of the bars shown here depicts a statistically significant relationship).

FLUCTUATING ASYMMETRY AND GOOD GENES. Over the last decade, a considerable literature has amassed on **fluctuating asymmetry**, which is random deviation from perfect symmetry of the body on its right and left sides. The work on fluctuating asymmetry is the result of a long search for generalizable traits that can be used as indicators of the overall health and vigor of an individual. Wouldn't it be wonderful, the argument goes, if there was a simple cue, or set of cues, that individuals could hone in on and use to determine the underlying genetic quality (with respect to health, resistance to disease, and so on) of potential mates? Symmetry may be such a cue.

Evolutionary biologists have focused their search for some measure of health and vigor by examining what is called **developmental stability** (Møller and Swaddle, 1998). Developmental stability is a measure of how well an organism handles changing environments as it matures. The underlying premise here is that organisms who can handle changing environments well may be those favored by the process of natural selection. Suppose that individuals do in fact differ in their ability to respond to the changing (and often adverse) conditions they face throughout their development. How would anyone else be able to tell who handles changing environmental conditions during development well and who doesn't? Bodily symmetry may be one way.

In animals, symmetry is a measure of the similarity of the left and right sides of an organism. For example, if your right and left arm were exactly the same length, your arms would be perfectly symmetrical. One way of measuring this would be to subtract the length of the left arm from that of the right. In that case, perfect symmetry would produce a score of zero, and any score greater or less than that would indicate some asymmetry. If others could assess your degree of asymmetry for a particular trait, they could presumably use that information in their mating decisions; and if symmetry is a good measure of developmental stability, using it as a cue should be favored by natural selection.

The "fluctuating" in fluctuating asymmetry refers to the fact that any deviation, positive or negative, from a score of zero is bad. In our example, it doesn't matter if you have too long a left or too long a right arm, just that you have too long an arm. The expectation is that positive or negative scores will fluctuate over evolutionary time, but the more an individual deviates from perfect symmetry, the more difficulty he will have in developing normally in his given environment (Figure 6.16).

The argument for symmetry being important in mate choice is that individuals who choose mates that are symmetric with respect to some trait are in reality choosing males with a high overall genetic quality, for such individuals appear to have a genetic constitution that allows them to weather the valleys and troughs of development better than individuals who are less symmetric (Møller and Thornhill,

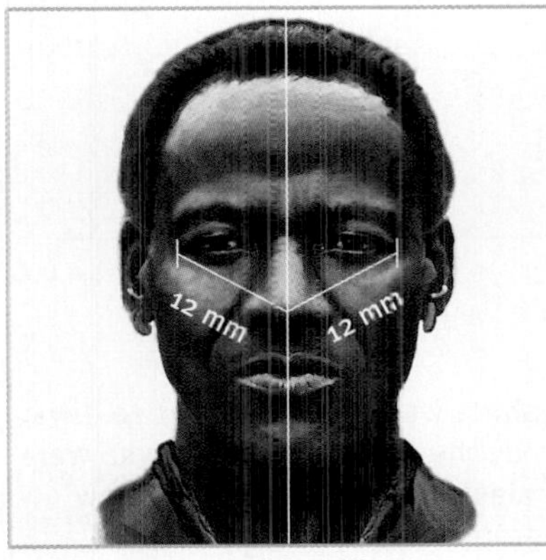

Symmetrical facial features

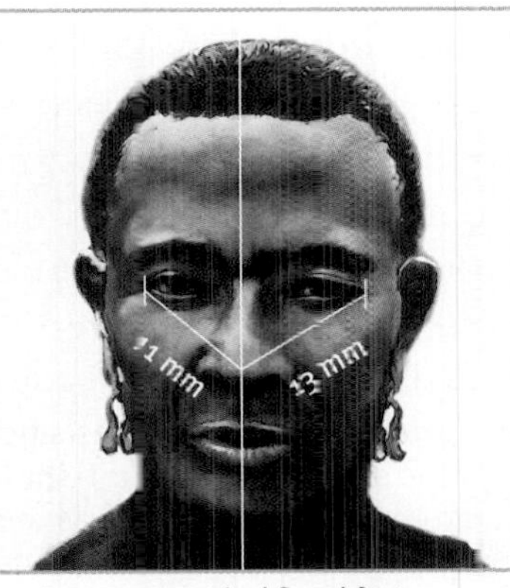

Less symmetrical facial features

FIGURE 6.16. Symmetry. The face on the left is more symmetrical than that on the right. Symmetry may be a cue of developmental homeostasis and is used as one cue in human mate choice.

1998). While this work has many critics (Swaddle and Cuthill, 1995; Palmer, 1996; Rowe et al., 1997; Whitlock and Fowler, 1997; Whitlock, 1998; David et al, 1999; Simmons et al., 1999; Bjorksten et al., 2000; Montgomerie, 2000), it is thought-provoking. Consider, for example, a 1998 study, in which Steve Gangestad and Randy Thornhill posed the following question: Do females change their reliance on symmetry as a function of their menstrual cycle? The hypothesis was straightforward: If symmetry is important to females as a way of assessing males, then they should be much more attuned to this when ovulating than at other times during the month (Gangestad and Thornhill, 1998). To test this, Gangestad and Thornhill began by giving women T-shirts worn by different males. Believe it or not, women can make a fairly good guess at male symmetry simply by smelling the odor that a male emits. Exactly why this is so is still not understood (although it probably harks back to our discussion of MHC and odor), but given that it is, T-shirts are a reasonable experimental tool.

Using forty-one college women as subjects, Gangestad and Thornhill distributed "smelly" T-shirts that had been worn by males for two days. Females were asked to rank the scent of a given T-shirt in terms of the attractiveness of the donor. The researchers also measured the symmetry of the male T-shirt donors and the reaction of females to the T-shirts at different periods during a normal ovulation cycle. Sure enough, women showed a stronger preference for the T-shirts of symmetric males just at the time when they were most fertile (Figure 6.17).

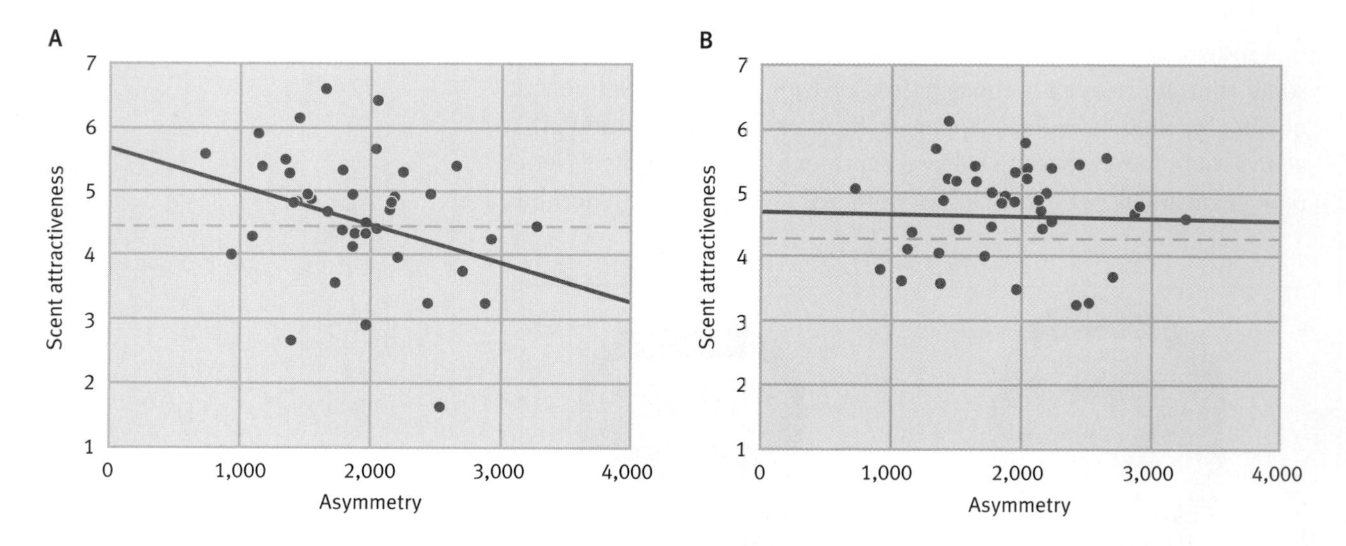

FIGURE 6.17. Smell, attraction, and fertility in women. College females smelled T-shirts worn by males for two days, and then the women rated the attractiveness of the scent of forty-one T-shirts worn by the men. **(A)** Women who were close to ovulation in their menstrual cycle preferred the scent of more symmetric males, while **(B)** women at a low fertility point in their cycle showed no such preference. The dashed line in the two graphs depicts the mean attractiveness rating, given a clean, unworn T-shirt. *(From Gangestad and Thornhill, 1998, p. 930)*

RUNAWAY SEXUAL SELECTION

Models of what is referred to as **runaway sexual selection** are probably the most difficult to understand of all those we have explored. These models assume the existence of two genes: the gene that codes for a particular trait in males, and the gene for mating preference in females. Such models can be expanded to consider numerous genes controlling for both male and female traits, but it will be easiest to follow the case of two genes. In runaway selection models, specific alleles from our two genes become *associated with each other through time*—when one allele is present in male offspring in a given clutch, the other allele will be present in female offspring from that clutch (Fisher, 1958; O'Donald, 1980; Kirkpatrick, 1982; Andersson, 1994).

To see how this runaway process works, imagine a population in which some fraction of the females have a heritable preference for brightly colored males, and that male coloration is itself a heritable trait. So, we have a group of females, some of whom prefer brightly colored males and some of whom don't, and a group of males, some of whom are more colorful than others. Now, we shall make an important assumption that is inherent in many good genes models—the genes coding for both male color and female preference are present in both males and females, but they are only "turned on" in the appropriate sex (preference genes in females, color genes in males).

Females that mate with colorful males should produce not only colorful sons but also daughters that possess their genetically coded preference for colorful males. Over time, the alleles that in females code for the *preference* for colorful males and the alleles that in males code for color become linked in the sense that, as the frequency of one changes, the frequency of the other changes as well. Once this positive feedback loop is set in motion, it can, under certain conditions "run away," like a snowball rolling down a snowy mountain. Across generations, selection may produce increasingly exaggerated male traits (for example, male color pattern) and stronger and stronger female preferences for such exaggerated traits.

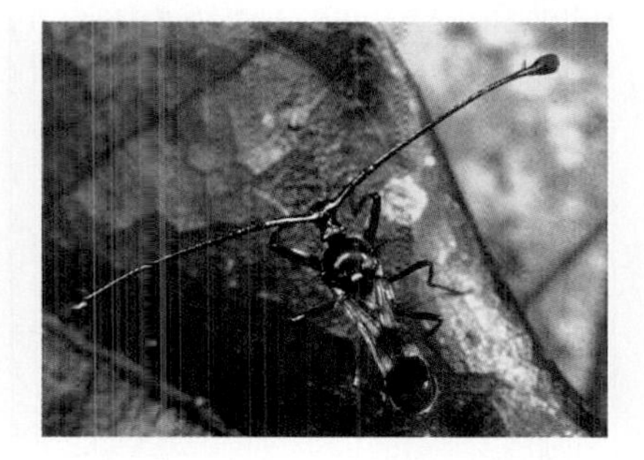

FIGURE 6.18. Stalk-eyed flies and runaway selection. Male stalk-eyed flies show marked variation in the length of their eye stalks. Wilkinson bred lines of flies with long and short eye stalks to test the runaway model of sexual selection. *(Photo credit: Mark Moffett)*

STALK-EYED FLIES AND RUNAWAY SELECTION. The most convincing case of runaway genetic selection comes from Jerry Wilkinson and his colleagues' work on the stalk-eyed fly (Wilkinson, 1993; Wilkinson and Reillo, 1994; Wilkinson et al., 1998a, 1998b; also see Bakker, 1993, for more on this in sticklebacks). In stalk-eyed flies, females prefer to mate with males possessing eyes that are at the end of long eye "stalks" (Figure 6.18). In one treatment of a meticulously well-done experiment lasting thirteen fly generations, males with the largest eye stalks were selected and allowed

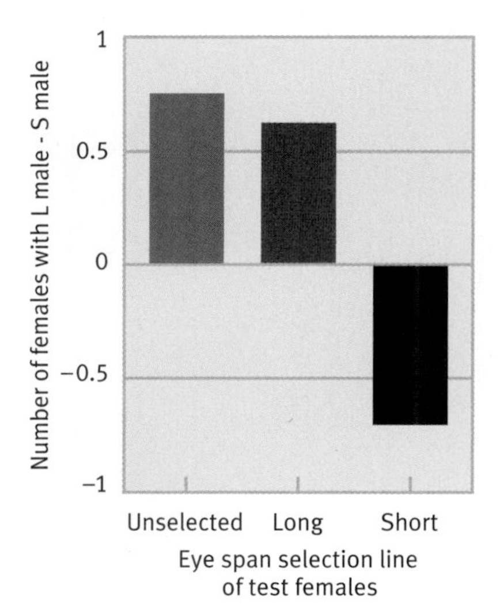

FIGURE 6.19. Genetic correlation of male trait and female preference in stalk-eyed flies. Over the course of a thirteen-generation experiment, Wilkinson and Reillo selected for males with long (L males) and short (S males) eye stalks. In both selection lines, females that mated were selected at random (that is, there was no artificial selection directly based on female choice). When females produced in the long and short eye treatments, however, were given a choice between males, they preferred males from their own selection treatment group, suggesting that female preference was genetically linked to the male trait being selected (stalk length). *(From Wilkinson and Reillo, 1994)*

to breed with females. In a second treatment, males with the shortest eye stalks were allowed to breed. In neither case were females chosen for any particular trait; that is, females were chosen randomly. Not surprisingly, in the treatment where long eye stalk length was selected for thirteen generations, the average male eye stalk increased in length, producing a more exaggerated version of this trait. Conversely, in the case where individuals with short eye stalks were selected, the size of the eye stalk decreased.

The critical finding in this study, however, did not center on the male trait. What goes to the heart of the runaway sexual selection model was the discovery of a positive link between the length of male eye stalks and the *female preference for this trait*. Recall that in both the long eye stalk and the short eye stalk male treatments, the females that the males mated with were selected at random—without any regard for their preference in male eye stalk length. All of the selection pressures were with respect to male eye stalk length. Yet, after just a few generations, when females were given the choice between mating with males with short and long eye stalks, females from the short eye stalk line preferred males with short eye stalks (Figure 6.19). This preference for short male eye stalk length by females in the short eye stalk line changed directly in response to genetic changes *in males*, in accordance with the prediction of the runaway selection model. In the long eye stalk line, females preferred males with long eye stalks, but then so did females who had not been subject to any selection treatment. While there are a number of possible explanations for what occurred in the eye stalk length treatment, Wilkinson and Reillo believe that female preference did in fact evolve in the direction of longer eye stalk length, but that males used in choice experiments did not differ enough in eye stalk length for the newly evolved female preference to be accurately measured.

Learning and Mate Choice

While genetic models of female mate choice have expanded our understanding of sexual selection greatly, these models do not incorporate the idea that individuals may *learn* various aspects of mate selection. Experimental psychologists have examined learning in the context of mate choice (Domjan and Hollis, 1988; Domjan, 1992; Domjan and Holloway, 1997). Although these studies seem much more removed from the normal environment of the animal than are studies on mate choice undertaken by behavioral ecologists, they are nonetheless quite interesting with respect to what they tell us about general aspects of animal mate choice. For example, work on conditioned stimuli and mating behavior has found that after exposure to a conditioned sexual stimulus, males are quicker to copulate (Zamble et al., 1985; Domjan et al., 1986), become better competitors with other males (Gutierrez and Domjan, 1996), display higher levels of courtship (Hollis et al., 1989), and produce more sperm (Domjan et al., 1998; Figure 6.20) and more progeny (Hollis et al., 1997).

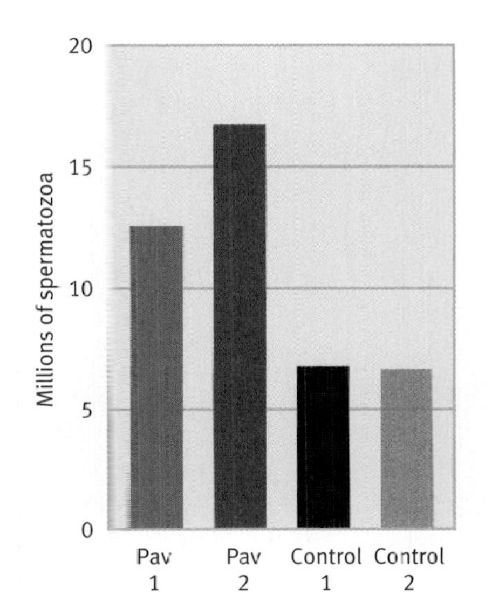

FIGURE 6.20. Pavlovian conditioning and number of sperm. Two different genetic lines of male quail (Pav 1 and Pav 2) learned to pair a distinctive experimental chamber with a chance to mate with a female. Males were then placed in the chamber with an artificial female. Subsequent to that, spermatozoa samples were taken. Males from both lines who had learned to pair the chamber with the chance to mate produced significantly more spermatozoa than control males (Control 1 and Control 2, which consisted of males from the two genetic lines who had not been given the chance to associate these two cues). *(From Domjan et al., 1998)*

SEXUAL IMPRINTING

One classic example of learning and mate choice is sexual imprinting (Lorenz, 1935; Bateson, 1978; Immelman, 1972; Shapiro, 1980; Marler and Terrace, 1984; ten Cate and Vos, 1999). In sexual imprinting young individuals "imprint" on what constitutes an appropriate mate from observing adults in their population. Normally, imprinting is restricted to some small time window during normal development, but the length of this window varies dramatically across species and behavioral contexts.

Cooke and his colleagues have evidence from both the lab and the field that imprinting plays a role in the mate choice of lesser snow geese (Cooke et al., 1972, 1976; Cooke and McNally, 1975). Lesser snow geese exist in two color morphs—blue and white—and prefer to mate with same-colored individuals (Figure 6.21). In a cross-fostering experiment in the laboratory, Cooke and his colleagues found that birds from families in which adopted parents were the same color as the young showed a stronger preference for same-colored mates than did individuals from families in which adopted parents were a different color than the young. These findings suggest that sexual imprinting plays a role in mate choice. To examine whether similar results on mate choice and imprinting

FIGURE 6.21. Lesser snow geese. These geese come in two color morphs and prefer to mate with same-colored individuals. *(Photo credit: Art Morris/Visuals Unlimited)*

occur in other species, studies on sexual imprinting have been undertaken in brown and white leghorned chickens (Lill and Wood-Gush, 1965), white and black king pigeons (Warriner et al., 1957), and different zebra finch morphs (Walter, 1973).

To further investigate what animals can learn about mate choice we turn now to Japanese quail, and then examine how aggressive behavior in fish can affect mate choice.

LEARNING AND MATE CHOICE IN JAPANESE QUAIL

Michael Domjan and his colleagues have been studying learning and mate choice in the Japanese quail for the better part of the last twenty years. One thing that Domjan and his colleagues have found is that male Japanese quail will quickly learn to stay by an area in which they have the opportunity to mate with a female (Domjan, 1987; Domjan and Hall, 1986; Figure 6.22). With this information in hand, Nash and Domjan (1991) examined whether this type of learning as an adult can override any effects of sexual imprinting as a juvenile.

Nash and his colleagues began by using different strains, or varieties, of Japanese quail. Each trial in their experiment had three subjects: a male quail from the standard "brown" strain, a female quail from the "brown" strain, and a female from a lighter-colored ("blond") strain of quail. In phase 1 of a trial, a brown male was allowed to see a blond female and then was given the opportunity to copulate with her. In the second phase of a trial, the same male could see a brown female quail, but was never in physical contact with her. In other words, a given male learned that in the laboratory, the presence of a blond female meant a mating opportunity, while this was not the case for brown fe-

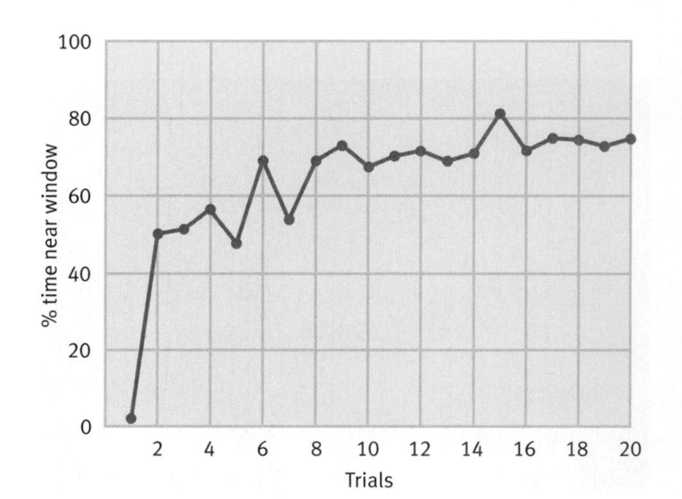

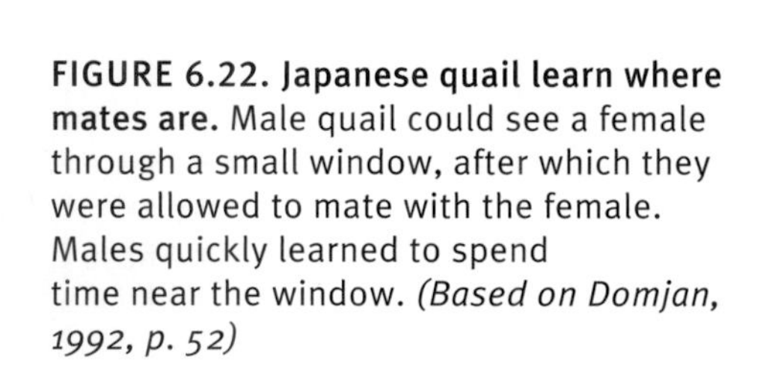
FIGURE 6.22. Japanese quail learn where mates are. Male quail could see a female through a small window, after which they were allowed to mate with the female. Males quickly learned to spend time near the window. *(Based on Domjan, 1992, p. 52)*

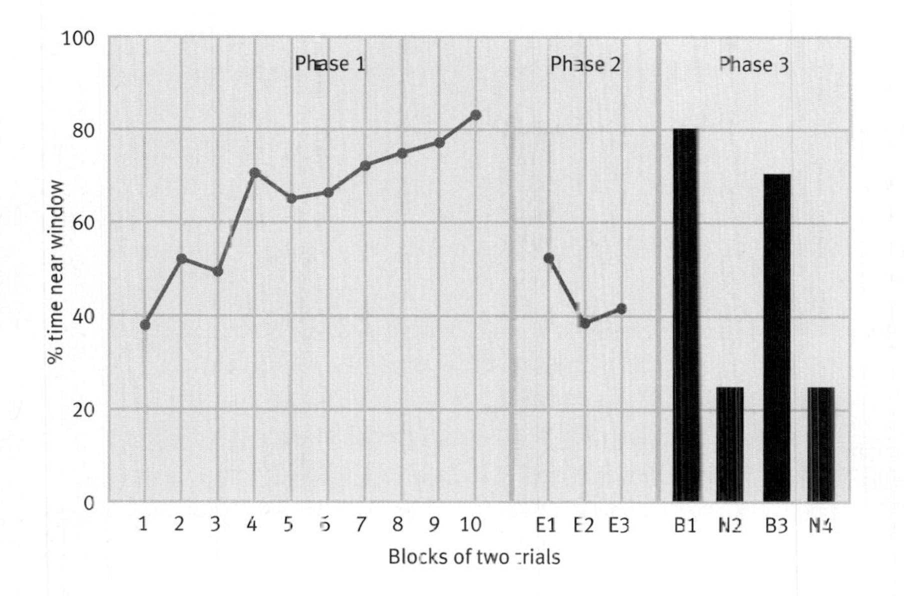

FIGURE 6.23. Overriding an imprinted sexual preference. Male Japanese quail were raised in an environment with normal brown-colored quail and should have imprinted on brown females as future mates. These males were then put through a battery of tests in which experimenters measured the percentage of time they spent near a window. First they saw a blond female on the other side of the window followed by numerous chances to mate with her (Phase 1). Then they saw through the window a brown female, but they were not allowed to mate with her (Phase 2). Lastly, they were tested to see how much time they spent near blond and brown females (Phase 3). In Phase 3, B = exposure to a blond female; N = exposure to a normal (brown) female. A significant preference for blond females was uncovered. *(Based on Domjan, 1992, p. 53)*

males. In the last stage of the experiment, males were tested to see how much time they spent near brown and blond females. The researchers found that blonds elicited a much stronger response (Figure 6.23).

To put Nash and Domjan's results into a larger mating/learning context, it is important to know that earlier work (Gallagher, 1976) had demonstrated that sexual imprinting plays a role in Japanese quail mate choice. The brown males used in Nash's study were raised in areas that contained other brown-strained individuals, and in which individuals had already imprinted on brown females. Despite this, however, learning as an adult who was likely to be a receptive mate overrode the effects of such sexual imprinting.

LEARNING AND MATE CHOICE IN BLUE GOURAMIS

Blue gourami fish are territorial, and when females first approach a male's territory, they are often attacked. Karen Hollis and her colleagues reasoned that if males could learn to anticipate the arrival of a female, they might reduce their initial aggression and thus mate more quickly and efficiently (Hollis et al., 1989).

Through Pavlovian conditioning, one group of males (PAV) was presented with a red light ten seconds before they were given access to a female (behind a partition in the red light section of the trial). Another group of males (UNP) was shown the red light a full six hours before they could interact with a female. After much training, males in both groups were presented with a red light and then given access to the female. Males from the UNP group

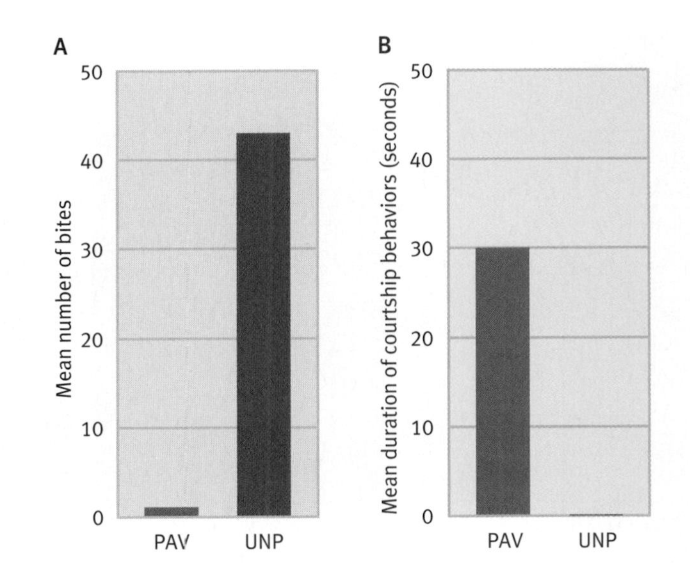

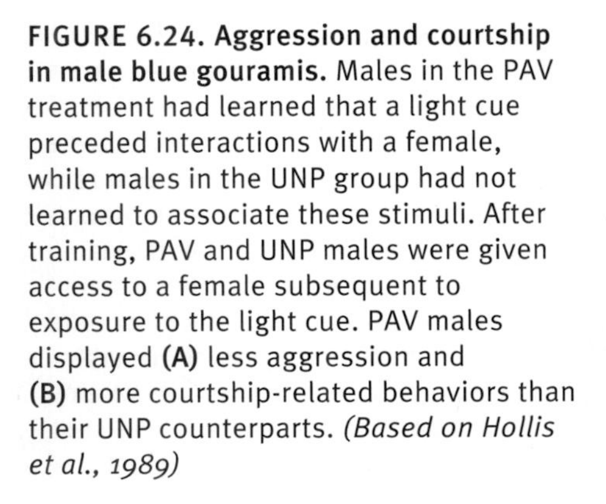

FIGURE 6.24. Aggression and courtship in male blue gouramis. Males in the PAV treatment had learned that a light cue preceded interactions with a female, while males in the UNP group had not learned to associate these stimuli. After training, PAV and UNP males were given access to a female subsequent to exposure to the light cue. PAV males displayed **(A)** less aggression and **(B)** more courtship-related behaviors than their UNP counterparts. *(Based on Hollis et al., 1989)*

responded with the typical aggression seen on encountering a female. In contrast, PAV males spent little time being aggressive and significantly more time courting females (Figure 6.24).

Cultural Transmission and Mate Choice

As we have seen earlier in this chapter, most work on female mate choice has assumed that a female's preference for a particular male trait is under some sort of genetic control (Fisher, 1958). It is, however, almost certainly true that social, nongenetic factors also play a role in mate choice (Boyd and Richerson, 1985). One area of sexual selection that has only recently been explored is to what extent a female's preference is affected by the preference of other females (Gibson and Hoglund, 1992; Dugatkin, 1996; Dugatkin, 2000; Westneat et al., 2000).

DEFINING MATE-CHOICE COPYING

We shall begin our discussion of cultural transmission and mate choice by defining mate-choice copying. We shall then ask whether females copy the mate choice of others, and if so, under what conditions. Subsequently we will examine some theoretical and empirical aspects of mate-choice copying.

Following Pruett-Jones (1992), we shall say that **mate-choice copying** has occurred when a male's probability of being preferred as a mate at time 2 increases (or decreases) as the result of being preferred by a female at time 1. In other words, if a male has an X percent chance of mating tomorrow if he has not recently mated, and a Y percent chance if he has recently mated, the effect of mate-choice copying is defined as the difference between Y and X. Mate-choice copying occurs if $Y - X > 0$; the greater $Y - X$, the stronger the effect of mate-choice copying. One nice feature of this operational definition is that these probabilities can be measured, and hence a mate-choice copying hypothesis can be supported or refuted by available data (Dugatkin, 1996).

Those interested in building mathematical models of how mate-choice copying might evolve have addressed three questions:

- What effect does female mate-choice copying have on the variance in male reproductive success (the "opportunity for selection" approach)?
- Under what circumstances will a copying strategy take over a population composed of noncopiers ("the game theory" approach)?
- How will female mate-choice copying affect the co-evolution of female preferences and the male trait that is the object of such preferences?

OPPORTUNITY FOR SELECTION MODELS. Wade and Pruett-Jones's (1990) "opportunity for selection" models have found that female mate-choice copying always increases the variance in male reproductive success. That is, mate-choice copying always creates a scenario in which both the number of males that never mate and the number of males with extraordinary mating success increase in frequency, thereby increasing the variance in male reproductive success (Figure 6.25).

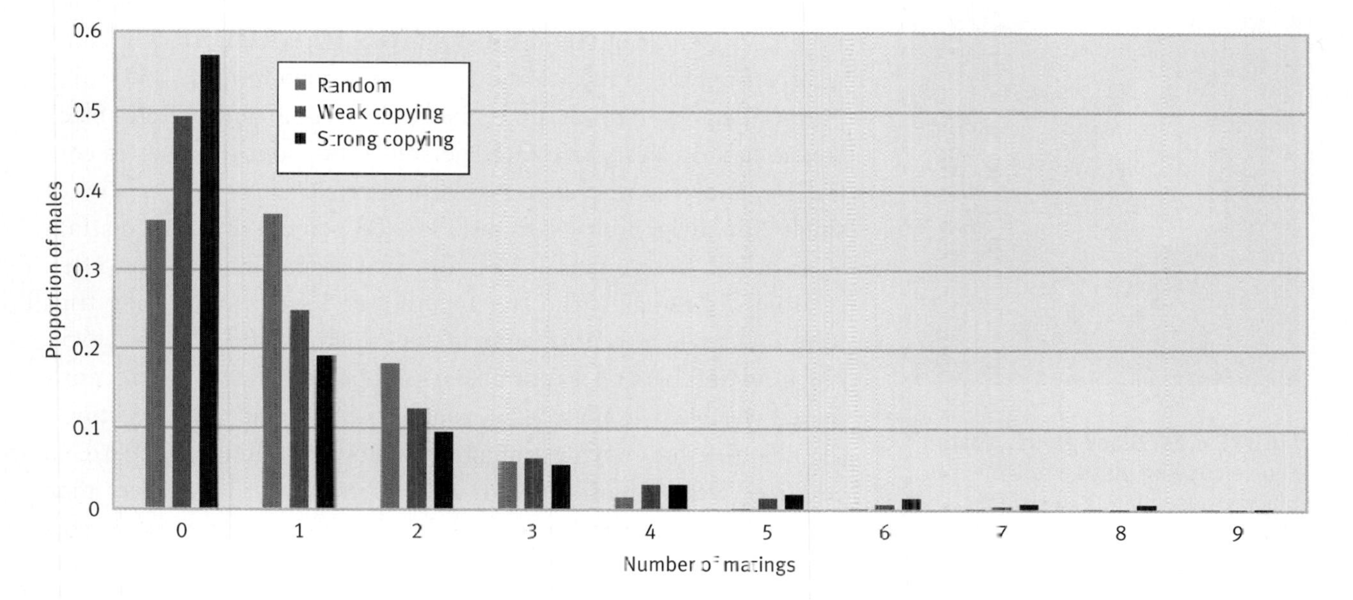

FIGURE 6.25. Mate-choice copying increases the variance in male reproductive success. Wade and Pruett-Jones built a computer simulation to examine the effect of mate-choice copying on the number of matings males obtained. What they found was that mate-choice copying increased both the proportion of males who obtained no mates and the proportion of males who had many mates. *(Based on Wade and Pruett-Jones, 1990, p. 5752)*

STRATEGIC MODELS. George Losey and his colleagues built a computer simulation that examined "copying" and "choosing" strategies in a structured environment that simulated a breeding area (Losey et al., 1986). What they found was that a mixture of copiers and choosers (with copiers in the minority) tended to be the optimal solution to the problem. When too many copiers were present, individuals playing this strategy were copying other copiers, rather than copying choosers, and hence the copiers were faring poorly. Losey and his colleagues also found that copying fares best when choosers are able to correctly identify high-quality males, when fecundity is low, and when females visit breeding areas for either long periods of time or on numerous occasions or both.

POPULATION GENETIC MODELS. Using population genetic models, investigators have examined the evolution of mate-choice copying, as well as other forms of cultural transmission of mate choice (Richerson and Boyd, 1989; Kirkpatrick and Dugatkin, 1994; Laland, 1994a, 1994b; Servedio and Kirkpatrick, 1996). In these models, males have a genetically based trait that females prefer, while a female's preference is affected by both a genetic and cultural component. In these models, traits "run away" to extreme values when mate-choice copying is allowed (Kirkpatrick and Dugatkin, 1994). This is similar to the process of genetic runaway selection we touched on earlier, but the critical distinction is that now the male and female traits become extreme, not because of genetic linkage, but because a female's preference is strongly affected by a culturally derived trait that co-evolves with the preferred genetic trait in males.

FIGURE 6.26. Black grouse. Mate-choice copying plays a role in black grouse mating decisions. *(Photo credit: niallbenvie.com)*

MATE-CHOICE COPYING IN GROUSE

Surrounded by Scotch pine, Norway spruce, and birch, black grouse mating arenas are interspersed throughout the bogs of central Finland (Figure 6.26). As with many lek breeding species, a single "top male" grouse obtains about 80 percent of all the matings at an arena. Before mating, females visit arenas many times, often in groups of females who stay together and synchronize their trips to various male territories at an arena (Höglund and Alatalo, 1995). Höglund and his colleagues observed that a male that had recently mated was likely to mate again sooner than by chance, suggesting a possible role for copying among females (Höglund et al., 1990). In addition, older females mated, on average, three days earlier than younger females, suggesting that copying, if it occurred, was most common among younger females. This is precisely what one would expect, as young and inexperienced females stand to gain much more from watching old pros than vice versa

Höglund and his colleagues subsequently undertook an ingenious experiment using stuffed dummy females placed on male territories within a lek (Höglund et al., 1995). In this experiment, seven males on a particular lek had stuffed black grouse dummies placed on their respective territories early in the morning before females arrived. The males courted these dummy females and even mounted them and attempted numerous copulations. Their results indicated that females were more interested in a male who had mated with other females on his territory, even if they were dummy females This finding is just what one would expect if copying, rather than some set of physical traits alone, explained the skew in male reproductive success among black grouse, with a few males obtaining a vastly disproportionate number of matings.

Mate copying likely plays a role in another species of grouse as well. In the early 1990s, Robert Gibson and his colleagues addressed the question of female mate-choice copying in sage grouse (Gibson et al., 1991). They examined the mating behavior of sage grouse females (*Centrocercus urophasianus*) in two different leks over a four-year period. One of their hypotheses regarding copying and mate choice was that the unanimity of female mate choice would increase as more hens mated on a given day, because more opportunities to observe and imitate would exist on such days. The data support this hypothesis (Figure 6.27). Further, their work reveals the "snowballing" effect often found in conjunction with imitation—not only did females copy each other, but copiers copied other copiers.

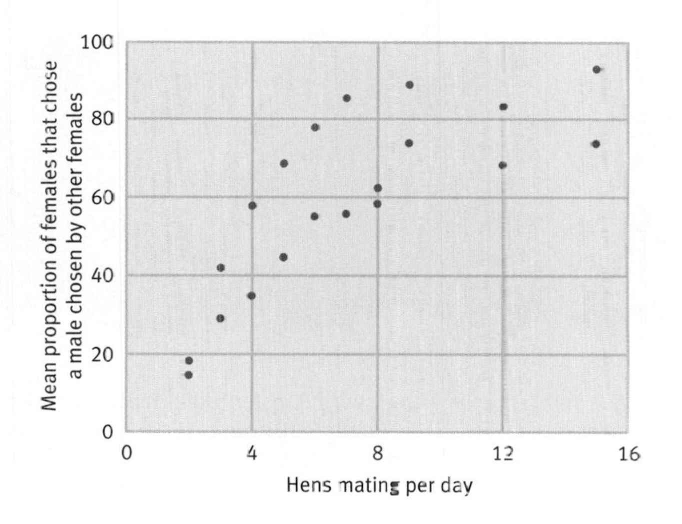

FIGURE 6.27. Female mate-choice copying in sage grouse. As predicted by models of mate-choice copying, when the number of hens mating per day increased, so too did the proportion of females choosing to mate with a male chosen by other females The green points are the observed matings, and the orange points are the expected matings if the females had not been copying the mate choices of other females. *(From Gibson et al., 1991)*

MATE-CHOICE COPYING IN GUPPIES

Some of the strongest evidence to date for female mate-choice copying comes from the guppy (*Poecilia reticulata*) (Dugatkin, 2000; Figure 6.28). Dugatkin (1992b) examined mate-choice copying using a ten-gallon aquarium situated between two end chambers constructed of clear Plexiglas. A single male was put into each of these end chambers. The "focal" female—the individual that potentially copies the behavior of other females—was placed in a clear Plexiglas canister in the center of the aquarium. A removable glass partition also created a section of the aquarium into which another female was placed. This individual will subsequently be referred to as the "model" female—the female whose behavior may be copied by the focal female.

At the start of a trial, males were put into the end chambers, and the focal and model females were placed in their respective positions in the aquarium. Fish were given ten minutes during which the focal female could observe the model female near one of the two males. The model female and the glass partition were then removed, and the focal female was released from her canister and given ten minutes to swim freely and choose whichever male she preferred. In these trials, the focal female chose the male that had been beside the model female seventeen out of twenty times.

FIGURE 6.28. Mate preferences in guppies. Female guppies from the Para River typically prefer mates with lots of orange color (Female C; Male A). If a female (D), however, sees a less colorful male (B) preferred by another female (E), she is much more likely to mate with that male (B) than with another male (C) who was not previously chosen as a mate. *(Based on Dugatkin and Godin, 1998)*

While the results of this experiment are consistent with the hypothesis that females copy the mate choice of others, there are several alternative explanations. For example, since guppies are schooling fish (Seghers, 1973; Magurran and Seghers, 1990), the focal female might simply be choosing the area that recently had had the largest group of fish (in this case, two). A control treatment was conducted to test this "schooling hypothesis." It was identical to the above protocol, except that females were placed in the end chambers. In this case, the focal female chose the female in the end chamber closest to the model in only ten out of twenty trials. As such, the tendency to school per se does not explain results of the first experiment; if it had, focal females would have chosen the end chamber closest to where the model had been placed.

Other controls in the guppy experiment found that: (1) When male courtship was removed from the experiment (via one-way mirrors), but all else was held constant, focal females chose randomly between males. As such, females were not simply going to the area that recently contained both a male and a female; the pair needed to be involved in courtship. (2) When choosing between males, focal females were not basing their decision on the activity patterns of males. (3) Focal females remembered the identity of the male chosen by the model—that is, if the focal female saw male 1 near the model female, but the position of the end chambers containing male 1 and male 2 was switched before that focal female herself chose, she still preferred male 1. Direct mate-choice copying by females is the only explanation to date that is compatible with all the results described. Furthermore, mate-copying guppies may even override their innate tendency to prefer colorful (orange) males (Figure 6.29). Two studies, however, found no evidence of mate-choice copying when replicating the guppy work (Brooks, 1996; Lafleur et al., 1997).

Using a protocol similar to that of the guppy experiments, mate-choice copying has been uncovered in sailfin mollies (*Poecilia latipinna*; Schlupp et al., 1994; Witte and Ryan, 1998), medaka (*Oryzias latipes*; Grant and Green, 1995), and quail (*Coturnix japonica*; Galef and White, 1998; White and Galef, 1999, 2000), but it does not appear to be a factor in fallow deer (*Dama dama*; Clutton-Brock and McComb, 1993; McComb and Clutton-Brock, 1994).

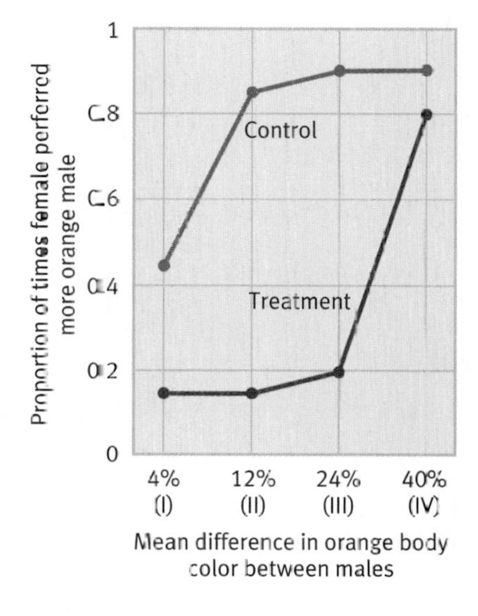

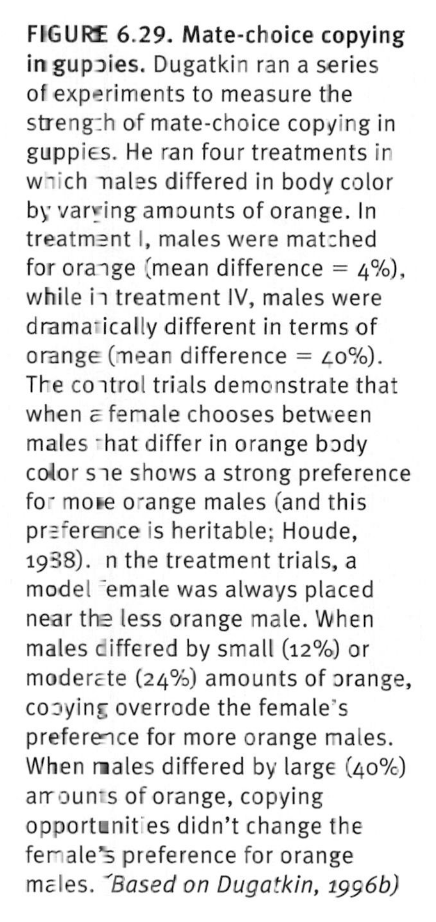

FIGURE 6.29. Mate-choice copying in guppies. Dugatkin ran a series of experiments to measure the strength of mate-choice copying in guppies. He ran four treatments in which males differed in body color by varying amounts of orange. In treatment I, males were matched for orange (mean difference = 4%), while in treatment IV, males were dramatically different in terms of orange (mean difference = 40%). The control trials demonstrate that when a female chooses between males that differ in orange body color she shows a strong preference for more orange males (and this preference is heritable; Houde, 1988). In the treatment trials, a model female was always placed near the less orange male. When males differed by small (12%) or moderate (24%) amounts of orange, copying overrode the female's preference for more orange males. When males differed by large (40%) amounts of orange, copying opportunities didn't change the female's preference for orange males. *(Based on Dugatkin, 1996b)*

SONG LEARNING AND MATE CHOICE IN COWBIRDS

Mate-choice copying is not the only means by which cultural transmission can affect mate choice in animals. Song learning in birds provides another classic example. One thing that is common to all songbirds is that they learn something about the songs they sing. In particular, much of the song-learning process involves learning songs from others, often referred to as tutors. Songs, in other words, are culturally transmitted.

In some ways, cowbirds (*Molothrus ater*) might seem an odd species to study the cultural transmission of behavior and its consequences for mate choice. Cowbirds are similar to cuckoo birds, in that they always lay their eggs in the nest of *another* species. So, in nature, cowbird young are never raised by adult cowbirds. Hence, it has long been thought that much of the behavior shown in this species must be innate. After all, you must "figure out" how to be a cowbird, and if you aren't learning it from your mom and dad, then "cowbirdness" must lay almost completely in the genes. Nonetheless, cowbirds do learn their songs, and even variations on such songs based on the slightly different song repertoires in different "subpopulations" of cowbirds. Furthermore, individuals from a given population prefer mating with others who sing the song native to their own area.

Todd Freeberg undertook a series of fascinating experiments with cowbirds to learn more about the cultural transmission of birdsong and its long-term consequences for mate choice in this species (Freeberg, 1998). He went about collecting birds from two different populations of cowbirds—one from South Dakota and one from Indiana. He chose these particular populations because the cowbirds in them are very different in terms of the behaviors they display and the songs they sing.

Freeberg ran what is referred to as a cross-fostering experiment. In rearing regime 1, he raised juvenile birds from the Indiana population with Indiana adults; in rearing regime 2, he had juvenile Indiana birds raised with South Dakota adults. He raised juvenile birds from South Dakota with South Dakota adults (rearing regime 3) or with adults from the Indiana population (rearing regime 4). Rearing regimes 2 and 4 had cross-fostered chicks, while regimes 1 and 3 served as controls (Figure 6.30). Cross-fostering experiments are a well-established technique for studying the effect of an animal's environment. If juveniles take on traits of the environment in which they are raised, regardless of whether or not it is their "natural" environment, this suggests a certain level of behavioral flexibility, and in this case, possibly a role for cultural transmission of birdsong.

After juveniles were raised with adults for a year, their own mating patterns were observed in a big aviary. All individuals were tested in an aviary that contained *unfamiliar* birds from both the Indiana and South Dakota populations. Even though they had never before interacted with the unfamiliar birds, the subjects preferred to mate with individuals that came from the same rearing regime they themselves came from. Freeberg's results were even more pronounced when the preference test was repeated after birds had been together longer.

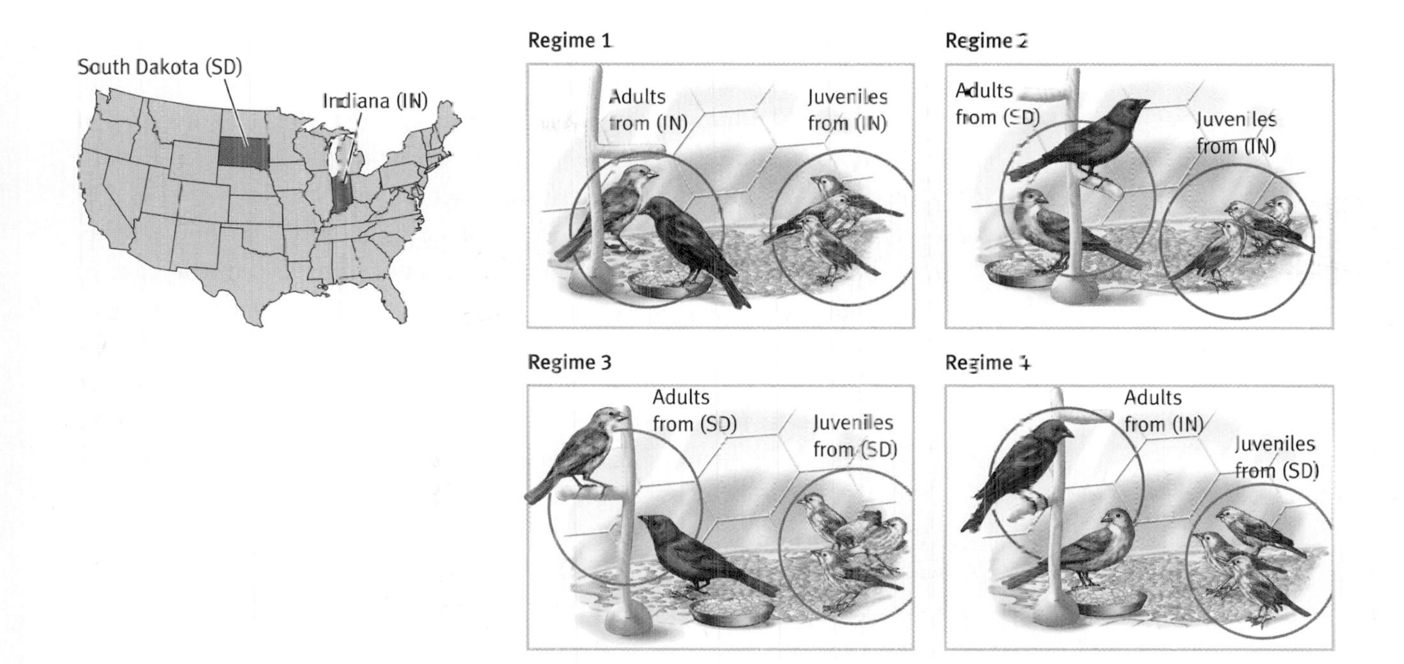

FIGURE 6.30. Cross-fostering. In Freeberg's cross-fostering experiments, juvenile cowbirds from Indiana were raised with adults from Indiana (regime 1) or adults from South Dakota (regime 2). South Dakota juveniles were raised with adults from South Dakota (regime 3) or adults from Indiana (regime 4).

Freeberg's findings show the importance of cultural transmission in shaping mate choice. First, they clearly demonstrate that information conveyed during early life may not have an impact until individuals mature, but the impact, when it occurs, is quite strong. Second, this work suggests that cultural transmission manifested itself in two different ways. Not only did males apparently learn what song to sing (and how to behave) by imitating the actions of the adults they grew up with, but females also appeared to learn which male traits to view as attractive by watching their mentors.

Male-Male Competition and Sexual Selection

As mentioned at the start of the chapter, once Darwin introduced the notion of sexual selection, much of the work in this area focused on intrasexual selection, which typically shows up in male-male competition for mates (although female-female competition does occur in "sex reversed" species such as the American jacana (*Jacana spinosa*; Jenni and Collier, 1972) and the pipefish (*Sygnathus typhle*; Berglund et al., 1986, 1988, 1989; Jones et al., 2000). This form of sexual selection is often more dramatic than female mate choice, as it typically involves some sort of direct competition between males, as in the classic case of various males bashing horns to determine access to a female.

Intrasexual selection need not be as dramatic as fierce fights between males. Competition may be less direct, as in the case of male cuckoldry, or as we shall see in Chapter 7, sperm competition. Naturally, many cases of sexual selection in nature will have elements of both intra- and intersexual selection, as when females choose to mate with males on the best territories, but only after males have first competed among themselves for that territory.

UNDERGROUND MATING AND MALE-MALE COMPETITION IN FIDDLER CRABS

Ideally, it is easiest to examine male-male competition when female choice is weak, if not completely lacking. This appears to be the case in fiddler crabs (*Uca paradussumieri*), a species in which females dig a hole underground and appear to accept the first male that enters their nests as a mate (Koga et al., 1999). Once a male finds such a nest, however, he spends a great deal of time and effort defending it from other males who attempt to usurp his position in that nest. When defending their nest, guarding males either fight intruders or use what is known as the "flat-claw defense" wherein a male stands in the entrance to a nest with his enlarged claw plain for all to see. This flat-claw defense is quite successful at deterring other males, even when intruders are larger than the guarding male.

Guarding behavior in a male lasts through the time a female ovulates, after which he abandons her nest.

RED DEER ROARS AND MALE-MALE COMPETITION

Some of the best-known studies of male-male competition are those of Tim Clutton-Brock and his colleagues on roaring in red deer (*Cervus elaphus*). Based on Zahavi's (1975) handicap principle, Clutton-Brock and Albon (1979) hypothesized that male red deer should use an honest indicator of each other's fighting ability to determine whether to compete for access to a female. Davies and Halliday (1978) used similar logic when examining depth of calls and body size in relation to fighting in toads (*Bufo bufo*). Red deer stags form harems during breeding season, and while serious

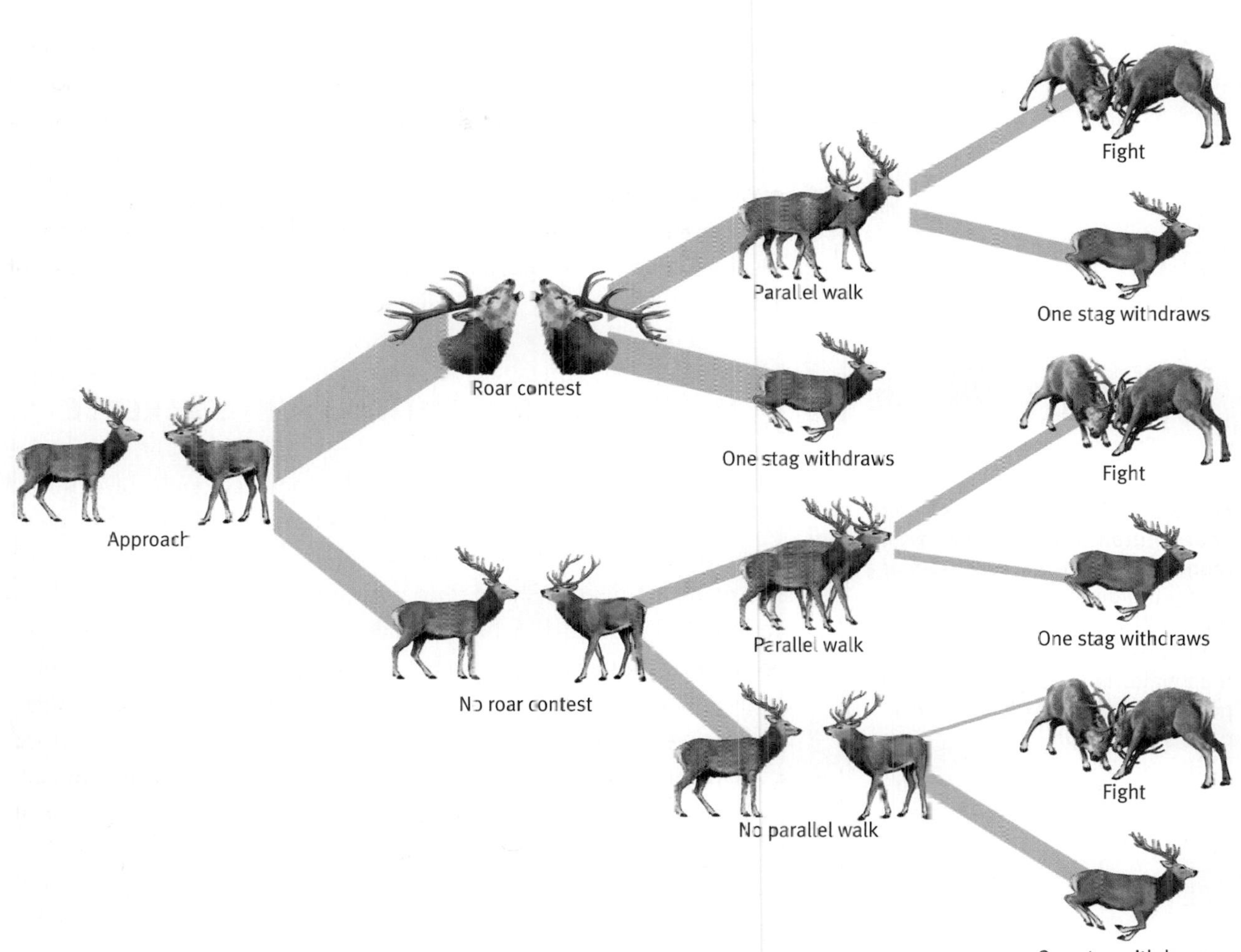

FIGURE 6.31. Approach and assessment in male red deer. When two male red deer approach within 100 m of each other during the mating season, any number of outcomes are possible. In this figure, thicker lines represent more likely outcomes. *(From Clutton-Brock et al., 1979)*

fighting is not common, 23 percent of harem holders show some sign of a fighting injury and 6 percent are permanently injured, suggesting strong selection pressure for accurately assessing an opponent's strength to avoid injury and possible death.

Clutton-Brock and Albon found numerous pieces of evidence suggesting that roaring was being used to assess an opponent's fighting ability in the context of harem holding (Clutton-Brock et al., 1979). To begin with, roaring and associated activities such as "parallel walks" were almost exclusively seen during the rut, that is, mating time (Figure 6.31). Furthermore, during the rut, harem holders roared more than those without a harem (Figure 6.32), and roaring increased when a harem holder was approached by another male (or heard taped playbacks of another male's roaring).

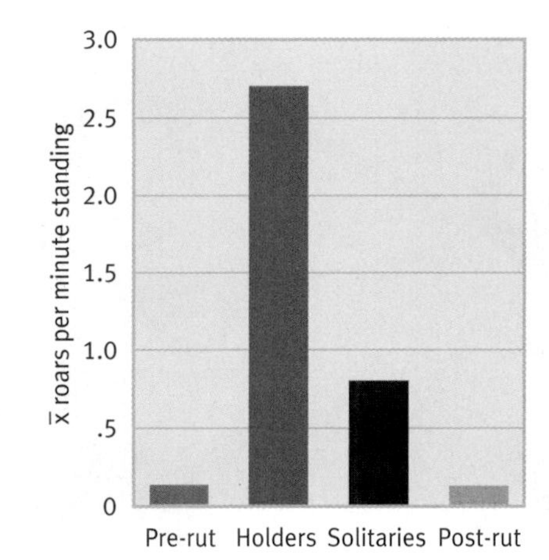

FIGURE 6.32. Roaring rates in red deer. Clutton-Brock and Albon found that males in possession of a harem (holders) roared much more than males not holding a harem (solitaries), as well as more often than males in the pre-rut or post-rut periods. *(Based on Clutton-Brock and Albon, 1979)*

Clutton-Brock and Albon also found that both the mean and the maximum roaring rate were positively correlated with stag fighting ability. Lastly, roaring contests were much more common between mature stags of approximately equal fighting ability, and roaring rarely escalated (through the "parallel walk" stage) to fighting except among the most closely matched stags, suggesting that roaring was a good indicator of fighting ability and one that stags often relied on.

MALE-MALE COMPETITION BY INTERFERENCE

Although many examples of male-male competition involve males fighting it out one way or another *before* a female is present, this need not be the case. An interesting subset of male-male interactions consists of males directly interfering with other males while the others are attempting to mate with a female. This sort of behavior is common during amplexus (in which males clamp on to females for prolonged periods of time) in amphibians (Duellman and Trueb, 1994) and during mating in insects. For example, in the European earwig (*Forficula auricularia*), heavier males defeat lighter males and gain access to females, with one earwig male often attempting to displace another male while he is copulating with a female (Forslund, 2000).

In at least some cases of male-male interference, females appear to solicit males to try to remove a rival during the actual mating event. Cox and Le Boeuf (1977) hypothesized that females may incite male-male competition, thereby increasing the probability that they mate with the highest-ranking male in a group. Cox and Le Boeuf tested their hypothesis on the elephant seal (*Mirounga angustirostris*), a species in which males form large harems of females (Figure 6.33). For example, harems of less than forty females are exclusively defended by a single dominant bull seal, and there is consequently huge variation in reproductive success among males. In one long-term study, only 8.3 percent of males mated, but some individual males inseminated 121 different females (Le Boeuf and Reiter, 1988).

FIGURE 6.33. Elephant seal fights. Male elephant seals are *much* larger than females and fight for access to females during the breeding season. *(Photo credit: François Gohier)*

Of the 271 estrous female elephants Cox and Le Boeuf had marked, approximately 87 percent of mounts were "protested" to some degree or another by females. Protesting included a large suite of behaviors, most prominently loud calls and constant back-and-forth movement to prevent the male from copulating. Such protests seem quite effective, as 61.4 percent of protested mounts were interrupted, as opposed to 25 percent of mounts that went unprotested. Female protests appeared to be designed to solicit more dominant males to interrupt mounts, as the probability that a

TABLE 6.1. Female seal elephants protest when mounted by nondominant males. Cox and Le Boeuf found that not only did alpha males mount females more often, but that the females protested their mounts less often than mounts by nondominant adults or subadults (SA_3, SA_4). *(From Cox and Le Boeuf, 1977, p. 324)*

AGE OR SOCIAL RANK OF MOUNTING MALE	MOUNTS OBSERVED* (*N*)	TOTALLY PROTESTED (%)	PARTIALLY PROTESTED (%)	NOT PROTESTED (%)
Alpha	74	37	43	20
Adult	70	49	34	17
SA4	9	78	22	0
SA3	4	100	0	0

Interrupted mounts are excluded.

female would protest a male mounting her was a function of that male's rank. Females rarely protested alpha male mounts, but they often protested mounts attempted by subordinates (Table 6.1). This resulted in females who protested mounts with subordinate males increasing their probability of copulating with the highest-ranking male in the vicinity.

The elephant seal example shows how difficult it can be to completely disentangle female mate choice from male-male competition. Clearly males interrupting the copulation attempts of others qualifies as male-male competition, but in the elephant seal case, this competition is initiated by a female to increase her probability of mating with the highest-ranking male in the area.

MALE-MALE COMPETITION VIA CUCKOLDRY

In many species of fish, one finds different male reproductive types—or morphs—that are distinct in structural, physiological, endocrinological, and behavioral traits (Gross and Charnov, 1980; Gross, 1985). In bluegill sunfish, three male morphs—parental, sneaker, and satellite—appear to coexist (Gross, 1982). Parental males are light-bodied in color (with a dark yellow-orange breast), build nests, and are highly territorial, chasing off any other males that come near their territory.

Sneaker males look and act very different from parental males. These males are smaller, less aggressive, and do not hold territories. Rather they lurk in hiding places near a parental male and zip into a territory while the parental male and female are spawning.

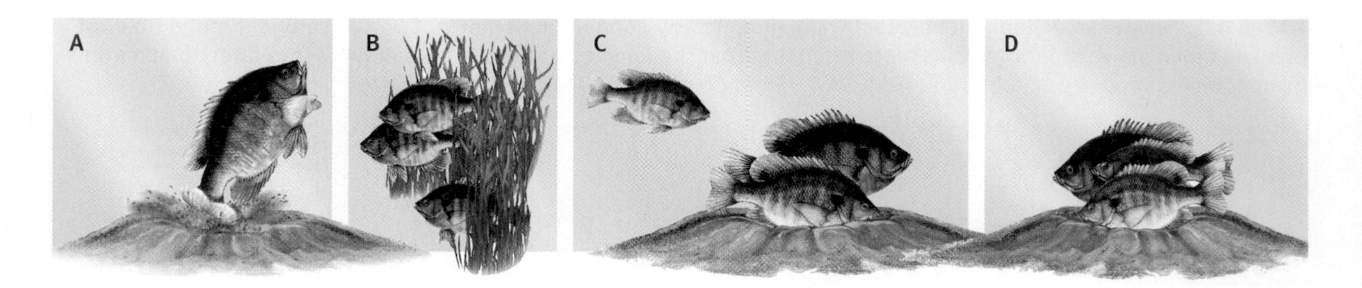

FIGURE 6.34. Bluegill morphs. (A) Bluegill parental male preparing nest. **(B)** Sneaker males hiding behind plants awaiting a chance to quickly sweep into parental nest. **(C)** A satellite male swimming over a nest containing a male and female. **(D)** A satellite male swimming between a parental male and a female. **(E)** A composite of A–D. *(Based on Gross, 1982)*

Sneakers then shed their sperm and swim away, with the whole process taking less than ten seconds (Gross, 1982). Male-male competition here then involves sneakers trying to outcompete others by a "hit-and-run" type strategy in which they sneak into the parental male's territory and cuckold the parental male (Figure 6.34). Without the parental male present, sneakers would have no chance to cuckold anyone, for without the parental male, females would not be drawn to the area where sneakers are lurking.

A third male morph, labeled satellite males, also exists in some bluegill populations. Satellite males tend to look like females. Disguised as such, they lower themselves down from the water column and position themselves between a spawning pair. If the parental male is duped, he will attempt to spawn with both the female and the impostor satellite male, at which point the satellite male will release his own sperm.

Neuroethology and Mate Choice

Neuroethology plays a large role in one hypothesis for the *initial* emergence of female mate preferences. This idea, put forth by Ryan (1990) and Endler and McLellan (1988), has been labeled the **sensory exploitation**, **sensory drive**, or **preexisting bias** hypothesis (West-Eberhard, 1979, 1983). Sensory exploitation argues that females may initially prefer male traits that elicit the greatest amount of stimulation from their sensory systems (Ryan and Keddy-Hector, 1992). What this often means is that neurobiological systems that are already in place in the female are "tapped into" and used in the context of sexual selection. This preference often leads to mating with males who do not provide any benefits, direct or indirect, to the female. For example, suppose that, for whatever reason, red berries are the most nutritious food source available to a fruit-eating songbird species. Females who are best able to search out and subsequently eat red berries survive and reproduce better (Rodd et al., 2002). Thus, natural selection works on the neurobiology of females such that they are acutely aware of the color red, and hone in on red things in their environment (Kirkpatrick and Ryan, 1991). If red feathers should suddenly arise in males of this normally blue-feathered species, they may be chosen as mates because the female's nervous system is already designed to preferentially respond to red objects. Males with red feathers are exploiting females in the sense that females prefer these males, not because the male trait has any fitness consequences, but because it draws the females' attention.

One thing to keep in mind when reading about the sensory bias hypothesis is that it was designed to elucidate the *origin*, not the long-term maintenance, of a trait. That is to say, the sensory bias hypothesis postulates how a trait initially arose in a population, not how it was maintained in a population over evolutionary time. Once a trait arises through some sensory bias, or through whatever means, natural selection will determine whether such a trait remains in a population and potentially increases in frequency or whether the trait decreases in frequency.

SWORDTAILS, PLATYFISH, AND SENSORY BIAS

One of the most convincing cases of the sensory exploitation of female mate choice is Alexandra Basolo's work on two related species of tropical fish, the green swordtail (*Xiphophorus helleri*) and the

Interview with Dr. MALTE ANDERSSON

Your book *Sexual Selection* is required reading for anyone in the field. How did you first get interested in this area?

As a bird-watching boy I was fascinated by some of the signs of sexual selection (although unaware of the process), in particular the conspicuous plumage of males in some species. The central tail feather of a cock pheasant was the crown jewel in my little collection of birds' feathers. Later I spent many summers in Lapland amidst birds such as lekking ruffs and role-reversed northern phalaropes, which further aroused my interest. As a graduate student I read some of the exciting theory, and on a journey to East Africa I saw breeding long-tailed and Jackson's widowbirds, which appeared to be made for testing the theory. Reading books by Williams (1966) and J. Merritt Emlen (1973) was very important. They showed that evolutionary theory was full of fascinating unsolved ecological and behavioral problems. They discussed Fisherian runaway sexual selection, and they independently outlined ideas similar to Fisher's (1915) index mechanism of sexual selection, ideas that were spread by Zahavi (1975) as the "handicap principle," provoking much discussion. These debates and unsettled problems were very inspiring.

Your work on tail length and mate choice in widowbirds is widely regarded as a classic, particularly with respect to the experimental manipulations you undertook. How did you get involved in studying widowbird mate choice? What prompted you to consider artificially changing the bird's tail length?

With a mathematics and physics background, modeling is one of my interests, and after finishing my thesis I took up modeling of sexual selection. I then realized that there was still much doubt as to whether female choice exists at all, so mate choice clearly needed to be tested empirically. My first plan was to study peacocks, and I briefly visited Sri Lanka, but the field situation seemed unfavorable. Instead, I went to Kenya with long-tailed widowbirds on my mind, and they turned out to be more accessible. Field experiments that I had done on other aspects convinced me of the power of this approach.

I was much inspired by Niko Tinbergen's writings and approach to science, which involved experimental testing in natural habitats to understand adaptive aspects of behavior and ecology. In my view Tinbergen is the main architect of behavioral ecology, pioneering research in this field long before it got its name. I cannot overstate how much I learned from eagerly reading his work early on—which was easy as his writing is so engaging. If you have not done so, you will probably enjoy reading his *Curious Naturalists*.

In addition, although far from infallible, controlled experimentation to me seems one of the most useful and also democratic human inventions ever made. It enables ordinary people like you and me to pose questions to nature, questions that she, if we are persistent enough, will usually answer in interesting and educating ways, increasing our understanding. Unless enemies of open-minded inquiry have too strong repressive power, insight gained from well-controlled experimentation and other empirical testing is likely to take precedence in the long term over authority, prejudice, cultural, political and religious correctness and taboo. It is a uniquely powerful tool for finding out about the natural world. This is a wonderful thing, not lessened because the ethics needed to make good use of the knowledge largely lie outside the realms of natural science.

So, there was no doubt in my mind that I ought to do controlled experimental manipulations to test the role of the male tail in the long-tailed widowbird. Once thought out, manipulation of tail length was not very difficult: scissors are an old invention, and rapidly hardening cyanoacrylate glue had just been developed as part of the space program. (So the widowbird study was sort of a spin-off from the Apollo project.)

Back in 1871 Darwin proposed that females might exert some choice in choosing their mates. Why do you think it took so much longer to study this form of sexual selection as compared to male-male combat?

For at least two reasons. First, male combat is much more easily observed than female choice. Mate choice may not be detected unless individuals are captured and marked and their behavior studied for long periods. Second, there was prejudice during Darwin's time and long after it against the idea that females are capable of choosing males based on "esthetic" traits.

For some time in the early 1980s it looked as if mathematical models of sexual selection were outpacing important empirical findings. Why do you think that was? Has the tide shifted, and if so, why do you think this shift has occurred?

In the early 1980s many suggested mechanisms of sexual selection created scope for mathematical modeling and logical testing, which attracted modelers to this field. During this theoretical Sturm und Drang period, modeling partly outgrew empirical testing. Field experiments in behavioral ecology then were still rather rare, and it took some time before suitable species and techniques for empirical work had been found. Toward the end of the decade, DNA-based methods of paternity determination began to revolutionize the field, enabling empirical tests that were not dreamt of ten years earlier, a revolution that is in full swing today. In addition, sexual selection theory has now been enriched and integrated with many other ideas and approaches, involving, for example, hormones, immune function, reproductive physiology, signaling and sensory mechanisms, which present attractive challenges to lab and field workers. The revolution in the study of sperm competition and other postcopulatory sexual selection has opened fascinating new areas, with vast opportunities for exciting empirical work. These are probably some of the reasons why the balance between modeling and empirical work has shifted toward the latter in recent years.

Many studies of sexual selection and mate choice don't actually measure direct fitness effects per se. Do you see this as a fundamental problem? Are there ways that more indirect measures can serve the same means?

Yes, this is a problem as there will always remain uncertainty about the total fitness effect of a trait in nature if only partial fitness effects have been studied, especially if only in the lab. This is one reason why long-term field studies that attempt to measure lifetime success over generations, such as the studies of red deer on Rhum and collared flycatchers on Gotland, are likely to remain important.

There have been thousands of studies of sexual selection in animals. What remains to be done?

I would be glad if I could give a hard-and-fast answer, but in my experience the formulation of a well-defined and interesting research question is half the job. When it is well formulated, the question often suggests what needs to be done, modeling-wise as well as empirically. Let me just say that I believe we are still some way from a general, empirically well-founded, broad understanding of the reasons for and consequences of mate choice. There is no lack of theoretical ideas, and there are many admirable empirical studies and exciting novel approaches and results, but critical testing is often difficult, and the hypotheses may not be mutually exclusive. Rather than refuting all but one hypothesis, the problem may involve estimating the relative magnitude of different mechanisms, which is often hard to do. But it should eventually be possible.

DR. MALTE ANDERSSON is a professor at the University of Gottingberg in Sweden. His early work on sexual selection in widowbirds is considered a classic study in the field, and his 1994 book Sexual Selection *(Princeton University Press) is regarded as the most important book written on this subject since Darwin's 1871 book* The Descent of Man and Selection in Relation to Sex *(J. Murray).*

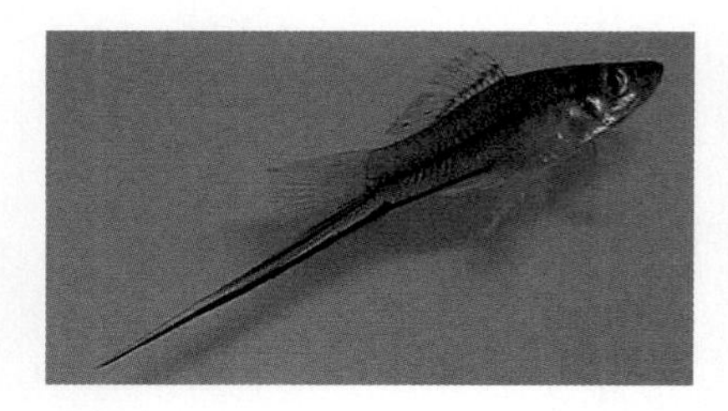

FIGURE 6.35. Artificial tails and sensory bias. In an attempt to test sensory bias models of sexual selection, Basolo used both swordtail fish (pictured here) and platyfish with an artificial sword sewed on. *(Photo credit: Ryan L. Earley)*

platyfish (*Xiphophorus maculatus*; Basolo, 1990, 1995). Female swordtails prefer males with long "swords" (a colorful elongation of the tail fin). Some male swordtails have long swords and some have short swords, but platyfish males lack swords completely. To experimentally examine sensory exploitation, Basolo sewed artificial (plastic) swords on platyfish males (Figure 6.35). In the treatment group of males, a yellowish sword was attached, and in a control group of males a clear, see-through sword was attached. Thus, the control males went through the same surgical procedure as those in the treatment group and had the same burden of swimming around with attached sword, but they had swords that were not visible to females (and were thus labeled "swordless").

Basolo examined how female platyfish responded to males in the treatment group as opposed to males in the control group. What she found was that female platyfish showed an immediate, strong, and consistent attraction to treatment males with newly acquired swords over the "swordless" males. This was despite the fact that there has been no evolutionary history of choosing males with swords in platyfish, since males in this species have never had swords. In fact, the ancestor to both swordtails and platyfish appears to have been swordless (Basolo, 1995; see Meyer et al., 1994, for a dissenting view). Basolo argues that female platyfish must have had a preexisting bias for long tails, because despite the fact that there had been no evolutionary history of choosing males with swords, this elaborate male trait was viewed as very attractive as soon as it appeared in the population. It thus appears that the neural system associated with vision in female platyfish has a built-in, preexisting bias for swords in males.

FIGURE 6.36. Frog calls and sensory bias. *Physalaemus pustulosus* males (one of whom is pictured here) add a unique "chuck" sound to the end of their advertisement calls. The calls of *Physalaemus coloradorum* males lack a whine. The calls of these two species have been used to test models of the sensory bias hypothesis. *(Photo credit: Michael Ryan)*

FROGS AND SENSORY BIAS

A second example of sensory exploitation comes from two species of frogs—*Physalaemus pustulosus* and *Physalaemus coloradorum* (Ryan and Rand, 1990; Ryan et al., 1990; Figure 6.36). Males in these species use advertisement calls to attract females and, as is the case for most *Physalaemus* species, males of both the *pustulosus* and *coloradorum* species begin such an advertisement with what is referred to as a "whine." *Pustulosus* males, however, are unique in that most add a "chuck" sound to the end of their call. When females choose between *pustulosus* males that chuck and those that don't, they prefer the former. The "chuck" part of the call is and has always been absent in the *coloradorum* species (that is, there is no evolutionary history behind chucks in *P. coloradorum*). Furthermore, as in the swordtail case, the ancestor of both *pustulosus* and *coloradorum* lacked a chuck as well. Yet, when modern acoustic gadgets are used to add a "chuck" sound to the end of *coloradorum* male calls, such calls are immediately preferred by females, suggesting that the auditory circuitry in these frogs was designed in such a way as to prefer a certain class of sounds—the ones that "chucks" fall into. Once again, the preference for a more elaborate call seems to have predated the actual appearance of the elaborate call, in accordance with the sensory bias hypothesis.

ZEBRA FINCHES AND SENSORY BIAS

Another fascinating possible example of preexisting bias comes from Nancy Burley's "leg band" experiments on zebra finches. Using both domesticated and natural populations of zebra finches (*Poephila guttata*), Burley found that the colored leg bands that she added to both male and female finches had significant effects on their fitness, even though the colors on the leg bands are not normally found on zebra finches. In a typical experiment, Burley would add red leg bands to one group of males, green leg bands to another group, and have a third group with no leg bands as a baseline control. Compared to controls, females showed a strong preference for red-banded males, and an aversion to green-banded males (Burley et al., 1982; Burley, 1988; Figure 6.37). Such preferences and aversions translate into red-banded individuals having greater differential reproductive success (Burley, 1986b) and lower mortality than green-banded (and control) individuals (Burley, 1985). Surprisingly, red-banded individuals also displayed lower levels of parental care and had broods with more same-sex offspring (Burley, 1986a).

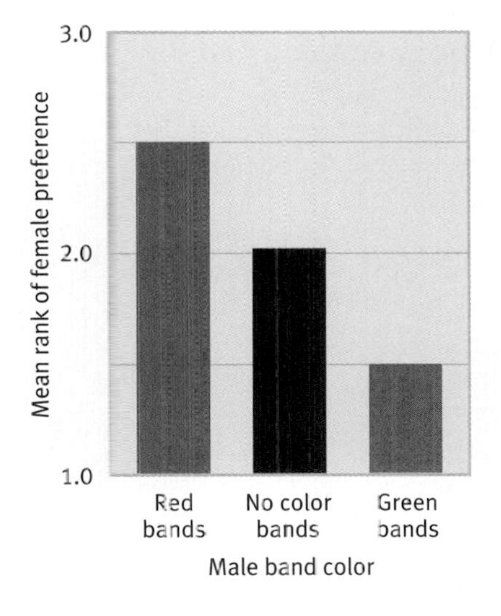

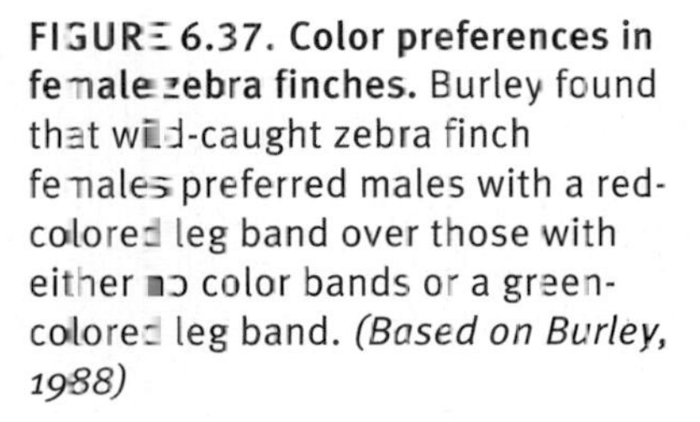
FIGURE 6.37. Color preferences in female zebra finches. Burley found that wild-caught zebra finch females preferred males with a red-colored leg band over those with either no color bands or a green-colored leg band. *(Based on Burley, 1988)*

Burley's work may not strictly qualify as sensory bias under some definitions, as we don't know the phylogenetic history of color preference in these birds and whether at some time in the past they may have had a preference for red. Nonetheless, this suggests that novel male traits can potentially tap into preexisting female biases in birds as well as in amphibians and fish.

SUMMARY

1. When developing his theory of sexual selection, Darwin outlined two important processes: intersexual selection and intrasexual selection.

2. Intrasexual selection involves competition among one sex, usually the males, for access to the other sex.
3. Intersexual selection involves mate choice, wherein individuals from one sex, usually the female sex, choose their mates from among members of the opposite sex.
4. Genetic models of female mate choice can be broken down into three classes: "direct benefits," "good genes," and "runaway selection" models.
5. Cultural transmission models of mate choice revolve around the issue of mate copying and address three related questions: What effect does female mate-choice copying have on the variance in male reproductive success? Under what circumstances will a copying strategy take over a population composed of non-copiers? How will female mate-choice copying affect the co-evolution of female preferences and the male trait that is the object of such preferences?
6. Sensory exploitation models of sexual selection argue that females may initially prefer male traits that elicit the greatest amount of stimulation from their sensory systems.

DISCUSSION QUESTIONS

1. Suppose that a group of males engaged in a series of fights and that male A emerged as the dominant individual. Now suppose that a female assessed all the males involved in fights and chose male A. Why might this example blur the distinction between intrasexual selection and intersexual selection?
2. Secure a copy Kirkpatrick and Ryan's 1991 paper "The evolution of mating preferences and the paradox of the lek" in *Nature* (vol. 350, pp. 33–38). Drawing from this paper, list the similarities and differences between sexual selection models in terms of both assumptions and predictions.
3. Sensory bias models of sexual selection examine the origins of female preference, not their subsequent evolution. Outline a system in which sensory bias establishes a preference, but direct selection or good genes models lead to the female preference evolving.
4. Why do you suppose it is so difficult to demonstrate mate-choice copying? Pick a species of your choice and design an experiment that would examine whether mate-choice copying is present. How many controls did you need to construct to rule out alternative hypotheses to mate-choice copying?

5. Pick any of the numerous dating game shows on television and watch a series of episodes. As you watch, imagine yourself as an ethologist studying mate choice. What sort of traits do males prefer in females? What traits do females prefer in males? Can you say anything about how your observations match up against current models of sexual selection?

SUGGESTED READING

Andersson, M. (1994). *Sexual selection*. Princeton: Princeton University Press. This book is considered the modern reference guide for work on sexual selection.

Darwin, C. (1871). *The descent of man and selection in relation to sex*. London: J. Murray. The book that started it all. It was in this book that Darwin laid out his ideas on "sexual selection."

Fisher, R. A. (1915). The evolution of sexual preference. *Eugenics Review*, 7, 184–192. One of the earliest models of the process of sexual selection.

Houde, A. E. (1997). *Sex, color and mate choice in guppies*. Princeton: Princeton University Press. A book-length case study of sexual selection in a model system: the guppy (*Poecilia reticulata*).

Kirkpatrick, M., & Ryan, M. (1991). The evolution of mating preferences and the paradox of the lek. *Nature*, 350, 33–38. This *Nature* article provides a nice view of the contrasting models in the sexual selection literature.

7

Different Mating Systems

- Monogamous Mating Systems
- Polygamous Mating Systems
- Promiscuous Mating Systems

Choosing Polygyny

- Polygyny and Resources
- The Polygyny Threshold Model

Surreptitious Promiscuity

- Extrapair Copulations
- Sperm Competition

Multiple Mating Systems in One Population?

- Dunnocks

Interview with Dr. Nick Davies

Mating Systems

Every few years in the United States a major headline notes, "Man Married to Two Women at Once." In the United States, of course, the law dictates a system of monogamy, wherein an individual can be married to only one person at a time. The controversy arises when religious dictates suggest otherwise. For example, while most modern-day Mormons do not support polygyny—where a male is married to numerous women at the same time—there are some who do have more than one wife because they hold that polygyny is allowed under their religious code (Figure 7.1). Controversy then erupts, partly for political and religious reasons, and partly because people are intensely interested in mating systems.

Mating systems can occur in many forms and gradations. In fact, in the above case, it is a bit disingenuous to say that U.S. law dictates monogamy. In fact, in terms of the mating systems literature, the United States can be characterized as a serially monogamous society. Because of the high divorce rate in the United States, many individuals will in fact be married to more than one person during their lifetime (serial monogamy), just not at the same time.

Different Mating Systems

Before examining a series of questions regarding the evolution of mating systems, let us do a quick survey of different forms that mating systems may take. In order to do so, we begin by tackling one definitional issue. There has been some debate in the mating systems literature as to whether "pair bonds" be-

FIGURE 7.1. Human polygyny. In many human cultures, polygyny is an acceptable form of pair bonding. This polygynous Mormon man is surrounded by his many wives and children. *(Photo credit: Nik Wheeler/Corbis)*

TABLE 7.1. Mating combinations. Various mating combinations and how they map onto potential mating success for males and females. *(From Davies, 1992, p. 29)*

	MATING SUCCESS	
MATING COMBINATION	FOR A MALE	FOR A FEMALE
Polyandry (e.g., 2 ♂ 1 ♀)	Share one female	Sole access to several males
Monogamy (1 ♂ 1 ♀)	Sole access to one female	Sole access to one male
Polygynandry (e.g., 2 ♂ 2 ♀)	Share several females	Share several males
Polygyny (e.g., 1 ♂ 2 ♀)	Sole access to several females	Share one male

tween males and females are important in classifying mating systems (Emlen and Oring, 1977; Thornhill and Alcock, 1983; Davies, 1991). Here we shall avoid having our broad-scale classification of mating systems depend on identifying pair bonds, and instead simply use the number of copulatory partners as a critical variable in defining our mating systems (Figure 7.2). Occasionally, however, when subdividing mating systems, it will be helpful to examine the role of pair bonding, and we shall do so when needed.

We now turn to a general classification of mating systems (see Table 7.1).

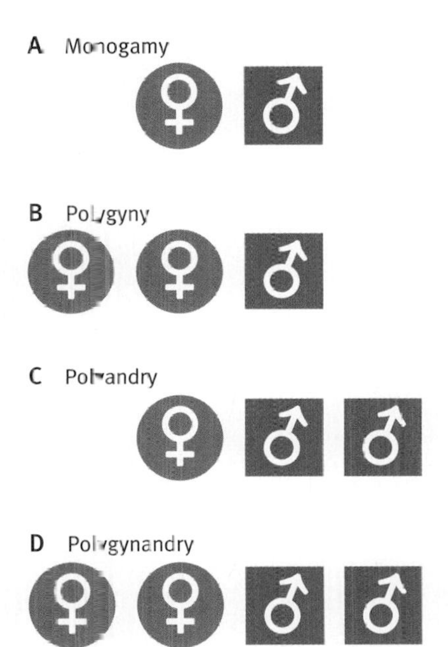

FIGURE 7.2. Four basic mating systems. These are: **(A)** monogamy (1 male, 1 female), **(B)** polygyny (1 male, more than 1 female), **(C)** polyandry (1 female, more than 1 male), and **(D)** polygynandry (more than 1 male, more than 1 female). Each mating system can be further subdivided in a number of ways.

MONOGAMOUS MATING SYSTEMS

A **monogamous mating system** is normally defined as a mating system in which a male and female pair mate only with one another during a given breeding season. As such, we can have animal societies in which a male and female mate only with each other during season 1, but in subsequent years have new mates (serial monogamy). In many territorial species, for example, a given male and female will mate exclusively with one another during the breeding season, but then they will switch mates during the next breeding season. Although it is comparatively rare, in some populations, a given male and female mate only with one another during their entire life span (or more precisely, the life span of the individual that

FIGURE 7.3. Long-term monogamy in the oldfield mouse. From data on 500 oldfield mouse burrows, Foltz found that 90 percent of the offspring in a family group were fathered by the male in their burrow. Behavioral observations also suggest that many females remained with the same mates across litters, suggesting long-term monogamy. *(Photo compliments of the Peromyscus Genetic Stock Center and Clint Cook)*

dies earliest). As an example of this form of lifetime monogamy, let us look at the oldfield mouse, *Peromyscus polionotus* (Figure 7.3).

Early work on monogamy suggested that it was relatively rare in mammals (Eisenberg, 1966). In the early 1980s, Foltz (1981) suggested that, based on the available data, monogamy in mammals was less common in large, conspicuous, diurnal species (those active in the daytime) and that monogamy was more common in smaller diurnal groups, particularly rodents. Indeed, behavioral work suggested that monogamy was fairly common in rodents (Kleiman, 1977), but the genetic evidence for monogamy in rodents was much weaker. To remedy this dearth of genetic support for monogamy in rodents, Foltz studied the breeding system of the oldfield mouse. Oldfield mice were known to have significant amounts of genetic variation (Selander et al., 1971) and were easily collected by excavating burrows.

Foltz excavated more than 500 oldfield mouse burrows, captured the individuals in each burrow, and then brought them into the laboratory. One hundred and seventy-eight families were collected from these burrows, and genetic analysis (in this case, a technique called gel electrophoresis) was conducted. From a subset of these families, Foltz calculated that 90 percent of the offspring found in a family group were fathered by the male in their burrow (Figure 7.4). Furthermore, behavioral observations of both males and females uncovered that most females remained with the same mates across litters, suggesting long-term monogamy to be the rule, rather than the exception, in this species. That being said, there may be a small proportion of females who adopt a different mating strategy in *P. polionotus*. Approximately 13 percent of the burrows excavated contained only an adult female (in most burrows there were both an adult female and an adult male). When this small sub-

FIGURE 7.4. Who fathers whom? In the oldfield mouse (*Peromyscus polionotus*), the male found in the burrow, on average, fathers 90 percent of the pups in that burrow.

set of females without live-in mates was analyzed, they were much more likely to switch mates across litters when compared to females that were already paired with a mate in the field.

POLYGAMOUS MATING SYSTEMS

A polygamous mating system is typically defined as one in which either a male or female has more than one mate during a given breeding season/cycle. Polygamy includes **polygyny**, wherein males mate with more than one female per breeding season, and **polyandry**, in which females mate with more than one male per breeding season. Polygamy can be also be subdivided temporally, in that polygamy can be simultaneous or sequential. Simultaneous polygamy refers to the case where an individual interacts with numerous mating partners in the same general time frame, whereas sequential polygamy involves individuals forming many short-term pair bonds in sequence during a given mating season.

One important thing to keep in mind with respect to polygamy is that this type of breeding system increases the variance in reproductive success in the sex that has more than one mate per season (see Chapter 6). For example, in polygynous systems there is often intense competition to mate with as many females as possible, and this typically produces a distribution of mating success in which a few individuals do extraordinarily well, but many males obtain no mates whatsoever (Figure 7.5). The converse holds true for the case of polyandry, although variance in reproductive success is usually less dramatic in this situation.

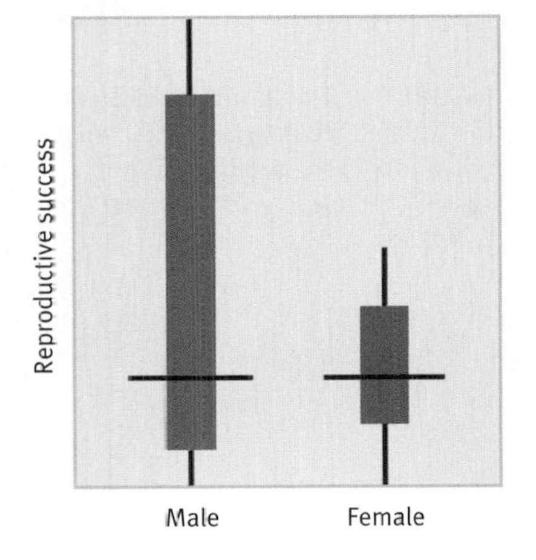

FIGURE 7.5. Variance and polygyny. Since each and every successful mating involves a male and a female, the average reproductive success must be equal in both sexes. In polygynous mating systems, however, the variance in reproductive success is often much higher in males than in females, with some males having very high reproductive success and others faring much more poorly (horizontal lines represent average reproductive success). *(Based on Low, 2000, p. 55)*

We shall return to the evolution of polygamy later in the chapter, but for the time being, let's examine one case of this very common mating system: the small Australian wasp in the aptly named genus *Epsilon* (Smith and Alcock, 1980). In this species, males mature earlier than females. They then proceed to search for unopened brood cells from which females will emerge. Such opened cells are often clustered together, and males are very territorial once they uncover such a potential gold mine of reproductive opportunities that can economically be defended as a unit. Once a virgin female emerges from her cell, the closest male present climbs on her back and mates with her. The rewards to males who guard against intruders may be mating opportunities with as many as two dozen virgin females. The mating system of these wasps is referred to as female defense polygyny.

Thornhill and Alcock (1983) argue that three characteristics are often present in insects displaying female defense polygyny: (1) females are short-lived and have low fecundity, receiving all the sperm they will ever use from a single male, (2) females mate shortly after becoming adults, and (3) females are grouped close together in space. These three characteristics apparently make female defense polygyny adaptive for males. Its relative effect on female fitness is less well understood.

Compared to polgyny, polyandry is not as common. In birds, polyandry takes two basic forms. In cooperative polyandry, several males will often defend a female's territory and mate with her. In resource defense polyandry, females themselves defend large multipurpose territories that include the small territories of a number of males. In birds, at least, both these forms of polyandry are often associated with some form of male paternal care of offspring (Davies, 1991).

FIGURE 7.6. Polyandrous birds. In the wattled jacana *(Jacana jacana)*, polyandry is the typical mating system. *(Photo credit: Ann and Rob Simpson)*

Polyandry has been particularly well studied in jacanas—a group of sex-role reversed shorebirds. Stephen Emlen and his colleagues studied one of these species, the wattled jacana (*Jacana jacana*; Figure 7.6) in Panama (Emlen et al., 1998). While polyandry in jacanas is often cooperative (Jenni and Collier, 1972; Tarboton, 1992), with females nesting with numerous males at one time, wattled jacana females display resource defense polyandry. In this species, males have small territories (40 meters in diameter) that abut one another. Female territories are considerably larger and contain anywhere from one to four male territories.

In the wattled jacana, a female lays clutches of eggs sequentially, after mating with males on her territory, with intervals of less than two weeks often separating the production of sequential clutches. Given that a female's territory encompassed a number of male territories, it was rather surprising that the probability was very high that a chick in a nest on a particular male's territory was his genetic offspring. That is, given the possibility that a female was also engaging in copulations with other males in her territory, the fact that there was a 92.5 percent probability that a given chick in a nest was the offspring of the male guarding that nest was rather remarkable (Oring et al., 1992, Owens et al., 1995; Delehanty et al., 1998; Emlen et al., 1998). Not surprisingly, this figure went down dramatically when males in surrounding nests were not incubating eggs or tending young chicks.

Although it is not relatively common compared to other mating systems, polyandry also occurs in other species besides birds. For example, in social insects, a single queen will often mate with many worker males. Compared to nests where a single male mates with a queen, polyandrous nests often have greater levels of within-group conflict. The reason that there is less within-colony harmony in the nest of polyandrous queens is the presence of numerous "patrilines," that is, offspring descended from different fathers. When polyandry is absent, all workers have the same mother and father, and hence the same genetic interest. With the creation of

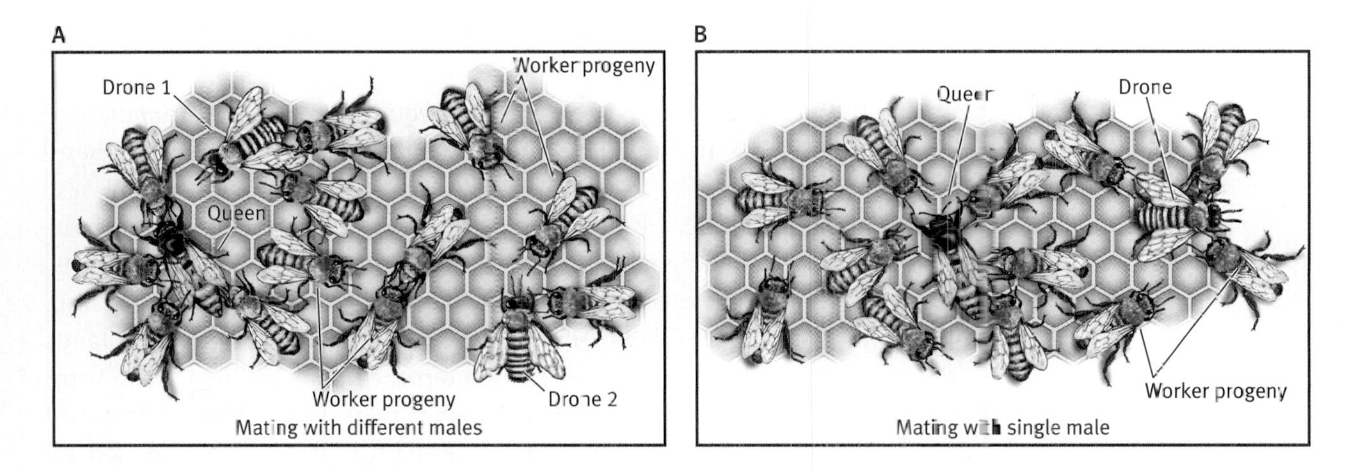

FIGURE 7.7. Polyandry and within-hive harmony. (A) When a queen mates with two different drones (denoted by red and blue), two patrilines (red and blue) exist in a single hive. Such hives have greater within-hive conflict than those in which **(B)** a queen mates with a single male (red).

patrilines, genetic interests are more divergent, with each patriline competing with the others for greater representation in the next generation (Seeley, 1995, 1997; Figure 7.7).

Thornhill and Alcock (1983) list a number of possible benefits that female insects accrue by adopting a polyandrous mating strategy. These benefits include:

1. Sperm replenishment
   - Female adds to depleted or low sperm supply
   - Female avoids the cost of storing sperm
2. Material benefits
   - Nutrients
   - Reduced predation
   - Protection from other males
3. "Genetic benefits"
   - Replacement of "inferior" sperm
   - Increased genetic variance in offspring
4. Convenience
   - Female avoids the costs of fending off copulation attempts by male

Of these, Thornhill and Alcock believe that material benefits account for most of the polyandry we see in insects, with such benefits ranging from the beneficial chemicals in the seminal fluid, nutritious spermatophores (sperm-encompassing packets filled with nutrients and produced by male insects such as moths and butterflies), "nuptial gifts" (as in flies and wasps), superior feeding sites and oviposition (egg-depositing) sites (as in dragonflies and dungflies), and male parental care in some waterbugs.

PROMISCUOUS MATING SYSTEMS

When polyandry and polygyny are occurring in the same population of animals, the breeding system is said to be **promiscuous**. There are, however, two very different kinds of promiscuity, which vary dramatically as a function of the presence or absence of pair bonds between mating individuals. In one form of promiscuity, both males and females mate with many partners and no pair bonds are formed. For example, a male may defend a territory that contains food, and females may visit such territories, obtain food, mate with a male, and then repeat this sequence many times (Davies, 1991).

The type of promiscuity outlined above need not be tied to male territoriality. In many primate species, when females are in estrous, both males and females mate repeatedly, often in rapid succession, with many opposite-sex partners. Take the case of the Barbary macaque (*Macaca sylvanus*; Taub, 1980, 1984). During their estrous period, female Barbary macaques "seek out new males after each ejaculatory copulation and do not allow any male sexual monopolization" (Whitten, 1987, p. 345). At the same time, of course, males also have numerous female sexual partners, hence producing a promiscuous breeding system.

In the second type of promiscuous breeding system, labeled **polygynandry**, several males form pair bonds with several females simultaneously. For example, in the dunnock (*Prunella modularis*), pairs of males will often jointly defend the territory(ies) of a pair of females (Figure 7.8). Nick Davies studied the dunnock mating system and the help provided by males to females. He observed cases in which one or two males were present with a female and found that females either received help from one, both, or neither of the

FIGURE 7.8. Versatile dunnocks. The dunnock mating system is extremely versatile and includes monogamy, polyandry, polygyny, and polygynandry. Here we see **(A)** a female with a newly hatched offspring (just visible in egg), and **(B)** a female feeding her brood. *(Photo credits: Nick Davies; W.B. Carr)*

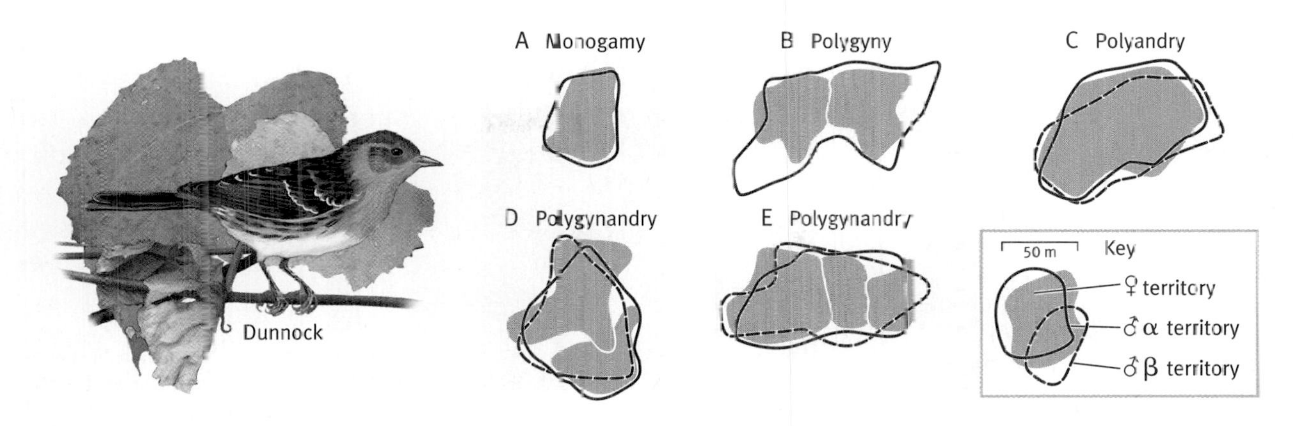

FIGURE 7.9. Incredible variation in dunnock breeding systems. Female territories are shown in green, while alpha male territories are depicted by solid red lines and beta male territories are shown by dashed red lines. In a single dunnock population, we can find mating systems from monogamy to polygamy, polyandry, and polygynandry. *(From Davies, 1992, p. 27)*

males present, but that on average, polygynandrous females received the equivalent help of one male (Davies, 1992; Figure 7.9). The more help the female received, the higher the mean nestling weight of the chicks in the brood and the lower the chick mortality rate due to starvation (Davies, 1986).

Margie Profet added an interesting new dimension to the study of promiscuous breeding by tying this type of breeding system to female reproductive health in mammals (Profet, 1993). Her basic argument goes as follows: Sperm are known to be vectors of various diseases (Ewald, 1994), some of which are quite dangerous to females. During insemination, various bacteria and other disease-causing agents from both male and female genitalia hitch a ride on sperm tails and work their way to the uterus.

Menstruation, Profet argues, is a defense that has evolved in females to rid the female reproductive tract of pathogens carried in by sperm. As such, menstruation should be most common in promiscuous breeding systems, as the probability of disease transfer via sperm should be highest in systems where both males and females engage in sexual activity with numerous partners (Figure 7.10). Promiscuous breeding systems not only expose females to greater quantities of sperm, but to a greater diversity of sperm, and hence to diseases that use sperm as vectors, both of which should increase a female's risk of infection. Furthermore, promiscuous breeding systems, in and of themselves, appear to favor the *evolution of* more virulent forms of a disease (Ewald, 1994) making the situation even more dangerous from the female perspective. While Profet suggests that a qualitative test of the data on breeding systems and menstruation in primates supports her hypothesis, a more quantitative analysis by Strassmann (1996) did not find any link between promiscuity and copiousness of menstruation in primates.

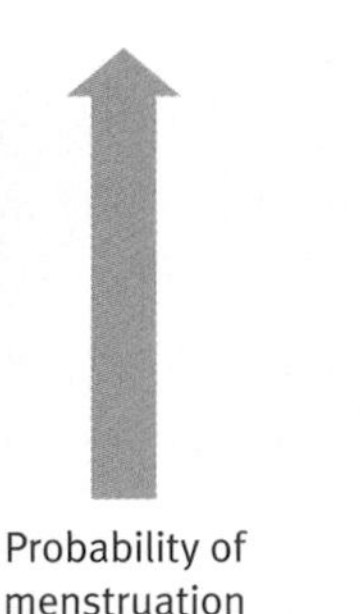

FIGURE 7.10. Sperm, breeding systems, and menstruation. Margie Profet has argued that promiscuous breeding systems deliver not only more sperm to females but a greater diversity of sperm (denoted by different colors). As the probability of sperm-transported disease increases, Profet predicts that menstruation should become more common.

Choosing Polygyny

With a survey of different mating systems in hand, we can now move on and address questions about the ecology and evolution of these different mating systems. We will first look at the role of resource dispersion on female mating decisions and then at a model to predict when polygyny is selected.

POLYGYNY AND RESOURCES

The question of whether a male should have one or many mates has long been tied to the question of dispersion of the sexes (Bradbury and Vehrencamp, 1977; Emlen and Oring, 1977). The basic argument goes as follows (Krebs and Davies, 1993): A female can often fertilize all her available eggs by mating with one or a very few males. That being the case, female fecundity is not so much tied to the availability of mates as it is to the availability of resources, such as food, defense, and so forth. The more resources available, the more offspring females can produce. Males, on the other hand, can potentially fertilize many, many females, and so their reproductive success is tied more to access to females than to access to resources. As such, female dispersion patterns should track the distribution of resources, whereas male dispersion patterns should track the dispersion of females.

If females track resources and males track females, then the mating system in a population is in fact intimately tied to the distribution of resources, as this distribution will affect whether males can defend more than one female at a time. If resources are dispersed fairly homogeneously, and/or if females must cover large areas to obtain enough resources to survive, it may not be possible for males to mate with and potentially defend more than one female. If resources are clumped, the economics shift, and males may be able to mate with and defend a number of females at once. For example, seal populations can be found aggregating on both ice packs and on beaches. On ice packs, females are fairly widely dispersed, and males typically guard and mate with anywhere from a single to a few females. When the seals are on land, females cluster in particularly safe areas on the beach, and males are able to defend herds of females against the approach of other females (Le Boeuf, 1978).

The effect of female dispersion on male mating patterns can be examined experimentally, as well as observationally. Take the case of the mating system of the grey-sided vole (*Clethrionomys rufocanus*). Ims (1987) found that when food distribution was experimentally manipulated, female dispersion patterns changed in the predicted manner—when resources became more clumped, so did females. Ims (1988) then undertook an ambitious experiment in which he introduced the grey-sided vole to a small island in Norway. In one treatment, Ims used caged female voles to simulate an environment in which females moved about a home range and were fairly spaced out. In the other treatment, caged females were clustered together. As a control, the reverse experiment was run in which males were in cages and their distribution was manipulated. As predicted, males tracked the distribution of caged females across treatments, but female dispersion was unaffected by the distribution of caged males.

THE POLYGYNY THRESHOLD MODEL

The idea that females often track resources and males track females does not mean that decisions regarding polygyny are strictly in the hands of males. For example, in systems where males have territories that house numerous mates simultaneously, females will often have to decide which territory they settle on by choosing between alternative male territories. Gordon Orians (1969) created the **polygyny threshold model (PTM)** to predict the behavior of females in just such a scenario (also see Verner and Willson, 1966; Downhower and Armitage, 1971).

HOW THE PTM WORKS. To see how the PTM works, imagine that males have settled on ten territories and that these territories vary with respect to some resource valuable to females. Let's say the resource in question is food, with territory 1 representing the area with the most food, and territory 10 that with the least food. The first female arriving to choose a territory can base her choice simply on what territory is optimal with respect to food intake, and so she should choose territory 1. A second female choosing between territories now faces the same landscape as the first, except that territory 1 is now occupied by a male and a female. If she chooses territory 1, she will only get some fraction of the food available there. If the food still available on that territory is greater, even though another female is present, than that available on any of the other nine territories, our second female should choose to settle there anyway. In so doing, she has passed the "polygyny threshold"—that is, she opts to stay on a territory with another female and thus be part of a male's polygamous relationship. Let's say, however, that the second female settles on territory 2 because the polygyny threshold has not been met.

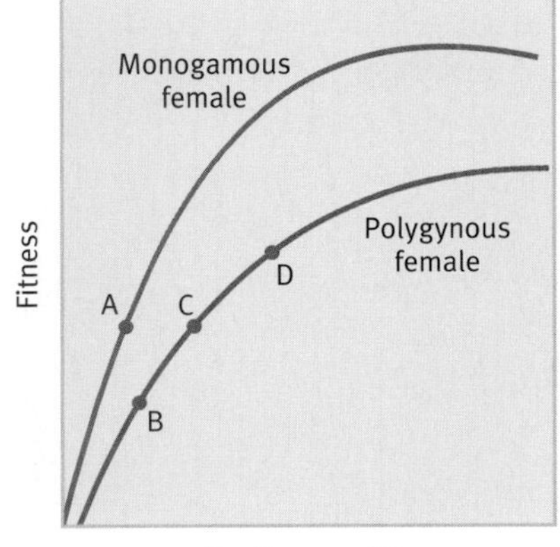

FIGURE 7.11. Female choice of territories. Imagine a female deciding among territories. She should choose the territory with the highest quality, that is, with the most food available or the most shade and so on, regardless of whether she would be the lone female (monogamous territory) or one of several females (polygynous territory), because it would provide her with the greatest fitness. If she had to choose between monogamous territory A and polygynous territory B, our female should choose to settle on territory A because she would have a higher fitness there. If polygynous territory D was available, however, she should opt for territory D over territory A, for now the polygynous territory (D) would provide her with greater fitness. If she had to choose between polygynous territory C and monogamous territory A, she should be equally likely to choose territory A or territory C, for each would provide her with the same fitness.

Our third female now comes in and chooses between territories 1 (other female present), 2 (other female present), and 3–10 (no females present). Her situation, however, is slightly different from that of female 2 when she arrived because the best open territory is now territory 3, and that isn't as profitable as territory 2. If female 3 chooses to settle on territory 1 or 2, then the polygyny threshold has again been crossed. This sort of logic can be applied to all subsequent females and allows us to make a few clear-cut predictions with respect to territory settlement (see also Brown, 1982, for a more general model of optimal group size in territorial animals). One such prediction is that polygyny should often occur in "patchy" environments, where males can defend valuable resources. In addition, the quality of a male's territory will affect his mating success, and when females settle about the same time, those on monogamous and polygamous territories should have about equal fitness (Krebs and Davies, 1987). It is this equivalency of fitness among monogamous and polygamous birds settling at the same time that makes the model stable. Because monogamous and polygamous birds settle in such a way as to produce approximately equal fitness, there is no temptation for females to move around once this state has been reached, as any such move would in fact lower an individual's reproductive success (Figure 7.11). Nonetheless, there are some limitations on the model (Borgerhoff-Mulder, 1990). These include the following:

- A female's reproductive success is affected not only by her choice, but also by the choice of others that *follow* her (Altmann et al., 1977).
- The PTM does not consider how factors independent of mating status affect female reproduction (Garson et al., 1981).
- The model assumes females have reliable information on territory quality and male mating status.
- Assessing territory quality is assumed to be a very low-cost activity (Christie, 1983; Slagsvold et al., 1988).

- The PTM makes no accommodation for the possibility that females on the same territory cooperate in some contexts (Altmann et al., 1977; Davies, 1989).
- The possibility that female aggression may curtail free choice is absent from the PTM (Lenington 1980).

THE PTM AND MATE CHOICE IN FEMALE BIRDS. One of the first experimental tests of the basic predictions of the PTM was Pleszczynska and Hansell's (1980) work with lark buntings (*Calamospiza melanocorys*). In this species, the resource that primarily determines settlement onto territories is shade cover. As the main cause of nestling mortality in lark buntings is overheating, the more shade on a territory, the better the territory. This can be shown experimentally by artificially increasing the shade in a given territory and demonstrating, as did Pleszczynska and Hansell, that such a treatment increases nestling survival.

When choosing a territory, female lark buntings often have a difficult decision to make. Since shade protection is such a critical resource, it will often be the case that the territories that are best suited to provide shade have already been settled by other females (Figure 7.12). Becoming a "secondary" female (Krebs and Davies, 1987) allows access to shade, but males only provide aid in rearing nestlings to a primary female—that is, the first female to breed on the male's territory. Despite this, not only did

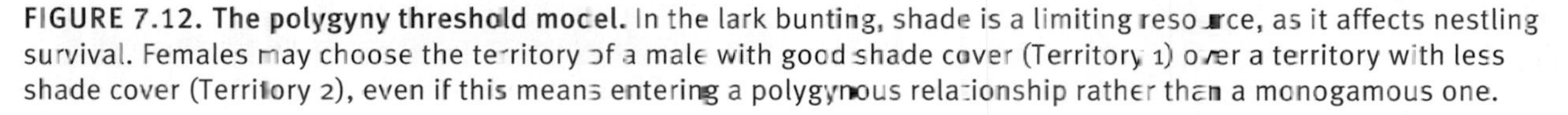
FIGURE 7.12. The polygyny threshold model. In the lark bunting, shade is a limiting resource, as it affects nestling survival. Females may choose the territory of a male with good shade cover (Territory 1) over a territory with less shade cover (Territory 2), even if this means entering a polygynous relationship rather than a monogamous one.

Pleszczynska and Hansell find that increasing shade availability on a territory made it much more popular, and that as the PTM predicts, secondary females, who often sacrificed male help on chick rearing for a shady area, had about the same reproductive success as did monogamous females who breed on territories with less shade cover.

FIGURE 7.13. Induced polygyny. Although prothonotary warblers are typically monogamous, polygyny can be experimentally induced. *(Photo credit: Ann and Rob Simpson)*

Rather than choosing a species that often displays polygyny, Petit (1991) worked with the normally monogamous prothonotary warbler (*Protonotaria citrea*) and attempted to induce polygyny—that is, to shift some females over the polygyny threshold (Figure 7.13). The warblers Petit examined have rarely been seen to stray from monogamy in most populations (Ford and McLauglin, 1983). Yet, despite this, twenty-three cases of polygyny occurred in Petit's long-term study. The question for Petit was why?

In her initial experimental manipulation, Petit supplemented male territories with extra nest boxes to examine whether such a manipulation might push some females over the polygyny threshold. Prothonotary warblers are "hole nesters," and as such, supplemental nest boxes should increase territory quality. Indeed, once such nest boxes were in place, numerous cases of polygamy were recorded (Figure 7.14). To see the pattern of polygamy, we must first recognize that habitat quality in the population Petit studied displayed marked spatial variance—that is, territories in the flooded plains were generally much preferred to territories in more inland areas. When examining the distribution of polygamy, it was found that all polygamous matings occurred on the flooded plains—that is, supplemental nesting allowed the polygyny threshold to be crossed in this habitat, but not in the inland habitats. In a logical follow-up experiment, Petit (1991) added nest boxes to *specific* male territories on the flooded plains and found that males on

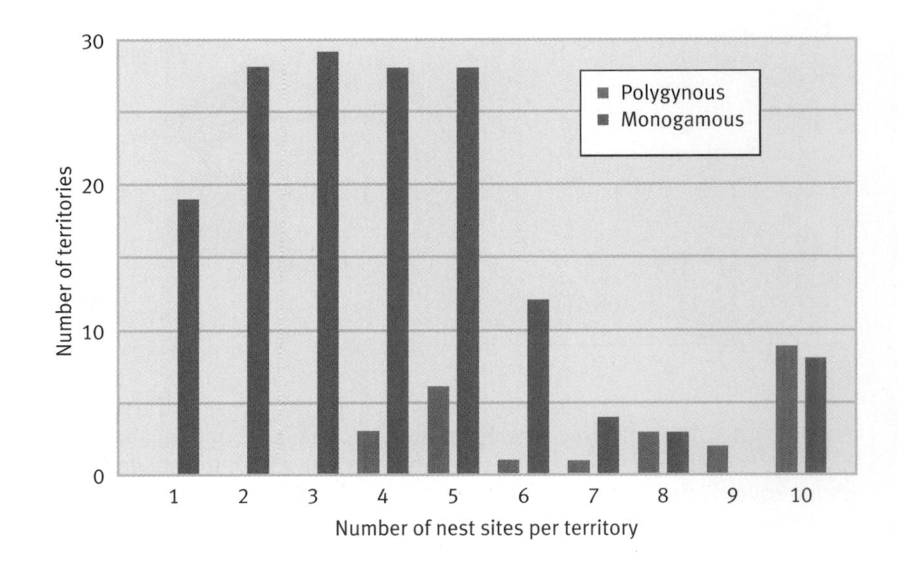

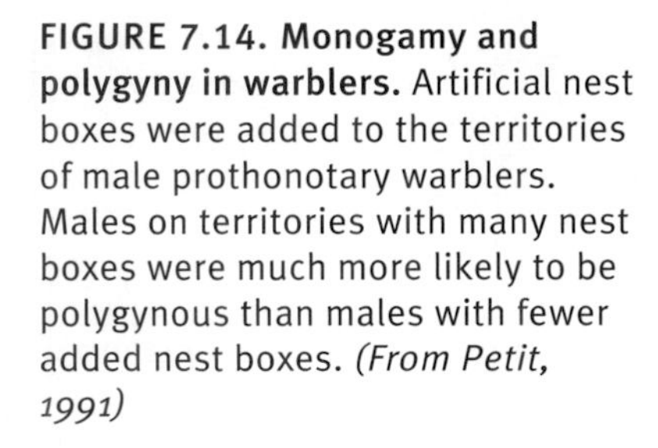

FIGURE 7.14. Monogamy and polygyny in warblers. Artificial nest boxes were added to the territories of male prothonotary warblers. Males on territories with many nest boxes were much more likely to be polygynous than males with fewer added nest boxes. *(From Petit, 1991)*

TABLE 7.2. Reproductive success and mating systems. The reproductive success of monogamous female warblers was compared to that of females that had a polygamous male. A primary female was almost always defined as the first female to breed on the territory of a polygamous male. *(From Petit, 1991)*

	MONOGAMOUS	PRIMARY	SECONDARY
Laying date	26 May	28 May	9 June
Overall nesting success			
Clutch size	4.0 ± 0.9	4.1 ± 0.4	4.1 ± 0.9
Number of young fledged	2.1 ± 1.9	2.9 ± 1.9	2.0 ± 1.9
Percent eggs fledged	51 ± 46	68 ±45	50 ±47
Nest success excluding predation			
Clutch size	4.2 ± 0.9	4.2 ± 0.4	4.1 ± 0.9
Number of young fledged	3.6 ± 1.3	4.0 ± 0.5	2.4 ± 1.8
Percent eggs fledged	36 ± 27	96 ± 10	61 ± 44
Fledging weight (g)	11.5 ± 0.8	11.7 ± 0.7	10.8 ± 0.9

such territories were much more likely to be polygamous. The general pattern appears to be that female warblers settle in the flooded plains first and only then will some females settle in the dryer inland areas. Once the flooded plain male territories all have a female resident, new females are faced with the option of monogamy on inland territories or polygamy in the flood plains.

Given that for some females the polygamy threshold is reached, we can ask whether the success of females on polygamous territories is in fact in line with that of monogamous females settling at approximately the same time. While Petit's data are not presented in a way to make this exact comparison, it does appear that, once predation is taken into account, polygamous flood plain females—that is, both the original female to settle on a territory and the "secondary" female who subsequently settled on the territory—are about as successful in terms of chick production and survival as are their inland monogamous counterparts (Table 7.2).

THE PTM AND HUMAN FEMALE MATE CHOICE. While much of the experimental work on the polygyny threshold model has occurred in birds (for example, Pleszczynska and Hansell, 1980;

Quinney, 1983; Bjorklund and Westman, 1986; Dhondt, 1987; Searcy and Yasukawa, 1996), Monique Borgerhoff-Mulder, an evolutionary anthropologist, has applied this model to mate choice in tribal humans (Borgerhoff-Mulder, 1990). On the surface, a test of the PTM in human female mate choice is reasonable, as wealth is still usually inherited through males (Whyte, 1980) and males who control resources often acquire lots of mates (Dickemann, 1979; Hartung, 1982; Irons, 1983; Betzig, 1986; Flinn and Low, 1986; Betzig, Borgerhoff-Mulder, and Turke, 1988). Borgerhoff-Mulder looked at polygyny in the Kipsigi tribesmen of Kenya, and she tested the basic prediction that when females are choosing between males with territories that differ in resource value, there should be a positive correlation between territory quality and female preference for males on superior territories. This sort of "settlement" experiment has been undertaken in redwinged blackbirds (Lenington, 1980) and pied flycatchers (Alatalo et al., 1984). Several studies have also tested the PTM in other human societies (Chisholm and Burbank, 1991; Roskaft et al., 1992; Perusse, 1994; Hames, 1996; Ezeh, 1997; Strassmann, 1997).

One consequence of the European colonization of parts of southwestern Kenya was the emigration of a pioneer group of Kipsigis (Figure 7.15) who moved to areas traditionally associated with the Masai tribe. Borgerhoff-Mulder obtained the settlement records of this migration and was able to determine when males settled on specific territories, how large each territory was, how many wives a male had at settlement time, and how many wives he obtained after settlement (Table 7.3). Somewhat surprisingly,

FIGURE 7.15. Polygynous Kipsigis. In the Kipsigi population, polygyny is the norm. Kipsigi "co-wives" both relax **(A)** and work **(B)** together. **(C)** A man with many wives will have numerous huts, grain stores, strips for maize, and so on. Here we see the plot of a man with three wives. *(Photo credits: Monique Borgerhoff-Mulder)*

TABLE 7.3. Polygyny in Kipsigi tribes. Kipsigi males are often polygynous. The number of wives a male obtains *after* settling in a new territory (pioneer wives) is a function of territory size. *(From Borgerhoff-Mulder, 1990)*

YEAR OF ARRIVAL	HUSBAND	SIZE OF PLOT SETTLED	PREVIOUS WIVES	PIONEER WIVES	HUSBAND'S TOTAL WIVES
1930	O	150	3	1	4
1932	D	100	2	2	4
	A	160	3	1	4
1933	C	100	0	2	4
	B	300	0	6	8
1935	F	50	2	1	3
1936	E	30	0	2	3
1937	I	70	2	1	4
1938	S	100	4	0	4
1939	H	60	1	1	2
	G	120	0	3	3
1940	P	60	1	1	3
1942	L	40	1	1	4
1943	J	130	0	2	4
	K	32	2	1	3
1944	M	40	1	1	3
	T	20	2	0	3
1945	U	20	2	0	3
1946	V	8	1	0	2
	N	36	1	1	2
1947	W	20	4	0	4
	X	20	2	0	3
1949	Q	29	0	1	2
	R	20	1	1	3
	Y	10	1	0	1

the number of wives a male had prior to settlement did not affect his subsequent territory size, suggesting that a male's prior success did not determine his territory size. That being said, the number of wives a male obtained subsequent to territory settlement was strongly affected by the size of his territory—with larger territories came an increase in the number of wives, as predicted by the PTM (Borgerhoff-Mulder, 1988, 1990). Female Kipsigis did pay a cost of polygyny, however, in that there was a negative relationship between the number of wives a male had and female reproductive success, even when controlling for territory size. This cost to polygyny may help explain why, in addition to preferring males with large territories, women also preferred bachelors to married men, all else being constant. As such, the ideal Kipsigi male was unmarried and in possession of a high-quality resource.

Surreptitious Promiscuity

Whereas some species are characterized by polygynous mating systems, others engage in "surreptitious promiscuity," that is to say, some individuals in species that are thought to be monogamous actually have been shown not only to mate with their partner but also to surreptitiously mate with others during the same mating season.

EXTRAPAIR COPULATIONS

In the early 1980s, ornithologists interested in bird mating patterns were starting to record more and more instances of what is now known as **extrapair copulations**, or **EPCs** (Ford and McLauglin, 1983; McKinney et al., 1984). Most species of birds nest during their breeding season, and some sort of pair bond is formed between a male and female at the nest. It was often thought that such pairings were truly monogamous, in the sense that partners in a given pair mated only with their nesting partner during a breeding season. More and more, however, it was discovered that males and females were leaving their territories and mating with other individuals, usually those in nearby territories, during the mating season. The occurrence of such EPCs prompted some ethologists to make a distinction between social monogamy and genetic monogamy. Most bird species in which EPCs were recorded formed pair bonds with just a single partner during a mating season, and as such displayed what is referred to as social pair bonding. Genetically, however, these systems resembled promiscuity more than monogamy, as matings occurred both with the partner and with other individuals during the mating season.

Among those studying mating dynamics in birds was David Westneat. Westneat found that behavioral observations in indigo buntings (Figure 7.16) suggested that about 13 percent of all matings were EPCs (Westneat, 1987b). Westneat was worried, though, that such observations might underestimate the actual percent of all offspring that were sired via extrapair interactions. This worry stemmed from the fact that the buntings were hard to observe (and hence a significant number of EPCs may have been overlooked) and because it was not clear how extrapair *matings* translated into extrapair *fertilizations*. For example, indigo bunting females generally resisted EPCs to a greater extent than mating with their nesting partner, and so not all EPCs would result in offspring (Westneat, 1987b).

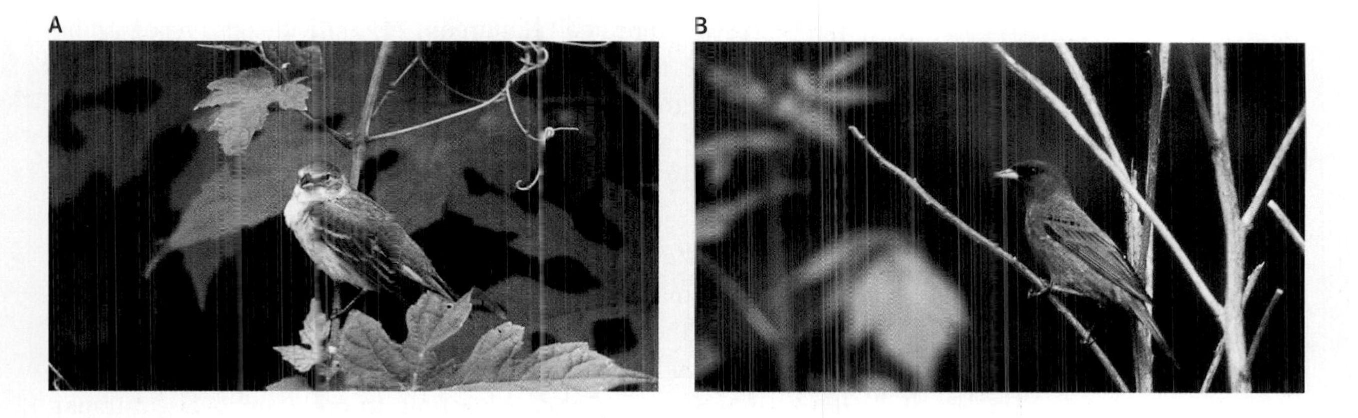

FIGURE 7.16. Indigo buntings. **(A)** A female indigo bunting. Although socially monogamous, female indigo buntings are often involved in extrapair copulations. **(B)** A male indigo bunting. Males defend their territories against intruders. Even so, an average of 13 percent of all matings in indigo buntings involve extrapair copulations. *(Photo credits: David Westneat)*

To examine what impact EPCs may have on mating dynamics in buntings, Westneat ran one of the first detailed genetic analyses of parentage done in conjunction with a detailed behavioral study (Westneat, 1987a; Figures 7.17). This study was conducted before DNA fingerprinting techniques were widely available, and instead relied on a technique called electrophoresis, which, although less powerful than DNA fingerprinting, does allow one to rule out a particular adult individual as the parent of a particular offspring.

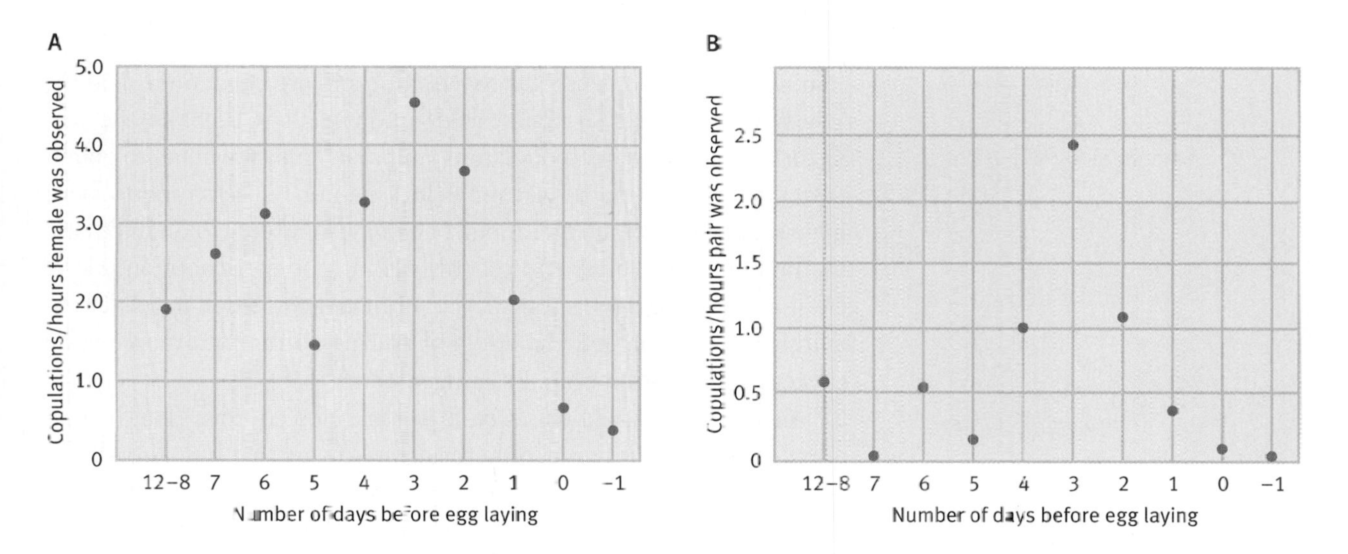

FIGURE 7.17. Copulations in indigo buntings. **(A)** Extrapair copulations occurred at the highest rate three days before egg laying (the day the egg was laid is 0 on the x-axis), just as was the case for within-pair copulations **(B)**. *(Based on Westneat, 1987b)*

Westneat biopsied hundreds of individually recognizable buntings over two years. Using electrophoretic comparisons, he found that of the 257 young biopsied, 37 had genotypes that were *not* consistent with the genotype of one of their presumed parents (Westneat, 1987a), and so at least 14 percent (37 out of 257) of all young were sired via an extrapair copulation—right in line with the 13 percent Westneat predicted based on his behavioral observations. Mathematical models, however, show that electrophoretic estimates of the percent of young fathered by extrapair fertilizations are an underestimate (Westneat et al., 1987). Plugging Westneat's numbers into the correction formula uncovers extrapair fertilization rates from 27.2 to 42.1 percent (depending on the year) in buntings. In addition, as mentioned earlier, females were much more likely to resist EPCs (34 of 43 attempts) than to resist mating with their pairmate (resistance on 72 of 320 attempts; Westneat et al., 1987).

Since Westneat's pioneering work, many studies have documented EPC frequency (Westneat et al., 1990; Webster and Westneat, 1998) in birds, with EPCs accounting for 76 percent of all young in one population of the superb fairy wren (*Malurus cyaneus*; Mulder et al., 1994). Given that monogamy was long considered to be the norm in birds, these are staggering numbers—numbers that are only available because of the revolution in molecular genetics that is still going full steam ahead.

SPERM COMPETITION

In many promiscuous breeding systems, there are big winners and big losers. With respect to males, a few individuals will obtain the vast majority of matings in a population, and many males will fail to obtain even a single mating opportunity. In Chapter 6, we saw how both male-male competition and female choice can affect which males are on the upper and lower ends of this mating curve distribution. Here, we shall look at a more subtle aspect of mate acquisition in relation to mating systems, namely the effect of **sperm competition** on mating success and the evolution of mating systems.

In some promiscuous (as well as some polyandrous) mating systems, males compete not only for access to mates, but directly for access to eggs. That is, a lot of competition in polygynous systems actually occurs *after* a female has mated with numerous males. If females store sperm from numerous matings, sperm from different males compete with one another over access to fertilizable eggs

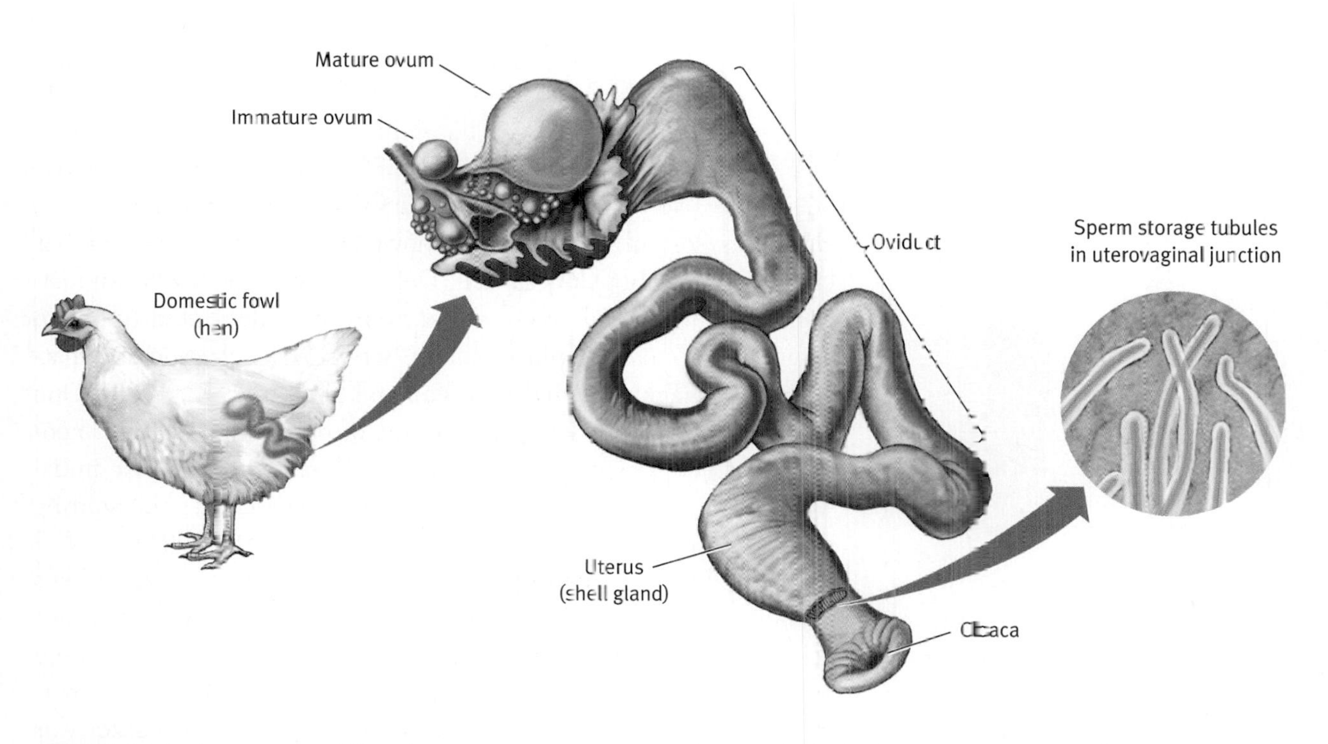

FIGURE 7.18. Sperm competition. The reproductive system of domestic fowl. Sperm storage occurs in sperm storage tubules (SST) at the uterovaginal junction. Females in many species can store sperm from multiple males, setting the stage for sperm competition. Only a small proportion of sperm makes it into the SST. *(Based on Birkhead and Møller, 1992)*

(Figure 7.18). When sperm competition exists, selection can operate directly on various attributes of sperm, such as sperm size and shape. In many ways, an adaptationist approach to sperm competition predicts that such competition should be under the strongest selection pressure imaginable. After all, if an individual wins a physical fight with another male, this *may* provide access to an estrous female, which *may* lead to a mating opportunity, which in turn *may* lead to fertilization. In sperm competition, if sperm hit the target—the ova—then from an evolutionary perspective, that's all that matters.

Sperm competition not only selects for a fascinating suite of behavioral strategies in males and females, and numerous sperm morphological characteristics (size, shape and so on), but it also has implications for the evolution of genitalia. For example, Sandy Harcourt and Paul Harvey examined the effect of polygyny and sperm competition on testes size in primates (Harcourt et al., 1981). In primates, even taking into account differences in absolute body size, testes increased as a function of promiscuity and presumably sperm competition, with testes size being larger in

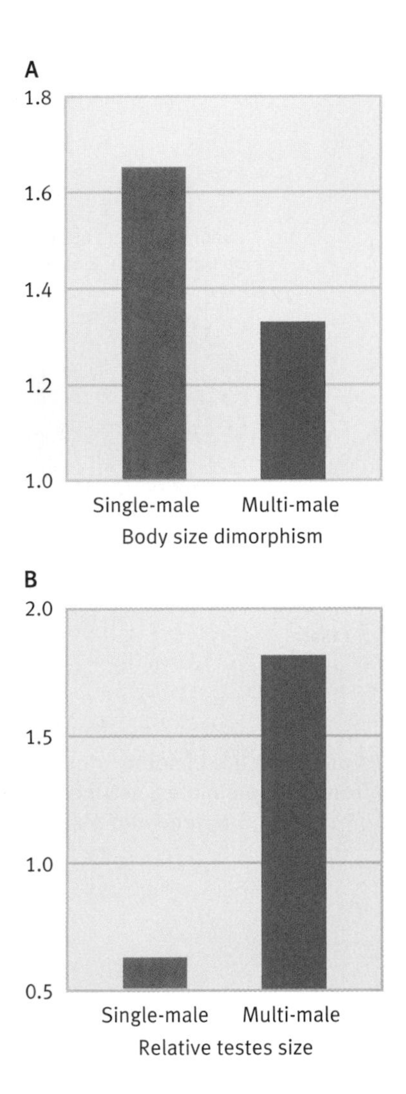

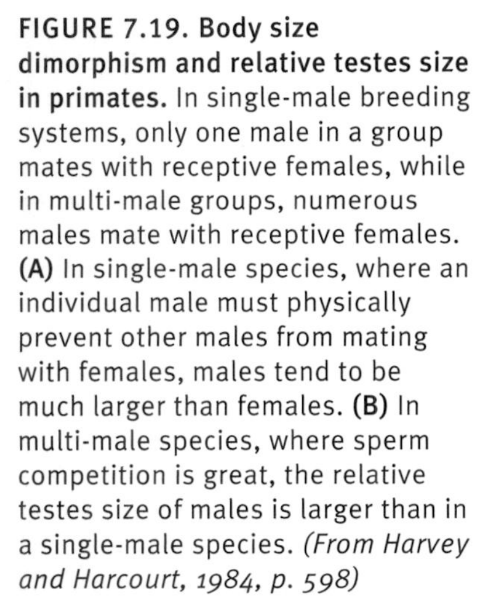

FIGURE 7.19. Body size dimorphism and relative testes size in primates. In single-male breeding systems, only one male in a group mates with receptive females, while in multi-male groups, numerous males mate with receptive females. **(A)** In single-male species, where an individual male must physically prevent other males from mating with females, males tend to be much larger than females. **(B)** In multi-male species, where sperm competition is great, the relative testes size of males is larger than in a single-male species. *(From Harvey and Harcourt, 1984, p. 598)*

primates in promiscuous versus monogamous systems (Harvey and Harcourt, 1984; Figure 7.19).

SPERM COMPETITION IN DUNGFLIES. Sperm competition has been extensively documented in many groups of animals, and an analogous sort of competition, known as "pollen competition" is known to occur in plants as well (Delph and Havens, 1998). In fact, numerous books and edited volumes are now dedicated to sperm competition and its implications for both sexual selection and mating systems (Smith, 1984; Birkhead and Møller, 1992, 1998). One of the acknowledged leaders, in terms of both empirical and theoretical aspects of sperm competition, is Geoff Parker, whose initial work on this subject in dungflies is rightly regarded as the starting point of sperm competition research (Parker, 1970b, 2001).

Dungflies use the droppings of large, often domestic, animals as their choice site for breeding. While in most insects, copulations generally last a matter of seconds, in dungflies they can last on the order of thirty minutes or more. After a detailed analysis of the natural history of mating in dungflies in British pastures, Parker was faced with a number of unresolved issues regarding dungfly mating (Simmons, 2001). For our purposes, we shall focus on two of these: (1) why did males remain with females during egg deposition?, and (2) why did dungflies copulate so very long? Sperm competition theory, first developed by Parker (1970b), provided the answer to these and many other questions surrounding the mating system of dungflies.

When a new dung pat is created and females begin to arrive, there ensues intense male competition for mating opportunities. A thousand or so males can descend on a single dung pat, all in search of females. Males that are lucky enough to find a female and begin copulating are under constant attack from other males trying to break up their pairing and start their own round of copulating (Figure 7.20). So strong is the competition for mates, that usurping males even attempt to mate with females as they are depositing their eggs on a pat of dung. To test for the role of sperm competition, Parker adopted a technique that entomologists of the day had been using in various biocontrol programs (Parker, 1970a). He irradiated the sperm of certain males, and such irradiated sperm would then fail to produce eggs that hatched. As such, Parker could take pairs of males, irradiate one of them, and then examine the relative success of each male by simply determining the proportion of eggs that failed to hatch (that is, the proportion of fertilizations attributable to the irradiated male).

What Parker found was that the number of eggs that were fertilized by the last male to mate with a female was proportional to

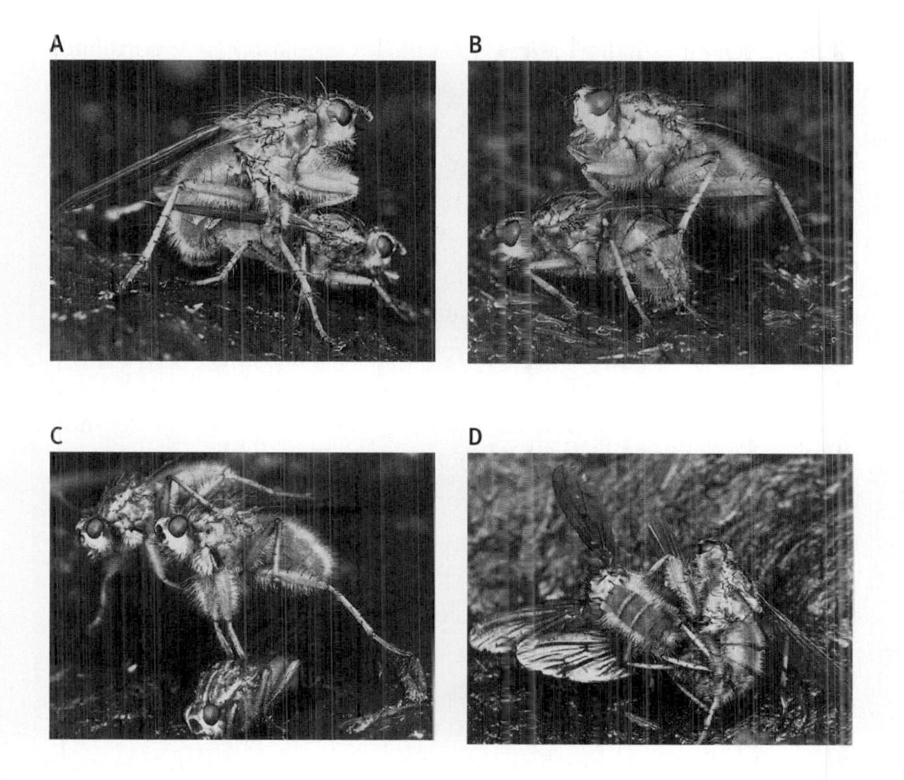

FIGURE 7.20. Dungfly mating. In dungflies, sperm competition can be intense, with the last male copulating with a female fathering up to 80 percent of her offspring. The images show: **(A)** dungflies copulating, **(B)** a male passively guarding a female against other males, **(C)** a male actively guarding a female from another male, and **(D)** a male struggling with an intruder attempting to displace him. *(Photo credits: Geoff Parker)*

how long such a mating lasted. The longer the last mating, the greater the reproductive success of the male. The longer such a copulation, the greater the extent to which the last male's sperm displaced the sperm of males who had copulated with the female earlier (Figure 7.21). Such "last male precedence" is common when sperm competition is in play, but it is not ubiquitous. In some mating systems, sperm competition appears to favor the first, rather than the last male to mate with a female.

In the dungfly system, the last male to mate with a female copulated on average for thirty-six minutes, and this resulted in his fathering approximately 80 percent of the young in the clutch of eggs

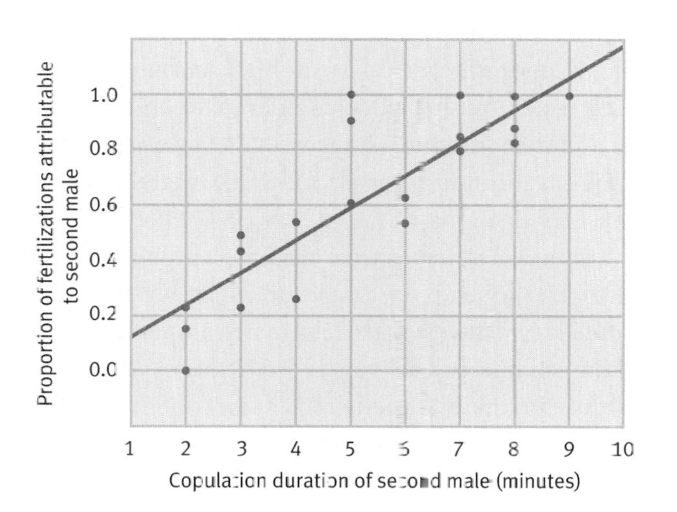

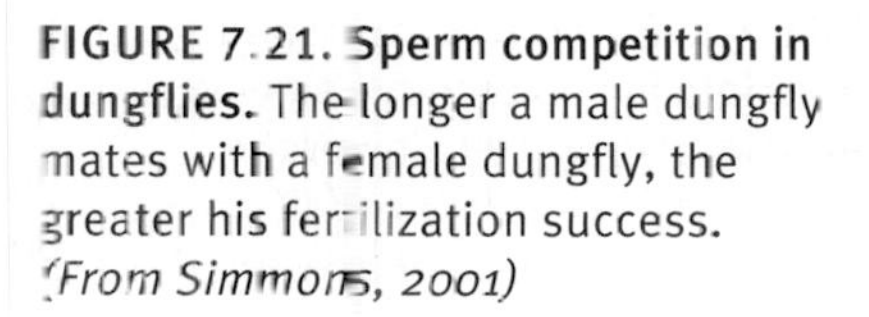

FIGURE 7.21. Sperm competition in dungflies. The longer a male dungfly mates with a female dungfly, the greater his fertilization success. *(From Simmons, 2001)*

deposited by a female (Parker, 1970a). If copulation time correlates with greater displacement of competitor's sperm, why don't males copulate for even longer periods and thereby attempt to displace 100 percent of a competitor's sperm? The answer appears to be that males must weigh such an option against what else could be done with the time in question. While increasing the time spent with female A will increase the displacement effect, it is also time that could have been used to find another female with whom to mate. Given that the rate of sperm displacement slows down with time, it will often pay a male to use such additional time to find other potential mates (Parker, 1974b; Parker and Stuart, 1976).

In any discussion of sperm competition, it is important to keep in mind that females are not simply "inert environments" that serve as receptacles of male sperm (Parker, 1970b). Rather, females themselves may play an active role in sperm competition via "cryptic choice"—that is, by selecting between sperm in a manner that would be very difficult to detect. Eberhard (1996) suggests that, among other things, such cryptic choice affecting sperm competition may affect how much sperm a female allows a copulating male to inseminate her with, how she goes about transferring such sperm to "sperm storage organs," and which sperm a female may select for actual fertilization.

SPERM COMPETITION IN SEA URCHINS. While sperm competition is often thought of in terms of its effect in utero, it can also play an important role in species that do not have internal fertilization. To get a better sense of the importance of sperm competition in such systems, let's consider Donald Levitan's work on sperm velocity and fertilization (Levitan, 2000). Levitan chose "primitive" sperm from the sea urchin and began by asking whether variation in the speed at which sperm traveled correlated with fertilization rate. Using sea urchin sperm makes this task a bit easier than, say, using human sperm, because sea urchins secrete their sperm and eggs into sea water. With a video camera that could tape sperm zipping along and a microscope to see which eggs were fertilized, Levitan was able to measure sperm swimming speed and fertilization success.

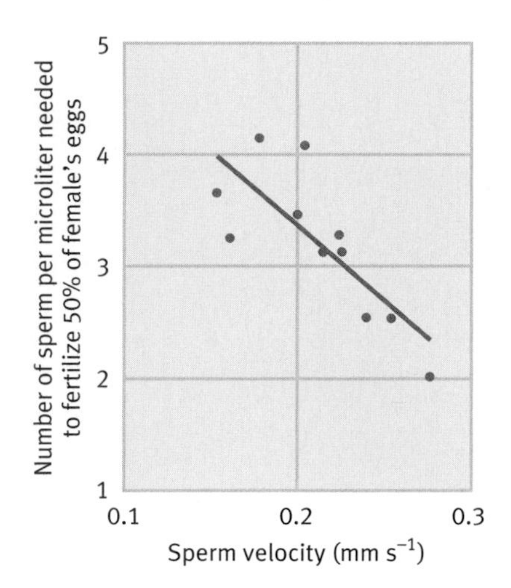

FIGURE 7.22. Sperm velocity and fertilization. In sea urchins, slower sperm fare poorly. The slower the sperm, the more sperm needed to fertilize a female's egg. *(Based on Levitan, 2000)*

Levitan found that males that produced slow-moving sperm required up to 100 times more sperm to fertilize a given number of eggs as quickly as males with fast-moving sperm (Figure 7.22). Levitan then examined what happens to sperm as they age. In so doing, he was examining a predicted tradeoff between the speed at which a sperm moves and how long that sperm survives. Basic natural selection thinking suggests that, since swimming fast and

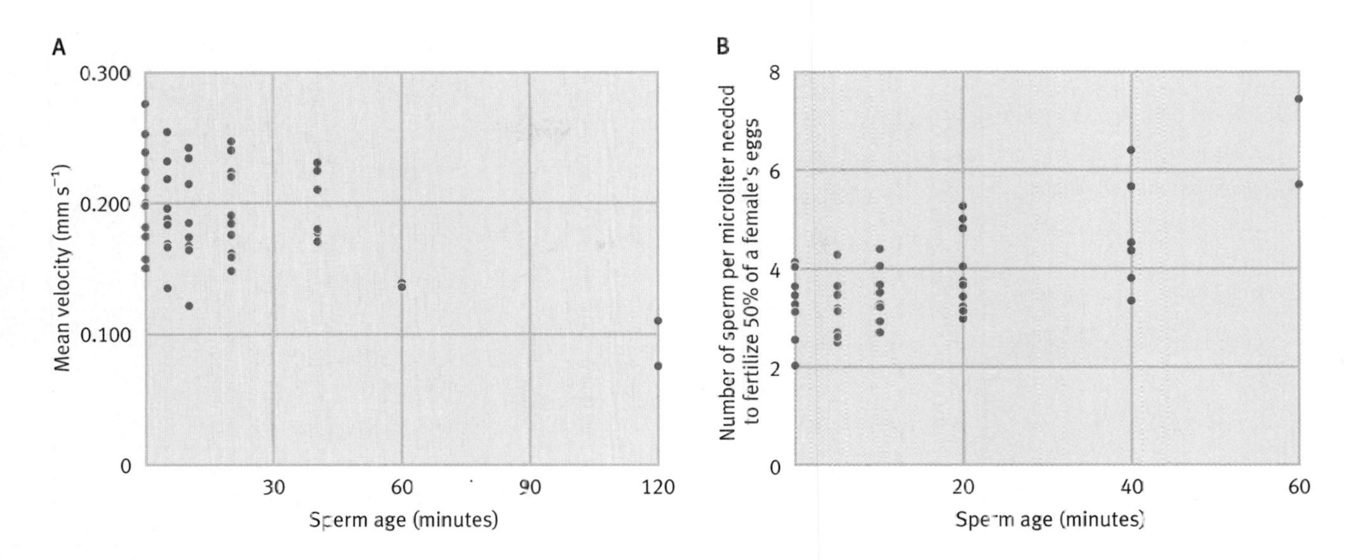

FIGURE 7.23. Older sperm fare poorly. (A) In sea urchins, older sperm swim slower, and **(B)** a greater quantity of such sperm is needed to achieve high fertilization rates. *(Based on Levitan, 2000)*

swimming for a long time both require energy, and since individuals only have so much energy, fast-moving sperm shouldn't live as long as slow-moving sperm. From a sperm's-eye view, you can't have your cake and eat it too.

When examining the expected tradeoff between speed and longevity, Levitan first found that *all* sperm slow down as they get older (Figure 7.23). Not only did sperm decrease their swimming speed with age, but they were much less likely to fertilize an egg, even when they encountered one. For example, sperm only an hour old could be as much as 100 times less likely to fertilize an egg as new sperm. After two hours, sperm were through—they fertilized no eggs at all. Clearly, then, time is of the essence when trying to fertilize sea urchin eggs.

With data on longevity and speed in hand, Levitan could return to the question raised earlier: Is there a tradeoff between sperm speed and sperm life span for individual sea urchins? The answer is a definitive yes, as Levitan found a negative correlation between velocity and endurance. Individuals who produced fast-moving sperm had their sperm become ineffective at much quicker rates than other individuals. The energy used up in swimming fast resulted in less energy for swimming for a long time, as well as a shorter life span.

OTHER EFFECTS OF SPERM COMPETITION. Sperm competition not only affects the speed at which sperm swim, but also the various shapes that sperm can take. For example, Baker and Bellis's **kamikaze sperm hypothesis** suggests that selection might

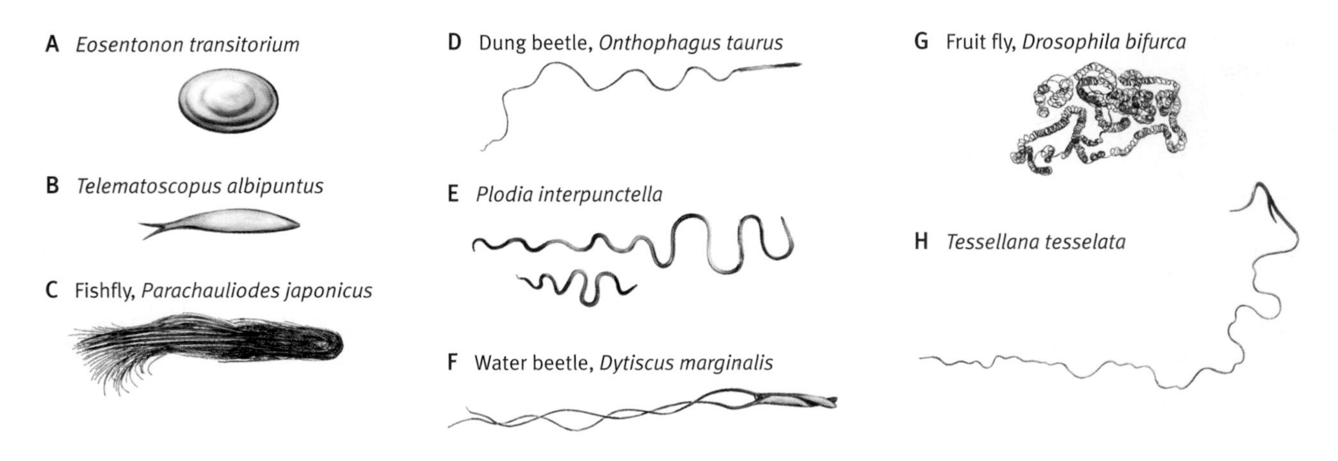

FIGURE 7.24. Variability in sperm morphology. Sperm competition is one of the many forces that have led to incredible variability in insect sperm morphology. Pictured here are: **(A)** sperm from *Eosentonon transitorium*, **(B)** fishlike sperm from *Telmatoscopus albipuntus*, **(C)** sperm bundle from the fishfly *Parachauliodes japonicus*, **(D)** 1 mm sperm from the dung beetle *Onthophagus taurus*, **(E)** short and long sperm from *Plodia interpunctella*, **(F)** paired sperm from the water beetle *Dytiscus marginalis*, **(G)** giant 58 mm sperm from *Drosophila bifurca*, and **(H)** hook-headed sperm from *Tessellana tesselata*. *(Based on Simmons, 2001)*

sometimes favor sperm that are designed to kill other males' sperm rather than fertilize eggs (Baker and Bellis, 1988). The evidence for this hypothesis is mixed, but in any case it demonstrates the myriad ways that selection *might* act via sperm competition (Bellis et al., 1990; Gomendio et al., 1998; Moore et al., 1999). While the evidence is equivocal in humans, sperm competition has had clear effects on sperm morphology (form and function; in this case sperm size) in insects (Pitnick et al., 1995; Figure 7.24), mammals (Gomendio et al., 1998), birds (Birkhead and Møller, 1992; Briskie et al., 1997), fish (Stockley et al., 1997), and nematodes (LaMunyon and Ward, 1999).

Another means by which sperm competition may operate is in relation to the number of sperm produced per mating. For example, Baker and Bellis (1993a) tested one basic prediction from sperm competition in humans, namely that the number of male sperm per ejaculate should be a function of the risk that a female has mated with other males. The greater the risk, the greater the sperm competition, and hence a predicted increase in sperm volume. Remarkably, Baker and Bellis found that not only did sperm number increase as a function of time since last copulation (Figure 7.25), but that even when absolute time was statistically removed from the equation, the relative amount of times couples spent together predicted sperm volume as well. When couples spent more time together, and hence the risk of extrapair copulations was low, sperm count was significantly less than when couples spent less time together.

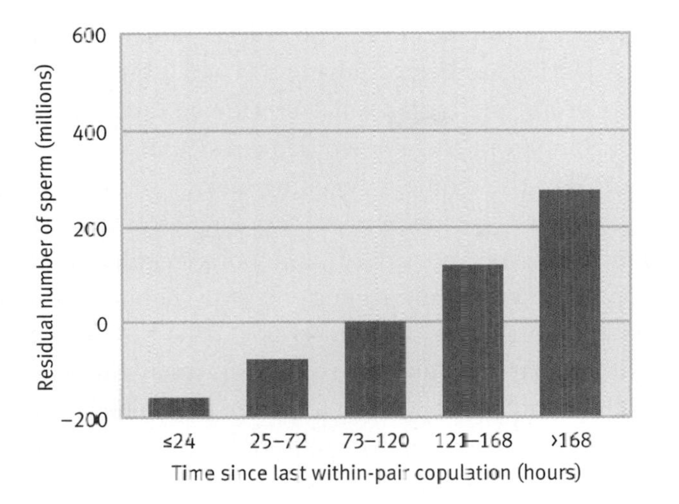

FIGURE 7.25 Sperm number in humans. In human males, the number of sperm ejaculated during a copulation is a function of the time since a pair last copulated. Note that the y-axis is a measure of "residuals"; hence negative values are possible. *(Based on Baker and Bellis, 1993a)*

Multiple Mating Systems in One Population?

There is some contention in the ethological literature over whether mating systems should be defined in terms of individual behavior or some population measure. For example, if some individuals, let's say some males, in a population have more than one mate, while others have just a single mate, how do we define the system? One way would be to argue that there are monogamous individuals and polygynous individuals. Alternatively, the mating could be defined at the population level. Here, if a significant proportion of males are polygynous, then the mating system is polygynous.

We shall generally adopt the second definition, primarily because in polygynous populations most individuals try and obtain more than one mate, even though some end up with only a single mate and some with no mate at all. Nonetheless, sometimes we must move away from a population-level analysis. In the case of dunnocks, we see monogamous, polygynous, polyandrous, and polygynandrous mating groups all in the same population.

DUNNOCKS

Let us return to the dunnock mating system, first touched upon during our discussion of polygynandry. While the dunnock has a fairly broad geographic range, it is most common in the British Isles, where Nick Davies has been studying a breeding population of about eighty individuals in the Botanical Gardens of Cambridge University. Davies has been studying this small brown bird for the past twenty years, and his work has provided a rare, detailed portal into a very complex mating system (Davies, 1992).

What makes the dunnock breeding biology so fascinating from a mating systems perspective is the long-term persistence of monogamy, polygyny, polyandry, and polygynandry in the same population. For Davies, unraveling the story behind the dunnock mating system is akin to detective work: "The puzzle is the dunnock's extraordinary breeding behavior and variable mating system. The job of the nature detective is to understand alternative options facing individuals, to assess their reproductive payoffs, and then to discover whether different mating strategies might emerge as a result of individuals competing to maximize their reproductive success" (Davies, 1992, p. 21).

Underlying much of the variance in mating systems is the fact that males and females have different interests—that is to say, there exists a conflict of interest between the sexes. For males, access to females increases from shared access to a single female (polyandry), sole access to a single female (monogamy), joint access to two females (polygynandry), through sole access to numerous individuals (polygyny). Indeed, a male's reproductive success over the course of a season is highest for the case of polygyny, and selection should favor male behavior that leads down this course. On the other hand, female reproductive success increases in precisely the opposite direction, with polyandrous and polygynandrous females having the highest reproductive success. As such, the sexes are clearly at odds with respect to what constitutes the optimal breeding system. This battle, though not completely one-sided, may be leaning in the direction of female dunnocks, as over the course of ten years, 75 percent of females and 68 percent of males were involved in either polyandrous or polygynandrous mating groups (Davies, 1992).

Examining the battle of the sexes, in conjunction with the dispersal patterns of dunnocks, we can begin to paint a clearer picture of the inherently complex breeding system of the dunnock. Early in the breeding season, females compete with each other to establish territories, and such territories are chosen independently of the position of males. Males then attempt to construct territories that overlay as many female territories as possible. Given this, Davies argues that the difference between monogamy and polygyny is a function of male territory size. Polygynous males had larger territories than monogamous males, but when comparing these two breeding systems, the territory size of females remained constant. On the other hand, the difference between polyandry and polygynandry was a function of female territory size. Male territory size

TABLE 7.4. Food and territory size. Supplementing the food on dunnock territories led to a decrease in female territory size but not male territory size. *(From Davies, 1992, p. 63)*

	TERRITORY SIZE (m^2)						
	WITH FEEDER			CONTROL			
	MEAN	(SE)	*n*	MEAN	(SE)	*n*	SIGNIFICANCE OF DIFFERENCE
Females	2,776	(379)	28	4,572	(456)	39	$P < 0.01$
Males							
Territory defended by one male	2,864	(340)	11	2,642	(416)	13	NS
Territory defended by two males	5, 276	(797)	14	6, 614	(674)	17	NS

remained constant across these systems, while female territory size was significantly larger in the former.

In Davies's population of dunnocks, because all individuals were marked (with color rings) and rarely did a bird stray more than two miles from the Botanical Gardens, Davies was able to construct experiments to test more precisely how resource defense and territoriality influenced the dunnock mating system. Davies and Lundberg (1984) reasoned that, since females were in strong competition with one another for territories with the best resources, if the resources available on any given territory were experimentally supplemented, territory size should shrink, since females would then be able obtain the same amount of resources without having to defend as large an area.

To test their hypothesis, Davies and Lundberg placed artificial feeders on a randomly selected area of female territories, and did indeed find that territory shrank as predicted. Moreover, they found that the male territories that overlay the manipulated female territories did not change in size (Table 7.4). What did occur, however, was a shift in the distribution of mating systems, in such a manner as to favor males. When female territory size shrank as a result of supplemental resources, males were better able to monopolize more than one female, and a shift away from polyandry and toward polygynandry occurred. This supports Davies's argument that in the dunnock, females track resources and males track females, and the resulting interaction helps us better understand the incredible variation in mating systems in this small brown bird.

Interview with Dr. NICK DAVIES

The dunnocks you work with have proven to be a model system for so many questions in behavioral ecology and animal behavior. Why did you choose to work with this species?

In 1979, when I started as a young University lecturer at Cambridge, I was excited by the new theoretical ideas of Bill Hamilton, Robert Trivers, John Maynard Smith, and Geoff Parker. They were beginning to explore the evolutionary consequences of individual conflicts of interest, not only the long-recognized conflicts among rival males, but also those between the male and female of a breeding pair and between parents and their offspring. These ideas led to a profound change in how we interpret individual adaptations. For example, the classical work on clutch size (pioneered by David Lack) had considered what would be optimal from a pair's point of view. Once genetic conflicts within families were recognized, a whole new world of possibilities was opened up, involving deception, manipulation, cheating, and compromise. Likewise, previous studies of mating systems had emphasized ecological pressures (food, nest sites, predation) that might lead to one system rather than another. But now we began to consider social conflicts as important selection pressures too.

So I was on the lookout for a field study which would allow me to take a fresh look at bird mating systems. While wandering in the University Botanic Garden, I noticed the dunnocks chasing around the bushes in pairs, threes, and fours and decided to color-band them for individual recognition to see what was going on. Once I had this initial excuse to begin the study, I then discovered more interesting new questions from the bird watching rather than from the theoretical literature.

Your work on the dunnock suggests these birds have an incredibly plastic mating system. Why do you think that is?

It is clear from watching behavior that males and females have different mating preferences. These conflicts make good sense in relation to individual reproductive payoffs from the different mating combinations. For a female, polygyny (one male plus two females) was the least successful option, because she had to share the help of one male with another female, and her chicks often starved. In polygyny, females often chased and fought each other. If one of them could drive the other away, then she could claim the male's full-time help and so enjoy the greater success of monogamy. A polyandrous female (one female with two males) was even more successful: if she shared copulations between both her males, then both helped to feed her brood. With this extra help, more chicks survived. This explained why a female encouraged both alpha and beta males to copulate with her. Our experiments showed that she maximized total male help by sharing copulations equally between the two males.

By contrast, a male fared least well in polyandry. Although more young were raised, there was the cost of shared paternity. Our DNA fingerprinting results showed that an alpha male did better with full paternity of a pair-fed brood. This explains why alpha males acted against the female's wishes, and tried to drive beta males away, or at least prevent them from copulating. A male did best of all in polygyny, the system in which a female did worst. Although each female was less productive, the combined output of two females in polygyny exceeded that of a monogamous female. So once again there was conflict—the male intervened in the squabbles between his two females in an attempt to retain them both.

The variable mating system thus reflects the different outcomes of sexual conflict. Where a male can prevent the conflicts between two females, we find polygyny. Where a female can escape the close guarding of her alpha male, and give a mating share to the beta male, we observe polyandry. Monogamy occurs when neither sex can gain a second mate. Polygynandry—for example, two males with two females can be viewed as a "stalemate": the alpha male is unable to drive the beta male off and so claim both females for himself, and neither female can evict the other and so claim both males for herself.

So the interesting question is: what determines whether particular individuals can get their best option despite the conflicting preferences of others? This is influenced by individual competitive ability and various practical considerations, such as vegetation density (which determines how easily an alpha male can follow a female). The main point is that the social conflicts are played on an ecological stage, which may affect the likely outcomes.

Sexual selection and mating systems are often presented as two distinct topics. Is it possible to understand one without the other?

Darwin's theory of sexual selection aims to explain the evolution of traits that increase an individual's mating success: size, weapons, ornaments, mate choice, and so on. Mating systems are the outcomes of this competition for mates. So you need to think about both process and outcome for a full understanding.

Why do you think polyandry is rare compared to polygyny?

When you look at social groups, it is certainly true that you more often see one male defending a group of females rather than the converse, namely one female defending a group of males. In theory, this is exactly what you would expect, because a male usually has a greater potential reproductive rate than a female. This means that an increase in the number of mates leads to a much greater increase in male reproductive success than it does in female reproductive success. Thus males often go for quantity, while females seek quality, when they search for mates. Nevertheless, until recent years we underestimated the frequency with which females seek matings with multiple males. They often do this surreptitiously, sometimes outside their social groups, to increase their access to resources, to increase male care, to reduce male harassment, or to improve the genetic quality of their offspring.

Recently there has been some debate as to who calls the shots when it comes to mating systems—males or females? Clearly any given mating requires both a male and female, but are there certain situations in which you expect males to be controlling the mating system, and others where mating systems are primarily under the control of females?

Darwin's theory of sexual selection aimed to explain how individuals competed for mates. He proposed two processes: people readily accepted the first, namely male-male competition (Darwin's "law of battle"), but were reluctant to believe the second, namely female choice. We now have good evidence for this, of course. It is interesting to see reactions to Parker's theory of sperm competition (sexual selection after the act of mating) follow the same history. At first, everyone focused on males—mate guarding, frequent copulation, mating plugs, sperm removal, testis size, and so on. Consideration of female roles ("cryptic female choice" of sperm) came later, and it is still controversial. My guess is that female roles will be crucial, simply because females should be better able to control events inside their own bodies.

In some cases, the act of mating seems to be under male control; for example, where female insects lay their eggs in localized patches (e.g., cow pats), powerful males can monopolize the laying sites, and a female is forced to mate in order to gain access. In other cases, females can more easily escape males; for example, insects that lay their eggs in more dispersed sites can often refuse matings and in birds (where females can often easily escape males by flying off) mating rarely occurs without female consent. It will be interesting to see whether cryptic female choice of sperm is more likely in cases where females have less control over the act of mating.

DR. NICK DAVIES is a professor at Cambridge University in England. His long-term work on dunnocks, summarized in Dunnock Behaviour and Social Evolution *(Oxford University Press), elegantly shows the complexity of animal mating systems and how one goes about testing fundamentally important hypotheses in such a system.*

SUMMARY

1. Animal mating systems are typically classified as monogamous, polygynous, polyandrous, polygynandrous, or promiscuous, depending on the number of mates that males and females take and the timing of such mating in relation to breeding season.
2. A female can often fertilize many of her available eggs by mating with one or a very few males. That being the case, female fecundity is not so much tied to the availability of mates as it is to the availability of resources, such as food, defense, and so on. The more resources available, the more offspring females can produce.
3. Males can potentially fertilize many, many females, and so their reproductive success is tied more to access to females than to access to resources. As such, female dispersion patterns should track the distribution of resources, whereas male dispersion patterns should track the dispersion of females.
4. The polygyny threshold model predicts under what conditions polygyny should occur in nature. In this model, females weigh the costs and benefits associated with being in a polygynous relationship on a good territory versus a monogamous (or more precisely, a less polygynous) relationship on a poorer territory.
5. While extrapair copulations and extrapair matings were once thought to be rare in birds, genetic evidence suggests this is far from true in many species.
6. In some mating systems, males compete not only for access to mates, but directly for access to eggs. If females store sperm from numerous matings, sperm from different males compete with one another over access to fertilizable eggs in what is known as sperm competition.

DISCUSSION QUESTIONS

1. Define and distinguish among serial monogamy, serial polygyny, simultaneous polygyny, promiscuity with pair bonds, and promiscuity without pair bonds.
2. Read Jenni and Collier's 1972 article, "Polyandry in the American jacana (*Jacana spinosa*)," in *Auk* (vol. 89, pp. 743–765). What selective forces favored polyandry in jacanas, and how did ecological factors affect selection pressure?

3. Why do you think that polygamous mating systems favor the evolution of more deadly diseases in animals and humans? Think about this from the perspective of the disease-causing agent.
4. Define an EPC. How does this differ from an extrapair mating? Why did it take ethologists so long to recognize the extent of EPCs in nature? How has molecular genetics revolutionized the way we think of mating systems in birds?
5. How has natural selection via sperm competition shaped both sperm morphology and male behavior? Also, create a list of potential ways in which females may affect sperm competition and its outcome?

SUGGESTED READING

Borgerhoff-Mulder, M. (1990). Kipsigis women's preferences for wealthy men: Evidence for female choice in mammals. *Behavioral Ecology and Sociobiology*, 27, 255–264. An early application of the PTM to humans.

Davies, N. B. (1991). Mating systems. In J. R. Krebs and N. B. Davies, *Behavioral ecology* (pp. 263–299). New York: Blackwell Scientific Publications. An overview of the behavioral ecology of mating systems.

Davies, N. B. (1992). *Dunnock behavior and social evolution*. Oxford, England: Oxford University Press. This is a delightful book about Davies's long-term work on behavior, with an emphasis on dunnock mating behavior.

Orians, G. (1969). On the evolution of mating systems in birds and mammals. *American Naturalist*, *103*, 589–603. The paper in which Orians lays out the idea for the polygyny threshold model.

Parker, G. (1970). Sperm competition and its evolutionary consequences in insects. *Biological Reviews of the Cambridge Philosophical Society*, 45, 525–567. The seminal paper (no pun intended) of sperm competition and animal behavior

8

Kinship and Animal Behavior
Kinship Theory
- Relatedness and Inclusive Fitness
- Family Dynamics
Conflict within Families
- Parent-Offspring Conflict
- Sibling Rivalry
Kin Recognition
- Matching Models
- Rule-of-Thumb Models
Interview with Dr. Stephen Emlen

Kinship

In an open field somewhere, a group of ground squirrels peacefully feed and go about their normal daily activities. Seemingly out of nowhere, a hawk begins its deadly dive from the air, targeting the squirrels in the field for its next meal. Suddenly a piercing shriek echoes through the valley—an alarm call given by one squirrel alerts others of the impending danger. The field comes to life with squirrels making mad dashes everywhere, doing whatever they can to reach their burrow, or at least some safe haven. Later, when the hawk has clearly set its sights elsewhere, the squirrels reappear.

A puzzle emerges in the above example. Why should an individual squirrel be the potential "sacrificial lamb" and give off an alarm cry? After all, screaming alarm calls as loud as possible, if nothing else, should make it the single most obvious thing in the entire field. Why would it do anything to attract the hawk in its direction and make itself the hawk's most likely next snack? Why not let someone else take the risks? Paul Sherman has been addressing these sorts of questions in a long-term study of alarm calls in Belding's ground squirrels (*Spermophilus beldingi*; Sherman, 1977, 1980, 1981, 1985; Figure 8.1). He has found that kinship (blood relatedness) affects animal behavior in important ways, playing a large role in whether or not a squirrel will emit an alarm call when a predator is detected.

FIGURE 8.1. Alarm calling squirrels. In Belding's ground squirrels, females **(A)** are much more likely than males to emit alarm calls when predators are sighted. Such alarm calls warn others, including female relatives and their pups **(B)**. *(Photo credits: George D. Lepp and Paul Sherman)*

In this chapter, after an introductory section demonstrating the power of kinship to affect animal behavior, we shall examine:

- The theoretical foundation underlying "inclusive fitness" or kin selection models of behavior.
- The evolution of the family unit.
- Parent/offspring conflict and sibling rivalry.
- How animals recognize kin.

Kinship and Animal Behavior

Belding's ground squirrels, like many other species—for example, prairie dogs (Hoogland, 1983, 1995)—give piercing alarm calls when a predator is spotted (Figure 8.2). These calls warn nearby individuals that a predator is in the vicinity, leading to a mass exodus toward places of safety.

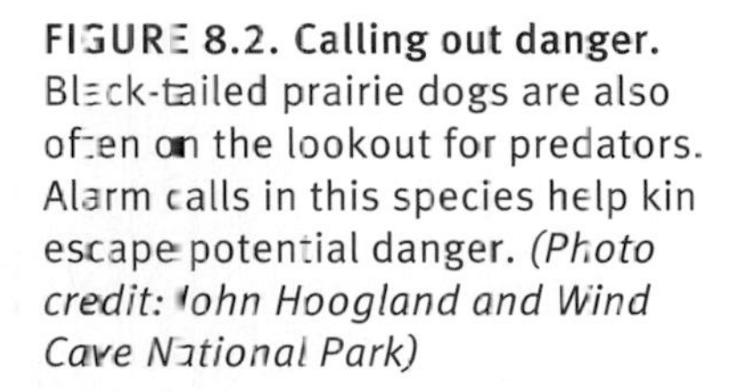

FIGURE 8.2. Calling out danger. Black-tailed prairie dogs are also often on the lookout for predators. Alarm calls in this species help kin escape potential danger. *(Photo credit: John Hoogland and Wind Cave National Park)*

To begin to answer the question about why individuals give alarm calls at the risk of their own lives, we need to recognize that alarm calls in Belding's ground squirrels are most often produced by females. That is, female squirrels give alarm calls when a predator is in the vicinity more often than expected by chance, while the converse is true for males (Figure 8.3). This is a critical piece of information, as it now allows the question to change from why

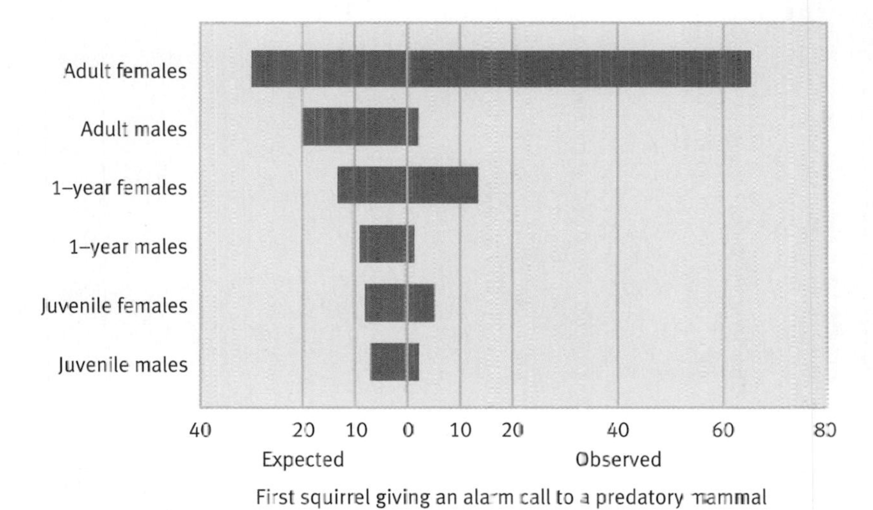

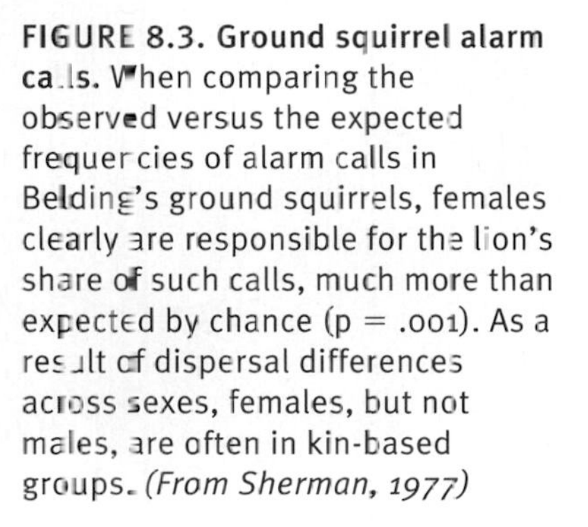

FIGURE 8.3. Ground squirrel alarm calls. When comparing the observed versus the expected frequencies of alarm calls in Belding's ground squirrels, females clearly are responsible for the lion's share of such calls, much more than expected by chance ($p = .001$). As a result of dispersal differences across sexes, females, but not males, are often in kin-based groups. *(From Sherman, 1977)*

alarm calls are emitted to why females give alarm calls so often. The answer lies in how population biology leads to asymmetries in kinship among members of a population.

In Belding's ground squirrels, males leave home and females mature in their natal area. This male-biased dispersal creates an imbalance in the way males and females are related to those individuals around them—females find themselves surrounded by relatives, while males are generally in areas with complete strangers (Figure 8.4). This asymmetry translates into females who warn close kin by emitting alarm calls, while males generally do not emit calls since their dispersal from their natal areas means their blood kin typically do not benefit from such a warning. Further support for the kinship-based alarm-calling hypothesis includes Sherman's finding that in the rare instances when females do move away from their natal groups and into groups with far fewer relatives, they emit alarm calls less frequently than do native females.

Kinship not only promotes prosocial behavior, but it also acts as a force in preventing antisocial behavior as well. As an extreme example, consider homicide in humans. Daly and Wilson (1988) examined 512 cases of homicide in Detroit, Michigan (in 1972). In

FIGURE 8.4. Kin selection and ground squirrels. Belding's ground squirrel groups are typically made up of mothers, daughters, and sisters who cooperate with one another in a variety of contexts. Males that emigrate into such groups cooperate to a much smaller degree. *(Based on Pfennig and Sherman, 1995)*

TABLE 8.1. Risk of homicide in cases where the victim and offender were cohabitants in Detroit in 1972. Observed values indicate the number of homicides that were actually committed. Expected values indicate the number of homicides in each category that we would expect if blood kinship were not playing a role. Relative risk rates were much higher for individuals who were not blood relatives. These numbers are underestimates since the "parent" and "offspring" categories include some stepfamily members and some in-laws. *(From Daly and Wilson, 1988)*

THE AVERAGE DETROITER ≥ 14 YEARS OLD IN 1972 LIVED WITH 3.0 PEOPLE	NUMBER OF VICTIMS		RELATIVE RISK (OBSERVED/EXPECTED)
	OBSERVED	EXPECTED	
0.6 Spouses	65	20	3.32
0.1 Nonrelatives	11	3	3.33
0.9 "Offspring"	3	29	0.27
0.4 "Parents"	9	13	0.69
1.0 Other "relatives"	5	33	0.15

the police records, 127 (25 percent) of these murders were classified as being committed by relatives. Nonetheless, Daly and Wilson found that if you consider only blood kin, rather than kin by marriage, only 6 percent of the murders can be pinned on relatives. Blood kin don't kill each other all that often

It can be argued, however, that the reason that homicide rates among blood kin are low is that in modern society people encounter nonblood kin much more often than blood kin. For example, if killers spent 94 percent of their time with unrelated individuals and 6 percent with blood kin, then the 6 percent murder rate for blood kin would be expected simply by chance, and would not indicate that blood relatedness reduces homicide. Daly and Wilson, however, found that even when the amount of time spent with blood kin and nonblood kin is taken into account, relatives rarely kill each other (Table 8.1).

Few forces have the power to shape animal behavior the way that kinship can. It could even be argued that the modern study of animal behavior and evolution began in the early 1960s, when W. D. Hamilton published his now famous papers on kinship and the evolution of social behavior (Hamilton, 1963, 1964). As we shall see in more detail in a moment, these papers formalized "inclusive fitness" or "kinship" theory, and revolutionized the way scientists understood evolution and ethology (Kuhn, 1962). Simply knowing which individuals in a population are blood relatives provides ethologists and

FIGURE 8.5. Helping offspring. One classic case of helping blood relatives is the mother-offspring relationship. In bank swallows, females remember the location of their nests and feed youngsters there. Since young chicks remain at the nest, this translates into mothers feeding their offspring. When chicks learn to fly, mothers learn to recognize their offspring's voices. *(Based on Pfennig and Sherman, 1995)*

behavioral ecologists with enough information to make some basic, but important, predictions about behavior (Figure 8.5).

Kinship Theory

The ground squirrel and human homicide examples give a flavor of the power of kinship to shape social behavior. With this in mind, we can now address the question of why kinship is so powerful an evolutionary force. W. D. Hamilton had this to say about kinship theory:

> In the hope that it may provide a useful summary we therefore hazard the following generalized unrigourous statement of the main principle that has emerged from the model. *The social behavior of a species evolves in such a way that in each distinct behavior-evoking situation the individual will seem to value his neighbors' fitness against his own according to the coefficients of relationship appropriate to that situation* [Hamilton's italics]. (Hamilton, 1964, p. 19)

It has been rumored that J. B. S. Haldane once remarked that he would risk his life to save two of his brothers or eight of his cousins. Haldane, who was quite versed in mathematics, made this rather bold statement by counting copies of a gene that might code for cooperative behavior. Such a gene-counting approach to kinship and the evolution of cooperation has been extended by theoreticians, but in its most elementary form, it is the heart and soul of kinship theory. So, let's see how it works.

RELATEDNESS AND INCLUSIVE FITNESS

The *Random House Dictionary* defines kinship as "family relationship," but the evolutionary definition is much more restrictive. In evolutionary terms, relatedness centers on the probability that individuals share genes that they have inherited from some common ancestor—parents, grandparents, and so on—a concept referred to as "identity by descent." For example, you and your brother are kin because you share some of the same genes, and you inherited them from common ancestors—in this case, your mother and father. In a similar vein, you and your cousins are kin because you share genes in common, only now your common ancestors are your grandparents. In general, common ancestors are the most recent individuals through which two (or more) individuals can trace genes that they share.

Once we know how to find the common ancestor of two or more individuals, we can calculate their relatedness, labeled r, which is equal to the probability that they share genes that are identical by descent. For example, two siblings are related to one another by an r value of 0.5. To see why, recall that all of the genes that siblings share come from one of two places—their mother or father. As such, there are two ways, *and only two ways*, that siblings can share gene X—via mother or father. If sibling 1 has X, then there is a 50 percent chance she received it from her mother; if sibling B has X there again is a 50 percent chance that her mother passed this gene to sibling 2. There is therefore a one in four chance that the siblings share gene X through the mother. The analogous argument can be made that there is a one in four probability that the father is the reason that siblings share gene X. To calculate the chances that you and your sibling share gene X through *either* mom or dad, we add the probabilities for each and obtain 0.25 + 0.25, or 0.5. This value, labeled r, can be calculated for any set of blood relatives, no matter how distant (Figure 8.6).

From a "gene's-eye" perspective, calculating relatedness starts the process that allows us to see why kinship is such a powerful force in animal behavior. The next step in the process is to recall that genes are selected based on the number of copies of themselves that they get into the next generation. To this point, we have been thinking

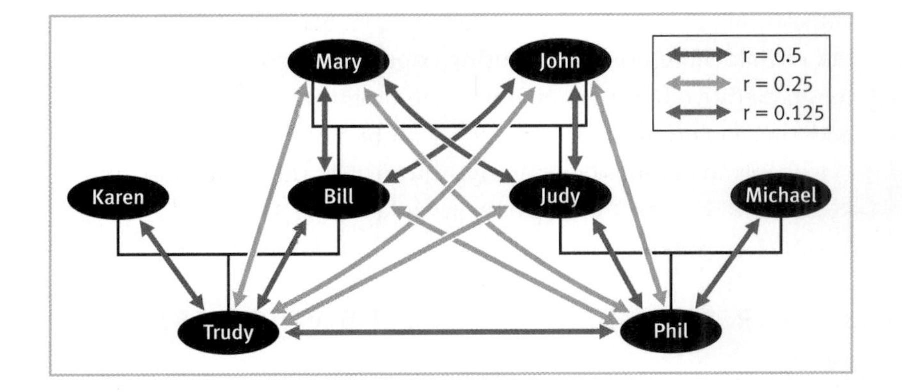

FIGURE 8.6. Genetic relatedness. Family trees can be used to calculate *r*, the genetic relatedness between individuals. Take this family tree of relatives. Mary and John are married. They have two children: Bill and Judy. Bill marries Karen, and they have one child: Trudy. Judy marries Michael, and they have one child: Phil. Orange arrows represent an *r* of 0.5, yellow arrows an *r* of 0.25, and the green arrow an *r* of 0.125.

about a gene in terms of the effect it has on the *individual* in which it resides, but kinship calculations suggest that this is a myopic view. Given that relatives, by definition, have a higher probability of sharing gene X than do strangers, then gene X may increase its chances of getting copies of itself into the next generation by how it affects not just the individual in which it resides, but that individual's blood relatives as well. When we depict fitness in this manner, we are talking about **inclusive fitness**. In the language of natural selection, kin are worth helping in direct proportion to their blood relatedness.

With an understanding of how *r* is calculated, we can now examine inclusive fitness theory in more detail. W. D. Hamilton tackled the question of kinship and animal behavior in his now famous pair of papers, "The Genetical Evolution of Social Behavior I and II" (1964). The essence of inclusive fitness models is that they add on to "classical" models of natural selection by considering the effect of a gene, not only on the individual that bears it, but on individuals sharing genes who are identical by descent (that is, blood kin). The equations in Hamilton's paper can be daunting even to those with a mathematical background. Fortunately, these equations can be captured in what is now referred to as "Hamilton's Rule," namely that a gene increases in frequency whenever:

$$\left(\sum_{1}^{A} rb\right) - c > 0$$

where b is the benefit others receive from the trait the gene codes for, c is the cost accrued to the individual expressing the trait, r is our measure of relatedness, and A is a count of the individuals affected by the trait of interest (Grafen, 1984). In other words, the decision to aid family members is a function of how related individuals are, and how high or low the costs and benefits associated with the trait turn out to be. When individuals are highly related and a gene codes for an action that provides a huge benefit at a small cost, selection strongly favors this trait. Conversely, the least likely trait to be selected is one in which those helped are not relatives, and where helping entails significant costs.

Hamilton's theory has had a profound impact on the work of ethologists, behavioral ecologists, and comparative psychologists. For example, although work had been going on in the field of cooperative breeding in birds for more than thirty years (Skutch, 1935, 1987; see also J. L. Brown, 1994), Hamilton's theory, in conjunction with Jerram Brown's (1970) empirical work on Mexican jays, caused a surge in studies in cooperative breeding that is still in full force today (J. L. Brown, 1987; we shall return to cooperative breeding below and in Chapter 9).

The impact of Hamilton's ideas was even greater than it might have been otherwise as a result of Jerram Brown's (1975) reformulation of Hamilton's Rule. Field-workers found the b and c terms of Hamilton's model difficult to measure in nature. Brown's "offspring rule" solved both these problems by using offspring as the currency of measure (for more on how and how not to measure inclusive fitness, see J. L. Brown, 1975; West-Eberhard, 1975; Charnov, 1977; Seger, 1981; Grafen, 1984, 1985; Creel, 1990; Queller, 1992). This formulation set up the possibility of field manipulations in which Hamilton's and Brown's ideas could be tested by counting the number of offspring born across treatments. For example, if one wanted to know the effect that young "helpers-at-the-nest" had on raising their siblings, one could examine the difference in the average number of chicks that survived in the presence and absence of such helpers (J. L. Brown et al., 1982; Figure 8.7).

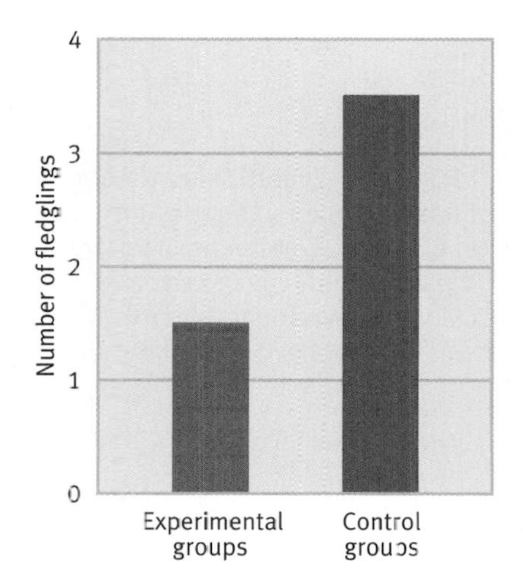

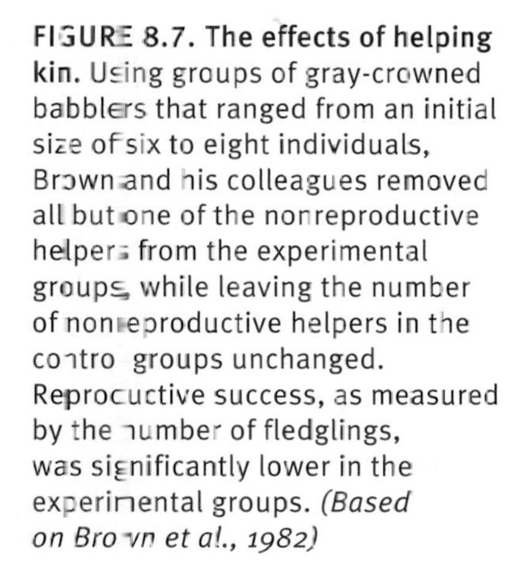

FIGURE 8.7. The effects of helping kin. Using groups of gray-crowned babblers that ranged from an initial size of six to eight individuals, Brown and his colleagues removed all but one of the nonreproductive helpers from the experimental groups, while leaving the number of nonreproductive helpers in the control groups unchanged. Reproductive success, as measured by the number of fledglings, was significantly lower in the experimental groups. *(Based on Brown et al., 1982)*

FAMILY DYNAMICS

While Hamilton's Rule makes some very general predictions about animal social behavior, subsequent work by behavioral ecologists and ethologists has generated a more specific list of predictions about family dynamics (Emlen, 1995b). In particular, Steve Emlen has been integral in the development of an "evolutionary theory of family" that sets as its goal testing specific predictions regarding

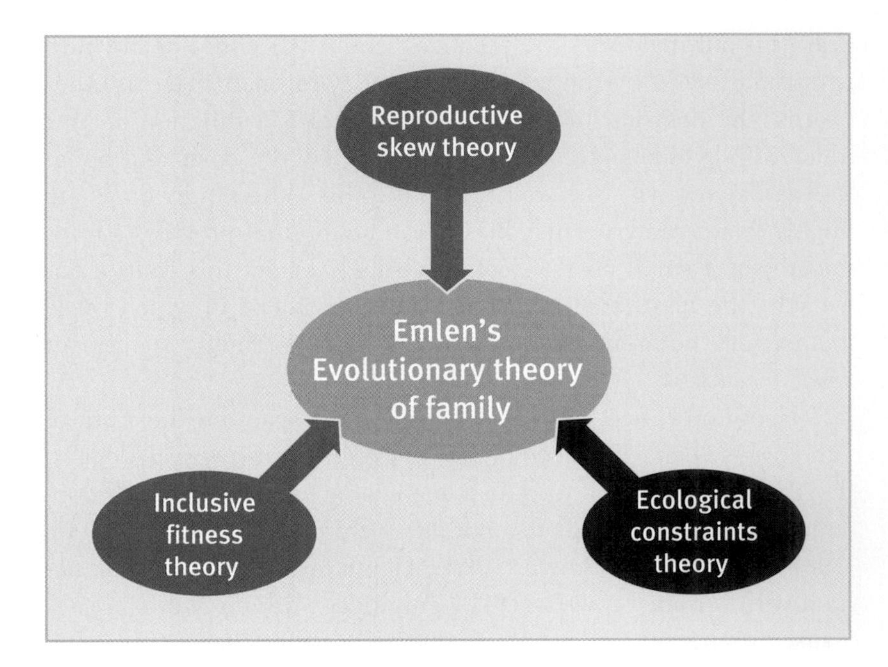

FIGURE 8.8. Evolutionary theory of family. Emlen's theories are generated by combining inclusive fitness, reproductive skew, and ecological constraints theory.

"the formation, the stability and the social dynamics of biological families" (Emlen, 1995b, p. 8092).

The building blocks for Emlen's predictions are: (1) inclusive fitness, which is also known as kin selection theory; (2) ecological constraint theory, which examines dispersal options of mature offspring (Koenig and Pitelka, 1981; Emlen, 1982a, 1982b; J. L. Brown, 1987; Koenig et al., 1992); and (3) reproductive skew theory, which seeks to understand how reproduction is divided among potential breeders by predicting conditions that should favor conflict or cooperation over breeding decisions (Figure 8.8).

Emlen puts forth fifteen specific predictions about animal family dynamics, and for each of these, he reviews the evidence from the animal literature, both for and against his predictions (Table 8.2). Subsequent to the publication of Emlen's paper, Davis and Daly (1997) tested Emlen's fifteen predictions, as they relate to human families. While Emlen's data are from a wide variety of animals, Davis and Daly's analysis is necessarily restricted to one species: *Homo sapiens*. Most of their data came from the 1990 Canadian General Social Survey (GSS), a telephone survey that amassed data from 13,495 households. While such a survey is probably reflective of modern Western society, it is important to recognize that it is not necessarily reflective of societies that live in more ancestral human conditions, such as hunter-gatherer societies.

Emlen and Davis and Daly provide us with a unique opportunity to examine and test evolutionary theories of family in both

TABLE 8.2. Biological families. The table lists the fifteen hypotheses associated with Emlen's evolutionary theory of the family. *(From Emlen, 1995b, p. 8093.)*

NO.	ABBREVIATED PREDICTION	EVIDENCE
1	Family groupings will be unstable, disintegrating when acceptable reproductive opportunities materialize elsewhere.	**Supportive:** 7 avian species, 2 mammalian species
2	Family stability will be greatest in those groups controlling high-quality resources. Dynasties may form.	**Supportive:** 5 avian species
3	Help with rearing offspring will be the norm.	**Supportive:** 107 avarian species; 57 mammalian species **Counter:** 5 avian species; 6 mammalian species
4	Help will be expressed to the greatest extent between closest genetic relatives.	**Supportive:** 5 avian species; 3 mammalian species **Counter:** 1 avian species
5	Sexually related aggression will be reduced because incestuous matings will be avoided. Pairings will be exogamous.	**Supportive:** 18 avian species; 17 mammalian species **Counter:** 1 avian species; 3 mammalian species
6	Breeding males will invest less in offspring as their certainty of paternity decreases.	**Supportive:** 1 avian species (many additional studies, some supportive, others counter, have been conducted on nonfamilial species)
7	Family conflict will surface over filling the reproductive vacancy created by the loss of a breeder.	**Supportive:** 6 avian species
8	Sexually related aggression will increase because incest restrictions do not apply to replacement mates. Offspring may mate with a step-parent.	**Supportive:** 4 avian species
9	Replacement mates (step-parents) will invest less in existing offspring than will biological parents. Infanticide may occur.	**Supportive:** 2 avian species; 2 studies summarizing mammalian data (Hausfater and Hrdy, 1984; Hrdy, 1999)
10	Family members will reduce their investment in future offspring after a parent finds a new mate.	**Supportive:** 2 avian species **Counter:** 2 avian species
11	Stepfamilies will be less stable than biologically intact families.	No data currently available
12	Decreasing ecological constraints will lead to increased sharing of reproduction.	**Supportive:** 2 avian species
13	Decreasing asymmetry in dominance will lead to increased sharing of reproduction.	**Supportive:** 2 avian species; 1 mammalian species
14	Increasing symmetry of kinship will lead to increasing sharing of reproduction.	**Supportive:** 4 avian species
15	Decreasing genetic relatedness will lead to increased sharing of reproduction. Reproductive suppression will be greatest among closest kin.	**Supportive:** 5 avian species; 2 mammalian species

humans and nonhumans alike, and we shall now examine a subset of five of their predictions in more detail.

PREDICTION 1. "Family dynamics will be unstable, disintegrating when acceptable reproductive opportunities materialize elsewhere."

In many ways, this is one of the most basic predictions that Emlen makes, in that it focuses clearly on the most fundamental costs and benefits of family life. Broadly speaking, individuals who have higher inclusive fitness when remaining with family should stay as part of the family unit, while those who have better opportunities for increasing their inclusive fitness elsewhere should depart. Evidence in support of this prediction in animals comes from seven long-term studies of birds and two long-term studies of mammals (Emlen, 1995b).

FIGURE 8.9. Superb fairy wren. In superb fairy wrens, young males often act as helpers-at-the-nest. When Pruett-Jones removed breeding males from their territories, almost all potential male helpers who could have dispersed to newly opened territories did so. *(Photo credit: Roger Brown/Auscape)*

The general technique for experimentally examining Prediction 1 is to create new, unoccupied territories and then examine whether mature offspring leave their natal area to live in such newly created areas (Pruett-Jones and Lewis, 1990; Walters et al., 1992; Komdeur, 1992). For example, consider Pruett-Jones and Lewis's (1990) work with superb fairy wrens (*Malurus cyaneus*), an insectivorous (insect-eating) Australian bird species (Figure 8.9). In this species, a breeding pair is often helped by their nonbreeding, young male offspring, while female offspring emigrate from their natal territory (Figure 8.10). To test the hypothesis that families will break down when suitable territories for helper young males emerge, Pruett-Jones and Lewis removed the breeding males from twenty-nine superb wren territories, for a total of forty-six removals over time, as some territories had a resident removed more than once.

By removing breeding males from their territories, new breeding opportunities arose for male helpers in nests near the area of the removals. All but one of the thirty-two potential male helpers who could have dispersed to the newly opened territories did so. Not only did male helpers leave their families when a breeding opportunity opened, they did so quickly—new territories were occupied by former male helpers within six hours (on average). Apparently, a shortage of females and breeding territories creates a scenario in which male helpers instantly seize the opportunity for a breeding territory and thus disband family life when the chance arises. As such, Pruett-Jones's work suggests that helping-at-the-nest may raise inclusive fitness of young males when territories are limited, but not otherwise.

Not surprisingly, the picture is not nearly as clear-cut when it comes to the data on humans and family stability. Davis and Daly found that married individuals were much more likely to live away

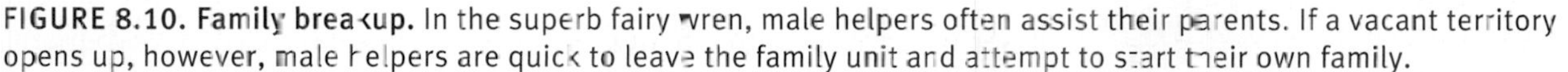

FIGURE 8.10. Family breakup. In the superb fairy wren, male helpers often assist their parents. If a vacant territory opens up, however, male helpers are quick to leave the family unit and attempt to start their own family.

from their parents than were single individuals in the same age/sex category. This suggests that new marriages cause existing family units to dissolve. It need not have turned out that way. Married individuals might very well have been more likely than single individuals to live with one set of parents. While the data on residence suggest that marriage does cause the dissolution of existing family units (and, of course, creates new such units), Prediction 1 is not supported if we use another set of relevant data. If we ask whether there exist differences between married and single individuals living away from their parents, in terms of contact with parents or grandparents, very few differences emerge. For most age/sex categories, married individuals living apart from either set of parents were just as likely to stay in contact (via phone, visits, letters) with parents and grandparents as were single individuals living away from home.

On the whole, Davis and Daly argue that the data from the Canadian GSS do not support Emlen's first prediction. Some differences were found in certain subcategories, but as Davis and Daly note "at best Emlen's first prediction holds only for some demographic groups with some types of relatives" (Davis and Daly, 1997, p. 415). Rather, Davis and Daly argue that, with some exceptions, it appears that human parents act as post-reproductive helpers to their own offspring, which may select for strong family bonds that do not easily dissolve when offspring get married.

PREDICTION 2. "Families that control high-quality resources will be more stable than those with lower-quality resources. Some resource-rich areas will support dynasties in which one genetic lineage continuously occupies the same area over many successive generations."

If the benefits associated with a natal territory are sufficiently great, then that, in conjunction with the indirect benefits of associating with relatives, produces a strong force to keep families intact. Individuals often leave their family and place of birth as a result of diminishing resources, including needed space for a new breeder. When dwindling resources are removed from the equation, as in the case of very high-quality territories, dynasties may develop over time. Emlen argues that offspring from families that control high-quality resources are likely to be much more reluctant to vacate the natal territory, as few alternatives provide what is available at home. Over the long run, this will create dynasties in families that occupy the very highest-quality territories. Not only are the offspring that remain on high-quality territories receiving a benefit, but their parents are as well, since they then pass down the best-quality territories to their blood kin (J. L. Brown, 1974).

FIGURE 8.11. Dynasty building in acorn woodpeckers. In cooperatively breeding acorn woodpeckers (*Melanerpes formicivorus*), the young not only survive better on territories with more storage holes, they are also more likely to remain on their natal territories throughout their life, creating a "family dynasty." Such dynasties enable kin groups to compete for superior territories. *(Photo credit: F.K. Schleicher/VIREO)*

Data from six species of birds support the dynasty-building hypothesis in that birds from high-quality family territories are indeed less likely to disperse from the natal territory than their counterparts from families with inferior territories. For example, in cooperatively breeding acorn woodpeckers (*Melanerpes formicivorus*), the critical measure of a territory's quality is the number of storage holes (Figure 8.11). In a New Mexican population of acorn woodpeckers studied by Stacey and Ligon (1987), territories varied from less than 1,000 to greater than 3,000 storage holes. Not surprisingly, individuals on territories with lots of storage holes produced a greater average number of offspring (Figure 8.12). More to the point, in areas with more than 3,000 storage holes, 27 percent of the young remained on their natal territories and helped their relatives, while only 2 percent of the young on territories with less than 1,000 storage holes stayed and helped, despite the fact that many open territories were available. The benefits of remaining on a high-quality territory appear to be real, as (male) birds who served as helpers had a relatively high probability of eventually entering the

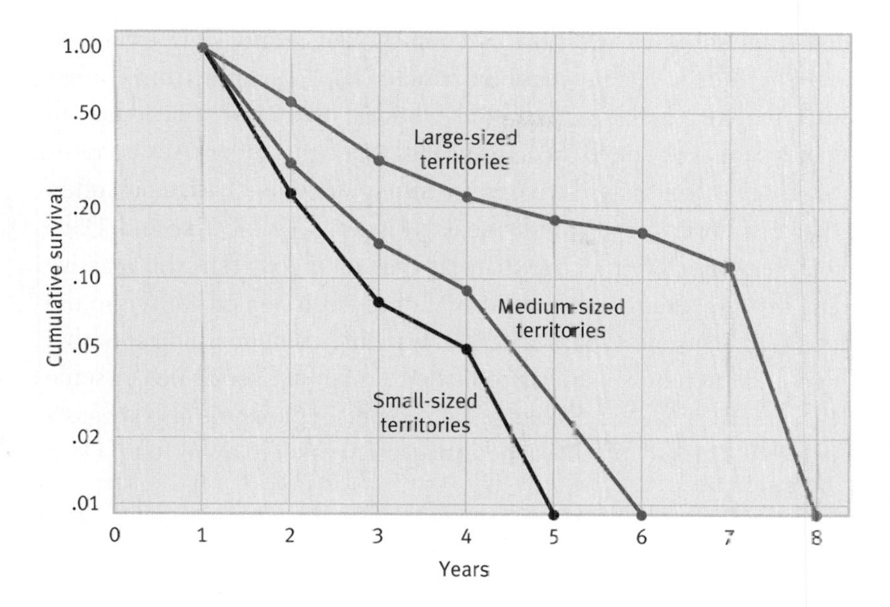

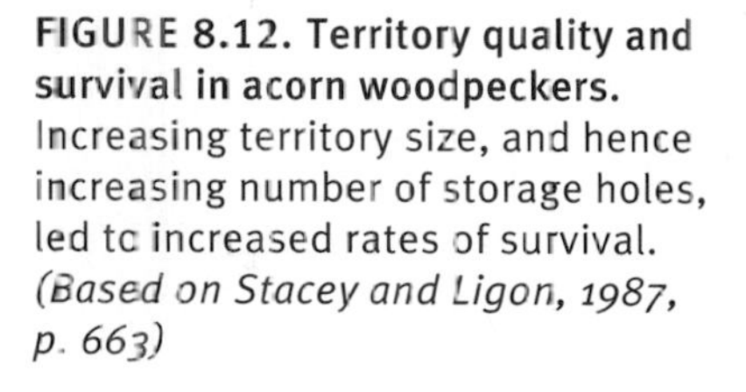
FIGURE 8.12. Territory quality and survival in acorn woodpeckers. Increasing territory size, and hence increasing number of storage holes, led to increased rates of survival. *(Based on Stacey and Ligon, 1987, p. 663)*

breeding population, often breeding in turn on their natal territory (which could hold numerous breeders simultaneously; Stacey and Ligon, 1987). In another cooperatively breeding bird species, the Florida scrub jays (*Aphelocoma coerulescens*; Figure 8.13), multigenerational dynasties are created on high-quality home areas via **territory budding**. Budding occurs when aggression on the part of the territory holder and his family produces a new expanded area that includes space for newly developing breeders

In terms of human family dynamics, Prediction 2 translates into well-to-do families being more stable than poorer families. Davis and Daly clearly demonstrate that if a stable family is defined in terms of co-residence (as in the nonhuman case), then this prediction is not supported. To cite just one of Davis and Daly's examples, White (1994a) has found that young adults from rich (First World) families tend to be less likely to be living with their parents than are same-aged individuals from poorer families. Nonetheless, since resources are much more mobile than ever in today's Western economies, it might be argued that familial co-residence is an inappropriate yardstick for measuring family stability. If the measure of stability is defined in terms of maintaining family contacts and providing social support during adulthood, the data are more supportive of Prediction 2.

FIGURE 8.13. Jay budding. Florida scrub jay territories often expand by "budding." *(Photo credit: Lynn M. Stone/Auscape)*

At the most general level, data suggest that such contact and support are indeed more often found in wealthy families (Taylor, 1986; White and Reidmann, 1992; Eggebeen and Hogan, 1990). Using GSS data, Davis and Daly addressed the more detailed question of whether contact with kin is not only more likely, but more frequent, as a function of wealth. Using letter, phone, or face-to-face conversations as a measure of contact, Davis and Daly examined whether individuals in wealthier families kept in contact more often with parents, grandparents, and siblings, than did individuals in poorer families. The GSS data suggest that for most age/sex cohorts, wealthier individuals did keep in touch with relatives more often than did lower-income individuals (Figure 8.14). Davis and Daly, however, offer an important caveat that emerged from the massive data set they examined (only a small fraction of which we see in the figure). They note that while "the available evidence indicates that high-income families are more likely to maintain social ties at some level and to engage in exchange . . . the present analysis also suggests that the strength of these ties may be weaker" (Davis and Daly, 1997, p. 420).

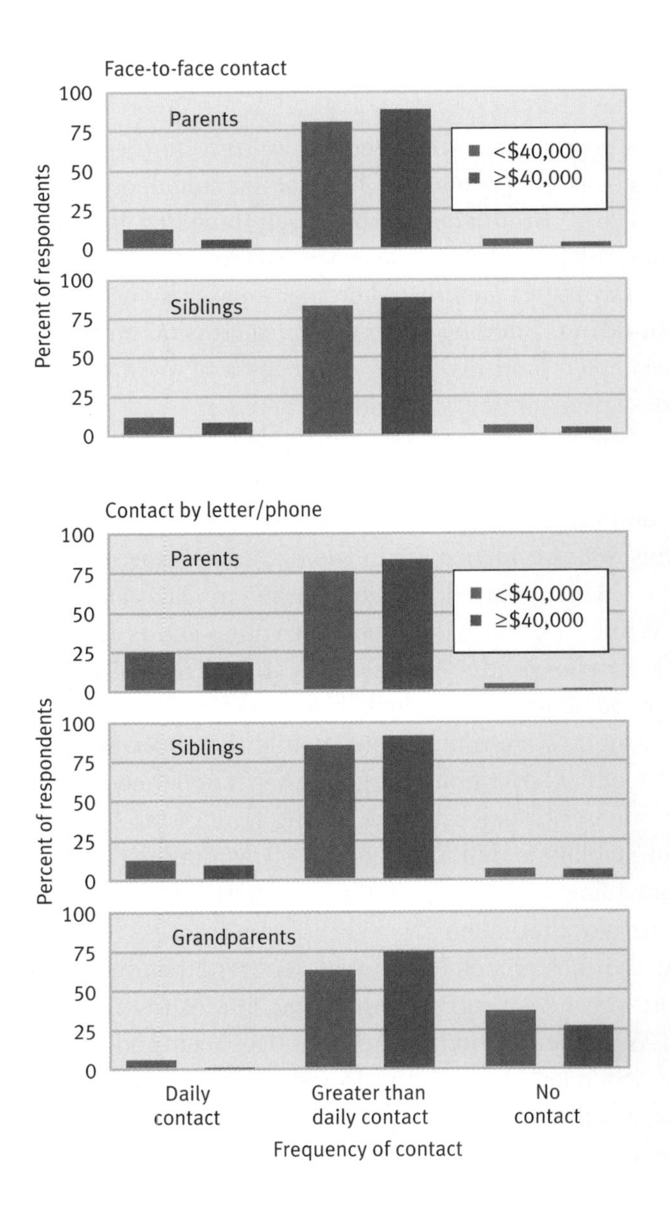

FIGURE 8.14. Wealth and familial contact. Based on Emlen's dynasty-building hypothesis, Davis and Daly hypothesized that familial contact would be greater in wealthy human family units. The data, from the 1991 Statistics Canada, appear to support this general prediction. *(From Davis and Daly, 1997)*

FIGURE 8.15. White-fronted kinship. Inclusive fitness models of behavior have been tested extensively in white-fronted bee-eaters. *(Photo credits: N. J. Demong)*

PREDICTION 4. "Assistance in rearing offspring (cooperative breeding) will be expressed to the greatest extent between those family members that are the closest genetic relatives."

Under Hamilton's Rule, increasing *r* has the effect of increasing prosocial interactions between individuals. As such, the higher the value of *r* between two individuals, the more likely they are to help one another. In addition, Hamilton's Rule suggests that, all else being equal, when given the choice between helping individuals who differ in *r*, aid should be differentially dispensed to individuals who are closest blood kin.

Emlen reports that in the nine studies published on cooperation in species of birds or mammals that live in extended families, eight studies uncovered individuals who preferentially extended aid as a function of blood relatedness. For example, in white-fronted bee-eaters (*Merops bullockoides*; Figure 8.15), helpers chose to aid the pair they themselves were most closely related to in 108 of 115 opportunities (Figure 8.16).

In many ways, kinship and aiding relatives is best exemplified by social insects, who show extreme levels of cooperation in conjunction with an odd genetic architecture that creates sisters that are "super relatives." Some social insects have a haplodiploid genetic system. Normally, we think of all individuals in a species as being either diploid (possessing two copies of each chromosome) or haploid (possessing only one copy of each chromosome). Haplodiploid species defy this convention in that males are haploid, while females are diploid. As a result of the genetics underlying haplodiploidy, sisters are related to one another on average by a coefficient of relatedness of 0.75, which has the bizarre effect of making females more related to their sisters than to their own offspring.

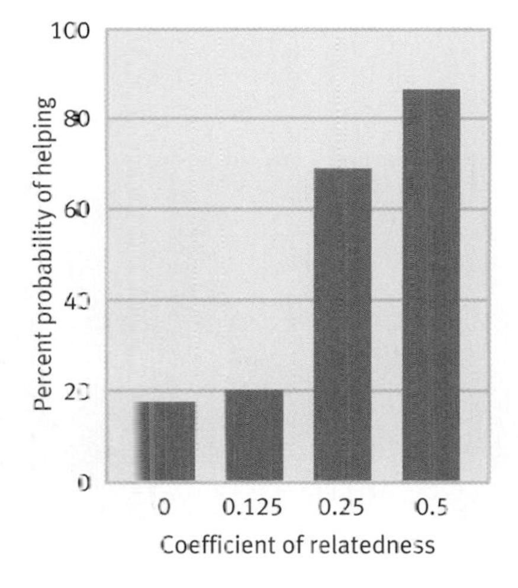

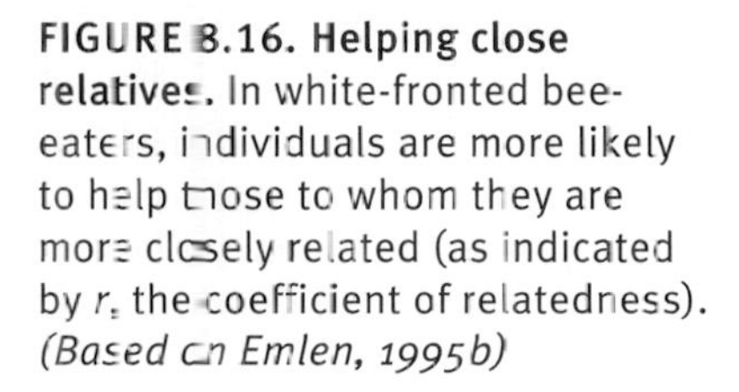

FIGURE 8.16. Helping close relatives. In white-fronted bee-eaters, individuals are more likely to help those to whom they are more closely related (as indicated by *r*, the coefficient of relatedness). *(Based on Emlen, 1995b)*

This value differs from the standard average relatedness of sisters in diploid species (r = 0.5), because in haplodiploids, full sisters inherit exactly the same genes from their father, while in diploid species, females have only a 50 percent chance that a gene that they inherited from their father is identical to a gene that their sister inherited from their father.

With an r of 0.75 between sisters, one would expect high levels of aid giving—just the sort of thing that social insects are well known for. It is, of course, the female workers in many species who go to suicidal lengths to defend a hive full of their sisters. A bee's stinger is often designed for maximal efficiency, and this often entails having the stinger ripped from the face of the bee doing the stinging in a way that causes the death of the stinging individual. Kinship need not, however, produce such ultra-altruism. If individuals are able to gauge their relatedness to others, then social insects may be influenced by kinship in any number of ways. Consider the fascinating case of worker "policing" in honeybees (*Apis mellifera*; Figure 8.17).

Using the mathematics of inclusive fitness theory, Ratnieks and Visscher (1989) found that in honeybee colonies with a single queen who mates one time, female workers are more related to their nephews (their sisters' sons, r = 0.375) than to their brothers (r = 0.25). This inequality, however, switches when the queen mates multiple times. In such a situation, workers may be more closely related to brothers than to nephews, with the exact values of relatedness depending on the number of different males with

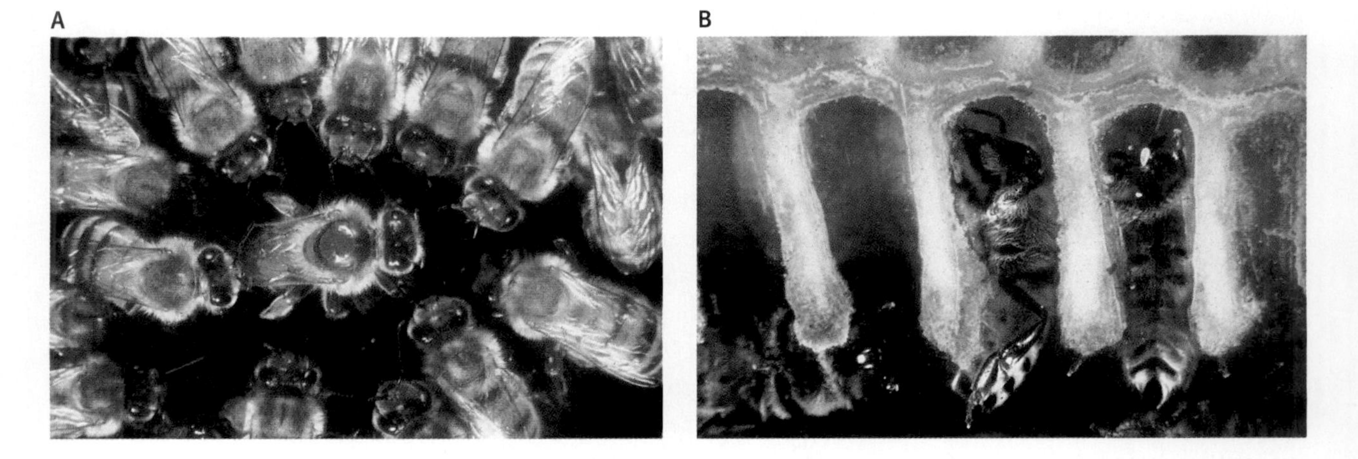

FIGURE 8.17. Honeybee policing. (A) While the queen (designated by the red dot on her back) typically lays the eggs in a honeybee colony, workers also attempt to lay unfertilized eggs. **(B)** When an egg laid by a worker is detected by worker police, it is eaten or destroyed. Workers are much more likely to destroy eggs produced by other workers than eggs produced by the queen. Such "policing" has inclusive fitness benefits associated with it. *(Photo credits: Francis Ratnieks)*

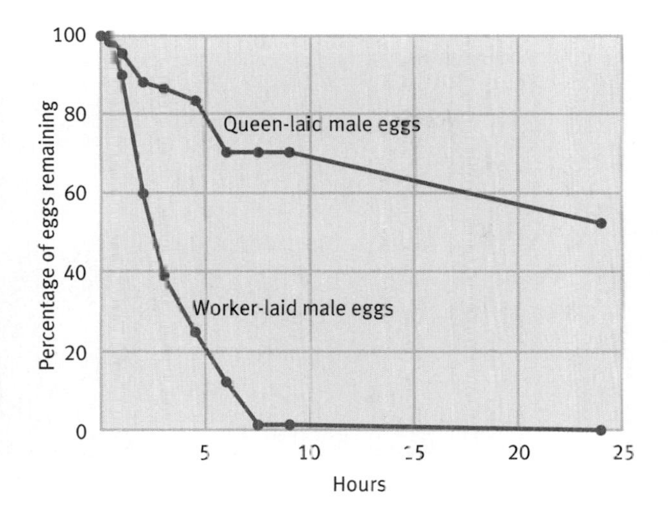

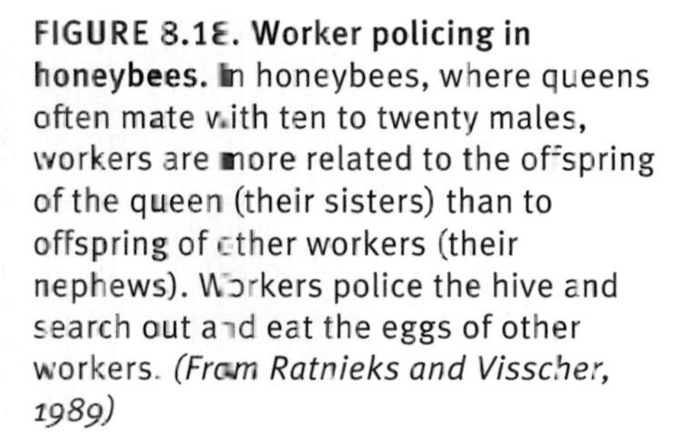
FIGURE 8.18. Worker policing in honeybees. In honeybees, where queens often mate with ten to twenty males, workers are more related to the offspring of the queen (their sisters) than to offspring of other workers (their nephews). Workers police the hive and search out and eat the eggs of other workers. *(From Ratnieks and Visscher, 1989)*

whom a queen mates. Under such conditions, wherein female workers may now be more related to brothers than to nephews, Ratnieks and Visscher (1988, 1989) argue that worker "policing" of honeybee reproduction may evolve. Such policing, for example, may take the form of workers favoring those eggs to which they are most highly related.

Ratnieks and Visscher examined the possibility that honeybee workers may favor brothers over nephews, where queens typically mate with ten to twenty different males. They found that honeybee workers showed remarkable abilities to discriminate between worker-laid eggs, which produce nephews, and queen-laid eggs, which produce brothers. After twenty-four hours, only 2 percent of the worker-laid eggs remained alive, while 61 percent of the queen-laid eggs remained alive (Figure 8.18). Workers appear to use a specific egg-marking pheromone produced only by queens to distinguish which eggs to destroy and which eggs to leave unharmed (Ratnieks, 1995), and in so doing, they police the hive in a manner that increases their inclusive fitness.

In accordance with the predictions of kin selection models, worker policing occurs in honeybees and common wasps (*Vespula vulgaris*; Foster and Ratnieks, 2001), where queens mate multiple times, but it is absent in stingless bees and bumblebees—species in which queens mate only once (Peters et al., 1999; Estoup et al.,1995). This interspecies comparison can also be made within a single species. In the vespine wasp, *Dolichovespula saxonica* (Figure 8.19), some nests have a single queen who mates only once, while others have a queen who mates multiple times. Foster and Ratnieks (2000) found that policing was the norm in multiply mated queen nests, but it was very rare in nests with a single-mated queen.

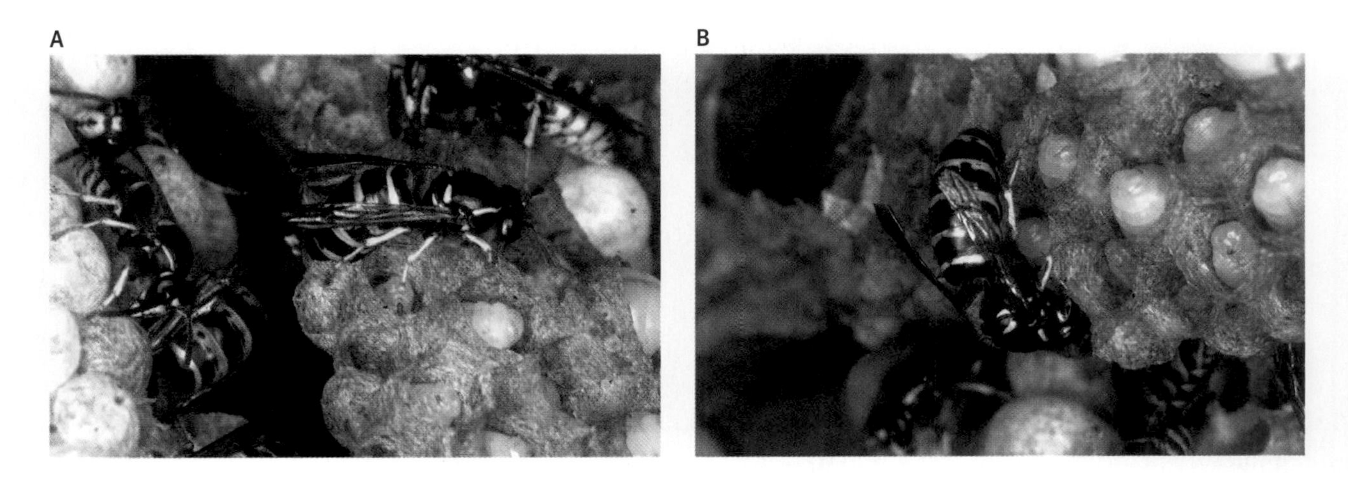

FIGURE 8.19. Wasp policing. Policing occurs in wasps, as well as in honeybees. In the wasp *Dolichovespula saxonica*, workers often lay (haploid) eggs, both in nests with single-mated and multiply mated queens. Such eggs are often eaten when detected by other workers. **(A)** Here a worker is eating another worker's egg. Policing is much more common in wasp colonies where the queen has mated with many males. **(B)** The wasp in the middle of the photo is a worker who has just laid an egg. *(Photo credits: Kevin Foster)*

Unfortunately, in Davis and Daly's examination of Prediction 4 in humans, the GSS data were not collected in a way to address this question. Individuals in the GSS study were related by an r value of either 0.5 or 0.0, and thus the distinction of how different relatives, that is, individuals with different positive values of r, are treated could not be addressed.

PREDICTION 9. "Replacement mates (step-parents) will invest less in existing offspring than will biological parents . . ."

This prediction stems from an extension of kinship theory to the case of stepfamilies. When step-parents are faced with decisions about how to distribute benefits to offspring, such parents should favor their biological offspring ($r = 0.5$) over their already existing stepchildren ($r = 0$). The extreme version of this hypothesis predicts that, under certain conditions, new step-parents may opt to harm, potentially in a fatal manner, their mate's current offspring, as such offspring are unrelated (in blood kinship terms), and as such are a drain on resources and return no kin-based benefits (Hrdy, 1979; Hausfater and Hrdy, 1984; Parmigian and Vom Saal, 1995; Hrdy, 1999). Moreover, as a result of dominance asymmetries that typically underlie male-female interactions, both Stephen Emlen (1995b) and Sarah Hrdy (1979) argue that such infanticide will be more common when the step-parent is a male.

Data from rodents, carnivores, and primates support the hypothesis that step-parents invest less in stepchildren. Moreover, infanticide on the part of male, as well as female, step-parents has been uncovered in numerous species in these groups (Hausfater and Hrdy, 1984). There is also growing evidence that avian step-parents destroy clutches containing potential stepchildren (Emlen, 1995b).

Compared with Emlen's other predictions, the human data relevant for Prediction 9 are both comprehensive and easier to interpret, though not necessarily pleasant to hear. Cross-cultural data clearly demonstrate that step-parents invest less in their stepchildren than in their biological children (Bray, 1988; Hetherington et al., 1989; White, 1994b), and children from stepfamilies leave home significantly earlier than children from biologically intact families.

Some of the strongest data on human behavior comes from comparisons of biological families versus stepfamilies. Stepchildren suffer child abuse at rates up to 100 times higher than those for biological children (Daly and Wilson, 1988, 1999; Figure 8.20).

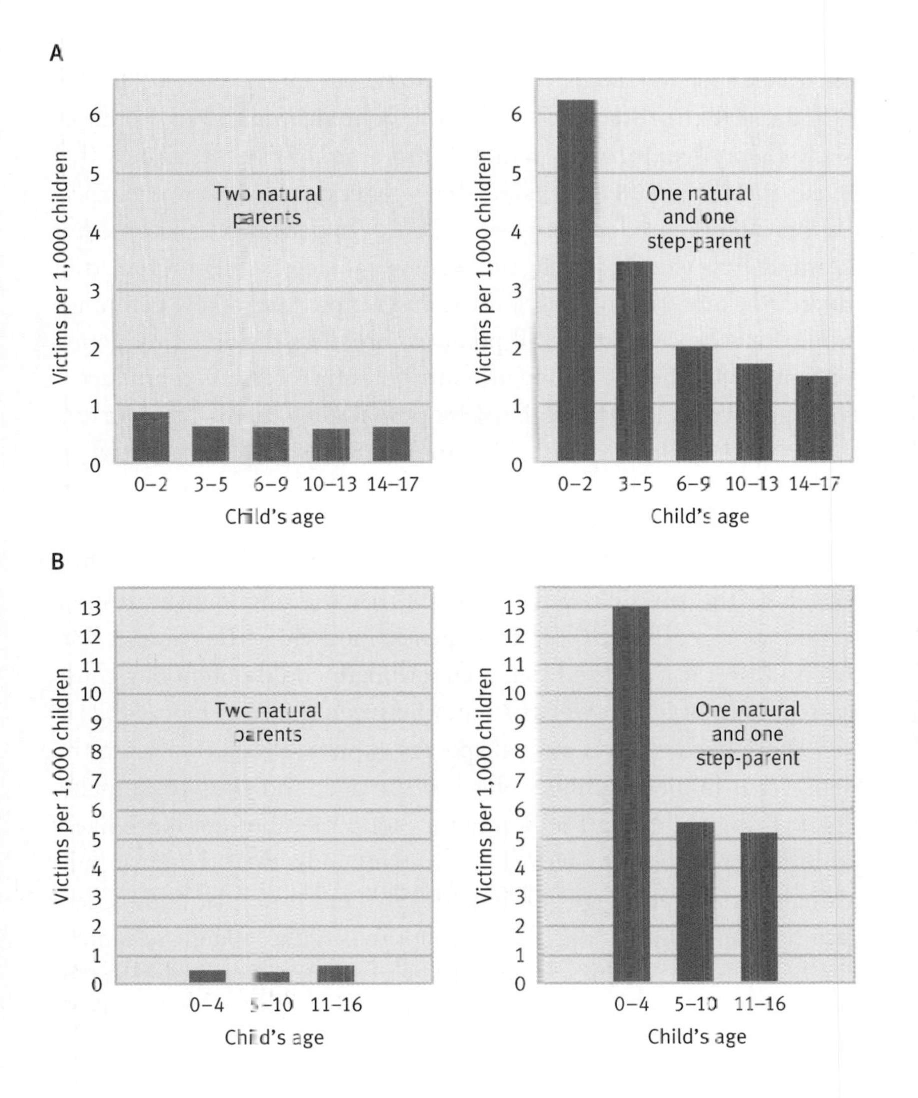

FIGURE 8.20. Child abuse and blood relatedness. (A) The per capita rates of child abuse reported in 1980 by the American Humane Association in the United States. Rates of child abuse were significantly greater in households where one parent was not blood kin. **(B)** The same relationship between child abuse and family type was also uncovered in Canada, as reported in 1983 by a provincial registry in Hamilton, Ontario. *(From Daly and Wilson, 1988)*

The extreme version of such abuse translates into stepchildren being murdered at a much higher rate than comparable numbers for biological offspring in most (Kaplun and Reich, 1976; Daly and Wilson, 1988, 1994, 1999), but not all large-scale studies (Temrin et al., 2000). Such abuse and murder are predominately at the hands of the step-parent, and because custom and law around the world favor a mother's rights, the abusing parent is almost always a stepfather.

The case for blood relatedness driving stepfamily violence is not constrained to interactions between parents and offspring. In addition to step-parents having no blood connection to their stepchildren, stepsiblings also have a lower r (0 or .25) than biological siblings ($r = 0.5$), and indeed some data suggest that conflicts are more common among half-sibs (Zill, 1988; Aquilino, 1991).

PREDICTION 13. "Reproduction within a family will become increasingly shared as the asymmetry in social dominance between potential co-breeders decreases."

To test this prediction often requires the use of what are known as optimal skew models. First introduced in models by Emlen (1982a, 1982b), Stacey (1982), and Vehrencamp (1983), optimal skew theory has recently matured into a sizable set of models that make specific predictions about how genetic, social, and ecological factors affect reproduction (Reeve, 2003). In particular, skew models examine how genetic, social, and ecological factors affect the distribution, or skew, of reproduction within groups. Such skew can range from 0, in the case where all potential breeders divide reproduction equally, to 1, where only one of many potential breeders reproduce.

With respect to Prediction 13, optimal skew theory predicts that since fighting is often most dangerous between equally matched individuals (Parker and Rubinstein, 1981; Archer, 1988; Riechert, 1998), the closer subordinate and dominant relatives are in fighting ability, the more likely the dominant individual will share reproduction with the subordinate. It predicts that the dominant individual would be more likely to offer a "peace incentive" (Reeve and Ratnieks, 1993) if it realized that a subordinate could potentially cause it great harm if a fight were to occur between the two (Figure 8.21).

Work in numerous animal species support Prediction 13 in the sense that family members close to the age, and thus presumably the fighting ability, of the dominant group member are more likely to breed concurrently with the dominant individual than are individuals dissimilar in age (Emlen, 1995b). For example, Scott Creel and his colleagues (1992) built an optimal skew model to examine the distribution of reproduction in dwarf mongoose populations. In this species, although most breeding is undertaken by a single

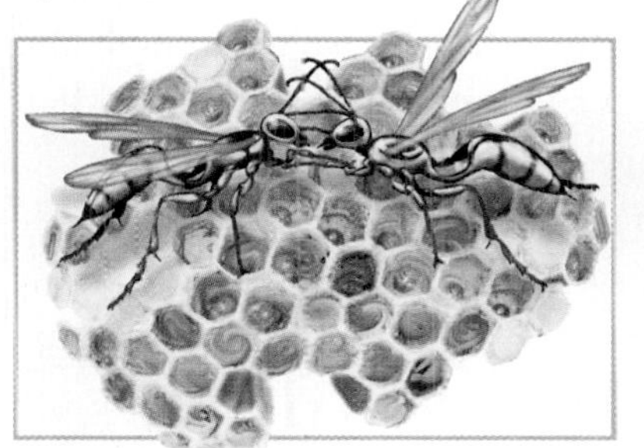

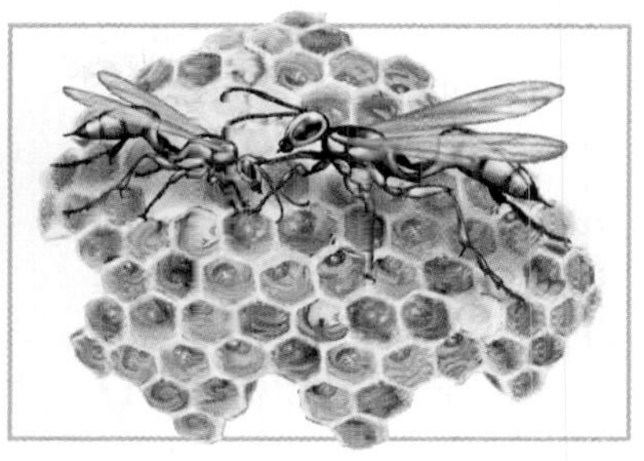

FIGURE 8.21. Reproductive skew theory. When two individuals are similar in size (left), the dominant individual is more likely to share reproduction with a subordinate than when size differences are greater (right). Such "peace incentives" are predicted by reproductive skew theory.

dominant pair of individuals, subordinates occasionally breed as well. Such breeding by subordinates had been hypothesized to be a breakdown of normal reproductive suppression mechanisms available to the dominant individuals. However, based on their skew model, Creel and his colleagues argued that, under certain conditions, occasional reproduction by subordinates is expected and is not the failure of some system of reproductive suppression. The skew model built by Creel predicted the frequency of subordinate females becoming pregnant quite accurately (Figure 8.22).

In their review of human behavior, Davis and Daly cite only one study—on the rule of marriage in Tibet—which provides data relevant to Prediction 13. In some Tibetan communities, concurrent polyandry—a mating system in which females have more than one male mate—is common. What is particularly interesting here is that females in the Tibetan system marry pairs of brothers. Why polyandry exists in these Tibetan communities is still debated, but from the perspective of Prediction 13, what really matters is that we have a system in which relatives (brothers) need to determine how much of the reproductive pie they will each receive. Crook

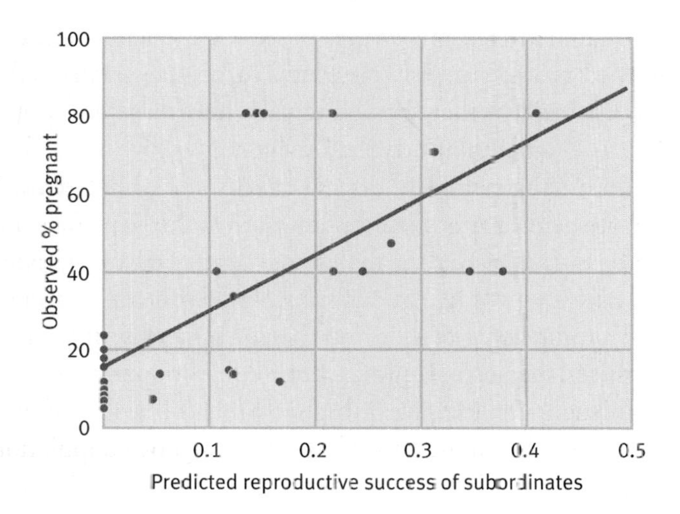

FIGURE 8.22. Optimal skew models and reproductive success in the dwarf mongoose. Creel and Waser built an optimal skew model of reproductive success in dominant and subordinate mongooses. The model predicted reproductive success in female subordinate mongooses remarkably well. *(From Creel and Waser, 1991)*

and Crook (1988) found that when brothers married to the same wife were close in age, and hence presumably close in strength and fighting skills, reproduction was more evenly divided than when brothers were of more disparate ages.

Conflict within Families

As is obvious from the above, blood relatedness plays a pivotal role in understanding how animal social behaviors evolve. Most often, inclusive fitness theory is used to better understand why relatives are more cooperative as a function of blood relatedness, or why individuals are often not particularly kind to those who are not their relatives. There is, however, another side to how relatedness shapes animal behavior, even when that behavior is being dispensed among blood relatives. Two ways in which this other side is manifest are **parent-offspring conflict** and sibling-sibling conflict.

PARENT-OFFSPRING CONFLICT

With a basic understanding of kinship under our belts, one thing should be perfectly clear—parents should be particularly willing to go to great lengths to help their offspring. After all, under standard diploid genetics (with no inbreeding), parents and offspring have an average r of 0.5. Furthermore, parents are almost always in a better position to help offspring than vice versa. As such, parental aid should be seen in many contexts. And indeed it is. Parents, especially mothers, provide all sorts of aid to their offspring (Clutton-Brock, 1991).

There are many reasons why it is the female who is most often the parent to provide aid to offspring. To begin with, females are almost always certain that the young they help raise are their own. Males, on the other hand, typically can only have some level of confidence in their paternity, as it is always possible that a given clutch of offspring were sired by another male. This lack of certainty reduces the resources males (as compared to females) are willing to invest in young. In addition, while females are often restricted to producing a certain number of offspring per unit time, males in many species can produce millions of sperm almost constantly. As such, males have a higher probability of producing a large number (or for the losers in the competition, a small number) of offspring per unit time (Clutton-Brock, 1988). The more a male mates, the more offspring he produces, and so males tend to invest more in multiple matings and less in raising offspring. Relatively speaking, in many species, females are more willing to invest in any given offspring than are males.

While it is true that a relatedness value of 0.5 creates strong selection pressure for parents (particularly mothers) to provide much aid to their offspring, there are limits to this aid, and parent-offspring theory sets forth these limits (Trivers, 1974). Understanding the fundamentals of parent-offspring conflict begins by recognizing that a parent's decisions about how much aid to give to any particular offspring are tempered by how much is available and by how many offspring such a parent is likely to have in the future.

In principle, a parent could dispense every ounce of energy it has to provide offspring X with all the benefits at its disposal. But if such an effort kills the parent or severely hampers future reproductive possibilities, it may not have been the wisest parental move in terms of the total number of offspring produced over the course of a lifetime. Every offspring has an r of 0.5, and natural selection should favor parents who raise as many healthy offspring over the course of their lives as is possible. So, there are limits on parental investment with respect to any given child (let's say child X).

Now, let us look at parental investment from child X's perspective. Child X will benefit from its parent aiding current and future siblings (who have an $r = 0.5$). Yet, child X is obviously more related to itself ($r = 1$) than to any of its siblings. As such, child X values the resources for itself more than the resources for its siblings. But to parents, all offspring are equally valued. This then sets up a zone of conflict between how much child X wants a parent to give to him, and how much a parent is willing to give (the former always being greater than the latter). This zone is where parent-offspring conflict takes place. Let's take a look at four cases of such conflict, two in animals and two in humans.

PARENT-OFFSPRING CONFLICT AND WEANING. One classic paradigm for studying parent-offspring conflict centers on weaning. In many mammals and birds, as young mature, they are "weaned" off a diet that is primarily (if not exclusively) provided by parents. In mammals, one measure of weaning is when young become independent of mother's milk. A zone of conflict exists for weaning from mother's milk in that the mother will provide only so much milk for a current brood of offspring, as she must consider her lifetime reproductive success. Yet, those in the current brood will be selected to beg for more than the maternal optimum, as each will weigh its own reproductive success more heavily than the reproductive success of other siblings (both current and future).

FIGURE 8.23. Weaning conflict. A classic example of parent-offspring conflict is reflected in the tension over when weaning should occur. *(Photo credit: Yves Lanceau/Auscape)*

Malm and Jensen (1997) examined parent-offspring conflict in mother-pup interactions in Swedish Dachsbracken dogs (Figure 8.23). Surprisingly, the weight of mothers stayed constant during the weaning period, suggesting that they were able to receive enough additional nutrition to maintain their body weight despite suckling their young. During this period, however, mothers spent less time with pups. In addition, as the pups matured, mothers decreased the number of times they urged the young to feed. Toward the end of the weaning period, while mother-initiated interactions decreased, there was a tendency for care-seeking and contact-seeking behavior from the pups to increase.

The period at the end of pup weaning may represent the zone of conflict between mother and pup. As pups reach seven weeks old, the maternal optimum in terms of mother's milk appears to have been reached, while the pup optimum has not yet been met. Hence, we see the mother's declining milk to her still begging pups as an outcome of parent-offspring conflict.

PARENT-OFFSPRING CONFLICT AND SEX RATIO IN SOCIAL INSECTS. The haplodiploid genetics underlying the social insects sets up a nice test of parent-offspring conflict over colony sex ratios (Trivers, 1974; Trivers and Hare, 1976; Alexander and Sherman, 1977). To see how parent-offspring conflict can be tested in hymenopteran—social insects (ants, bees, wasps)—we must recall that a queen's relatedness to all her offspring, both male and female, is equal to 0.5 (in that both male and female offspring contain copies of 50 percent of the queen's genome). For female workers, who tend the young as they mature, however, relatedness is skewed. Remember that haplodiploidy creates a scenario in which sister workers can be related to one another by an r of 0.75, but only share

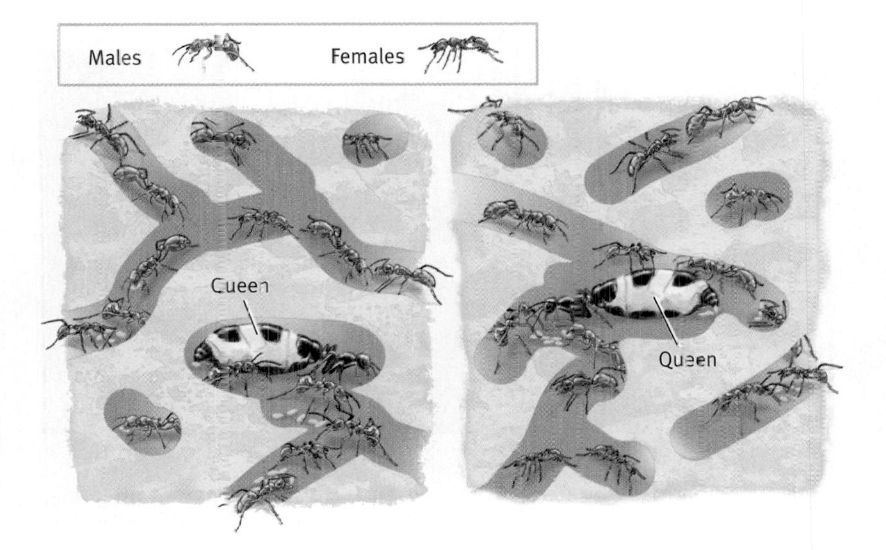

FIGURE 8.24. Control of the colony. If a queen mates with just one male, kin selection predicts a 1:1 sex ratio of females to males if queens are in charge (left), but a 3:1 sex ratio of females to males if workers dictate sex ratios (right). Males are in blue, females in red.

an r of 0.25 with their haploid brothers. As such, the queen should be favored to produce a sex ratio of 1 female:1 male, where both sexes are valued equally. If workers (females) control the sex ratio, however, through policing eggs, destroying other workers' eggs, or some other means, then this ratio should be closer to 3:1, as sisters are three times as valuable to the workers as brothers (Figure 8.24). Hence, there is a parent-offspring conflict over sex ratio.

This idea has been modeled mathematically numerous times (Trivers and Hare, 1976; Oster et al., 1977; Charnov, 1978), and two lines of empirical evidence suggest that workers may be in charge of determining the sex ratio. First, sex ratios in social insects are indeed closer to 3:1 than 1:1 (Trivers and Hare, 1976). Second, experimental work has examined what happens when a queen dies and is replaced by one of her daughters (Yanega, 1988, 1989; Mueller, 1991). After such a replacement, workers are equally related to the new queen's sons and daughters, and hence they should favor a 1:1 sex ratio. In such cases, the frequency of males produced indeed rises, and the sex ratio moves toward 1:1—precisely what one would expect if the workers were controlling the sex ratio.

IN-UTERO CONFLICTS IN HUMANS. Parent-offspring conflict is not restricted to postbirth interactions between mothers and offspring (Figure 8.25). David Haig makes the case that in humans, and likely in many other species as well, parent-offspring conflict is occurring between a pregnant mother and her developing fetus (Haig, 1993). That is, despite the fact that pregnancy has typically been viewed as the ultimate set of cooperative interactions between mother and fetus, evolutionary conflicts seep in even at this level. What makes

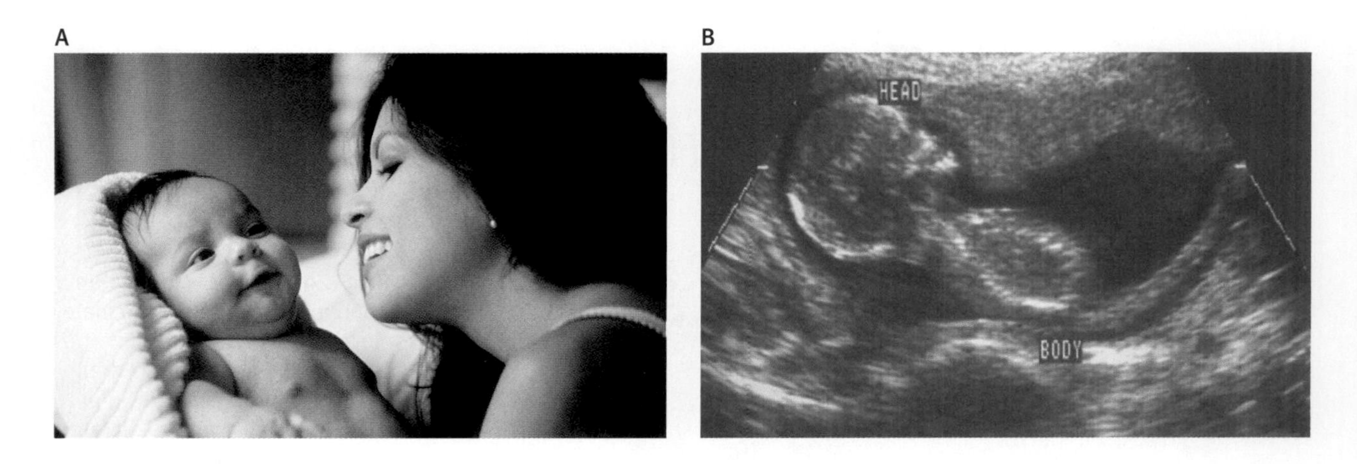

FIGURE 8.25. Mothers and babies. While the parent-offspring relationship is usually cooperative **(A)**, parent-offspring conflict can occur, even in utero as shown by an ultrasound **(B)**. *(Photo credits: Ariel Skelley/Corbis; Visuals Unlimited)*

in-utero parent-offspring conflict particularly fascinating is that it shifts the balance of power between parent and offspring. In most cases of parent-offspring conflict, a parent has the upper hand, as she is almost always behaviorally dominant to her offspring. When the offspring is still in utero, however, mothers can't deprive offspring of resources without depriving themselves as well, thus shifting the balance of power away from the mother and toward neutral.

Haig summarizes his general argument succinctly: "Fetal genes will be selected to increase the transfer of nutrients to their fetus, and maternal genes will be selected to limit transfer in excesses of some maternal optimum" (Haig, 1993, p. 495). Again, if we envision two sets of genes, one turned on inside the fetus during the fetal stage and one turned on inside the mother during pregnancy, we see that these genes do not have identical interests. From the perspective of maternal genes, all else being equal (and we will see in the next example, all else need not be equal), each fetus, current and future, is valued equally. From the perspective of the fetal genes, the current fetus (the one they reside in) is worth more than future siblings (r of 1.0 versus 0.5).

Using published medical literature, Haig argues that fetally derived cells that have "invaded" the maternal endometrium (the membrane lining the mother's uterus) during implantation are designed to change maternal spiral arteries in such a way as to make constriction of the arteries much more difficult. Such an invasion benefits the fetus in two ways: (1) by providing the fetus direct access to maternal arterial blood and allowing the fetus to release hormones and other substances directly into the maternal bloodstream, and (2) by putting the volume of blood (and hence the nutrients it contains) under fetal, rather than maternal, control.

Haig argues that placentally produced hormones, such as human placental lactogen and human chorionic gonadotropin, manipulate the in-utero environment in a manner that benefits the fetus at the cost of the mother. For example, a fetus may use human placental lactogen to increase maternal resistance to insulin. If this action goes unopposed by the mother, it would normally cause sugar to remain in the blood a longer time, thus providing a longer period of time for the fetus to access such sugar for itself. The maternal counterresponse is, however, to increase insulin. If this countermeasure is unsuccessful, we have the fetus obtaining extra sugar, but the mother suffering from gestational diabetes. Similar arguments can be made regarding gestational high blood pressure.

Gestational diabetes and high blood pressure are not only clear cases of parent-offspring conflict, but may have serious implications for treatment of pregnancy-related medical conditions. If, for example, medical doctors viewed gestational diabetes as a "disease" that needed to be cured, they would presumably act differently than if they viewed gestational diabetes as an evolutionary measure selected by fetal genes to increase sugar flow to the fetus.

PARENT-INDUCED INFANTICIDE IN HUMANS. Under certain extreme conditions, parent-offspring conflict can lead to infanticide at the hands of a parent. Imagine the following scenario, which has no doubt occurred numerous times during our evolutionary history. A woman gets pregnant at a time when resources are relatively plentiful. As the pregnancy develops, the resource base degenerates and is very poor by the time the woman gives birth. Further suppose that the near future seems bleak with respect to resources as well.

Let's examine the above scenario from three perspectives—those of the child, a young mother, and an older mother. In no case do we need to assume that the individuals are consciously making any cost/benefit decisions, only that they act as if such decisions have been made at some level. This is an important statement, as I would not want the argument made below to be interpreted as a cold-hearted, conscious decision on the part of the participants. We will simply assume that natural selection has favored certain behaviors over others, and we will leave unspecified precisely how these behaviors manifest themselves cognitively.

From the child's perspective, future siblings (with an $r = 0.5$) are important, but unless there is a very high probability that future siblings will be born in good health, their survival may not be as important as the child's own survival. In such cases, the offspring will do whatever it can to get as many resources from the parent as possible.

Now, switch perspectives to that of a young mother who has just given birth in this resource-poor environment. Such a female will likely have many more reproductive opportunities in her future,

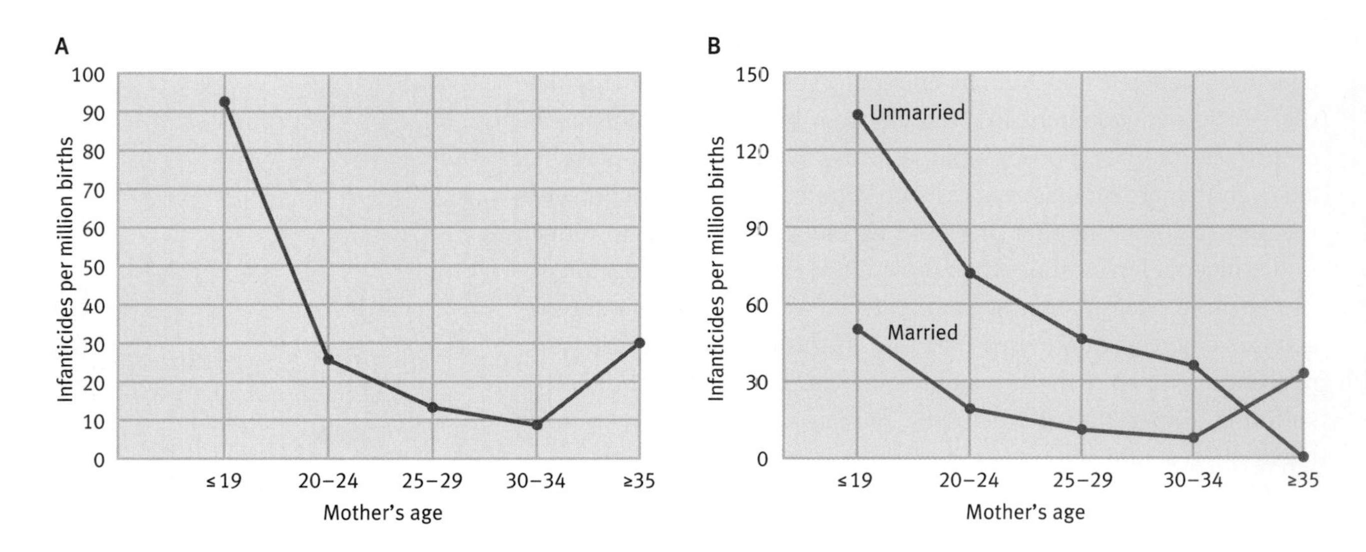

FIGURE 8.26. Risk of infanticide. (A) Since younger women have greater residual reproductive value, infanticide is expected to be higher in younger women than in older women. Data from Canada (1973–1984) support this prediction. **(B)** The relationship between infanticide and mother's age holds for both married and unmarried women, as supported by data from Canada (1974–1983). *(From Daly and Wilson, 1988)*

and she is said to have a high "residual reproductive value." Some of these offspring may be born in times when resources are more plentiful, and our young mother at some level must now weigh the future value of such offspring against expending a great deal of energy to provide resources for her current child. When the costs and benefits to future offspring outweigh the costs and benefits to the current offspring, infanticide may be favored by natural selection.

If we imagine that an older mother is in the exact same situation as that described above, she may come to a different decision. For the sake of argument, imagine our hypothetical mother is certain that because of her age, the child she just had will be her last. Her residual reproductive value is zero (or close to it), so we might expect her to put all her effort into raising her child, regardless of the state of the environment. This translates into a clear age-based prediction about parent-offspring conflict as it relates to infanticide. Younger women with higher reproductive value should be more likely to commit infanticide than older women. Data from three populations suggest that this prediction is supported (Figure 8.26A). Furthermore, this relationship between infanticide and the age of the mother holds true whether the mother is married or unmarried (Figure 8.26B).

SIBLING RIVALRY

Most readers will be familiar with sibling rivalry from basic psychology classes, but evolutionary biologists have been fascinated with such rivalries as well, and they have developed a substantial

FIGURE 8.27. Sibling competition. In this "hierarchy model" of sib-sib competition developed by Mock and Parker, the dominant chick (A) decides how much food to take, given that it has two siblings (B and C). Chick B, the next most dominant chick, decides what proportion of the remainder of the food it will take. Chick C receives whatever remains. *(Based on Mock and Parker, 1997)*

empirical and theoretical literature on this subject (Mock and Parker, 1997). The logic underlying the models of sibling rivalry is in many ways similar to that of parent-offspring conflict (Figure 8.27). Instead of imagining the world from the viewpoint of genes in parents versus genes in offspring, however, we now shift perspectives and focus on genes in siblings. Mathematical models are constructed to consider sibling rivalry among many different sibs, but for the sake of simplicity, we can make the case by just considering the evolution of sibling rivalry when two siblings are involved.

From the viewpoint of sib 1, sib 2 is indeed a valuable addition to family life, as sibs 1 and 2 share an r of 0.5, which means that what is good for sib 1 is *usually* good for sib 2. Imagine, however, an extreme environment in which there is only enough food for one sibling to survive. In such a world, each sibling values the other at a level of half of what it values itself. We might then expect extreme, perhaps even lethal, competition among siblings to emerge. If we make the environment less harsh, we expect sib-sib interactions to be less competitive, but since each values itself more than the other, some level of competition should be the norm, rather than the exception (Figure 8.28). We simply expect the rivalry to emerge in less lethal ways when resources are not as limited.

Sibling rivalry is illustrated nicely in the long-term work of Douglas Mock and his colleagues studying egrets (Mock and Parker, 1997). As idealic as downy chicks in a bird nest may seem to the casual observer, interactions in egret bird nests resemble

FIGURE 8.28. Sib-sib conflict. Kin selection theory predicts that individuals generally should not be aggressive toward kin such as sibs. This is especially true when there are abundant resources **(A)**. But if there are limited resources **(B)**, conflict over the resources will increase as each individual is more related to itself (r = 1) than to its sib (r = 0.5).

prize fights more closely than anything out of *Bambi* (Figure 8.29). Consider the following summary of work on sib-sib interactions:

> Sibling fights take many forms, depending mainly on how the loser concedes and how quickly it does so. The simplest fights, which usually occur while the participating dyad has had a series of increasingly one-sided battles, are those in which the attack inspires no retaliation. At the next level, return fire is brief until the loser is tagged with several unanswered shots and crouches low. From there, the severity of the beating is left largely to the victor's discretion. Sometimes it continues to jab at its opponent, causing the latter to screech and hide its face. As an alternative to jabbing, a dominant chick may seize the cowering victim by its head or neck, lift that part a few centimeters and then slam it down forcefully against the nest cup. If the attack persists for more than a few extra blows, the loser is likely to flee, sometimes squawking loudly and racing about the nest dodging behind the other nest occupants while being hit. During such chases, the primary target is the back of the head. Frequently bullied chicks soon develop a characteristic baldness, dotted with fresh and crusted blood, where the nape feathers have been plucked forcibly during fights. (Mock and Parker, 1997, pp. 103–104)

FIGURE 8.29. Sib-sib competition in birds. In nests of egrets, sib-sib competition can be intense and result in the death of smaller, less dominant chicks. *(Photo credit: Gary Carter/Visuals Unlimited)*

When egret chicks first hatch, parents bring back enough food to fill all the chicks' guts, thus removing any impetus for sib-sib aggressive interactions. As the chicks grow, however, a point is quickly reached at which the food brought to the nest is not enough to feed everyone, and an intense battle scene emerges when a parent returns to the nest with a juicy bolus (a soft mass of chewed, partially digested fish). Parents regurgitate such boluses in the nest, and most are grabbed before they hit the ground. The key to obtaining access to a bolus is positioning within the nest, and specifically vertical positioning (the higher the better when mom returns). Even a chick tilting its head up above horizontal is a cue likely to spark aggression in egrets' nests.

Egrets, like most birds, hatch eggs asynchronously, that is, at different times. Thus, hatching order produces chicks that can differ in age by a significant number of days. Such age differences play a critical role in determining who emerges the victor in sib-sib interactions. A very clear age-related dominance hierarchy exists in chicks. First-hatched individuals are often much heavier than second-hatched chicks, who are themselves heavier than those hatched later (Figure 8.30A). In sib-sib interactions, large size means better fighting ability and that translates into significantly more food (Figure 8.30B).

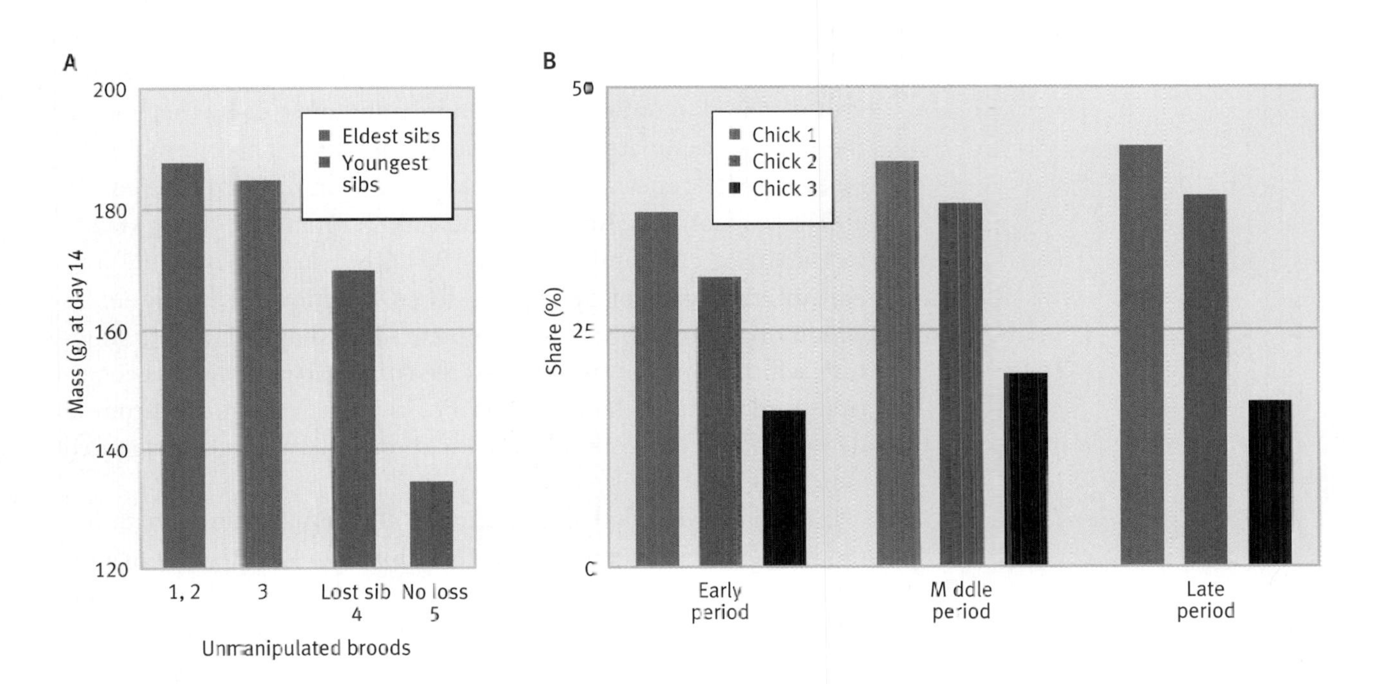

FIGURE 8.30. Birth order and food intake. (A) Normal broods of little blue herons include four to five chicks that are hatched asynchronously. The three eldest chicks (chicks 1, 2, and 3) are heavier than the youngest chicks (chicks 4 and 5). If a younger chick is part of a brood in which one of its older sibs has died (lost sib), its average weight is less than that of its older sibs but more than it would be if it were in a brood in which no sibs had died (no loss). **(B)** In egret broods, the oldest, dominant chick (1) receives more food than the middle chick (2), who in turn gets more than the youngest chick (3). This holds for the early period after hatching (1–13 days), the middle period (14–21 days), and the late period (21–30 days). *(Based on Mock and Parker, 1997)*

Kin Recognition

Given the power of kinship to affect social interactions, it should come as no surprise that ethologists and behavioral ecologists have a long-standing interest in how kin recognize one other (Fletcher and Michener, 1987; Hepper, 1991; Pfennig and Sherman, 1995; Hepper and Cleland, 1998). Early on, we went through a general procedure to show how to calculate relatedness (r). Ethologists are not so foolish as to assume that animals are able to calculate relatedness in the manner we described. We need only assume that natural selection favors individuals who act in a manner that makes it appear as though they are able to make such calculations.

MATCHING MODELS

Many models of kin recognition center on individuals having some "internal template" against which they match others and gauge relatedness (Waldman, 1987; Reeve, 1989; Fletcher and Michener, 1987; Hepper, 1991). These "matching" models differ in their specifics, but the basic idea is that individuals determine if others are kin or nonkin, depending on how closely they match the internal template of some focal individual. The internal template may be generated genetically (Crozier, 1987; Boyse et al., 1991), via learning (Robinson and Smotherman, 1991), or via social learning (Alexander, 1979, 1991), but in all cases, the animal infers degree of kinship as some function of the extent to which others match its own template. The shape of such a function will depend on how fine-scaled the particular kin recognition under study happens to be, and it can range from a dichotomous "kin/nonkin" classification system to a more graded system of kinship.

The most detailed and predictive of template kin recognition models is Reeve's conspecific threshold model. Reeve's (1989) model uses sophisticated mathematics to determine how individuals establish acceptance thresholds—the level at which the other's phenotype is considered a match to the focal individual's internal template. Reeve analyzes the evolution of acceptance thresholds in six different contexts (for example, individuals guard against nonkin entering an area, individuals search out kin, and so forth) and finds that the evolutionarily stable acceptance threshold is a function of the relative frequency at which individuals interact with different classes of conspecifics and the

FIGURE 8.31. Tadpole cannibals. As in the spadefoot toad, two different tadpole morphs—a carnivorous cannibal and an herbivorous omnivore—exist in a number of amphibian species. Here a tiger salamander (*Ambystoma tigrinum*) cannibal morph (right) is eating an omnivore morph (left). *(Photo credit: David Pfennig)*

fitness consequences of rejecting kin by mistake or accepting an individual as kin when it isn't.

The specific evolutionary function of kin recognition varies dramatically, from facilitating various forms of kin-biased altruism to increasing the probability of optimal outbreeding. Pfennig and his colleagues (1993) have uncovered a fascinating case of putative template matching in the cannibalistic behavior (Elgar and Crespi, 1992) of spadefoot toad (*Scaphiopus bombifrons*) tadpoles. In nature, two feeding morphs of spadefoot toads exist: juveniles that feed on detritus (small, often drifting vegetative clumps of food) typically develop into herbivorous omnivores, while those that feed on shrimp fairies tend to mature into carnivorous cannibals (Figure 8.31).

Pfennig and his colleagues (1993) examined kin recognition abilities in both the herbivore and cannibal spadefoot morphs. Individuals were then tested with unfamiliar siblings or unfamiliar nonrelatives. When visual and chemical cues are both in play, herbivores preferred associating with their siblings (rather than with unrelated individuals). Carnivorous individuals (taken from the same sibship as the omnivores tested), however, spent more time near unrelated individuals (Figure 8.32). This is precisely what one would predict if individuals were able to recognize kin based on some internal template.

Pfennig and his colleagues also offered carnivores a choice between unfamiliar siblings and unfamiliar nonrelatives in a protocol that allowed carnivores to actually eat other tadpoles. Carnivores were not only more likely to eat unrelated individuals, but they were able to distinguish between relatives and nonrelatives by some sort of taste test. That is, carnivores were equally likely to suck relatives and nonrelatives into their buccal cavities, but relatives were

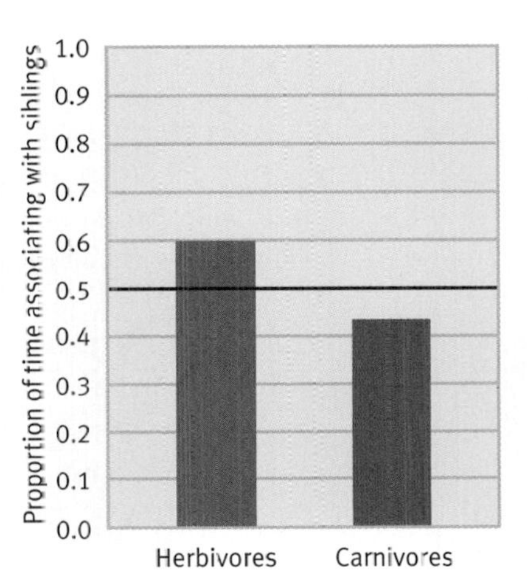

FIGURE 8.32. Kin recognition in spadefoot toads. Spadefoot toad tadpoles come in two morphs: carnivorous and herbivorous. Individuals from each tadpole morph were placed between two groups of tadpoles, one of which contained sixteen unfamiliar siblings, while the other was composed of unfamiliar nonsiblings. Herbivorous morphs preferred unfamiliar sibs, while carnivorous morphs (who cannibalize other tadpoles) preferred nonkin. The solid horizontal line represents the value expected if tadpole behavior was random with respect to kinship. *(Based on Pfennig et al., 1993)*

Interview with Dr. STEPHEN EMLEN

You've worked on a whole array of questions in animal behavior. Why have you chosen to make the study of kinship in bee-eaters a primary focus?

My initial interest in the bee-eaters was to understand the evolution of altruism. Scientists of all persuasions have long been fascinated by the paradox of altruism: Why do individuals engage in behaviors that benefit others, yet are costly to themselves? It was known that some species of bee-eaters are cooperative breeders, which means that adults forego breeding to help to rear the offspring of others. It was also known that some species of bee-eaters are colonial breeders, meaning that many pairs of birds live and breed in close proximity. So I felt that an intensive study of a cooperative species of bee-eater would shed light on the important question of altruism, and by choosing to study a species that was also colonial, I would be examining altruism in a more complex society than had been looked at previously. The intuition paid off. White-fronted bee-eaters proved to be an excellent species for studying not only the cooperative rearing of offspring but also the decision rules that govern which individuals become helpers and whom, and how much, they help.

Soon, however, my interest in the bee-eaters came to extend far beyond the initial question of altruism. I learned that these birds live in a society that is remarkably similar in its structure to the societies of early humans. This society is comprised of several "layers." First, there are parents who, often with the help of genetic relatives, attempt to rear offspring each year. Second, each set of parents is part of a larger, extended family that typically comprises their parents, siblings, nieces and nephews, and cousins. Finally, a dozen or so different extended families all live together in a single colony, much like the extended families of early humans which banded together to live in villages. The parallels to human society extend even further since bee-eaters form socially monogamous pair bonds that usually last for the lifetime of the individuals involved. Further, both parents, the mother and the father, play essential roles in child care. When these facts became clear, I realized that the bee-eaters provided an excellent system for studying the complexities of family social life—what I call family dynamics. I currently believe that species such as bee-eaters, species that live in societies based upon extended family memberships, will provide us with our best model systems for understanding the social dynamics of family living.

Do the terms "kinship" and "family" differ in meaning when discussed in ethology as opposed to when they are used by nonscientists?

To the evolutionary biologist, kinship is defined strictly by genetic relatedness. A parent is an individual from whom one has obtained one-half of one's genes. A sibling is an individual with whom one shares one-half of one's genes by virtue of common descent. A grandchild is an individual with whom one shares one-quarter of one's genes by virtue of common descent through an offspring. And so on. Anthropologists use the term very differently. Kin are those individuals that a particular society refers to as their relatives. Some societies have broadly inclusive terms for "brother" or "cousin" that do not correspond with genetic usage. Thus the anthropological meaning of kinship categories is culturally, rather than genetically, based. In our own Western society, nonscientists typically use kinship terms that reflect blood lines. Thus, when a sibling marries, her/his spouse is referred to as joining the family, but the term "in-law" is retained to distinguish the spouse from the blood-linked, or genetically related, family members. Similarly, when an individual with children remarries, the new spouse becomes an integral part of the family unit, but the term "step-parent" is used. Thus our terminology retains information about the assumed degree of genetic relatedness among different individuals. Scientists and nonscientists alike instantly know the meaning of such terms and can relate them to their personal experiences and their own families.

Is there something special about the breeding systems of birds that makes them so amenable to long-term kinship and cooperation studies?

There are several special features of birds. First, many persons are aesthetically attracted to birds. Birds are active during the daytime, and most of their behaviors are visually mediated and thus easy to observe. As a result, a myriad of people have studied birds and there is a vast background literature on them. Second, the breeding systems of birds make them logical candidates for studies of cooperative breeding. The majority of bird species are socially monogamous and exhibit biparental care of their young. Several hundred species are known to be cooperative breeders—more than in any other class of vertebrate. Third, the life spans of birds are relatively short. Thus data on the fitness consequences of behaviors such as helping can be collected over the full lifetimes of birds without subsuming the full lifetimes of the researchers. Finally, birds show great complexity in their social interactions, yet their behaviors are largely devoid of cultural variability. For these reasons, birds provide some of the best animal models that we have for deciphering basic, natural (noncultural) rules of social engagement associated with family interactions and child-rearing behaviors.

Many persons assume that because humans are primates, studies of primate social behaviors will yield the most useful information about ourselves. But this is not always the case. Very few primates form long-term pair bonds, and fewer yet have males that contribute regularly to the care of young. The "family" structure of most primates thus shares much less in common with humans than does that of many birds. Because cooperative breeding appears to have evolved primarily in family-based societies, it is not surprising that bird studies have played such a prominent role in studies of kinship and cooperation.

What has been the general reaction to your "evolutionary theory of the family" models? Do natural and social scientists view your work differently?

The response has been excellent. Whenever one tries to bridge the gap between the social and the life sciences, one must remember that each group speaks its own language. So it takes time for each to understand precisely what the other is saying. Most psychologists and sociologists who study human family dynamics are concerned primarily with the proximate causes of behaviors—for example, "How does the early developmental history of a child influence her/his behavior toward others as an adult?" or "What situations trigger conflict behavior with other family members?" The biologist is more interested in the ultimate causes of the same behaviors—for example, "How has the evolutionary history of family interactions shaped the decision rules that influence our social behavior toward other family members?" or "Can understanding the fitness consequences of behaviors allow us to predict why certain situations are likely to elicit family conflict?" At first it might seem like the sociologist, the psychologist, and the evolutionist have little in common. They are asking different questions. But they are all trying to understand the same behaviors, for example, "When, where, and between whom is cooperation, as opposed to conflict, to be expected?" Once the different levels of analysis are sorted out, the approaches of the sociologist, the psychologist, and the evolutionary biologist can be seen to complement one another. Each provides additional information for the others. More and more social and life scientists are coming to understand this.

DR. STEPHEN EMLEN is a professor at Cornell University. He has made important contributions to the subjects of migration, kinship, and the evolution of cooperation (to name just a few). His paper "An Evolutionary Theory of Family" integrates numerous theories in behavioral ecology to generate fifteen hypotheses about family life in nonhumans and humans.

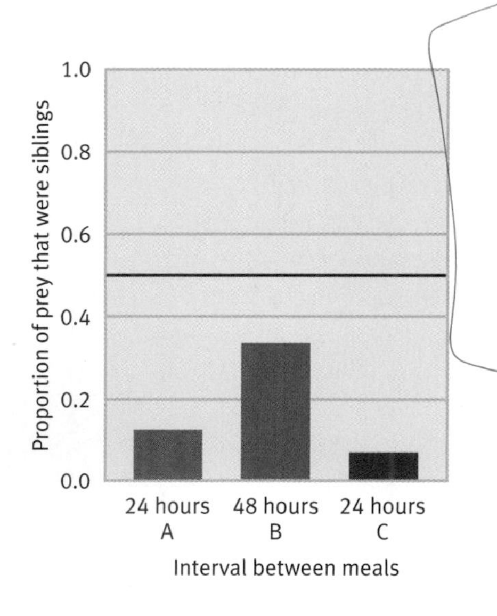

FIGURE 8.33. Hunger and carnivorous toads. The carnivorous morph of spadefoot toads prefers to eat nonkin over kin. When starved for 24 hours (A), only a little more than 10 percent of individuals eaten were kin. If starved for 48 hours, this figure rises. As a control, toads were again starved for 24 hours (C), and results were similar to the original 24-hour deprivation treatment (A). The solid horizontal line represents the value expected if cannibalism were random with respect to kinship. *(Based on Pfennig et al., 1993)*

released much more frequently than unrelated individuals. Being able to recognize kin has clear advantages in the spadefoot toad, as ingesting kin would generally be selected against whenever alternative food sources were available. In accordance with this, the researchers found that cannabilistic toads were much less picky when they had been starved for twenty-four hours (Figure 8.33).

RULE-OF-THUMB MODELS

In species where kin groups are spatially segregated from one another over relatively long periods of time, a second, and in some sense simpler, form of kin recognition may evolve. In such scenarios, natural selection might favor a kin recognition rule of the form "if it lives in your nest/cave/territory, then treat it like kin" (Holmes and Sherman, 1983; Sherman and Holmes, 1985; Blaustein, 1983; Waldman, 1987; but see Tang-Halpin, 1991, for arguments that such rules should not be considered as true kin recognition).

To see how such a rule of thumb might work, imagine a population of animals in which family units live in a fixed area, set apart from other such areas, such that there is not much interaction across the areas. In such a population, all of the machinery (cognitive, genetic, sensory) necessary to distinguish kin from nonkin may be superfluous. It may be that a rule that instructs individuals to treat all individuals in their living area as kin works just as well in terms of kin recognition as the more complicated rules associated with matching. After all, if everyone in an area is almost always kin, selection should favor the simplest possible kin recognition rule. Of course, such rules are subject to cheating (Figure 8.34), as shown in

FIGURE 8.34 Kin recognition breakdown. While adopting an "if it's in your nest, it is your offspring" rule often works for mothers, the system can be sabotaged by "nest parasites." Here a mother dunnock is feeding a baby cuckoo that has been dumped into her nest. *(Photo credit: Eric and David Hosking/ Corbis)*

the nest parasitism that cuckoos undertake, leading a bird of another species to feed the cuckoo nestling that hatched in its nest as if it were its own offspring (Ortega, 1998). Nonetheless, these rules are generally quite effective.

Spatial cues and kin recognition rules can often change through the lifetime of an individual. For example, John Hoogland and Paul Sherman (1976) have found that in bank swallows (*Riparia riparia*), parents *initially* feed any chick in their nest, thus adopting an "if it is in my burrow, it is likely kin" rule. This makes good sense, because for the first three weeks of life chicks cannot fly, and it is therefore extremely likely that any chick in a burrow is kin. At three weeks, however, chicks learn to fly and there is consequently much more mixing among young. Michael Beecher has found that when their chicks are about twenty days old, bank swallow mothers switch from their rule of thumb (feed what is in your nest) to using distinctive vocal cues (that is, a template-based system) to recognize and feed their offspring (Beecher et al., 1981a, 1981b, 1986).

SUMMARY

1. W. D. Hamilton's "inclusive fitness" or "kinship" theory has revolutionized the way scientists understand evolution and ethology.
2. In evolutionary terms, relatedness centers on the *probability* that individuals share genes that they have inherited from some *common ancestor*—parents, grandparents, and so on—a concept referred to as "identity by descent." The essence of inclusive fitness models is that they add on to "classical" models of natural selection by considering the effect of a gene not only on the individual that bears it, but on those sharing genes that are identical by descent (that is, blood kin).
3. The decision to aid family members is a function of how related individuals are, and how high or low the costs and benefits associated with the trait turn out to be. When individuals are highly related and a gene codes for an action that provides a huge benefit at a small cost, selection strongly favors this trait.
4. Parents should be particularly willing to go to great lengths to help their offspring. But a zone of parent-offspring conflict is also predicted under basic kinship theory, as in weaning situations.
5. While sibling rivalry is most often associated with the field of psychology, basic kinship theory also defines the conditions under which sibling rivalry should be favored, including competition for food.

6. Many models of kin recognition center on individuals' having some "internal template" against which they match others and gauge relatedness.
7. In species where kin groups are spatially segregated from one another over relatively long periods of time, a second, and in some sense simpler, form of kin recognition may evolve. In such scenarios, natural selection might favor a kin recognition rule of the form "if it lives in your nest/cave/territory, then treat it like kin."

DISCUSSION QUESTIONS

1. Why would it be an advantage for animals to gauge very small differences in blood kinship relationships? Why, for example, would it often be better if animals could distinguish relatives at the level of cousins ($r = 0.125$) rather than simply distinguishing siblings ($r = 0.5$)? What sorts of benefits might be possible when small differences in relatedness could be gauged?
2. Read Emlen's 1995 article, "An evolutionary theory of the family," in *Proceedings of the National Academy of Sciences, U.S.A.* (vol. 92, pp. 8092–8099). List the evidence for and against three of the predictions that we did not examine in this chapter.
3. Try and find some statistics on the rate of child abuse in families where children are adopted. What might you expect based on the logic adopted by Daly and Wilson? What do your data show?
4. Based on the parent-offspring conflict model, what differences in weaning behavior would you expect to see between younger and older mammalian mothers?
5. How might both kin selection and kin recognition rules be useful in understanding cases of "adoption" in animals?

SUGGESTED READING

Crozier, R. H., & Pamilo, P. (1996). *Evolution of social insect colonies: Sex allocation and kin selection*. Oxford: Oxford University Press. A book-length treatment of kin selection in social insects.

Emlen, S. T. (1995). An evolutionary theory of the family. *Proceedings of the National Academy of Sciences, U.S.A.*, *92*, 8092–8099.

Here Emlen lays out all fifteen predictions derived from his "evolutionary theory of family."
Hamilton, W. D. (1963). The evolution of altruistic behavior. *American Naturalist, 97*, 354–356. Hamilton's ideas on kinship, boiled down to their main ingredients.
Reeve, H. K. (1998). Game theory, reproductive skew and nepotism. In L. A. Dugatkin and H. K. Reeve (Eds.), *Game theory and animal behavior* (pp. 118–145). New York: Oxford University Press. An overview of how kinship is incorporated into a family of models known as "reproductive skew" models.
Sherman, P. W. (1977). Nepotism and the evolution of alarm calls. *Science, 197*, 1246–1253. This paper on kin selection and alarm calls in Belding's squirrels is perhaps the most well-cited paper in all the inclusive fitness literature.

9

The Range of Cooperative Behaviors

- Helping in the Birthing Process
- Social Grooming
- Group Hunting
- Nest Raiding

Three Paths to Cooperation

- Path 1: Reciprocity
- Path 2: Byproduct Mutualism
- Path 3: Group Selection

Phylogeny and Cooperative Breeding in Birds

Hormones, Reproductive Suppression, and Cooperative Breeding

Coalitions

- Coalitions in Baboons
- Alliances and "Herding" Behavior in Cetaceans

Interspecific Mutualisms

- Ants and Butterflies—Mutualism with Communication?

Interview with Dr. Hudson Kern Reeve

Cooperation

In many streams of the Northern Mountains of Trinidad, the water is crystal clear, and during the dry season, the behavior of guppies can be seen from the bank. If you sit beside such streams for a bit, you will likely see some guppies, perhaps a pair of them, break away from their group and inspect a potentially dangerous predator, such as a pike cichlid, *Crenicichla alta*. The guppies could just as easily have headed for cover, but rather than choose this course of action, they opt to inspect and approach the threat (Pitcher et al., 1986). Not only do some fish take the risks inherent in approaching a predator, they do so in coordinated pairs, and they may even return to their group and somehow pass on to others the information they just obtained (Figure 9.1).

How can such cooperative behavior emerge among animals? The risk-taking guppies described above are in what appears to be a no-win situation. In a pair of fish, the temptation for each guppy to stop inspecting and let the other fish take the risk is always present. After all, the best thing that could happen for guppy 1 is that guppy 2 takes the risks and passes the information it receives on to everyone else for free. But, of course, the same holds true for guppy 2, as its highest payoff comes when guppy 1 approaches the predator alone. Yet, if both individuals opt to wait for the other to inspect the predator, each may be worse off than if they had inspected the predator as a pair.

Is there a solution to this dilemma? Is any sort of cooperation possible in such a scenario? The answer is yes, and we shall examine exactly how later in this chapter. But before we do, we need two things under our belt: a definition and a broader overview of cooperation.

The word **cooperation** typically refers to an outcome—an outcome that despite potential costs to the individual provides

FIGURE 9.1. Risk taking and cooperation in guppies. Two male guppies (lower left and lower center of photo) inspect a pike cichlid predator. Guppies cooperate during such risky endeavors. *(Photo credit: Michael Alfieri)*

some benefits to group members. But the phrase "to cooperate" has two common usages. "To cooperate" is sometimes used to convey the idea of achieving cooperation—something the group does. But "to cooperate" also may mean to behave in a way that makes cooperation possible, even though such cooperation may not be achieved (Mesterton-Gibbons and Dugatkin, 1992; Dugatkin, 1997a). Here "to cooperate" will mean the latter.

Cooperation occurs in many species and is manifested over a wide variety of behaviors. In Chapter 8, we examined cooperation and altruism among kin. To better understand the origins and the benefits of cooperation among unrelated individuals, we will answer the following questions in this chapter:

- What paths leading to cooperation have been identified by ethologists? What theory lies behind each? What empirical evidence supports each of these paths?
- What role does phylogeny—the evolutionary history of an organism—play in explaining the distribution of animal cooperation?
- What roles do hormones play in unraveling the mysteries of animal cooperation?
- What role does cooperation play in coalition formation?
- How can we explain *interspecific* cooperation among animals?

The Range of Cooperative Behaviors

Although cooperation is far from ubiquitous in the animal kingdom, it occurs in almost every imaginable context: foraging, predation, antipredator behavior, mating, play, aggression, and so forth. To give you a sense of the sorts of cooperation found in animals, we will briefly discuss helping in the birthing process, social grooming, group hunting, and nest raiding, all of which involve cooperative behavior.

HELPING IN THE BIRTHING PROCESS

Kunz and Allgaier (1994) describe an interesting case of cooperation in the Rodriques fruit bat (*Pteropus rodricensis*). In this bat, unrelated females assist pregnant individuals in the birthing process

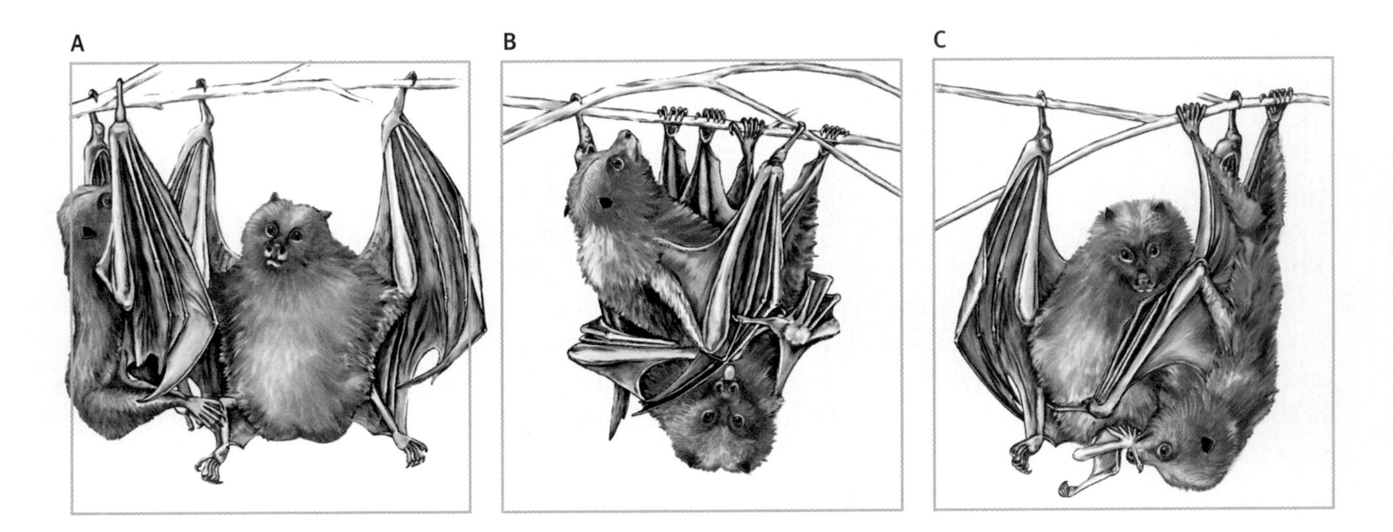

FIGURE 9.2. Bat midwives. (A) A nonpregnant female helper bat assumes the "feet down" birthing position as it "tutors" the pregnant female. **(B)** A pregnant female in a cradle position during birth is being groomed by a helper. **(C)** As the wing and foot of the pup emerge, the helper continues to groom the female that is giving birth. *(Based on Kunz and Allgaier, 1994)*

(Figure 9.2). Kunz and Allgaier found that those assisting in the birth were involved in licking the pregnant individual, as well as wrapping their wings around expectant mothers. Remarkably, mothers-to-be didn't go into the typical feet down birthing position until helpers took this position. Other anecdotal evidence for assistance during birth has been recorded in marmosets (*Callithrix jacchus;* Lucas et al., 1937), Indian elephants (*Elephas maximus;* Poppleton, 1957), African hunting dogs (*Lycaon pictus;* Gaffrey, 1957), raccoon dogs (*Nyctereutes procyonoides;* Yamamoto, 1987), and bottle-nosed dolphins (*Tursiops truncatus;* Essapian, 1963).

SOCIAL GROOMING

Social grooming, or **allogrooming,** wherein one individual grooms another, is one of the most obvious and frequently observed cooperative behaviors observed by primatologists (Figure 9.3). From the time of the earliest experimental work on primates, grooming has been considered a large part of the glue that holds primate troops together (Zuckerman, 1932; Carpenter, 1942).

The benefits associated with grooming include removal of ectoparasites (Sparks, 1967; Hutchins and Barash, 1976; but see Dunbar, 1991) and "tension reduction" (Terry, 1970; Boccia, 1987; Schino et al., 1988; Keverne et al., 1989). In addition, grooming can be "exchanged" for other currencies such as coalition formation; reduced aggression from other individuals, particularly dominant group members (Fairbanks, 1980; Silk et al., 1981; Silk,

FIGURE 9.3. Primate grooming. Primates of many different species engage in various forms of grooming behavior. Such behavior may serve numerous functions simultaneously. *(Photo credit: Chris Crowley/Visuals Unlimited)*

1982); access to scarce resources such as water (Weisbard and Goy, 1976; Cheney, 1977) or food (Kummer, 1978; de Waal, 1989b); entrance into new groups (Hauser et al., 1986); aid in chasing potential predators away (Kummer, 1978); and future association with individuals who have "special skills" that others do not possess (Stammbach, 1988). Grooming others, however, has it costs. Maestripieri (1993), for example, found that adult female macaques involved in social grooming were less vigilant with respect to the whereabouts of their offspring, thereby subjecting the offspring to increased rates of aggression from other troop members.

GROUP HUNTING

Group hunting has been documented in many raptors (Bent, 1938; Cade, 1982), including Harris's hawks *(Parabuteo unicinctus*; Figure 9.4). Consider James Bednarz's description of cooperative hunting in this species:

> The most common tactic can be described as a surprise pounce (seven kills) involving several hawks coming from different directions, and converging on a rabbit from far away. . . . When a rabbit found temporary refuge or cover, a flush-and-ambush strategy was employed. Here, the hawks surrounded and alertly watched the location where the quarry disappeared while one or possibly two hawks attempted to penetrate the cover. When the rabbit flushed, one or more of the perched birds pounced and made the kill. . . . The final tactic used, "relay attack," was the least common (2 of 13). This technique involved a nearly constant chase of a rabbit for several minutes while the "lead" was alternated among party members. (Bednarz, 1988, p. 1526)

FIGURE 9.4. Harris's hawk foraging. Groups of Harris's hawks sometimes hunt for prey in a cooperative manner. *(Photo credit: Jack Milchanowski/Visuals Unlimited)*

A quite impressive array of hunting tactics, to say the least.

FIGURE 9.5. Nest raiding by female sticklebacks. In three-spined sticklebacks, males (left; red on abdomen) guard nests containing eggs. In some populations, females (right) form large "marauding" groups, which attack such nests and eat the eggs within them. *(Photo credit: Dwight R. Kuhn)*

NEST RAIDING

Perhaps because of our experiences in human society, we often do not picture "gangs" of females cooperating to overpower males. Yet, this is just what occurs in the context of nest guarding in three-spined sticklebacks (*G. aculeatus*). Stickleback parental care is undertaken by males alone (Wooton, 1976), and females appear to form large groups (of up to a few hundred) that raid and destroy the nests of males guarding eggs (Figure 9.5), often eating the eggs contained in such nests (Abdel-malek, 1963; Rohwer, 1978; Kynard, 1979; Snyder, 1984; Whoriskey and FitzGerald, 1985).

The possible benefits accrued from nest raiding include: (1) the nutritional value of the eggs eaten, (2) mating with the male whose nest has been destroyed—a benefit accrued by a single female from the raiding group, and (3) the "spiteful" payoff of eliminating potential competitors for your own offspring. The little data available suggest that males are able to defend their nests against small groups of females, but they are helpless against onslaughts by larger groups. Although only a single female—usually the one who initiates a raid (FitzGerald and van Havre, 1987)—will mate with the male whose nest was destroyed, this probability of mating, in addition to the nutritional value of the eggs (no matter how slight), might make being part of a cooperative nest-raiding group profitable, particularly in environments that are limited in terms of either good nest sites or high-quality males.

Three Paths to Cooperation

Within ethology and evolutionary biology, work on the evolution of cooperation can be traced at least as far back as Darwin's interest in cooperation and altruism in insects and other groups (Darwin, 1859). Darwin was well aware that social insects were often hypercooperative. Although he was unable to completely explain why, Darwin did come close to explaining the "riddle" of such behavior via kinship. After Darwin, the study of cooperation was primarily (but not exclusively) kept alive first by the Russian prince and naturalist Petr Kropotkin (1902) and later on by the likes of W. C. Allee, A. E. Emerson, and "the Chicago School of Animal Behavior" (Banks, 1985; Mitman, 1988, 1992). Although these natural historians, ethologists, and population biologists amassed a good deal of data that documented cooperation among

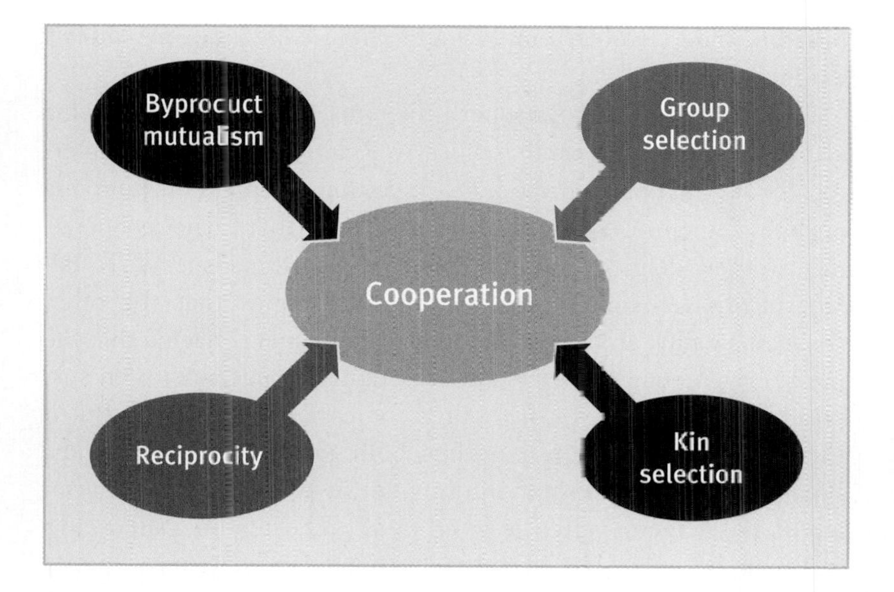

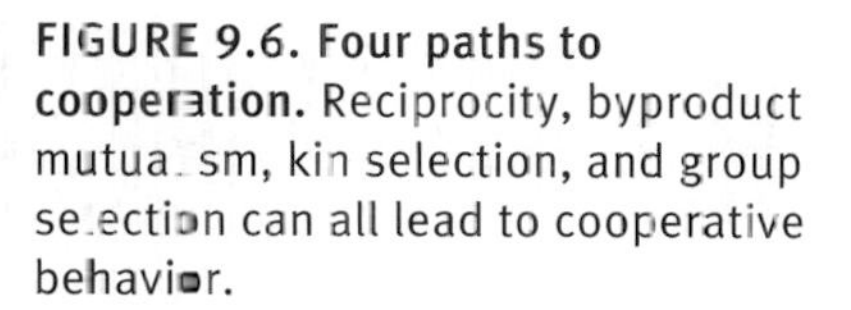

FIGURE 9.6. Four paths to cooperation. Reciprocity, byproduct mutualism, kin selection, and group selection can all lead to cooperative behavior.

animals (Allee, 1951), they did little in the way of advancing any theory on how cooperation evolved. A theoretical foundation for the study of cooperation has, however, emerged over the last forty years. From Hamilton's innovative work in 1963 (see Chapter 8) to the present, four paths to the evolution and maintenance of cooperation in animals have been developed: kin selected cooperation (Hamilton, 1964), reciprocity (Axelrod, 1984), byproduct mutualism (J. L. Brown, 1983), and group selection (D. S. Wilson, 1980) (Figure 9.6). There is now ample empirical evidence available to put the models underlying these paths to the test (Dugatkin, 1997a). As we focused on kin selection in Chapter 8, we shall only consider the remaining three paths to cooperation here.

PATH 1: RECIPROCITY

Robert Trivers (1971) tackled the problems associated with the evolution of cooperation between unrelated individuals in his seminal paper, "The Evolution of Reciprocal Altruism." Trivers argued that genes for cooperative and altruistic acts may be selected if individuals differentially allocate such prosocial behaviors to those who have already been cooperative and altruistic toward the donor. Trivers addresses the question of cooperation via reciprocal altruism in a theoretical framework called "game theory." Game theory is a mathematical tool that is used when the "payoff" an individual receives for undertaking an action is dependent on what behavior others adopt. In particular, Trivers, with some help from W. D. Hamilton suggests that the evolution of cooperation may best be understood by employing a game called the **prisoner's dilemma.**

To understand the prisoner's dilemma, imagine the following scenario: While in separate rooms, two criminal suspects are interrogated by the police in the hope of getting a confession from one or both of them. Cooperation and defection (that is, not cooperating) are now defined from the perspective of the suspects. To defect means to "squeal" and tell the authorities that the other suspect is guilty, and to cooperate is to stay quiet. Imagine that the police have enough circumstantial evidence to put away both suspects for one year, even without a confession from either. But if each suspect informs on the other, both go to jail for three years. Finally, if only one suspect informs on his partner, such behavior allows the defector (often referred to as a cheater) to walk away a free man, but causes his partner to go to jail for five years.

Table 9.1 depicts a payoff matrix that demonstrates that player 1 will receive a higher payoff if he cheats, regardless of what player 2 does, since he will serve fewer years in prison if he defects, regardless of what his partner chooses. As such, player 1 should always defect. The same holds true for player 2, and he should also always defect. The *dilemma* in the prisoner's dilemma is that while each player receives P (three years in prison) when they mutually defect, both players would receive a higher payoff (R, which is only one year in prison) if they had both cooperated (Poundstone, 1992).

TABLE 9.1. The prisoner's dilemma game. In this game, each player can either cooperate or defect. For the matrix to qualify as a prisoner's dilemma game, it must be true that T > R > P > S. T = "Temptation to cheat" payoff, R = "Reward for mutual cooperation" payoff, P = "Punishment for mutual defection" payoff, and S = "Suckers" payoff.

		Player 2	
		Cooperate	Defect
Player 1	Cooperate	R 1-year prison term	S 5-year prison term
	Defect	T 0-year prison term	P 3-year prison term

Math Box 9.1. Evolutionarily Stable Strategies (ESS)

An evolutionarily stable strategy (ESS) is defined as "a strategy such that, if all the members of a population adopt it, no mutant strategy can invade" (Maynard Smith 1982). Here "mutant" refers to a new strategy introduced into a population, and successful invasions center around the relative fitness of established and mutant strategies. If the established strategy is evolutionarily stable, the payoff from the established strategy is greater than the payoff from the mutant (new) strategy. To see this more formally, let's consider two strategies, I and J. We shall denote the expected payoff of strategy I against strategy J as E(I, J), the payoff of J against I as E(J, I), the payoff of I against I as E(I, I), and the payoff of J against J as E(J, J). Strategy I is an ESS, if the following two conditions are met:

Either

$$E(I, I) > E(J, I) \tag{1}$$

or

$$E(I, I) = E(J, I), \text{ but } E(I, J) > E(J, J). \tag{2}$$

If condition (1) holds true, then I does better against other I's than J does. Since we start with a population full of I's (except for our one J mutant), this condition ensures that I is an ESS.

Condition (2) addresses what happens when I and J have the same fitness when interacting with I. If this is true, then J may now reach higher frequencies by chance, since when it occurs as a mutant it does just as well as I. I is then an ESS if it does better than J when paired up against J, that is, when E(I, J) > E(J, J).

Robert Axelrod and William Hamilton (1981) used both analytical techniques and computer simulations to examine the success of an array of behavioral strategies in an *iterated*, or repeated, prisoner's dilemma game. In the iterated version of the game, players are trapped in this prisoner's dilemma for many interactions, and the exact end point is unknown. Thus, this type of game tends to mimic certain natural situations, since social, group-living animals generally encounter each other more than once. Axelrod and Hamilton searched for the **evolutionarily stable strategy (ESS)** to such a game (Maynard Smith, 1982; see Math Box 9.1). They demonstrated that if the probability of meeting a given partner in the future was above some critical level, then in addition to the success of a simple strategy of "always be noncooperative" (labeled "always defect" or ALLD), a conditionally cooperative strategy called **tit for tat** (TFT) was a robust solution to the iterated prisoner's dilemma (see Math Box 9.2).

Math Box 9.2. An ESS Analysis of TFT and the Prisoner's Dilemma Game

Axelrod and Hamilton (1981) examined whether tit for tat (TFT) and the primitive "always defect" (ALLD) strategy were evolutionarily stable—that is, whether they could resist invasion from mutants if they themselves were at a frequency close to 1. They began this task by proving that if the strategy ALLD or a strategy that alternates D and C (ALTDC) could not invade TFT, then no single pure strategy could invade. They then considered whether ALLD or ALTDC could in fact invade TFT.

Recall the payoffs of the prisoner's dilemma (T, R, P, and S). If we let w equal the probability of interacting with the same player on the next move of a game, then moves of the game are a geometric series, and the expected number of interactions with a given opponent is equal to $1/(1-w)$. So, for example, if $w = 0.9$, the expected number of interactions with a given partner is 10 $[1/(1-0.9)]$. This being the case, when TFT is close to being at a frequency of 1, virtually all TFT players meet other TFT players, and their payoff is:

$$R + wR + w^2R + w^3R + \ldots, \text{ which sums to } R/(1-w) \tag{1}$$

An ALLD mutant would have all its interactions with TFT, and its payoff would be:

$$T + wP + w^2P + w^3P + \ldots, \text{ which sums to } T + wP/(1-w) \tag{2}$$

and so TFT can resist invasion from ALLD when:

$$R/(1-w) > T + wP/(1-w) \tag{3}$$

Solving the inequality for w, ALLD fails to invade TFT when:

$$w > (T-R)/(T-P) \tag{4}$$

Now, when playing TFT, ALTDC gets a payoff:

$$T + wS + w^2T + w^3T \ldots, \text{ which sums to } (T + wS)/(1-w^2) \tag{5}$$

TFT is thus resistant to invasion when:

$$R/(1-w) > (T + wS)/(1-w^2) \tag{6}$$

Solving the inequality for w, ALTDC fails to invade TFT when:

$$w > (T-R)/(R-S) \tag{7}$$

Hence Axelrod and Hamilton concluded that TFT is resistant to any invasion when:

$$w > \text{maximum of these two values: } (T-R)/(T-P) \text{ and } (T-R)/(R-S). \tag{8}$$

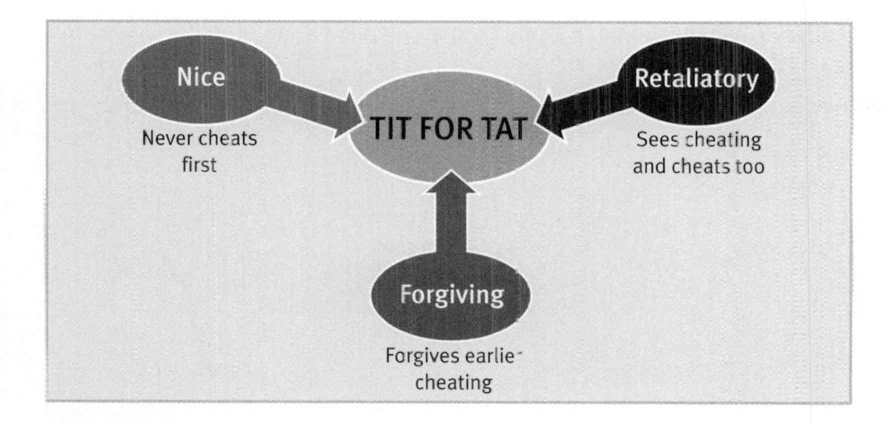

FIGURE 9.7. Tit for tat. The tit-for-tat strategy has three fundamental characteristics. The individual using TFT is (1) nice—it never cheats first, (2) retaliatory—it always responds to a partner that is cheating by cheating itself, and (3) forgiving—it only remembers one move back in time, and hence is capable of "forgiving" cheating that is done early in a sequence.

TFT is a rule that instructs a player to cooperate on the initial encounter with a partner and to subsequently copy its partner's previous move. This behavioral rule then incorporates social learning into our cooperation paradigm, as individuals copy what their partners do, and such behavior can potentially ripple through an entire population. Axelrod (1984) hypothesized that TFT's success is attributable to its three defining characteristics: (1) "niceness"—the player using TFT is never the first to defect, as it initially cooperates with a new partner, (2) swift "retaliation"—a TFTer immediately defects on a defecting partner, as it copies its partner's previous move and so, if its partner defects, a TFT individual defects, and (3) "forgiving"—TFTers remember only *one move back in time*. As such, the player using TFT forgives prior defection if a partner is currently cooperating. That is, TFTers do not hold grudges (Figure 9.7). Since the original models of TFT have appeared, dozens of variants of this strategy have been examined (Dugatkin, 1997a), but most share the essential characteristics we have just discussed.

Numerous studies examine whether animals apparently trapped in a prisoner's dilemma game use TFT and TFT-like strategies, but here we shall focus on two of them. The first examines the use of TFT during antipredator behavior in guppies and the second addresses reciprocity (but not necessarily TFT) in the context of vampire bat food sharing.

PREDATOR INSPECTION AND TFT IN GUPPIES. Let's return to the scenario outlined at the start of the chapter—that of **predator inspection,** wherein guppies inspect a potentially dangerous predator. Before addressing whether guppies use TFT while inspecting a predator, we first need to examine whether a pair of fish inspecting a predator are indeed trapped in the payoffs associated with the prisoner's dilemma game (see Table 9.2; Dugatkin,

TABLE 9.2 Inspection and game theory. The payoffs to predator inspection qualify as a prisoner's dilemma when $T > R > P > S$.

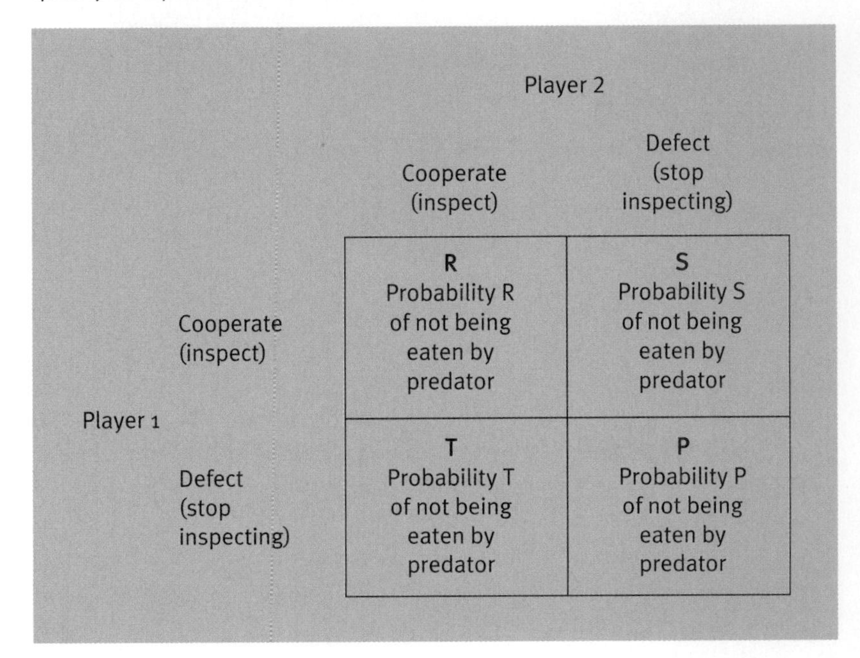

		Player 2	
		Cooperate (inspect)	Defect (stop inspecting)
Player 1	Cooperate (inspect)	R Probability R of not being eaten by predator	S Probability S of not being eaten by predator
	Defect (stop inspecting)	T Probability T of not being eaten by predator	P Probability P of not being eaten by predator

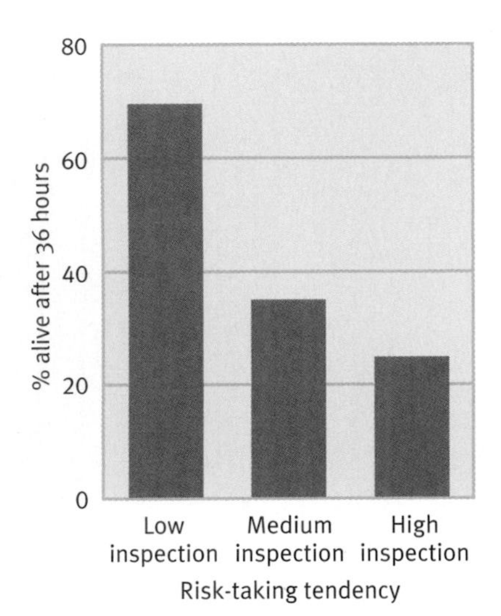

FIGURE 9.8. The risk of inspecting predators. Ten groups of six guppies each—two low inspectors, two medium inspectors, and two high inspectors—were placed with a predator in a pool that was one meter in diameter. The probability of surviving thirty-six hours was a function of inspection tendencies, with those inspecting most often suffering the highest mortality. *(From Malinski et al., 1997)*

1997a). As we see in Tables 9.1 and 9.2, for a payoff matrix to be a prisoner's dilemma, it must be true that $T > R > P > S$, where T is the payoff for defecting against a cooperator, R is the payoff to mutual cooperators, P is the payoff to mutual defectors, and S is the payoff to a cooperator matched with a defector. What evidence is there that these inequalities hold true for inspecting guppies?

Is $T > R$? Milinski (1987) has argued that any fish that trailed behind its partner while inspecting the predator would do better than its co-inspector, as it could observe whether the lead fish was attacked. In such a scenario, the trailing fish would receive a payoff of T if it stayed behind, but a payoff of R if it swam beside its partner. Evidence that inspectors are more likely to get eaten the closer they approach a predator (Dugatkin, 1992c; Milinski et al., 1997; Figure 9.8), coupled with the evidence that inspectors transfer the information they receive (Figure 9.9; Magurran and Higham, 1988), provides some empirical support that T is in fact greater than R for predator inspection.

Is $R > P$? If $P > R$, and the payoff for inspecting in a group is less than the payoff when no one inspects, it would not pay for any individual to inspect and the phenomenon of inspection would be rare and maladaptive when it occurred. Given that inspection is a well-known phenomena (Dugatkin and Godin, 1992b; Pitcher,

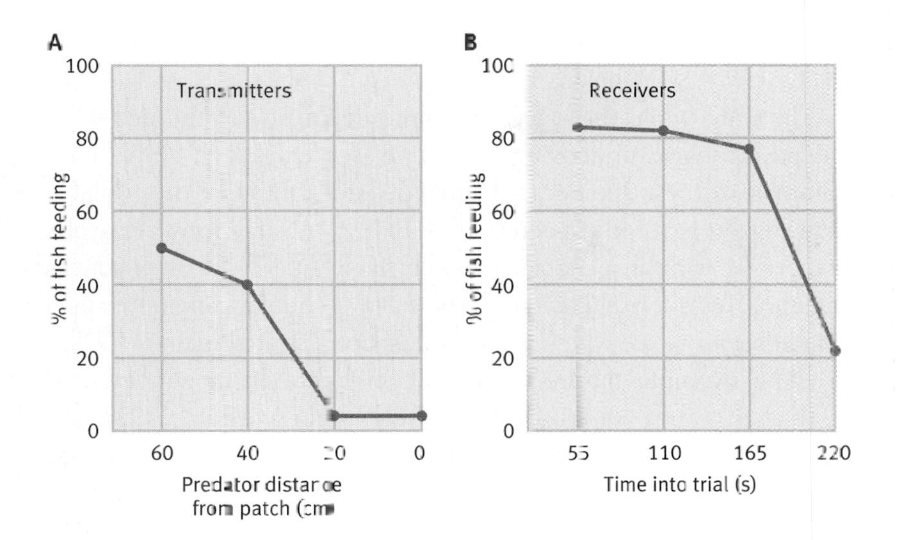

FIGURE 9.9. Information transfer in minnows. Magurran and Higham designed an experimental protocol in which some fish (transmitters) were allowed to inspect, while others (receivers) could see the transmitters, but not the predators. As the predator approached the transmitters, the latter fed less often **(A)**, as did the receivers **(B)**, suggesting that receivers were getting information about danger from the transmitters. *(Based on Magurran and Higham, 1988, p. 157)*

1992), this is unlikely to be the case, and so R is likely greater than P (Figure 9.10).

Is P > S? If predator inspection is to qualify as a prisoner's dilemma, it must also be true that the payoff to mutual noncooperation (P) be greater than that associated with inspecting alone (S). While it appears to be the case that having no inspectors in a shoal is dangerous, the most perilous situation for a single fish is to be the lone inspector in a group. Evidence from a number of experiments indicates that

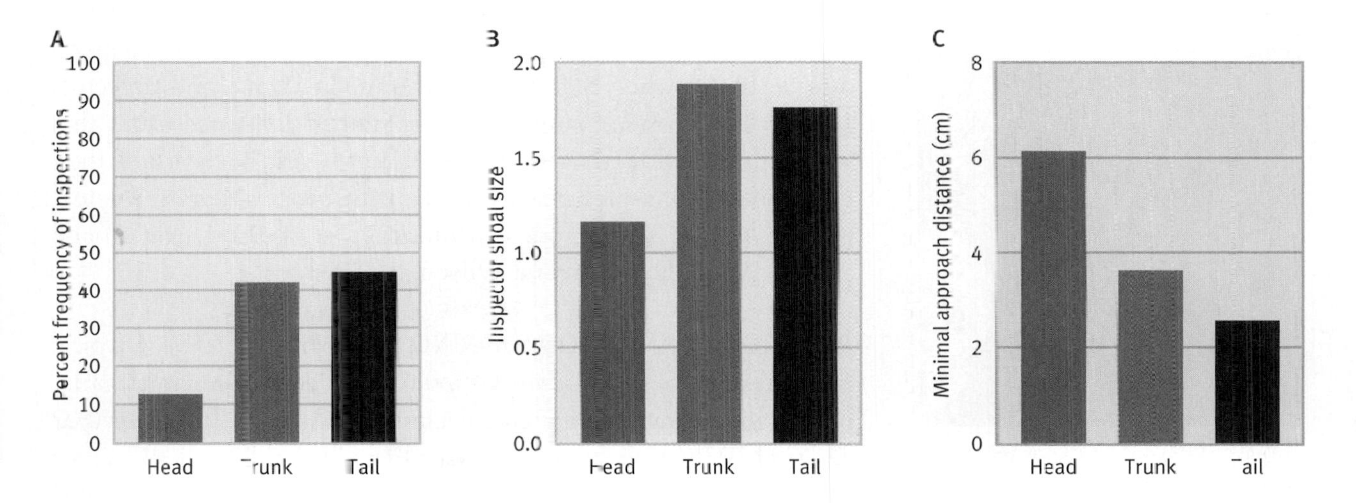

FIGURE 9.10. Inspection behavior in the wild. Researchers placed a model predator into a tributary of a river in Trinidad, and they recorded inspection behavior. Inspectors clearly recognize the head region of a predator as most dangerous. **(A)** There were fewer inspections of the predator's head than of its trunk and tail. **(B)** Inspector group was smallest when inspecting the head region of a predator. **(C)** Approach distance was a function of the part of the predator's body that was being inspected, with inspectors staying the furthest away when they were inspecting the predator's head. *(Based on Dugatkin and Godin, 1992b)*

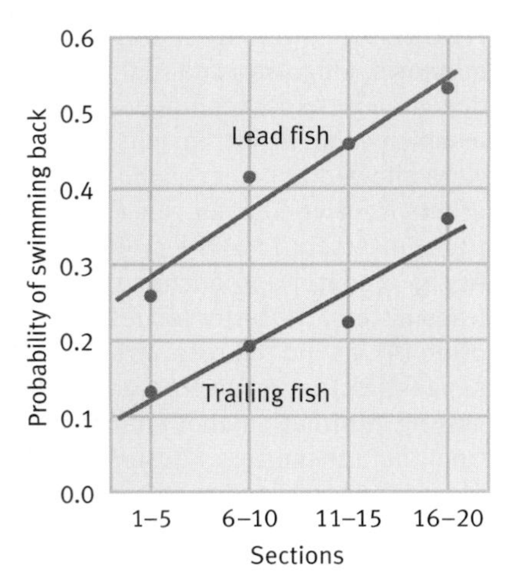

FIGURE 9.11. "Retaliation" in guppies? When pairs of guppies inspect a predator, lead fish in a given section of a tank are more likely to turn back to safety than are trailing fish (at about the same distance). This might be interpreted as lead fish "retaliating" against trailing fish, who fail to stay by their side. *(Based on Dugatkin, 1991, p. 130)*

single fish (stragglers) suffer very high rates of predation, suggesting that P > S (Milinski, 1977).

In sum, while there are still some debates over the details of the payoff matrix underlying inspection (see Dugatkin, 1997a, for a summary of these sometimes heated debates), much of the available data suggest that for inspectors, T > R > P > S. Hence, the payoffs associated with inspection behavior likely qualify as a prisoner's dilemma. As such, theory suggests that we should examine inspection behavior to see whether individuals use a strategy akin to TFT.

The dynamic nature of inspection behavior in guppies and sticklebacks (but not mosquitofish; see Stephens et al., 1997) supports the idea that inspectors do, in fact, use the TFT strategy when inspecting potential predators (but again see Dugatkin, 1997a, for more on the question of inspection and cooperation, including a lengthy discussion of the critiques of this work). As predicted by TFT, inspectors appear to be

- "nice," as each starts off inspecting at about the same point in time,
- "retaliatory," as inspectors cease inspection if their partner stops (Figure 9.11),
- "forgiving," if inspector A's partner has cheated on it in the past, but resumes inspection, A then resumes inspection as well (Dugatkin, 1991).

If the payoffs of inspection behavior match those of the prisoner's dilemma, then the payoffs of Table 9.2 demonstrate that everyone should prefer to associate with cooperators, as cooperators do better when paired with other cooperators, and defectors also do better when paired with cooperators. Evidence exists that inspectors do in fact remember the identity and behavior of their co-inspectors and prefer to associate with cooperators over defectors, at least in small groups (Milinski et al., 1990; Dugatkin and Alfieri, 1991a; Dugatkin and Wilson, 2000).

FIGURE 9.12. Blood-sucking reciprocators. To survive, female vampire bats require frequent blood meals. Individuals often regurgitate part of their blood meals to others, but they are much more likely to do so for those who have shared a meal with them in the past. *(Photo credit: Jerry Wilkinson)*

RECIPROCITY AND FOOD SHARING IN VAMPIRE BATS. A typical group of vampire bats (*Desmodus rotundus*) is composed largely of females, with a low average coefficient of relatedness (between 0.02 and 0.11; Wilkinson, 1984, 1990; see Figure 9.12). Females in a nest of vampire bats regurgitate blood meals to other bats that have failed to obtain food in the recent past (Wilkinson, 1984, 1985). This sort of food sharing can literally be a matter of life or death, as individuals may starve if they don't receive a blood meal every sixty hours (McNab, 1973). Jerry Wilkinson examined whether relatedness, reciprocity, or some combination of the two best explained the evolution and maintenance of blood sharing in this species (Figure 9.13).

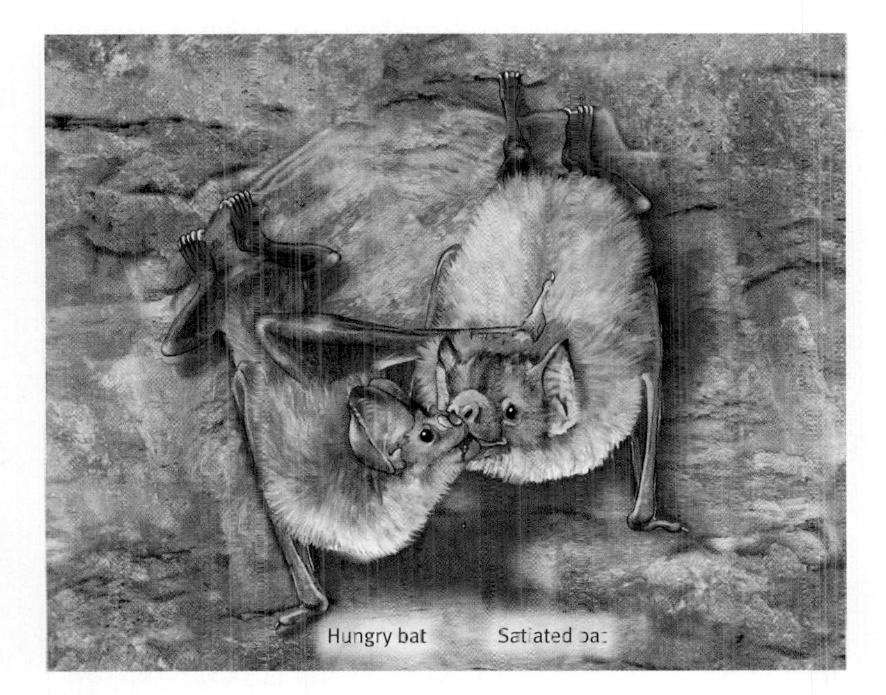

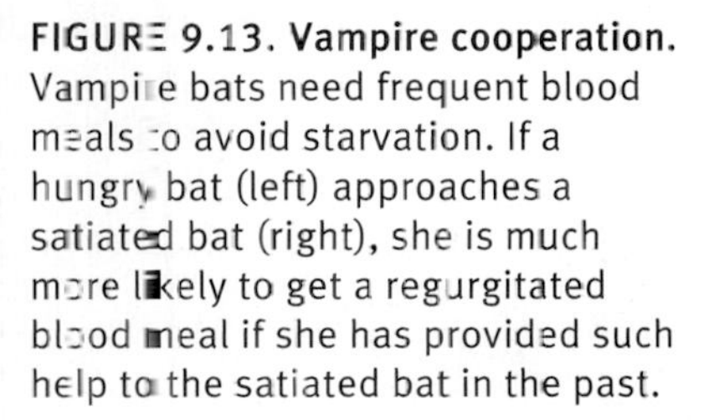

FIGURE 9.13. Vampire cooperation. Vampire bats need frequent blood meals to avoid starvation. If a hungry bat (left) approaches a satiated bat (right), she is much more likely to get a regurgitated blood meal if she has provided such help to the satiated bat in the past.

While relatedness does play a role in regurgitating food, Wilkinson also created what he calls an "index of opportunity for reciprocity." When analyzing the vampire data with this sort of index, Wilkinson suggested three lines of evidence that reciprocity is important in this system: (1) the probability of future interaction is high as predicted by TFT models, (2) the blood meal obtained is critical, while the cost of giving up some blood may not be that great, and (3) vampire bats are able to recognize one another and are more likely to give blood to those that have donated blood to them in the past (Figure 9.14). So, while the vampires don't necessarily play a strict version of TFT, they are clearly employing some sort of reciprocal altruistic strategy when it comes to blood sharing.

PATH 2: BYPRODUCT MUTUALISM

Another path to the evolution of cooperation is **byproduct mutualism** (J. L. Brown, 1983; Connor, 1995; see also West-Eberhard, 1975; Rothstein and Pirotti, 1987). Cooperation via byproduct mutualism occurs when an individual pays an immediate cost or penalty for not acting cooperatively, such that the immediate net benefit of cooperating outweighs that of cheating. So, for example, if you and a friend are trapped in a cave and the only way out is for you both to push the large boulder blocking the entrance, there is no temptation for you not to do so, as you will both

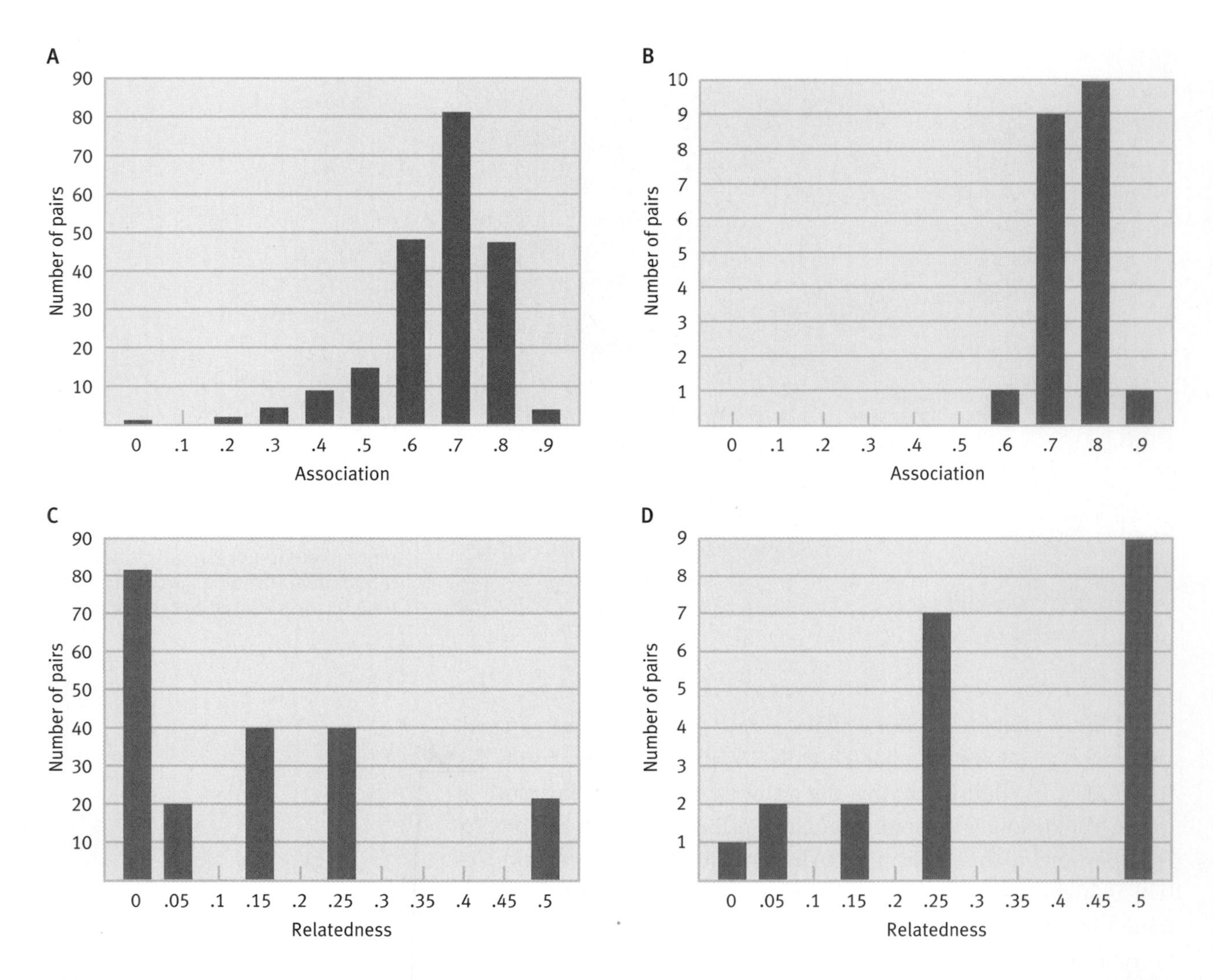

FIGURE 9.14. Vampire bat blood meals. Wilkinson used twenty-one regurgitation events not involving mothers and their offspring to examine the role of relatedness and reciprocity in blood sharing. The graphs show: **(A)** all possible association patterns (a measure of when the recipient bat and some other bat are together) that could be found between the recipient bat and others in the nest, **(B)** the actual association patterns found between donors and recipients, **(C)** the genetic relatedness between recipients and all other roost members, and **(D)** the actual relatedness between a recipient and a donor. Bats were much more likely to regurgitate a meal to close kin and to those with whom they associated more often. Follow-up laboratory work found that bats were capable of keeping track of who fed them in the past and who didn't. *(From Wilkinson, 1984)*

pay a large cost if either of you fails to cooperate (Figure 9.15). Michael Mesterton-Gibbons and I have argued that a prerequisite for byproduct mutualism is the boomerang factor—any uncertainty that increases the probability that a defector will be the victim of its own cheating (Mesterton-Gibbons and Dugatkin, 1992).

Byproduct mutualism differs from reciprocity in two fundamental ways. First, as alluded to in the cave example, there is no temptation to cheat under byproduct mutualism. That is, the environment favors either everyone cooperating or no one cooperating. In addition, while most forms of reciprocity require some form of

FIGURE 9.15. Byproduct mutualism. In the case of byproduct mutualism, there is an immediate cost paid for not cooperating. Here, neither person gains by not helping to move a stone that neither can budge alone.

scorekeeping (tracking your partner's behavior over time), no scorekeeping is required for cooperation via byproduct mutualism.

SKINNERIAN BLUE JAYS AND BYPRODUCT MUTUALISM. One critique of the work on cooperation and the prisoner's dilemma is that it is very difficult to obtain precise payoffs to enter into the payoff matrix. A potential way around this problem is to use Skinner boxes (as in operant psychology experiments) which allow precise control over the payoffs that animals encounter in the lab. Clements and Stephens (1995) did just this in their study of blue jay (*Cyanocitta cristata*) cooperation.

Clements and Stephens tested pairs of blue jays, each of whom could peck one of two keys—the cooperate key or the defect key (Figure 9.16). Two payoff matrices were used. The first was

FIGURE 9.16. Blue jays and food payoffs. The experimental set-up used by Clements and Stephens involved two birds pecking at food. The food rewards that the birds received after pecking at cooperate or defect keys were based on whether a prisoner's dilemma or byproduct mutualism payoff was used. *(Photo credit: Jeff Stevens)*

a prisoner's dilemma matrix (P matrix), while the second was a byproduct mutualism matrix (M matrix) (Table 9.3). In each case, bird 1 would begin a trial by pecking one of the keys and bird 2 would end the trial (again by pecking either the cooperation key or the defect key). Birds were given food according to the payoffs assigned to the game they were playing. For example, if a pair was in the P matrix part of the game and bird 1 cooperated when bird 2

TABLE 9.3. The P matrix and the M matrix. These two payoff matrices were used to examine the relative importance of reciprocity (a possible outcome of the P matrix) and byproduct mutualism (a possible outcome of the M matrix) in bluejays. In the M matrix, there was no temptation to defect, as cooperating always fared better than defecting (that is, 4 > 1, 1 > 0). The first number in each cell represents the payoff to bird 1, and the second number the payoff to bird 2. *(Data from Clements and Stephens, 1995)*

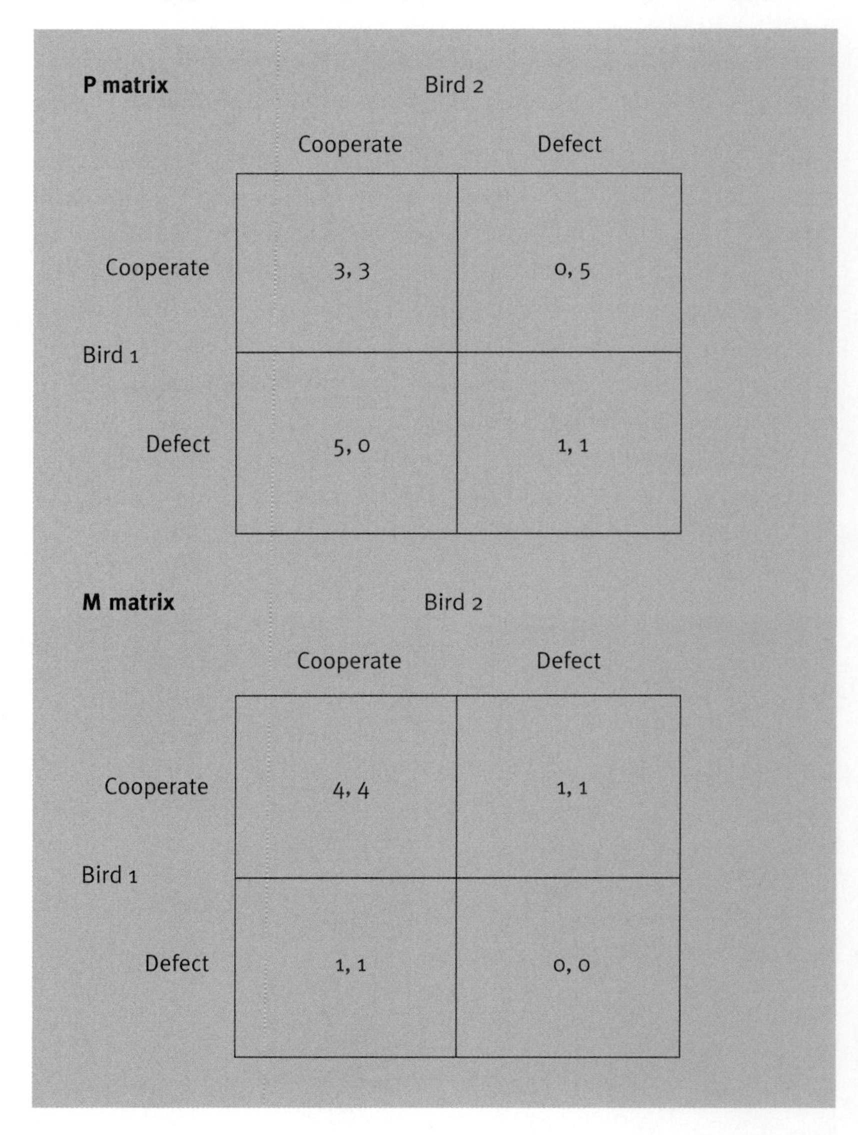

P matrix

		Bird 2	
		Cooperate	Defect
Bird 1	Cooperate	3, 3	0, 5
	Defect	5, 0	1, 1

M matrix

		Bird 2	
		Cooperate	Defect
Bird 1	Cooperate	4, 4	1, 1
	Defect	1, 1	0, 0

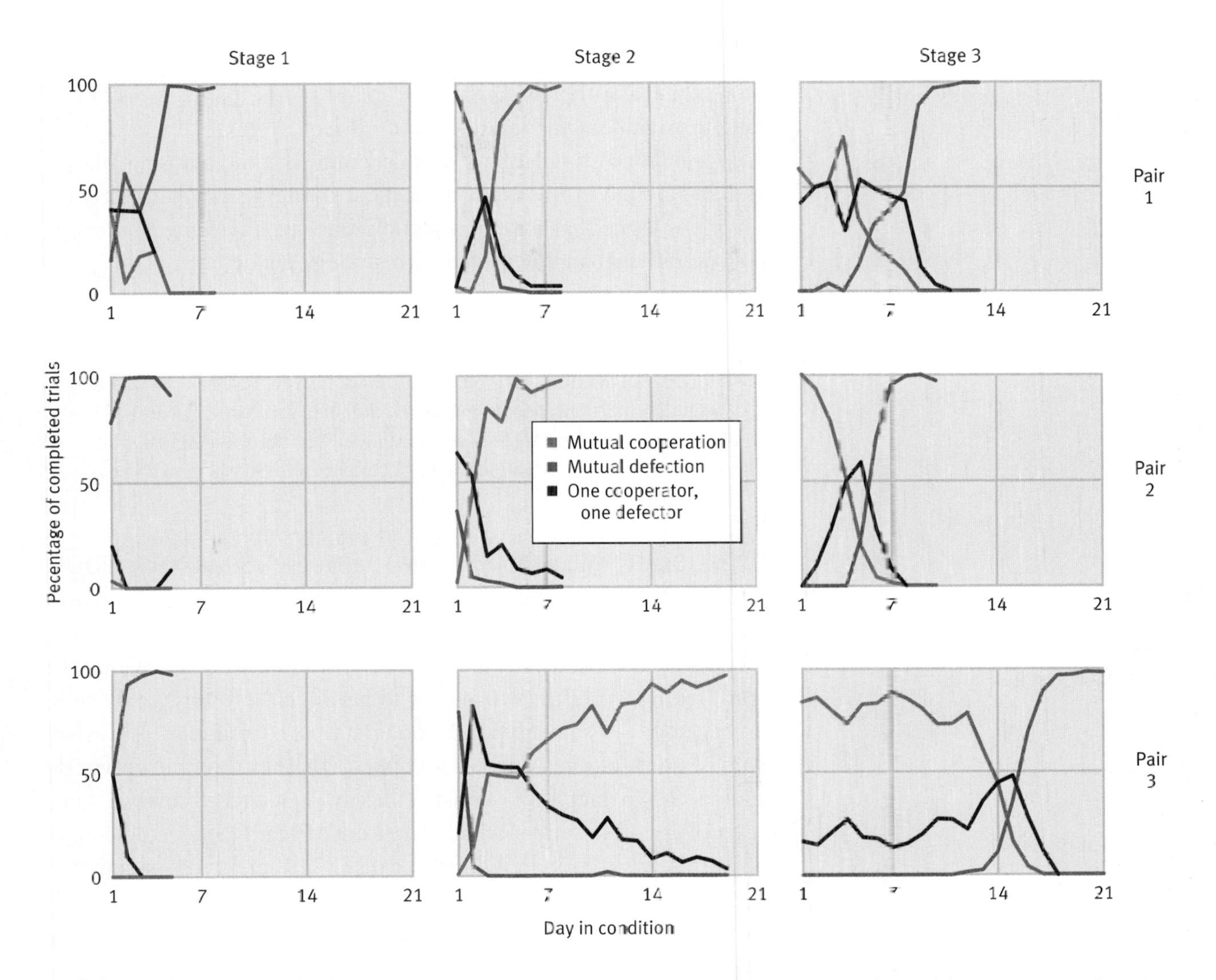

FIGURE 9.17. Byproduct mutualism and blue jays Three pairs of blue jays were tested in a three-stage experiment: Stage 1 = prisoner's dilemma, Stage 2 = byproduct mutualism, and Stage 3 = prisoner's dilemma. Jays cooperated when the payoff matrix matched byproduct mutualism, but not when it matched the prisoner's dilemma. *(Based on Clements and Stephens, 1995)*

defected, bird 2 obtained five food items, while bird 1 received no food items.

Birds were exposed to the P matrix, then to the M matrix, and finally to the P matrix once again, and on any given day a pair of birds would play these games with each other more than 200 times. Clements and Stephens found that, regardless of whether or not the jays could see each other, birds defected in the first P matrix, cooperated in the M matrix, and reverted to defection the second time they encountered the P matrix (Figure 9.17). As such, blue jays appeared to cooperate via byproduct mutualism and not reciprocity, even when a payoff matrix (the prisoner's dilemma) that should promote reciprocity was presented. Reciprocity then failed to emerge as a stable solution, even when pairs of bird had more than a sufficient number of plays of the game to establish long-term mutual cooperation.

One interesting aspect of this study is that learning played such a key role in the establishment of cooperation. Clements and Stephens argue that natural selection may favor *learning to be cooperative,* rather than any cooperative strategy per se.

> We do not suppose that natural selection has favored key-pecking in Skinner Boxes. Indeed, we offer no argument that blue jays ever faced Prisoner's Dilemmas in their evolutionary history. Rather . . . natural selection has equipped animals with the ability to recognize and implement cooperative strategies in novel situations, when economic circumstances favor cooperation. (Clements and Stephens, 1995, p. 531)

BYPRODUCT MUTUALISM AND HOUSE SPARROW FOOD CALLS. House sparrows *(Passer domesticus)* produce a unique "chirrup" call (Summers-Smith, 1963) when they come upon a food resource. Such calls may attract other birds to a newly discovered bounty, and if this turns out to be the case, chirrup calls may be regarded as some type of cooperation. To examine just what type of cooperation, Marc Elgar tested whether the chirrup vocalization did in fact bring conspecifics to a newly discovered food source, and if so, under what conditions (Elgar, 1986). To accomplish this, Elgar recorded chirrup calls at artificial feeders containing pieces of bread. The patches contained either bread that was divisible among sparrows, or bread that was just enough food for a single bird.

To begin with, Elgar did find some evidence that those sparrows arriving at a patch of food first were the most likely to produce chirrup calls. Furthermore, chirrup call rates were higher when the food resource was divisible (Figure 9.18). Elgar argues that food items in one of his treatments were small enough that sparrows could pick them up and fly away, which is what the sparrows did, without producing chirrup calls. Chirrup calls were, however, associated with larger food items—that is, those that were too big to remove from the experimental area. It may be that given that sparrows needed to remain at a feeder, and that it is safer to do so in the company of other sparrows, the benefits associated with predator detection outweighed the costs of inviting other foragers to share one's food at the site. If this proves to be the case, then chirrup calls are most easily understood in the context of byproduct mutualism. The calls are emitted when the immediate net benefit for calling (predator detection) is greater than the net benefit for not calling (more to eat).

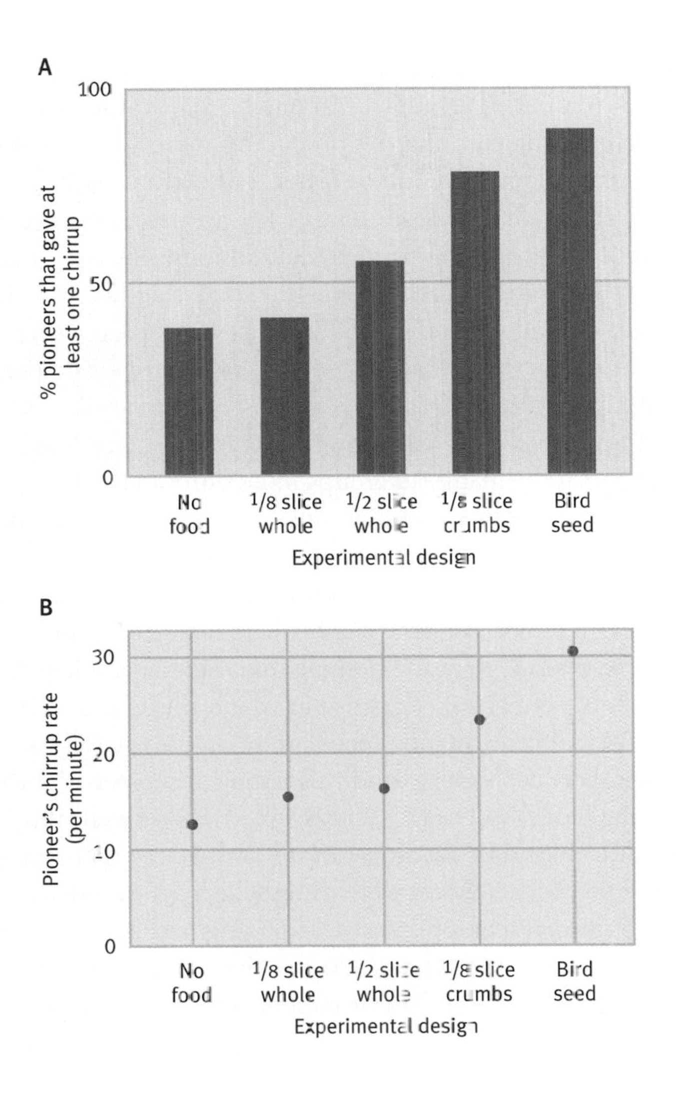

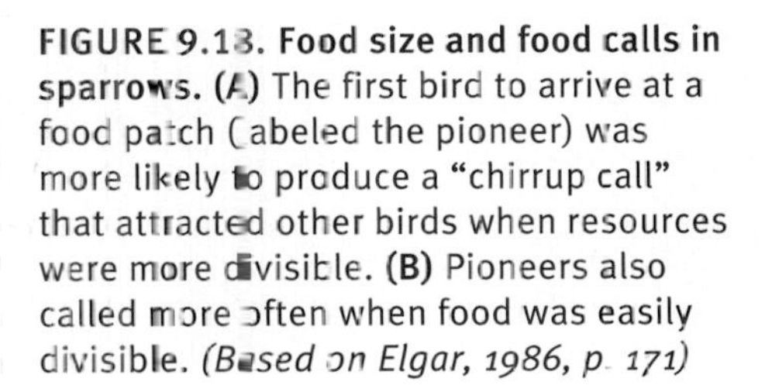

FIGURE 9.13. Food size and food calls in sparrows. **(A)** The first bird to arrive at a food patch (labeled the pioneer) was more likely to produce a "chirrup call" that attracted other birds when resources were more divisible. **(B)** Pioneers also called more often when food was easily divisible. *(Based on Elgar, 1986, p. 171)*

PATH 3: GROUP SELECTION

Group selection has a long history, but we can only touch on the salient aspects of it here (Williams, 1966, 1992; Alexander and Borgia, 1978; D. S. Wilson, 1980; Sober and Wilson, 1998). Modern group selection models of cooperation, although still controversial, are quite straightforward (Wilson and Sober, 1994).

The kernel of trait-group selection models is that natural selection operates at two levels: within-group selection and between-group selection. In the context of cooperation and altruism, within-group selection acts *against* cooperators and altruists, since such individuals, by definition, take on some cost that others do not. Selfish types are always favored by within-group selection since they receive any benefits that accrue because of the actions of cooperators and altruists, but they pay none of the costs.

As opposed to within-group selection, between-group selection favors cooperation if groups with more cooperators outproduce other groups—for example, by producing more total offspring or being able to colonize new areas faster. For example, when individuals in a group of unrelated animals give alarm calls, they pay a cost within the group, as they may be the most obvious target of a predator honing in on such a call. But their sacrifice may benefit the group overall, as other individuals—including other alarm callers, as well as those that don't call—are able to evade predators because of the alarm call. Thus, groups with many alarm callers may outproduce groups with fewer cooperators. For such group-level benefits to be manifest, groups must differ in the frequency of cooperators within them, and groups must be able to "export" the productivity associated with cooperation.

Group selection models remain controversial, as many ethologists believe that because they can be mathematically translated into selfish gene models, they add nothing to our understanding of selection and behavior. There is a sense in which this is true. Yet, group selection models do instantly focus attention on what is happening within and between groups, and this is not necessarily the case for selfish gene models (Dugatkin and Reeve, 1994). Thus, group selection models may spur investigators to construct experiments that would not be obvious if they were using selfish gene models.

To get a more comprehensive view of group selection, following D. S. Wilson (1990), let's focus on the evolution of specialized foraging behavior in an environment containing many insect colonies.

GROUP SELECTION IN ANTS? Cooperative colony foundation occurs in a number of species of ants (Holldobler and Wilson, 1990), where cooperating co-foundresses are *not* closely related (Rissing and Pollock, 1986, 1988; Rissing et al., 1989).

Cooperative colony foundation has been well studied in the desert seed harvester ant *Messor pergandei*. Adult ants are very territorial (Wheeler and Rissing, 1975; Ryti and Case, 1984), and brood raiding—wherein brood captured by nearby colonies are raised within the victorious nests and colonies that lose their brood in such interactions die—is seen among young starting colonies in the laboratory. All queens in a nest cooperate in establishing their living quarters by excavating the ground for a nest, and each produces approximately the same number of offspring (Table 9.4). Furthermore, there exists a positive correlation between the number of cooperating foundresses in a nest and the number of initial workers (these workers are also the brood raiders) produced by that colony (Rissing and Pollock, 1991). Nests with more workers—that is, those with more cooperating foundresses—are more likely

TABLE 9.4. Cooperating co-foundresses. In the ant *Messor pergandei*, unrelated queens (here W, Y, B, and O) co-found nests. The reproductive output of queens within a nest tends to be approximately equal. Differences in all three measures across the four queens were not statistically significant. *(From Rissing and Pollock, 1986)*

		% LAID BY			
NEST NO.	NO. OF EGGS	W	Y	B	O
1	22	—	32	41	27
2	24	38	25	38	—
4	32	41	28	25	6
5	11	45	—	55	—
8	29	28	34	38	—
9	44	34	16	27	23
10	36	31	19	50	—
11	21	43	—	24	33
15	29	38	—	21	41

to win brood raids (Rissing and Pollock, 1987). Until workers emerge (through eclosion), queens within a nest do not fight, and no dominance hierarchy exists (Figure 9.19). After workers emerge and the between-group benefits of multiple foundresses are already set in place with the presence of brood raiders, all that remains is within-group selection, which favors being noncooperative. It is at this juncture that queens within a nest often fight to the death.

The scenario depicted above is precisely what is necessary for group selection to operate, as group selection requires the differential

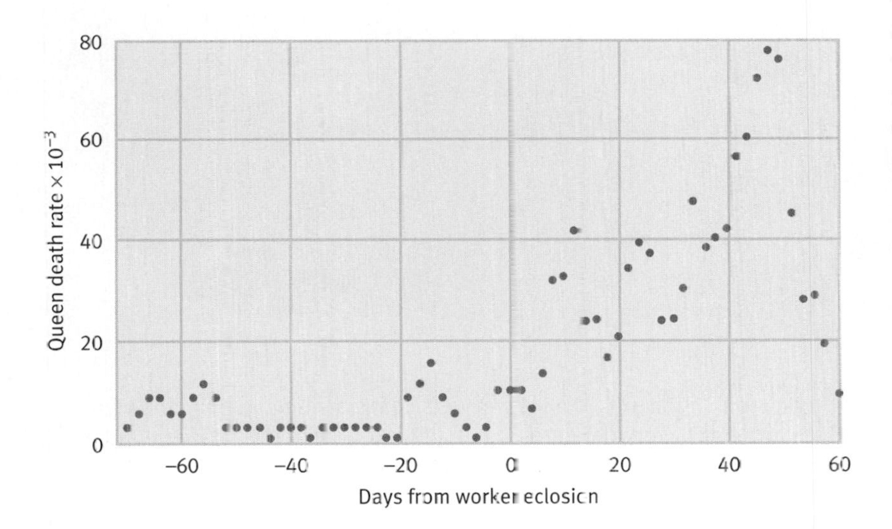

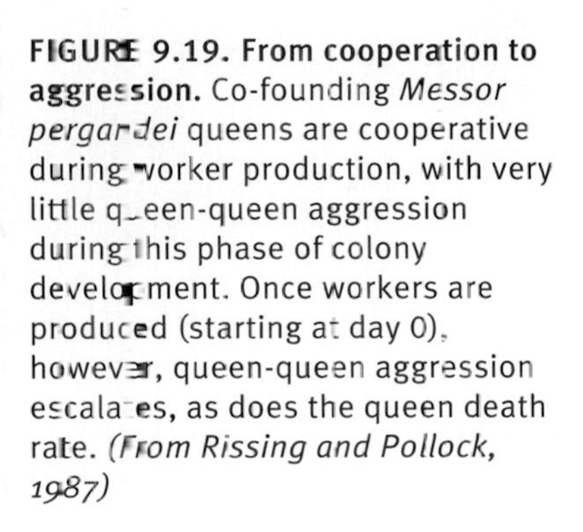

FIGURE 9.19. From cooperation to aggression. Co-founding *Messor pergandei* queens are cooperative during worker production, with very little queen-queen aggression during this phase of colony development. Once workers are produced (starting at day 0), however, queen-queen aggression escalates, as does the queen death rate. *(From Rissing and Pollock, 1987)*

productivity of groups based on some trait. In the case of *M. pergandei,* the trait is queen-queen cooperation. Such cooperation is selected against *within* groups (since cooperators pay a cost not born by noncooperators), but it may be selected for *between* groups, since groups with many cooperators differentially survive brood raiding, which is the between-group component of group selection. Yet, these studies have recently been challenged. Pfennig (1995) constructed a field experiment contrasting single and double foundress associations in *M. pergandei*. Not only did double foundress nests not outlive single foundress nests in Pfennig's experiment, but no brood raiding at all was observed, calling into question all of the elements necessary for group selection to operate in the wild.

One of the strongest cases to date of group selection comes from Rissing and his colleagues' (1989) work on another ant, *Acromyrmex versicolor*. In this species, many nests are founded by multiple queens, no dominance hierarchy exists among queens, all queens produce workers, and brood raiding among starting nests is common. Furthermore, the probability of the nest surviving the brood raiding period is a function of the number of workers produced.

A. versicolor differs from *M. pergandei*, however, in that queens in the former species forage after colony foundation (Figure 9.20).

FIGURE 9.20. Extreme cooperation by foraging queen. In the ant *Acromyrmex versicolor*, a single queen (shown in the blowup circle) is the forager for a nest. Such foraging is very dangerous, but all food collected is shared equally among (unrelated) queens.

TABLE 9.5. Harmony among *Acromyrmex versicolor* queens. In *A. versicolor*, unrelated queens co-found nests, with a single queen taking on the dangerous role of forager for everyone in the nest. All indications are that reproduction within nests is equal between foragers and nonforagers. *(Based on Rissing et al., 1989)*

	FORAGER	NONFORAGER
Mean number of primary eggs	8.6	8.5
Mean primary egg length	0.52	0.54
Mean number of total eggs	20.37	18.94

As a result of increased predation pressure and parasitization, foraging is a very dangerous activity for a queen. Yet, once a queen takes on the role of forager, she remains in that role. After a queen becomes the sole forager for her nest, she shares all the food brought into her nest with her co-foundresses. In other words, the forager assumes both the risks and the benefits of foraging, while the other queens in her nest simply assume the benefits but not the costs (Table 9.5). Once again, however, cooperation—in this case, extreme cooperation on the part of the forager—within nests appears to lead to more workers. This in turn affects the probability that a given nest will be the one to survive the period of brood raiding, thus providing the between-group component necessary for cooperation to evolve (Seger, 1989; Rissing et al., 1989).

THE HUTTERITES. A second fascinating example of how trait-group selection can facilitate cooperation is a human one. The Hutterites are a fundamentalist religious group that bears some resemblance in lifestyle to the Amish and other Anabaptists. Between-group selection so strongly favors cooperation among the Hutterites that they themselves refer to their groups as the equivalent of human bee colonies (Ehrenpreis, 1650; Figure 9.21).

Hutterites live in small farmlike communities, primarily in Alberta and Saskatchewan, Canada. Within communities, private property is rare, and all behaviors are geared toward increasing group productivity by stressing an allegiance to God, rather than to man. A remarkable division of labor exists in Hutterite groups, and this has allowed Hutterites to prosper in what others consider marginal lands. Hutterite cooperation translates into extreme productivity—as Hutterites have the highest birthrate of any known human society

FIGURE 9.21. Cooperation in humans. The Hutterites, a religious sect of the Anabaptists, are one of the most cooperative human societies ever studied. *(Photo credit: Annie Griffiths Belt/Corbis)*

and, at the same time, are very efficient at transforming new land areas into productive Hutterite farms. They have been so productive, in fact, that throughout their five-hundred-year history, laws have been specifically enacted to stop their communities from spreading too quickly.

The creation of a new Hutterite community is an excellent example of group-level efficiency:

> The Hutterites' passion for fairness is perhaps best illustrated by the rules that surround the fissioning process. Like a honeybee colony, Hutterite brotherhoods split when they attain a large size, with one half remaining at the original site and the other half moving to a new site that has been pre-selected and prepared. In preparation for the split, the colony is divided into two groups that are equal with respect to number, age, sex, skills and personal compatibility. The entire colony packs its belongings and one of the lists is drawn by lottery on the day of the split. (Wilson and Sober, 1994, p. 604)

The Hutterites have also developed an elaborate system of rules to handle violations by cheaters. Hutterites take cheaters as a serious problem—a threat to their very being—and the rules they have devised are designed to suppress cheating within groups. These rules minimize the cost of cooperation and maximize the between-group component of trait-group selection models—the component that fosters cooperation.

While it is true that many Hutterites in a colony are related to one another, it is also true that most individuals in a colony are not related. What we really have are many family units acting together to further the good of the community. So, while relatedness is high within families, it is not necessarily that high between families in a community. Furthermore, with respect to blood relatedness, group selection models assume that kin groups are simply one type of group and hence do not make as big a distinction between related and unrelated individuals as do inclusive fitness models.

Phylogeny and Cooperative Breeding in Birds

The three paths to cooperation outlined above, in conjunction with inclusive fitness theory, provide the theoretical foundations upon which evolutionary work on cooperation rests. With this under our belt, we can now move on to other aspects of cooperative behavior, including a possible phylogenetic component, which provides a window to examine how evolutionary history may affect cooperative behavior.

Phylogenetic analysis allows us to examine whether a trait may be common in a group of animal species as a result of common descent rather than independent selection regimes. In other words, phylogenetic analysis lets us ask whether one reason that a trait is common in a taxa is that all members of that taxa share that trait because their common ancestor did. For example, if we find cooperation in species 1, 2, 3, and 4, and all of these species evolved from species 5, which we find to cooperate, then species 1–4 may not represent independent data points with respect to cooperation. Natural selection may very well be maintaining a trait in many different species, even when such a trait originated in a common ancestor, but phylogenetic logic suggests that to count all such cases as originating independently is questionable.

Cooperative breeding in birds (Chapter 8) is common enough that numerous adaptationist hypotheses have been put forth to explain its selective value and distribution (J. L. Brown, 1987; Stacey and Koenig, 1990). In fact, early studies of cooperative breeding in birds were quite different from the current adaptationist approach, and they gave much more weight to phylogeny's role in cooperative breeding (D. Davis, 1942; Hardy, 1961; J. L. Brown, 1974; Fry, 1977). With many new phylogenetic tools in hand, Edwards and Naeem (1993) have reexamined cooperative breeding from a phylogenetic perspective.

Using 166 species of cooperatively breeding passerine birds in ninety-seven genera (J. L. Brown, 1987), Edwards and Naeem began their work by testing whether the distribution of cooperatively breeding species was random. To do this, they created a computer simulation to predict what the distribution of cooperative breeding species would be if they distributed into genera simply based on the number of species in that genera. They found that the distribution in nature differed significantly from the random distributions generated by the computer simulations, with some genera having too many cooperatively breeding species and others too few (Figure 9.22). Edwards and Naeem followed this analysis by using already published phylogenetic trees to examine the distribution of cooperative breeding. For example, their phylogenetic analysis of jays, Australian songbirds, Australian treecreepers, and New World wrens suggests that cooperative breeding may have arisen a limited number of times in some common ancestor(s) of modern-day species and has simply been lost by those species that do not cooperate today. In other words, common ancestry, rather than natural selection alone, may explain the distribution of cooperative breeding in some modern groups of animals.

The phylogenetic approach does not necessarily conflict with an adaptationist view of cooperative breeding, although it does emphasize the importance of history (phylogeny) in our understanding of the distribution of cooperative breeding. For example, it may be that a set of species derived from a single ancestor initially "inherits" the tendency to cooperate (a phylogenetic component) but that, subsequent to this, selection favors continued cooperation in some species but not others (a selective component).

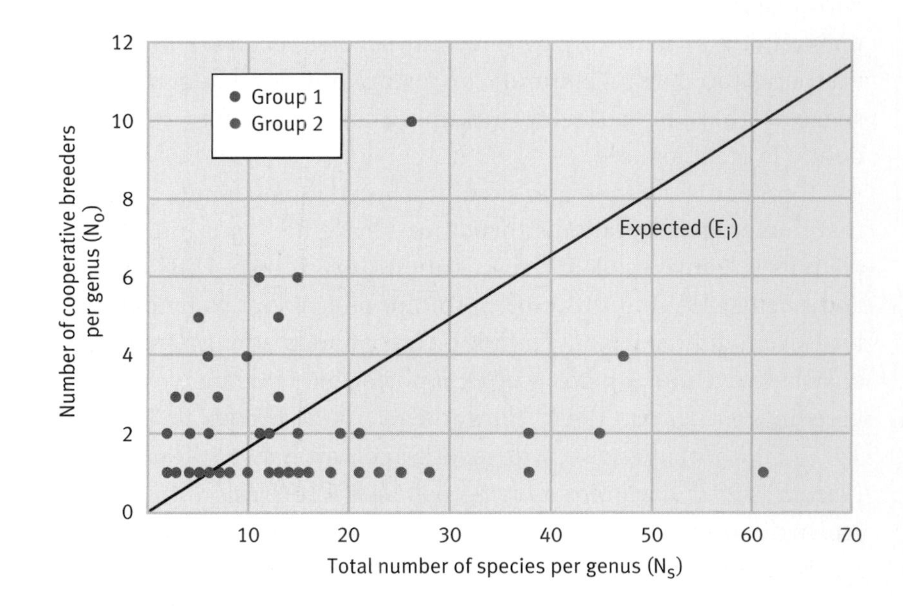

FIGURE 9.22. Phylogeny and cooperative breeding. The solid line represents the slope expected if the number of cooperators per passerine genus (seventy-one) were proportional to genus size. In group 1, cooperative breeding was overrepresented; in group 2, cooperative breeding was underrepresented. *(Based on Edwards and Naeem, 1993, p. 761)*

Hormones, Reproductive Suppression, and Cooperative Breeding

Much of our discussion so far has focused on ultimate issues relating to cooperation. This reflects the fact that much of the work on cooperation has been done from an evolutionary perspective, focusing on natural selection and, to a much lesser degree, phylogeny. But there is also a considerable literature that examines proximate factors in relation to cooperation. In Chapter 8, we examined the evolution of cooperative breeding in light of kin selection models; in this chapter, we shall examine the possible hormonal correlates associated with cooperative breeding (as such, this section involves cooperation among both kin and nonrelatives).

Cooperative breeding systems are generally broken down into two subcategories: **plural breeders,** which are multiple pairs of breeders that nest together and help raise each other's offspring, and **singular breeders,** which are a single pair in a group that sires virtually all of the offspring, with other group members helping to raise the young (J. L. Brown, 1987). Here we shall specifically focus on the hormonal underpinnings of cooperative singular breeding systems. In such systems, the norm is for the single pair to breed, with dominant individuals enforcing reproductive suppression in subordinates by displaying aggression toward any subordinate that prepares to breed (Creel, 1996; see also J. L. Brown, 1974; Stacey, 1979; Emlen, 1991).

The question of why subordinates "accept" reproductive suppression is an evolutionary one, and numerous models have been designed to address this fascinating issue (Vehrencamp, 1983; Keller and Reeve, 1994; Johnstone et al., 2000; Reeve, 2003). Such models generally ask how ecological, demographic, energetic, and genetic parameters interact to produce an evolutionarily stable level of reproductive suppression (see Chapter 8). Rather than looking at *why* reproductive suppression can evolve in cooperative animal breeding systems, we will ask what is known about the role of hormones in leading subordinates to cooperate—that is, to help raise the offspring of others in their group.

Scott Creel and his colleagues have been examining both proximate and ultimate questions surrounding singular cooperative breeding in the dwarf mongoose (*Helogale parvula*) since the

FIGURE 9.23. Cooperation among dwarf mongooses. (A) In this dwarf mongoose group, we can see dwarf mongooses exhibiting many kinds of cooperative behaviors—feeding, grooming, and "baby-sitting" young offspring. **(B)** Another kind of cooperative behavior is transporting young dwarf mongooses from place to place. This may be done for both an individual's own offspring and for other young. *(Photo credits: Scott Creel)*

mid-1980s (Figure 9.23). Dwarf mongooses live in small social groups in sub-Saharan Africa (Kingdon, 1977). Sociality in this species likely evolved in response to the availability of sufficiently abundant prey (Waser, 1981; Waser and Waser, 1985), as well as from the need to defend against aggressive predators, as mongooses suffer high rates of predation (Rasa, 1986).

Cooperation on the part of reproductively suppressed dwarf mongooses takes many forms, including guarding (baby-sitting), grooming, and carrying food (Rood, 1980), and spontaneous lactation by nonbreeders (Creel et al., 1991), and such aid is effective in increasing the reproductive success of dominant individuals. Dominant pairs produce the great majority of all offspring sired in a group, yet they are aided by reproductively suppressed, often genetically related, individuals in raising such young. Given that, Creel and his colleagues asked what means were used to suppress reproduction in some mongooses and, in so doing, to produce a class of individuals that not only fails to reproduce, but may also help raise the young of others.

Creel and his team set out to test how both behavioral and endocrinological factors operate to produce reproductive suppression in a wild population of dwarf mongooses. This alone was an important undertaking. While some studies had examined similar questions in captive primates and rodents, until Creel undertook his work, little was known about how both behavioral and endocrinological factors operate to create suppression in the wild (for exceptions, see Reyer, 1986; Faulkes et al., 1990). Using focal animal sampling of aggressive interactions (Altmann, 1973), measurements of body mass, and endocrinological analysis of urine samples, Creel and his colleagues (1992) were able to piece together an interesting scenario that explains how behavioral and endocrinological factors promote both reproductive suppression and subsequent cooperation, but do so differently across dwarf mongoose sexes.

Not surprisingly, Creel and his team found that in dwarf mongooses the number of matings an individual obtained was highly correlated with rank in both sexes. High-ranking (older) members of a group were much more likely to obtain matings than were subordinate individuals (Figure 9.24), and high-ranking males and females preferentially mated with each other. Females in estrous

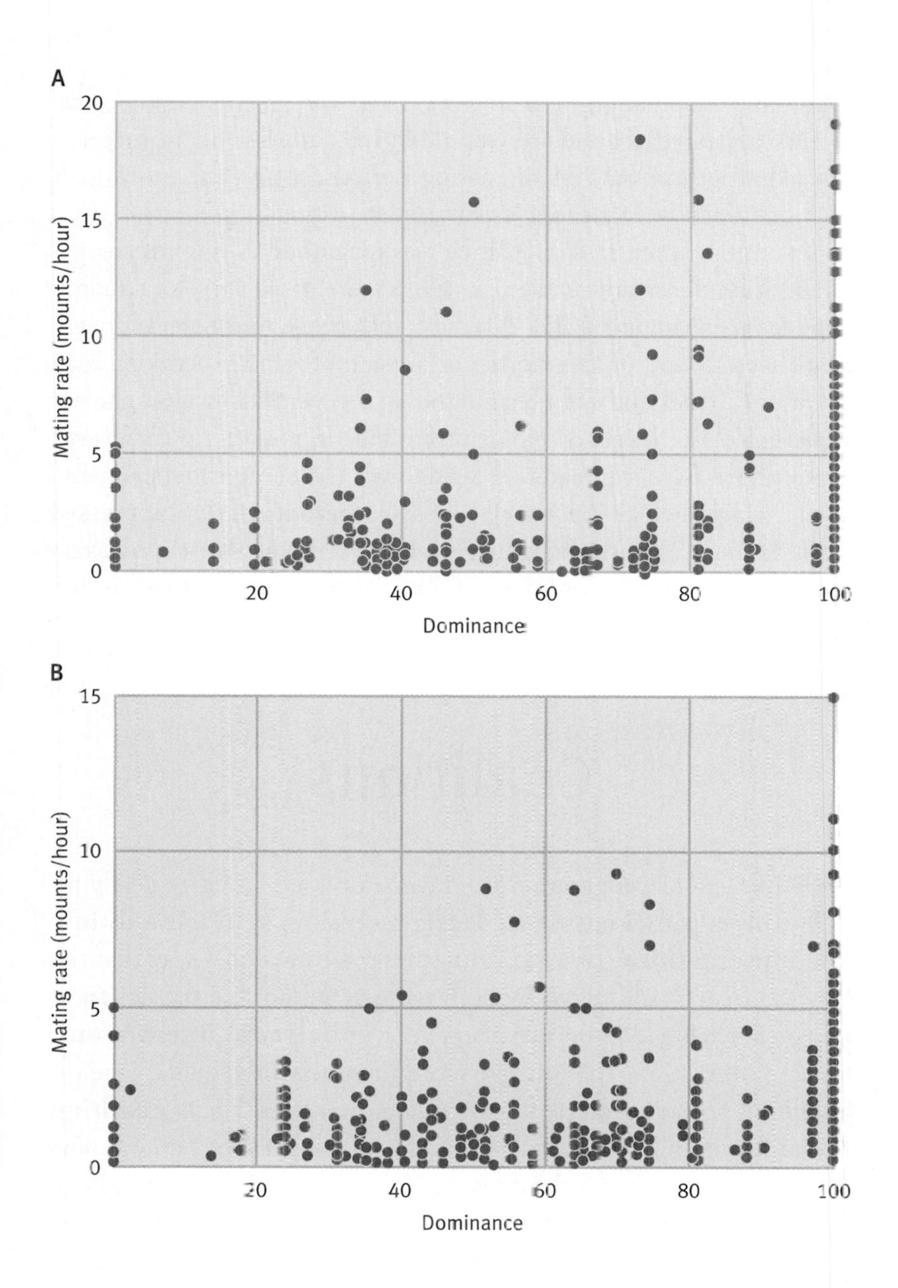

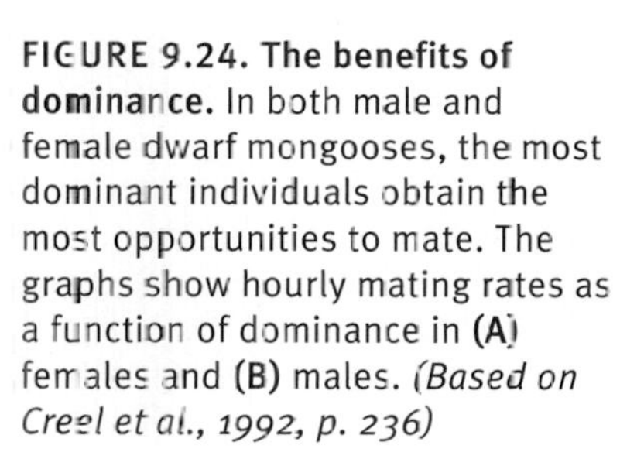
FIGURE 9.24. The benefits of dominance. In both male and female dwarf mongooses, the most dominant individuals obtain the most opportunities to mate. The graphs show hourly mating rates as a function of dominance in **(A)** females and **(B)** males. *(Based on Creel et al., 1992, p. 236)*

rarely refused mounting by a male, even a subordinate male. But such attempted mounts by subordinate males were primarily curtailed by male-male interactions. During mating periods (which were synchronized, as females tended to be in estrous at the same time; Rood, 1980), male-male aggression tripled, with most of the increase due to alpha males attacking subordinate males in fights over access to females. For males, then, reproductive suppression, and eventual helping behavior, was primarily due to behavioral interactions with other males.

The situation in female dwarf mongooses was quite different from that of their male counterparts. Although there is a strict dominance hierarchy among females throughout the year, female-female aggression did not increase during mating periods. Rather than behavioral interactions alone producing reproductive suppression (and eventually cooperation in raising others' offspring), endocrinological factors appeared to underlie reproductive suppression in females. Both during and outside of mating periods, dominant individuals had significantly higher levels of urinary estrogen conjugate (EC) concentrations, but this difference was magnified during mating periods, when dominant female EC levels were triple those of subordinate females (Figure 9.25). No such differences in male androgen levels were found (either during or outside of mating periods). Furthermore, Creel and his team found that several days after the estrous peak, EC levels of pregnant females increased up to seventy times above baseline readings, while over the same time span, EC levels went back to baseline values in reproductively suppressed subordinates. This strongly suggests that reproductive suppression was a function of subordinates' failing to become pregnant, rather than aborting fetuses from successful copulations.

Coalitions

The type of cooperation we have addressed so far typically involves pairwise, that is, dyadic interactions. In these **dyadic interactions,** two individuals interact in such a way that the fitness of each is affected by both its own action and the action of its partner. Cooperation can also occur in **polyadic interactions,** that is, interactions that involve more than two individuals. One example of polyadic interactions involving cooperation is **coalition behavior,** which is typically defined as a cooperative action taken by at least two other individuals or groups against another individual or group. When coalitions exist for long periods of time, they are often referred to as **alliances** (Harcourt and de Waal, 1992).

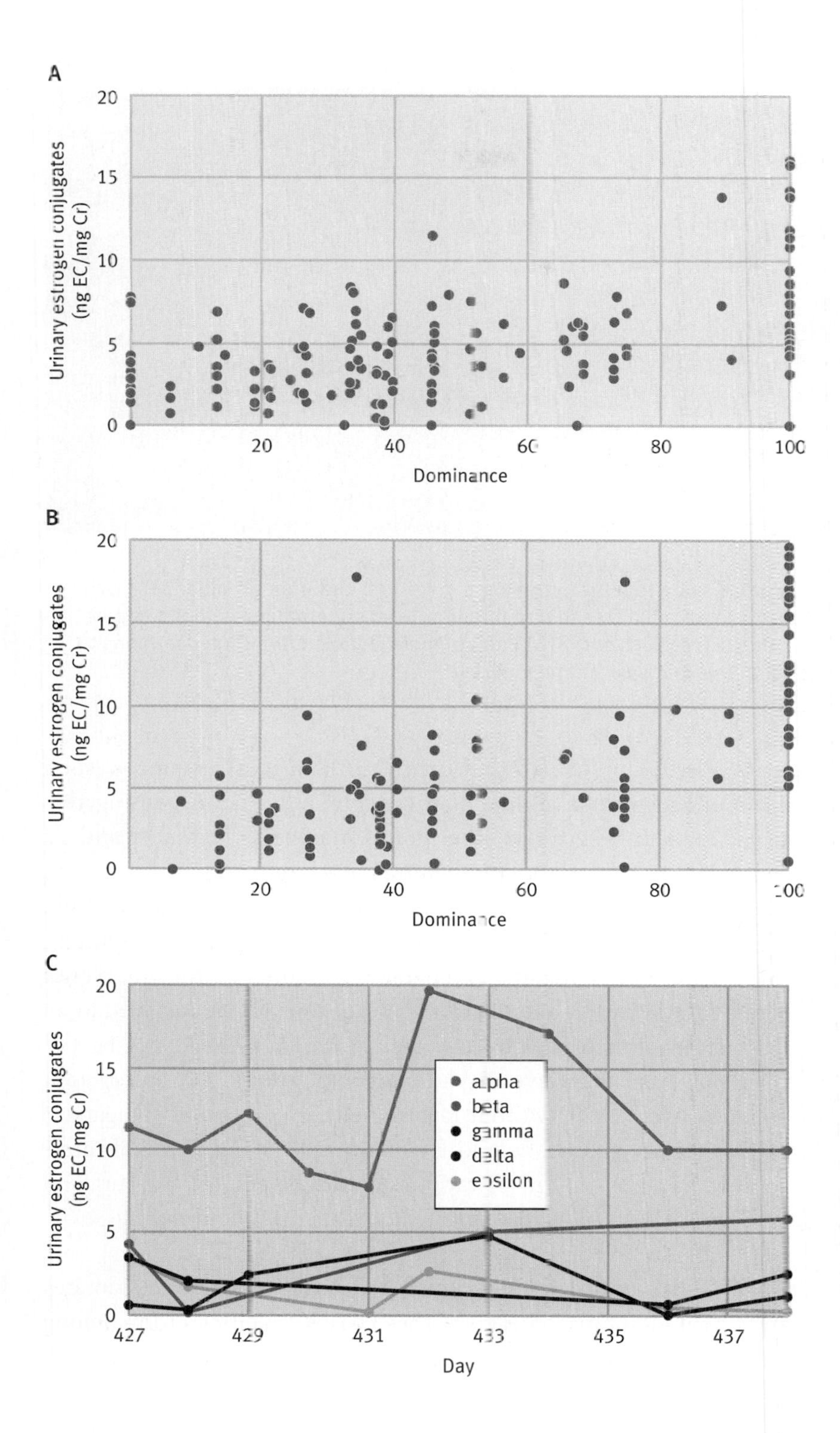

FIGURE 9.25. Suppression of mongoose reproduction. In female dwarf mongooses, urinary estrogen levels are higher in dominant individuals both **(A)** during nonmating periods and **(B)** during mating periods. **(C)** Urinary estrogen levels are shown in five females. Mating occurred on days 431–435, and it was during this period that urinary estrogen levels differed most between the dominant female and the other four females. *(Based on Creel et al., 1992, p. 240)*

Coalitions have been documented in numerous primate species (Harcourt and de Waal, 1992; Chapais, 1992), hyenas (*Crocuta crocuta*; Zabel et al., 1992), wolves (*Canis lupus;* Fentress and Ryon, 1986), lions (*Panthera leo;* Packer and Pusey, 1982), cheetahs (*Acinonyx jubatus;* Caro, 1994a), coatis (*Nasua narica;* Russell, 1983), and dolphins (*Tursiops truncatus;* Connor et al.,

FIGURE 9.26. Coalitions. (A) Three male dolphins swim together, forming a long-term coalition (or alliance). Such male coalitions "herd" females. Occasionally different alliances join together to form superalliances that compete against other such superalliances. *(Photo credit: Richard Connor)* **(B)** Pairs of male chimps often form coalitions to act against larger, more dominant, individuals. *(Photo credit: Frans de Waal)*

1992, 1999; Connor, 1992; Figure 9.26). In most instances, such coalitions involve an animal *intervening* in a dyadic, usually aggressive, interaction between other group members (Harcourt and de Waal, 1992). Often, the intervening individual is dominant to others involved in the interaction, but this is not always so (see Silk, 1992a, 1992b, for a review of rank and intervention). In addition, in primates, interventions often take the form of the intervener coming to the aid of one of the two other individuals involved in an interaction (Harcourt and de Waal, 1992). This need not be the case, however, as intervening animals may break up an interaction between two others, without favoring either combatant (Dugatkin, 1998a, 1998b; Johnstone and Dugatkin, 2000). This type of intervention has been found in various primate species (Harcourt and de Waal, 1992), as well as in the cichlid fish (*Melanochromis auratus*; Nelissen, 1985).

We shall examine two examples of coalitions, one in primates and one in dolphins. In each of these cases, coalitions form among males to gain access to reproductively active females.

COALITIONS IN BABOONS

One of the most well-cited examples of coalition formation is Craig Packer's (1977) work on male reproductive coalitions in baboons, *Papio anubis*. In *P. anubis*, an "enlisting" male solicits a coalition partner by rapidly turning his head between the solicited animal (that is, the individual he is requesting aid from) and his opponent, while at the same time continuously threatening his opponent (Figure 9.27). In his study of coalition formation, Packer observed ninety-seven solicitations that resulted in coalitions being formed. On twenty of these occasions, the opponent was consorting with an estrous female, and this increased the probability of a coalition being formed between the other two individuals (the enlisting male and the solicited male). On six of these twenty occasions, the estrous female deserted the opponent and went to the enlisting male, creating a clear benefit to coalition formation, at least for the enlisting individual.

What about the costs and benefits to the animal that is solicited into a coalition? Joining a coalition may in fact be fairly costly to solicited individuals, who rarely ever obtain access to the estrous female but who risk attack from the opponent. Packer's study suggests that solicitors may overcome such costs by having the individual who enlisted them respond when they themselves need help (reciprocal coalitions). In fact, Packer discovered that baboons had "favorite partners," and that favorite partners solicited each other more often than they solicited other group members, suggesting that "alliance" rather than coalition may be the correct label for this sort of baboon interaction

Other studies of coalition and alliance formation in baboons have generally supported Packer's (1977) findings (Smuts, 1985; Noe, 1986). Bercovitch (1988), however, found that in olive baboons

FIGURE 9.27. Baboon coalitions. A male baboon (middle) involved in an aggressive interaction (with male on left) will often solicit others to aid him by turning his head in the direction of a potential coalition partner (male on right).

(Papio cyanocephalus anubis), males who attempted to enlist others in coalitions were no more likely to obtain the females than any other coalition member, and that baboons who declined to join a coalition were again solicited in the future. The reason for the discrepancy between Packer's and Bercovitch's studies are as yet not clear, but Hemelrijk and Ek (1991) suggest that it may be tied to the presence of a clear alpha male in the Bercovitch study, but not in the Packer study.

ALLIANCES AND "HERDING" BEHAVIOR IN CETACEANS

Richard Connor has uncovered an interesting case of alliance formation in bottleneck dolphins *(Tursiops truncatus).* What makes this case particularly unique is that dolphins are notoriously difficult to track for long periods of time, and while ethologists have generally thought dolphins capable of sophisticated behavior in the wild, this has been very difficult to document in any scientific manner. Connor and his colleagues examined pairs and trios of males forming close associations in the bottleneck dolphin population of Shark Bay in Western Australia. He and his colleagues found not one, but two different types of alliances between male dolphins, with both forms of alliances focusing on males who were "herding" reproductive females to keep them close.

So-called "first-order" alliances are composed of pairs or trios of males acting in a coordinated fashion to keep females by their side, presumably for the purposes of mating. Males in first-order alliances stay very close to one another, and alliances remain stable for many years. When females herded by an alliance try to swim away (which they often do), males act in a very coordinated, aggressive manner (they angle off to different sides) to prevent the females from leaving (Connor et al., 1992).

What makes alliance formation in dolphins unique is that dolphins in different first-order alliances also join together in "second-order" superalliances and aggressively attack and steal females from other alliances. Connor showed that on two occasions, a defending alliance was assisted by another alliance in its attempts to maintain the female it was herding, creating a battle of second-order alliances. Second-order alliances have been documented in only one other species—humans—and Connor (1992) argues that the complex social interactions inherent in dolphin superalliances, as well as in other aspects of dolphin society, may explain the evolution of large brain size in dolphins.

Interspecific Mutualisms

We have been focusing on intraspecific cooperation. There is, however, much evidence that *interspecific* cooperation is important in shaping animal social behavior as well (Thompson, 1982; Vandermeer, 1984; Boucher, 1985; Law and Koptur, 1986; Kawanabe et al., 1993; Bronstein, 1994; Connor, 1995). Such cooperation *between species* is usually referred to as **mutualism** (instead of byproduct mutualism), and so we will use that label here. A classic case of mutualism is that between ants and butterflies. Below we shall consider one aspect of this mutualism.

ANTS AND BUTTERFLIES—MUTUALISM WITH COMMUNICATION?

In numerous species of butterflies and ants, a mutualistic relationship has developed in which butterfly pupae and larvae produce a sugary secretion that ants readily consume, and ants protect the larvae from parasitoid wasps and arthropod predators. Such a mutualistic relationship constitutes a win-win scenario, with individuals in both species better off than they would be otherwise.

Naomi Pierce has been studying one such mutualistic relationship between the imperial blue butterfly (*Jalmenus evagoras*) and the ant (*Iridomyrmex anceps*) (Figure 9.28). The benefits to both parties in this mutualism are enormous. Pierce and her colleagues

FIGURE 9.28. Butterflies and ants in a mutualistic relationship. In the mutualism between the butterfly *Jalmenus evagoras* and the ant *Iridomyrmex anceps*, butterfly larvae cannot survive in the absence of ants, and ants receive some of their food from the nectar produced by the butterfly larvae. *(Photo credit: Naomi Pierce)*

Interview with Dr. HUDSON KERN REEVE

You've worked on cooperation in both wasps and naked mole rats. Isn't that a strange combination? How did you settle on these species?
Superficially, it is indeed a very strange combination—social paper wasps are aerial insects and naked mole rats are subterranean mammals! But there is a deep evolutionary connection between them that has stimulated my interest in both: I think that both manifest common principles of social evolution. The ways that the societies are organized are actually quite similar, with high reproductive skews (near reproductive monopoly by one or a few individuals) occurring among cooperating relatives. In both, the highest-ranking reproductive female behaviorally enforces her reproductive dominance.

I know you spend a good chunk of time engaged in fieldwork each year. Do you find that long hours of behavioral observation and tracking lead you toward new work on cooperation? Could you provide an example?
Yes, I think ultimately that all of the best evolutionary theories are rooted in field observation. (For example, Bill Hamilton, the architect of kin selection theory, was a superb naturalist.) My own view that social organisms, including social insects, engage in reproductive transactions (as embodied in "transactional skew theory") grew out of repeated observations of resource exchanges and *restraint* of aggression in social wasp colonies. Social wasps are constantly exchanging food and water and are curiously nonaggressive

when each is laying an egg, as if they had the equivalent of a "social contract" over the division of resources, and, ultimately, reproduction. I don't think that I ever would have seriously considered the idea that social wasps are reproductively paying each other to cooperate, had I not watched nearly a thousand hours of videotapes of field colonies.

Perhaps more theory is devoted to the evolution of cooperation than to any other issue in ethology. Why do you think that is?
I think that it has attracted a great deal of theoretical interest because it represents an evolutionary puzzle that even had Darwin fretting: How could an organism ever achieve a reproductive advantage by enhancing the reproductive output of another organism at the expense of its own? Hamilton provided one important answer to this puzzle through his kin selection theory. But this solution left unanswered the question of how cooperation could evolve between unrelated organisms—the latter, of course, has been the focus of intense theoretical activity. No doubt, we humans are especially interested in cooperation among nonrelatives, because each of us observes and experiences it regularly (to varying degree!) every day.

Do you ever encounter the sense that some in the scientific community simply believe animals are not cognitively sophisticated enough to undertake acts of cooperation and altruism?
Yes, unfortunately, this is still a pervasive view. I think that this view is highly questionable. An organism needn't have a huge brain in order to engage in high-order social interac-

tions (I refer back to the notion that social wasps can have "social contracts" over reproduction!). I think that many behavioral biologists currently underestimate the cognitive complexity of their organisms, and the situation isn't helped by the fact that most theorists find it easiest to model simply behaving organisms!

In fact, there is now a major push to think of social organisms (especially social insects) as self-organizing robots whose rules of social interaction are very simple. I think that the latter view is headed in the wrong direction, because it ignores the evidence that organisms are highly conditional (context-specific) in their social behavior. I think that the evidence will eventually reveal that most social organisms are best viewed as proactive inclusive-fitness maximizers, i.e., behave according their cognitive *projections* of the inclusive fitness consequences of alternative social actions—and are not anything like toasters with a few settings determining simple input-output (stimulus-response) relationships. An inclusive fitness projector will always win evolutionarily over a self-organizing robot, provided that the neural machinery costs are not too great, and it is precisely the latter that I think has been systematically overestimated. Another way to put this is that the difference between a large brain and a small brain is not that the latter results in less sophisticated behavior; rather, the smaller brain still enables very sophisticated behavior, but over a somewhat narrower range of conditions (the ones regularly encountered by its bearers).

How have the fields of mathematical economics and political science contributed to our understanding of the evolution of cooperation?

The influence of economics and political science on evolutionary biology is immense, in large part because the theoretical apparatus of game theory was developed and refined in these two disciplines by people such as von Neumann and Nash and many others. Now, game theory is *the* central theoretical tool in understanding the evolution of social behavior.

Ironically, I think that evolutionary biology, now enriched by game theory, eventually will absorb human economics and political science as part of itself, because the latter two disciplines are just two subfields of the study of human social behavior, and we evolutionary biologists believe that the only satisfying theory of any organism's social behavior ultimately must be evolutionary! I am sometimes criticized for thinking that social wasps have reproductive transactions on a human economic analogy—surely that is too anthropomorphic! But this criticism gets things backwards: If human economic behavior and wasp social behavior are evolved responses to similar selection pressures, then the connection between them is much deeper than a simple analogy, i.e., if they really are manifestations of the same evolutionary principles, then anthropomorphism is the correct stance!

Do you think that the rules governing cooperative behavior are sometimes transmitted culturally in nonhuman primates?

There is growing evidence that this is so, and the important consequence is that, for such primates (and certainly for humans), we need to understand how cultural and biological evolution will interact. Do they proceed independently of each other, as some believe, or does one somehow entrain the other? This will be a hugely important focus for theoretical and empirical research in the years ahead.

How close are we to having a comprehensive understanding of animal cooperation? What, if anything, remains to be done?

My view is that we are still very far away. There is no shortage of theories, but I suspect the best ones have yet to be developed. What is most limiting are the data that cleanly discriminate among alternative theories. The latter is true in part because theorists have not always been clear about which predictions separate alternative theories and in part because empiricists have not always been good at deriving the right theoretical predictions for their study organisms (or for the contexts in which the latter are studied). This is not to sound pessimistic. On the contrary, it is an extremely exciting time to be a sociobiologist, as we are on the brink of beginning to solve many, many puzzles! In my view, what we cannot afford to lose is the conviction that these puzzles have general, elegant solutions.

DR. HUDSON KERN REEVE is an associate professor at Cornell University. His work integrates theoretical, empirical, and conceptual approaches to ethology to understand cooperation, kinship, aggression, and the distribution of reproductive opportunities within nonhuman and human groups.

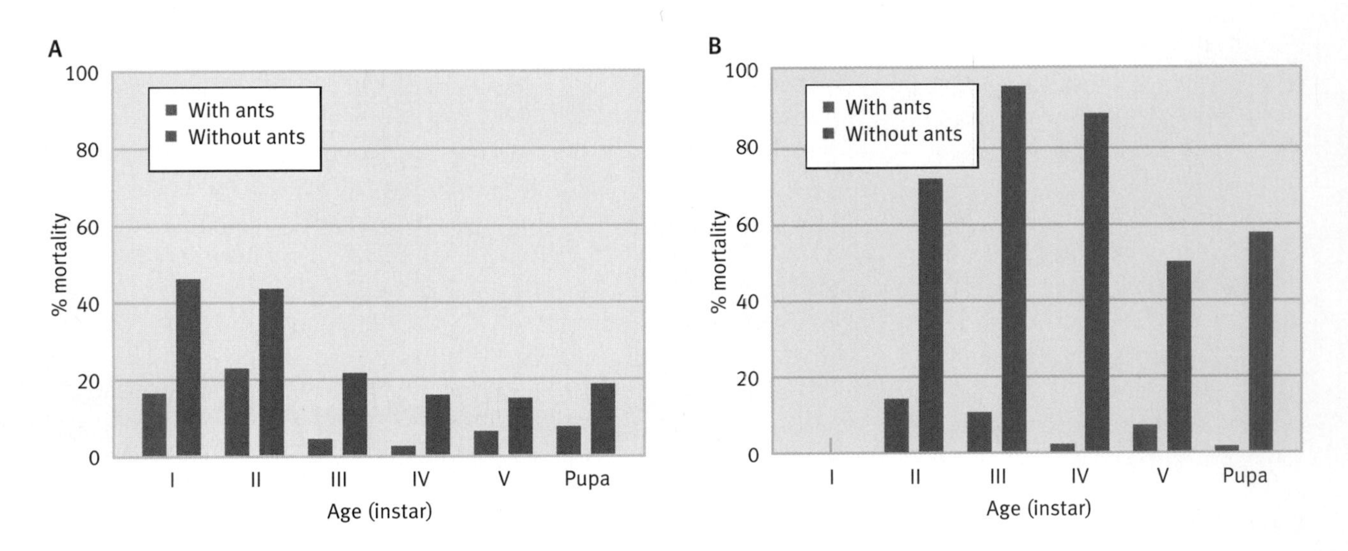

FIGURE 9.29. Butterflies need their ant partners. The probability of survival of *Jalmenus evagoras* larvae and pupae when faced by predation was much higher when ants were present than when they were experimentally excluded at two Australian field sites: **(A)** Mt. Nebo and **(B)** Canberra. *(From Pierce et al., 1987, p. 242)*

(1987) have found that, in nature, when faced by predation, butterfly larvae cannot survive when ants are experimentally removed from their environment (Figure 9.29). While ants can survive in the absence of the nectar they consume from larvae and pupae, they obtain a significant portion of their nutrients from their butterfly larvae partners (Pierce et al., 1987; Fiedler and Maschwitz, 1988).

This ant-butterfly mutualism involves costly investment in the other by both parties. Larvae raised in a predator-free laboratory environment pupate later and at a much larger size (Pierce et al., 1987), as they are able to modify the amount they secrete and use the nutrients normally distributed to ants for their own development. Size in both male and female *J. evagoras* is correlated with reproductive success (Hill and Pierce, 1989; Elgar and Pierce, 1988; Hughes et al., 2000) and as such pupating early represents a significant cost (smaller body size as a result of pupating early and hence less reproductive success) to the butterflies. Although the cost to ants for protecting butterfly larvae has not been quantified, it is likely associated with an increased risk of detection by their own predators and parasitoids, as well as the metabolic costs associated with defense (Pierce et al., 1987).

Travasso and Pierce (2000) extended the study of the *J. evagoras/I. anceps* mutualism one step further by examining whether there was interspecific communication in this system. It turns out that ants are almost deaf when it comes to airborne sounds, but they are quite sensitive to vibrational cues. Given this, Travasso

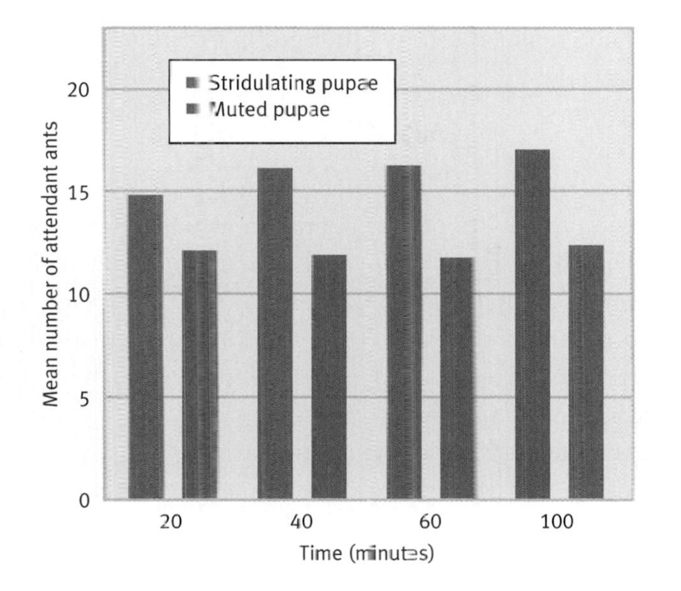

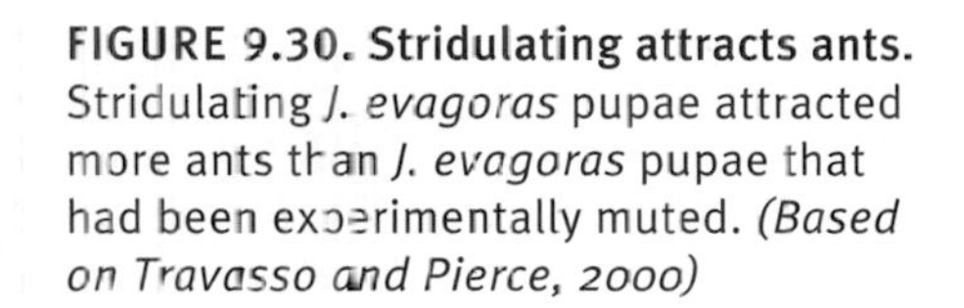

FIGURE 9.30. Stridulating attracts ants. Stridulating *J. evagoras* pupae attracted more ants than *J. evagoras* pupae that had been experimentally muted. *(Based on Travasso and Pierce, 2000)*

and Pierce began by examining the role of vibrational cues in communication between ants and butterflies. They found that larval sound production (stridulation) was higher when ants were in the vicinity, suggesting that vibrational cues are "intended" to be used as a way to communicate with ant guards.

In a follow-up experiment, pairs of butterfly pupae were tested together, with one of the pair having been "muted" by applying nail polish to its stridulatory organs. Then, using an elaborate "preference" testing device that included two bridges on which the ants could march, Travasso and Pierce examined whether ants were more attracted to the muted individual in a pair, or the individual who was free to produce vibrational communication. What they uncovered was that ants demonstrated a clear preference for associating with the pupae that could (and did) produce vibrations, providing further evidence that vibrational communication plays a real role in this ant-butterfly mutualistic relationship (Figure 9.30).

SUMMARY

1. Cooperation is an *outcome* that—despite *possible* costs to the individuals involved—provides some benefits to the members of a group. To behave cooperatively means to behave in such a manner that renders the cooperation possible (something the individual does)—even though the cooperation will not

actually be realized unless other group members also behave cooperatively.

2. Cooperation has been studied in taxa ranging from insects and fish to birds and mammals, including in the contexts of foraging, predation, antipredator activities, and aggression.
3. In addition to kin-selected cooperation, which we discussed in Chapter 8, three paths to the evolution and maintenance of cooperation in animals have been developed—reciprocity, byproduct mutualism, and group selection. Many mathematical tools have been employed to model these paths, and of these, game theory is the most often used tool.
4. Work on cooperation via reciprocity has centered on the prisoner's dilemma game and a strategy called tit for tat.
5. Group selection models of cooperation have both a within- and a between-group component. Within-group selection favors cheating, while between-group selection favors cooperation.
6. Phylogeny can be used to better help understand the distribution of cooperation among related species. The phylogenetic approach does not necessarily conflict with an adaptationist view of cooperative breeding, although it does emphasize the importance of history (phylogeny) in our understanding of the distribution of cooperative breeding.
7. In addition to examining *why* reproductive suppression can evolve in cooperative animal breeding systems, we can also address what is known about the role of hormones in leading subordinates to cooperate.
8. Cooperation can also occur in interactions that involve more than two individuals. One example of this type of cooperation is coalition behavior, which is typically defined as a cooperative action taken by at least two other individuals or groups against another individual or group.
9. In addition to cooperation between members of the same species (that is, *intraspecific* cooperation), there is a great deal of evidence that *interspecific* cooperation is also important in shaping animal social behavior.

DISCUSSION QUESTIONS

1. Read G. Wilkinson's (1984) article, "Reciprocal food sharing in vampire bats," in *Nature* (vol. 308, pp. 181–184). Then outline how Wilkinson was able to separate out the effects of kinship and reciprocity in his study of vampire bats.

2. Run a small prisoner's dilemma experiment with a few other students. In one group, using coins as payoffs, test pairs of subjects (who cannot communicate with each other in any manner) and tell them beforehand that they will only play this game once. In a second treatment, use pairs of subjects (who cannot communicate with each other), but inform them that they will play this game many, many times together, but do not tell them exactly how many times. In a third and fourth treatment, repeat treatments 1 and 2, but allow the subjects to communicate with each other before the game starts. What sort of differences and similarities do you predict across treatments? What do the data say?
3. How would you respond to the following statement: Animals aren't capable of humanlike thought processes and therefore they cannot be cooperating.
4. Why do you suppose that work on animal behavior and cooperation draws more attention from other disciplines, such as mathematics, political science, and psychology than any other area in ethology? What might we learn about human cooperation from studies of animal cooperation? What sorts of things would be difficult to glean about human cooperation by studying animal cooperation?

SUGGESTED READING

Axelrod, R., & Hamilton, W. D. (1981). The evolution of cooperation. *Science, 211,* 1390–1396. A classic paper that essentially introduced most ethologists to using game theory to address cooperative behavior.

Connor, R. C. (1995). The benefits of mutualism: A conceptual framework. *Biological Reviews of the Cambridge Philosophical Society, 70,* 427–457. A conceptual overview of work on mutualism and byproduct mutualism.

Kropotkin, P. (1902). *Mutual aid* (3rd ed.). London: William Heinemann. One of the first important books on cooperative behavior in animals.

Packer, C. (1977). Reciprocal altruism in *Papio anubis. Nature,* 265, 441–443. A highly cited study that was one of the first to experimentally tackle alliances and cooperation.

Wilson, D. S., & Sober, E. (1994). Re-introducing group selection to the human behavioral sciences. *Behavioral and Brain Sciences, 17,* 585–654. An overview applying group selection thinking to human communities.

10

Optimal Foraging Theory

- Basic OFT: What to Eat and Where to Eat It
- Specific Nutrient Constraints
- Risk-Sensitive Foraging

Learning and Foraging

- Foraging, Learning, and Brain Size in Birds
- Learning and "Work Ethics" in Pigeons

Foraging and Group Life

- Group Size
- Social Learning and Foraging
- Public Information and Foraging

Molecular, Neurobiological, and Hormonal Aspects of Honeybee Foraging

- The Period Gene, mRNA, and Foraging
- Juvenile Hormone, "Mushroom Bodies," and Foraging

Interview with Dr. John Krebs

Foraging

If the number of copies of genes making it into the next generation is the currency of natural selection, then it is an inescapable fact that getting food is a critical part of every animal's existence. Depending on the circumstances, individuals can survive to reproductive age without being aggressive, they can survive to reproductive age without play, they can survive to reproductive age if they don't cooperate or find a territory, but they can almost never survive to reproduce if they don't eat. As such, we expect that animals should spend a fair share of their available time looking for food—foraging—and that natural selection should favor efficient foraging strategies. All evidence indicates that this is in fact the case (Figure 10.1).

Consider the following incredible story of ant foraging. About fifty million years ago, ants began cultivating their own food by entering into a mutually beneficial relationship with certain species of fungi. The ants promote the growth of the fungi (good for the fungi), while also feasting on the vegetative mycelium produced by their fungal partners (good for the ants). Aside from humans, ants are one of the few groups on the planet that grow their own food, and they do so in this rather spectacular way.

Cameron Currie and her colleagues (1999a, 1999b) have found that the ant-fungus relationship is even more complicated than scientists first believed. Researchers who study fungus-growing ants have long known of a whitish-gray crust found on many ants involved with fungus. It turns out that the substance is actually a mass of *Streptomyces* bacteria. This group of bacteria produces many antibiotics. Currie and her colleagues hypothesized that ants actually use the bacteria to produce antibiotics that kill parasites

FIGURE 10.1. Foraging. Many animals spend a good deal of their waking hours foraging. Here **(A)** a black bear and **(B)** a Richardson's ground squirrel forage on a simple meal. *(Photo credits: Ron Erwin)*

that grow in their fungal gardens (Figure 10.2). That is, they hypothesize that not only have ants evolved a complex relationship with their food source (fungi), but they have even evolved a means to protect this food source from destruction.

Four lines of evidence support the researchers' claim. First, all twenty species of fungus-growing ants they examined had *Streptomyces* bacteria associated with them. Second, ants actually transmit the bacteria across generations, with parents passing the bacteria on to offspring. Third, when male and female reproductive ants are examined (before their mating flights), only females possess the bacteria. This is critical, as only females start new nests that will rely on the bacteria to produce antibiotics, and only females are involved in "cultivating" fungus gardens. Fourth, and most important, the bacteria found on fungus-growing ants produce antibiotics that wipe out only *certain* parasitic diseases. When Currie and her colleagues tested the antibiotics produced by the bacteria, they found they were potent only against *Escovopsis*, a serious parasitic threat to the ants' fungus garden. Other parasitic species (those not a danger to fungus-growing ants) were unaffected by *Streptomyces* antibiotics.

FIGURE 10.2. Tending the garden. A worker of the leaf-cutter ant *(Acromyrmex octospinosus)* tending a fungus garden. The thick whitish-gray coating on the worker is the mutualistic bacterium *(Actinomycetous)* that produces the antibiotics that suppress the growth of parasites in the fungus garden. *(Photo credit: Christian Ziegler)*

What makes the work on foraging behavior so fascinating is that the examples are often as incredible as the ant story above. As another case in point, consider the phenomenal ability of some birds and mammals to remember where they have stored literally thousands of food items. Watching a squirrel, for example, finding the nuts it has stored for winter all over your garden often inspires a sense of awe in the observer. And it isn't only squirrels. Many species, especially many bird species in the family *Corvidae* (for example, crows, magpies, jays, jackdaws), possess such uncanny abilities. How is that possible? How can an animal find scores of food items that are scattered across its environment and often hidden over the course of months and months?

To address the relation between foraging and spatial memory, Susan Healey and John Krebs (1992) examined hippocampal volume and food-storing abilities in seven species of corvid birds. The hippocampal region in birds is known to be associated with food retrieval (Krushinskaya, 1966, 1970; Krebs et al., 1989; Sherry et al., 1989; Sherry and Vaccarino, 1989), and the corvids are an ideal group in which to examine the relationship between hippocampal volume and food storage in more detail, since so much variation in food-storing behavior exists within this group (Goodwin, 1986; Balda and Kamil, 1989; Vander Wall, 1990). Some corvids store no food, while others rely on the food they have stored over the course of nine months.

Healey and Krebs studied two species—jackdaws (*Corvus monedula*) and Alpine choughs (*Pyrrhocorax graculus*)—that rarely if ever

FIGURE 10.3. Spatial memory in corvids. Healey and Krebs examined hippocampal volume and food-storing abilities in seven species of corvid birds, including **(A)** jackdaws and **(B)** European magpies. *(Photo credits: Glenn Oliver/Visuals Unlimited; William Grentell/Visuals Unlimited)*

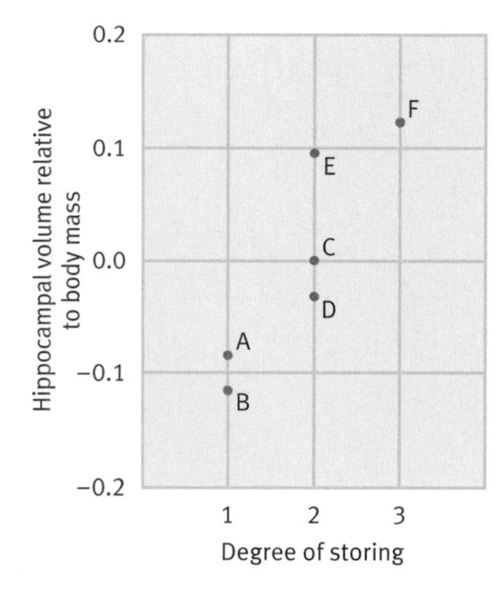

FIGURE 10.4. Foraging and brain size. The volume of the hippocampal region relative to body mass was positively correlated with the extent of food storing in six species of birds: (A) alpine chough; (B) jackdaw; (C) rook and crow combined; (D) red-billed blue magpie; (E) magpie; (F) European jay. *(Based on Healey and Krebs, 1992)*

stored food, four species—rooks (*Corvus frugilegus*), European crows (*Corvus corone*), European magpies (*Pica pica*), and Asian red-billed blue magpies (*Cissa erythrorhyncha*)—in which food storing plays some role, and one species—European jays (*Garrulus glandarius*)—in which not only does food storing play an important role, but the location of 6,000 to 11,000 seeds must be remembered for nine months. When food-storing behavior was examined in relation to hippocampal volume in these seven species, a strong positive relationship was uncovered (Figure 10.3). The more food-storing behavior in a species, the greater the hippocampal volume, strongly supporting the hypothesis that hippocampal volume is a critical variable in the evolution of spatial memory and foraging (Figure 10.4).

In this chapter, we shall touch on the following foraging-related questions:

- How does foraging theory predict where and what animals eat?
- What role does learning play in foraging decisions?
- How does group social dynamics affect foraging?
- How can we incorporate neurobiology, molecular biology, and endocrinology into our study of animal foraging behavior?

Optimal Foraging Theory

In this section, we shall examine a class of models that falls under the umbrella known as **optimal foraging theory (OFT)** (Stephens and Krebs, 1986; Kamil et al., 1987; Sih and

Christensen, 2001). As can be inferred from the name, these models use a form of mathematical analysis known as optimization theory. In general, optimization techniques are used to predict the best possible solution to a problem within a given set of constraints. For instance, if we want to know what the optimal diet is for species X, we begin by making a set of assumptions—for example, we might assume that foragers can't search for more than one item at a time—and then proceed to see which types of food a forager should take, given our constraints.

Optimal foraging theory began in earnest in 1966 with back-to-back papers by MacArthur and Pianka (1966) and John Emlen (1966) in *The American Naturalist*. These papers set the stage for ethological and behavioral ecological models to come (Emlen, 1966, 1968; Schoener, 1971; Charnov, 1973, 1976; Maynard Smith, 1974a; Pulliam, 1974), and were followed by an exponential rise in the number of ethological studies conducted on animal foraging.

While there are many optimal foraging models in the animal behavior literature, we shall examine four such models, which address the following questions:

- What food items should a forager should eat?
- How long should a forager stay in a certain food patch?
- How is foraging affected when certain nutrient requirements are in place?
- How does variance in food supply affect a forager's decision about what food types to eat?

BASIC OFT: WHAT TO EAT AND WHERE TO EAT IT

Optimal foraging theory is first concerned with the basic questions of what to eat and where to eat it. It proposes mathematical models to determine what food foragers will eat and how long to remain in a food patch, given that they could move elsewhere at some cost.

WHAT TO EAT. One of the most basic foraging problems faced by animals is deciding which type of food items should be in a diet and which should be excluded (Stephens and Krebs, 1986). For example, imagine that a forager could search for, and potentially consume, food types 1, 2, and 3. Should the forager eat all three? Only one? Two of the three? If so, which subset? To tackle this question, ethologists and behavioral ecologists have developed optimality models designed specifically to handle these issues. We shall now examine the underpinnings of one of the first optimal foraging models and then review a classic test of it.

FIGURE 10.5. Foraging decision. Here a female cheetah (the forager) has killed a hare (the prey). In making the decision whether to take hares rather than some other prey into the diet, the animal will compare the energy value (the amount of calories provided to the forager by eating this prey), encounter rate (how often the prey is encountered by the forager), and handling time (time for the forager to kill and ingest the prey) for each putative prey. *(Based on Caro, 1994a)*

Let's first consider the simplest possible case—choosing between two different types of food (Figure 10.5). This might be a choice between two prey species (for carnivores), two different species of seeds (for granivores), or even two different size classes of the same food. Each prey item will have an energy value, encounter rate, and handling time associated with it. For example, let's say one prey type is encountered every three minutes (encounter rate), and once it is encountered, such a prey item takes two minutes to kill and ingest (handling time), and provides the forager with 300 calories (energy value). Let's define the profitability of a prey item as its energy/handling time. The greater the energy/handling ratio, the greater the profitability of a prey type. We shall assume that the prey type with the highest profitability—let's call it prey type 1—is always taken by a forager. This means that our optimal diet problem boils down to two simple questions: Should prey type 2 be taken? If it should, under what conditions? We can now examine the assumptions of this optimal prey choice model, which can always be modified, and then proceed to its main conclusion. Our model assumes:

- We can measure our prey types in some standard currency (for example, calories).
- Foragers can't simultaneously handle one item and search for another.
- Prey are recognized instantly and accurately.
- Prey are encountered sequentially.
- Foragers are designed to maximize their rate of energy intake.

With these assumptions in place, a bit of mathematical analysis (Math Box 10.1) produces a fascinating, and somewhat counterintuitive, prediction. The encounter rate with the less profitable prey item (prey type 2)—that is, how often a predator encounters the less profitable item—does *not* affect whether that item should be added to the diet or not. Rather, a critical encounter rate with the most profitable item is calculated; if that rate is high enough, only prey item 1 is taken and, if it isn't, then both prey item 1 and prey item 2 are taken. That is, the decision of whether to add prey type 2 to the diet is not dependent on how many prey type 2 items are around for our forager; rather, it is related to how many prey type 1 items are around. This basic prediction of optimal diet choice has been tested many times.

One nice example of how to experimentally examine the predictions of this optimal diet model can be found in work by Krebs and his colleagues (1978) on great tit birds (*Parus major*). In an ingenious laboratory experiment, great tits were placed in front of a moving conveyer belt (Figure 10.6). The researchers

FIGURE 10.6. Great tit foraging. One classic early experiment using optimal foraging theory had mealworms of different sizes presented on a conveyor belt to great tits.

Math Box 10.1. The Optimal Diet Model

Imagine a forager has only two prey types that it consumes (for example, two types of seeds, two types of herbivores, and so on). Within our model, a forager must eat one of these prey type items (or it will starve), and so our question becomes: Which item does a forager always take, and under what conditions does a forager take both types of prey? To begin, let us define the following terms:

e_i = energy provided by prey type i
h_i = handling associated with prey type i (how long does it take to eat prey type i once it is caught)
λ_i = encounter rate with prey type i
T_s = amount of time devoted to searching for prey
T = total time

We shall assume that an animal always takes the prey type that has a higher e_i/h_i value (called the profitability of a prey type), and we will label this prey type as prey 1. The question then becomes whether a forager should take prey 1 alone, or whether a forager should take both prey 1 and prey 2 upon encountering them.

We begin by calculating the total energy (E) associated with prey 1 divided by the total time associated with this prey (T).

For prey type 1:

$$\frac{E}{T} = \frac{T_s\lambda_1 e_1}{T_s + T_s\lambda_1 h_1} \quad (1)$$

The numerator here takes the total number of prey type 1 captured ($T_s\lambda_1$) and multiplies it by the energy value (e_1) of each prey, producing the total energy obtained from prey type 1. The denominator adds together the total search time (T_s) and the total handling time ($T_s\lambda_1 h_1$), given such a search. This can be simplified to:

$$\frac{E}{T} = \frac{\lambda_1 e_1}{1 + \lambda_1 h_1} \quad (2)$$

Now, we can ask whether this value is greater than the E/T associated with taking both prey types. To find the E/T of taking both prey type 1 and prey type 2, we calculate the following:

$$\frac{E}{T} = \frac{T_s(\lambda_1 e_1 + \lambda_2 e_2)}{T_s + T_s\lambda_1 h_1 + T_s\lambda_2 h_2} \quad (3)$$

The numerator represents the total energy obtained from prey types 1 and 2, while the denominator adds together the total search time (T_s) and the total handling time for prey types 1 and 2. This can be simplified to:

$$\frac{E}{T} = \frac{\lambda_1 e_1 + \lambda_2 e_2}{1 + \lambda_1 h_1 + \lambda_2 h_2} \quad (4)$$

Our question then boils down to when is the following inequality true:

$$\frac{\lambda_1 e_1}{1 + \lambda_1 h_1} > \frac{\lambda_1 e_1 + \lambda_2 e_2}{1 + \lambda_1 h_1 + \lambda_2 h_2} \tag{5}$$

When this inequality holds true, our predator should take only prey type 1. When it does not hold true, our predator should take prey type 1 and prey type 2. Equation (5) can be shown to hold true when:

$$\lambda_1 > \frac{e_2}{e_1 h_2 - e_2 h_1} \tag{6}$$

From (6) we can derive two important predictions:

1. Once a critical encounter rate with prey type 1 is reached, it alone should be taken.
2. The decision about whether to take prey type 2 does *not* depend on how common prey type 2 is (that is, on prey type 2's encounter rate). This can be seen in (6) by the absence of λ_2 from our inequality.

used two different-sized pieces of mealworm as the two prey item types and, by using a conveyor belt to present the prey, they had precise control of the rate at which these two prey types were encountered by the tits, the exact energy provided by both types of prey, and the precise handling time associated with each size of mealworm. With these parameters measured, it was possible to predict when the birds should take only the most profitable prey types and when they should take both types of prey. The researchers found that while the birds' behavior was not perfectly predicted by the optimal choice model, the model did a fairly good job of predicting how the tits would forage, in that it was the encounter rate of the most profitable prey, not the least profitable prey, that determined whether tits took the least profitable items (Figure 10.7A). Similar results have been found in foraging experiments using bluegill sunfish (Werner and Hall, 1974; Figure 10.7B).

WHERE TO EAT. Another critical decision a forager must make is how long it should stay in a patch of food. For example, how long should a hummingbird spend sucking nectar from one flower, given that there are other flowers available, or how long should a bee spend extracting pollen from one flower before moving on to the next one? To address such questions, Eric Charnov (1976) developed an

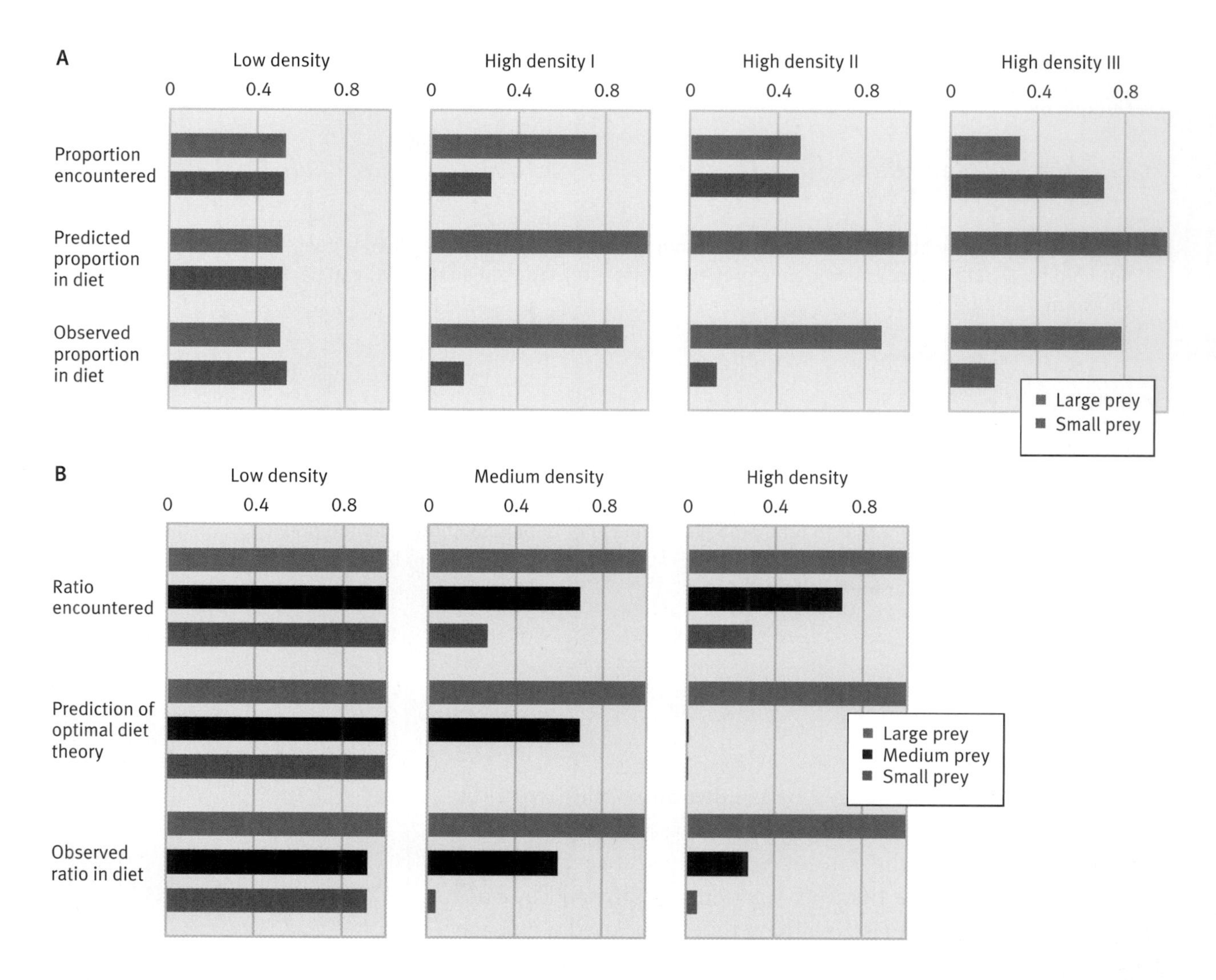

FIGURE 10.7. Optimal choice of diet. (A) Optimal foraging in great tits was examined by Krebs and his colleagues (1977) in four density conditions. With a knowledge of exact encounter rates, handling times, and energy values, they were able to predict the birds' optimal diet of large and small prey. **(B)** Optimal foraging in bluegill sunfish was studied by Werner and Hall (1974). They were able to predict the optimal foraging of the fish on three different size classes of *Daphnia*. The fit between expected and observed foraging is quite good, although the fish tended to oversample medium and small *Daphnia* in the high-density treatment. *(Based on Krebs, 1978, p. 31).*

optimality model that has become known as the **marginal value theorem** (this idea was also independently derived by Parker and Stuart, 1976).

To see how this model works, imagine a forager feeding in an area that contains different patches of food. A patch of food for a chimp might be a tree full of fruit, but for a bee it might be a flower (Figure 10.8). More generally, a patch is defined as a clump of food that can be depleted. Once a forager begins foraging in a patch, the rate at which it takes in new food slows down, as the more the forager eats, the more the patch is depleted. Other less

FIGURE 10.8. Patch choice. For a bee, different flowers in a field of flowering plants might represent different patches.

depleted patches will then be attractive, but in order to get to these patches, the forager must pay some cost, such as lost time foraging while traveling, increased rate of predation during travel, and so forth. The question then becomes: How long should a forager stay in a patch it is depleting before moving on to other patches?

The marginal value theorem makes a few clear, testable predictions regarding patch residence time. First, if we know the rate of intake within a patch, a forager should stay in that patch until the marginal rate of intake—that is, the rate of intake associated with the next prey item in its foraging patch—is equal to that of the average rate of intake across all patches available. In other words, a forager should stay in a patch x time units, where x is that point in time when its marginal rate of food intake is equal to what it could get in other patches, given that it has to pay a cost to get to such other patches. Second, the greater the time between patches, the longer a forager should stay in a patch. This makes sense, since increased travel time leads to an increase in the costs associated with such travel, and such costs need to be compensated. Remaining in a given patch longer is one means by which such compensation can be achieved. Third, for patches that are already of generally poor quality when the forager enters the patch, individuals should stay longer in such patches than if they were foraging in an environment full of more profitable patches. Think of it this way: In order to make up for the travel costs associated with a move from a

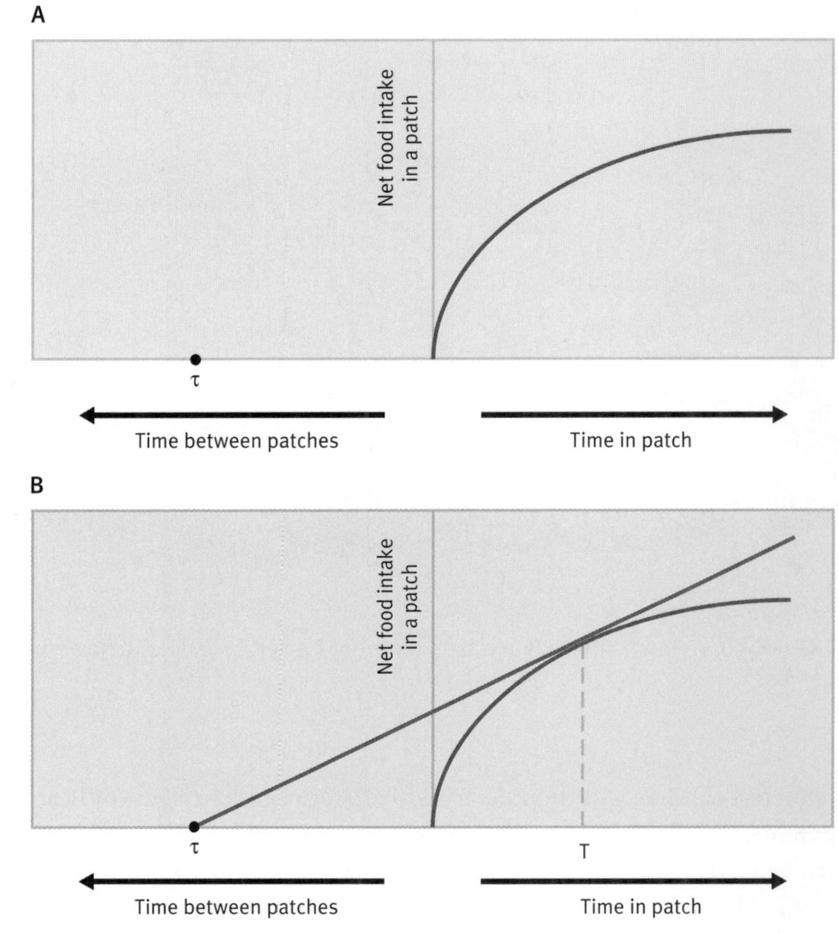

FIGURE 10.9. Graphical solution to marginal value problem. (A) To calculate the optimal time for a forager to remain in a patch, we begin by drawing a curve that represents the cumulative food gain in an average patch in the environment. Then, going west on the x-axis we find the average travel time between patches (τ). **(B)** We then draw a straight line from τ that is tangent to the food gain curve. From the point of tangency, we drop a perpendicular dashed line to the x-axis, which gives us an optimal time (T) for the forager to stay in the patch.

patch, a forager has to stay longer in a poor patch than in a good patch to obtain a fixed amount of energy (Figure 10.9).

One aspect of the marginal value theorem has also been tested using great tits as subjects. In a large aviary, Cowie built a series of artificial trees, each of which contained numerous branches (Cowie, 1977). Attached to some of the branches were sawdust-filled potting baskets that had mealworms hidden in them (Figure 10.10A). Cowie was able to calculate the rate of gain a tit would encounter in such sawdust-filled patches. In addition, Cowie could manipulate the travel time between patches, as each pot had a lid, and the lids could be made to be easily dislodged, providing a short travel time from leaving one patch to beginning to forage at another, or very difficult to open, simulating a long travel time. Based on all this, the optimal time to stay in a patch as a function of travel time between patches was calculated. Results clearly demonstrate that the amount of time birds spent in a patch matched the optimal time predicted by the marginal value theorem (Figure 10.10B).

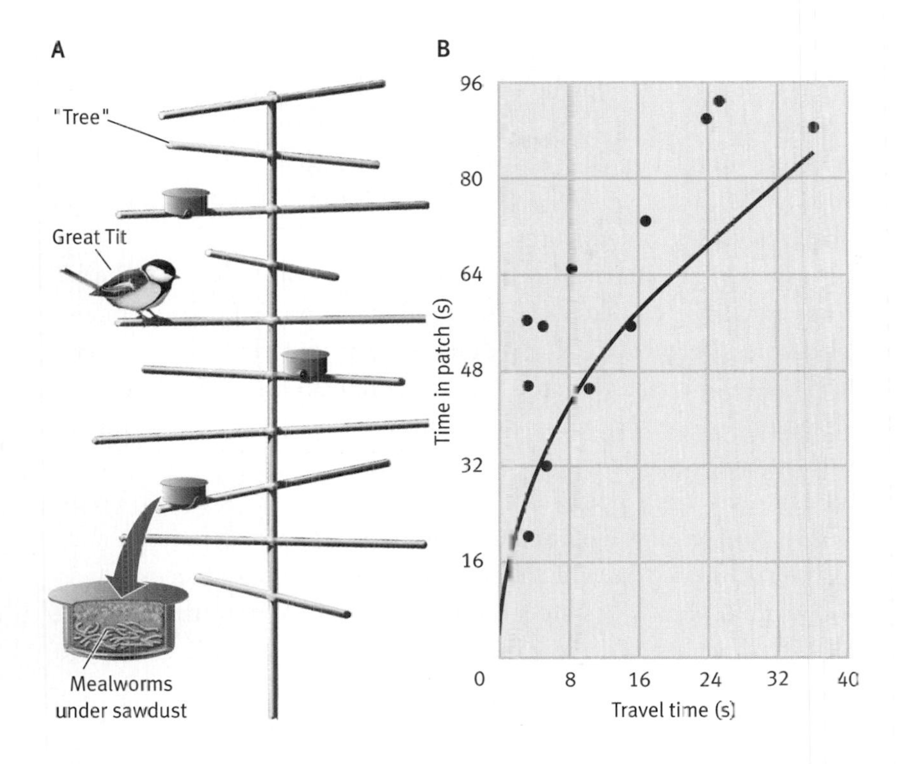

FIGURE 10.10. Optimal time in patch and travel time. In a test of the marginal value theorem, Cowie (1977) constructed **(A)** an artificial tree that allowed him to control both patch quality and travel time. **(B)** The solid line is the predicted optimal time in a patch plotted against the travel time, which was calculated based on the marginal value theorem, while the data points are the observed times the birds stayed in the patch plotted as a function of travel time between patches. *(Based on Krebs, 1978, p. 44.)*

Before moving on to other models from optimal foraging theory, it is worth noting that the marginal value theorem is not restricted to cases of foraging. Any time a resource depletes as a function of use, and costs associated with traveling between patches of that resource are present, one can use the marginal value theorem to solve for optimal patch time. This approach has, for example, been used to calculate how long a male should search for and mate with a female (where the female is now the "patch"; Parker and Stuart, 1976), and how long a cheater should stay in a group of cooperators (where a cheater "forages" on the cooperative nature of its groupmates; Dugatkin and Wilson, 1991; Dugatkin, 1992a).

SPECIFIC NUTRIENT CONSTRAINTS

In addition to maximizing their rate of energy intake, animals must also attempt to minimize the probability that they fail to ingest particular nutrients. For example, many animals need to take very small quantities of trace elements like zinc in order to remain healthy, and this requirement may compromise the general foraging rules they follow to maximize energy intake per unit time. If looking for trace element X takes time away from other activities (searching for other foods, antipredator behavior, mating, and so forth), fulfilling minimum requirements may be very costly. These constraints appear to be particularly severe in herbivores, who often must take large amounts of low-energy food to survive (Crawley, 1983; see Roguet et al., 1998, for a review, and Fritz and Simms, 1992, for more on plant responses to herbivores).

FIGURE 10.11. Moose foraging on a salt budget. Linear programming models of foraging have been used to predict the diet of salt-limited moose. *(Photo credit: Joe McDonald/Visuals Unlimited)*

One classic study of nutrient constraints on foraging in herbivores is Gary Belovsky's (1978) study of sodium intake in moose (*Alces alces*; Figure 10.11). Sodium is a particularly good candidate for a nutrient constraint study, as vertebrates require large amounts of sodium (since sodium is lost in urine), sodium is scarce, and besides water, sodium is the only nutrient for a which a "specific hunger" has been documented in animals (Stephens and Krebs, 1986). Moose have a particular sodium problem in certain habitats where sodium is restricted to aquatic plants that are inaccessible in the winter, when lakes are frozen. This, coupled with the fact that terrestrial plants provide much more energy per unit time than aquatic plants, presents the difficult problem of how much of a moose's diet outside the winter should be based on aquatic plants. Foraging on such plants provides much needed sodium (which can be stored by the moose), but it takes time away from foraging for energy-rich terrestrial plants (Figure 10.12). What's a moose to do?

To predict foraging behavior in the above circumstances, ethologists and behavioral ecologists use a mathematical technique called linear programming. Linear programming models are designed to handle certain optimality problems, specifically those in which some optimum must be achieved in the face of a particular constraint. In our case, natural selection should favor any behavioral strategy that provides moose with the most energy/time, subject to the constraint that a certain amount of sodium must be ingested. Again, in nature, this question translates into how much time during nonwinter periods a moose should devote to foraging for energy-poor, but relatively sodium-rich, aquatic plants.

Belovsky built a linear programming model to predict moose foraging behavior. Into this model, Belovsky incorporated the minimum amount of food a day that a moose needs to survive, how quickly a moose digests its food, the energy value of aquatic and terrestrial plants, and the sodium constraint. A moose needs to take in about 2.57 grams of sodium a day in the summer to get enough sodium to make it through the winter. Somewhat incredibly, the model's prediction—that a moose should spend about 18 percent of its summer foraging time on aquatic plants—matches the moose's behavior almost exactly. Belovsky (1984) has extended his use of linear programming techniques to other herbivore foraging problems as well.

FIGURE 10.12. Specific foraging constraints. Moose need salt, and they acquire it from energy-poor plants. This takes time away from foraging on energy-rich terrestrial plants. Linear programming models can help predict the moose's foraging behavior.

RISK-SENSITIVE FORAGING

One aspect of foraging that we have ignored so far is *variability*, known in statistical parlance as "risk" (not to be confused with risk in the sense of danger). Consider the following scenario: An animal can choose to forage for six hours in one of two patches. Further imagine that both patches have the exact same type of food item, and animals know what to expect in each patch. In patch 1, each forager will receive eight items with certainty. In patch 2, there is a 50 percent chance a forager will receive sixteen food items and a 50 percent it will get nothing. The mean number of food items that the foragers can expect in each patch is identical (eight), but the variance (risk) in food intake per hour is greater in patch 2 (Figure 10.13). Should our foragers take the differences in risk into account when deciding between patches? This is the sort of question that **risk-sensitive optimal foraging models** address (Oster and Wilson, 1978; Caraco, 1980; Real, 1980; Real and Caraco, 1986; McNamara and Houston, 1992).

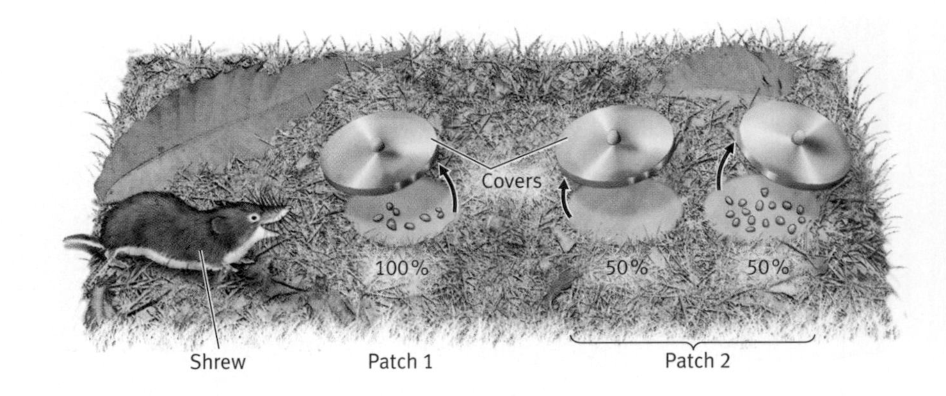

FIGURE 10.13. Risk-sensitive optimal foraging model. Imagine a shrew that must decide between a patch (1) that always yields eight pellets once the cover is removed and a patch (2) in which half the time there are no pellets and half the time there are sixteen pellets. Both patches have the same mean (eight), but the variance is greater in patch 2. If our forager takes variance into account, it is foraging in a risk-sensitive manner.

For a basic understanding of risk-sensitive foraging, let's consider foragers in three different states. Forager A values every new food item equally. Whether it is eating its first or fiftieth item, each item provides it with the same reward value. Forager B is stuffed, and although every additional item it takes in has some value, each additional item is worth less and less. Think about the value of a slice of cake to you after you have just had two ice cream sundaes. It might be worth something, but not what it would have been worth before consuming the sundaes. Forager C is starving, and every additional item it eats is worth more and more (to a limit), perhaps providing the difference between surviving or starving to death (Figure 10.14).

Risk-sensitive foraging models, though mathematically complex, make a very straightforward, but powerful, prediction. Returning to our two patches that have the same mean number of food items but different variances, risk-sensitive models predict that the *hunger state* of a forager will determine whether it prefers the patch with more or less variance. Our satiated predators are predicted to be *risk-averse*—that is, they should choose to forage in patch 1. On the flip side of the coin, our starving foragers should opt to be *risk-prone*, and to select foraging in patch 2. All other foragers should be indifferent with respect to variance and should not show preference for one patch type or another.

To see why variance is favored in starving birds, but avoided by relatively satiated birds, we need to think about survival probabilities. For our relatively filled forager, each additional piece of food

FIGURE 10.14. Utility of food. A forager may value each and every food item the same **(A)**, it may value each additional food item less and less **(B)**, or each additional food item more and more **(C)**. The bird on the left **(A)** is predicted to be risk insensitive, the middle bird **(B)** is predicted to be risk averse, and the bird on the right **(C)** is predicted to be risk prone.

isn't worth all that much more, and so it should opt for a consistent food source, since sixteen pieces of food isn't worth all that much more than eight when it is stuffed. For our starving bird, eight pieces of food may not be enough to provide it with enough energy to survive the night, but sixteen food items might. In that case, it is worth risking getting no food at all for the chance of getting sixteen items, since eight food items, even if they are a certainty, wouldn't be enough to pull the starving forager through to the next foraging period. Naturally, as with all the mathematical models we analyze, we are not suggesting that animals make the mental calculations that we just went through, but rather that natural selection favors any "rule of thumb" behavior.

One of the first, and best, studies of risk-sensitive foraging was that of Thomas Caraco and his colleagues (1980). They presented yellow-eyed juncos with two seed trays, and once a bird made a choice to go to one tray, the other was immediately removed (Figure 10.15). One tray had a "fixed" amount of food—for example, a fixed tray might always have five seeds—while the other tray had a "variable" amount of food—for example, either no seeds half the time, or ten seeds half the time. Keep in mind that the risky tray had a mean number of seeds equal to that on the fixed tray—a mean of five seeds. The researchers calculated the value of every new item of food for both hungry birds (negative energy budget) and those birds that were meeting their daily food requirements

FIGURE 10.15. Optimal juncos. Junco foraging behavior has been used to test numerous optimal foraging models. *(Photo credit: Gary Meszares/Visuals Unlimited)*

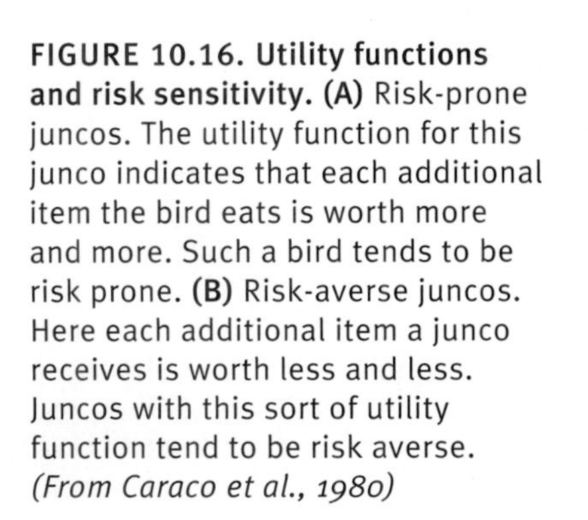

FIGURE 10.16. Utility functions and risk sensitivity. (A) Risk-prone juncos. The utility function for this junco indicates that each additional item the bird eats is worth more and more. Such a bird tends to be risk prone. **(B)** Risk-averse juncos. Here each additional item a junco receives is worth less and less. Juncos with this sort of utility function tend to be risk averse. *(From Caraco et al., 1980)*

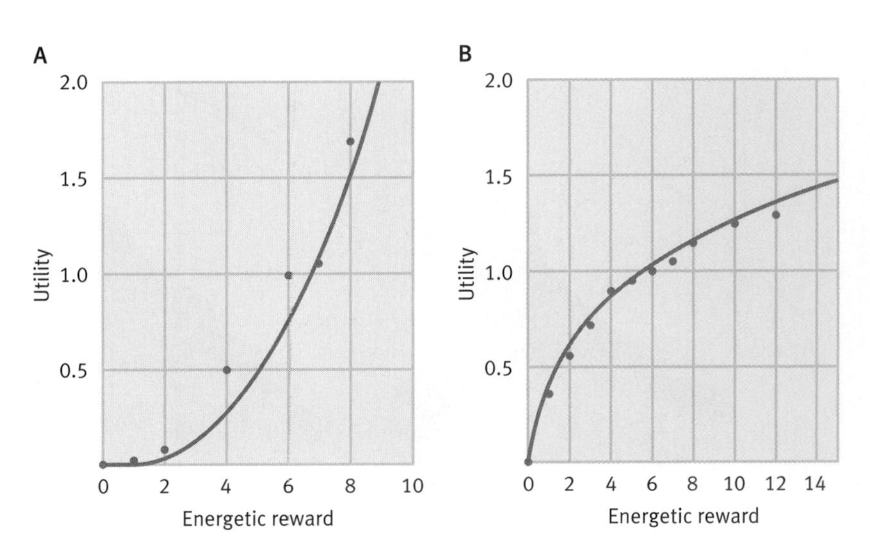

(positive energy budget). Birds on a negative energy budget valued every additional food item much more than those on a positive energy budget (Figure 10.16).

When choosing between trays, risk-sensitive theory predicts that birds on a positive energy budget should choose the fixed trays, while very hungry birds should choose the variable trays. Caraco's juncos behaved in a fashion remarkably similar to that predicted from theory. Yellow-eyed junco foraging is indeed risk-sensitive, shifting between risk-averse foraging and risk-prone foraging according to energy budgets. Soon after Caraco's findings were published, similar results on risk sensitivity were uncovered in other seed-eating birds (Caraco, 1981, 1983) and shrews (Barnard and Brown, 1985).

Learning and Foraging

While the studies described so far often involve animals learning something about their foraging environment, they were not designed as experiments on foraging and learning per se. To even touch on the tip of the iceberg of studies done on learning and animal foraging would require a series of books. After all, these sorts of studies have been a mainstay in psychology over the last fifty years, and there have been thousands of studies examining learning in the context of foraging (for a nice review from a particularly evolutionary mindset, see Shettleworth, 1998, and for more on the relationship between foraging and learning experiments in biology and psychology, see Shettleworth, 1984,

1998; Pietrewicz and Richards, 1985; Stephens and Krebs, 1986; Kamil et al., 1987; Yoerg, 1991; Krebs and Inman, 1994; Kacelnik and Abreu, 1998; Balda et al., 1998). Rather than even trying to summarize this work, we shall look at a tiny, but important, subsection of it in two studies that examine:

- Foraging, learning, and brain size in animals.
- How animals value what they have learned about food.

FORAGING, LEARNING, AND BRAIN SIZE IN BIRDS

A number of ethologists, behavioral ecologists, and evolutionary biologists have hypothesized a link between forebrain size and learning in animals (Rozin, 1976; Riddell and Corl, 1977; Eisenberg and Wilson, 1978; Clutton-Brock and Harvey, 1980; Mace et al., 1981; Johnstone, 1982; Wyles et al., 1983; Jolicoeur et al., 1984; Dunbar, 1992; Byrne, 1993). The premise underlying this link is as follows: Since the forebrain appears to be associated with behavioral plasticity—including learning—animals with larger forebrains should often be better learners. While this approach is susceptible to numerous potential biases, Louis Lefebvre and his colleagues used its basic principles to test for a relationship between "foraging innovations" and forebrain size in North American and British Isle bird groups (Lefebvre et al., 1997).

Lefebvre defined a "foraging innovation" as "either the ingestion of a new food type or the use of a new foraging technique." For our purposes here, since we are looking at the role of learning in foraging *behavior*, we are mostly concerned with the latter types of innovations (which also include examples of social learning). Lefebvre and his colleagues used the "short notes" section of nine ornithology journals to gather data on 322 foraging innovations—126 in British Isle birds and 192 in North American birds (Table 10.1). Innovations included behaviors ranging from herring gulls' "catching small rabbits and killing them by dropping on rocks or drowning" (Young, 1987) to common crows' "using cars as nutcrackers for palm nuts" (Grobecker and Pietsch, 1978). The researchers then calculated how these innovations were distributed across different orders of birds, at the same time taking into account how common or rare a particular bird order was in Britain or North America.

With the information about foraging innovation in hand, Lefebvre and his colleagues obtained data on relative forebrain size in the bird species of interest (Portmann, 1947; Scott, 1987; Holden and Sharrock, 1988). In both North American and British Isle birds, relative forebrain size correlated with foraging

TABLE 10.1. Examples of foraging innovations in birds. Lefebvre and his colleagues gathered data on 322 foraging innovations, including those in this list. *(Based on Lefebvre, 1997, pp. 552–553)*

SPECIES	INNOVATION
Cardinal (Florida subspecies)	Nipping off nectar-filled capsule on flower and eating it
Herring gull	Catching small rabbits and killing them by dropping on rocks or drowning
Ferruginous hawk	Attracted to gunshot, preys on human-killed prairie dog
Magpie	Digging up potatoes
Storm petrel	Feeding on decaying whale fat
Great skua	Scavenging on road kill
House sparrow	Using automatic sensor to open bus station door
House sparrow	Systematic searching and entering of car radiator grilles for insects
Galápagos mockingbird	Pecking food from sea lion's mouth
Common crow	Using cars as nutcrackers for palm nuts
Osprey	Opening conch shells by dropping them on concrete-filled drum
Turnstone	Raiding gastric cavities of sea anemones
Red-winged + Brewer's blackbirds	Following tractor and eating frogs, voles, and insects flushed by it
Sparrowhawk	Drowning blackbird prey
Carrion crow	Landing on floating sheep corpse and feeding from it
Downy woodpecker	Using swaying caused by wind to catch meat hung from a branch

innovation. Bird orders that contained individuals that possessed larger forebrains were more likely to have high incidences of foraging innovation (Figure 10.17).

Although many of the cases of foraging innovation were based on a single observation, the fact that a larger-scale analysis spanning 322 cases found a striking relationship between forebrain size and one measure of learning (foraging innovation) suggests that this is an important but relatively understudied area in ethology (also see Riddell and Corl, 1977; Eisenberg and Wilson, 1978; Clutton-Brock and Harvey, 1980; Mace et al., 1981; Wyles et al., 1983; Jolicoeur et al., 1984; Byrne, 1993). In support of this, Lefebvre and his colleagues

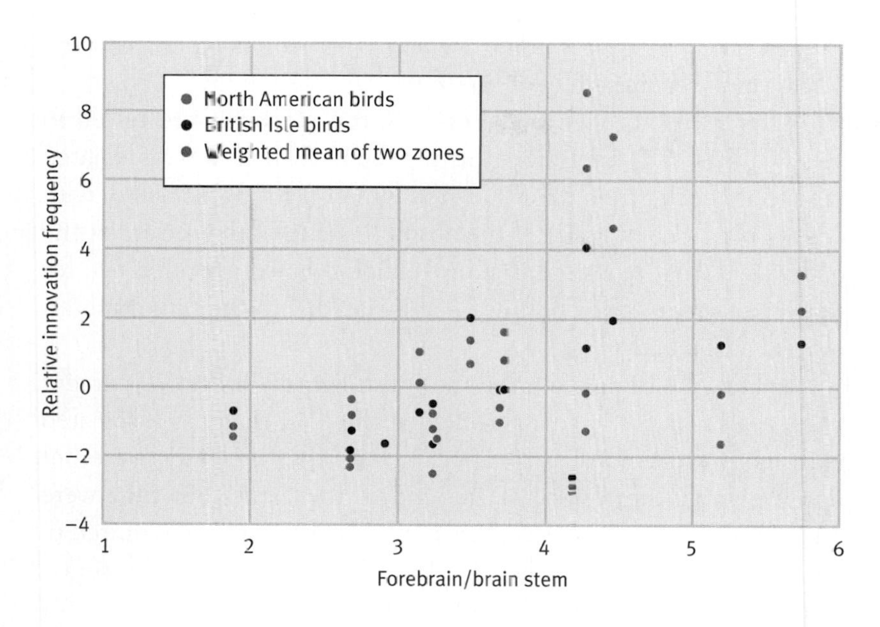

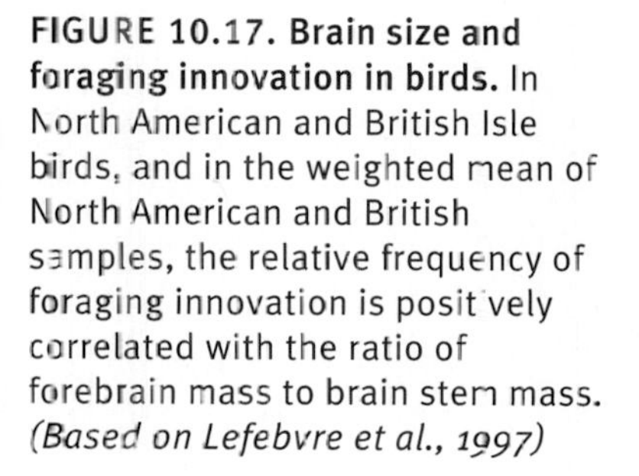
FIGURE 10.17. Brain size and foraging innovation in birds. In North American and British Isle birds, and in the weighted mean of North American and British samples, the relative frequency of foraging innovation is positively correlated with the ratio of forebrain mass to brain stem mass. *(Based on Lefebvre et al., 1997)*

(1998) ran a similar analysis to those with the North American and British Isle birds and uncovered very similar results in Australasian birds. In addition, Sol and Lefebvre (2000) found evidence that brain size and foraging innovation were good predictors of which birds were likely to invade new areas of New Zealand.

LEARNING AND "WORK ETHICS" IN PIGEONS

One particularly robust finding in behavioral work is that if two tasks produce the same reward, animals, including humans, will choose the easier of the tasks. In other words, if I give you the choice between having to climb a mountain to get a steak dinner, or walking down the block to get the same dinner, chances are you'll choose the walk.

What if, however, I change the scenario, and make you walk up the mountain or down the block first, and only then give you a great steak dinner? If I inquire which steak tasted better (provided a greater reward), odds are you'll say the mountaintop steak, because you had to work harder for it. Somewhat surprisingly, this sort of phenomenon also appears to be displayed by pigeons (Clement et al., 2000).

Tricia Clement and her colleagues undertook an experiment in which pigeons were trained to one of two different learning regimes. In regime 1, if a bird pecked a white circle once, a red light would go on, and the bird would be rewarded with a food item. In regime 2, pigeons were trained to peck the white circle twenty times, at which point a green light would go on, and the birds would receive a food item.

After going through both regimes, the pigeons were tested to see whether they preferred the red light (which was associated with one peck, then food) or the green light (associated with twenty pecks, then food). If the absolute value of the reward at the end was all that mattered, the birds should have no preference for red versus green, since both were associated with the same amount of food at the end.

In fact, the birds preferred the green light in almost 70 percent of all trials. That is, pigeons pecked at the light that was associated with *more* work, not less. It appears then that the birds were not only weighing the reward at the end of the game, but they were taking into account the actual amount of work that was required to get the reward as well. When the birds had to work hard for something, they valued it more (Clement et al., 2000).

Clement and her colleagues' findings are inconsistent with "traditional" theories of learning that presume that animals base their decisions on the consequences of a behavior, rather than on what precedes a particular action (Skinner, 1938; Hull, 1943). Rather, their results are more in line with the "relational theory of learning" (Lawrence and Fetsinger, 1962), which argues that it is the *change* in value that occurs from the initial stimulus (the white light) to the positive value associated with the discriminating stimulus (the red or green light) that predicts preference on the part of the birds. That is, it is the relative change from what immediately preceded the colored light to what came after the light that really matters. Pigeons should treat the twenty pecks preceding the green light as a more aversive stimulus than the one peck preceding the red light. Since the change from a very averse situation to a reward (green) is greater than the change from a mildly averse situation to a reward (red), the relational theory of learning predicts that birds should favor the green cue.

Foraging and Group Life

For species that spend all or part of their lives in groups, various aspects of group life can impact the foraging behavior, as well as the foraging success, of individuals. For example, in Chapter 1 we touched on Galef's excellent work on the role of imitation and rat foraging. The behavior of other rats in a group clearly affects how any given individual behaves. We shall explore other cases of foraging and social learning (Galef and Giraldeau, 2001), but the effect of group life on foraging can be manifest in a wide variety of forms. With that in mind, in addition to examining the role of social learning in foraging behavior, we shall tackle the following questions in this section:

- What is the role of group size per se on animal foraging?
- What role does cooperation within groups play in foraging, and how can this be distinguished from the effect of group size?
- How does work on "public information" shed light on animal foraging?

FIGURE 10.18. Bluegill foraging. Bluegills were among the first species used to test optimal prey choice models. The bluegills' foraging patterns approximated those predicted by theory. *(Photo credit: Gary Meszares/ Visuals Unlimited)*

GROUP SIZE

In many species that live in groups, simply increasing the foraging group size increases the amount of food each forager receives (Eggers, 1976). This may be because more foragers flush out proportionally more prey or because cooperative hunting can mean a division of labor between different group members.

FORAGING IN BLUEGILLS. Consider foraging success in bluegill sunfish. These fish primarily eat small aquatic insects that live in dense vegetation (Figure 10.18). Such items need to be flushed from the vegetation, and flushing may be more efficient as a function of group size (Morse, 1970; Bertram, 1978; Mock, 1980). That is, while bluegill prey are quite difficult to catch in dense vegetation, increasing foraging group size may force more prey out of the vegetation. While such prey may not be taken by the specific bluegill chasing them, they may be ingested by other bluegill group members. Gary Mittlebach (1984) examined this hypothesis by experimentally manipulating bluegill foraging group size in a controlled laboratory setting.

Mittlebach placed 300 amphipods (small aquatic prey) into a large aquarium containing juvenile bluegill sunfish. He studied the success of bluegills foraging alone, in pairs, and in group sizes of three to six. Mittlebach measured the number of prey captured per bluegill, and he uncovered a clear positive relationship between foraging group size and individual foraging success (Figure 10.19). It appears that the increased uptake rate was due to two factors. First, proportionally more prey were flushed when group size rose. In addition, however, prey were apparently clumped together and,

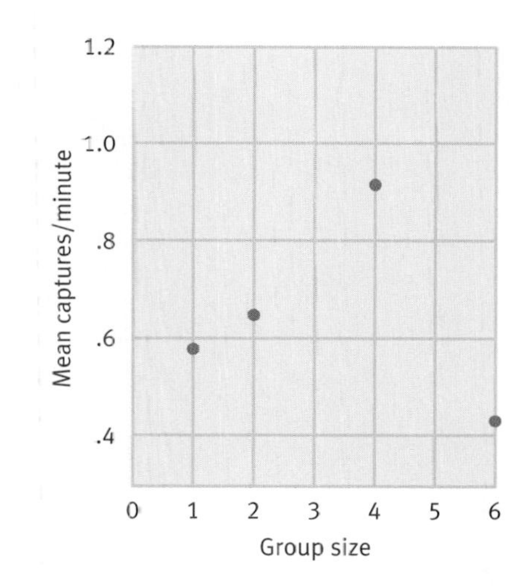

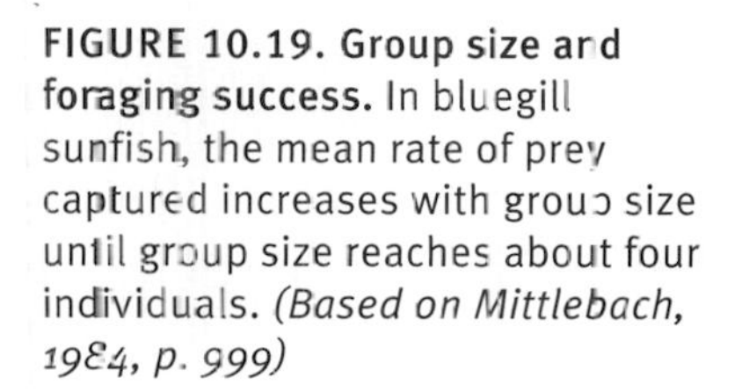

FIGURE 10.19. Group size and foraging success. In bluegill sunfish, the mean rate of prey captured increases with group size until group size reaches about four individuals. *(Based on Mittlebach, 1984, p. 999)*

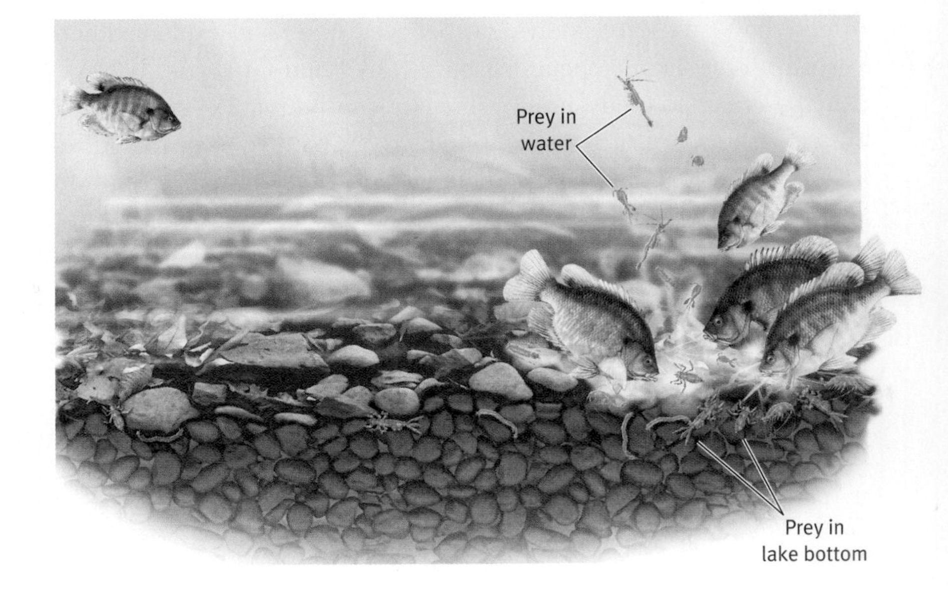

FIGURE 10.20. Bluegill group foraging. When bluegills forage in groups, they flush out more prey and attract other fish to the foraging site.

when one bluegill found amphipods, others swam over to this area and then often found food themselves (Figure 10.20).

While increased group size in bluegills does benefit each individual forager in its intake of food (its per capita foraging success), there is no evidence to suggest that bluegill foragers hunt in any coordinated fashion—the flushing effect is merely a byproduct of having more individuals searching for food (see below for more on cooperative foraging). Furthermore, the means by which bluegills increase foraging efficiency in larger groups (that is, by flushing) is only one means of increased foraging efficiency. For example, in a number of species, increased group size reduces the amount of time that any given individual needs to devote to antipredator activities, and this often (Caraco, 1979; Barnard and Sibly, 1981; Sullivan, 1984), but not always (J. N. M. Smith, 1977; Ens and Goss-Custard, 1984), increases per capita foraging success.

The general relationship between group size and foraging success uncovered by Mittlebach has been found over and over in animal studies. For example, Creel (2001) ran a meta-analysis on foraging success and group size in seven different species that hunt in groups. Overall, Creel found a strong positive relationship between individual foraging success and group size (Packer and Rutton, 1988; Caro, 1994a, 1994b). Though the meta-analysis covered only seven species, this set included stalkers and coursers that hunt by swimming, climbing, running, and flying.

DISENTANGLING THE EFFECT OF GROUP SIZE AND COOPERATION. Often, when hunting occurs in groups, there is some element of cooperation (Chapter 9) involved. For example, when wild

FIGURE 10.21. Cooperative hunting in wild dogs. (A) Wild dogs typically hunt in large groups. Different members of the hunting pack play different roles during a hunt—flushing out prey, undertaking the initial attack, disemboweling the prey, and so on. **(B)** After a successful hunt, wild dogs capture a prey and disembowel it. *(Photo credits: Scott Creel; Paul Souders/Worldfoto)*

dogs hunt down a prey item, it is a coordinated effort. Different members of the hunting pack play different roles in the hunt—flushing the prey, making the initial attack, disemboweling the prey, and so forth (Figure 10.21). In such cases, it is useful to separate out the effects of coordinated cooperation from that of group size per se. To see how ethologists disentangle such effects, let us examine hunting behavior in chimps.

Cooperative hunting or the lack of it has been examined in chimp populations in the Gombe Preserve in Tanzania (Lawick-Goodall, 1968; Telecki, 1973; Busse, 1978; Goodall, 1986; Boesch, 1994a, 1994b, 1994c), the Mahale Mountains in Tanzania (Nishida et al., 1983, 1992; Uehara et al., 1992), and the Tai National Park in the Ivory Coast (Boesch and Boesch, 1989; Boesch, 1994b; Figure 10.22). The most comprehensive studies of cooperative hunting

FIGURE 10.22. Cooperative chimps. While we often think of chimps as primarily vegetarian, meat, when it is available, is readily added to their diet. In the Tai Forest (Ivory Coast), chimps cooperate in both capturing and consuming prey. Once a prey is caught, subtle rules for food distribution are invoked. *(Photo credit: Christopher Boesch)*

among chimps are those of the Boesches, who compared hunting patterns across the Gombe and Tai chimp populations. Major differences in hunting strategies emerged between these populations.

In Tai chimps, hunting success was correlated with group size in a nonadditive fashion. That is, adding more hunters did not simply increase intake by an additional x units/hunter. Rather, the additional units of food added to each hunter's diet increased more with each new hunter (up to a limit). In addition to these group-size effects, Boesch (1994b) also uncovered clear evidence of cooperation in Tai chimp hunting behavior. Very complex, but subtle, social mechanisms exist that regulate access to fresh kills, and assure hunters greater success than those who fail to join a hunt.

The situation was quite different with Gombe chimps. There, the success rate for solo hunters was quite high (compared with the Tai population), and no correlation between group size and hunting success was found. That is, no group-size effect was uncovered in the context of foraging. In addition, behavioral mechanisms limiting a nonhunter's access to prey were absent, and such individuals received as much food as those who hunted cooperatively (Goodall, 1986; Boesch, 1994b).

SOCIAL LEARNING AND FORAGING

Information about foraging can be culturally transmitted. Individuals observe others to see how they find food and what they eat. To understand more about social learning, we will consider foraging in pigeons and food choice in preschool children.

FIGURE 10.23. Urban foragers. Pigeons are scavengers, coming across novel food items all the time. Such a species is ideal for studying foraging and cultural transmission. *(Photo credit: Bob Newman/Visuals Unlimited)*

PIGEONS. Pigeons (*Columbia livia*) are an ideal species in which to examine cultural transmission of feeding behavior (Figure 10.23). Being primarily scavengers feeding on human garbage, pigeons face the same dilemma faced by rats: Which new food items are safe, and which are dangerous? Over the last fifteen years, Louis Lefebvre and his colleagues Luc-Alain Giraldeau and Boris Palameta have run an intriguing series of experiments that attest to the strength of cultural transmission in shaping diet in the pigeon. This work has focused on three related issues: (1) What type of information do pigeons transfer about food? (2) How does such information spread or fail to spread through a population of pigeons? (3) What factors favor the cultural transmission of information over alternative means of acquiring information?

Palameta and Lefebvre (1985) set out to examine cultural transmission in a three-part experiment that used observer and demonstrator animals. The task that observer pigeons needed to master was piercing the red half of a half red/half black piece of paper covering a box. Under the paper was a bonanza of seeds for the lucky bird who made it that far.

An observer pigeon was placed in an arena with such a food box (half red/half black paper cover) and exposed to one of four scenarios. In the first group, birds saw no model on the other side of a clear partition. None of the pigeons in this group learned how to get at the hidden food. In a second group, observers saw a model pierce the red side of the paper but get no food. Once again, pigeons in this treatment failed to learn how to get food from the multicolored box. In the final two treatments of the experiment, birds either saw a model eating from a hole that had been pierced by Palameta and Lefebvre (the birds did not see the model solve the hidden food puzzle), or they saw a model both pierce and eat. Although birds in both of these treatments learned to solve the food-finding dilemma, those in the latter treatment did so much faster (Figure 10.24).

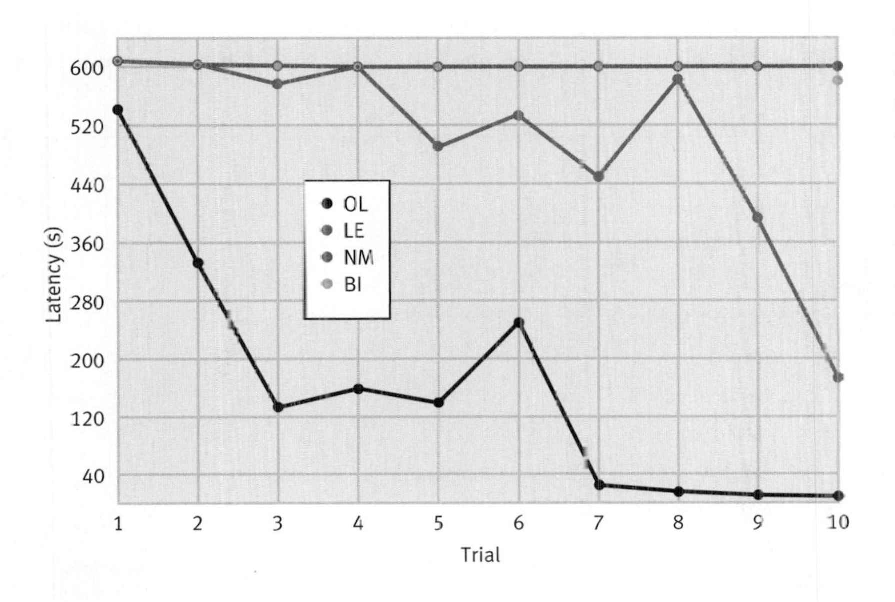

FIGURE 10.24. Social learning and foraging in pigeons. Pigeons in this experiment need to learn to pierce the red half of a paper covering a box of seeds. The graph shows average latency to eating for four groups: NM (no model) group—these birds were never exposed to a model; BI ("blind" imitation) group—these birds observed a model bird pierce a paper cover but never eat; LE (local enhancement) group—these birds observed a model bird eat through a hole in the paper made by the experimenters; and OL (observational learning) group—these birds saw a model bird pierce the paper and eat the seeds in the box. Pigeons in the NM and BI treatments never learned to feed in the experimental apparatus. The quickest learning occurred in the OL treatment. *(Based on Palameta and Lefebvre, 1985)*

There is an interesting twist to the pigeon story. In many animals that live in groups, there are two strategies that individuals can use for foraging: producing and scrounging. **Producers** find and procure food, while **scroungers** make their living parasitizing the food that producers have uncovered (Barnard, 1984). Layered onto the social learning and foraging story we have seen in pigeons, we find producers and scroungers in this species as well (Giraldeau and Lefebvre, 1986). And it is the unusual way that producing and scrounging interact with social learning that makes the pigeon story particularly useful in furthering our understanding of cultural transmission and foraging.

Despite Palameta and Lefebvre's work demonstrating social learning in pigeons in cages, when birds are tested in groups, only a few birds learn new feeding behaviors by observing others. This apparent paradox caused Giraldeau and Lefebvre to examine whether scrounging behavior somehow inhibited cultural transmission (Giraldeau and Lefebvre, 1987). To do so, they used a different set-up than the one described above (Figure 10.25). Now, flocks of pigeons were allowed to feed together. Forty-eight little test tubes were placed in a row and five of these tubes had food. Which five, of course, was unknown to the birds. To open a tube, an individual had to learn to peck at a stick in a rubber stopper at

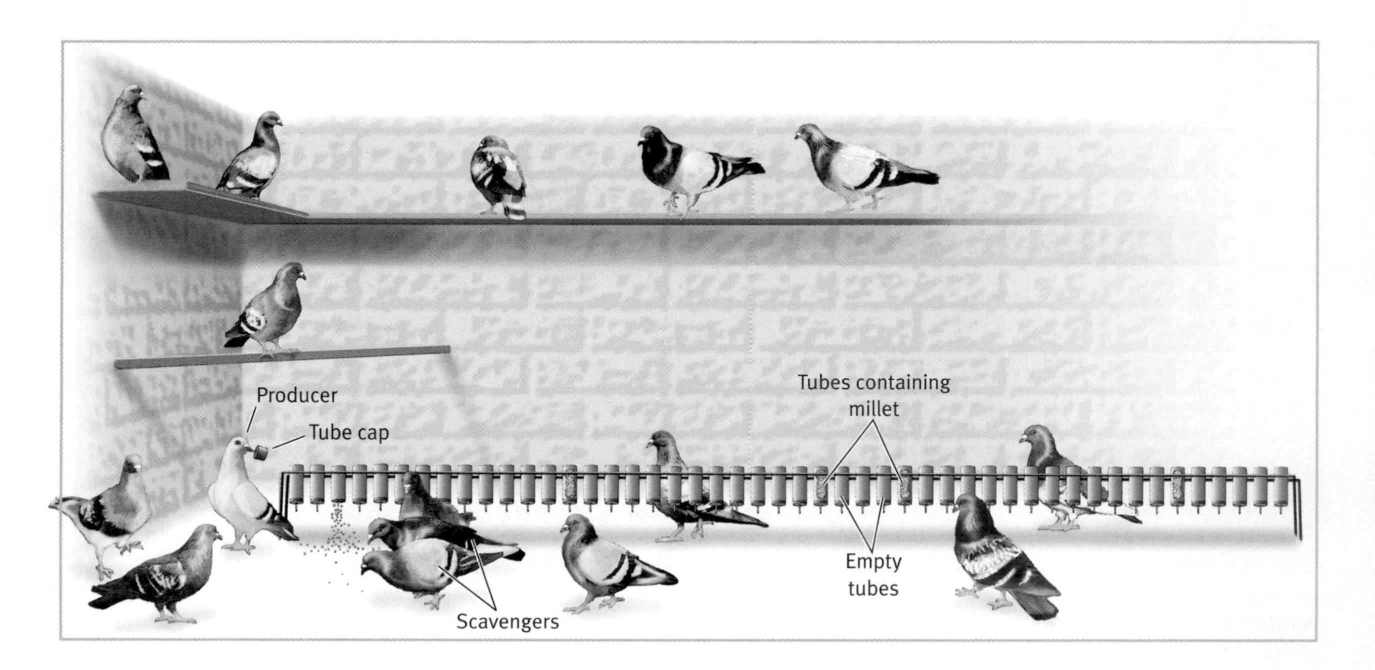

FIGURE 10.25. Producing and scrounging. When a group member finally opens a tube with food in it, the food spills on the floor and is accessible to all. In Giraldeau and Lefebvre's study using such a scenario, out of sixteen pigeons, only two learned to open tubes (these were the producers), while fourteen acted as scroungers. Labels show where millet was hidden, but the birds were not privy to this information.

the top of the tube. This caused the test tube to open and the contents to spread over the floor below. Once the food was out, any bird in the vicinity, not just the one that opened the tube, could eat it.

The results of their experiment were striking. As predicted based on earlier work, only two of the sixteen pigeons in their group learned to open tubes—that is, the flock was composed of two producers and fourteen scroungers. Two additional findings suggested to Giraldeau and Lefebvre that scrounging inhibited an individual from learning how to open tubes via observation. To begin with, scroungers followed producers and seemed more interested in where producers were than in what producers did to get food. Second, by removing the two producers from the group, Giraldeau and Lefebvre were able to ascertain that scroungers not only didn't display the tube-opening trait, but they also didn't know how to open these tubes. That is to say, it was not as if scroungers could open the tubes, but opted not to; they truly never learned from observing the producers.

In order to get a firm handle on just how scroungers blocked cultural transmission of foraging skills, Giraldeau and Lefebvre (1987) ran a second set of more controlled experiments (Figure 10.26). Here, they paired a single observer with a single demon-

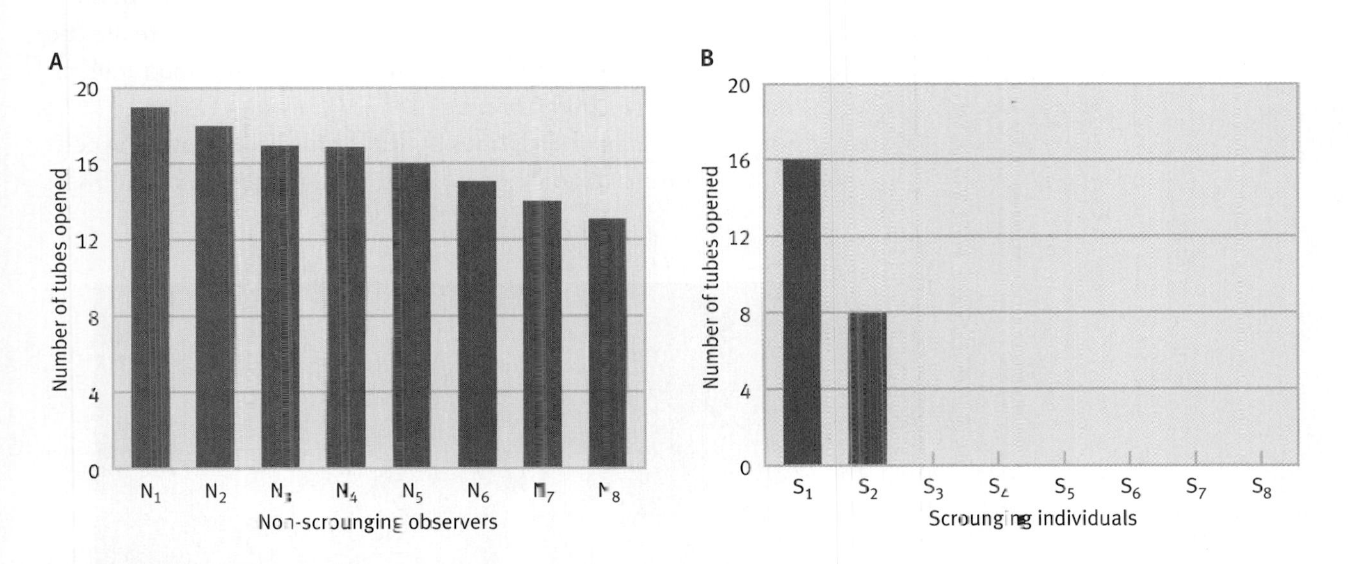

FIGURE 10.26. Scrounging prevents social learning In an ingenious experiment, Giraldeau and Lefebvre trained two groups of pigeons. (A) One group of birds saw a model bird peck at a stick in a rubber stopper at the end of the tube. This provided food to the model. These birds learned to peck at the stick to obtain food after observing the model, and so they were considered non-scroungers. (B) In the second group of birds, when the model pecked at the stick, food was released, but to the observer. In this group, observers were scroungers. Scroungers fared much worse than non-scroungers when faced with the task of pecking at the stick. *(Based on Giraldeau and Lefebvre, 1987)*

strator (who already knew how to obtain food). If an observer had the chance to view a demonstrator open tubes and obtain food, in time the observer learned how to open tubes. That is, all birds were capable of learning the foraging task. In an ingenious manipulation, Giraldeau and Lefebvre then set up the experimental cages so that every time the demonstrator opened a tube, the food in that tube slid over to the observer's side of the cage. In these treatments, the observers rarely learned how to open the tubes themselves. Their scrounging on the food found by others interfered with the transmission of information. Giraldeau and Lefebvre's finding remains one of only a few cases in which researchers knew enough about a specific case of cultural transmission to figure out when it is effective and when it isn't.

YOUNG CHILDREN, SOCIAL LEARNING, AND FOOD CHOICE. Back in 1938, Duncker examined the food preferences of preschoolers. After establishing what a child liked and disliked, that child was given the opportunity to view another individual choose one of the foods that he himself did not rank particularly highly (Figure 10.27). While the effects that Duncker uncovered lasted only a week or so, he did in fact find that children shifted their food preference toward that of the individual they observed. The effect was especially strong just where one might expect—when the child was observing a friend or a powerful child who was not a friend (Duncker, 1938). In addition, Duncker read stories to children in which a hero figure ate the food that the child ranked low, and sure enough, that food took on new meaning to the child subject.

Some forty years after Duncker, Birch ran a fascinating experiment on children's acquisition of food preferences (Birch, 1980).

FIGURE 10.27. Preschool imitators. The specific food preferences of preschoolers are significantly affected by what they believe specific other children prefer. This effect is fairly short-lived. *(Photo credit: Nancy Alexander/Visuals Unlimited)*

Birch began by recording the vegetable preferences of children before the start of a trial. Such children were then placed at a table where three or four other youngsters chose what vegetable they would eat before the child who was the subject (the observer) chose. Slowly, but surely, subjects began to shift their choice toward that of their peers. Given that none of the models in either Birch's or Duncker's work was actively teaching the subjects (that is, explaining why their choice was a good one), these studies not only show how cultural influences on food choice are developed, but how strong such forces can be, at least with respect to children. Thus, cultural transmission is strong enough that foods that initially produce a strong negative effect (such as chili peppers; see Rozin and Schiller, 1980) can eventually come to be part of a child's diet through enculturation.

PUBLIC INFORMATION AND FORAGING

Tom Valone and his colleagues have suggested that another potential foraging-related benefit associated with group dynamics is the acquisition of what they refer to as "public information" (Valone, 1989; Valone and Benkman, 1999). The idea behind public information models is that a forager needs to be assessing a whole host of environmental variables (food availability, predators, and so on). One way to update such information is to use the actions of others as a cue to changes in environmental conditions. This differs from social learning models, in which individuals learn something specific (a new behavior, the preference of others). In public information models, individuals merely use the actions of others as a means of assessing the condition of the environment, and as such, public information allows group members to reduce environmental uncertainty.

While public information models are general in nature and can apply to numerous environmental parameters, here we shall focus on how these models have been employed in the context of foraging behavior. Solitary foragers can reduce environmental variability by acquiring knowledge about foraging patches before arriving at a patch (Green, 1980; Valone, 1991) and/or by keeping track of how long they have been in a patch and how much they have eaten (Valone and Brown, 1989; Valone, 1991; Valone and Giraldeau, 1993; Ollson et al., 1999). Foragers in a group, however, can acquire knowledge about patches via public information. In particular, a forager in a group can use the foraging success of others as an estimate of patch quality.

FIGURE 10.28. Public information. Social foragers such as starlings have been used to test "public information" models of foraging behavior. Starlings in this public information experiment were tested using an array of food placed into cups. *(Photo credit: Jennifer Templeton)*

With public information available, social foragers should have access to quicker and better estimates of the productivity of a patch of food than solitary foragers. For example, these public information models predict that social foragers in poor patches should leave such patches earlier than solitary individuals would, since social foragers may use the failed foraging attempts of their groupmates as additional information about when they themselves should leave a patch of food. Jennifer Templeton and Luc-Alain Giraldeau tested this prediction of public information in the starling (*Sturnus vulgaris*). Starlings in this experiment fed at an artificial feeder with thirty cups that were either empty or contained a few seeds (Templeton and Giraldeau, 1996; Figure 10.28). A given bird (B1) fed from such a feeder either alone or paired with a second bird (B2). Prior to being paired with B1 partners, B2 birds had been given the chance either to sample a few cups in this feeder or to sample all such cups. Two results support the predictions of public information models. First, when tested on completely empty feeding patches, B1 birds left such patches earlier when paired with any B2 bird than when foraging alone. Even more importantly, B1 birds left patches earliest of all when paired with B2 birds that had complete information about the patches (as compared with those B2 birds with only partial information); similar results have been found in red crossbills (*Loxia curvirostra*; Smith et al., 1999).

Molecular, Neurobiological, and Hormonal Aspects of Honeybee Foraging

As a result of its ethological, evolutionary, agricultural, and commercial value, honeybee foraging has been the subject of intense investigation. Here, we shall use aspects of this model organism to examine some molecular, neurobiological, and hormonal underpinnings of foraging.

THE PERIOD GENE, mRNA, AND FORAGING

In many social insects, including honeybees individuals typically spend the early part of their lives working in the hive, while the latter half of their lives is often devoted primarily to foraging outside the nest (Robinson, 1992; Figure 10.29). To examine the possible molecular biological underpinnings to this developmental shift to forager (Robinson et al., 1999), Toma and his colleagues (2000) capitalized on findings that suggest that the period (*per*) gene is known to influence circadian rhythms and development time in fruit flies (Konopka and Benzer, 1971; Kyriacou et al., 1990). As such, they focused on the *per* gene as a "candidate" gene involved in control of the developmental changes to foragers that occur in honeybees. More specifically, they examined how *per* messenger RNA (mRNA) levels might influence such changes.

Toma and colleagues measured brain *per* mRNA levels in three sets of laboratory-raised bees. The three groups of bees were composed of individuals that were four to six, seven to nine, and twenty to twenty-two days old. In addition, researchers marked and added

FIGURE 10.29. The *per* gene and bee foraging. Many adult worker bees spend a significant proportion of their time foraging for food away from the hive. Ethologists have examined the effects of the *per* gene on such foraging behavior. *(Photo credit: George D. Lepp/Corbis)*

a group of one-day-old bees to a natural colony of bees in the field. Marked individuals were recaptured at day 7 and day 24, and the researchers measured their brain *per* mRNA levels. Results demonstrate a clear relationship between *per* mRNA level, age, and foraging in honeybees. In both laboratory and natural populations, *per* mRNA was significantly greater in older individuals who foraged for food and brought such food to their colony, when compared to younger bees that remained at the nest (Figure 10.30).

Toma also examined whether *per* mRNA increases were strictly associated with age per se, or whether *per* mRNA and foraging were linked, even if age could be removed as a factor. Luckily, "precocious" foragers are sometimes found in beehives. These foragers begin searching for food outside of the nest at about seven days old. Thus, precocious foragers provided the chance to remove age effects from the *per* mRNA/foraging connection, as the researchers now had a sample of bees that foraged but were much younger than typical food-gatherers.

Precocious foragers had *per* mRNA levels that did not differ from those of typical (older) foragers, suggesting a fundamental link between *per* mRNA and foraging, as opposed to a more general connection between *per* mRNA and development. Of course, the arrow of causality here is still somewhat clouded, as this research can't tell us whether increased *per* mRNA leads to foraging or foraging produces high *per* mRNA levels. To determine causality,

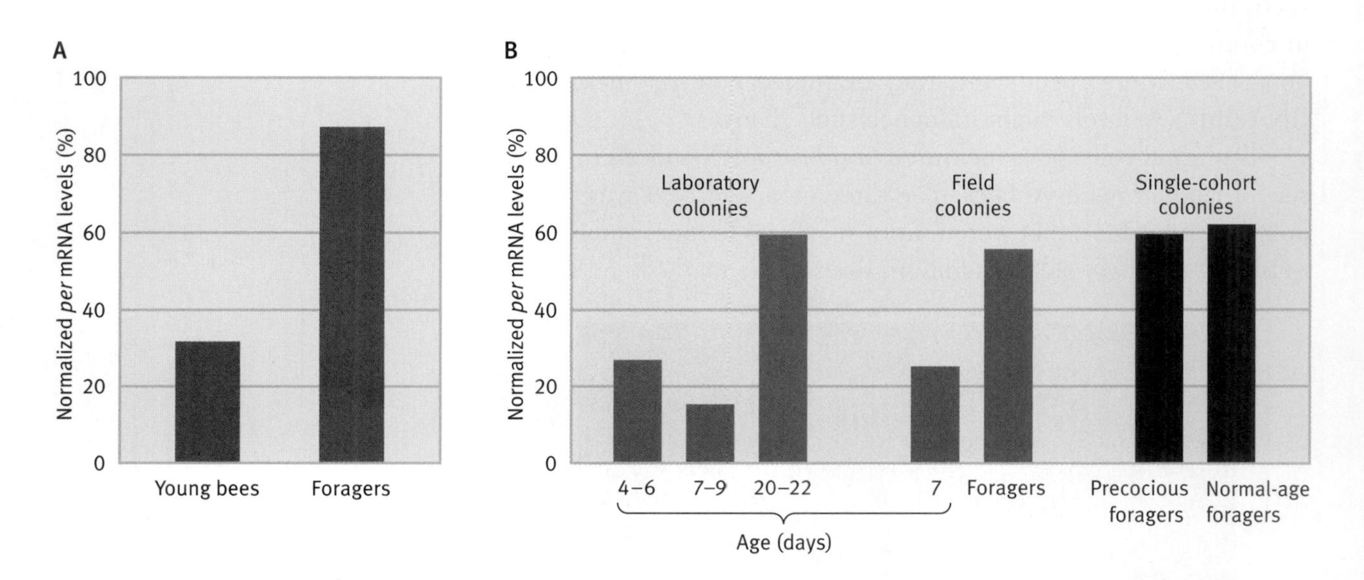

FIGURE 10.30. Foraging, age, and mRNA. (A) Foraging bees have significantly higher levels of *per* mRNA than younger, nonforaging bees. This difference could be due to age, behavior caste (forager versus nonforager), or both. **(B)** Some individual bees developed into precocious foragers who began searching for food much earlier than usual. When ten-day-old precocious foragers were compared to normal foragers (twenty-two-day-old foragers), Tuma and his colleagues found no statistical differences in *per* mRNA levels. *(From Toma et al., 2000)*

researchers would have to manipulate *per* mRNA levels in bee brains, which has not yet been attempted.

JUVENILE HORMONE, "MUSHROOM BODIES," AND FORAGING

As bees mature into foragers who leave the nest an associated increase occurs in levels of juvenile hormone (JH) (Riddiford, 1994 Wyatt and Davey, 1996), particularly JH III (Hagenguth and Rembold, 1978; Huang et al., 1991). To better understand the relationship between juvenile hormone and foraging behavior, Sullivan and his colleagues (2000) began by removing the corpus allatum—the glandular source of JH—in experimental groups of bees in four colonies (Figure 10.31). This group of allatectomized bees was then compared to two control groups, one in which the bees went through a similar surgical procedure but did not have their corpus allatum removed, and the other in which bees were anesthetized only.

While the presence of JH is not a necessary precondition for bee foraging, in three of the four colonies observed, bees that had been allatectomized, and thus had no JH, began foraging significantly later than bees in the control groups, suggesting a role in foraging-related development in honeybees. No other major behavior shifts resulted from removing the corpus allatum. In addition, Sullivan and his colleagues found that when allatectomized bees were given a dose of a methoprene—a JH analog (Robinson, 1985, 1987; Robinson et al., 1989)—they showed no differences in age-related foraging when compared to controls, providing even stronger evidence for the role of JH in foraging behavior in honeybees.

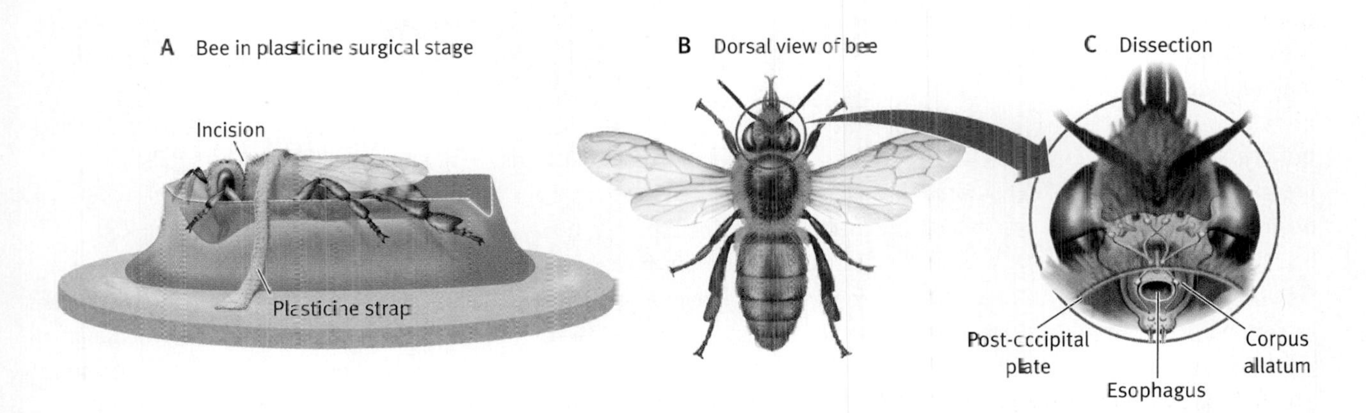

FIGURE 10.31. Bee surgery. To examine the effect of juvenile hormone (JH) on honeybee foraging, Sullivan removed the corpus allatum—the gland that produces the hormone. The inset shows a view through the incision. *(Based on Sullivan et al., 2000)*

Interview with Dr. JOHN KREBS

How did it come to be that someone whose father won a Nobel Prize for coming up with the Krebs cycle made his initial mark in science by studying animal behavior?

No single factor, but a concatenation of many events, led me to a career in animal behavior. I came to zoology as a bird-watcher; as a high school pupil I spent two summers working in the laboratory of the famous ethologist, Konrad Lorenz; and as an undergraduate at Oxford, I was lucky enough to be taught by two giants of their generation, David Lack and Niko Tinbergen. Having said all this, my father dissuaded me from my first choice of a university course, namely archaeology, and I ended up as a zoologist, having failed to get in to study medicine! My Ph.D. at Oxford was unusual in that I studied population ecology in Niko Tinbergen's animal behavior group. This meant that right from the start I saw the linkages between ecology and behavior that led to the emergence of "behavioral ecology."

One of the most innovative aspects of your work on foraging in great tits was the use of a conveyer belt to present food to the birds. How did you come up with that idea? What did this allow you to do that otherwise would have been impossible?

Throughout my career I have been extraordinarily fortunate in the exceptional colleagues, students, and co-workers I have had. The "conveyor belt" was not my idea. A graduate student in the animal behavior group at Oxford, Jon Erichsen, had devised it for psychophysical studies of reaction times in birds. Together we saw its potential for

foraging studies. It would enable us to control encounter rates very accurately by having the prey move past the bird, rather than vice versa. I remember vividly posting to (no e-mail in those days) Ric Charnov the first results from the conveyor belt. These showed that the inclusion or exclusion of the less profitable prey type depended exclusively on the encounter rate with the more profitable prey. The theory actually worked!

Psychologists have been studying the feeding behavior of animals for more than fifty years. Why the lag in ethological studies in this area? Why did they start up in earnest in the 1970s?

Ethologists had in fact been studying feeding behavior for many years. The important concept of "searching image" was well established, and there were many studies of the cues used by animals for detecting food, the development of feeding skills during the life of an individual, as well as of predator avoidance strategies, such as crypsis, mimicry, and warning colors. The new approach in the 1970s, known as "optimal foraging theory," resulted from a confluence of ideas from ethology, ecology, evolution, psychology, and economics. As so often happens when new ideas are beginning to crystallize, more than one person independently came to similar conclusions. The seeds of foraging theory were more or less independently sown by J. T. Emlen, R. H. MacArthur and E. R. Pianka, J. D. Goss-Custard, and G. A. Parker. The attraction of foraging theory for me at the time, and I think for many others, was the juxtaposition of a broad theoretical framework, testable predictions from mathematical models, and both field and laboratory data.

How do classical psychologists and ethologists differ in the way they study animal foraging?

Traditionally, psychologists studied the mechanisms of behavior in the laboratory whilst ethologists and behavioral ecologists studied function or adaptation in the field. In the last fifteen years, these boundaries have become blurred, to the benefit of both disciplines. For example, ethologists have adopted some of the experimental and theoretical tools of operant and classical conditioning as well as of cognitive psychology. Psychologists have increasingly taken on board the potential value of an evolutionary framework for understanding behavior.

Many optimal foraging models involve complex mathematics. Does that mean the animals have to know how to solve mathematical equations to forage efficiently?

Mathematical models of flight are complex, but birds do not need to be mathematicians in order to fly. Nor do they need to do sums in order to forage according to the predictions of optimal foraging models. Mathematics is the universal language used by a scientist to describe and analyze what goes on in nature. There is no implication that nature itself uses mathematics.

A number of foraging models focus on what are called "rules of thumb." Could you explain the basic idea here?

Traditional foraging models were concerned with the end result (or adaptive significance) of foraging—for example, maximizing energy intake, rather than about the mechanisms that control feeding behavior. The notion of "rules of thumb" is that the end result, analyzed in a foraging model, might be achieved by a mechanism that approximates the "right" answer. If, for example, you were trying to find the best textbook of animal behavior, the rule of thumb "read if it is by Dugatkin" might bring you close enough to the optimal solution to be virtually indistinguishable from a more complex search strategy. If students following a different rule of thumb did better in exams, over time, the "Dugatkin rule" would be replaced by another rule that came closer to the optimal solution. In other words, rules of thumb are themselves subject to evolution by natural selection.

One potential critique of the foraging literature in ethology is that so many of the experiments undertaken are run in the laboratory and not in the field. In your estimation, how serious a problem is this?

If you are studying fundamental properties of animals—for example, decision-making rules—then they can be elucidated in the laboratory as well as in the field. A potential danger is that in the laboratory you may end up finding out more about the properties of your apparatus and experimental set-up than about the animal! This is a criticism that has sometimes been leveled at earlier studies of "schedules of reinforcement" by operant psychologists. Probably the ideal approach is to use a combination of field and laboratory work. Laboratory work certainly has the advantage that it is possible to control the variables more precisely. For laboratory work, intelligent design of the experimental set-up will help to ensure that the essence of the natural situation is captured.

What's "the big question" to be tackled next in the study of animal foraging behavior?

Foraging theory, like any other branch of inquiry, will continue to evolve and change. I do not see a single "big question" on the horizon, but I see several potentially important ways in which the study of foraging behavior could develop. One of these is to apply ideas from foraging behavior to conservation. The pioneering work of J. D. Goss-Custard and W. J. Sutherland has shown how foraging models can be used to predict the effects of habitat loss and disturbance on threatened populations. The key has been to link the foraging of individuals to density-dependent influences on the population as a whole. Another area of recent, and likely continuing, growth consists of the links between behavioral economics, decision theory, and foraging behavior. This is a development of the links between animal psychology and foraging behavior that began about fifteen years ago. New insights are flowing in both directions. More generally, I do not think we should regard "foraging behavior" as an ineluctable, distinct field of inquiry. The basic ideas may be reconfigured, remolded, and, reinverted so that the categories used to describe behavior today might seem outmoded in ten years' time.

DR. JOHN KREBS is a professor at Oxford University in England. Dr. Krebs's work on foraging behavior was seminal to the development of optimal foraging theory. Dr. Krebs (along with Dr. Nick Davies; see Chapter 7 interview) is the editor of Behavioural Ecology (Blackwell Science)*, the most well-read book in the field of behavioral ecology.*

SUMMARY

1. Animal foraging theory encompasses a class of models that predicts what and where animals should eat, as well as nutrient constraints and the effect of hunger state on risk-sensitive foraging.
2. Somewhat surprisingly, optimal foraging models predict that the encounter rate with the less profitable of two prey items does not affect whether that item should be added to the diet or not. Rather, a critical encounter rate with the *most* profitable item determines what prey items should be added to a diet.
3. The marginal value theorem predicts how long a forager should stay in a given patch, given that it can leave and travel to feed in more undisturbed patches. Moving between food patches, however, is assumed to entail some cost, including lost time foraging while traveling and the increased rate of predation while traveling.
4. In addition to maximizing their rate of energy intake, in some species individuals must simultaneously attempt to minimize the probability that they fail to ingest particular nutrients, such as trace minerals.
5. Hungry foragers should be risk prone, that is, willing to assume greater variance (risk) in food intake. Less hungry individuals tend to be more risk averse.
6. Animals often learn how and on what to forage; indeed, work in this area has been a mainstay in the psychological literature for years. Learning sometimes manifests itself in foraging innovations; brain size often correlates with the frequency of such foraging innovations.
7. In group-living species, social interactions within a group can impact foraging success via social learning, public information, or some form of cooperative foraging behavior.

DISCUSSION QUESTIONS

1. Read R. Pulliam's 1973 article, "On the advantages of flocking," in the *Journal of Theoretical Biology* (vol. 38, pp. 419–422). Based on the article, outline the "many eyes" hypothesis and discuss how it relates to foraging behavior.
2. Read E. L. Charnov's 1976 article, "Optimal foraging, the mar-

ginal value theorem," in *Theoretical Population Biology* (vol. 9, pp. 129–136). How might you modify the model to examine other behaviors displayed by animals?

3. Using Math Box 10.1 as a starting point, construct a foraging model with three prey types. Imagine that you already know that it pays for a forager to eat prey type 1 and prey type 2. What are the conditions under which it should add prey type 3?
4. Why do you suppose it took so long for ethologists and psychologists to recognize the larger literature on foraging behavior that exists in the other field? What do you think were the biggest differences in the way foraging was studied across these disciplines?
5. Using the graphs in Figure 10.9 as a starting point, examine what happens to the time an animal should spend in the patch as a function of how profitable that patch is. This will involve changing the shape of the curve that describes food intake as a function of patch residence time and examining what this change does to optimal time in a patch.

SUGGESTED READING

Belovsky, G. E. (1978). Diet optimization in a generalist herbivore: The moose. *Theoretical Population Biology, 14,* 105–134. One of the first optimal foraging studies that incorporated specific food constraints (salt).

Capaldi, E., Robinson, G., & Fahrbach, S. (1999). Neuroethology of spatial learning: The birds and the bees. *Annual Review of Psychology, 50,* 651–682. This review of the neurobiology of spatial learning has nice examples drawn from the bee foraging literature.

Giraldeau, L. A., & Caraco, T. (2000). *Social foraging theory.* Princeton: Princeton University Press. A comprehensive overview of social foraging theory, with many empirical examples to lead the reader through the math.

Krebs, J. (1978). Optimal foraging: Decision rules for predators. In J. Krebs & N. Davies (Eds.), *Behavioral ecology: An evolutionary approach* (2nd ed., pp. 23–63). Sunderland, MA: Sinauer Associates. A good overview of early work on foraging.

Stephens, D., & Krebs, J. (1986). *Foraging theory.* Princeton: Princeton University Press. Regarded by most as the definitive book on foraging theory and animal behavior.

11

Behavioral Tradeoffs Associated with Predation

- Predation and Foraging
- Predation and Hatching Time in Wasps

Alarm Signals

- Vervet Alarm Calls
- Tail Flagging

Prey Approaching Their Predators

- Costs and Benefits of Thomson's Gazelles Approaching a Predator

Interpopulational Differences

- Interpopulational Differences in Antipredator Behavior in Minnows
- Predator vs. Prey Arms Races

Learning and Antipredator Behavior

- The Direct Fitness Consequences of Learning about Predators

Social Learning and Antipredator Behavior

Interview with Dr. Manfred Milinski

Antipredator Behavior

At the California site where Richard Coss, Donald Owings, and their colleagues have been studying predator-prey interactions, gopher snakes and ground squirrels have both been present for approximately a million years and snakes rely heavily on squirrels as a main staple of their diet. Not surprisingly, after so great a period of evolutionary time, natural selection has strongly selected for squirrels to be able to identify snakes and to respond to them with behavior intended to ward off predators (Owings and Coss, 1977; Coss and Owings, 1985; Coss, 1991). For example, groups of squirrels will often mob, or converge on, a snake predator, biting and harassing it until the snake is forced to leave the area (Owings and Coss, 1977). In addition, squirrel antipredator behavior includes throwing dirt, pebbles, and roots at putative predators, as well as emitting alarm calls that are specifically made when snakes, but not other predators, are present (Owings and Leger, 1980; Figure 11.1). But there is more to it than that. It seems that the squirrel's immune system has also evolved in concert with the effects of snake predation. To see this, let us look at some behavioral immunological work undertaken by Naomie Poran and Richard Coss.

Poran and Coss (1990) focused on maturing young California ground squirrels and examined both antipredator behavior and immunological defenses against snakes. With respect to the latter, they studied serum-to-venom binding levels—a measure of immunological defenses against snake bites—in six ground squirrels. To do this, they employed radioimmunoassays of serum-to-venom binding levels in squirrels at ages fourteen, twenty, forty-eight, and eighty days. At very young ages, pups have very low serum-to-venom binding levels, but this measure shoots up to

FIGURE 11.1. Snakes and squirrels. (A) Ground squirrels recognize snakes as predators upon emergence from their burrow. **(B)** Confrontations with rattlesnakes are common, and **(C)** they sometimes lead a squirrel to throw dirt and roots at the snake (note the snake's head at the left) to defend itself. *(Photo credits: Richard Coss)*

almost adult levels somewhere between days fourteen and thirty (Figure 11.2). This occurs even when pups are not exposed to snakes during the experiment. Why is there a spike in serum-to-venom binding levels?

The answer appears to be tied to the natural history of the squirrels. Squirrel pups don't emerge from their natal burrows until they are about forty days old (Linsdale 1946). In their burrows, the pups are relatively safe from snake predation, at least compared to the threats they face when emerging from their burrow (Figure 11.3). It seems then that selection has favored a spike just prior to when the threat of snake predation will increase dramatically.

One thing for certain about animal behavior—if an individual slips up with respect to antipredator tactics, its future fitness may well be zero. This obvious, but striking, fact suggests that natural selection should operate very strongly on antipredator behavior and that, in addition, we might expect fine-tuned learning in this behavioral venue as well. Perhaps because interactions between predators and their prey are often spectacular to observe, ethologists have a long history of studying antipredator behavior.

For example, consider Tony Pitcher's work on antipredator behaviors in schooling (or shoaling) species of fish (Pitcher, 1986). In addition to the potential hydrodynamic and foraging benefits accrued by living in groups, fish in schools display a wide assortment of antipredator tactics. When a predator is sighted, schooling fish often increase "elective group size" by schooling much more tightly

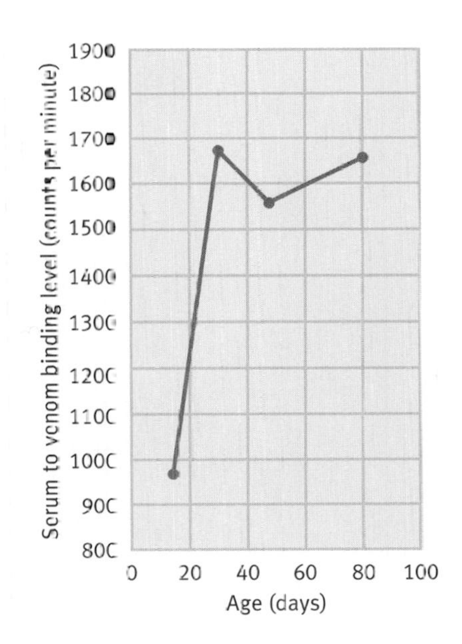

FIGURE 11.2. Immunology and predation. Ground squirrel pups emerge from their burrows at about forty days, at which time there is also an increase in their immunological defenses (their serum-to-binding levels) against snakes. *(From Poran and Coss, 1990, p. 237)*

FIGURE 11.3. Danger at emergence. Ground squirrel pups often face serious predation threat on their first emergence from their burrow.

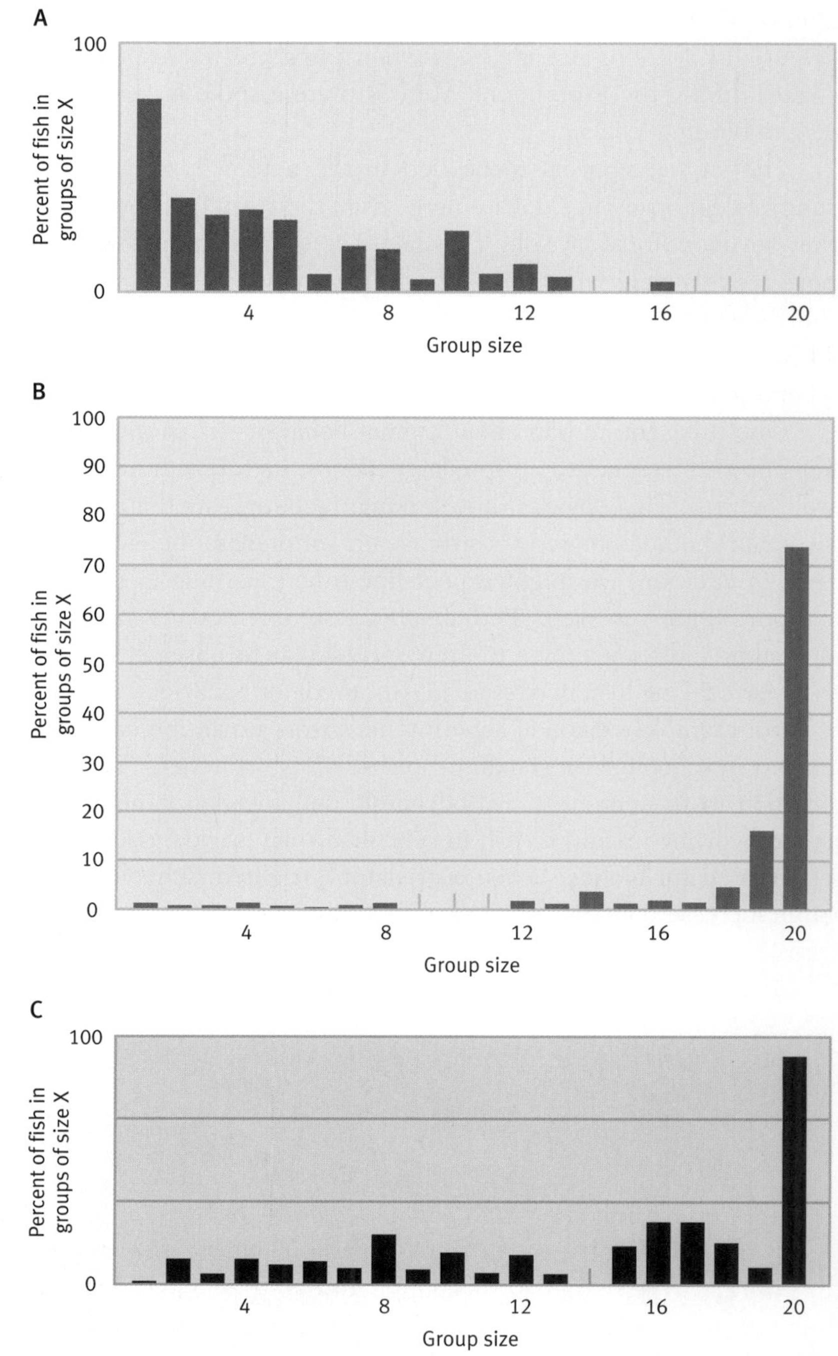

FIGURE 11.4. Group size and predation. Pitcher examined elective group sizes in a population of twenty fish. Elective group size was measured as fish within four body lengths of each other. The histograms show elective group size during **(A)** pre-exposure to the predator, **(B)** two hours after a one-hour exposure to the predator, and **(C)** one day after exposure to the predator. *(Based on Pitcher, 1986)*

(Figure 11.4). Cohesive schooling allows for the following antipredator tactics (Figure 11.5):

- Fountain Effect—Schools maximize their speed, split around a stalking predator, and then reassemble behind the putative danger (Potts, 1970; Nursall, 1973; Radakov, 1973; Pitcher and Wyche, 1983).

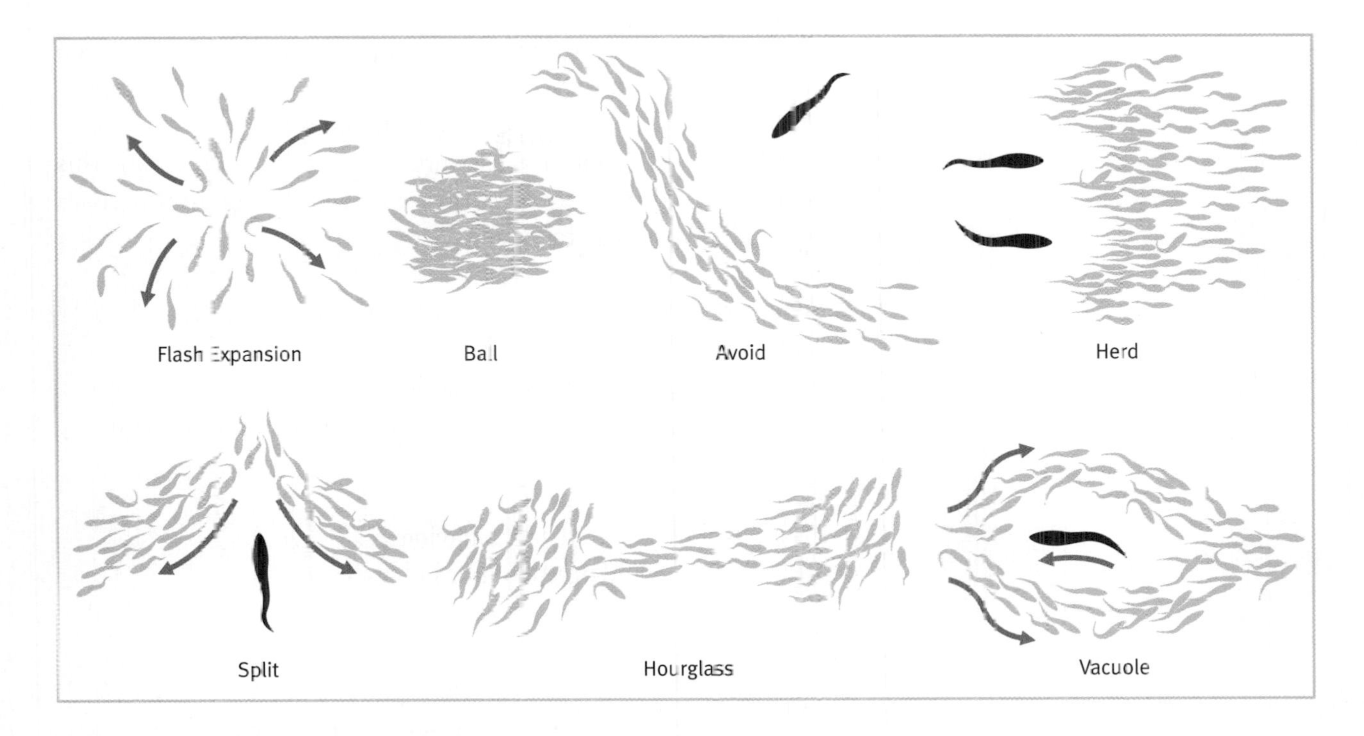

FIGURE 11.5. A myriad of antipredator defenses. Schooling fish (in green) can use a whole array of antipredator strategies. (Predators are shown in red.) *(Based on Pitcher and Wyche, 1983)*

- Trafalgar Effect—In tight groups, information about a predator spreads from individual to individual more quickly than in loose schools (Webb and Skadsen, 1980). Treherne and Foster (1981) coined this the Trafalgar Effect because it resembled the quick transfer of information through the use of battle signals by Lord Nelson's fleet at the battle of Trafalgar.
- Flash Expansion—Tightly packed schools of fish "explode," with the fish quickly swimming off in all directions. This can temporarily confuse a predator and allow for escape.
- Confusion Effect—Large groups of fish, in and of themselves, can "confuse" a predator, making it difficult to hone in on a single target and follow it (Milinski, 1979a). Schooling fish often add to this effect by moving around in very erratic patterns within groups. This effect differs from flash explosions in that the school remains a unit.
- Predator Inspection—Here, a few individuals break away from a school, approach a predator to gain information, and then return to the school, where this information may be spread across individuals (Pitcher et al., 1986; Pitcher, 1992). Later on in the chapter, we shall go into this behavior in depth.

As we shall see, a wide array of antipredator activities are seen in many species of animals. In addition to the many types of antipredator tactics animals employ, antipredator activities also interact with other behaviors we examine in chapters throughout this book—mating, foraging, cooperating, communicating. In this chapter we shall explore the following:

- Behavioral tradeoffs associated with predation
- Alarm signals
- When prey approach their predators
- Interpopulational studies, natural selection, and antipredator behavior
- The predator-prey arms race
- Plasticity and antipredator behavior
- Social learning and antipredator behavior

Behavioral Tradeoffs Associated with Predation

Ethologists who study antipredator tactics focus on "tradeoffs" more than on any specific behavior. That is, when animals spend time engaged in antipredator activity, they could, in principle, be doing something else—foraging, mating, resting, playing, and so forth (Lima and Dill, 1990; Figure 11.6). Or, rather than totally curtailing alternative behavior, antipredator behavior could be creating pressure to perform other behaviors in a different manner—for example, to forage in the vicinity of a refuge, to mate at times when predation is minimal, and so on. In either case, tradeoffs between some other behavior and antipredator tactics are often a fact of life when it comes to animal behavior.

Ethologists have performed hundreds of experiments to examine the tradeoffs between antipredator behavior and almost every other conceivable behavior. For example, Mouget and Bretagnolle (2000) found that the loud mating calls emitted by blue petrel birds (*Halobaena caerulea*) attract brown skua (*Catharacta antarctica lonnbergi*) avian predators. Conversely, blue petrels use the territorial calls of the brown skua as a cue to cease their mating vocalizations, demonstrating that the relationship between predation and other behaviors can be both fascinating and complex. Here we shall

FIGURE 11.6. Tradeoffs. When animals are being vigilant, it is often at the cost of other activities. The starlings here can't be foraging for insects while they scan the sky for hawks.

examine predation-based tradeoffs in case studies of (1) squirrels and foraging, and (2) wasps and offspring maturation.

PREDATION AND FORAGING

Predation pressure affects virtually all aspects of foraging—everything from when a forager begins feeding (Clarke, 1983; Lima, 1988a, 1988b), to when it resumes feeding after an interruption (De Laet, 1985; Hegner, 1985), where it feeds (Dill, 1983; Schneider, 1984; Ekman and Askenmo, 1984; Lima, 1985), what it eats (Hay and Fuller, 1981; Dill and Fraser, 1984; Lima and Valone, 1986), to how it handles its prey (Krebs, 1980; Valone and Lima, 1987). As an example, let us consider Lima and Valone's (1986) work on predation and foraging in the gray squirrel (*Sciurus carolinensis*).

Early work by Lima had demonstrated that squirrels alter their foraging choices as a result of predation pressure from redtailed hawks (*Buteo jamaicensis;* Lima, 1985). Squirrels who could either eat their food items where they found such items or carry the food to cover were more likely to carry items to an area of safe cover, particularly as the distance to safe cover decreased. That is, the closer the refuge from predation, the more likely they would use

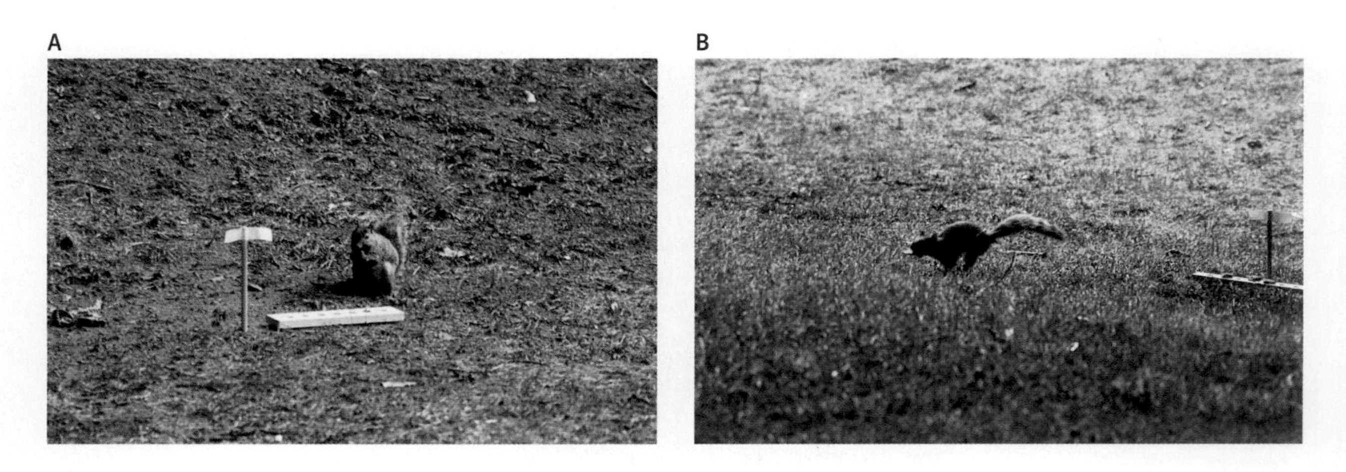

FIGURE 11.7. Squirrel balancing act. (A) When squirrels forage in open fields, they need to balance the risk of predation (a function of distance to cover) against the benefits associated with various food items. Here a squirrel is foraging at a feeding station like that used in Steve Lima's work. **(B)** A squirrel in Lima's study heads for cover with a food item (part of a cookie) in its mouth. *(Photo credits: Steven Lima)*

such a shelter when foraging; when it was a quick run to reach safety, squirrels generally chose to do so (Figure 11.7). Furthermore, squirrels were much more likely to carry larger (rather than smaller) items to safe areas before continuing to forage.

Lima and Valone followed up the above study with one in which they presented squirrels with two types of food—a large chunk of cookie (associated with long handling times) or a small chunk of cookie (associated with short handling times). Cookie chunks, rather than nuts, were used to avoid the confounding variable of food storage, as nuts are often buried, but cookies are always eaten. A combination of large and small items was placed either close (8 m) to an area of cover or farther from safety (16 m).

In order to make sense of Lima and Valone's results, we need two critical pieces of information. First, the profitability (see Chapter 10) of small food items was greater than the profitability of large food items. Thus, in the absence of predation, we would expect that squirrels would always take any small food item that they encountered. Second, the total handling time associated with larger food items was great enough that, using optimal foraging theory, Lima reasoned that larger items should be brought to cover, where it is safe, before being eaten, particularly when the distance to cover was not great. This was not the case for smaller items. Lima and Valone reasoned that if faced with predation, squirrels might sometimes pass up the smaller, more profitable food items and continue to search for larger morsels to bring back to cover. This, in fact, is just what the squirrels did. Smaller items were indeed rejected in favor of larger items that needed to be brought back to cover. In other words, predation pressure redefined which food items should be taken and which should be passed up.

PREDATION AND HATCHING TIME IN WASPS

While adults, and to some extent juveniles, have a variety of means to combat predation, antipredator options available to embryos are clearly more limited. The situation, however, is not hopeless (Werner and Gilliam, 1984; Werner, 1986, 1988), as embryos can alter the time at which they hatch, and this can act as an effective antipredator tool (Sih and Moore, 1993; Warkentin, 1995; Moore et al., 1996; Vonesh, 2000). As a case study of this phenomenon, let us examine Karen Warkentin's work on red-eyed treefrogs (*Agalychnis callidryas*) and their predators to see how it nicely demonstrates the way natural selection can produce behaviors on the part of embryos that can, in fact, lessen their risk of predation (Warkentin, 2000).

Red-eyed treefrogs are native to the forests of Panama, where Warkentin undertook her field study of the antipredator tactics available to embryonic treefrogs. These frogs attach their eggs to the various types of vegetation that hang over water, and once tadpoles hatch, they immediately drop down and take to their aquatic habitat. The terrestrial habitat of the egg and the aquatic habitat of the tadpole have a set of dangerous, but different, predators that feed on treefrogs. If terrestrial predation from snakes and wasps is weak, embryos hatch late in the season (Warkentin, 1995, 1999; Figure 11.8). Such late hatching allows the frogs to grow to a size

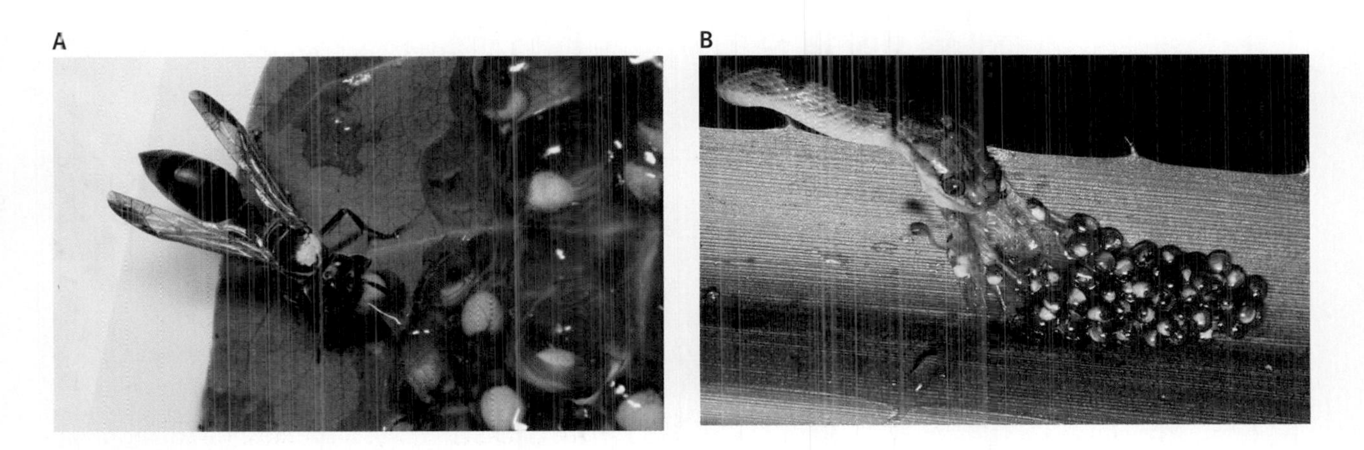

FIGURE 11.8. Predators that feed on treefrogs. Red-eyed treefrogs have numerous predators that specialize in feeding on their eggs. **(A)** Here a wasp forages on treefrog eggs. **(B)** Snakes are another dangerous predator on red-eyed treefrog eggs. *(Photo credits: Karen Warkentin)*

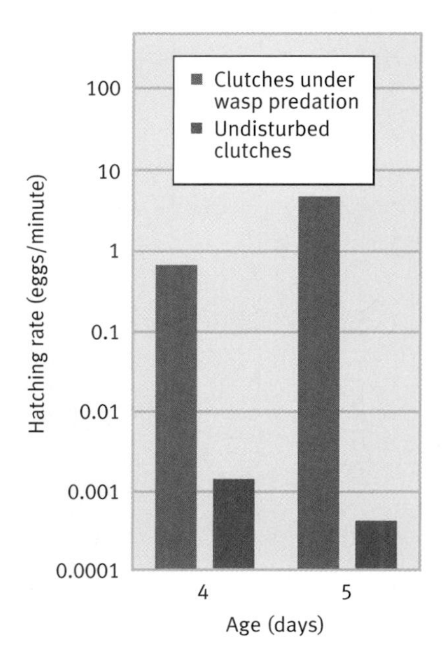

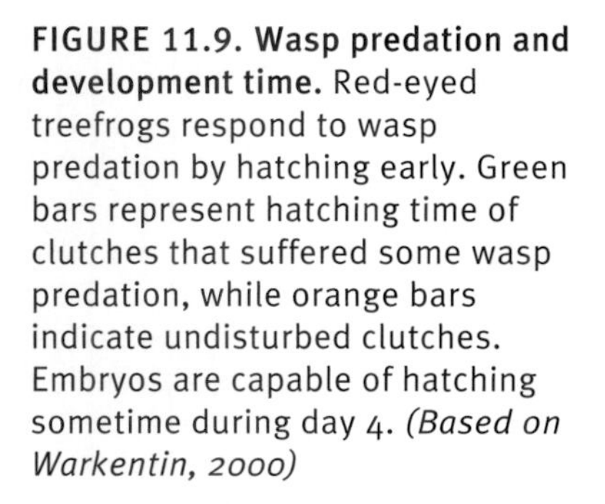

FIGURE 11.9. Wasp predation and development time. Red-eyed treefrogs respond to wasp predation by hatching early. Green bars represent hatching time of clutches that suffered some wasp predation, while orange bars indicate undisturbed clutches. Embryos are capable of hatching sometime during day 4. *(Based on Warkentin, 2000)*

that lowers the levels of fish predation once they hatch and fall into the water.

Based on mathematical models that were designed to predict development times when embryos faced a dramatically different environment than hatchlings (as in the red-eyed treefrog), Warkentin predicted that treefrog eggs would hatch sooner if predation in the terrestrial environment increased. Both snakes and wasps are terrestrial predators on treefrog eggs, with the latter taking one egg at time, while the former is capable of much more damage per attack. When predation from snakes and wasps is high, it often pays to mature early and drop into the water, away from heavy terrestrial predation.

Warkentin marked 123 egg clutches and found that less than half the eggs she kept track of survived to hatch, the majority falling prey to wasp attacks. The question of interest then is whether the eggs that survived responded in any way to the high rate of predation by wasps. Warkentin (2000) found that eggs in clutches that were not disturbed by predators often hatched at about six days. When comparing eggs from these undisturbed clutches to clutches that had already suffered some predation by wasps, hatching rates were dramatically different. Eggs hatched at a much quicker rate when their clutch had been the victim of some wasp predation, with most eggs from attacked clutches hatching at four or five days (as opposed to six; Figure 11.9). Remarkably similar results were found when examining the effect of snake predation on hatching rates (Figure 11.10). In fact, in the case of snake-induced early hatching, Warkentin found that while early hatching provided a temporal refuge from snakes (that is, developmental shifts from eggs to tadpoles stopped predation by snakes), it led to increased rates of predation by other predators on (the now relatively small) treefrog tadpoles in their aquatic environ-

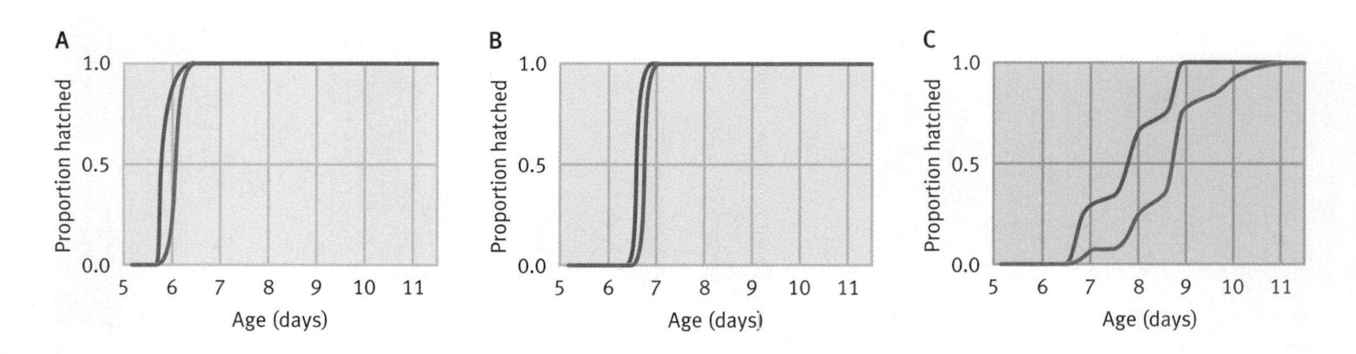

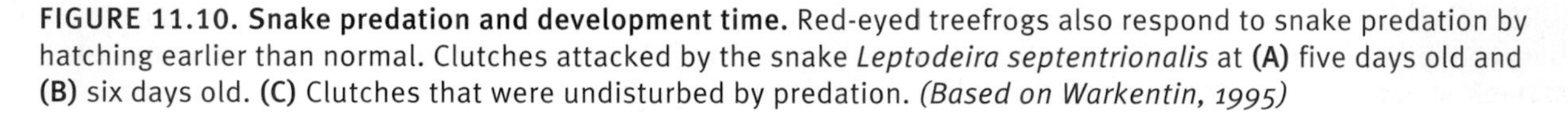

FIGURE 11.10. Snake predation and development time. Red-eyed treefrogs also respond to snake predation by hatching earlier than normal. Clutches attacked by the snake *Leptodeira septentrionalis* at **(A)** five days old and **(B)** six days old. **(C)** Clutches that were undisturbed by predation. *(Based on Warkentin, 1995)*

ment, demonstrating a clear and significant, but often complicated, impact of predation on hatching times.

Alarm Signals

Some forms of antipredator behavior can be much more dramatic than the more subtle forms we have seen above. One of the most obvious forms of antipredator activity is the alarm call, wherein some sort of signal is given after a predator is sighted (see Chapter 8).

In principle alarm calls can be employed as a means of deception (Matsuoka, 1980; Munn 1986; Møller, 1988, 1990), or as a means to manipulate others in a group (Perrins, 1968; Charnov and Krebs, 1975). For example, in some bird species, individuals may use alarm calls to quickly empty an area of potential competitors, and then forage in an unimpeded manner (Munn, 1986). Deception, however, may be the exception rather than the rule, as alarm calls most often function simply as a defense against predators.

VERVET ALARM CALLS

Alarm calling in vervet monkeys is serious business (Figure 11.11). Many researchers believe that vervets make very distinct and different calls in response to leopard, snake, and eagle predators (Struhsaker, 1967a, 1967b). While there is still some debate as to whether monkeys actually use specific alarm calls to refer to specific predators (Marshall, 1970; S. M. Smith, 1977), Seyfarth and his colleagues (1980) have found evidence that vervets respond in adaptive ways to different calls. For example, vervets run to the trees when they hear a leopard alarm call, or they hide in bushes when a fellow vervet utters an eagle alarm call (Figure 11.12). This suggests that vervets are at least using calls to refer to the *mode* of predator attack (Macedonia, 1990; Evans, Evans, and Marler, 1993).

FIGURE 11.11. Vervet monkey. Because of their complex social structure, vervet monkeys have proven a model system for examining the evolution of alarm calls. *(Photo credit: Adam Jones/ Visuals Unlimited)*

One interesting question regarding primate alarm calls concerns intentionality (Dennett, 1978, 1987; Hauser and Nelson, 1991; Bekoff and Allen, 1992): Do alarm callers actually *intend* to warn others of potential danger? Cheney and Seyfarth (1990) addressed the intentionality of alarm calling with data from vervets and Japanese macaques (*Macaca fuscata*). In support of the notion that alarm callers intend to notify those who might hear the call,

FIGURE 11.12. Vervet alarm calls. Vervet monkeys give different alarm calls depending on the predator sighted. When other monkeys hear specific alarm calls, like that for an eagle, they behave in adaptive ways (for example, hiding in bushes).

they found that lone vervets do not give alarm calls. Furthermore, vervets give more alarm calls when their offspring are in the vicinity than when the offspring of others are present, and males give more calls when a female, as opposed to a dominant individual, is nearby (Cheney and Seyfarth, 1985). Both the failure of lone individuals to call and context specificity are necessary, but not sufficient, to demonstrate intentionality.

A strong test of intentionality would examine whether the state of those potentially being warned affected the alarm caller's behavior. For example, if alarm calls are intended to warn others in a group, such calls need not be emitted if the caller recognizes that its groupmates are already aware of any putative dangers. In a similar vein, once others in the group hear an alarm call, the individual emitting it should stop. Field data indicate that vervets continue to give alarm calls after everyone in the group has heard them (Cheney and Seyfarth, 1981, 1985). To test whether this indicates a lack of intentionality, Cheney and Seyfarth (1990) ran a controlled experiment on alarm calls in a laboratory colony of Japanese macaques. In this experiment, mother-daughter pairs were tested in two treatments, "knowledgeable" and "ignorant." In the knowledgeable treatment, mother and daughter sat next to each other and could both see a "predator" (in this case, a human with a net). In the ignorant treatment, only the mother could see the predator. If mothers intended to inform their daughters of danger, then alarm calls and other protective behaviors should have been greater in the ignorant treatment. This was not the case, and the evidence did not support the notion that mothers intended to warn offspring.

TAIL FLAGGING

In ungulates, individuals are known to "flag" their tails after a predator has been sighted (Figure 11.13). Such flagging is part of a sequence of antipredator behaviors, and often involves an individual lifting its tail and "flashing" a conspicuous white rump patch. Flagging often, but not always, occurs when a predator is at a relatively safe distance from its potential prey (Hirth and McCullough, 1977). Tail flagging has been postulated to: (1) warn conspecifics (kin and nonkin) of potential dangers (Estes and Goddard, 1967); (2) "close ranks" and tighten group cohesion (McCullough, 1969; Kitchen, 1972; P. S. Smith, 1991), perhaps to ensure group-related foraging and antipredator benefits in the future (Trivers, 1971; R. J. F. Smith, 1986); (3) announce to the predator that it has been sighted and should therefore abandon any attack (Woodland et al., 1980; Caro, 1995a; Caro et al., 1995); (4) entice the predator to attack from a distance that is likely to result in an aborted attempt (Smythe, 1977); (5) cause other group members to respond, thereby confusing the predator, and making the flagger itself less likely to be the victim of an attack (Charnov and Krebs, 1975); or (6) serve as a sign for appeasing dominants, and only secondarily play a role in antipredator behavior (Guthrie, 1971).

FIGURE 11.13. Tail-raising event. In white-tailed deer, individuals often display a white patch while running from danger. At least a half-dozen hypotheses have been put forth to explain the function of the patch. *(Photo credit: Gary Carter/ Visuals Unlimited)*

Hirth and McCullough (1977) and Smith (1991) found support for the hypothesis that tail flagging may cause an increase in group cohesion; that is, one function of tail flagging appears to be to keep group size large. Larger group size in and of itself dilutes the probability that any given deer will be taken when a predator attacks. As such, tail flaggers benefit themselves, as well as others in the group, by their behavior.

The story with respect to tail flagging in white-tailed deer is even more complex. When Bildstein (1983) also examined tail flagging in this species, he could not rule out the cohesion hypothesis, but he argued that tail flagging was also directed at the predator, not others in the group, and was probably a signal to the predator that an attack was not likely to succeed. This "pursuit-deterrence" argument (Woodland et al., 1980) is supported by Tim Caro and his colleagues' (1995) work on white-tailed deer. They found no evidence for a time cost to flagging (LaGory, 1987), and no evidence that tail flagging was aimed at conspecifics in the context of cohesion. Rather, they showed that white-tailed deer that run fast flick their tails and are using this signal to communicate to the predator that an attack would probably be unsuccessful.

Prey Approaching Their Predators

Although it may appear paradoxical at first, animals sometimes approach predators when they initially encounter them. This sort of behavior has been extensively documented for vertebrates, particularly in fish (Pitcher et al., 1986; Dugatkin and Godin, 1992a, 1992b), birds (Altmann, 1956; Curio, 1978; Curio et al., 1978a), and mammals (Walther, 1969; Kruuk, 1972; Cheney and Seyfarth, 1990). The phenomenon of prey approaching predators has been referred to as boldness (Huntingford, 1976), investigative behavior (Loughry, 1987), predator inspection behavior (Pitcher, 1986; Pitcher et al., 1986), predator harassment (Hennessy and Owings, 1978; Curio and Regelmann, 1986), and mobbing (Altmann, 1956; Owings and Coss, 1977; Curio, 1978), depending mainly on its outcome and presumed functions(s).

Among vertebrates, the dynamics of approach toward a potential predator are quite similar. Prey typically approach a putative predator from a distance in a tentative, jerky manner. The approach is characterized by a series of moves toward the predator interrupted by stationary pauses and sometimes alternating with moves away from the predator (Curio et al. 1983; Pitcher et al., 1986; Milinski, 1987; Dugatkin, 1997a). In birds and mammals in particular, the prey may emit alarm signals or exhibit distraction or threat displays (Dominey, 1983; Donaldson, 1984; Brunton, 1990) during an approach toward the predator. The approach may culminate in a number of possible outcomes along a continuum, ranging from the prey simply retreating, perhaps to rejoin a social group of conspecifics nearby, to an escalation of prey attacking the predator. Approaches with the latter outcome are generally referred to as predator harassment or mobbing.

For this sort of approach behavior to evolve and to be maintained in a population, the average fitness benefits accrued must equal or exceed the associated average fitness costs. Therefore, a resolution of the apparent paradox of prey approaching their predators should be based on a consideration of this behavior's fitness-associated benefits and costs to individual prey.

Given this, we shall focus on outlining the possible costs and benefits to individual prey animals when approaching predators. (A similar analysis could be made of fleeing from predators; Ydenberg and Dill, 1986.)

COSTS AND BENEFITS OF THOMSON'S GAZELLES APPROACHING A PREDATOR

An individual prey presumably incurs an increased risk of mortality by approaching a potential predator at a distance, and particularly by rendering itself more conspicuous during its approach by alarm signaling or threat/distraction displaying. A systematic investigation of this potential direct fitness cost has been difficult, mainly because predator attacks on approaching prey are infrequently observed in nature. Nevertheless, numerous anecdotal field observations strongly indicate that this cost is real (but see Hennessy, 1986).

Using natural populations of Thomson's gazelles (Figure 11.14) as her experimental subjects, Claire Fitzgibbon has undertaken one of the most thorough analyses of the costs and benefits of approach behavior (Fitzgibbon, 1994). In the Serengeti National Park (Tanzania), gazelles live in groups that can vary from fairly small (<10 individuals) to fairly large (>500) and interact with four main predators: lions, cheetahs, spotted hyenas, and wild dogs. While we shall focus on the predators that gazelles in fact do

FIGURE 11.14. Gazelle antipredator behavior. (A) Gazelles are constantly vigilant for potential predators. **(B)** Many different species, including the cheetah (pictured here chasing a gazelle) hunt gazelles. *(Photo credits: Gerald and Buff Corsi/Visuals Unlimited; Joe McDonald/Visuals Unlimited)*

approach (lions and cheetahs), it is interesting to note that gazelles rarely approach wild dogs and hyenas. This difference appears to be primarily due to the fact that cheetahs and lions rely on surprise and short, fast chases, while wild dogs and hyenas rely on stamina. Approaching the former group may be wise, while approaching the latter is often a surefire invitation to a hunt.

With respects to the benefits of approaching a predator, Fitzgibbon speculated on three possible scenarios:

- Approach behavior might actually *decrease* the current risk of predation (Figure 11.15).
- Approach behavior might allow gazelles to gather information about a potential threat.
- Approach behavior might serve to warn other group members of the potential danger associated with predators.

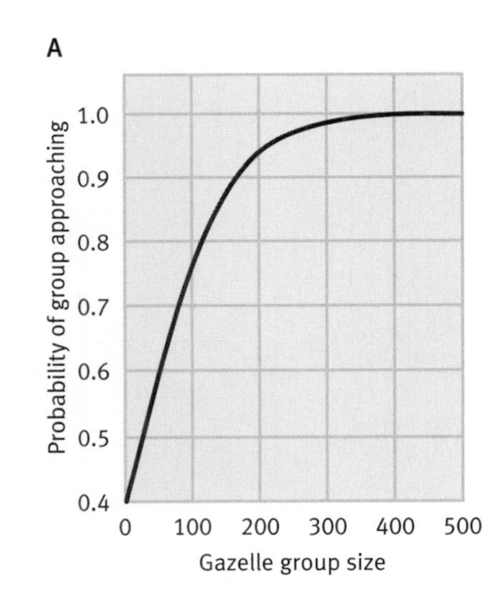

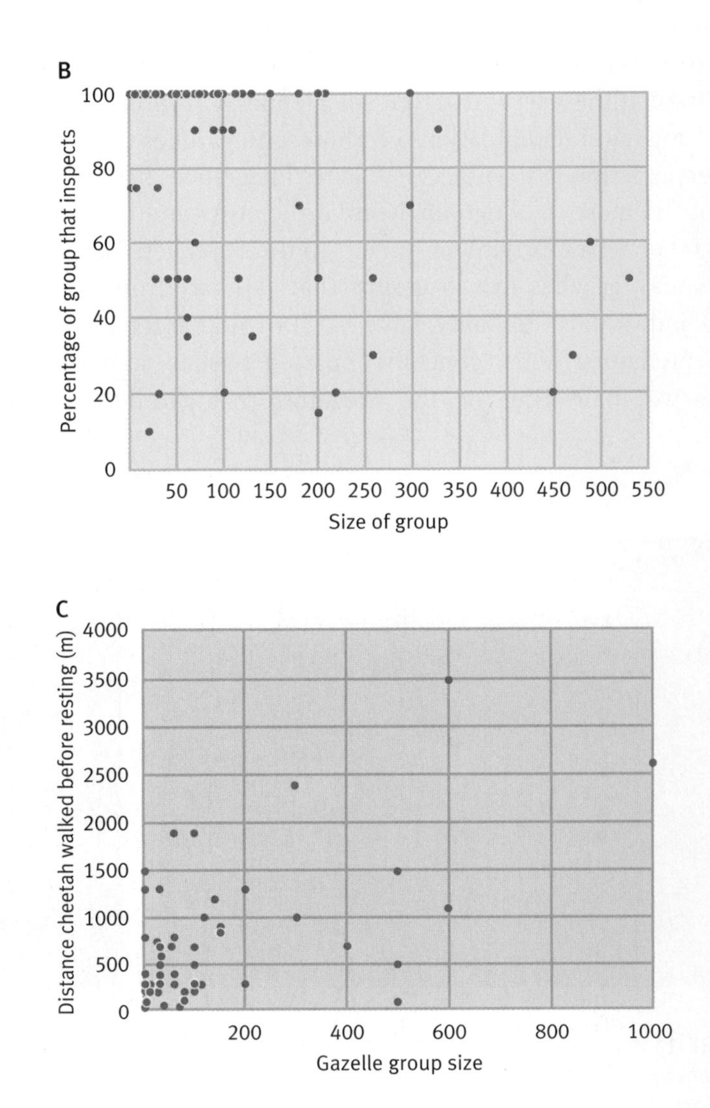

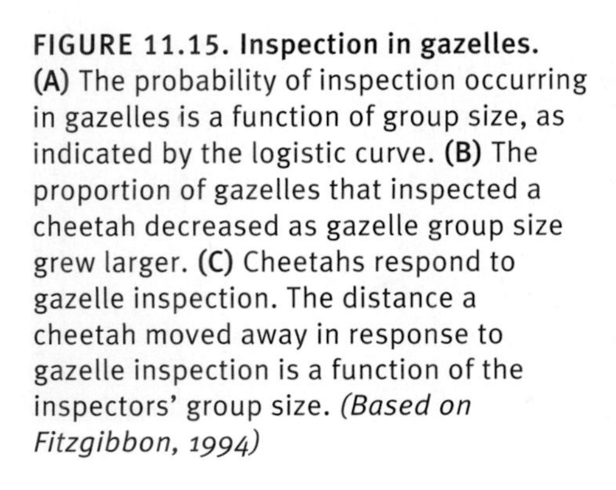

FIGURE 11.15. Inspection in gazelles. **(A)** The probability of inspection occurring in gazelles is a function of group size, as indicated by the logistic curve. **(B)** The proportion of gazelles that inspected a cheetah decreased as gazelle group size grew larger. **(C)** Cheetahs respond to gazelle inspection. The distance a cheetah moved away in response to gazelle inspection is a function of the inspectors' group size. *(Based on Fitzgibbon, 1994)*

In the course of her two-year field study, Fitzgibbon found some evidence for benefit 1. In particular she found that cheetahs responded to gazelle inspection behavior, which is most common in large gazelle groups (Figure 11.15A and B), by moving further between rest periods and between hunting periods. This in turn could cause cheetahs to leave a particular area sooner than normal as a result of gazelle approach behavior, leading to decreased rates of mortality among potential prey (Figure 11.15C).

The cost of gazelle approach behavior is manifested primarily in terms of lost time/energy and increased risk of predation. Gazelles actually spend approximately 4 percent of their waking hours involved in approach behavior. This 4 percent could otherwise be devoted to other activities (foraging, mating, resting) and thus represents a real "opportunity cost" to the animals. In terms of more direct costs, while the odds of an approaching adult being killed by a cheetah are very low (on the order of 1 in 5,000), the probability of younger individuals being taken during an approach are an order of magnitude greater (about 1 in 400).

Interpopulational Differences

One way to examine how strongly natural selection has operated on antipredator behavior is by comparing the behavior of individuals that live in populations that differ in terms of the predation pressure they suffer. The between-population approach generally works as follows: Find two (or more) populations of the same species that live in environments that are similar but differ in one significant way—predation. In population 1, we might find the species we are studying under strong predation pressure, while in population 2, we might have individuals under little, perhaps even no, such predation pressure. We would then look for differences in the antipredator behaviors across these populations. If such differences were uncovered, it would suggest that selection had been operating on antipredator behaviors, and that this might produce heritable differences across our populations. Of course, it is always possible that such between-population differences could be due to differences in individual experiences per se, and hence in learning, but even then we would need to examine the possibility that populations differ in their ability to learn, which itself could be under selection pressure.

The population comparison method for studying selection and antipredator behavior has been employed in many animal systems, including (but not limited to) ground squirrels (Owings and Coss, 1977), guppies (Magurran et al., 1995; Houde, 1997), sticklebacks (Giles and Huntingford, 1984; Bell and Foster, 1994), and minnows (Magurran and Pitcher, 1987; Magurran, 1990).

INTERPOPULATIONAL DIFFERENCES IN ANTIPREDATOR BEHAVIOR IN MINNOWS

FIGURE 11.16. Interpopulational differences in minnows. Minnows from high-predation and low-predation areas respond very differently when confronted with a potential predator. For example, Gwynedd minnows (those from the low-predation site) totally ceased foraging and eating in the presence of a predator. In contrast, Dorset minnows (those from the high-predation site), who are used to foraging in the face of danger, curtailed their foraging activity, but not nearly to the extent of Gwynedd minnows. *(Photo credit: Johnny Jensen/Visuals Unlimited)*

Anne Magurran and her colleagues have examined antipredator behavior in two different populations of minnows *(Phoxinus phoxinus)*. Minnows from the Dorset area of southern England and the Gwynedd area of northern Wales were chosen (Figure 11.16), as the Dorset population is under strong predation pressure from pike predators, while pike are absent from the Gwynedd population of minnows (and have apparently never colonized this area). The basic protocol employed in Magurran and Pitcher's (1987) study was to test to see if antipredator behavior across these populations differed as a function of their different predation pressures.

While fish from southern England and northern Wales are all members of the same species, their antipredator repertoires are quite different. To begin with, in the laboratory, before exposure to a predator, Dorset minnows (those from high-pike-predation areas) were found to swim around in larger groups than Gwynedd minnows. Individuals in such large groups often have greater safety from predators. Dorset minnows also seemed to have more stable groups, with less movement of individuals from group to group than the Gwynedd minnows. Once a predator was added to the protocol, both minnow populations dramatically increased their group sizes. What is particularly interesting, however, is that once the predator was removed, it took the Gwynedd minnows significantly longer to adjust their group size back to normal (Figure 11.17). That is, not only did the high-predation Dorset minnows have generally stronger antipredator responses, but they were also quicker to respond to the removal of danger by resuming normal, nonpredator-based activities.

Returning to the foraging/predation tradeoff we discussed earlier, Gwynedd minnows completely ceased eating once a predator was presented, while Dorset minnows, who are accustomed to foraging in the face of danger, curtailed their foraging activity, but not nearly to the extent of Gwynedd fish. With re-

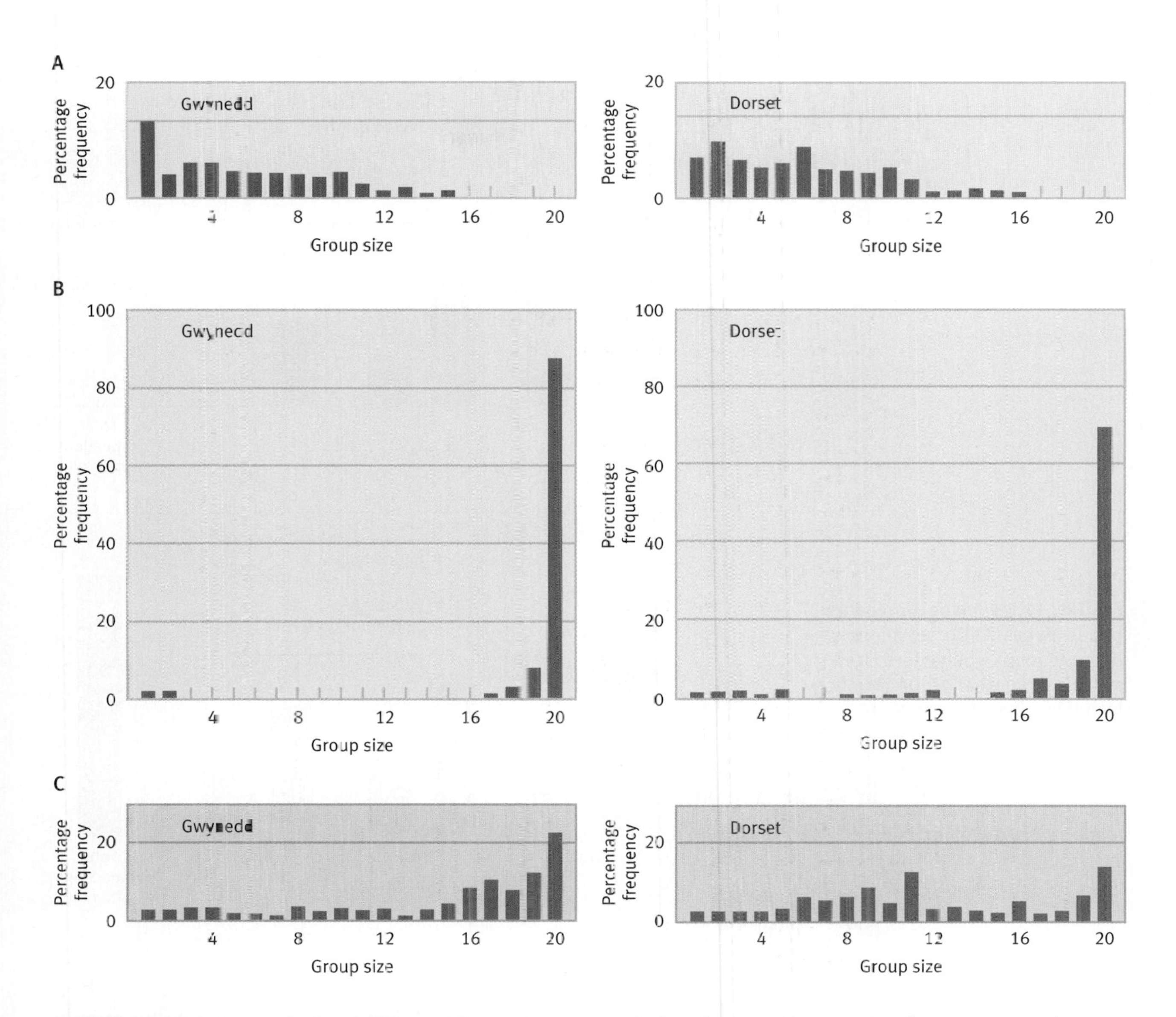

FIGURE 11.17. Interpopulational difference in response to predation. Dorset minnows come from an area with many predators, including pike, while Gwynedd minnows come from an area that does not have pike and hence they do not have experience with pike predators. **(A)** The two graphs indicate group size in Gwynedd and Dorset minnows before exposure to a pike predator. Dorset minnows normally swim in larger (and more stable) groups than do Gwynedd minnows. **(B)** The two graphs indicate group size in Gwynedd and Dorset minnows two hours after exposure to a predator. Both populations responded strongly to the presence of the pike predator. **(C)** The two graphs indicate group size in Gwynedd and Dorset minnows one day after exposure to the pike predator. Dorset minnows were quicker to return to normal group size than were Gwynedd minnows. *(Based on Magurran and Pitcher, 1987)*

spect to inspection behavior, Dorset minnows inspected more often than did Gwynedd minnows, but they were also much more likely to stop inspecting if a conspecific was eaten by a pike (Figure 11.18).

Are these discrepancies between the antipredator behavior of the Dorset population, which lives with pike, and the Gwynedd

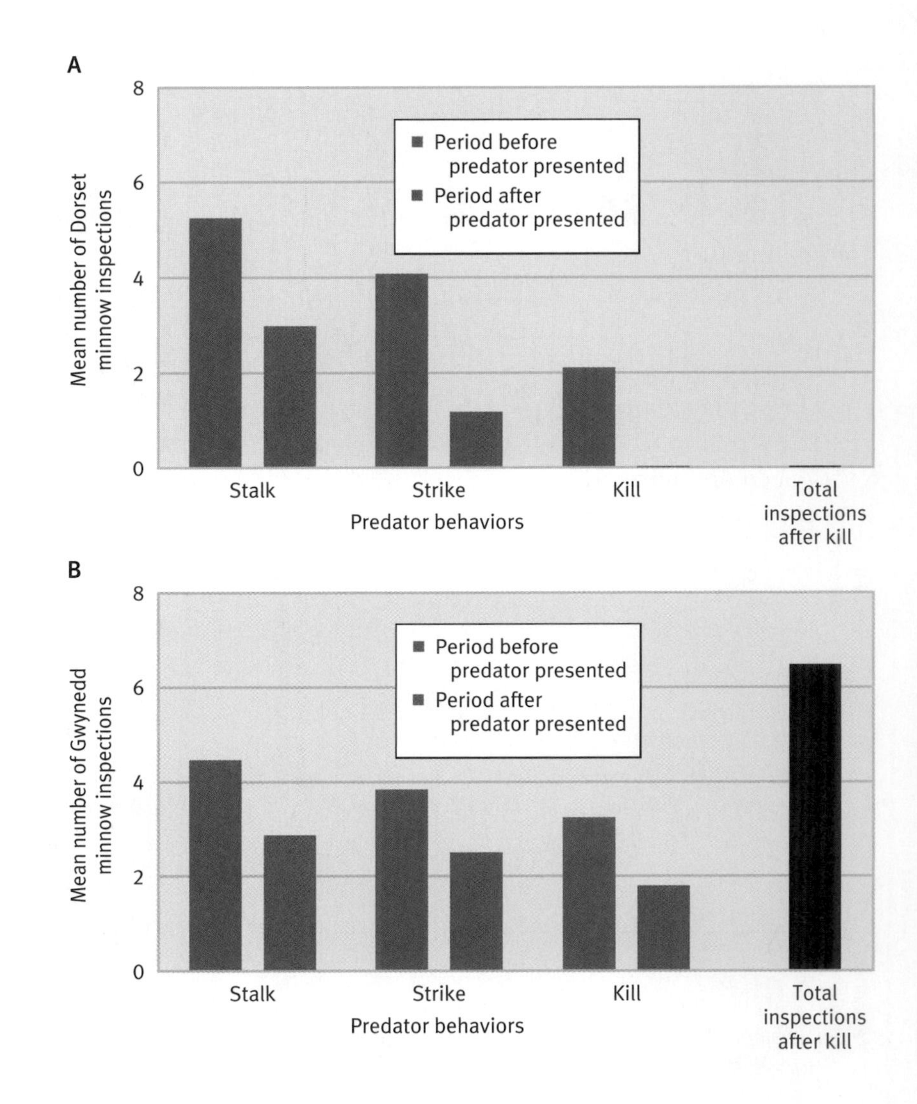

FIGURE 11.18. Inspection across populations. (A) Dorset minnows respond to pike behaviors (stalk, strike, kill another group member) by decreasing their predator inspection behavior, while **(B)** no statistically significant decrease in inspections occurs in the Gwynedd minnow population. *(Based on Magurran and Pitcher, 1987)*

population, which does not, a result of selection operating differently across these groups? To address this question, adult minnows from Gwynedd and Dorset populations were captured in the wild, and their offspring were raised in the laboratory (Magurran, 1990). Four treatments were undertaken. In treatments 1 and 2, two-month-old offspring from Gwynedd and Dorset minnows were exposed to a "model" pike predator (constructed from wood, but made to be a realistic representation of a predator) five times during development, and their antipredator behaviors were recorded. In treatments 3 and 4, rearing conditions were identical, except that no minnows in these treatments were exposed to the model predator.

When comparing the antipredator behavior of two-month-old fish exposed to predators, Magurran found that young Dorset minnows differed from young Gwynedd minnows in the same manner uncovered in her earlier work (using wild fish). Dorset minnows inspected more often and tended to be found in larger groups (Figure 11.19), suggesting that natural selection has produced "hardwired" differences in antipredator behavior across these two populations. The real story, however, is more complex, and hence more interesting.

After raising these minnows in the laboratory for a total of two years, Magurran examined the antipredator tactics of minnows in these four groups. The differences found in two-month-old Dorset and Gwynedd populations were also found in two-year-old-fish, but experience also played a key role. That is, while all Dorset fish (experienced and inexperienced) inspected more and in larger groups than all Gwynedd minnows, there were also interesting between-population differences in the role that experience played in shaping antipredator behavior. In both populations of minnows, experienced adults inspected more often than inexperienced adults, but this difference was most pronounced in Dorset fish. In addition, Dorset fish tended to be found in larger groups than Gwynedd fish, but again experience played a role, with experienced Dorset fish schooling in larger groups than inexperienced individuals from this population. No such experience effect for group size was found in Gwynedd fish.

Natural selection seems to have operated in one, and perhaps two, ways on antipredator behavior in minnows. First, as a result of differences in predation pressure, Dorset and Gwynedd fish have genetically based differences in the antipredator tactics they employ. Second, it may be that natural selection has actually selected

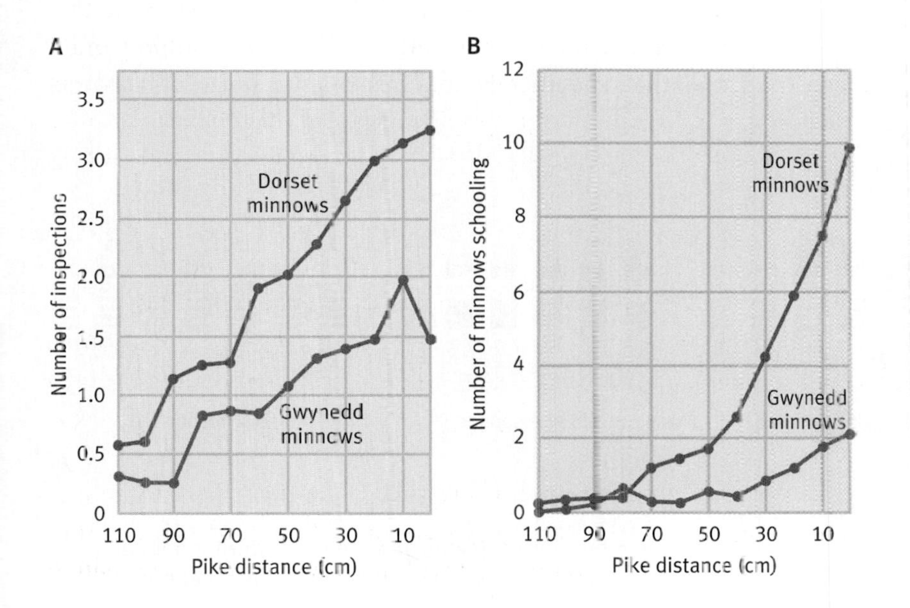

FIGURE 11.19. Reaction to predators. (A) Population differences in inspection behavior were found when comparing two-month-old minnows, who had never before been exposed to a predator, from high-predation (Dorset) and low-predation (Gwynedd) sites. **(B)** Population differences in schooling behavior were also found in two-month-old inexperienced fish originating from high-predation (Dorset) and low-predation (Gwynedd) sites. *(Based on Magurran, 1990)*

for stronger responses based on experience in the Dorset population of minnows from high-predation areas.

PREDATOR VS. PREY ARMS RACES

So far we have essentially been focusing on the antipredator activities of prey but have not discussed the behavior of the predators themselves in much detail. Clearly, if selection operates on prey to avoid predators, then selection should also act on predators to overcome whatever antipredator behavior their prey have evolved. One good example of how this might operate is the so-called predator-prey "arms race." The concept of an arms race became common parlance during the Cold War between the United States and the former Soviet Union (Figure 11.20). At its core, the idea of the Cold War arms race was that each time one side increased its defensive or offensive capabilities, the other side needed to respond, or else it would be at a significant disadvantage. In the end, this produced a spiraling effect that made each side more and more dangerous to the other.

Jerison (1973) has applied this concept of an arms race to mammalian predator-prey interactions over evolutionary time (Vermej, 1987; Ridley, 1996). When one looks at the fossil record at any given point in time, it appears that carnivore predators normally possess larger brains than the herbivore prey that they hunt. It may be that hunting is a complicated task that simply requires much in the way of brain power. Conversely, though, one could argue that avoiding being eaten is also a complicated task, and so perhaps prey should have larger brains. While we can't settle this debate, we can focus on another aspect of predator and prey brain size, and one that specifically touches on the issue of an arms race.

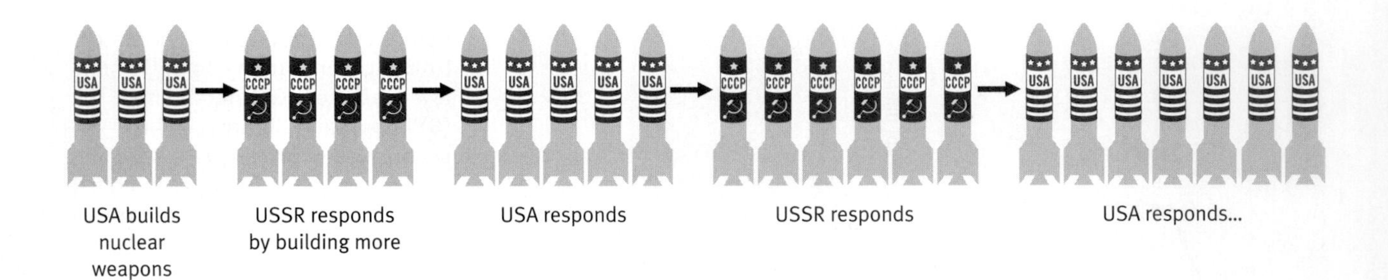

FIGURE 11.20. Arms race. A classic arms race occurred between the United States and the USSR during the Cold War. Each side responded to the increase in the other's nuclear arsenal by increasing its own nuclear weapons. A similar type of phenomenon occurs in predator-prey arms races.

From a historical perspective, it has been found that brain size in both predators and prey has increased through time (Figure 11.21). There are, of course, numerous possible reasons why brain size in both groups has increased over time. One interesting hypothesis comes from Jerison (1973), who interprets this pattern as evidence of a predator-prey arms race (where the weapon here is brain power). Jerison's argument goes like this: If at time 1 predators have a larger brain size than prey, then selection should favor larger brain size in prey. Such selection on prey brain size should not only produce prey with larger brain sizes, it should also cause additional selection on predators to maintain their initial brain size advantage. As a result, we might see natural selection favoring even larger brains in predators, which again creates selection pressure on brain size in prey. Through time, we see the spiraling up in brain size that one might expect if the arms race analogy was a good one (Vermej, 1987).

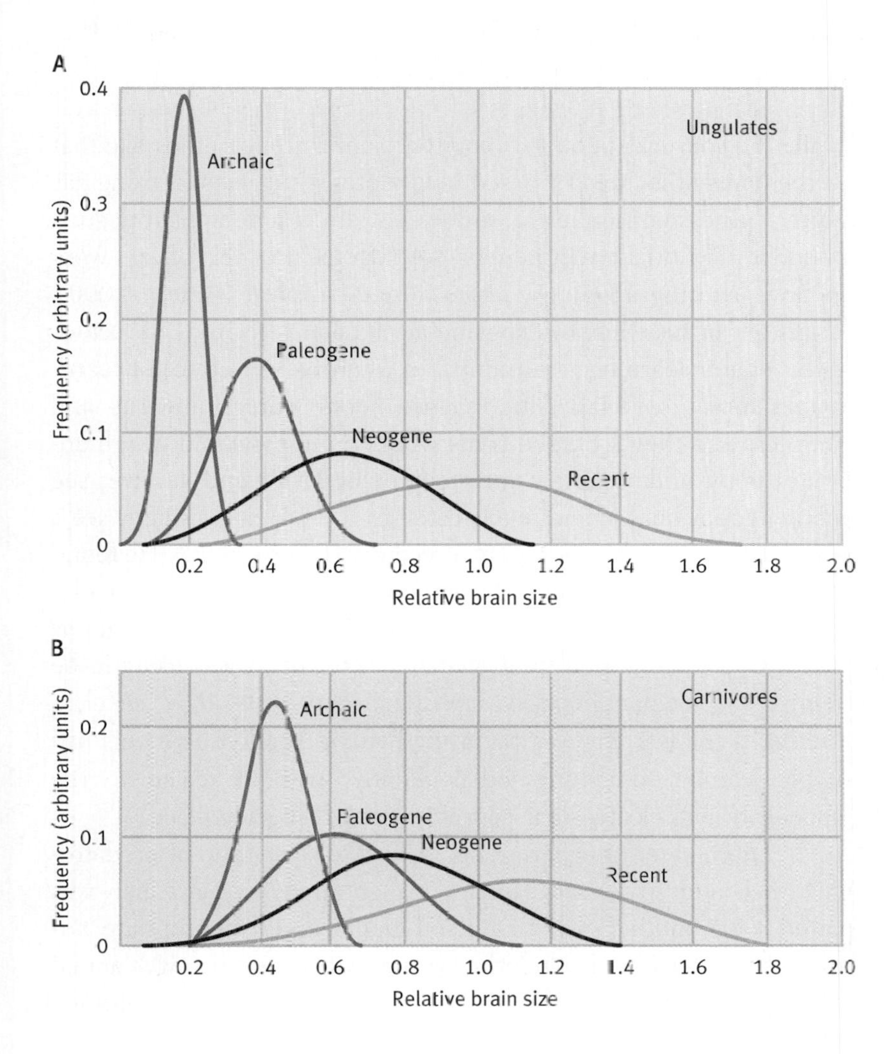

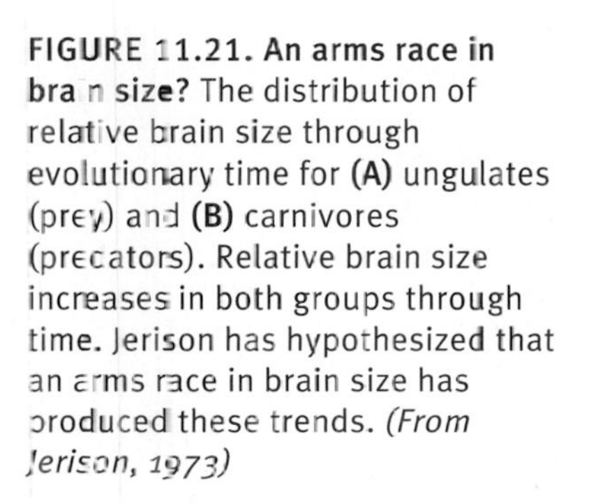
FIGURE 11.21. An arms race in brain size? The distribution of relative brain size through evolutionary time for **(A)** ungulates (prey) and **(B)** carnivores (predators). Relative brain size increases in both groups through time. Jerison has hypothesized that an arms race in brain size has produced these trends. *(From Jerison, 1973)*

Learning and Antipredator Behavior

In Chapter 4, we argued that learning could be conceptualized as one type of phenotypic plasticity that animals have in their behavioral arsenal. Recall from Chapter 4, for example, the bryozoan species, *Membranipora membranacea* (Harvell, 1998), which remains spineless when living in predator-free environments, but grows spines relatively quickly when exposed to predatory cues (Harvell, 1994; Tollrian and Harvell, 1998). Here, we shall examine how learning—a form of behavioral plasticity—provides prey with the ability to recognize what constitutes a predator, as well as looking at direct evidence for some of the precise fitness benefits associated with learning and antipredator behavior.

THE DIRECT FITNESS CONSEQUENCES OF LEARNING ABOUT PREDATORS

While the sort of data on learning that comes from studies like that on the damselflies we discussed in Chapter 4 suggests learning has a direct fitness consequence, it does not directly demonstrate such a benefit. To find a direct benefit, we turn to Michael Alfieri's work on how learning about predators affects survival (Alfieri, 2000). Alfieri began his work by exposing adult (male) guppies to various "antipredator learning treatments" to examine the effect of learning on direct survival. Alfieri's treatments were numerous and complex, and they included trials where a guppy would watch another guppy undertake an antipredator behavior and survive, see another guppy chased and finally eaten by a predator, or simply see a guppy dropped into the mouth of a predator (Figure 11.22). He found that the type of information that an observer guppy obtained had a significant impact on its survival. Males that saw another guppy chased and finally eaten by a predator were much less likely to be eaten by a predator themselves in subsequent tests. It is not clear whether social learning was occurring here or whether observers had simply learned something independently (without regard to the model fish), but clearly some sort of learning had taken place.

In another set of experiments, Alfieri took a group of predator-naive fish and split them up. In one group, each naive fish was paired with another fish that itself had learned appropriate antipredator behaviors. In the other group, each naive fish was paired with a guppy like itself—that is, an individual with no predator ex-

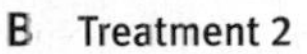

FIGURE 11.22. Learning about danger. Michael Alfieri examined the role of learning in guppies during antipredator activities. **(A)** In one treatment, a guppy simply observed a predator from behind a partition, **(B)** in a second treatment, an observer saw the predator eat a guppy, and **(C)** in a third treatment the observer saw another guppy returning safely. The observer's behavior was then tested in the presence of a predator.

perience. He found that fish that were paired with experienced partners survived interactions with a predator with the same probability as their partners, while naive fish paired with inexperienced partners suffered much higher mortality rates (Figure 11.23). That is, not only does personal experience have a direct effect on survival in the face of a predator, but the experience level of one's partner does as well.

Exactly how naive fish benefit from their partner's experience is not yet known. That being said, Alfieri took the next logical step and asked whether guppies preferred to associate with partners that had learned how to avoid predators (see also Magurran et al.,

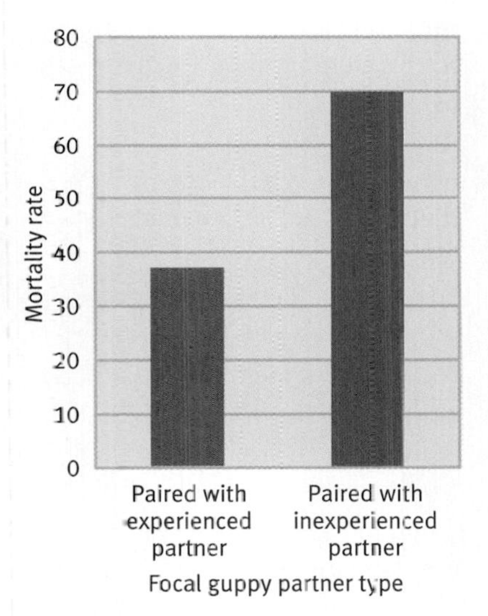

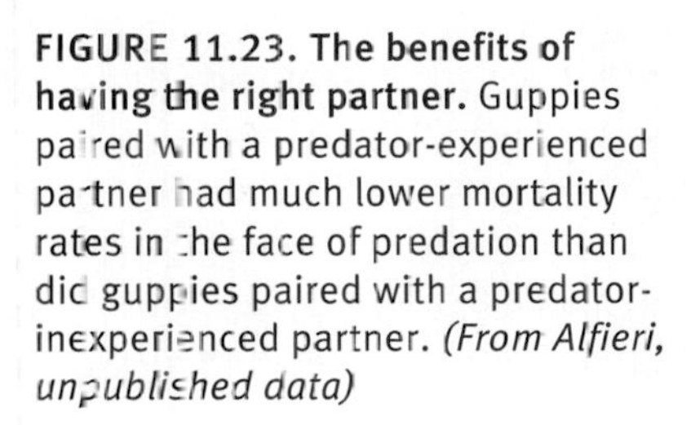

FIGURE 11.23. The benefits of having the right partner. Guppies paired with a predator-experienced partner had much lower mortality rates in the face of predation than did guppies paired with a predator-inexperienced partner. *(From Alfieri, unpublished data)*

Interview with Dr. MANFRED MILINSKI

You've made fundamental contributions to the study of cooperation, foraging, and mate choice. What drew you to study predation?
The animal behavior group that I joined quite early as a student at Bochum University in Germany studied the function of mobbing behavior of birds. I found this behavior fascinating if not paradoxical: a small bird that has detected a predator of its own, e.g., an owl, approaches it as closely as 30 cm, flipping its tail and wings and giving noisy calls. How can the little bird profit from this strange behavior?

A good deal of your work has used the stickleback as a model system. Why?
I decided to focus on three-spined sticklebacks for several reasons: (1) they were plentiful in the ponds around my university, (2) much of their behavior was known via Tinbergen's classical papers, (3) they are found everywhere in the Northern Hemisphere, from Canada to Switzerland, under a broad range of ecological conditions, so that many questions can potentially be answered with them (no expensive and adventurous fieldwork in nice parts of the world is necessary, which I regret now and then), (4) they are easy to breed in large numbers in the lab, (5) they are only a few centimeters long, and thus do not need much lab space, (6) they have spines which are ideal for individual marking, (7) they become quickly accustomed to tanks also when kept singly, and (8) they are extremely interesting animals—it is still a pleasure to work with them.

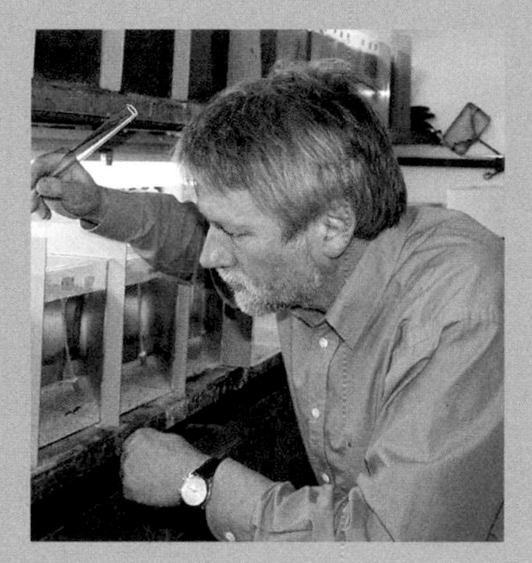

Interpopulational comparisons have proved particularly useful in terms of understanding the evolution of antipredator behavior. Why is that?
This is a kind of natural experiment: there are populations which have a specific predator and others in which it has been missing for many generations. We expect that predation risk has selected for specific antipredator behavior and can test for it with offspring from high-predation populations and use the offspring from predator-free populations as controls. Ideally a population is the statistical unit, so you need several populations of either type. Especially convincing are cases in which a predator has been introduced recently, and you can show that the antipredator behavior evolves under your eyes.

Initially a great deal of work on predation focused on foraging/predation tradeoffs. How would you summarize the findings from these studies?
We expect that animals optimally balance their needs, both of finding and selecting food efficiently and of avoiding being eaten themselves. We can demonstrate this balance by increasing one need while keeping the other constant, e.g., hungrier animals "pay" more attention to their foraging task at the expense of vigilance for predators, whereas under increased predation risk they watch out for predators and slow down foraging or simply hide. Many studies have shown that animals are surprisingly good in finding almost the optimal tradeoff.

Most animals are both predator and prey. That being said, some animals are primarily predators and others are primarily prey. Is there some

fundamental difference between these last two groups?

The famous "life-dinner principle" comes to my mind: a fox that is chasing a rabbit runs for its dinner, whereas the rabbit runs for its life. This means that the fox may give up if the task appears to be too demanding and wait for an easier prey; the rabbit, however, should invest all its reserves to escape. This suggests that selection is stronger on prey to escape a predator than on predators to subdue a prey. However, a predator has to catch prey on a regular basis to avoid starving, whereas a prey animal might not meet a predator for long periods of time, or may just be lucky because shoal mates have been selected by the predator. So it's difficult to predict who should be ahead in the arms race.

It's been suggested that predation has favored an "arms race" in brain size between predators and prey. Where do you come down on this question?

This idea is quite attractive. There are many examples showing that both predators and prey are limited in their ability to process information by unit time while hunting prey or escaping a predator. They suffer from information overload because the "channel capacity" of their brain is too limited. A larger brain would indeed be advantageous for either side.

WE EXPECT THAT ANIMALS OPTIMALLY BALANCE THEIR NEEDS, BOTH OF FINDING AND SELECTING FOOD EFFICIENTLY AND OF AVOIDING BEING EATEN THEMSELVES.

More fieldwork has been undertaken on predation than is the case for many other areas in animal behavior. How has that helped us better understand predation pressure?

A well-designed lab experiment can prove that a specific cause-effect relationship exists, e.g., you may find that a prey animal can detect a predator by smell, all other influences being carefully excluded. However, only under natural conditions can one find out how important this relationship is; odor cues might be much less important than visual cues (or vice versa).

DR. MANFRED MILINSKI is a professor at the Max Planck Institute in Germany. He has made fundamental contributions to the areas of antipredator behavior, sexual selection, and the evolution of cooperation.

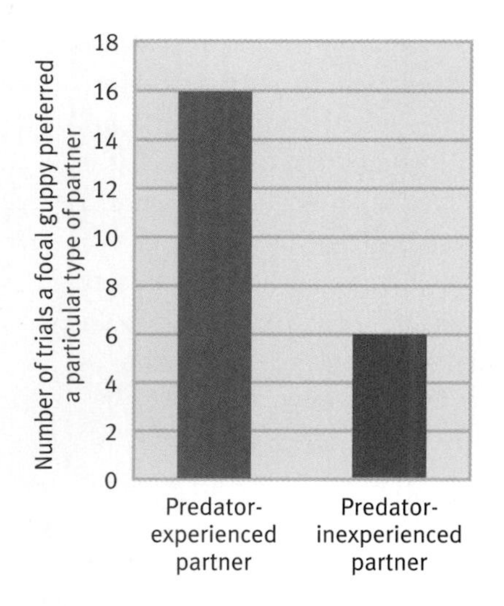

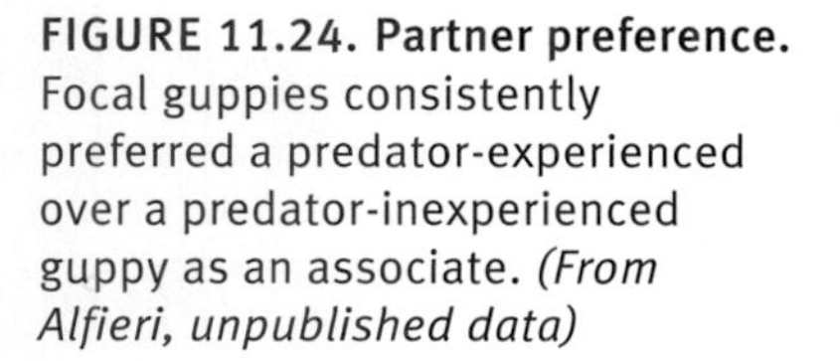

FIGURE 11.24. Partner preference. Focal guppies consistently preferred a predator-experienced over a predator-inexperienced guppy as an associate. *(From Alfieri, unpublished data)*

1994; Griffiths and Magurran, 1997, 1999). To test this, a guppy was given the choice of spending its time near one of two other fish—one inexperienced, one experienced. Results showed clearly that guppies, who had never interacted with either the experienced or inexperienced potential partners before a choice test, demonstrated a clear preference for partners that had learned how to avoid being eaten. Presumably such a choice increased the chooser's rate of survival if a predator suddenly appeared (Figure 11.24).

Social Learning and Antipredator Behavior

Motmots, a tropical bird species, instinctively fear poisonous coral snakes. The particular coral snakes that are dangerous to motmots have a specific color pattern—red and yellow bands. When baby motmot chicks are presented with a wooden dowel with red and yellow bands painted on it, the chicks *instantly* fear it. But if green and blue bands or even red and yellow stripes—neither of which resemble snakes dangerous to the motmot—are painted on a dowel, motmot young no longer treat it like a danger (S. M. Smith, 1975; J. N. M. Smith, 1977).

The motmot solution to knowing who the enemy is—hardwiring the answer into the genetic code—works well under certain conditions, namely when the predatory species involved are few and constant through time. Yet, if there are lots of predators to handle and/or if the kinds of predators are constantly changing, innate fears may be an inadequate or inappropriate solution to the "know your enemy" problem. Under such conditions, it might pay to learn who the enemy is by observing how others respond to potential threats. Blackbirds, it turns out, are a good species in which to examine this hypothesis.

Blackbirds, like many other bird species, undertake a fascinating antipredator defense mechanism called "mobbing" (Sordahl, 1990). Once a flock of blackbirds spots a predator, some of them join together, fly toward the danger, and aggressively attempt to chase it away. Such attacks often work well enough to force predators to leave the blackbirds' area.

Eberhard Curio and his colleagues (1978b) examined whether one function of mobbing behavior might be to help predator-naive blackbirds identify what constitutes a predator (Altmann, 1956). In each trial of their experiment, Curio's team began with a "model" and "naive" bird, each in its own aviary. The experimental apparatus was designed so that each of the birds could see another bird—

a noisy friarbird (*Philemon corniculatus*). The friarbird was a novel species that neither the model nor the naive bird had seen before. Furthermore, it looked nothing like any known predator of blackbirds. The friarbird was presented in such a way that the naive blackbird saw the friarbird alone, but the model saw both a friarbird and, adjacent to it, a predator of blackbirds (a little owl, *Athene noctua*). From the viewpoint of the naive subject, the little owl was out of sight. In other words, when the model mobbed the little owl, the naive individual saw it mobbing a friarbird.

Curio and his team found that once they had seen a model apparently mobbing a friarbird, naive birds themselves were much more likely to mob this odd new creature than if they had not been exposed to a teacher. The researchers then took the experiment one step further and essentially asked whether the subject (now not so naive) could then act as a model for a new naive bird. And if that worked, how many times could they get a former naive bird to successfully act as a model? In other words, how long is the "cultural transmission chain" in blackbirds? The longer such chains are, the more powerful cultural transmission may be in spreading novel behaviors through a population. Though their sample was small, Curio and his colleagues found that the blackbird cultural transmission chain was at least six birds long. That is, after the initial model (bird 1) believed that friarbirds were dangerous, a new model (bird 2) then saw bird 1 respond to friarbirds. If bird 2 then responded as if friarbirds were dangerous, a new model (bird 3) then observed bird 2 respond to the friarbird as a predator. This procedure went on until six different birds acted as if the friarbird was a predator, and then the "chain" was broken.

SUMMARY

1. Predation affects almost every aspect of animal life—from foraging to mating to habitat selection.
2. One well-studied tradeoff involving predation occurs in the context of foraging. Animals alter their foraging choices in response to predation pressure. Classic optimal foraging models need to be, and have been, amended to incorporate predation as a variable affecting numerous aspects of foraging, including what to eat, the size of food items, and when and where to eat.
3. Predation pressure not only affects how an animal behaves, but can also have a dramatic effect on such life-history variables as development time—for example, when eggs will hatch.

4. Alarm signals of various forms are a very common type of antipredator activity. A large literature exists on how alarm signals operate, including the possibility that they can be used in deceptive ways.
5. In some primate species, and perhaps in other species as well, alarm calls can vary in response to the presence of different types of predators.
6. Although it may appear paradoxical at first, animals sometimes *approach* predators when they are initially encountered. Ethologists have examined the costs and benefits associated with such actions.
7. Brain-size evolution in predator and prey may be the result of a predator-prey arms race.
8. Comparing a single species in populations with and without a particular predator is a powerful means for studying how selection operates on antipredator behaviors.
9. Some prey species use both individual and social learning to identify predators and to learn the appropriate way to react when in danger's path.

DISCUSSION QUESTIONS

1. Pick your animal of choice, and sketch what a normal time budget (how much time it spends feeding, sleeping, mating, and so on) might look like for this animal. Now, besides the direct time spent looking for predators, examine how predation might directly or indirectly affect all the behaviors on your time budget.
2. Jerison interprets an historical increase in the brain size of predators and their prey as evidence of a predator-prey arms race, where the weapon is brain power. Can you think of other alternative hypotheses that might explain this finding?
3. Some prey, particularly birds, actually mob their predators and harass them until the predators leave. List some of the costs and benefits associated with such mobbing, and construct a hypothesis for what sorts of environments might favor mobbing.
4. Abrahams and Pratt (2000) used a thyroid hormone to manipulate growth rates of the fathead minnow (*Pimephales promelas*). Thyroid treatment stunted growth rates, and the researchers found that such stunted individuals were less likely to risk exposure to predators to gain access to food. Why might that be? Is it possible to construct an argument that would predict the

exact opposite of what was found? Also, what does this study tell you about the relationship between proximate and ultimate factors shaping antipredator behavior?

5. A number of studies have found that laboratory-raised animals can learn what constitutes danger by watching other animals respond to potential predators. How might such cultural transmission be employed by those interested in wildlife reintroduction programs?

SUGGESTED READING

Caro, T. M. (1986). The functions of stotting: A review of the hypotheses. *Animal Behaviour, 34,* 649–662. This paper systematically reviews hypotheses regarding one type of antipredator behavior—stotting. For a companion paper that tests these hypotheses in Thomson's gazelles, see Caro, T. M. (1986). The functions of stotting in Thomson's gazelles: Some tests of predictions. *Animal Behaviour, 34,* 663–684.

Caro, T. M. (1994). Ungulate antipredator behavior: Preliminary and comparative data from African bovids. *Behaviour, 128,* 189–228. A systematic study of the antipredator behavior of one class of large African mammals.

Kerfoot, W. C., & Sih, A. (Eds). (1987). *Predation: Direct and indirect impacts on aquatic communities.* Hanover, NH: University Press of New England. This book looks at the ecological and behavioral implications of predation in aquatic communities.

Lima, S. L. (1998). Stress and decision making under the risk of predation: Recent developments from behavioral, reproductive, and ecological perspectives. *Advances in the Study of Behavior, 27,* 215–290. A very nice review of decision making in the context of antipredator activities.

Milinski, M. (1977). Do all members of a swarm suffer the same predation? *Zeitschrift fur Tierpsychologie, 45,* 373–378. One of the first well-controlled studies examining group behavior and predation.

Smith, R. J. F. (1992). Alarm signals in fishes. *Reviews in Fish Biology and Fisheries, 2,* 33–63. A good introduction to the role of chemicals and alarm signals in one group of animals (fish).

Stanford, C. B. (1998). *Chimpanzee and red colobus: The ecology of predator and prey.* Cambridge, MA: Harvard University Press. A book-length case study of predator-prey interactions among two species of primates.

12

Communication and Honesty

Communication Venues

- Foraging
- Play
- Mating
- Aggression
- Predation
- Songs

Interview with Dr. Amotz Zahavi

Communication

When we see a chimp bare its canine teeth at another chimp, we and the chimps in the vicinity know that trouble is brewing. Similarly, when ravens hear an individual yell, they as well as any ornithologists around know that food has been found. Clearly, animals communicate with each other. In fact, without some communication system involving signalers and receivers, much of animal social behavior as we know it would simply cease to be. As such, we shall focus on many aspects of communication in this chapter (Hauser, 1997; Bradbury and Vehrencamp, 1998), but before doing so, we need to delineate what sort of animal communication we shall deal with in depth, and what sort we shall only touch upon.

In his book *If a Lion Could Talk*, Stephen Budiansky (1998) expounds on a quote made famous by philosopher Lutwig Wittgenstein: "If a lion could talk we would not understand him." While Budiansky's book is primarily about consciousness and intelligence, the Wittgenstein title has implications for communication—namely, that animals may see and experience the world in a fundamentally different way than we do as humans, that even if a lion could talk, we wouldn't understand what he was saying. This would not be because lions would speak a different language—the Wittgenstein quote implies this problem would be overcome somehow—but because lions would be communicating about the world in a fundamentally different way. Budiansky sums it up nicely by noting ". . . if a lion could talk, we probably could understand him. He just wouldn't be a lion anymore; or rather his mind would no longer be a lion's mind."

Budiansky is interested in communication as a means of understanding animals and animal behavior. This is clearly the approach we are going to take here. That is, **communication**—defined as the transfer of information from a signaler to a receiver—will be analyzed here in light of understanding animal behavior. But there is a somewhat different reason that some people study animal communication—namely, to understand how humans can communicate with animals. There are legitimate reasons for doing this—for example, in order to better interact with our pets or domesticated livestock. Here, however, we are not interested in animal-human communication, except in terms of how it helps us better understand the evolution of animal communication.

There are a number of reasons not to focus on animal-human communication per se. First, chances are that this phenomenon has only *evolved* with respect to a small number of animal species—dogs, cats, horses, and other species with which humans have long-term relationships. In other species, there is no a priori reason to assume that animal-human communication systems have evolved, as there is no evolutionary history of interaction to begin with. Second, animal-human communications are notoriously difficult to study. That is to say, it is very hard to know exactly what is being communicated from the animal side. A classic case is the study of "Clever Hans," which is usually paraded as a cautionary tale about learning, but is also illustrative with respect to communication (Pfungst, 1911; Sebeok and Rosenthal, 1981).

Around the turn of the twentieth century, a schoolmaster named William von Osten became famous because of the apparently prodigious ability of his horse Hans. Hans was able to do mathematical puzzles, identify music, and answer questions regarding European history. He did this, of course, not by actually speaking, but by using his hoof to count out the right answer to a math problem, or shaking his head "yes" or "no" to a verbal query. When the Prussian Academy of Science put Hans to the test in a controlled environment, they found something interesting. Hans could give the right answer only when some person in the room knew the right answer. If two people each gave Hans part of a question, but each was ignorant of what the other told Hans, the horse *never* got the correct answer. The reason was that Hans was indeed clever, but not in the way people thought (Figure 12.1). His cleverness, it turns out, was that he could pick up on very subtle body cues and facial cues that investigators in the room were (unconsciously) emitting while they were providing correct and incorrect answers for Hans to select from. So, for example, if the problem was "What is 10 + 10?" Hans was able to pick up on subtle cues people emitted when he was approaching twenty hoofbeats on the floor.

The cautionary lesson with respect to animal intelligence is clear, but there is also a lesson about human-animal communica-

FIGURE 12.1. Clever Hans. Clever Hans, the horse, was thought capable of incredible mental feats. In fact, Clever Hans was picking up very subtle cues from the individual who asked him a question (and consequently knew the answer to that question).

FIGURE 12.2. Animal communication. Communication can take many forms, including **(A)** tactile (as in the nose touching by yellow-bellied marmots), **(B)** auditory (as in tail slapping by beavers), and **(C)** auditory and visual (as in the "drumming" by ruffed grouse). *(Photo credits: Elinor Osborn; Michael Quinton)*

tion in the story of Hans. Human-animal communication can be very subtle and extremely difficult to pin down. Of course, the fact that something is difficult to study is no reason not to study it, but this, in conjunction with the first caveat about animal-human communication, leads me to focus on evolved animal communication systems.

There are a number of ways we could break up the study of animal communication. One way is to examine different *modes* of communication. For example, animals employ a wide variety of modalities—visual, vocal, chemical, vibrational, electrical, and so forth—to communicate with one another (Figure 12.2), and our discussion could revolve around these different modalities. We could have had sections on how different chemicals are invoked in both sending and receiving messages, or on how we might expect auditory vibrational communication to be more common in environments like dense forests, where vision is limited, and so forth. In fact, we do touch on many communication modalities, but they will not be the fulcrum on which this chapter rests. Instead, we will take a different approach (Figure 12.3).

After addressing a nagging question in animal communication—is communication honest?—we will examine case studies of communication in different behavioral venues, including foraging, play, mating, fighting, and predation. While we have brought up communication while discussing these behaviors in earlier chapters, we will expand on the subject here, and in so doing, touch on the different communication modalities. So, for example, we shall see how foragers might communicate about food sources and how individuals might let each other know that it is "playtime." We shall examine how communication affects mating, how it provides much-needed information about predation threats, as well as how it plays a fundamental role during aggressive interactions and the

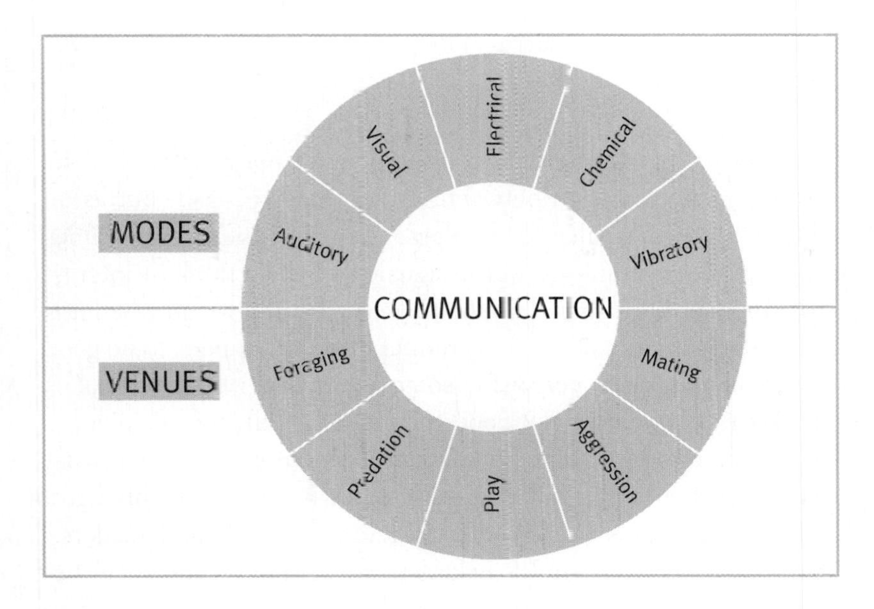

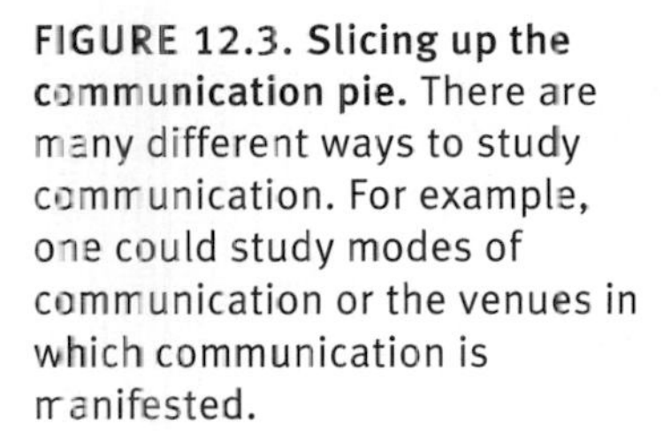
FIGURE 12.3. Slicing up the communication pie. There are many different ways to study communication. For example, one could study modes of communication or the venues in which communication is manifested.

reconciliations that sometimes follow such interactions. We shall also see that natural selection with respect to communication operates not only on sender and receiver behavior, but on the morphological and neurobiological traits that underlie communication.

Communication and Honesty

If we adopt the definition of communication given above—the transfer of information from signaler to receiver—then an evolutionary slant on this topic immediately raises the question of honesty. That is, should the signaler be honest in the information it conveys (Dawkins and Krebs, 1978), or should it attempt to deceive the recipient in some way or another?

Consider the following analogy of communication from Richard Dawkins and John Krebs:

> When an animal seeks to manipulate an inanimate object, it has only one recourse—physical power. A dung beetle can move a ball of dung only by forcibly pushing it. But when the object it seeks to manipulate is itself another live animal there is an alternative way. It can exploit the senses and muscles of the animal it is trying to control, sense organs and behavioural machinery which are themselves designed to preserve the genes of that other animal. A male cricket does not physically roll a female along the ground and into his burrow. He sits and sings, and the

> female comes to him under her own power. (Dawkins and Krebs, 1978, p. 282)

Dawkins and Krebs view communication not so much as the exchange of information between signaler and receiver, but as an attempt by the signaler to manipulate the recipient. Sometimes what is in the best interest of the signaler is also in the best interest of the recipient. This need not be the case, however, and when it is it's usually just an incidental byproduct of what happens to be good for the signaler (see below for some exceptions). When what is good for the signaler is not good for the recipient, the recipient is then selected for the ability to unscramble what is honest and what isn't, and act in ways that maximize its own fitness. In this light, Krebs and Dawkins (1984) refer to recipients as "mind readers." This view leads Krebs and Dawkins to speak of an "arms race" between signaler and recipient, in which the signaler is selected to better manipulate the receiver, who then is selected to better filter out only that information that benefits it, and so on.

This is quite a different view from what Dawkins and Krebs refer to as "the classic ethological approach" of animal behavior (Tinbergen, 1964; Marler, 1968; W. J. Smith, 1968, 1977). Implicit in this classic approach is that both parties benefit from the information exchange, and thus there is little selection pressure for either to be deceitful. In the classic ethological view, the signaler and receiver are "on the same page," and selection here is thought to favor the most economical way to share information.

Krebs and Dawkins recognize that the sort of cooperative signaling that the classic ethological approach viewed as predominant may in fact be occurring in some systems, particularly those involving kin or reciprocal exchanges (Chapters 8 and 9). They offer an ingenious way to distinguish between those systems in which there is co-evolution in the manipulator/mind-reader sense and those in which cooperation dominates. They argue that when communication is of the manipulator/mind-reader type, the signals employed should be exaggerated, as one might expect from a salesman attempting to convince you his product is the top of the line. But when cooperative signaling is in play, Krebs and Dawkins believe that natural selection should favor "conspirational whispers." Since signaling often involves some costs (for example, energy costs, drawing attention from predators), selection should favor minimizing these costs, and hence reducing the conspicuousness of the communication itself (Johnstone, 1998a). For example, take the case of ptarmigans (*Lagopus mutus*). Both male and female rock ptarmigans are a stark white against the snow of winter. Once spring arrives, females molt quickly, while males stay white, and as such are very conspicuous against the dark brown background. Such conspicuousness attracts females, but it also attracts predators. As the breeding season progresses, males, rather than molting, soil their plumage to become less conspicuous (Montgomery et al., 2001). In other words, males communicate information to females via coloration but when the period during which communication is useful comes to end, they shut down the communication system in a cost-efficient manner—by soiling their white plumage.

While formal mathematical models of the "conspirational whispers" vs. conspicuous display hypotheses have shown that this dichotomy oversimplifies communication systems a bit (Johnstone, 1998b), it is nonetheless a very useful heuristic tool. There is, however, another means besides cooperative signaling by which we might expect communication to be honest. Even within the manipulator/mind-reader view of communication, honesty might evolve if the signals being sent are either impossible or, at the very least, difficult to fake. To see how this might work, imagine that females produce more offspring when mating with larger males. All males, even small ones, would be selected to produce signals that made a female believe they were large. But selection should favor females that were paying attention only to those cues that are in fact true indicators of large size. As such, we should end up with a system in which females pay heed only to cues that can't be faked (that is, honest cues). This appears to be the case in toads, where deep croaks can be produced only by large males (because of the design of the vocal system), and females use croaks as one major cue in choosing among males (Davies and Halliday, 1978; Figure 12.4).

Amotz Zahavi has suggested that honesty is also possible in the case where traits are not impossible, but merely very costly, to fake (Zahavi, 1975, 1977, 1997; see Grafen, 1990a, 1990b, for a mathematical analysis). Under the handicap principle, if a trait is costly to produce, it may be adopted as an honest signal, because only

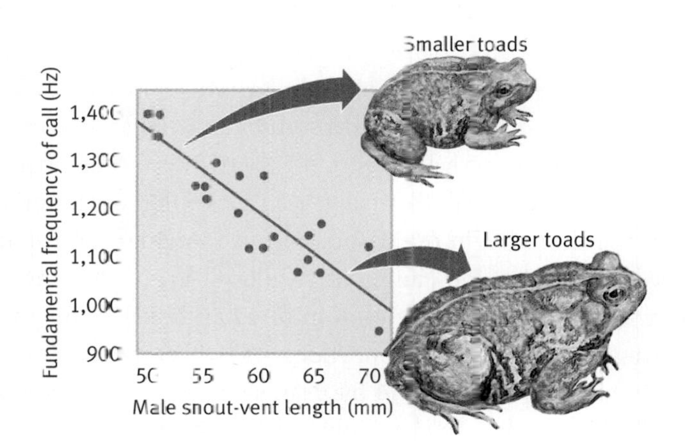

FIGURE 12.4. Toad size and croaks. Male toads *(Bufo bufo)* use the frequency of other males' calls to gauge their potential opponents' size. The relationship between a male's size (as indicated by his snout-vent length) and the frequency of his call is strong, suggesting that call frequency is an honest indicator of size, and hence of fighting ability. *(Based on Davies and Halliday, 1978).*

those individuals who can pay the cost will typically adopt the signal in question. For example, imagine that females use the length of a male's energy-costly song as a cue for the amount of resources a potential mate is able to garner. While a male that is not particularly good at garnering resources could use virtually all of his resources to sing, and thus give the female a false impression of his resource-garnering skills, most of the time only males that are genuinely good at gathering resources will be able to afford to sing, as singing is an energy-costly activity. Hence, honest communication may be an outcome even when deception is possible in principle, as long as deception is very costly in nature (Johnstone, 1995, 1998a; Adams and Mesterton-Gibbons, 1995; Zahavi and Zahavi, 1997; Mesterton-Gibbons and Adams, 1998). As Johnstone (1998a) notes, this idea bears a striking resemblance to the idea of conspicuous consumption in humans, where "consumption of valuable goods is a means of reputability to the gentleman of leisure" (Veblen, 1899). Thus, conspicuous consumption is a trait not easily faked, and is an honest indicator of who possesses resources.

Communication Venues

As we are about to see, communication between animals occurs in a wide variety of venues. It is hard, but not impossible, to imagine how it could be otherwise. *What* the particular communication skills might be will depend on both the species and the behavioral venue in question, but *that* good abilities to communicate should be the focus of selection seems inevitable.

FORAGING

In Chapter 9, we examined food calls in house sparrows as a potential case of cooperation via byproduct mutualism. Aside from the potentially cooperative aspects of "food calls," these behaviors are interesting in and of themselves as examples of communication. We therefore begin our discussion of foraging and communication by examining food calling in swallows. We then continue to explore the role of communication in foraging by describing the famous "waggle dance" that honeybees use to convey information about the location of food, and by discussing the intricate communication system underlying foraging in leaf-cutter ants.

SWALLOW "SQUEAKS" AND FOOD PATCHES. Colonial breeding cliff swallows (*Petrochelidon pyrrhonota*) live in nests that serve as "information centers" (Ward and Zahavi, 1973; C. R. Brown, 1986). For some time, researchers believed that individuals living in nests only received "passive" information—that is, it was thought that group members simply observed their nest mates and followed them to potential resources.

The idea that individuals were *recruited* to food sites was first suggested by Charles Brown and his colleagues (C. Brown et al., 1991). Using both playback experiments (playing tape-recorded bird calls) and provisioning experiments (putting food out to entice birds), Brown and his team found that cliff swallows gave off "squeak" calls, which alerted conspecifics that a new food patch had been found (Table 12.1). In addition, such squeak calls were emitted only in the context of recruiting others to a food site (no squeak calls were given at the colony nest); these calls did not appear to be used in any other context.

In terms of the costs and benefits of this form of foraging-based communication, those recruited certainly obtain a resource—food. Recruiters also obtain benefits from calling, however, since with increasing group size, it is more likely that one group member will track the insect swarm, and thus provide further foraging opportunities (C. Brown, 1988; C. Brown et al., 1991). This tracking behavior may be especially critical, as individual swallows must

TABLE 12.1. Squeak calls attract others. The mean number of birds and squeak calls heard two minutes before and two minutes after insects were flushed by foraging birds. For each two-minute period there were significantly more birds after insects were flushed than before, and there were significantly more squeak calls that were heard after the insects were flushed than before. *(From Brown et al., 1991, p. 559)*

PERIOD	MEAN NO. OF BIRDS		MEAN NO. OF CALLS/BIRD/2 MINUTES		NO. OF TRIALS (*N*)
	BEFORE	AFTER	BEFORE	AFTER	
8 May 1990	37.8	240.3	0.062	0.267	11
9 May 1990	9.5	20.5	0.000	0.137	4
24 May 1990 A.M.	9.7	22.7	0.000	0.100	6
24 May 1990 P.M.	30.8	88.7	0.053	0.171	9
27 May 1990	7.9	45.4	0.031	0.133	7
29 May 1990	8.8	54.8	0.000	0.098	5
30 May 1990	8.5	38.5	0.033	0.109	2
15 June 1990	25.9	102.1	0.032	0.134	7

often return to the colony to provision young, and hence they would likely have great trouble relocating an insect swarm without the help of others.

HONEYBEES AND THE WAGGLE DANCE. Food collecting in honeybees often involves thousands of workers covering large areas of ground, at least from a honeybee's perspective (Visscher and Seeley, 1982, Figure 12.5). Given this, how do individual foragers monitor their own intake rate, the changing distribution of resources through time, and the colony's needs? The now famous **waggle dance**, wherein bees learn some aspects of foraging by following other foraging bees in a dance that provides some information on the spatial location of food, provides part of the answer.

FIGURE 12.5. Bee foraging. Honeybee foraging involves a complex communication system, including waggle dances. This dance, along with other informational cues, gives bees in a hive information about the relative position of newly found food sources. *(Photo credit: Leroy Simon/Visuals Unlimited)*

The waggle dance of the honeybee was first studied experimentally by Karl von Frisch (1967). Seeley eloquently sums up the waggle dance as

> a unique form of behavior in which a bee, deep inside her colony's nest, performs a miniaturized reenactment of her recent journey to a patch of flowers. Bees following these dances learn the distance, direction and odors of these flowers and can translate this information into a flight to specified flowers. Thus the waggle dance is a truly symbolic message, one which is separated in space and time from both the actions on which it is based and the behaviors it will guide. (Seeley, 1985, pp. 84–85)

The waggle dance has an almost mystical flavor to it. Fortunately, it is possible to maintain that air of the amazing, while at the same time studying this behavior in detail, as ethologists have now been doing for more than half a century (Seeley, 1985). To see this, imagine that a worker bee has just returned from a bountiful cluster of flowers that are approximately 1500 meters from her nest, and that these flowers are located 40 degrees to the right of an imaginary straight line running between the worker's nest and the sun (Figure 12.6).

Upon returning to the nest, our worker bee would quickly start "dancing" up and down a vertical comb within the hive, with as many sisters as possible trying to keep up with her (and making as much physical contact as they can). While dancing vigorously (waggling), our worker would be conveying crucial information to her relatives in the hive. That is, her dance would provide topographical information (north, east, south, west, northwest, and so on) for finding the food source from which she has just returned. When compared to a straight up-and-down run along a comb, the angle at which our forager danced would describe the position of the food source of interest in relation to the hive and to the sun.

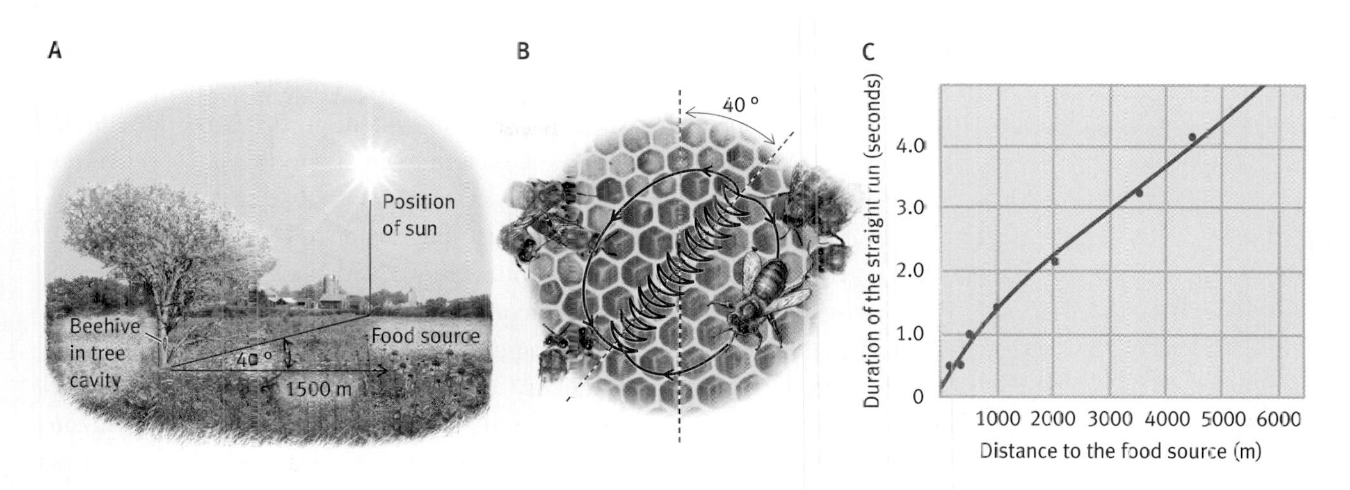

FIGURE 12.6. Honeybee waggle dances. (A) Imagine a patch of flowers that is 1500 meters from a hive, at an angle 40° to the right of the sun. **(B)** When a forager returns, the bee dances in a figure eight pattern. In this case, the angle between a bee's "straight run" (up and down a comb in the hive) and a vertical line is 40°. **(C)** The length of the straight run portion of the dance translates into distance from the hive to the food source. *(Based on Seeley, 1985)*

Moreover, the longer the bee danced (in a part of the waggle dance known as the "straight line"), the further away the bounty. In fact, every extra 75 milliseconds of dancing would translate into the resource being about an additional 100 meters from the hive.

In addition, the scent of the returning forager, as well as any food that might be stuck to her body, would provide would-be foragers with even more information about the food source. While experimental work suggests limits to the accuracy of the waggle dance (Seeley, 1985, 1995), overall it is a remarkable communication system that is breathtaking to observe.

CHEMICAL AND VIBRATIONAL COMMUNICATION IN FORAGING ANTS. When it comes to communication in the animal world, few groups can challenge the ants for mastery. Almost every aspect of ant life revolves around communicating through chemical cues (E. O. Wilson, 1971; Holldobler and Wilson, 1990; Billen and Morgan, 1997; Vander Meer et al., 1997). Here, we shall examine chemical communication in the leaf-cutting ants of the genus *Atta*. Subsequent to this, we shall discuss vibrational communication in this same species. Ants in this genus are one of the dominant lifeforms in the neotropics, consuming more vegetation there than any other group. Remarkably, leaf-cutters subsist entirely on the fungus they grow on leaves and on the sap produced by the plants whose leaves they harvest (Littledyke and Cherrett, 1976).

Leaf-cutters live in huge nests containing hundreds of thousands of workers (plus a single queen) and have evolved an elaborate, caste-based system for obtaining the leaves on which they subsist

(E. O. Wilson, 1980a, 1980b). Leaves must first be cut, then carried to the nest, ground up, chewed and treated with enzymes, placed into the "fungus garden," and subsequently cultivated. Holldobler and Wilson (1990) liken the system to an assembly line where different castes handle different tasks. Here we shall examine two components of foraging in leaf-cutters: long-distance foraging, which relies heavily on chemical communication, and short-distance foraging, which relies on vibrational communication.

When it comes to long-distance foraging communication, two chemicals are particularly important in *Atta* species—methyl 4-methylpyrrole-2-carboxylate and 3-ethyl-2,5dimethylpyrazine (Morgan, 1984). These substances are produced in the poison gland of leaf-cutter ants and are used to recruit fellow workers to foraging sites that are relatively long distances from their nest. Recruitment pheromones are incredibly powerful:

> . . . the discoverers of methyl 4-methylpyrrole-2-carboxylate . . . estimated that one milligram of this substance (roughly the quantity in a single colony), if laid out with maximal efficiency, would be enough to lead a column of ants three times around the earth. (Holldobler and Roces, 2001, p. 94)

Recruitment pheromones, which fade slowly, are placed along the trails leading to trees where leaves are being harvested. These pheromones are also deposited along branches and twigs, bringing recruited foragers very close to the leaves they need to harvest. When they are close, however, leaf-cutters also rely on a second mode of communication—vibrations (stridulations)—to determine which precise leaves to work on.

FIGURE 12.7. Devastating leaf-cutters. Leaf-cutter ants can devastate foliage in their path. The ants don't attack all the leaves, however, but instead they often strip some leaves to the stalk (for example, those that are most tender or have fewer secondary compounds present), while leaving other leaves untouched. *(Photo credit: Bert Holldobler)*

When *Atta* workers are cutting leaves, the process seems systematic. Sections of leaves are not randomly cut throughout a tree (or even a branch), but rather certain leaves are cut and cut until there is virtually nothing left (Figure 12.7). Holldobler and Roces (2001) hypothesized that *Atta* workers were cutting the most desirable leaves, where desirability might be measured in terms of leaf tenderness or the amount of secondary plant compounds present (the fewer, the better). They further hypothesized that because *Atta* workers appeared to raise and lower their gasters in a manner similar to the way they created stridulatory vibrations (Markl, 1968), they were using vibrational cues to recruit other workers to the best leaves in the vicinity.

To begin their work on vibrational communication, Holldobler and Roces used noninvasive Doppler vibrometry to test whether workers in *Atta cephalotes* stridulate while they are cutting leaves. Indeed, workers were stridulating while they cut, and the vibrations were sent along the length of a leaf in a long series of vibrational "chirps" (Figure 12.8). Next, Holldobler and Roces offered workers leaves that differed in toughness (tough or tender). In addition, in a follow-up experiment, both tough and tender leaves were dipped in sugar water and offered to ants. The results were clear-cut. While only 40 percent of the ants stridulated when cutting tough leaves (no sugar water), the number increased to 70 percent when the ants were cutting tender leaves (no sugar water) and to almost 100 percent when either type of leaf was dipped in sugar water (Roces, Tautz, and Holldobler, 1993). Ants, then, clearly stridulate while cutting, and they vary this behavior, depending on the value of the leaf in question.

The above two experiments, although they are powerful, do not demonstrate that ant workers are recruited by stridulation. To test for the potential recruiting nature of stridulatory communication, Holldobler and Roces hooked one leaf to a vibrator and used a similar ("silent") leaf as a control. When given the choice to cut on either of the leaves, ant workers clearly preferred the vibrating leaf, demonstrating that these cues were not only produced by, but used to recruit, workers. Further, through an ingenious set of experiments, they were able to demonstrate that communicating to other workers was the primary function of stridulation. Interestingly, though, stridulation does serve another role in communication. In *A. cephalotes*, a small-sized caste, called minim workers, exists. These minim workers cannot cut leaves, but they are often found hitchhiking rides on the backs of leaf-cutters. Minims protect these leaf-cutting workers from attack by parasitic flies (Eibesfeldt and Eible-Eibesfeldt, 1967; Feener and Moss, 1990), and so it is certainly beneficial for leaf-cutting ants to have minims find them. Hitchhiking minims apparently use the vibrational cues created by stridulating leaf-cutting nest mates to locate the leaf-cutters (Figure 12.9).

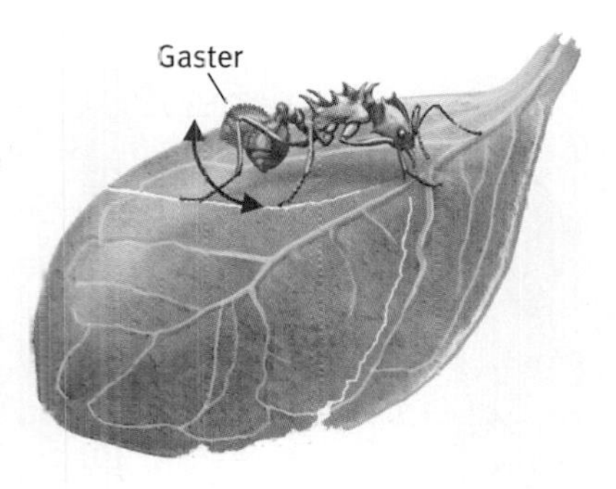

FIGURE 12.8. Stridulating communication. A schematic of a leaf-cutter ant cutting a leaf and stridulating its gaster up and down. *(Based on Holldobler and Roces, 2001)*

FIGURE 12.9. Minim workers hitchhiking on a leaf. The stridulating signals emitted by leaf-cutters are used in numerous contexts. One such venue is between leaf-cutters and "minimum" workers (minims), who use these signals to eventually hitch rides on cut leaves that are carried on leaf-cutters' backs. *(Photo credit: Bert Holldobler)*

PLAY

Play takes on a fundamental role in the development of many canid species (see Chapter 15). One challenge associated with play is for the individuals undertaking it to realize that what is about to occur is play and not some associated "real" behavior (aggression, predation, mating, and so on). Without knowing this, individuals may attempt behaviors that are completely inappropriate for play. The question of how animals communicate that they are about to enter a bout of play is one that researchers take seriously. There are a number of ways that this might take place, but for now, let us focus on what have been labeled **play markers**, or play signals—

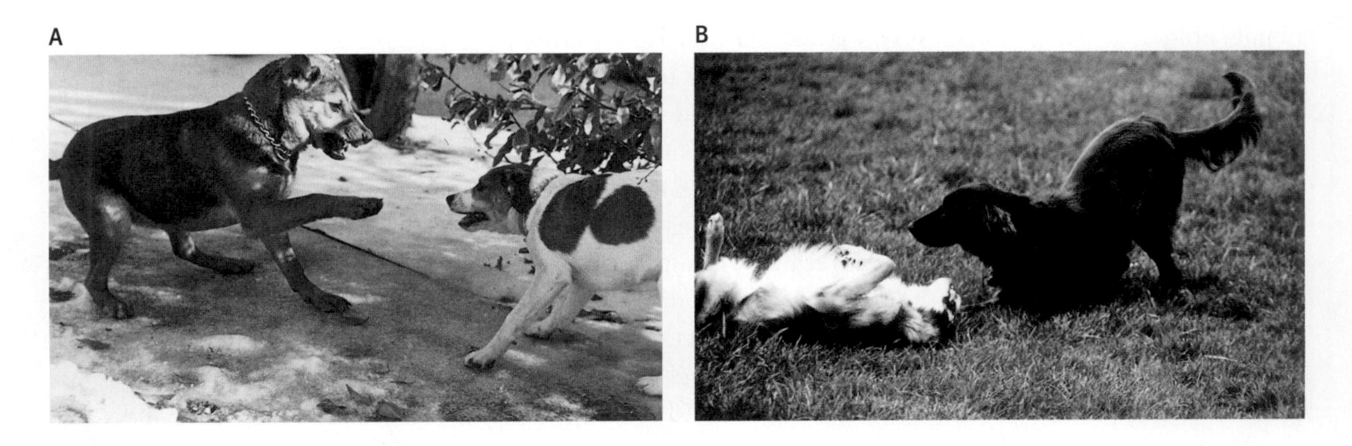

FIGURE 12.10. Play markers. (A) Play signals override canine aggression. The dog on the right growls, while the dog on the left "paws." Pawing is a play signal that can turn a potentially dangerous encounter into a playful one. **(B)** Play bows are also often used as signals that the bowing individual wants to play. The dog in this photo is play bowing to another individual. *(Photo credits: Marc Bekoff)*

indicators that a behavior is part of play (Bekoff, 1977, 1995; Berger, 1980).

In canids, biting/shaking is usually performed during dangerous activities such as fighting and predation. Yet, this behavior is also part and parcel of play in young canids. Bekoff (1995) found that play markers such as a "bow" would precede biting and rapid side-to-side shaking of the head to indicate that they were not dangerous behaviors (Figure 12.10). The bow would communicate that these actions should be viewed in a new context—that of play.

MATING

Most aspects of animal mate choice involve one form of communication or another. When animals choose among potential mates, they are often selecting their mate based on traits that are designed to communicate something to the chooser. As such, much of our Chapter 6 discussion could be recast in the language of communication, and as we saw earlier in this chapter, a good chunk of it could be conceptualized in terms of honest and dishonest communication. In this section, we will examine the role of communication during mate choice in two extremely different systems: water striders and electric fish. In so doing, we shall also be touching on two very different but equally fascinating modalities associated with communication.

RIPPLE COMMUNICATION AND MATE CHOICE IN AQUATIC INSECTS. In 1972, Stim Wilcox discovered a theretofore unknown form of communication—ripple communication in water striders

(Figure 12.11). This form of communication has now been demonstrated in numerous species of water striders, and it has also been found in giant water bugs (Bottger, 1974; R. L. Smith, 1979; Kraus, 1989). In water striders, ripples are usually produced by an up-and-down movement of the legs, with both right and left legs in synchrony and in constant contact with the water surface. Ripple signals vary from 2 to100 Hertz (Wilcox, 1995), with most falling in the 20 to 50 Hertz range. Such signals can range from 0.2 second (courtship behavior of *Rhagadotarsus anomalus*; Wilcox, 1972) to 30 seconds (Polhemus, 1990). Here we shall touch on the little that is known about ripple communication and courtship, but it is important to recognize that ripple communication has been found to play a role in an incredible array of behaviors, including copulation, postcopulation behavior, sex discrimination, mate guarding, induction of oviposition, territoriality, and food defense (Table 12.2).

In *R. anomalus*, males produce ripple signals that can travel more than 60 cm, and females find such ripple signals very attractive. Wilcox (1972) undertook a series of ingenious playback experiments that demonstrated that females from as far away as 60 cm were attracted to mating ripple signals—ripples that are very different from those produced between aggressive males. Females would often grasp males and even begin to oviposition in response to such

TABLE 12.2. Contexts of ripple communication. In water striders, ripple signals are used in many different contexts, including calling (CALL), courtship behavior (CSP), copulatory behavior (COP), postcopulatory behavior (PCOP), territorial/spacing behavior (TERR), and sex discrimination (SEX). Within the table, M = male signal; F = female signal. *(Based on Wilcox, 1995, p. 112)*

SPECIES	CALL	CSP	COP	PCOP	TERR	SEX
Aquarius remigis			M		M F	M
Gerris elongatus	M	M			M	
Gerris conformis			M			
Limnoporus notabilis						
Limnoporus dissortis		M	M		M	M
Limnoporus rufoscutellatus		M			M	M
Rhagodotarsus anomalus	M	M F	M	M	M	M
Rhagadotarsus kraepelini	M	*	*	*	*	*
Rhagadotarsus hutchinsoni	M	*	*	*	M	*
Tenogogonus albovittatus		M F				
Microvelia longipes					M	

**Signals not reported; however, the signal systems of these three* Rhagadotarsus *species appear to be very similar, so the signal contexts indicated by * appear likely.*

A

B

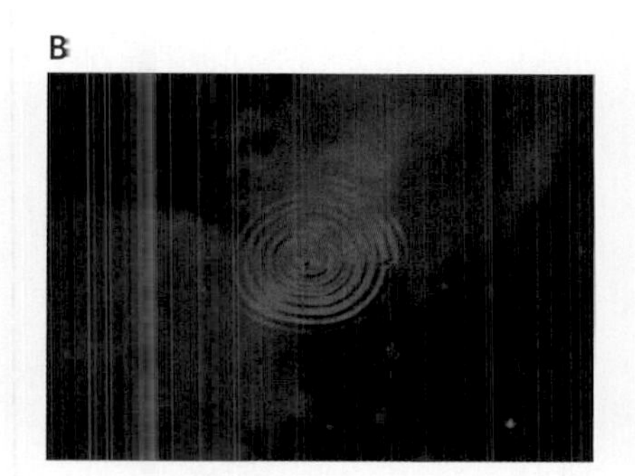

C

FIGURE 12.11. Ripple communication by water striders. **(A)** The concentric circles of these ripples in a pond are part of the communication used by the water strider, *R. kraepelin*. **(B)** A closeup of the male water strider (*R. kraepelin*) and the ripples he is making to communicate with other water striders. **(C)** An experimental set-up to study ripple communication, in which a dead *A. remigis* female (mounted with silicon rubber) is making a computer-generated signal via a magnet glued to her leg. In nature, female water striders don't emit such signals. *(Photo credits: Stim Wilcox)*

signals, once the signaler was detected. In addition, Wilcox (1995) speculates that ripple calls designed to attract mates also serve as a means for species identification, as *R. anomalus* are often found in the same streams and ponds as other water striders (Polhemus and Karunaratne, 1993).

Ripple signals are also employed by giant water bugs in the context of mating (Bottger, 1974; R. L. Smith, 1979; Kraus, 1989). While performing courtship displays, male giant water bugs pump up and down with their legs, thereby creating ripples, which then attract female water bugs. The roots of such pumping behavior may lie in the motion that giant water bugs use to aerate eggs (which are laid on their backs) (R. L. Smith, 1979).

FIGURE 12.12. Shocking communication. The electric catfish, *Malapterurus electricus*, along with many other species, uses electric current to communicate with nearby individuals. *(Photo credit: E. R. Degginger/Color-Pic, Inc.)*

A SHOCKING EXAMPLE OF MATE CHOICE. Humans tend to communicate by speech, touch, sight, sound, and odors. As such, we focus on these modalities and forget that there are other ways to communicate. In particular, communication through electric current is not uncommon in fish. Carl Hopkins has spent the better part of his career studying all aspects of communication in the African electric fish (Figure 12.12), and along the way, he has made some incredible discoveries (Hopkins, 1986, 1999a, 1999b).

Electric organ discharges (EODs) are usually species specific (Hopkins, 1981) and are often used for species recognition (Hopkins and Bass, 1981). EODs are used in a variety of social contexts (Hopkins, 1986), but a relatively consistent finding is that during the mating season, male EODs are longer than those of females. This sex difference suggests that males are using EODs to attract mates. Although direct experiments on mate selection and EODs have not been undertaken to date, this hypothesis is further supported by the consistent finding that androgen tends to increase EODs in juveniles, females, and nonreproductive males (Bass et al., 1986; Bass and Hopkins, 1983,1985; Bass and Volman, 1987; Freedman et al., 1987).

To examine the link between androgens, mate choice, and communication, Carlson and colleagues (2000) examined the nature of social interactions on EOD discharge in males and any possible correlates with circulating levels of androgens in *Brienomyrus brachyistius*. To begin their experiment, the researchers housed each of eighteen male electric fish with a single female electric fish. Subsequent to this, all fish were transferred to larger aquaria, where each tank held three male and three female fish. Prior to being placed in these larger tanks, the relative dominance of the three males was determined by a pairwise competition of males.

When a male was alone with a female, male EODs were significantly higher than female EODs, but male EODs did not predict rank in the second part of the experiment. When placed in group

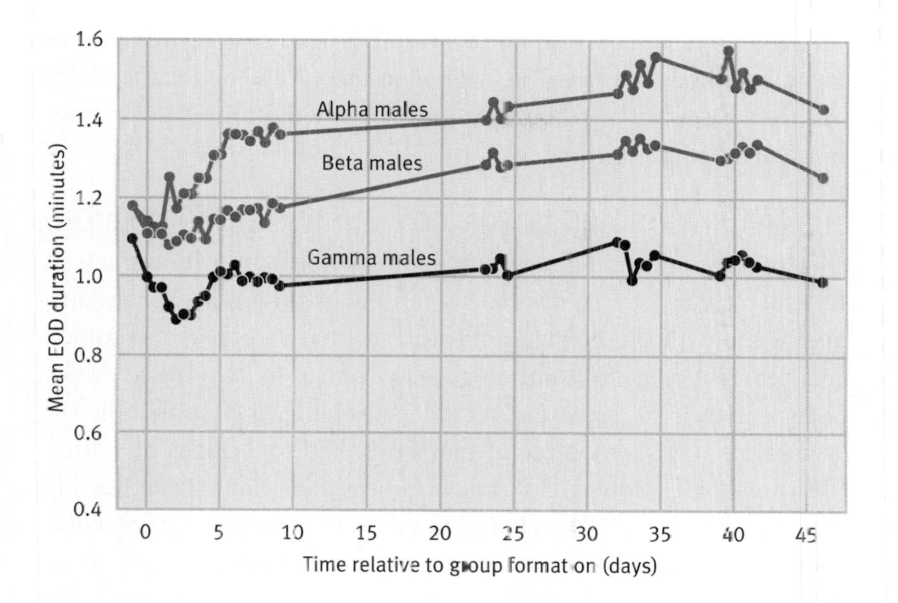

FIGURE 12.13. The effects of rank on electric organ discharge. Three male fish were first housed individually with a female fish. All six fish (three male, three female) were then placed together in a social group. The EODs of the alpha male fish increased greatly, the EODs of the beta male fish increased somewhat, while the EODs of the gamma fish decreased. *(Based on Carlson et al., 2000, p. 181)*

tanks, female EODs remained relatively constant, but male EODs changed markedly as a function of rank (Figure 12.13). The EODs of alpha males increased greatly, the EODS of beta males increased somewhat, and the EODs of gamma males dropped (these differences remained throughout the forty-plus-day study). With respect to hormones, by the end of the experiment (when males were housed together), rank correlated positively with levels of 11-ketotestorone (no differences in 11 K-T existed at the start of group formation). It thus seems that androgens are not the cause of differences in dominance, but rather a consequence of differences in rank. In short, rank appears to be communicated to both other males and females via differences in EODs, and females may use such EOD differences in selecting mates (Hopkins, personal communication).

AGGRESSION

In his classic, but often underappreciated book, *The Expression of Emotions in Man and Animals*, Darwin (1872) described the often ritualized sequence of behaviors that some animals undertake during aggressive interactions. Many of these behaviors communicate the level of danger associated with a step in the sequence leading to severe fighting. For some time, ethologists built "ethograms" that detailed the aggressive sequence in their species of interest. Although these ethograms are less common today, and are often discussed in the context of some mathematical model of aggression (see Chapter 14), they are still valuable in placing aggression within the behavioral repertoire of whatever species is being studied.

Here we shall examine communication and aggression in salmon territorial defense, as well as in raven food acquisition. We will then discuss the reconciliation behavior that often follows aggression in primates.

SUBORDINATION AND COLOR COMMUNICATION IN SALMON. One obvious means for communicating information in animals is via color change. If color change can occur on a time scale commensurate with the behavior of interest, then it represents a quick and efficient way to communicate information to others. Color change may be a particularly good communication vehicle in aggressive contests, where color may be linked to "badges of status" (Rohwer, 1982; Roper, 1986), and hence color *change* can quickly indicate an individual's relative rank in a hierarchy. Quick color change may help individuals avoid the costs associated with fighting an opponent that is likely to defeat them anyway (Geist, 1974b; Enquist and Leimar, 1990; Hurd, 1997).

Because fish have tight hormonal and neuronal control over the expansion of pigment cells (Waring, 1963; Fujii and Oshima, 1986; Nelissen, 1991; Hulscher-Emeis, 1992), they are particularly adept at quick color change over short time periods. Consequently, they have been the subject of numerous experiments on color change and aggression (Barlow, 1963; Baerends et al., 1986; Hulscher-Emeis, 1992; Guthrie and Muntz, 1993). For example, researchers have studied Atlantic salmon (*Salmo salar*), in whom aggression is most often associated with territorial defense. Their early work suggested that dominant males develop dark vertical eye bands (Keenleyside and Yamamoto, 1962) and that subordinate individuals develop darker body color (Figure 12.14).

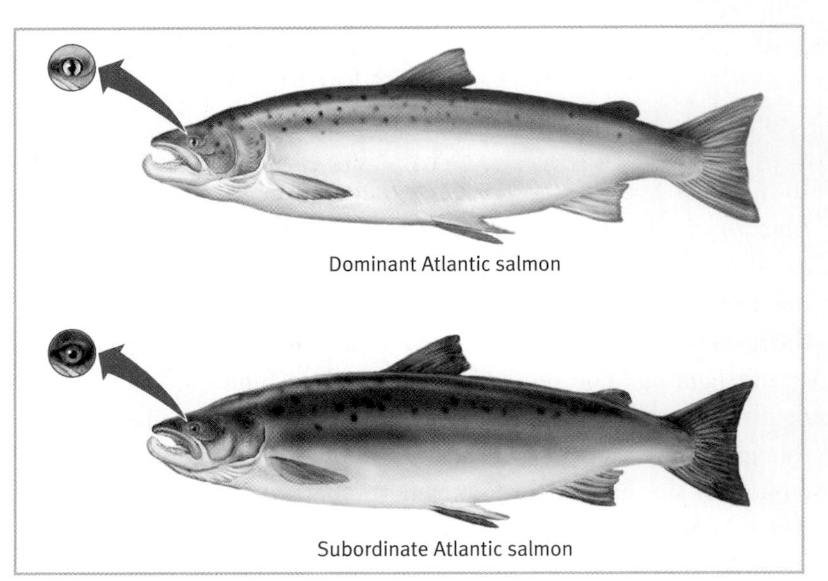

FIGURE 12.14. Color as a signal. In Atlantic salmon, subordinate individuals often assume a much darker body color. Dominants' body color remains light, but they develop dark vertical eye bands.

If change to a darker body color occurs as a result of subordination, then it may be viewed as a means of communication on the part of the subordinate fish, and this in turn may lead to behavioral changes on the part of the dominant fish. O'Connor and colleagues (1999) tested these hypotheses using size-matched pairs of juvenile Atlantic salmon placed in an aquarium that served as a fighting arena. One salmon was placed on either side of a partition, and both were allowed time to acclimate to their side of the tank, and subsequently to behave as if that area were their territory. At that point, one salmon was removed from its side of the tank and became an "intruder" on the territory of the other fish. As aggressive interactions or "contests" commenced, the researchers gathered data on body color to test their hypothesis regarding color and communication.

At the beginning of the experiment, when each salmon was in its respective half of the aquarium, and before one fish was chosen as an "intruder," the vast majority of fish (94 percent) had very pale body color and light gray eyes. Once the researchers moved "intruders" into the other side of the tank, they found that in six cases neither fish became aggressive and there was no color change in such fish. In twenty-four other cases after the move, however, displaced fish developed dark body color during contests, and there was a tight link between dark body color and subordination, with nineteen of the twenty-four dark fish being subordinate (Figure 12.15; the same pattern was found with respect to eye color, with subordinate fish developing darker eye color).

Based on these results, a link appears to exist between color and rank. Nonetheless, the above data don't tell us whether color change communicated anything to the dominant individual. To test their hypothesis about communication, O'Connor's team examined the behavior of the winner subsequent to the loser's color change.

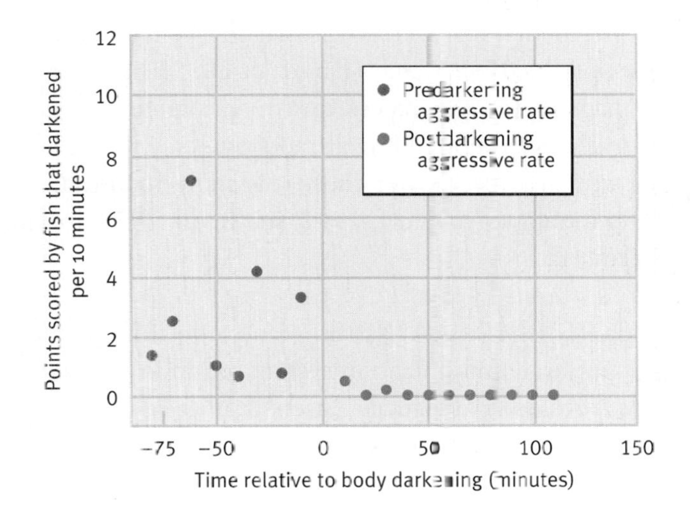

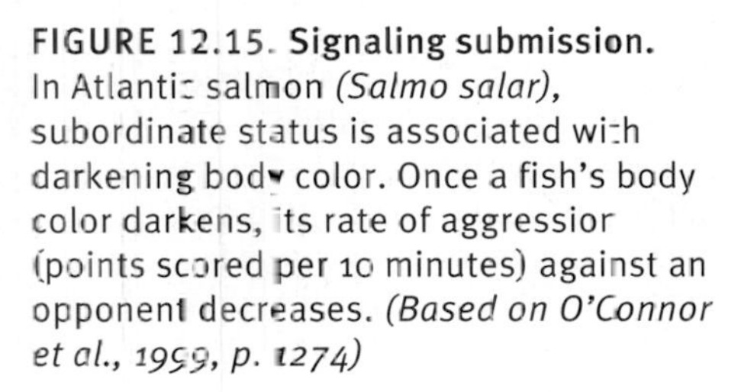
FIGURE 12.15. Signaling submission. In Atlantic salmon *(Salmo salar)*, subordinate status is associated with darkening body color. Once a fish's body color darkens, its rate of aggression (points scored per 10 minutes) against an opponent decreases. *(Based on O'Connor et al., 1999, p. 1274)*

While dominant individuals continually increased their aggression toward subordinates before the latter's color change, once the color change occurred there was an equally quick decrease in aggression toward the subordinate. This temporal shift suggests that the subordinate's color change acted as a signal to the dominant fish. It should be noted, however, that once the subordinate fish became darker, its aggressive behavior also decreased (Figure 12.15). As such, it is difficult to say for certain that color alone acts as a signal of submission (as decreased aggressive behavior may also signal this). Work in other species (Abbott et al., 1985), however, does suggest that the link between color, communication, and subordination is well worth exploring in more depth in Atlantic salmon.

FIGURE 12.16. Raven yells. Under certain conditions, ravens emit a loud "yell" upon uncovering a new food source. Such yells attract other birds. *(Photo credit: Bernd Heinrich)*

RAVEN "YELLS," FOOD PATCHES, AND AGGRESSION. Ravens are scavengers that can often survive for days on end if they uncover a large food patch (for example, a rotting carcass). When a raven uncovers a new food source, it often emits a very loud "yell," that among other things, attracts other ravens to the caller's newly discovered bounty (Figure 12.16).

Bernd Heinrich and his colleagues address both proximate and ultimate questions surrounding why common ravens (*Corvus corax*) call when discovering food patches (Heinrich, 1988b, 1989; Heinrich and Marzluff, 1991). On the proximate end, it appears that yelling is both a response to hunger level and an indication of rank in a group hierarchy (Heinrich and Marzluff, 1991). That is, hungry birds call more often than satiated birds (Figure 12.17). In addition, dominant individuals suppress the yelling behavior of those below them (Figure 12.18). So, if we want to know who yells, we need to know something about hunger level and rank in a group. Status via rank may also play a role in our ultimate explanation of yelling, as it may translate into increased reproductive success. That being said, the predominant ultimate causal factor underlying yelling appears to be attracting others to a food source in order to overpower resident ravens and increase the yeller's fitness. Most ravens are unrelated "vagrants" (Parker et al., 1994), and the only way for a vagrant to gain access to a food source that is already being defended is to yell. Yelling attracts others, who together with the yeller can then overpower those originally found at the food source (Heinrich and Marzluff, 1991; see Roell, 1978, for more on this in jackdaws).

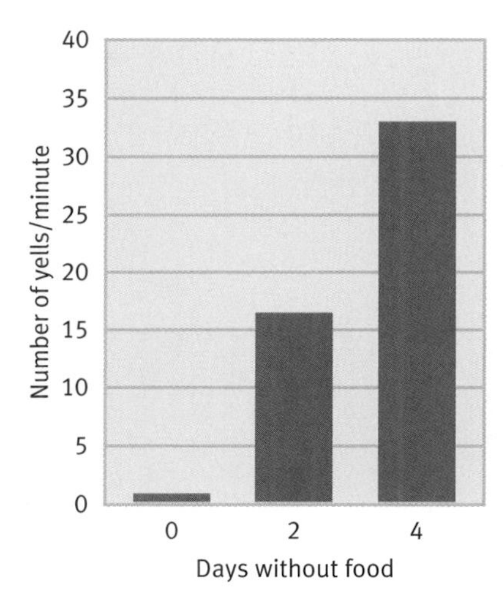

FIGURE 12.17. Yellers are hungry. In ravens, "yelling" is often associated with foraging—in particular, calling others to a food bonanza. Immature ravens yell progressively more as a function of hunger. *(Based on Heinrich and Marzluff, 1991)*

"PANT HOOTS" AND LEARNING. One question that emerges from the complex communication system found in primates is how individuals develop a "language" such that they may be able to gauge one another's intentions (Pinker, 1991, 1994; Nowak and

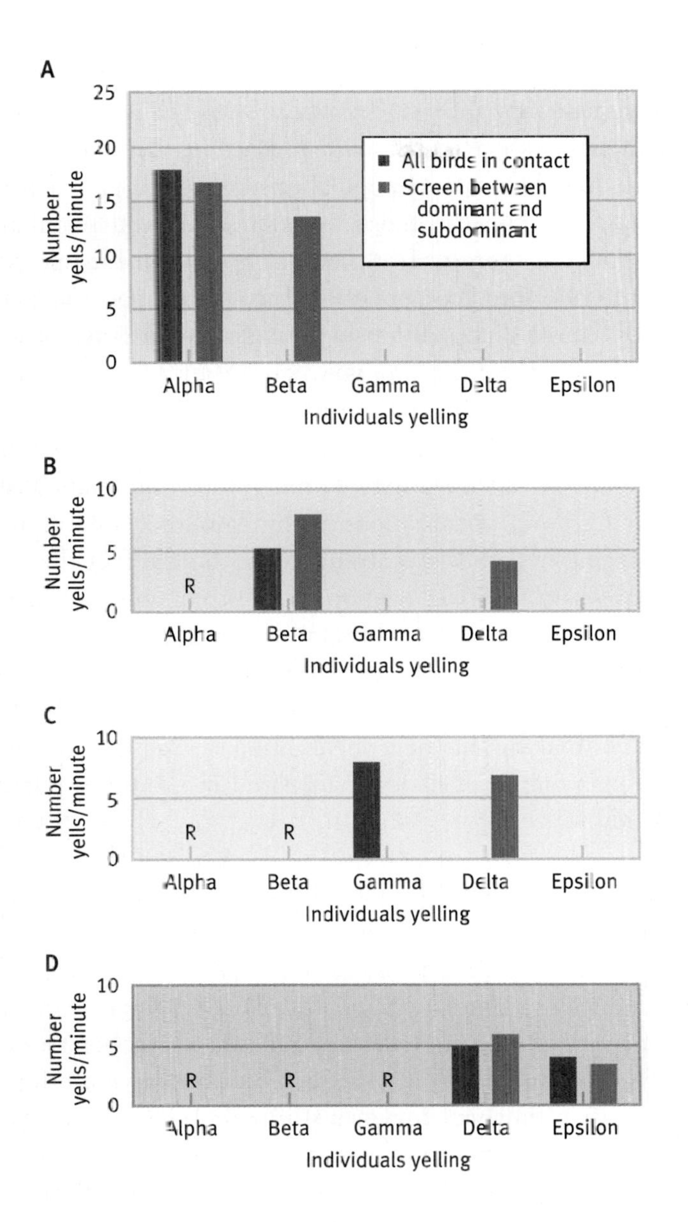

FIGURE 12.18. Dominance and yelling in ravens. In addition to hunger level, dominance also affected yelling rates. **(A)** Five birds, ranked alpha to epsilon, were first placed together, and yelling rates were measured. **(B)** Yelling rates when the alpha bird was removed (indicated by R). **(C)** Yelling rates when the alpha and beta birds were removed (indicated by R). **(D)** Yelling rates when the alpha, beta, and gamma birds were removed (indicated by R). The green bar indicates all birds in contact, and the orange bar indicates a screen between the dominant and subdominant birds. *(Based on Heinrich and Marzluff, 1991, p. 18)*

Krakauer, 1999; Nowak et al., 2000). This might be particularly useful in aggressive encounters, where gauging the intentions of others may affect strategic behavior. Much has been done on the conceptual and theoretical issues of language development in nonhumans. But aside from work on birdsong, empirical studies on the development of complex communication systems have lagged behind theory to some degree (Janik and Slater, 1997; Seyfarth and Cheney, 1997; Husberger and Snowdon, 1997).

Marshall and his colleagues (1999) tackled complex communication by looking at what role learning played in structuring the vocalizations of two chimpanzee zoo populations, one from Florida and one from North Carolina. They examined the four sequential components to chimp pant hoots (Goodall, 1986; Mitani et al.,

FIGURE 12.19. Male hoots. Pant hoots, such as the one emitted by the male chimp shown here, are part and parcel of the chimpanzee communication system. *(Photo credit: Frans de Waal)*

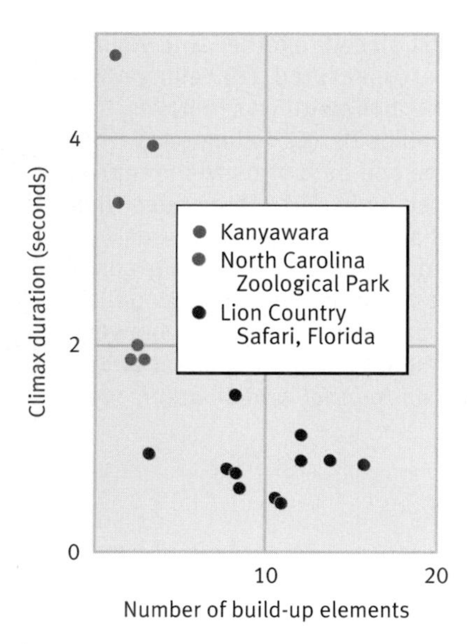

FIGURE 12.20. Pant hoots vary across chimp populations. Using two measures of pant hoots (number of build-up elements and climax duration), different geographic populations clustered together. *(Based on Marshall et al., 1999)*

1992). Pant hoots are given in many behavioral contexts, including both aggression and foraging (Figure 12.19). When Marshall and his colleagues compared the two populations they were working with, they found that both populations produced pant hoots that were relatively similar to those found in a natural population of chimps, but that there were differences between the zoo populations in the first three components of the pant hoot. For example, the North Carolina population had longer "introductory" and "climax" sections to their pant hoots, while the Florida population had more "build-up" elements in their hoots (Figure 12.20).

Marshall and his team argue that the differences they uncovered were due to social learning within groups that led to a convergence on different vocalizations within groups. Support for this social learning hypothesis is strengthened by the fact that in only one group was a new vocalization (the "Bronx cheer") uncovered. Apparently, the Bronx cheer sound was introduced into the Florida population sometime in the last ten years by a single male who was making this vocalization before he joined this population. Subsequent to his making this new vocalization, it was learned by other males in his group, but it never appeared in the other population under study.

COMMUNICATION AND RECONCILIATION IN PRIMATES. "Making up" or reconciliation after bouts of aggression is an integral part of post-conflict interactions in numerous primate species. Because of their genealogical relationship to man and their sophisticated behavioral repertoire, reconciliation in chimps has been studied extensively. As Frans de Waal (1989a) notes, chimpanzee fights are more bark than bite, but they are still potentially dangerous. Reconciliation follows about 40 percent of aggressive interactions in the Arnhem Zoo population of chimps that de Waal initially worked with.

Reconciliation in chimps is clearly delineated and contains behavioral elements that are hard to miss. One such element is the outstretched arm and open hand gesture. In addition, kissing plays a fundamental role in chimpanzee reconciliation, as do eye contact, yelping, and soft screaming, all of which increase in frequency (Figure 12.21). Consider de Waal's description of one bout of aggression and reconciliation:

> . . . Nikkie, the leader of the group, has slapped Hennie during a passing charge. Hennie, a young adult female of nine years, sits apart for awhile feeling with her hand the spot on her back where Nikkie hit her. Then she seems to forget the incident; she lies in the grass, staring in the distance. More than fifteen minutes later Hennie slowly gets up and walks straight to a group that includes Nikkie and the oldest female, Mama.

FIGURE 12.21. Rhesus reconciliation. In many primate species including rhesus monkeys (pictured here), reconciliation behavior often follows acts of aggression. *(Photo credit: Frans de Waal)*

> Hennie approaches Nikkie, greeting him with a series of soft pant grunts. Then she stretches out her arm to offer Nikkie the back of her hand for a kiss. Nikkie's hand kiss consists of taking Hennie's whole hand rather unceremoniously into his mouth. This contact is followed by a mouth-to-mouth kiss. Then Hennie walks over to Mama with a nervous grin. Mama places a hand on Hennie's back and gently pats her until the grin disappears. (de Waal, 1989a, p. 39)

Given the ubiquity of cultural transmission of behavior in chimps (Whiten et al., 1999; Whiten and Boesch, 2001), it would not be surprising to learn of consistent culture-based differences in reconciliation across chimp groups, although it should be noted that in a study examining sixty-five categories of chimp behavior across six natural populations, reconciliation is not listed as a culturally transmitted behavior.

Reconciliation has not only been studied in well-known primates such as the chimpanzee, but also in species about which we are just starting to learn. For example, in 1987 Ren and his colleagues captured nine golden monkeys in the Chinese province of Shanzi. Little was known of this monkey's behavior, but the researchers built an ethogram of golden monkey communication based on 10,000 social interactions while the monkeys were kept in small social groups in captivity (Ren et al., 1990a, 1990b). It became clear from this ethogram that reconciliation plays a large role in post-conflict interactions in golden monkeys.

In golden monkeys, reconciliation often follows one of the four most common aggressive interactions, which in order of severity (from least to most dangerous) are threats/lunges, chases, hitting, and biting. After 130 such interactions, Ren and his team gathered data on the behavior of the recipient of the aggression. Among other things, they simply asked whether two individuals who had just been in an aggressive interaction were more, less, or equally likely to be found near each other in subsequent nonaggressive interactions. When compared to matched controls, individuals who were just in an aggressive interaction were significantly more likely to be near one another in a nonaggressive context. These nonaggressive interactions were often reconciliatory in nature and included embracing, grooming, holding on to one another, and crouching.

PREDATION

Communication can also be used to warn conspecifics about predators. Part of the value of foraging in groups derives from the protection of numbers. If one individual recognizes the approach of danger, it can communicate this to others through alarm calls.

WOODPECKER AND CHICKADEE ALARM CALLS. In Chapter 8, we saw the very powerful role that blood kinship plays in shaping alarm calls. Recall that in Belding's ground squirrels, alarm calls are primarily driven by the degree of relatedness between caller and recipient. But kinship is not a prerequisite for alarm calling. In both downy woodpeckers (*Picoides pubescens*) and black-capped chickadees (*Parus atricapillus*), individual alarm calls are emitted to protect *mates*, but not other conspecifics (Figure 12.22). Kimberly Sullivan (1985) found that downy woodpeckers never gave alarm calls when they were foraging alone (0/46), when in a flock with no other conspecifics (0/23), or when with a same-sex downy woodpecker (0/6). In seven of nine instances, however, they emitted calls when in the presence of a woodpecker of the opposite sex. Three of these calls were made by females and four by males, suggesting that both sexes use this form of communication to attract opposite-sex partners. The situation is similar in black-capped chickadees. In these chickadees, mated individuals stay close to one another in wintering flocks (Ficken et al., 1980), and alarm calls appear to be emitted to increase a mate's probability of surviving the winter (Witkin and Ficken, 1979). This "mate investment" form of communication has also been suggested to be a causal factor selecting for alarm calls in other birds as well (Morton, 1977; Leopold, 1977; East, 1981; Curio and Regelmann, 1985; Hogstad, 1995).

FIGURE 12.22. Protecting a mate. While downy woodpeckers don't give alarm calls when they are paired with same-sex partners, they emit such alarm calls when they are paired with a member of the opposite sex. *(Photo credit: E. R. Degginger/Color-Pic, Inc.)*

ALARM CALLS AS DECEPTIVE COMMUNICATION. Alarm calls are a powerful sort of communication. Failure to listen for such calls might lead to death, and so natural selection should favor paying close attention. This, however, sets up the possibility for using alarm calls in a deceptive manner. In Chapter 11, we mentioned that using alarm calls in a deceptive manner is probably the exception rather than the rule. That being said, there is, in fact, some evidence that this sort of deceptive communication exists for vervets and two species of birds, and it is worth examining these cases for what they tell us about communication.

Cheney and Seyfarth (1990) argue that there is anecdotal evidence that vervets use deceptive alarm calls during some intergroup encounters or when an interloper approaches a troop. Intergroup encounters, as well as the appearance of interlopers, are dangerous events in the lives of vervets, as they often lead to serious aggression. Occasionally, male vervets give an alarm call when encountering a new troop, even though no predator is in the vicinity. While it is possible, it is also unlikely that such calls are mistakes, as vervets are quite sophisticated when it comes to alarm calls, and their behavior seems to indicate that they know when a predator is around and when it isn't (see Chapter 11). Part of the reason that ethologists doubt that such calls are given in error is that the response to alarm calls emitted when a new group is encountered is that everyone heads for cover, and so a potentially dangerous intergroup encounter is avoided, at least for the moment.

Cheney and Seyfarth state that in some 264 intergroup interactions they observed, males gave false (leopard) alarm calls about 2 percent of the time, and it was almost always a low-ranking male—one who stands the most to lose from intergroup interactions—who gave the call. In fact, one might expect that false alarm calls should be relatively rare, because if they were too common, they would simply be ignored, which could prove fatal when a real predator was around. On the other hand, vervets don't have the whole deception package that, say, a person might have at their disposal. For example, on a number of occasions one individual (Kitui) who emitted an alarm call when an interloper from another group was near his troop subsequently behaved in a way that seemed out of place.

> As if to convince his rival of the full import of his calls, Kitui twice left his own tree, walked across the open plain, and entered a tree adjacent to the interloper's, alarm calling all the while. Kitui acted as if he got only half the story right; he knew that his alarm calls had caused others to believe there was a leopard nearby, but he did not seem to realize that other aspects of his behavior should be consistent with his calls. To leave his tree and to walk toward the other male simply betrayed his own lack of belief in the leopard. (Cheney and Seyfarth, 1990, p. 215)

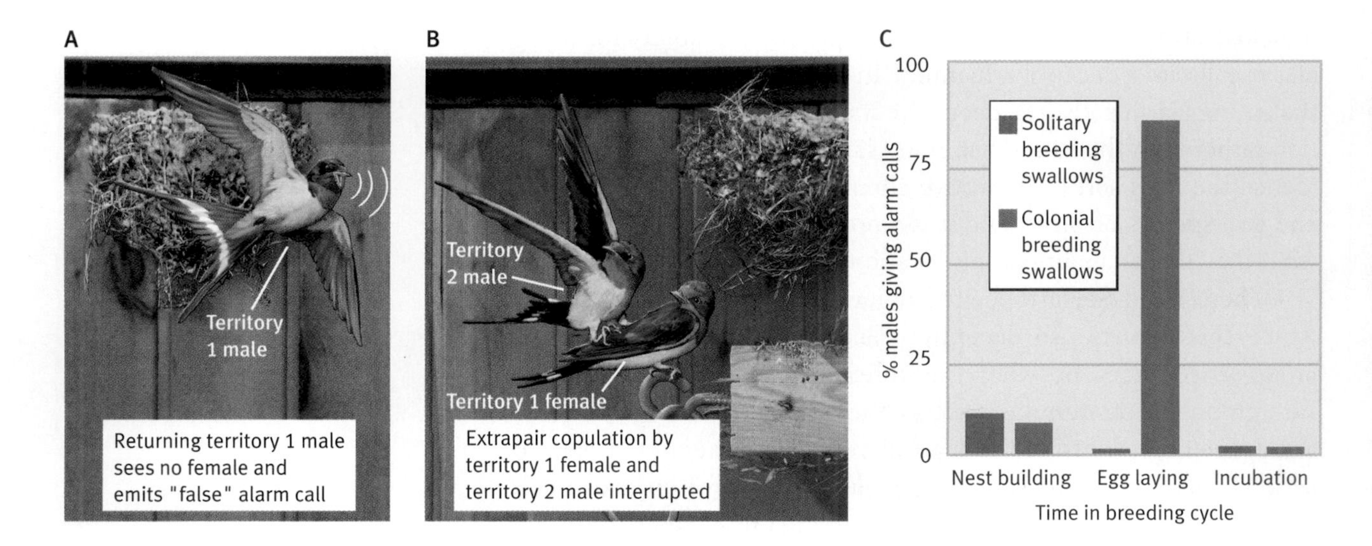

FIGURE 12.23. Dishonest alarm calls in swallows? (A) Male barn swallows often give false alarm calls when their fertile mates leave the nest vicinity. **(B)** These false alarm calls sometimes disrupt extrapair copulations (EPCs). **(C)** Møller hypothesized that male swallows would give false alarm calls when they were at the greatest risk of EPCs to disrupt the EPCs. To test this, Møller removed a female from the nest at different stages in the breeding cycle for both solitary breeding swallows and colonial breeding swallows. Solitary breeding males almost never emitted alarm calls when their mate was temporarily gone. Colonial breeding males emitted false alarm calls during the period in which EPCs were most likely (during egg laying). *(Based on Møller, 1990)*

Deceptive alarm calls are not unique to primates and have been found in at least three studies on birds (Munn, 1986; Møller, 1988, 1990). Møller (1990), for example, hypothesized that male swallows emitted false alarm calls when their paternity was threatened by extrapair copulations (see Chapter 7). Such alarm calls break up extrapair matings, as all swallows take flight in response to alarm calls. To test this idea, when males were away from their nests, Møller chased their presumptive mate from the nest as well. He did this both during the period when extrapair copulations were most likely (during egg laying) and when such pairings were least likely (during incubation periods). When females were absent during egg laying, the males who returned to their nests almost always gave false alarm calls, which might serve to disrupt any extrapair copulation that their mate was undertaking. In the same situation, during periods when extrapair copulations were unlikely, males almost never emitted alarm calls (Figure 12.23).

SONGS

Some forms of animal communication are more pleasing to humans than others. Go into your local music or book shop, and you are bound to find dozens of CDs of animal "song." Most likely, these will be birdsongs, but since about the 1970s, whale songs have also been in vogue. In addition to their practical, sedative effects on human listeners, these two forms of song are inherently interesting in and of themselves and have been studied by naturalists, ethologists, behavioral ecologists, comparative psychologists, and evolutionary biologists for hundreds of years.

THE MOTOR BASIS UNDERLYING BIRDSONG. Few communication systems in vertebrates have been studied to the extent of birdsong (Thorpe, 1961; Catchpole and Slater, 1995; Hauser and Konishi, 1999; Todt and Naguib, 2000; see Weilgart and Whitehead, 1997, for more on whale song, and Bailey, 1991, for more on insect song). The literature on birdsong is huge and spans from the developmental (Kroodsma, 1999), to the molecular and anatomical (Nottebohm, 1999), the physiological (Suthers, 1999), and neurological (Mundinger, 1999), through to the role of learning in the development of song (Marler, 1999). It would be impossible to attempt to review all the work in these areas, so here we narrow our focus considerably and zoom in on one aspect of birdsong—the motor basis for birdsong in oscine birds (Suthers, 1999).

Birdsong is incredibly diverse in structure, pattern, tempo, frequency, and repertoire size (Searcy, 1986, 1992; Searcy and Brenowitz, 1988; Read and Weary, 1992; Hauser, 1997). This diversity, however, is somewhat mystifying, as the **syrinx**, the vocal organ used in birds, varies little between different species in this group. Our question here will be: How it is possible that morphological (structural) invariance in the syrinx can translate into great diversity in song? The answer sheds light not only on the proximate factors that underlie birdsong, but also on the incredible power of natural selection to tinker along the edges when necessary.

To unravel this mystery, we begin by recognizing that the avian syrinx has two compartments—left and right—and that the two sides of a bird's brain can control these compartments independently (Suthers, 1997). This one piece of information sets the stage for moving from invariance in the structure of the syrinx to variance in the song output. Now, although we have a syrinx that varies little in structure, its *parts* can be manipulated to create new permutations that may give rise to new sounds (Suthers, 1999).

Over a hundred years ago, it was suggested that two different sides of the syrinx contribute differentially to the frequency component of song, but it was not until Greenewalt (1968) and Stein (1968) that stronger evidence of this phenomenon arose. This independence of the two sides of the syrinx allows songbirds to switch off one side at any time, and these sorts of switches can occur between song syllables, or even within them (Table 12.3). This allows for song variation in that it permits some species to:

- Operate both sides of the syrinx independently throughout their song without one side being dominant, as in brown thrushes (*Toxostoma rufum*) and gray catbirds (*Dumetella carolinensis*) (Suthers and Hartley, 1990; Suthers et al., 1994, 1996).
- Have one side of the syrinx dominate song generation, as in canaries (*Serinus canaria*) (Nottebohm and Nottebohm, 1976).
- Alternate which side of the syrinx dominates during a song, as in brown-headed cowbirds (*Molothrus ater*) (Allan and Suthers, 1994).
- Have one side of the syrinx dominate for certain frequencies, and the other side dominate for the remainder of the frequencies used in a song (sequential lateralization), as in the northern cardinal (Suthers and Goller, 1996; Suthers, 1997).

TABLE 12.3. The different ways to sing. The costs and benefits of different song lateralization patterns. *(Based on Suthers, 1999)*

SONG LATERALIZATION PATTERN	SPECIES	ADVANTAGES	DISADVANTAGES
Independent bilateral phonation	Brown thrasher and gray catbird	Two independent voices increase spectral and phonetic complexity.	Expensive in use of air supply. Best suited for a low syllable repetition rate.
Unilateral dominance	Waterslager canary	Conserves air, favoring shorter minibreaths and longer phrases with pulsatile expiration. Separation of phonatory and inspiratory motor patterns to opposite sides of syrinx. Both of these may facilitate higher syllable repetition rates.	Use of one voice limits frequency range and certain kinds of spectral and temporal complexity. Minibreath may be smaller than tracheal deadspace.
Alternating lateralization	Brown-headed cowbird	Enhances spectral contrast between notes. Efficient use of air supply. Extended frequency range for overall song.	Two-voice complexity limited to note overlap.
Sequential lateralization	Northern cardinal	Extended frequency range for continuous FM sweeps. Conserves air supply.	Lacks spectral complexity of two voices.

In addition to all of the above, the respiratory muscles that supply air to the syrinx can also be employed to achieve the variance in song output across oscine birds. Varying the respiratory muscles can have dramatic effects on the structure and timing of birdsong. So, again, the syrinx itself is invariant, but variation around other anatomical structures can produce variation in the output (birdsong).

WHALE SONG CODAS. Sperm whales (*Physeter macrocephalus*) are simply magnificent creatures to behold (Figure12.24). Females can live up to seventy years (Rice, 1989), and they live in very stable matrilineal groups (Whitehead et al., 1991; Richard et al., 1996). Whales communicate with each other through a series of short clicks known as codas (Watkins and Schevill, 1977). When sperm whales are in close contact, "conversations," or exchanges of codas, are common. Moreover, such conversations are often linked with some visual activity occurring at the same time (Whitehead and Weilgart, 1991). Given this social form of communication and the well-known cognitive sophistication of cetaceans (Rendell and Whitehead, 2001), Weilgart and Whitehead (1997) undertook a large-scale study of codas in sperm whales in different locations, with the primary goal being to determine whether there were coda "dialects," as had been found in killer whales (Ford, 1991; Strager, 1995).

In their study of sperm whale populations in the Caribbean Sea and in the vicinity around the Galápagos Islands, Weilgart and Whitehead found 3,644 codas. They characterized these codas by the number of clicks and the time interval between clicks and

FIGURE 12.24. Sperm whales. A group of sperm whales *(Physeter macrocephalus)* seen from above. These whales communicate through a series of short clicks known as codas. Populations of sperm whales from different locations use different codas to communicate with each other. *(Photo credit: Hal Whitehead)*

Interview with Dr. AMOTZ ZAHAVI

How did your long-term work on Arabian babbler birds lead you to focusing on communication?

I started the study of the babblers three years before I had the idea of the handicap principle. I wanted to know how a group functions (as distinct from a flock, which I had studied earlier in wagtails). I did not start the study to test a theory. Later on, the theory of the handicap principle helped me understand the communication and the cooperation of babblers.

Communication can be a dicey term to define. How do you define it?

Communication is the active transfer of information through signals. I define a signal as a trait, or part of a trait, that evolved solely in order to convey otherwise unknown information to other individuals. For example, the long tail of a peacock is used as a signal to attract mates; peacocks also use their tails for navigation in flight, like any other tail. However, the tail's extra length, the characteristic that handicaps its bearer, functions only as a signal. Information may also be collected by observation. This is not communication.

How did you come up with the handicap principle?

I developed the idea of the handicap principle (Zahavi and Zahavi, 1997) as a consequence of a remark by a student of mine who rejected the logic of Fisher's model to explain waste in mate choice. It took me a few months of thinking to come up with the handicap principle as a stable alternative to Fisher's model. Almost immediately I realized that the handicap principle could explain the deterrence of rivals and predators

through signals. It could also explain the makeup of other types of signals: vocal, movement, and colors. I also realized that "specific" signals, which were supposed to convey information about the identity of signalers, may be interpreted as standards in a competition between individuals. These signals convey information about the differences between individuals rather than their similarities.

The big finding in the babblers study, that connected it with communication and with the handicap principle, came when I realized that their altruistic behavior evolved as a signal. I found that helper babblers who brought food to the nest did not actually increase the fitness of the group, and that dominants often rejected subordinates' help. This did not make sense when one viewed the help as evolving in order to assist others, but it made perfect sense when one viewed the act of helping as a signal conveying quality. These findings also led me to examine the logic of kin selection and of reciprocity. I realized that these two theories were as unstable as group selection models, and should be rejected for the same reasons that group selection is rejected.

The handicap principle suggests that acts of altruism should be viewed as acts of advertisement by which an individual advertises its ability or intention. In other words, they are signals. Thus, I found the handicap principle to be not only a key for understanding reliability in communication, but also a factor that stabilizes all social systems. Any social system, including a monogamous pair, is faced with one key question: Why should an individual partner invest in the system rather than exploit its partner or partners? The assumption that altruism functions as a signal can stabilize the functioning of all social systems.

In short, I think that all signals involve handicaps, and that many social adaptations are best explained as evolving thanks to their use as signals loaded with handicaps.

Is it true that you told theoreticians interested in the handicap principle that if they just could understand Hebrew, you could easily explain this idea to them? Could you share a story about that?

My remark to Alan Grafen about Hebrew was a joke. I had gone over my simple verbal argument to him, about the handicap principle, several times, and he did not accept it. I saw no point in repeating yet again, and ended our discussion with that joke. My verbal argument was simple: If one is of a better quality than another, and its quality is not known to potential partners or adversaries, the individual may benefit from investing a part of its advantage in making its superior quality known and still retain an advantage over the other. Every business does that.

How much of communication do you view as honest communication?

Most communication is honest. The basic idea of the handicap principle is that signals are selected by their receivers to be honest. Dishonest signals may occur in the relatively few cases when it does not pay the receiver to check the reliability of the signal. There is no reason to believe that natural selection should favor signalers over receivers.

Do you think communication is more important in some behavioral venues (for example, foraging, aggression, cooperation) than others? If so, why?

Communication is a cooperative act. All communication evolves to facilitate cooperation. Since all social behavior involves cooperation, communication is important for understanding any social behavior.

Do you think of human communication via language as fundamentally different from all other forms of communication?

Human verbal communication is not reliable. It developed from nonverbal communication when man had to carry out transactions that could not be described with nonverbal signals. Once verbal communication evolved, it evolved handicaps for the skillful use of language: poetry, for example, handicaps poets by restrictions imposed by style, thus enabling them to demonstrate their superiority through the skillful use of words. But language of itself does not contain any element that can ensure the truthfulness of the information conveyed. One has to assess the reliability of the information by other means.

I FOUND THE HANDICAP PRINCIPLE TO BE NOT ONLY A KEY FOR UNDERSTANDING RELIABILITY IN COMMUNICATION, BUT ALSO A FACTOR THAT STABILIZES ALL SOCIAL SYSTEMS.

How do you think the current work in brain science and cognition will change the way that animal behaviorists think about communication?

I am not well-acquainted with the studies of cognition and brain sciences. However, I do not think results from brain sciences will affect the basic logic of communication. After all, developments in computer science did not change the basic statement that two and two are four.

What do you think of as the most important unanswered question in the area of evolution and communication?

I did not intend to develop the handicap principle as a general principle for all communication systems. I started out using it to understand mate choice, and gradually found that I could apply it in my studies of the babblers for understanding all their signals and their altruism. One of the most important implications of the handicap principle is that the pattern of a signal is related to the message encoded in it. Wealth is advertised by waste; courage is advertised by taking a risk; strength is advertised by feats that require strength. I am now applying the same logic to understanding chemical signaling: in other words, I investigate the relationship between the form of a signal molecule and the message it carries. I am also applying the logic of the handicap principle to the evolution of signals within the multicellular organism. This may well become a very important field for discoveries in signaling in the near future, and can also be of practical use in understanding hormones, neurotransmitters, ligands, and receptors.

DR. AMOTZ ZAHAVI is a professor at Tel Aviv University in Israel. His handicap hypothesis is summarized in his book The Handicap Principle: A Missing Piece of Darwin's Puzzle *(Oxford University Press).*

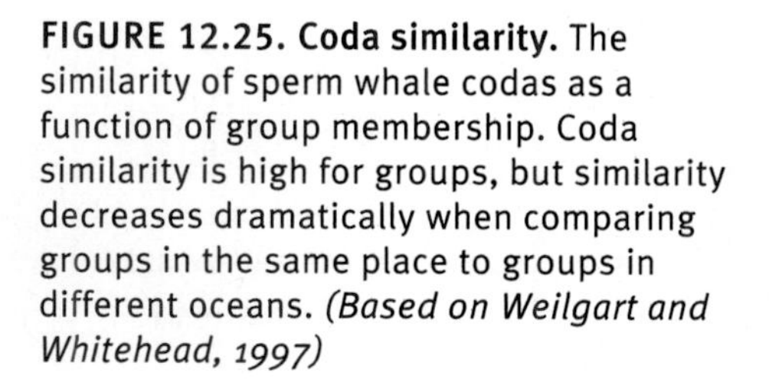
FIGURE 12.25. Coda similarity. The similarity of sperm whale codas as a function of group membership. Coda similarity is high for groups, but similarity decreases dramatically when comparing groups in the same place to groups in different oceans. *(Based on Weilgart and Whitehead, 1997)*

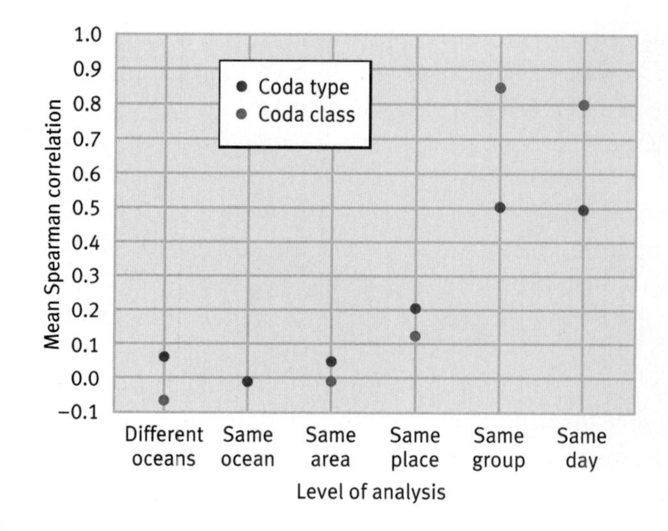

grouped them into thirty coda types. The researchers analyzed the coda types in terms of the individual whales who emitted them, the day they were emitted, the specific locations where the coda types were sung, and the broad geographic areas associated with the coda type.

Coda types from a given group turned out to be much more similar than codas from different groups in the same general locality (Figure 12.25). In addition, there seems to be little degradation of codas—when the same groups were studied over a two-year period, the same codas were generally found to be sung within groups. These two findings suggest that groups develop their own stable dialect (Weilgart and Whitehead, 1997). Although most of the variation in coda type could be accounted for by comparisons of groups in the same area, it was also true that smaller chunks of variance were explained by groups in the same general area being more similar than groups in other areas, and by groups in the same ocean being more similar than groups in different oceans.

Weilgart and Whitehead argue that dialect development within groups can best be explained by a simple social learning process, where younger individuals gradually develop the coda associated with their group by imitating their mother's coda. Young sperm whales start off singing much simpler codas than older whales, and this is consistent with a mother-offspring social learning process. Of course, this in itself is not evidence of social learning, as such a pattern could also be due to a developmentally encoded program that simply changes coda complexity with age. But DNA analysis of groups throughout the area sampled by Weilgart and Whitehead shows little genetic variation between the groups, and hence it is unlikely that genetic differences explain differences in dialect. In addition, humpback whales are known to learn their songs (Payne et al., 1983), and so it is not unreasonable to suggest a similar song-learning process in sperm whales.

SUMMARY

1. Rather than structure the study of communication according to *modes* of communication (visual, vocal, chemical, vibrational, electrical, and so on), we have broken down communication by behavioral venues (foraging, play, mating, aggression, predation).
2. One question that consistently rises to the surface when studying communication is whether it is "honest." One framework to study this question begins by assuming that the exchange of information between signaler and receiver is an attempt by the signaler to *manipulate* the recipient. Recipients are then selected to be good "mind readers."
3. Zahavi's handicap principle hypothesizes that honest signaling can evolve when traits are very costly to fake.
4. In the context of foraging, communication often works to the benefit of the signaler (who may lose some food, but gain from group-level protection from predators) and the recipient (who obtains food at a low cost).
5. Work on water striders, ants, and electric fish demonstrates that new modes of communication are often being discovered—"ripple" communication in the water striders, "stridulation" in ants, and electric discharge in electric fish
6. Researchers have addressed the question of how animals communicate that they are about to enter a bout of play. There are a number of ways that this might take place, one of which is a "play marker," which indicates that a behavior is part of play.
7. Various forms of communication (for example, color change) are employed during aggressive interactions, often to signal relative fighting status.
8. In primates, and perhaps in other taxa, subtle communication tactics allow for the evolution of complex behaviors such as "reconciliation."
9. Deceptive communication, such as fake alarm calls, has been uncovered in primates and birds, suggesting that animals may indeed adopt communication tactics designed to trick others in a manner beneficial to the deceiver.

DISCUSSION QUESTIONS

1. Bacteria often release chemicals that affect other bacteria in their vicinity. Would you consider this communication? If so, why? If not, would you consider the chemical trails that ants use to direct one another to a food source communication? How does that differ from the bacteria case, if at all?
2. One problem in examining whether a communication system more resembles exaggerations or "conspirational whispers" is that it is difficult to know how to define those terms for any given animal social system. How might you overcome this problem? Consider using a comparative study involving many related species.
3. Imagine you are studying a group of amphibian species that vary in their habitats, some living in dense, murky water, and others living in very clear ponds. What sorts of differences in communication systems would you expect to see across such species?
4. Suppose you are studying a heretofore unexamined species of primates. During your observations you note that individuals are often throwing heavy rocks against trees, causing a large "booming" sound. You speculate that individuals are communicating to one another using this technique. How might you go about testing this hypothesis?

SUGGESTED READING

Bradbury, J. W., & Vehrencamp, S. (1998). *Animal communication*. New York: Oxford University Press. A well-done overview of animal communication systems.

Cheney, D. L., & Seyfarth, R. M. (1990). *How monkeys see the world*. Chicago: University of Chicago Press. A wonderful introduction to how vervet monkeys perceive the world. Contains probably the best work on animal deception in the field.

Dawkins, R., & Krebs, J. R. (1978). Animal signals: Information for manipulation? In J. R. Krebs & N. B. Davies, *Behavioural ecology* (pp. 282–315). Sunderland, MA: Sinauer Associates. Also Krebs, J. R., & Dawkins, R. (1984). Animal signals: Mind reading and manipulation? In J. R. Krebs and N. B. Davies, *Behavioural ecology* (pp. 380–401). Sunderland, MA: Sinauer Associates. A pair of book chapters that nicely outline the question of manipulation and mind reading.

Hauser, M. (1997). *The evolution of communication*. Boston: MIT Press. Also Hauser, M., & Konishi, M. (Eds.) (1999). *The design of animal communication*. Cambridge, MA: MIT Press. A pair of books that synthesize biological and psychological perspectives on communication.

Zahavi, A. (1997). *The handicap principle*. New York: Oxford University Press. A book-length treatment on the evolution of traits that are hard to fake, including traits related to communication.

13

Models of Habitat Choice

- The Ideal Free Distribution Model and Habitat Choice
- The IFD Model and Foraging Success

Territoriality

- Territoriality and Learning
- Territory Owners, Satellites, and Sneakers
- How to Keep a Territory in the Family
- Conflict in Family Territories

Migration

- The Challenges of Migration
- The Heritability of Migratory Behavior
- Learning and Migration in Fish

Interview with Dr. Judy Stamps

Habitat Selection, Territoriality, and Migration

Few things affect predation pressure, foraging, mating opportunities, and aggression more than where an animal lives. To see why, let's imagine a male bird who starts off his day near a forest track we have divided into 1,000 zones. Let's say our avian subject starts his day in zone 1 (of 1,000 zones). Where should he spend his time that day? To answer that, our bird must balance any number of factors. If it is the mating season, one question would be: Where in our forest are the females? But before expending the effort both to find a potential mate and to court her, perhaps a meal would be useful. The question would then be: Where is the food? Keep in mind that the prey that our subject eats may also be in the diet of other species, and some of those other species may also prey on the type of bird species we are watching, and so the decision-making process would be even more complicated, as it would now include mates, food, and predators.

As our brisk morning starts to warm up a bit, many factors may strongly influence our bird's choice of where to spend time. Perhaps the best place to find food would be in an open area, where desiccation in the midday sun would be an issue. Perhaps parasites would frequent some of the zones in our field more than others. The point of all this being that the choices faced by animals are complex and multidimensional (Figure 13.1).

Let's imagine that in deciding where to spend its time, our bird ended up often flying between zones 1 and 50, not spending all that much time in any one zone, and not attempting to stop others from using these areas either. In such a case, we might speak of zones 1 and 50 as being its **home range** for that day. In some species, it might be that individuals were nomads; they were constantly wandering and never returning to the same place with any regularity at all. But let's return to our hypothetical bird. If over the course of a longer timescale, our subject frequented zones 1–50 every other day, and in between spent his time in zones 51–100, we might then speak of its home range as being in zones 1–100.

Suppose that, in addition to spending his time in zones 1–100, our male bird did not like having other males in zones 1–100 and actively defended this area from usurpers. Now we can speak of a **territory**—an area occupied and defended by the bird. It might be that while he frequented zones 1–100, our bird only kept intruders out of zones 1–50. In that case, his home range would be twice as big as his territory.

Some species stay in the same territory and home range for long periods of time. Naturally, what constitutes a long period of time will vary dramatically across species, but as an extreme example, some long-lived birds can maintain the same territory for years upon years and even "bequeath" their territory to their offspring.

FIGURE 13.1. Habitat choice. Imagine a redwinged blackbird deciding where to form a territory. All sorts of factors—mates, temperature, predators—play a role in the decision making.

Even animal species with defined home ranges and territories, however, can make dramatic shifts in the habitats in which they live; that is, they can periodically move from one region to another through migration. The most salient examples of this are the large-scale migrations that are common in insects, mammals, and birds. Let's imagine that we continue studying our bird, who wanders about his home range (zones 1–100) and defends his territory along the way. Six months after watching our subject, we note that he actually leaves the forest we have been studying and spends the next six months 2,000 miles south of where we first saw him. At the end of these six months, our bird friend returns back to zones 1–100 and starts back where he left off. In this case, we have home ranges and territories mixed in with large-scale migration. We might, in fact, get any number of permutations here. It may be that 2,000 miles south, our subject has another home range or territory or neither.

Ethologists and behavioral ecologists have made considerable progress in understanding the many issues raised above. For example, through long hours of sometimes backbreaking observation, we know much about the natural history of many species, and in such knowledge lies information about which variables seem most critical to habitat choice in a particular species. In some cases, abiotic (nonliving) factors dominate habitat choice; in other cases, biotic (living) factors are the major influence. The abiotic factors that affect habitat choice may encompass heat, availability of water, wind, "cover" from danger, availability of specific nutrients, and so on. The biotic factors affecting habitat selection may include the location of potential mates, potential food, and potential predators and parasites.

Ethologists have gone far beyond simply creating a laundry list of variables that affect habitat choice. Far more interesting are the studies we shall examine that delve into how such variables interact in determining habitat. A respectable literature on both theoretical and empirical studies of the tradeoffs involved in habitat choice now exists and is rapidly growing. For example, as we shall see below, a series of studies has compared these effects—say, the availability of food and the presence of predators—to see how these two variables fare in relation to one another. Further, researchers are now explicitly incorporating learning into habitat selection and finding fascinating new trends that may pave the way for future work in this area.

In terms of the other focus of this chapter—migration—behavioral ecologists and ethologists have also made considerable strides in the last few decades. Animal behavior researchers have moved beyond simply observing and recording migration. This is not meant to denigrate the fantastic and labor-intensive work still under way documenting monumental migrations like those of the monarch butterflies, but only to suggest there is a lot more migration work these days that focuses on hypothesis testing per se. For example, we now have a much better handle on the physiology and resultant behavioral changes associated with switching from a "nonmigratory" to a "migratory" mode, as well as the costs and benefits of migration. What seemed impossible to study experimentally a generation ago is now commonplace in the study of migration.

Having introduced habitat selection, territoriality, and migration, we will take the rest of the chapter to examine the following topics in more depth:

- Models of habitat choice and territoriality. Ethologists and behavioral ecologists have developed a family of models that make specific predictions regarding habitat choice and territoriality. Such models often examine the various tradeoffs associated with choosing one area over another.
- Territoriality and learning. Until recently, the role of learning in the acquisition and maintenance of territories was fairly under-

studied. We shall examine a model of territoriality and learning and a test of that model.

- Territory ownership, satellites, and sneakers. Once an individual establishes a territory, selection pressures then exist for others to parasitize the benefits built up by territory owners, without paying any of the associated costs. "Sneakers" and "satellites" are two such parasitic strategies we shall examine in detail.
- Family dynamics and territoriality. Oftentimes relatedness plays a role in habitat selection and territoriality. In Chapter 8, we reviewed some evidence that kinship creates "dynasties," wherein territories are passed down from generation to generation. Here we shall expand on this and examine how relatedness can create a conflict regarding territoriality.
- Migration. We shall examine not only the incredible means that animals use to migrate long distances, but the potential costs of migration as well.

Models of Habitat Choice

Before building an explicit model of territoriality per se, it might be useful to examine a more general phenomena, that of a **habitat choice**. While territoriality implies the defense of a set area, habitat choice models examine how animals distribute themselves in space and time with respect to some resource in their environment (Rosenzweig, 1981, 1985, 1990, 1991; Brown and Rosenzweig, 1985; Bateson, 1990; Kacelnik et al., 1992; Kennedy and Gray, 1993; Morris, 1994). The resource in question could range from food to mates to safety, and we shall envision each habitat as having some level of the resource under consideration (for example, habitat X has five food items per second of search time, two potential mates per acre, and so on).

THE IDEAL FREE DISTRIBUTION MODEL AND HABITAT CHOICE

The **ideal free distribution (IFD) model** (J. L. Brown, 1969a; Orians, 1969; Fretwell and Lucas, 1970; Fretwell, 1972; Parker and Stuart, 1976) was developed to predict how animals will distribute themselves among habitats with varying levels of resource availability (Parker and Sutherland, 1986; Korona, 1989; McNamara and Houston, 1990; J. S. Brown, 1998). To see how the ideal free distribution model works, let us consider the simplest rendition of this model. Consider a habitat choice model, where individuals choose between only two habitats, H1 and H2, which have two resources, R1 and R2, respectively. For example, we might consider two food patches as habitats: one provides five food units/minute, and the other allows foragers access to three food units/minute. Imagine that we are studying a population that has N individuals and that individuals can move freely from habitat 1 to habitat 2, and that such moves are cost free. How many individuals should end up in H1 and how many in H2?

The ideal free distribution model predicts that the equilibrium distribution of individuals into patches should be that distribution at which, if any individual moved to the patch it was not in, it would suffer a reduced payoff. That is, at the ideal free distribution equilibrium, any individual that moved from H1 to H2, or vice versa, would obtain fewer resources as a result of its move. It can be shown that this translates into individuals settling in habitats in proportion to the resources available in that patch. The equilibrium proportion of individuals in H1 and H2 should be reached when $R_1/N_1 = R_2/N_2$—that is, when the per capita intake rate of individuals in both patches is equal. This is also known as the **resource matching rule**, as the distribution of individuals at equilibrium matches the distribution of resources across patches.

THE IFD MODEL AND FORAGING SUCCESS

The IFD model has been applied to everything from dispersal (Wahlstrom and Kjellander, 1995) to predation (Kacelnik et al., 1992) to mating decisions (Parker, 1974b; Hoglund et al., 1998) to the interaction of Icelandic fishing trawls (Rijnsdorp et al., 2000a, 2000b). Early work on this model generally focused on foraging behavior, however, and in particular addressed whether: (1) animals distributed themselves in a manner matching resource input, and (2) at such a distribution, all individuals in fact received about the same amount of food.

The first direct test of IFD models of foraging was conducted by Manfred Milinski (1979b). In an elegantly simple experiment, Milinski arranged two feeders that distributed water fleas (*Daphnia magna*) at opposite ends of a tank. Foraging behavior of six sticklebacks was then observed when the feeders produced fleas in a 5:1 or a 2:1 ratio. After some initial sampling, the fish in each treatment distributed themselves under feeders in a ratio similar to the input ratio at the feeders (Figure 13.2). While Milinski's work was the first experimental work demonstrating the resource

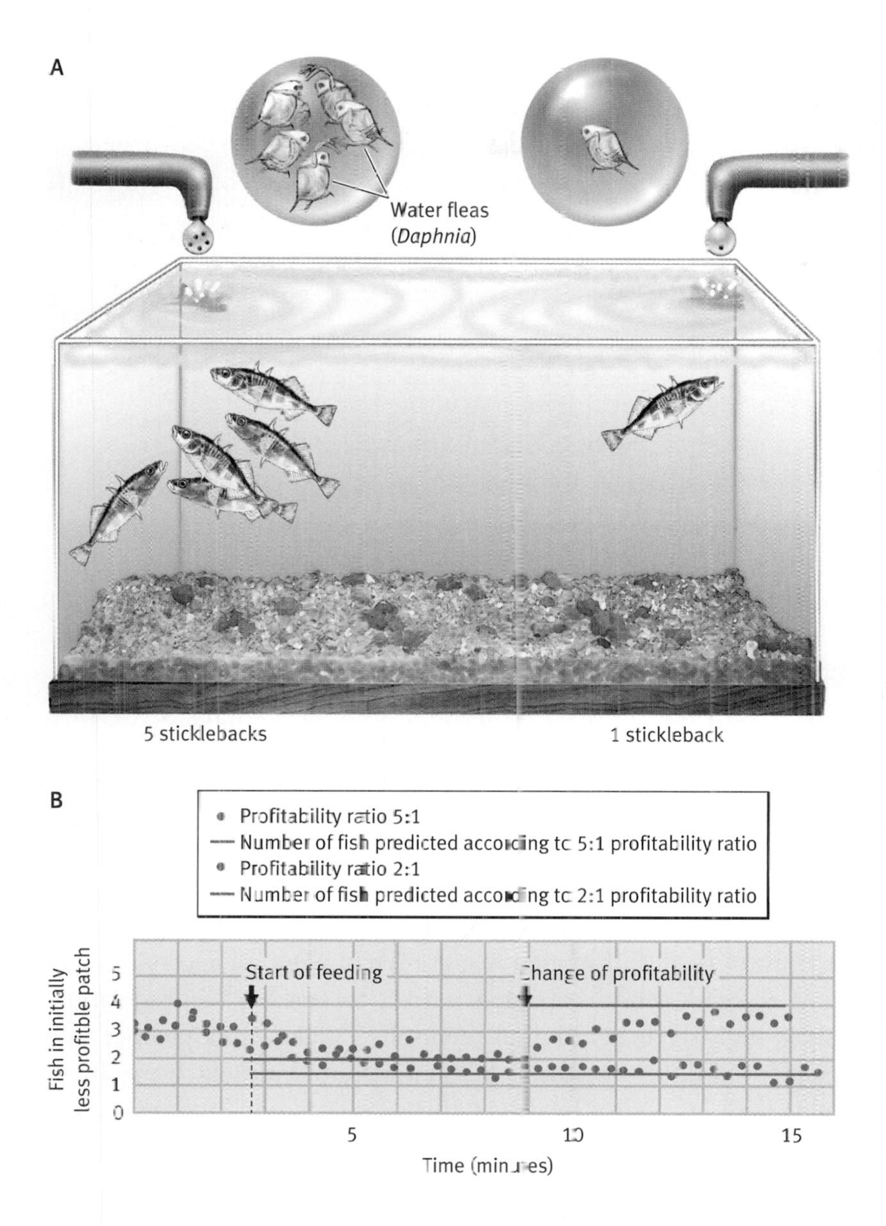

FIGURE 13.2. Ideal free fish. **(A)** When sticklebacks are presented with two foraging patches that produce food at a 5:1 ratio (5 water fleas being dropped into the tank from the left feeder versus 1 water flea being dropped into the tank from the right feeder), they distribute themselves in accordance with predictions from the ideal free distribution model. **(B)** Milinski tested the ideal free distribution model using sticklebacks foraging from two feeders on opposite sides of an aquarium. In his first experiment, one feeder was five times as profitable as the other. The arrow on the left indicates the start of feeding. The green points indicate the number of fish in the less profitable patch, and the green line shows the predicted number of fish at the less profitable patch under a profitability ratio of 5:1. In his second experiment, Milinski shifted the profitability of stickleback foraging patches to a ratio of 2:1. The orange points indicate the number of fish in the less profitable patch, and the orange lines show the predicted number of fish in the less profitable patch under a profitability ratio of 2:1. The arrow on the right indicates the point at which the profitabilities of the more and less profitable patches were reversed. *(Based on Milinski, 1979b)*

matching rule, it was not designed to examine the feeding success of individuals, and hence it could not determine whether individual foraging success was approximately equal across foraging patches. For that, we must turn to David Harper's (1982) work on mallard ducks.

Using individually recognizable mallards (*Anas platyrhynchos*) in a lake at Cambridge University, Harper ran an IDF experiment similar to that of Milinski. Two observers were stationed 20 meters apart, and each threw pre-cut, pre-weighed pieces of food into the lake. Thus, the observers acted as data takers *and* as the food stations themselves. Harper varied the profitability of a patch by varying either the number of pieces of food thrown or the weight of each piece (Figure 13.3).

FIGURE 13.3. Ducks feeding at a pond. (A) One of the first controlled experiments on the ideal free distribution model involved ducks feeding at a pond. **(B)** To test the ideal free distribution model, food was thrown into a pond from two locations, and the distribution of ducks to each feeding station was recorded. *(Photo credit: Marc Epstein/ Visuals Unlimited)*

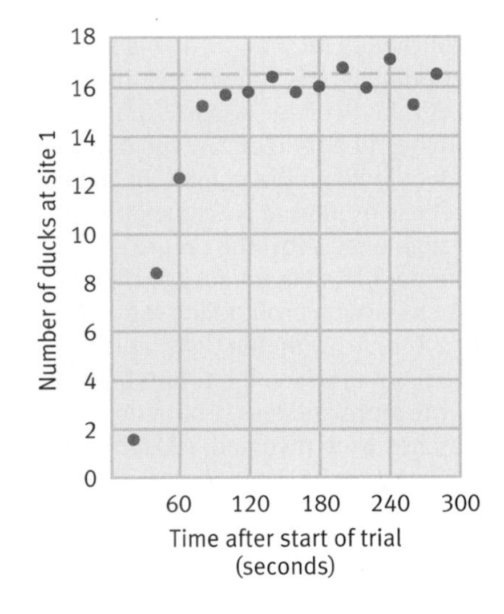

FIGURE 13.4. Ideal free ducks? Harper stationed two individuals along a pond to create foraging patches for ducks. Bread was thrown into the pond at each station. Here the patches had equal profitability. The dashed line represents the predicted number of ducks at site 1. *(Based on Harper, 1982)*

When equal amounts of food were thrown in by both observers, ducks quickly distributed themselves in a 1:1 ratio (Figure 13.4), and similarly they distributed themselves in a 2:1 ratio when one observer threw in twice as much bread as the other. During these trials, however, Harper noted that some individuals seemed to be very aggressive and tended to receive a disproportionate amount of food. Subsequent experimental work on aggression and foraging success in these ducks indeed confirmed this observation. While individuals distributed themselves across food patches in a manner similar to the input of each resource into each patch, all individuals were *not* receiving the same amount of food across patches, and as such, the ducks' behavior might be better represented by "despotic" models of patch distribution (Sutherland, 1983; Sutherland and Parker, 1985; Parker and Sutherland, 1986).

Territoriality

Having discussed the distribution of individuals among habitats, we can now move on to examine territoriality—the occupation and defense of particular areas. Most of our discussion of territoriality will focus on territories that contain either a single individual or a family. Territories are sometimes defended by groups of individuals, however, and this sort of group territoriality can often result in dramatic intergroup interaction.

Groups may attempt to enlarge a territory or take over another group's territory. For example, consider the "between-group raiding" that is occasionally seen in natural populations of chimps. Between-group interactions often appear to be "warlike," and in fact they resemble the raiding behavior so common among many tribes of humans (Boehm, 1992). During raids, all-male chimpanzee "patrol groups" often travel into areas that border their territorial boundaries (Bygott, 1979; Nishida, 1979; Goodall, 1986). In contrast to excursions for food, in which vocalizations are common, patrols move in a wary fashion and remain silent (Goodall, 1986). These raids often involve the killing of a small number of members of the raided group and the capture of females. Occasionally, raiding parties from two groups will meet one another. Rather than all-out aggression, both groups will engage in hostile vocalizations and then withdraw (Goodall, 1986). Despite the fact that all-out warfare does not emerge when two raiding parties meet, raiding can, in the long run, amount to the extinction of one group. For example, Nishida and his colleagues (1985) provide evidence that raiding behavior in the Mahale Mountains of Tanzania amounted to a larger group extinguishing a smaller group of chimps.

TERRITORIALITY AND LEARNING

While most models of territory acquisition assume that animals are capable of assessing various characteristics about a potential territory, they do not explicitly consider how learning (as defined in Chapter 4) affects the establishment and maintenance of a territory per se. Below we discuss two models that examine how learning affects decisions regarding both the acquisition and subsequent maintenance of territories.

TERRITORIALITY AND LEARNING DURING SETTLEMENT. Judy Stamps has been studying territoriality in juvenile *Anolis aeneus* lizards for more than twenty years (Stamps, 2001). These tiny lizards form territories early in life, and Stamps was interested in, among many other things, where and how juveniles decided to stake out a territory (Figure 13.5). Her initial inclination was to test for the importance of food availability in structuring territoriality in *A. aeneus*. Yet, despite numerous elegant experiments manipulating food availability, she did not uncover a clear-cut primary effect of food availability on territory formation. Rather, in subsequent experiments, Stamps found that protection from predators and suitable temperature were the most important attributes of a desirable territory.

FIGURE 13.5. Territorial lizards. (A) Juvenile *Anolis aeneus* are very territorial, and their territory formation has been studied in the context of habitat choice and learning. **(B)** *Anolis aeneus* stake out territories in areas such as those depicted here. *(Photo credits: Judy Stamps)*

Stamps next turned to the question of *how* juvenile lizards determined what territories were suitable with respect to temperature and predation pressure. More specifically, given that these factors affect which areas lizards choose as territories, do lizards *learn* what areas are best from their interactions with other lizards? Stamps noted that juvenile lizards appeared to be very interested in what other lizards were doing and reasoned that they might be determining territory quality as a result of their interactions with conspecifics. If another individual has determined that a territory is safe and doesn't overheat, then it probably is safe and doesn't overheat. This led Stamps to develop field experiments to examine this conspecific cueing hypothesis.

Stamps (1987a) began testing her conspecific cueing hypothesis by examining whether a territory that previously had an owner would be viewed as more attractive to other lizards. A juvenile would be allowed to observe two equally good territories, one currently occupied, one vacant. When given the choice between these two areas, with the territory owner now removed, juveniles showed a strong preference for the previously inhabited area (Figure 13.6). Furthermore, juveniles who entered the test arena but did not observe the territories during the initial part of the experiment displayed no preference for the previously occupied territory, suggesting a strong visual component to conspecific cueing in *A. aeneus* (Stamps, 1987b).

Based on her conspecific cueing hypothesis and her subsequent tests of this idea, Stamps developed a learning-based model of territory formation that produces patterns remarkably similar

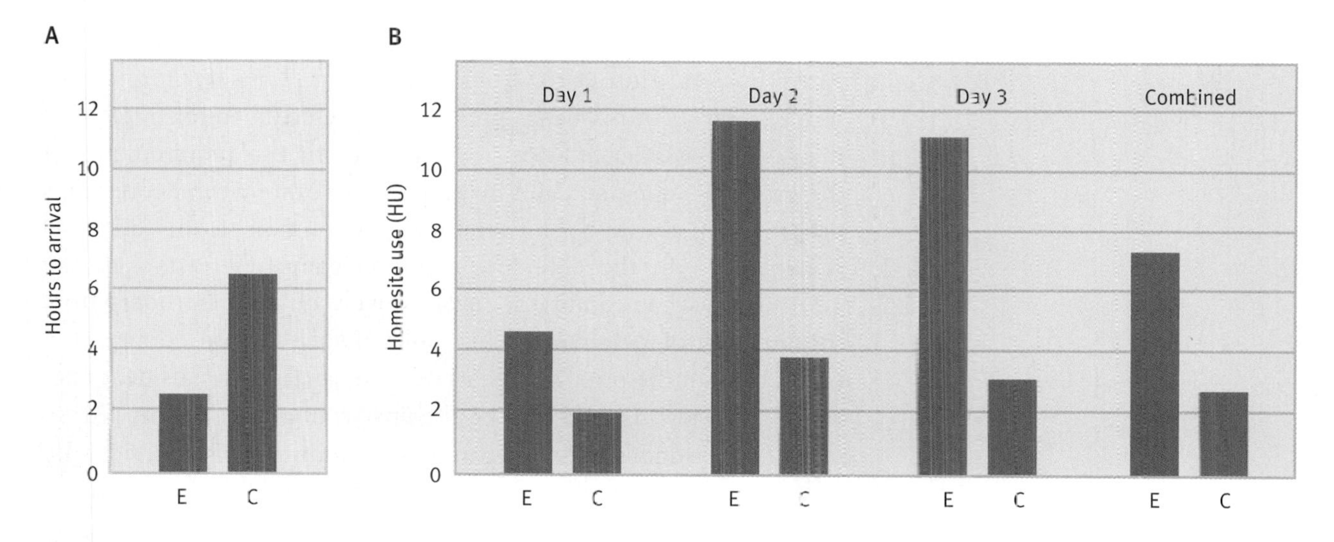

FIGURE 13.6. Territorial lizards. (A) Focal juveniles not only spent more time on experimental (E) versus control (C) homesites, they also arrived at experimental homesites more quickly. **(B)** Juvenile lizards were drawn to experimentally manipulated homesites (E) over control (C) homesites. Control territories remained empty prior to the focal's choice, while experimental territories had formerly contained a territorial juvenile. *(Based on Stamps, 1987a)*

to those seen during territory formation in *A. aeneus* (Stamps and Krishnan, 1999).

TERRITORIALITY AND LEARNING AFTER SETTLEMENT. The role of learning in territory-related decisions does not end after a territory has been acquired. For example, once an individual has settled on a territory, it might pay for it to learn who lives as its neighbors in the adjacent areas. Territoriality may foster cooperative bonds between neighbors, and it may even lead to what has been called the **dear enemy effect** (Fisher, 1954). Neighbors, though competitors with respect to certain variables, may have common enemies, such as other unknown individuals attempting to take over a territory in the area of interest. Krebs (1982) portrays the dear enemy phenomenon as one in which territorial neighbors are at first involved in a costly battle over setting up territory boundaries. Once these boundaries are established, however, neither party stands to gain by having to "renegotiate" such boundaries with a new neighbor, and so a new cooperative, nonaggressive component is added to neighbor relations—producing dear enemies (Falls, 1982; Ydenberg, 1988). In this light, the dear enemy strategy has been defined as: (1) not challenging your neighbor for portions of its territory, and (2) not being vigilant against your neighbor, on the assumption that it can be trusted (Getty, 1987).

Getty employs the prisoner's dilemma, and the economic metaphor of the oligopoly to model: (1) the dear enemy phenomenon, and (2) active coalitions among neighbors to inhibit third-party "floaters" (intruders) from challenging either territory holder (Getty, 1987). In his dear enemy model, Getty demonstrates that the payoffs associated with dear enemies fit the requirements of the prisoner's dilemma, and he proceeds to examine the conditions under which the tit-for-tat strategy is a solution to this dilemma (Chapter 9). In the defensive coalition game, Getty tackles the question of what conditions favor actively chasing intruders from the territory of neighbors. This game involves more complicated behavior than the prior model, as not only are individuals dear enemies, but they *aid* one another against a common enemy. Again using the prisoner's dilemma game, he examines two strategies: "Be a dear enemy, but don't help your neighbor" and "Be a dear enemy and help your neighbor." Results indicate that tit-for-tat can be an evolutionarily stable strategy to this game.

FIGURE 13.7. Hooded warblers. Economic models of the dear enemy phenomenon have been tested in hooded warblers. *(Photo credit: Rob Simpson/Visuals Unlimited)*

Hooded warblers (*Wilsonia citrina*; Figure 13.7) appear ideal to test Getty's dear enemy ideas, as individuals in this species recognize their neighbors and are involved in many interactions with them over a long, but indeterminate period of time (Godard, 1991). As such, Godard (1993) examined whether hooded warblers used a tit-for-tat, or TFT-like strategy in the context of neighbor interactions (see Chapter 9). Using a taped playback system, she simulated an intrusion deep into male A's territory (Figure 13.8). Intrusions were playbacks of either a territorial neighbor (male B) or of a stranger. Male A's responses to a playback of male B's song were examined before and after such an intrusion. Using seven different behavioral measures, Godard (1993) found that male territory owners were much more aggressive to neighbors after an intrusion if the neighbor's song was played deep within that male's territory.

Godard noted that heightened aggression among territorial neighbors was evident on the day of a neighbor intrusion, but not on subsequent days, suggesting the possibility that warblers forgive neighbors if they do not repeat the same offense. Of course, an equally plausible argument is that hooded warblers simply forget about an intrusion over the course of time (Healey, 1992). Which of these explanations, if any, best explains Godard's results remains an open question.

TERRITORY OWNERS, SATELLITES, AND SNEAKERS

As we have seen, territory ownership requires constant vigilance against intruders. But the relationship between vigilance and territoriality can be even more complex. Here we shall examine this complexity by addressing two issues. First, do some individuals

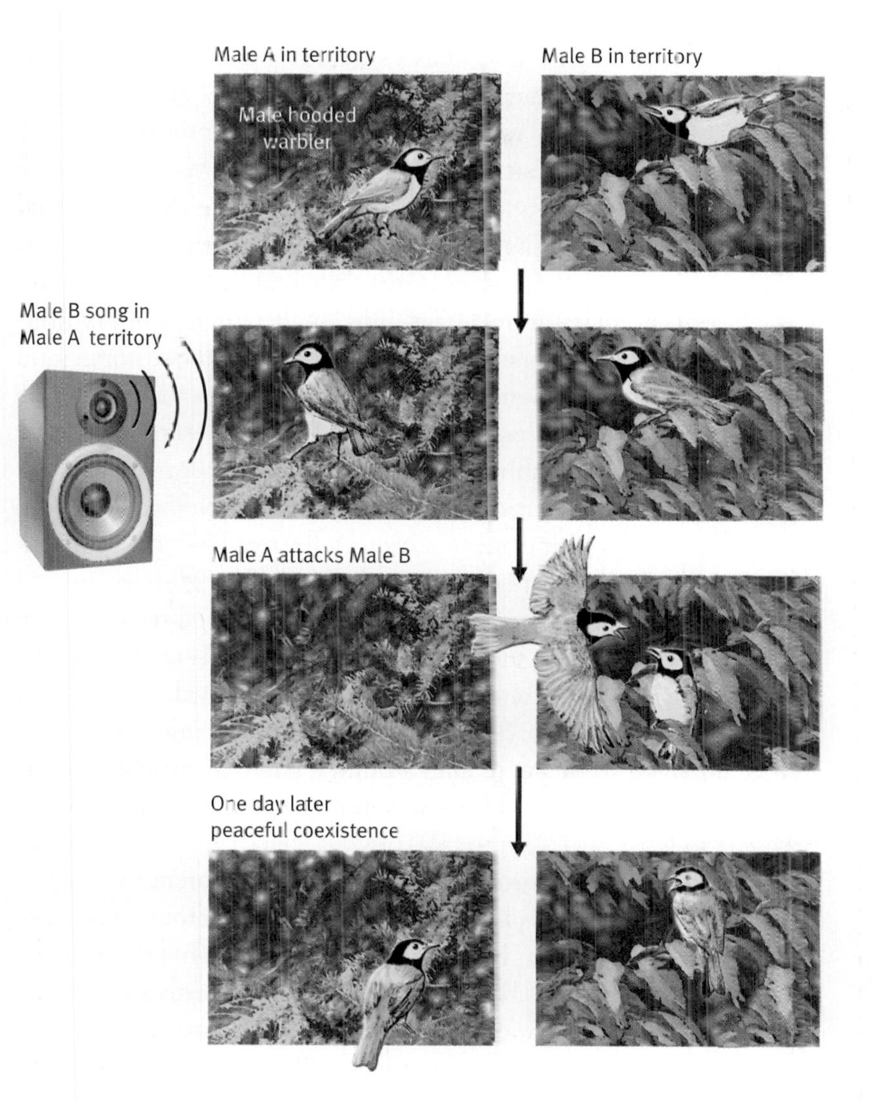

FIGURE 13.8. Dear enemies. The dear enemy phenomenon was tested using a playback experiment in male hooded warblers.

lurk in the mist and parasitize the hard work that others have devoted to establishing a good territory? In this context, we shall examine "sneakers" and "satellites," which don't attempt to take over a territory but rather use more sophisticated strategies to parasitize territory holders (see Chapter 6). Second, the dear enemy work just outlined demonstrates that under some conditions neighboring territory owners tolerate one another, and may even form active coalitions against third parties. One can extend the logic of the dear enemy scenario a bit and ask whether territory owners may temporarily allow other individuals on their territories. That is, do the costs and benefits ever play out such that it is in a territory holder's interest to temporarily allow other individuals onto its territory?

SNEAKER AND PARENTAL MALES IN SUNFISH. It is one thing for a male territory owner to have to repel challenges from other individuals trying to take over his territory, but it's another thing when an entire class of males has its behavior and morphology designed to usurp the benefits of territoriality. Yet, this is precisely what the male bluegills we first discussed in Chapter 6 face on a constant basis. Recall that Mart Gross found three types of bluegill sunfish males: parental, sneaker, and satellite males (Gross and Charnov, 1980; Gross, 1982). "Parental" males are older individuals who aggressively defend territories in which females shed their eggs. "Sneaker" males are younger individuals who do not hold territories. Instead, they swim in the area of parental territories and dart into such areas when a ripe female appears to mate with the parental male in the area of his nest. Sneakers then quickly shed their sperm. A remarkable 60 percent of bluegill egg releases involve some sort of intrusion, almost exclusively by sneakers, with 16.9 percent of the intruders successfully reaching the females' eggs. Not surprisingly, parental males are very aggressive toward sneakers when they discover such individuals. "Satellite" males are intermediate in age to parentals and sneakers. They mimic the color pattern of female bluegills, and they descend slowly down in a parental male's territory when a female enters. Satellites mimic female behavior while swimming around a parental male and a real female, and then they shed their sperm at the same time as the parental male.

Neither satellites nor sneakers ever become parentals. Rather, there appear to be two distinct life-history paths that males can take. One path involves diverting resources from growth to reproduction early on. This path leads to smaller individuals designed to deliver sperm at a very young age—that is, sneakers. Sneakers, should they live long enough, become satellites, but never parentals. The alternative male life-history path is to divert resources toward growth instead of reproduction. This in turn produces larger, but older, parental males, who use their size to secure territories, thereby attracting females (this phenomenon has also been observed in other species—for example, in salmon; Gross, 1985).

FIGURE 13.9. Territorial wagtails. Economic models of territorial ownership, and the acceptance of intruders, have been put to the test in pied wagtails. *(Photo credit: Charles McRae/Visuals Unlimited)*

OWNERS, SATELLITES, AND TERRITORY DEFENSE IN PIED WAGTAILS. From the perspective of a territory holder, intruders need not always be strictly a liability. To see an example of how this might work, let's look at pied wagtail (*Motacilla alba*) behavior with respect to territorial economics (Figure 13.9). Pied wagtails defend riverside winter territories, foraging on insects that wash up on the banks (Davies, 1976; Figure 13.10). This food source is "renewable" in the sense that, after a period of time, prey abundance on a territory again increases after being depleted.

Davies and Houston (1981) demonstrated nicely how pied wagtails systematically search their territories, providing time for de-

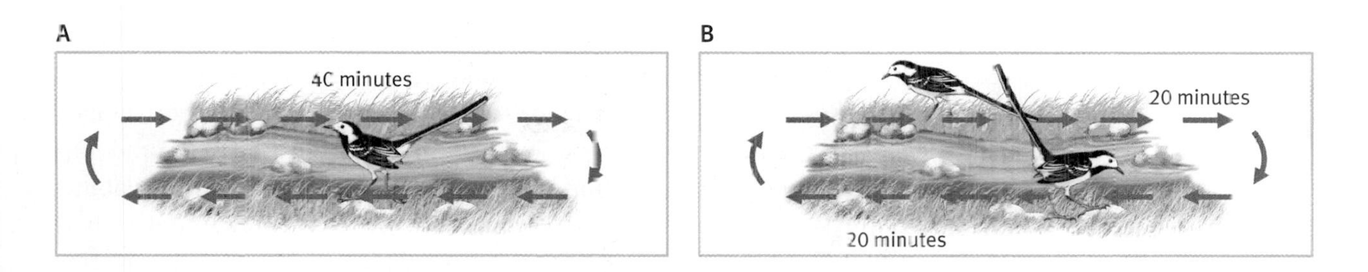

FIGURE 13.10. Pied wagtails and food search. Pied wagtails systemically search for food on their territories along riverbanks. **(A)** A single bird can complete a circuit of the riverbank in 40 minutes and gets all the food it finds. **(B)** When a territory is shared by two birds, however, the circuit is divided up as well, and so each bird primarily gets the food that it finds in its 20 minutes of walking the riverbank. *(Based on Davies and Houston, 1984)*

pleted resources to again increase. But life on a pied wagtail's territory isn't as easy as simply waiting for a new crop of insects to be washed ashore, as territory owners must often deal with intruders. At times, such intruders are tolerated, at which point they are called satellites, and at times, they are aggressively chased off a territory. Davies and Houston wanted to know what determines whether the satellites are tolerated or defended against. When answering this, it is important to note that a territory is worth more to a pied wagtail owner than to an intruder because intruders do not know which areas have recently been cropped, while owners do. If an owner permits an intruder to stay, it loses some foraging-related benefits, but it also gains assistance in territory defense against further intruders (Table 13.1). Having a satellite around makes a territory owner's residence a bit safer, but not quite as profitable in terms of food.

TABLE 13.1. Owners and satellites. Pied wagtails form winter foraging territories along a river. Owners often allow "satellite" individuals to forage on such territories. Some territorial defense is provided by the satellites. *(Based on Davies and Houston, 1981)*

TERRITORY	NUMBER OF DEFENSES BY		PERCENT DEFENSES BY SATELLITE
	OWNER	SATELLITE	
1	13	5	27.8
2	2	1	33.3
4	8	2	20.0
5	6	6	50.0
7	44	22	33.3
8	8	7	46.7
9	13	15	53.6
Total	94	58	38.1

Davies and Houston devised a model, called the owner's dilemma, and they used this model to predict when owners would allow satellites on their territories. As predicted by their model, they found that when the benefits of having a satellite around outweighed the costs—for example, on days when food availability was particularly high—territory owners tolerated satellites; otherwise, they evicted such intruders. This is a particularly interesting case of conditional cooperation because it suggests that the costs and benefits of byproduct mutualism (see Chapter 9) can vary depending on territory-holding status. Would-be satellites, who have no foraging territories of their own, always face a harsh environment with respect to food allocation and are willing to pay the costs of territory defense in order to have access to food on territories. When food becomes more sparse for owners, however, they do not allow intruders in. So, intruders are always willing to undertake cooperative territory defense in exchange for food, but the costs and benefits of byproduct mutualism do not always select for owners to permit this cooperative act.

HOW TO KEEP A TERRITORY IN THE FAMILY

FIGURE 13.11. Scrub jays and territories. In Florida scrub jays, territories are "inherited" across generations, leading to the establishment of "family dynasties." Territory size is increased through a process called budding. *(Photo credit: Charlie Heidecker/Visuals Unlimited)*

Rather than having to establish a territory from scratch, territory ownership is "inherited" in Florida scrub jays (*Aphelocoma coerulescens*; Figure 13.11). Unlike closely related species in the western United States in which the young disperse from their natal territories before their first birthday, Florida scrub jays have an elaborate system of helpers-at-the-nest, resulting in an interesting territorial inheritance system and sometimes leading to "family dynasties" (Fitzpatrick and Woolfenden, 1984, 1988; Woolfenden and FitzPatrick, 1990; Quinn et al., 1999).

For more than thirty years, Glen Woolfenden, J. Fitzpatrick, and their colleagues have been studying Florida scrub jays at the Archibold Biological Station to "record all events pertinent to survival, movement and reproduction," with the goal of testing various theories in ecology, evolution, and behavior (Woolfenden and FitzPatrick, 1990). In this regard, they have been able to generate vast amounts of data regarding territory ownership and have been able to put together the following picture.

Each Florida scrub jay territory is initially "owned" by a monogamous pair of birds. In addition, territories often contain up to six helpers-at-the nest (see Chapters 3 and 8), and all individuals, including helpers, defend the territory. When territories are formed, they may be relatively small. But as family size increases, so does territory size (up to a point). When one territory increases in size, it is almost always at the cost of an adjacent territory. The way this typically occurs is through a process known as territorial budding (Woolfenden and Fitzpatrick, 1978). At about two and a half years of age, a male, with the help of his family members, will expand the family territory and actively defend this new area as his own. While there are some aggressive interactions between the male who has acquired this "budded" territory and the family members in adjacent territories, these are rare when compared to the aggressive acts between such a male and any nonfamily members that border his territory. As such, the budding procedure serves to increase a family's total territory size, provides a male with his own smaller territory, and essentially creates a form of inheritance. This inheritance system, with larger families outcompeting smaller families, may lead to very large territorial family dynasties.

CONFLICT IN FAMILY TERRITORIES

Under some conditions, different family members on a territory may come to different conclusions about when individual X should breed. For example, it may be in individual X's best interest to breed, while it may be in the best interest of other family members on the territory for X not to breed. Conversely, it may be in X's interest not to breed, but in the interests of other family members for X to breed. In fact, there is a burgeoning literature on **optimal skew theory**, which studies the distribution of breeding within a group and whether there will be cooperation or conflict over reproductive activities (see Chapter 8). For a good example of conflicting interests in terms of territoriality and breeding, we turn to Steve Emlen's work on parent-offspring conflict over breeding in white-fronted bee-eaters (Emlen and Wrege, 1992).

Like the young of many communally breeding species (J. L. Brown, 1987), young male white-fronted bee-eaters (*Merops bullockoides*) often forego the opportunity to breed on their own. Such males often remain on their natal territory and aid their relatives, usually their parents, in raising young (usually their siblings). When breeding opportunities for young males are rare, no conflict exists between young males and their parents—it is in the best interest of both parties for the male to remain at home and help his parents. The situation gets more complex when breeding opportunities away from the natal nest are more readily available. Under such conditions, it may be in the best interest of a young male to breed on his own, but in the best interest of such a male's parents for him to remain at home and help raise his siblings. This leads to conflict between offspring and parent.

The conflict over breeding arises out of the benefit structure associated with helping. Pairs without helpers raise on average

0.51 offspring. Every helper a pair has adds approximately 0.47 offspring. Keeping in mind that young males who breed for the first time will rarely have a helper, let's start with the perspective of a young male's parents. If their son attempts to breed, he will, on average, produce 0.51 offspring. If he helps them, he will on average add 0.47 offspring to their next clutch. But parents are twice as related to the offspring they produce as they are to the grandoffspring their son will produce. Hence, parents have an incentive to keep offspring around to help, and they will try to block any attempt by their son to mate. The accounting is slightly different from the young male's perspective. Since he is equally related ($r = 0.5$) to his own offspring and his siblings, the incentive to resist attempts to suppress breeding may be minor. Hence, we end up with parental suppression of offspring breeding met by little resistance on the part of their son. This interpretation is bolstered by the fact that when a breeding pair tries to suppress the reproductive efforts of more distant kin, or even nonkin, their actions are met with much stiffer resistance.

Migration

While we often think of choosing a habitat as a means by which a more sedentary lifestyle is achieved, it need not always be the case. For animals that migrate, particularly those that migrate long distances, choice of the appropriate habitat throughout a year may require vast amounts of energy to be spent on traveling.

> Twice each year, billions of birds, entire species, swarm across the globe, travelling thousands of miles as they follow the sun to populate regions that are habitable for only part of the year. The spatial scope of these migrations exceeds all other biological phenomena. So fantastic are they that ancient civilizations devised a host of myths to explain the periodic appearance and disappearance of such vast numbers of animals. Those apocryphal stories were concocted in part because what we know to be true seemed then so completely beyond the pale. It seemed more likely that swallows buried themselves in the mud at the bottom of ponds than that they flew all the way from Europe to Africa and back twice each year. But the truth turned out to be more amazing than the myth. (Able, 1999, p. vii)

From wildebeests swarming across the African plains to monarch butterflies heading by the tens of millions to Mexico each year, migration is certainly one of the more spectacular of all animal behaviors (Figure 13.12). In some species, annual migration is obligatory; in some, it occurs only when conditions become poor (irruptive mi-

FIGURE 13.12. Animal migration. In some species of birds and mammals, massive yearly migrations take place. Here, we see migration in **(A)** geese and **(B)** gnu. *(Photo credits: William J. Weber/Visuals Unlimited; Joe McDonald/Visuals Unlimited)*

gration); in some cases, only a portion of a population will migrate and the rest will stay put (Able, 1999). No matter where a species falls on this migration continuum, questions abound—how do they know where to go, when to go, how to go, how to prepare?

THE CHALLENGES OF MIGRATION

The literature on animal migration is huge (Heape, 1931; Griffin, 1964; R. R. Baker, 1978; Gauthreaux, 1980, 1997; Rankin, 1985; Alerstam, 1990; Able, 1999). Rather than attempt to summarize this work in a concise form (an impossible task), we shall briefly touch on two ethological questions regarding migration: (1) How do animals prepare for long migrations and what awaits them at the end of such migrations?, and (2) What are some of the means by which animals manage to find the far-off places to which they migrate?

DEFENSE AGAINST PARASITES. Migratory birds face not only extreme levels of exercise that can reduce immune responsiveness (Hoffman-Goetz and Pedersen, 1994) and increase the level of potentially dangerous and mutagenic free radicals (Leffler, 1993), but long-distance migrants also face new parasites and diseases upon arrival at their migratory end point. That is, while nonmigratory birds must combat parasites in one environment, migratory birds face that challenge in two very different environments and are thus combating a more diverse array of potential parasites. As such, Møller and Erritzøe (1998) hypothesized that migratory birds should invest more heavily in immune function compared to related nonmigratory relatives.

Møller and Erritzøe tested their hypothesis by comparing the size of immune defense organs in pairs of bird species. One member of each pair was a migratory species and the other pair member was a closely related species that was nonmigratory. The immune defense organs that were compared were the bursa of Fabricius and the spleen. The assumption was that a larger size in either of these defense organs would provide better immunological resistance to parasites (Toivanen and Toivanen, 1987; John, 1994). What Møller and Erritzøe found was that in nine of ten pairwise comparisons, the bursa of Fabricius was larger in the migratory species, while in nine of thirteen pairwise comparisons, the spleen was larger in the species that migrates.

Are there other variables besides the tendency to migrate or not that might explain differences in the size of immune defense organs when comparing the pairs of bird species that Møller and Erritzøe tested? The authors themselves note that differences in sexual selection pressure, mating systems, and the pattern of nest use and reuse are also known to affect investment in immune defenses (Møller and Erritzøe, 1996). As such, they reanalyzed their data, using only pairs of species that were known not to differ on any variable (besides migratory tendencies) that might affect the size of immune defense organs. While this comparison lowered their sample size to six pairwise comparisons for both the bursa and the spleen, their initial findings remained unchanged, suggesting an important relationship between migratory habits and the immune system.

FIGURE 13.13. Bunting migration. Among other cues, indigo buntings use the stars as a navigation tool while migrating at night. *(Photo credit: Maslowski/Visuals Unlimited)*

INDIGO BUNTINGS AND NAVIGATING THE STARS. The challenges posed by large-scale migration during the daytime hours seem magnified at night. Yet, the vast majority of passerine birds migrate almost exclusively at night (Able, 1999). What cues might such birds use when traveling after the sun has set? One of the earliest and best-known studies of how migrating individuals might use the stars to guide them was undertaken by Stephen Emlen. Building on prior ethological work by Kramer and by Sauer (1957), Emlen undertook an ingenious set of experiments involving indigo buntings (*Passerina cyanea*), a nocturnal migratory species that travels 2,000 miles each winter from the northeastern United States to the Bahamas, Mexico, and Panama (Emlen, 1967, 1970, 1975; Figure 13.13).

Emlen began by constructing funnel-shaped test cages for buntings, at the bottom of which he placed an ink pad. He constructed cages such that each time a bunting tried to fly out, the lo-

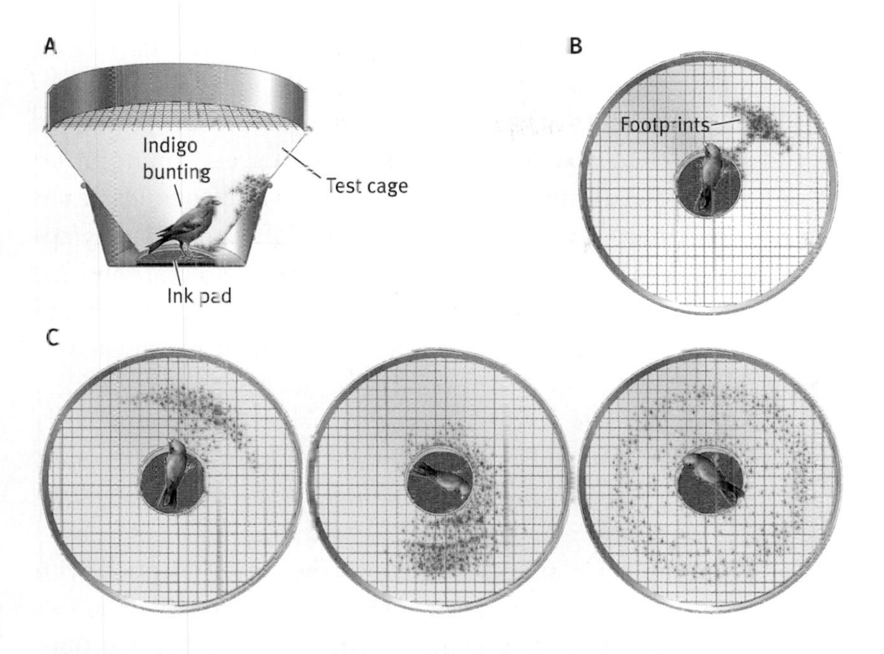

FIGURE 13.14. Migration and orientation. (A) A cross-sectional view and **(B)** a top view of the circular test cage used by Emlen in his orientation/migration work in indigo buntings. The funnel portion of the cage was made of white blotting paper, with an ink pad at the bottom. The entire apparatus was placed in an outdoor cage. The migration tendencies of the buntings were recorded as the bird tried to hop out of the funnel. **(C)** Each time the bunting hopped in one direction, it appeared as black footprints on the blotting paper. The left footprint records show a bunting during spring (when it orients north), the middle footprint records show a bunting during fall (when it orients south), and the right footprint records show a bunting on nights when the stars are obscured (and it hops randomly with respect to direction). *(Based on Emlen, 1975.)*

cation of its footprint was marked by ink, and so its orientation pattern was easily recorded (Figure 13.14). When such cages were placed under a star-lit sky, buntings clearly oriented their attempts in the direction they would normally migrate—southerly in September and October, and northerly in April and May. In addition, these patterns all but disappeared on cloudy nights when stars were not visible, suggesting a strong role for some sort of star-based navigational system.

Emlen repeated his experiments inside a planetarium, where he could have a lot more control over exactly what the birds were exposed to. Results were very similar to those obtained in the field, but Emlen was now able to artificially shift the position of the North Star. In response to this manipulation, buntings shifted to the "new south" or "new north" (depending on the season). Emlen was able to further demonstrate that buntings appear to use the geometric patterns of stars in northern skies as one guide on their long journey. In fact, buntings appear to employ a remarkably adaptive system for using the stars around the North Star to lead them on their migrations.

There are numerous constellations that fall within 35 degrees of the North Star, and buntings appear to be able to cue in on more than one of these constellations. Removal of one star from the normal pattern does not throw the birds off track, as they apparently fall back to relying on the stellar information that remains. Emlen sums it up nicely by noting "since birds frequently migrate on nights when there is variable cloud cover, such redundancy is obviously adaptive."

MAGNETIC BUTTERFLIES. Few natural spectacles are as magnificent as seeing tens of millions of monarch butterflies migrating each year from North America to the mountain ranges of central Mexico. During their annual migration, the air is so thick with monarch butterflies that branches of trees have been known to collapse from the weight of too many butterflies (Figure 13.15). Monarchs traverse up to 6,000 miles on their migratory trip, and they do so the first time out. How is that possible? How can a creature that weighs half a gram navigate thousands of miles on the first try?

Based on earlier work, Jason Etheredge and his colleagues already knew that monarch butterflies use the sun as a compass of sorts when heading down to Mexico (Etheredge, 1999; Perez et al., 1997, 1999). But there are lots of cloudy and rainy days during their migrations. On those days, the butterflies don't sit put and wait for the sun, and they don't get lost. How do they get their bearings then?

Given what was already known about migration in birds and fish, Etheredge and his colleagues hypothesized that monarchs use a magnetic compass to fly southwest to Mexico each year (Wiltschko and Wiltschko, 1995). This is as not as strange a hy-

FIGURE 13.15. Monarch migration. Migration of monarch butterflies can involve tens of millions of individuals. *(Photo credit: Fritz Polking/Visuals Unlimited)*

pothesis as it might seem on the surface, as monarchs are thought to contain magnetite, a mineral suspected of being involved in magnetic orientation in other species (MacFadden and Jones, 1985; Jungreis, 1987).

In an attempt to test their hypothesis, Etheredge and his team captured some monarchs during their migration and brought them back to the laboratory. They divided the butterflies into three groups. They tested one group under normal magnetic conditions, one in the absence of a magnetic field, and the last group under a reverse magnetic environment. Those tested in a normal magnetic field flew southwest (that is, in the direction of Mexico); those tested in an environment with no magnetic field flew in no particular direction; and those exposed to a reverse magnetic field flew northeast—the exact opposite direction migrating butterflies normally head (Figure 13.16). This is just what one might expect if magnetic field played a role in monarch butterfly migration. Nonetheless, there is a cautionary lesson here about unexpected experimental problems. About a year after this article was published in the prestigious *Proceedings of the National Academy of Sciences*, the authors retracted the paper. The retraction read as follows:

> The positive response to magnetic fields in two experiments cannot be repeated. Further experiments show the false positives in these tests result from a positive taxis by the butterflies to the light reflected off the clothing of the observers. We therefore retract our report. We regret the inconvenience that publication of this study may have caused.

This provides an important lesson about how controlled one's experiments must be when studying something as complex as migration.

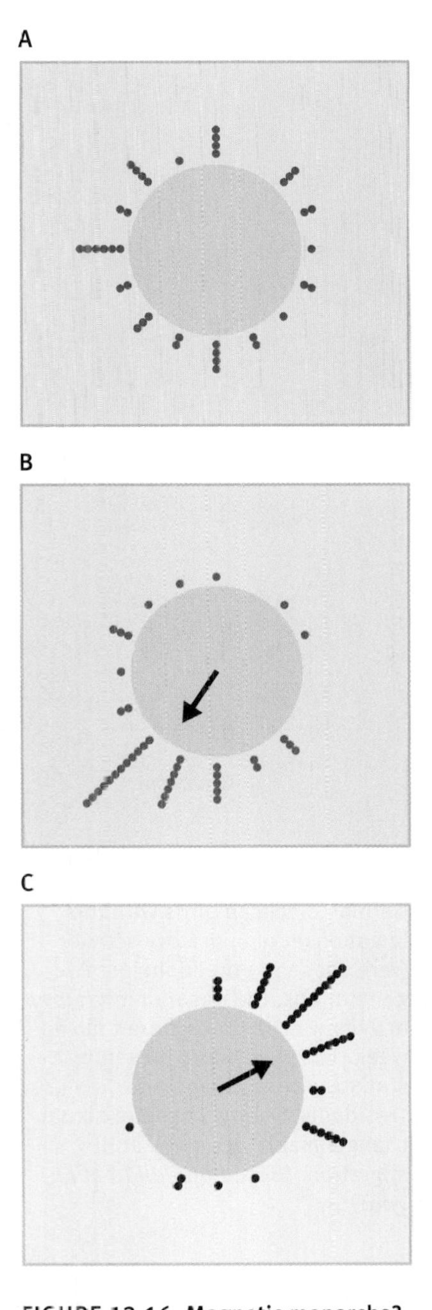

FIGURE 13.16. Magnetic monarchs? To test whether monarch butterflies migrate via magnetic orientation, three treatments were conducted. **(A)** In a nonmagnetic environment, monarchs oriented randomly. **(B)** In a normal magnetic field, the butterflies oriented in a southwestern direction, as they do in nature. **(C)** When the magnetic fields were reversed, monarchs reversed their normal orientation (and flew in a northeastern direction). See text for caveat on these results. *(Based on Etheredge et al., 1999)*

THE HERITABILITY OF MIGRATORY BEHAVIOR

Given the broad and varied consequences that various aspects of migration have on an individual's fitness, it is not that surprising that evolutionary biologists and ethologists have examined the possibility that migratory behavior is a heritable trait. Whether migratory behavior is heritable is of particular interest nowadays because of the unprecedented increase in global temperature over the last 100 years. Such environmental changes are hypothesized to affect various aspects of migration, such as time and length of migration, and hence it is important to understand what proportion of the variance in migratory behavior is genetic and what proportion is not.

Using an array of different techniques, Berthold and his colleagues have been studying the evolution of migratory activity in birds for the better part of thirty years. For example, Berthold

(1990) compared the onset of migratory activity in laboratory and wild birds of eighteen different species. He found a strong correlation between onset of migratory activity in both groups of birds, suggesting that the timing of departure for migration may be under genetic control. Such studies cannot, however, directly measure heritability. To do this, Pulido and Berthold and colleagues (2001) ran a selection experiment on autumn migration behavior in the German blackcap bird (*Sylvia atricapilla*).

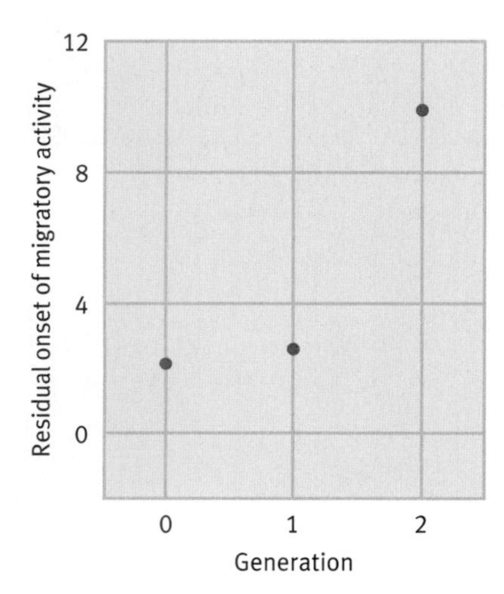

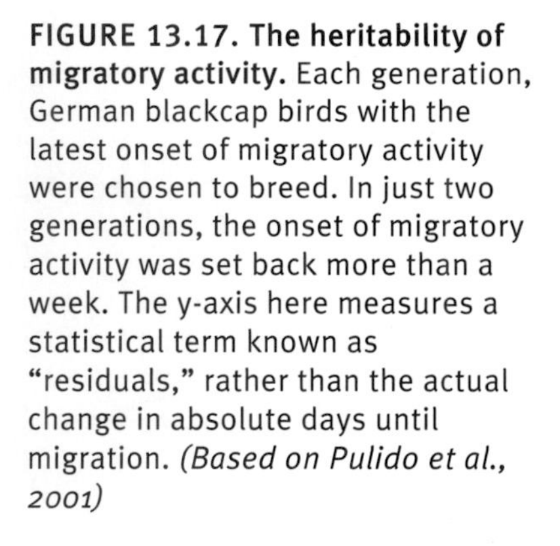

FIGURE 13.17. The heritability of migratory activity. Each generation, German blackcap birds with the latest onset of migratory activity were chosen to breed. In just two generations, the onset of migratory activity was set back more than a week. The y-axis here measures a statistical term known as "residuals," rather than the actual change in absolute days until migration. *(Based on Pulido et al., 2001)*

Pulido and his colleagues brought blackcaps in from the field and measured their "zugunruhe" or migratory restlessness during the autumn migratory season. Migratory restlessness in the lab correlates well with the onset of migration in the wild, and they defined the onset of such restlessness as the first night on which a bird was active during at least five thirty-minute periods. While they ran a large number of different treatments, we shall concentrate on a classic selection experiment that they conducted.

In 1994, the researchers collected forty blackcaps from the field and brought them back to the lab. The ten birds with the latest dates for migratory restlessness were then allowed to mate. They produced a total of twenty-six offspring, and from that group four breeding pairs with late-onset migratory restlessness were again allowed to breed, producing fourteen second-generation birds in a line selected for late migratory activity. Somewhat remarkably in only two generations, Pulido and his team selected for migratory activity being delayed for an average of 7.65 days (Figure 13.17), and they calculated a heritability of 0.72 for the onset of migratory behavior (Pulido et al., 2001). In conjunction with their other findings, Pulido and his colleagues interpreted this result as evidence that changes in migratory patterns as a result of global warming may indeed be due to genetic changes in migratory behavior rather than due to some sort of phenotypically plastic response on the part of birds.

LEARNING AND MIGRATION IN FISH

There are numerous theories on how some species of fish, such as coho salmon, make the often very long migration back to their natal stream as adults (Figure 13.18). For example, there is some evidence that fish use celestial and magnetic compasses (Quinn, 1984), while at the same time other studies have found that fish may use drifting currents (Miller et al., 1985) or some physiological change in response to their current position (Barkley et al., 1978) to guide them back home. There is, however, more and more evidence that demonstrates that learning may play a role in fish migration (Dodson, 1988).

FIGURE 13.18. Coho migration. During migration by coho salmon, olfactory cues are central for guiding individuals back to their home areas. *(Photo credit: Science/Visuals Unlimited)*

Juvenile salmon appear to learn the odors associated with their natal streams, and subsequently use such information to guide them home again (Dittman and Quinn, 1996). Dittman and Quinn examined the age at which young salmon learn about their home streams, and whether such learning is age-dependent in hatchery-reared coho salmon (*Oncorhynchus kisutch*). At different stages during development, young salmon were exposed to either natural odors or to an artificial odor (beta-phenylethyl alcohol), and then tested in the presence of such odors when they matured into adults. Results from the beta-phenylethyl alcohol (PEA) treatment show clearly that learning the odor of the natal stream was taking place when the young matured from the parr to the smolt stage of salmon development, and not earlier. This age-dependent learning effect is particularly interesting, as learning occurs at the same time as a surge in plasma thyroxin levels, suggesting a possible endocrinological correlate to migratory learning (Figure 13.19).

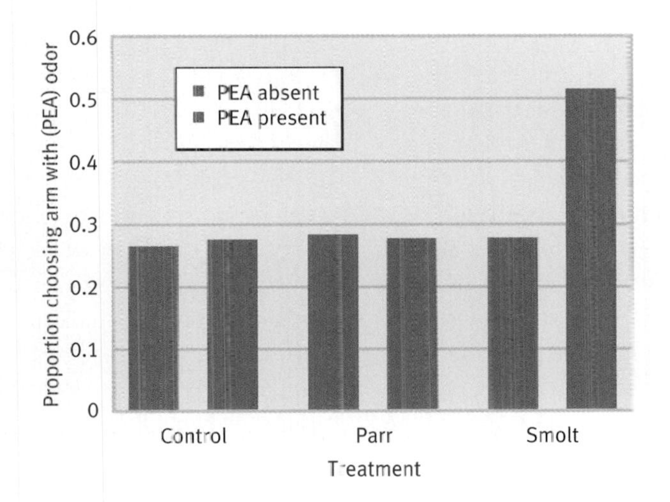

FIGURE 13.19. Learning the smell of home. To test whether coho salmon learn the smell of their natal stream, three groups of coho salmon were compared, and beta-phenylethyl (PEA) was chosen as the chemical imprinter of choice. Fish were exposed to PEA at the parr stage, the smolt stage, or never (control), and they were then given a choice between PEA odor or no odor in a two-arm maze. When comparing across groups, only salmon in the smolt stage learned to associate PEA with their stream. *(Based on Dittman and Quinn, 1996)*

Interview with Dr. JUDY STAMPS

Of all the systems you could have chosen to study territoriality, how did you end up working with small lizards?
I originally began studying social behavior in lizards because of a comment made by my professor in an undergraduate animal behavior class. He suggested that lizards might be a good choice for studying stereotyped behavior patterns (displays) because many lizards use stereotyped behavior—headbobbing patterns—to communicate with each other. So, I began by studying the displays of male *Anolis* lizards in captivity. After several years of studying lizards in the laboratory, I became curious about their behavior under natural conditions, and made a preliminary field trip to their native habitat, in the West Indies, to observe their behavior there. This first trip not only convinced me that these lizards were doing some pretty interesting things in nature, but also that the place they were doing them (the West Indian island of Grenada) had obvious attractions of its own. So, my first fieldwork focused on the social behavior of adult lizards in the West Indies.

While observing adults in the field, I noticed that tiny hatchlings exhibited many of the same behavior patterns as adults, including headbob displays, chasing, fighting, and the defense of territories. In addition, since territory size typically scales with body size, hatchlings that weighed only 2 grams had commensurately small territories, on the order of only .5 square meter in size. As a result, it was possible to study an entire "neighborhood" of territories in a small area in the field and to manipulate structural fea-

tures of habitats under natural conditions to determine the features that free-living juveniles preferred in their territories. After several months of playing around with juvenile lizards, it was obvious that they were much more amenable to both laboratory and field studies of territorial behavior than adults, and from that point on, I focused on the juveniles.

You've recently started modeling territoriality in terms of learning. What spurred you to take on this approach?
Many hours spent observing juveniles in the field convinced me that I needed to consider the role of learning in territorial behavior. For instance, it was quite apparent that a juvenile who entered a neighborhood for the first time had no idea where any of the territory boundaries were, and that it only became aware of their locations as a result of being chased from one territory to the next by the residents. Similarly, two juveniles of similar size meeting for the first time engaged in an extended, strenuous fight, involving lengthy exchanges of a wide array of headbob patterns, and culminating in physical combat, in which the two parties grabbed one another by the jaws, and then banged one another against the perch. It seemed that this "knock down, drag out" fight made a strong impression on both combatants, because when they encountered each other again the next morning, both seemed reluctant to venture near that opponent, and instead they began to exchange headbob displays from adjacent perches, across an intervening "no-man's-land" that eventually became the border between their two territories. Other observations indicated that if an individual was summarily attacked the first time it ventured into a

novel area, it almost never returned to that area. In contrast, if an individual received a comparable attack in an area it had used for an extended period of time, its first response was to flee, but typically it returned to the area within an hour or so, and persisted in using its familiar area in the face of repeated attacks by its opponent. These and other observations suggested that salient experiences, both positive (a snooze on a familiar, comfortable perch with a full stomach) and negative (being attacked by an opponent) had important effects on the subsequent social and spatial behavior of these animals.

Inspection of the psychology literature revealed that scientists studying the effects of positive and aversive stimuli on space use had already described very general behavioral phenomena that could account for the behavior of these lizards. For instance, rodents who receive aversive stimuli (e.g., electric shock) as soon as they first venture into an unfamiliar area are unlikely to return to that area, whereas comparable individuals who receive the same stimulus in an area in which they have previously been rewarded are much more likely to return to that area. Translated into territorial terms, these studies suggested that territorial animals may be behaving like little psychologists, delivering punishment to one another in an attempt to dissuade members of their own species from using a particular area.

Does being bigger usually mean you are guaranteed a good territory? If not, what else plays a role in territory acquisition?

In some situations, large body size can be helpful in acquiring a good territory, because if two individuals are competing for an area that is novel to both of them, and if one is larger than the other, the larger one is capable of delivering more punishment to its opponent during aggressive interactions than vice versa. However, size is not everything, because individuals who are already in possession of an area typically retain possession of that area, even when competing with larger opponents, a phenomenon called the "prior residency advantage." The prior residency advantage may be attributable to the fact that a resident knows more about an area than does a newcomer, as a result of which the resident is likely to fight more vigorously for that area, and persistently return to that area even after losing one or more fights to a larger newcomer.

Regardless of the reason, the prior residency advantage means that animals can acquire space by being the first to settle in an area, rather than by being bigger than their opponents. This means that in many species, the ability to find newly vacant territories, and the ability to settle in them before anyone else finds them, plays an important role in territory acquisition. Thus, the prior residency advantage may explain why some birds establish territories weeks or months in advance of the time that the territory will be needed for reproduction: the early bird gets the territory.

Are there any general patterns in territoriality that emerge when you look across taxa, or perhaps across the sort of medium that animals live?

There are a number of general patterns that occur in territorial species from a wide range of taxa. I have already mentioned the prior residency advantage, which occurs in virtually all territorial animals. Another generalization is that individuals in territorial species exhibit site tenacity, meaning that a given individual tends to remain in the same area for an extended period of time. Site tenacity is observed in all territorial species, but the reverse is not the case (e.g., all species with site tenacity do not defend areas in which they live). The basic elements of territorial behavior (stay in an area and use aggressive behavior to discourage other individuals from remaining in or returning to that area) are exhibited by a very large array of species, ranging from sea anemones to primates, but taxa with a long evolutionary history of territoriality typically add certain refinements to this basic pattern. For instance, territory owners in birds, mammals, frogs, lizards, and insects may produce songs, olfactory signals, conspicuous visual displays, or other conspicuous signals using other sensory modalities. These broadcast signals indicate to other inhabitants of the area that the territory is occupied, as well as providing additional information about the sex, condition, and (sometimes) identity of the territory owner. Territory owners in species with a long evolutionary history of territoriality are also likely to be able to recognize different categories of conspecifics, and tailor their aggressive behavior accordingly.

DR. JUDY STAMPS is a professor at the University of California at Davis. Her long-term work on lizards and territoriality has produced fundamentally new ideas on both territoriality and the role of learning in territory formation.

SUMMARY

1. A suite of biotic and abiotic factors affect habitat choice in animals. Abiotic factors include, but are not limited to, heat, availability of water, wind, "cover" from danger, and the availability of specific nutrients, while biotic factors encompass the location of potential mates, food, predators, and parasites.
2. The ideal free distribution model is a simple, but powerful, predictor of habitat choice as a function of resource distribution. The ideal free distribution model predicts that the equilibrium distribution of individuals into patches (or habitats) should be that distribution at which, if any individual moved to a patch (or habitat) it was not in, it would suffer a reduced payoff.
3. Territorial neighbors, though competitors with respect to certain variables, may have common enemies, such as other intruders that attempt to take over a territory in their general area. The "dear enemy" phenomenon is one in which territorial neighbors are at first involved in a costly battle over setting up territory boundaries. Once boundaries are established, neither party stands to gain by having to reestablish such boundaries with a new neighbor, and so a new cooperative element is added to neighbor relations—producing dear enemies.
4. Not all adults in a population necessarily opt to be territorial. In certain species, like bluegill sunfish, some individuals (that is, "parentals") defend territories in order to attract mates, while others opt for a "sneaker" strategy. Sneakers do not form territories, but rather hover on the outskirts of parental territories and attempt to intercept females on their way into territories.
5. Ethologists studying migration not only focus on the obvious costs (for example, the energy needed for travel) and benefits (breeding in a warmer climate) of migration, but also on more subtle factors, such as the immunological costs of being a migrant, the various cues that migrants use to locate their breeding and nonbreeding grounds, and the interaction between genetic and environmental factors in shaping migratory behavior.

DISCUSSION QUESTIONS

1. Make a general list of the costs and benefits of territoriality. Using that list, determine what sort of environments would generally favor the formation of long-term territories.

2. The ideal free distribution model predicts that animals will distribute themselves among patches in proportion to resources. What sort of cognitive abilities, if any, does this assume on the part of animals? Would bacteria potentially distribute themselves in accordance with the predictions of IFD models?
3. Suppose that young individuals watch older conspecifics choose their territories and subsequently use such information in their habit-choice decisions about valuable resources. Outline one scenario by which such observational learning could increase competition for prime habitat sites, and one in which it would decrease competition for such sites.
4. In the literature on "sneaker" and "parental" male bluegill sunfish there has been some debate as to whether there are distinct genetic morphs or whether each individual is capable of becoming either a sneaker or a parental male and uses environmental and social cues to make this decision. Can you sketch out an experiment to distinguish between these hypotheses?
5. Consider Møller and Erritzøe's work on immune defense organs and migration behavior. Can you make any predictions regarding how a migrating species might fare against local parasites (in both its habitats) as compared to resident species? What is the logic underlying your hypotheses?

SUGGESTED READING

Able, K. P. (1999). *Gatherings of angels: Migrating birds and their ecology*. Ithaca, NY: Comstock Books. A wonderful, easy-to-read book on migration in birds.

Brown, J. L. (1969). Territorial behavior and population regulation in birds. *Wilson Bulletin, 81*, 293–329. A classic paper examining the effects of territoriality on population dynamics.

Emlen, S. T., & Wrege, P. H. (1992). Parent-offspring conflict and the recruitment of helpers among bee-eaters. *Nature*, 356, 331–333. A crisp and concise experiment examining parent-offspring dynamics and territory establishment.

Milinski, M., & Parker, G. (1991). Competition for resources. In J. Krebs & N. Davies (Eds.), *Behavioral ecology: An evolutionary approach* (pp. 137–168). Sunderland, MA: Sinauer Associates. A nice overview of the ideal free distribution model.

Stamps, J. (2001). Learning from lizards. In L. A. Dugatkin (Ed.), *Model systems in behavioral ecology* (pp. 149–168). Princeton: Princeton University Press. The most up-to-date review of learning and territory establishment.

14

Game Theory Models of Aggression

- The Hawk-Dove Game
- The War of Attrition Model
- The Sequential Assessment Model

Winners, Losers, Bystanders, and Aggression

- Winner and Loser Effects
- Bystander Effects

Endocrinology, Neurotransmitters, and Aggression

- Corticosterone and Aggression
- Testosterone and Aggression
- Neurotransmitters and Aggression

Interview with Dr. John Maynard Smith

Aggression

In the eyes of Thomas Henry Huxley, one of the predominant intellectual figures of the nineteenth century and Darwin's most vociferous defender, interactions in the animal world resulted in a bloodbath. Huxley seemed to think that aggression, often extreme aggression, was the norm for everyday life among the animals:

> From the point of view of the moralist, the animal world is on about the same level as the gladiator's show. The creatures are fairly well treated, and set to fight; whereby the strongest, the swiftest and the cunningest live to fight another day. The spectator has no need to turn his thumb down, as no quarter is given . . . the weakest and the stupidest went to the wall, while the toughest and the shrewdest, those who were best fitted to cope with their circumstances, but not the best in any other way, survived. Life was a continuous free fight, and beyond the limited and temporary relations of the family, the Hobbesian war of each against all was the normal state of existence. (Huxley, 1888, p. 163)

In contrast, Russian Prince Peter Kropotkin, a well-respected naturalist, whose book *Mutual Aid among Animals* (Kropotkin, 1902) is rightly regarded as a classic, looked around and saw a world that seems almost antithetical to that of Huxley's:

> . . . in all these scenes of animal life which crossed before my eyes, I saw mutual aid and mutual support carried on to an extent which made me suspect in it a feature of the greatest importance for the maintenance of life, the preservation of each species and its further evolution. (Kropotkin, 1902, p. 18)

The point of these quotes is not to try to debate whether animals are aggressive or not. To some extent, toned-down versions of Huxley's and Kropotkin's views are both correct. In Chapter 9, we discussed the role of cooperation in animal societies, and here we shall turn to what is known about aggression and animal behavior. Before progressing, though, it is important to note that cooperation and aggression are not the flip sides of a coin—individuals in groups often cooperate with each other in order to compete, often aggressively, with other groups of animals (Dugatkin, 1997a; Gadagkar, 1997). Aggression is part and parcel of even those most cooperative of units—the social insect group. In wasps, for example, while cooperation is often the order of the day within hives when "foreign" individuals from other hives try to enter a nest, wasp guards often respond with vigorous and sometimes deadly defensive behaviors (Gamboa et al., 1986; Figure 14.1).

Since ethologists and naturalists have been fascinated with animal aggression for hundreds of years, the literature in these areas is huge (for example, Scott, 1958; Lorenz, 1966; Huntingford and Turner, 1987; Archer, 1988; Wrangham and Peterson, 1996; Mock and Parker, 1997; Moynihan, 1998). Rather than attempt to sum-

FIGURE 14.1. Intruder aggression. When a wasp (left) approaches a nest, guards at the nest determine whether it's a hive mate or an intruder. Intruders are aggressively repelled.

marize this literature, which would require a multivolumed series itself, we shall look at three general aspects of aggression in animals:

- Models of aggression, including the hawk-dove game, the sequential assessment game, and the war of attrition,
- The implications of winner, loser, and bystander effects on aggressive interactions and hierarchy formation,
- Endocrinology, neurotransmitters, and aggression.

Before progressing, note that while this chapter focuses on aggression per se, the role of aggression as it relates to other aspects of animal behavior—for example, cooperation, kinship, foraging, habitat selection, and mating—can be found throughout this book.

Game Theory Models of Aggression

As we have already discussed in other chapters, game theory models are set up to examine behavioral evolution when the fitness of an individual depends both on its own behavior and on the behavior of others. What better behavior than aggression for developing and testing game theory models of behavior? If, for example, an individual is willing to fight for a contested resource, clearly its fitness is going to depend on whether its opponent opts to fight or simply to flee.

Evolutionary game theorists have built a suite of models that examine the evolution of fighting behavior (Riechert, 1998). Here we shall focus on the three most well-developed game theory models of aggression—the hawk-dove game, the war of attrition game, and the sequential assessment game. All three of these game theory models have certain elements in common. For example, a *cost to fighting* is assumed either explicitly or implicitly in hawk-dove, war of attrition, and sequential assessment games. Such a cost can take many different forms, ranging from opportunity costs (the cost associated with not doing something else), to the cost of physical injury, up to and including mortality costs. Taking into consideration a cost of fighting in all game theory models of aggression seems more than justified, as opportunity costs must exist (by definition), and fights clearly often cause injuries, ranging from minor to fatal (particularly in insects and arachnids; Enquist and Leimar, 1990).

A second critical variable in all game theory models of aggression is the value of whatever resource is being contested. In some cases—for example, a carcass—the value should be fairly easy to calculate. In other cases, such as a male's *access* to reproductively active females, resource value can be much more difficult to calculate. That being said, clearly the value of a resource affects an animal's decision to fight, including how long and/or how hard to fight (Figure 14.2).

The value assigned to a resource need not be the same for both of the contestants deciding on whether to fight for such a resource. For example, imagine that two animals are going to contest who gets a ten-pound prey item that has just been discovered. While ten pounds of meat is ten pounds of meat, it may be *valued* differently by putative fighters. To a starving animal, ten pounds of food might have a greater value than its value to an animal that just re-

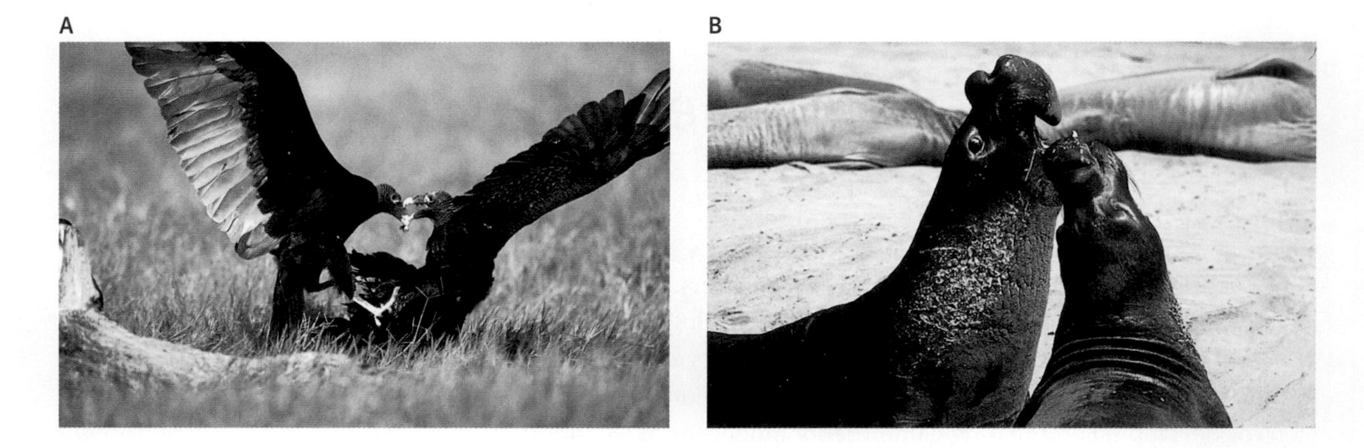

FIGURE 14.2. Deciding to fight. (A) One of the many resources animals will fight over is food, as shown here by these vultures that are fighting over a carcass. **(B)** Males also fight over females. Here, male elephant seals are fighting over access to reproductively active females. *(Photo credits: Tom Vezo; Jeff Mondragon/mondragonphoto.com)*

FIGURE 14.3. Value estimation. When two animals contest a resource, the hungrier animal may fight harder or longer to obtain it. If one of these cats were hungrier than the other, it might be willing to risk more to obtain the remains of the fish.

cently ate (Figure 14.3). As another example, consider the value of territory to a potential intruder and to a territory holder. Above and beyond the fact that the territory holder might be on a territory because it is a good fighter to begin with, it may well be that a territory holder will value a contested area (that is, its territory) more because it has already invested time and energy in learning where the resources in such a territory are located (Table 14.1). These

TABLE 14.1. Resource value and fighting. The value that an individual animal assigns to some resource affects how long contests last and to what extent they escalate. *(Based on Riechert, 1998)*

RESOURCE	CONTEST DURATION	LEVEL OF ESCALATION
Duration of food deprivation		
▸ Crayfish		increases
▸ Domestic cat		increases
Limited mating opportunities		
▸ Marine isopod crustacean		increases
▸ Bowl-and-doily spider	increases	
▸ Dungfly	increases	
▸ Red-spotted newt	increases	
▸ Red deer	increases	
Limited food/site availability		
▸ Funnel-web spider	increases	increases
▸ Digger wasp	increases	
▸ Iguana		increases
▸ Fulmar		increases
▸ Bald eagle		increases

sorts of asymmetries in value have been documented many times in the ethological literature (Barnard and Brown, 1984; Ewald, 1985; Verrell, 1986).

Let us now examine some specific evolutionary game theory models in a bit more detail.

THE HAWK-DOVE GAME

The most well-known game theory model of the evolution of aggression is the **hawk-dove game** (Maynard Smith and Price, 1973; Maynard Smith, 1982). In historical terms, this game holds a special place in the hearts of ethologists and behavioral ecologists because it was in its creation that evolutionary game theory truly came into being. Prior to the description of this game, only a few papers had discussed how game theory might be used to address evolutionary issues, particularly behavioral evolution (Lewontin, 1961; Hamilton, 1967; Trivers, 1971). Maynard Smith and Price's publication of the hawk-dove game is rightly regarded as a watershed event in theoretical ethology, as it was *the* paper that made animal behaviorists recognize the potential utility of game theory to their discipline.

Given the impact of the hawk-dove game on how ethologists and behavioral ecologists think about aggression, it is remarkable how basic (yet brilliant) a game this really is. Imagine individuals can adopt one of two behavioral strategies when contesting some resource: (1) hawk—wherein a player will escalate and continue to escalate until either it is injured or its opponent cedes the resource, or (2) dove—wherein a player displays as if it will escalate, but retreats and cedes the resource if its opponent escalates (this strategy was originally labeled "mouse").

If we let V = the value of the contested resource, and C = the cost of fighting, we can create a payoff matrix for the hawk-dove game (see Table 14.2). There are two assumptions underlying this matrix. First, we are assuming that if two doves meet, on average they will receive half the value of the resource (for example, they may opt to split the resource). Second, when it comes to hawk-hawk interactions, we are assuming that the only cost involved is paid by the loser, who is injured. Hence, hawks have a 50 percent chance of obtaining the resource (V/2) and a 50 percent chance of being injured and not receiving the resource (C/2), for a total expected payoff of (V−C)/2). One could, for example, change the payoff in the hawk-hawk cell to (V/2)−C, which would imply that both hawks would pay a cost of fighting. Here, however, we will stick to the original game.

TABLE 14.2. The payoff matrix for the hawk-dove game. Both player 1 and player 2 choose between the hawk (always be aggressive) strategy and the dove (bluff, but retreat if opponent escalates) strategy. V = value of resource, C = cost of fighting. The payoff to player 1 is shown here.

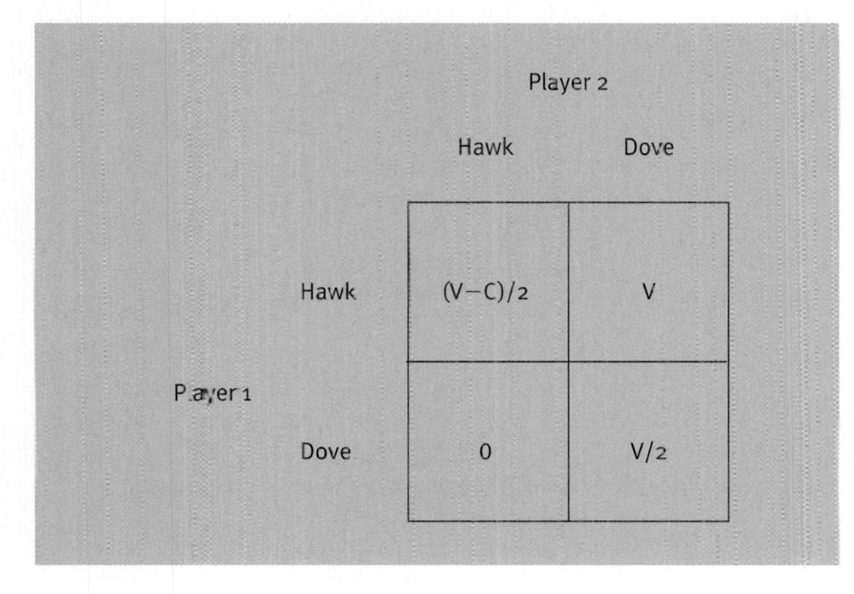

		Player 2	
		Hawk	Dove
Player 1	Hawk	(V−C)/2	V
	Dove	0	V/2

When calculating the evolutionarily stable strategy (Chapter 9) to this game, the ESS depends on whether the value of the resource or the cost of fighting is greater. If V > C, then hawk is an ESS, as the hawk-hawk payoff, (V−C)/2, is positive, and hence greater than the payoff dove obtains when it meets hawk (0). Dove, however, is not an ESS, as the dove-dove payoff (V/2) is less than the payoff hawk obtains when it meets dove (V). Hence, when V > C, hawk is the only ESS. This makes intuitive sense, since V > C implies that the cost of fighting, paid only by hawks, is low compared to the prize that awaits the winner of any contest (V), and hawks do well. The situation is a bit more complicated when the cost of fighting is greater than the value of the resource. Now, neither hawk nor dove is an ESS. But it can be shown that some mixture of hawks and doves is in fact evolutionarily stable in this case (see Math Box 14.1).

There are many more complicated varieties of the hawk-dove game than the one we examined above (Riechert, 1998; Mesterton-Gibbons and Adams, 1998). One particularly interesting version of this game adds in two new strategies: bourgeois and antibourgeois. Now, in addition to hawks and doves, we have a bourgeois strategy that instructs an individual to play hawk if it is a territory holder, and dove if it does not own a territory. An antibourgeois strategy codes for the opposite—play dove if you are a territory holder, and hawk if you are not.

Math Box 14.1

Let's look at the case of the hawk-dove game, when $C > V$, in a bit more detail.

If $C > V$, then hawk is not an ESS, as hawk's payoff against other hawks, $(V-C)/2$, is now negative, while dove's payoff against hawk is 0. Using the same calculations as we did for the case of $V > C$, we can see that dove is not an ESS either.

To see if some combination of hawks and doves is an ESS, let p = the frequency of hawks, and $1-p$ be the frequency of doves. Hawk's payoff is then:

$$p(V-C)/2 + (1-p)\,V \tag{1}$$

where the first term is hawk's payoff against other hawks and the second is hawk's payoff against dove.

Dove's payoff can be calculated as:

$$p(0) + (1-p)V/2 \tag{2}$$

where the first term is dove's payoff against hawk and the second term is dove's payoff against other doves.

When (1) = (2), the fitness of hawks is equal to the fitness of doves, and we then have our equilibrial frequency of hawks and doves. A little algebra shows that this occurs when $p = V/C$.

While bourgeois as an ESS makes some intuitive sense, the same cannot be said for antibourgeois, which also is an ESS. And yet solutions that are not intuitive are often the most interesting, as they sometimes explain otherwise inexplicable phenomena. To see this, let's take a look at Mexican spiders.

ANTIBOURGEOIS MEXICAN SPIDERS. *Oecibus civitas* is a small, territorial, Mexican spider that shows what appears to be a very odd set of behaviors when it comes to aggression and home range defense (Burgess, 1976). These spiders set down their territories under rocks and play the antibourgeois strategy. When an intruder approaches a territory, a territory holder flees rather than fights. The former territory holder then searches for a new abode, and if he comes upon such a prize, the territory owner *there* vacates (Figure 14.4). This produces a potentially long, bizarre series of events in which territory owners leave, then challenge someone still in possession of another territory, causing that territory owner to leave, and so on.

How can such a seemingly strange cascade of events come to be? Usually, home court advantage is a real phenomenon in animals, with territory holders winning fights with an intruder. What factors then caused this antibourgeois strategy to evolve in Mexican spiders (as well as other species; Komers, 1989; Hodge and

FIGURE 14.4. Reverse bourgeois strategy. In *Oecibus civitas,* territory holders relinquish their territories when challenged by an intruder. This can have cascading effects throughout a cluster of *O. civitas* territories. *(Based on Mesterton-Gibbons and Adams, 1998)*

Uetz, 1995)? The answer, at least in part, probably centers around the cost of fighting versus the number of territories available and the danger in moving between territories. Imagine that the cost of fighting is relatively high, but numerous territories are available, and it is safe to move to such territories. Under such circumstances, fleeing when an intruder approaches—that is, playing the antibourgeois strategy—may in fact be the most profitable option available to individuals (Mesterton-Gibbons, 1992).

Looking at the Mexican spider case from an ESS perspective, we see that an antibourgeois strategist runs off in search of a new territory if a hawk strategist appears at its abode (Mesterton-Gibbons and Adams, 1998). If a hawk strategist owns a territory and another hawk strategist approaches, however, the territory owner will fight, attempting to win as quickly as possible and thereby retain its territory. If predation is generally low, and the odds of a territory holder winning a fight are also low—both of which may be the case for our spiders—the antibourgeois territory owner is more likely to move safely to a new area than the hawk territory holder is to win his fight. As such, antibourgeois strategists will invade and spread in a population of hawk strategists.

BOURGEOIS BUTTERFLIES. On the flip side of the coin from the strategy employed by Mexican spiders is the bourgeois strategy displayed by some territorial butterflies. In the speckled wood butterfly

(Pararge aegeria), territories are not, in fact, set in space. That is, rather than having a territory with a set place in three dimensions, the male has a territory that is an open patch where the sun happens to break through the clouds. When a male comes upon an empty patch (of sunlight), he immediately occupies it and secures a mating advantage compared to males who are not in sunlit territories.

When a male speckled wood butterfly comes upon a territory that has another male in it, the contest is *always* settled as follows: resident wins, intruder retreats. In fact, there is very little aggression at all when males come upon occupied territories, perhaps because a prolonged fight over a short-lived resource such as a sun patch is not worth the costs. Rather, once a male is aware that a territory is occupied, he simply leaves (Figure 14.5). What makes the "resident wins" rule even more dramatic is that an individual need only be the resident of a sun patch for a few seconds to secure victory over an intruder. In a study by Davies (1978), for example, a male, let's call him M1, is experimentally made a territory owner. M1 then always defeats M2, an intruder male. Yet, if M1 is removed from his territory and M2 is resident for only a short period of time, M1 now defers to M2 when he is reintroduced into his original sun patch. The only escalated contests occur when two

FIGURE 14.5. Conventional rules. The set-up for Davies's experiments on territory ownership and contest rules in the speckled wood butterfly *(Pararge aegeria)*. *(Based on Dawkins and Krebs, 1978, p. 299)*

males both "believe" that they are the rightful owner of a sun patch. This happens naturally when two butterflies come to a sun patch at about the same time but do not notice the other individual for some period (this can also be simulated experimentally). In such cases, real fights can and do occur between the speckled wood butterfly males.

THE WAR OF ATTRITION MODEL

In the hawk-dove game, the only choice an individual has is to fight or not fight (flee). While it is true that the hawk-dove game can be made more complicated by adding strategies that code for rules like "fight if *x*, don't fight if *y*," in all cases the behavioral choice is dichotomous (fight or flee). In order to examine aggression when the choice available to individuals is more continuous—for example, "fight for *x* seconds, then stop," the war of attrition model was created (Maynard Smith, 1974b; Parker, 1974a; Maynard Smith and Parker, 1976; Bishop et al., 1978; Hammerstein and Parker, 1982). The war of attrition model has three important underlying assumptions (Riechert, 1998): (1) the strategy set—how long to contest a resource—is continuous; (2) the aggressive behavior displayed is relatively mild, in that it does not lead to severe injury—the cost of fighting is not huge; and (3) there are no clear cues such as size, territory possession, and so forth, that contestants can use to settle a contest. If we again let V = the value of the resource, and let x = contest length, we can demonstrate that at the ESS, the probability that a contest lasts x units of time is equal to $(2/V)e^{-2x/V}$ (Parker and Thompson, 1980). So rather than getting a set time for fighting, the war of attrition predicts an ESS *distribution of contest lengths* (Figure 14.6).

This predicted distribution of contest lengths matches certain contest durations in nature. In particular, the duration of dungfly male mating behavior (Parker and Thompson, 1980) and damselfly aggression over feeding sites (Crowley et al., 1988) appear to match the predicted ESS distribution of the war of attrition very nicely.

For example, recall the dungflies we discussed in Chapter 7. Females arrive at fresh dung pats to lay eggs, and hence males aggregate at such pats for access to females. The question, in terms of war of attrition models, is how long should a male stay at a given pat. If he stays too long, he will encounter fewer and fewer females, and per unit time he will also increase the number of aggressive contests with other males at the pat. If he leaves too quickly, he will pay the cost of moving and may miss the opportunity to mate with females at the patch he left. Parker and Thompson (1980), using

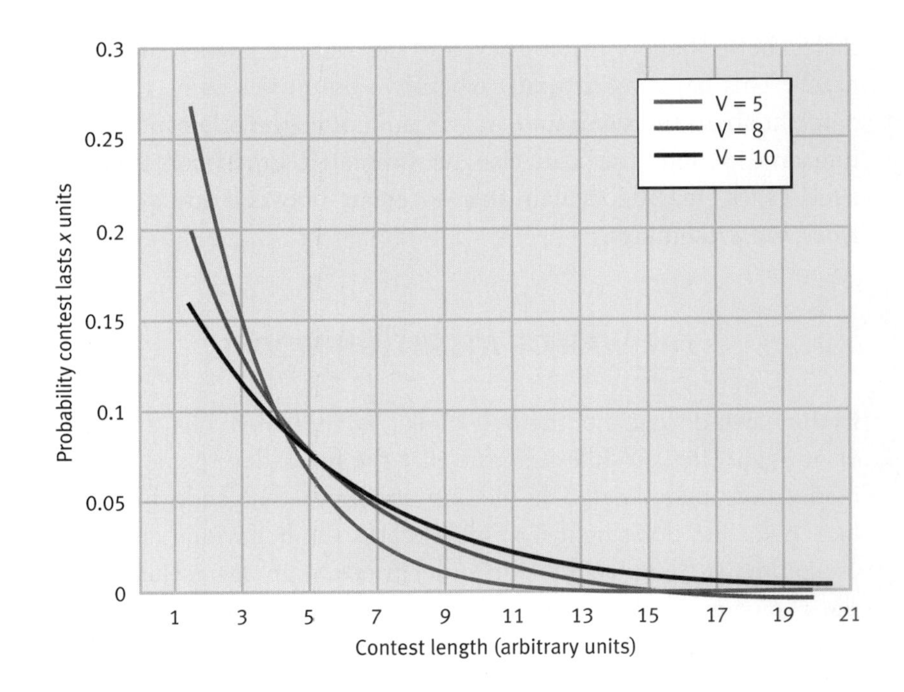

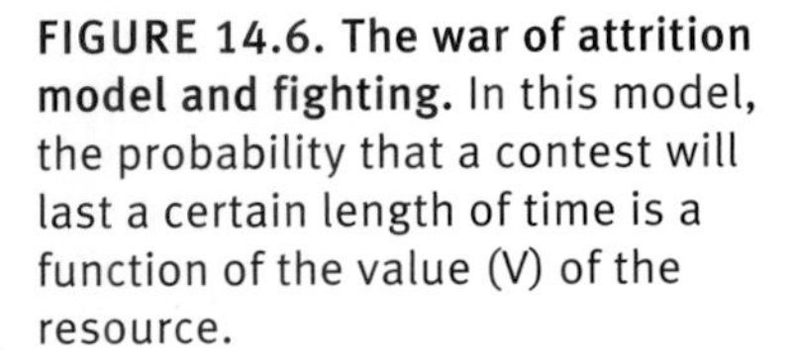

FIGURE 14.6. The war of attrition model and fighting. In this model, the probability that a contest will last a certain length of time is a function of the value (V) of the resource.

detailed measures of the costs and benefits of dungfly mating, found that male stay times (the time they remained at a patch) were exponentially distributed, as one might predict from a war of attrition model. More to the point, if one assumes that it takes about four minutes to find a new pat, then male stay times all yield approximately equal fitness, again in accord with the predictions of a war of attrition model (Figure 14.7). When Parker calculated the mean time to find a new patch from his dungfly fieldwork, it was indeed approximately four minutes (see Curtsinger, 1986, and Parker and Maynard Smith, 1987, for more on dungflies and the war of attrition).

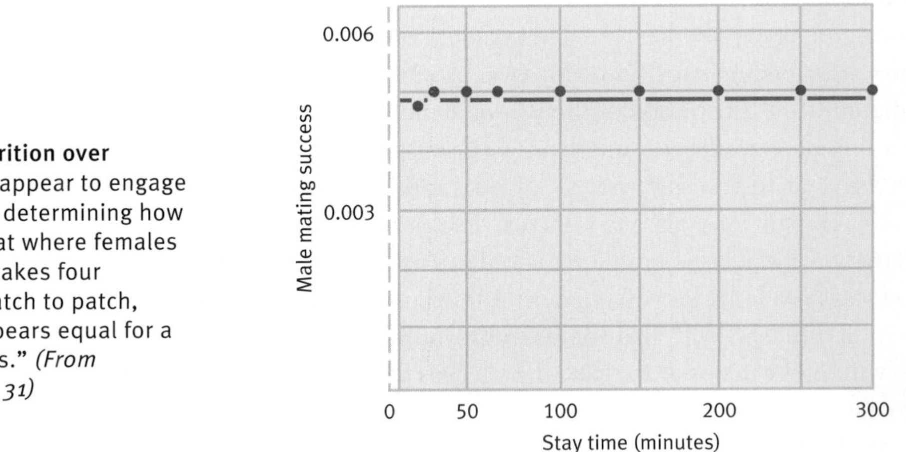

FIGURE 14.7. War of attrition over females. Male dungflies appear to engage in wars of attrition when determining how long to stay on a dung pat where females may alight. Assuming it takes four minutes to move from patch to patch, male mating success appears equal for a wide range of "stay times." *(From Maynard Smith, 1982, p. 31)*

THE SEQUENTIAL ASSESSMENT MODEL

One of the more complicated of the game theory models of aggression is the sequential assessment model (Enquist and Leimar, 1983, 1987, 1990; Leimar and Enquist, 1984; Enquist et al., 1990). In presenting the case for developing a new set of game theory models of aggression, Enquist and his colleagues argue:

> . . . the hawk-dove game, the war of attrition and similar games are not based on realistic behavioral mechanisms, i.e. mechanisms that produce sequences of behaviour patterns as a result of internal and external stimulation. This means that much of the behavioural diversity found in fighting behaviour, like the use of different behavioural elements, cannot be analyzed with such models. (Enquist et al. 1990, p. 1)

In response to this supposed lack of realistic models, the sequential assessment game was created. What separates the sequential assessment game from other models of aggression is that it is built to analyze fights in which individuals continually assess one another in a series of "bouts." Since this is arguably the manner in which most animal fights take place, let us examine how this model works, what predictions it makes, and what experimental evidence is available to test its basic predictions.

The underlying foundation of the sequential assessment model rests on the idea that assessing an opponent's fighting abilities is critical to contest behavior. Assessing one's opponent, Enquist and his colleagues argue, is analogous to statistical sampling. A single sample (assessment) introduces significant random error; the more sampling (assessment) one does, the lower the error rate and hence the more confident one can be in whatever is being estimated—in our case, the opponent's fighting ability.

Most sequential assessment models examine contests that contain different types of aggression that can vary from relatively mild to very dangerous. At the evolutionarily stable solution to the sequential assessment game, the following type of scenario is predicted: Individuals should begin with the least dangerous type of aggressive behavior (let's call it behavior A), and "sample" each other with respect to A for some period of time. Soon, however, all the information about an opponent with respect to A will be garnered. At that point, the next most dangerous behavior (B) should be the dominant action among protagonists. Again, at some point, information about an opponent and behavior B will reach saturation. Depending on the behavioral repertoire of the animals in question, more and more dangerous behaviors will be tacked on to the sequence. The more evenly matched opponents should engage in the more dangerous behaviors.

The sequential assessment game allows predictions not only with respect to who will win a fight, but also as to what the *entire sequence* of aggressive interactions leading up to a victory should look like. The model allows one to map a "giving up" line, such that once an animal crosses this line, it should stop fighting and hence end an aggressive interaction.

Studies testing various predictions of the sequential assessment game have produced mixed results. Most studies find partial support of the model's basic predictions (Koops and Grant, 1993; McMann, 1993; I. C. Smith et al., 1994; Jennions and Backwell, 1996; Hack, 1997; DiMarco and Hanlon, 1997; Jensen and Yngvesson, 1998; Molina-Borja et al., 1998; Brick, 1999). Other studies more definitively support the great majority of the basic predictions of the sequential assessment game (Leimar et al., 1991). One such study examined contest behavior in *Nannacara anomala.*

FIGURE 14.8. Fighting fish. *Nannacara anomala* has proven to be an ideal species for testing various predictions derived from the sequential assessment game. *(Photo credit: Magnus Enquist)*

SEQUENTIAL ASSESSMENT IN *NANNACARA ANOMALA*. *Nannacara anomala* is an ideal fish species in which to test the sequential assessment model (Figure 14.8). Unlike many other cichlid species, males are not territorial but rather form hierarchies starting with a clear top-ranked individual who searches for females with whom to mate. Aggressive interaction in males of this species ranges from "changing color" (least dangerous) and "approaching" through "tail beating," biting and mouth wrestling (Figure 14.9), up to "circling," which is the most dangerous of the aggressive activities, wherein fish repeatedly attempt to bite each other while they swim in a circular pattern.

Enquist and his collaborators staged pairwise interactions among male fish to test predictions of the sequential assessment game (Enquist et al., 1990). All trials were videotaped so as to allow a detailed analysis of subtle behavioral changes through time. Overall, interactions between aggressive males matched the predictions of the sequential assessment game very well.

One of the most elementary predictions of the sequential assessment game is that the more evenly matched the opponents, the

FIGURE 14.9. Sequential assessment. Much work on the sequential assessment game has been done in the fish *Nannacara anomala*. Here two individuals are engaged in a "mouth wrestle." *(Based on Enquist et al., 1990)*

longer the fights and the more phases of a fight. The closer individuals are in fighting ability, the more sampling is needed to assess who in fact would likely win a dangerous fight. In *N. anomala*, as in many fish, individuals are able to assess weight asymmetries (Enquist et al., 1987), and weight differences appear to have a very large effect on contest outcome, with heavier fish more likely to emerge from a contest victorious. In accordance with the predictions of the sequential assessment game fights do indeed take longer when fish are more closely matched for weight than they do when large asymmetries in weight exist.

The sequential assessment game also predicts that when numerous behaviors are used in aggressive contests, they should be used in approximately the same order in all fights. That is, while some fights are predicted to last longer and contain more elements than others, *the order* in which new aggressive behaviors appear in a fight should be similar in all contests, and hence shorter contests should simply have fewer types of behavior. Again, the behavior of *N. anomala* matches the model's prediction quite nicely. Males typically begin aggressive interactions with some sort of visual assessment, progress to tail beating, and then to biting and mouth wrestling, and occasionally to circling. How many acts in this sequence are played out depends on differences in the opponent's weight, but *the order in which these acts are displayed* tends to be the same regardless of weight differences.

Winners, Losers, Bystanders, and Aggression

As we saw in the sequential assessment game, prior experience can have a serious impact on the outcome of aggressive interactions. While the sequential assessment game focuses on experience during a contest, other recent work (both empirical and theoretical) has examined experience across many aggressive encounters. These studies examine three phenomena we shall address in turn: winner effects, loser effects, and bystander effects.

WINNER AND LOSER EFFECTS

Anyone who follows sports knows that individual athletes, as well as teams, appear to have "winning streaks" and "losing streaks." That is, above and beyond their physical talent per se, it appears that the act of winning itself begets more winning, and the act of losing begets more losing. A similar set of phenomena exists in the animal literature, wherein winning a fight increases the probability of future wins—so-called **winner effects**—and vice versa for losing an aggressive interaction—so-called **loser effects** (Landau, 1951a, 1951b).

Winner and loser effects are usually defined as an increased probability of winning at time T, based on victories at times T-1, T-2, and so on, and an increased probability of losing at time T, based on losing at times T-1, T-2, and so on, respectively. Winner and loser effects are *not* necessarily two sides of the same coin, as one can be present without the other. For example, A may be more likely to defeat B if it has just defeated C (winner effect), but this does not necessarily mean that C is more likely to be defeated (loser effect) during its next interaction.

The tendency to win or lose a fight after a prior victory or defeat has been documented in a wide variety of species. To get a flavor for such studies, let us examine winner and loser effects in birds, fish, and snakes.

WINNER AND LOSER EFFECTS IN BLUE-FOOTED BOOBIES. As we saw in our discussion of sibling rivalry, hierarchies encompassing a great deal of aggressive interactions often form among young related nest mates in many species of birds (Mock and Parker, 1997). In a long-term study of nest-mate aggression in blue-footed boobies *(Sula nebouxii)*, Hugh Drummond and his colleagues have been investigating aggressive behaviors commonly found during nest-mate hierarchy formation (Figure 14.10). Early work on the

FIGURE 14.10. Blue-footed boobies. For blue-footed booby chicks involved in aggression, both winner and loser effects are in play. Winner effects, however, are much shorter lived. *(Photo credit: Hugh Drummond)*

role of experience in shaping hierarchy formation found that when older chicks became aggressive early on and younger chicks became submissive, older chicks would defeat younger chicks that were considerably larger in size (Drummond and Osorno, 1992). While this work clearly hints at the possibility of winner and loser effects in blue-footed boobies, it is not possible to definitively say whether loser effects (in younger chicks), winner effects (in older chicks), or both played a role in establishing hierarchies in this species.

In order to experimentally uncouple winner and loser effects, Drummond and Canales (1998) followed up the study mentioned above by pitting each dominant and subordinate individual against *inexperienced individuals*. Thus, in any given treatment, only one of the animals had prior experience, and it was possible to examine winner and loser effects in isolation from one another. Specifically, Drummond and Canales predicted that when pitted against inexperienced individuals, initially subordinate animals will have a decreased probability of winning aggressive bouts, and initially dominant individuals will have an increased probability of winning aggressive interactions. To be conservative, they pitted subordinates against inexperienced partners that were slighter smaller (thus *increasing* a subordinate's chance of victory), and dominants against slightly larger individuals (*decreasing* the dominant's probability of victory).

Drummond and Canales uncovered winner and loser effects early on (at about four hours) in their experiment. At this critical initial stage, subordinates were much less likely to be aggressive than were their inexperienced pairmates, and dominants were much more likely to be aggressive (Figure 14.11). The winner effect waned with the passage of time, however, while the loser effect stayed intact throughout the ten days of observation, showing that young blue-footed boobies learn to accept the subordinate role.

It is interesting to note that, as in the Drummond and Canales study, most studies that search for winner and loser effects uncover the latter, with only some fraction finding the former. This suggests a more general role for loser effects than for winner effects in structuring aggressive interactions.

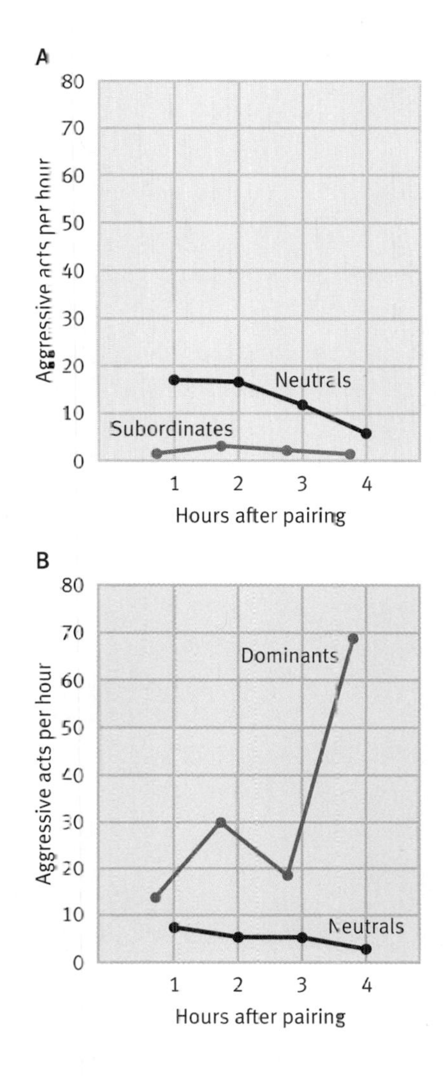

FIGURE 14.11. Winning and losing booby chicks. (A) When subordinate chicks (green curve) who have experienced many losses are paired against neutral singleton chicks (red curve), they display very little aggressive behavior. **(B)** When dominant chicks (orange curve) who have recently won many aggressive interactions are paired against neutral chicks (red curve), they are very aggressive. The pattern begins to dissipate after four hours. *(Based on Drummond and Canales, 1998, p. 1672)*

WINNER AND LOSER EFFECTS IN *RIVULUS MARMORATUS*.

Somewhat surprisingly, more controlled experimental studies of winner and loser effects have been undertaken in fish than in any other group. Why this is the case is hard to know for certain, but in all likelihood it is related to the fact that: (1) aggression is common in fish, (2) aggression in fish is easily quantified in controlled laboratory settings, and (3) the endocrinology of aggression in fish has been well documented in numerous species.

Hsu and Wolf (1999) examined winner and loser effects in the hermaphroditic fish, *Rivulus marmoratus*. In their comprehensive and well-designed experiment, they studied not only the effect of recent wins and losses (at time T-1) on subsequent aggressive behavior (at time T), but also the effect of penultimate wins and losses (at time T-2) on aggressive behavior (at time T).

Fish were presented with different combinations of wins (W), losses (L), or "neutral" (N; no win, no loss) events (Table 14.3). A group of fish that were not used as experimental subjects in the winner/loser trials were allowed to fight, and the individual that defeated all others was used to fight with the experimental fish and provide them with a loss. Similarly, when experimental fish were provided a win, it was by being paired with another fish that had just been defeated by many others (none of whom were used as subjects in the winner/loser/neutral treatments).

By comparing fish in the WW versus LW and LL versus WL treatments (see caption of Table 14.3). Hsu and Wolf were able to document the first experimental evidence that the penultimate (next to last) aggressive interaction a fish experiences also affects its current probability of winning or losing. To see this, let us focus on the effect of penultimate interactions on winner effects. Here, fish experiencing WW are significantly more likely to win a fight

TABLE 14.3. Penultimate and ultimate contests. Hsu and Wolf examined many combinations of winner and loser effects on contest outcomes through time. WW = win at time T-1, win at Time T; LL = loss at time T-1, loss at time T; WL = win at time T-1, loss at time T; NN = "neutral," no loss, no win. Outcomes of (1) and (2) directly measured the effects of penultimate winner and loser experiences. Outcomes of (3) measured the importance of penultimate versus ultimate experience on the probability of winning. Outcomes of (4) showed the difference between LL-WL and LN-WL, measuring immediate loser effects. Outcomes of (5) showed the difference between LW-WL and LN-WL, measuring immediate winner effects. Lastly, outcomes at (6) showed the difference between WW-LW and NN-LW, measuring the effects of penultimate and ultimate winning experiences. *(From Hsu and Wolf, 1999)*

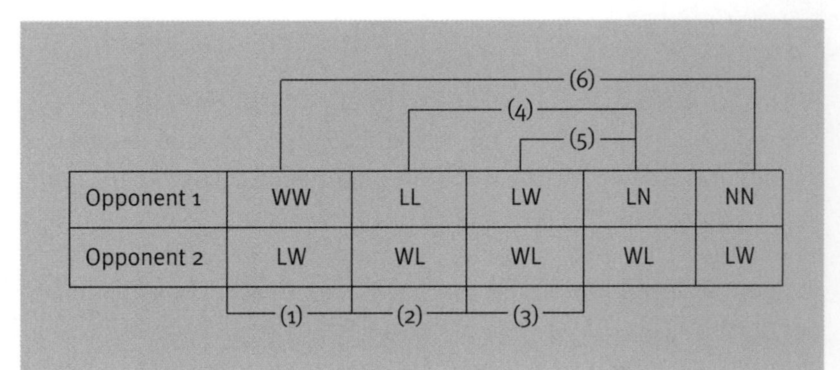

Opponent 1	WW	LL	LW	LN	NN
Opponent 2	LW	WL	WL	WL	LW

than fish experiencing LW. If the penultimate interaction had no effect on current aggressive interactions, one would expect no such difference across the WW versus LW treatments. Comparing numerous other treatments, Hsu and Wolf were able to demonstrate that, while penultimate interactions were important, they were not as important as the interaction that immediately preceded an outgoing interaction (that is, interactions one time unit in the past). In addition, unlike many other species, *Rivulus marmoratus* showed no asymmetry in winner and loser effects—that is, the loser effect was not stronger than the winner effect, nor vice versa.

WINNER AND LOSER EFFECTS IN COPPERHEAD SNAKES. By and large, when ethologists study experiential effects, they do little work on reptiles. Luckily, this asymmetry in choice of study organism does not hold for studies of winner and loser effects. Gordon Schuett (1997) has done the most elegant work on these phenomena in copperhead snakes (*Agkistrodon contortrix*; Figure 14.12). In this species, male-male aggression has a significant impact on mating

FIGURE 14.12. Winner and loser snakes. (A) In copperhead snakes, losses can have a significant effect on future contest outcome. *(Photo credit: Gordon Schuett)* **(B)** Snake fights in *Agkistrodon contortrix* include such aggressive behaviors as: (1) ascend, (2) ventrad-to-ventrad sway, (3) laterad-to-laterad sway, (4) ventrad-to-dorsad sway, (5) hook, and (6) stiffen. *(Based on Schuett and Gillingham, 1989, p. 248)*

success. As such, Schuett examined whether any winner and loser effects impacted aggression (and indirectly mating success).

The basic experimental protocol for examining aggression in copperheads involved a trial arena that had a female in the center of the arena and one male on each end. To begin with, Schuett used males that had no aggressive interactions for six to twelve months prior to his study, and he pitted against each other two males that differed in size by approximately 10 percent. In all thirty-two trials, the larger male was dominant to the smaller male and gained reproductive access to the female.

Once the size contests were complete, ten winners and ten losers from such contests were chosen, and each was matched against a same-sized male copperhead that had no prior experience. Prior winners were not more likely to win again, nor were they more likely than opponents who had no experience to obtain access to a female. Nonetheless, as in blue-footed boobies, the loser effect did have a significant impact on subsequent interactions, with losers more likely to lose again and to cede access to reproductively active females to winners. In an interesting twist, Schuett than examined how individuals who had now lost twice (in the size and loser effect treatments) fared against opponents who had no experience and who were about 10 percent smaller. Would the loser effect outweigh the positive size advantage, or vice versa? Results were crystal clear and pointed to the magnitude of the loser effect in copperheads—two-time losers lost all contests with smaller opponents.

MATHEMATICAL MODELS OF WINNER AND LOSER EFFECTS. In addition to empirical work on winner and loser effects, there are a number of mathematical models that examine the implications of winner and loser effects on the formation of animal hierarchies (Landau, 1951a, 1951b; Dugatkin, 1997b; Hsu and Wolf, 1999; Mesterton-Gibbons, 1999).

A dominance hierarchy is usually defined as a rank ordering of individuals in a group based on their aggressive interactions with each other. For example, if individual A wins the majority of fights with B, C, and D, then A is the top-ranked (or alpha) individual in a hierarchy. If B beats C and D, C beats D, and D beats no one, then a linear hierarchy exists (A > B > C > D) and that hierarchy is said to be transitive, in that if A defeats B and B defeats C, then A defeats C as well (and the same holds true for B defeating C, C defeating D, and B defeating D).

Ethologists' interest in hierarchies can be traced back at least as far as Schjelderup-Ebbe's (1922) work on this subject in chickens. Part of the interest in hierarchy formation no doubt arises from the difficulty in understanding why an animal would ever "accept" a role other than that of alpha individual in the group (Allee, 1951; Vehrencamp, 1983; Reeve and Ratnieks, 1993; Dugatkin, 1995). Initial work on the impact of winner and loser effects on hierarchy formation was undertaken by Landau over fifty years ago (Landau, 1951a, 1951b). Landau was troubled that when he modeled hierarchy formation mathematically, the linear, transitive hierarchies often found in nature did not seem to emerge from his mathematical work, wherein he simply considered inherent differences in fighting abilities across individuals. As such, Landau added winner and loser effects to his models to examine if linear, transitive hierarchies then emerged. Sure enough, they did. Landau then was able to demonstrate an important role for winner and loser effects in shaping our understanding of dominance hierarchies in nature.

Despite the importance and impact of Landau's papers on research in the area of dominance hierarchies, a number of critical questions surrounding winner and loser effects and how they interact remain unanswered more than fifty years after the original papers were published. In particular, Landau did not examine winner and loser effects independently, but only considered the effect on hierarchy formation when both were present. Furthermore, Landau did not take into consideration the fact that animals assessed each other's fighting ability, technically referred to as **resource holding power (RHP).** Is it possible that when RHP is assessed, winner and loser effects can, when separated, have different implications for hierarchy formation?

To address these questions, Dugatkin (1997b) developed a simple computer simulation that independently examined winner and loser effects when individuals assessed each other's fighting abilities and could choose to fight or "flee." The most salient result of these models is that the type of hierarchy predicted depends critically on whether winner effects exist alone, loser effects exist alone, or some combination of these forces are acting in a system.

Winner effects alone produce hierarchies in which the rank of individuals all the way from top rank to bottom rank can be unambiguously assigned. Loser effects alone produce hierarchies in which a clear alpha individual exists, but the relationship among other group members remains murky (Table 14.4). The difference between the hierarchies generated by the winner effect and by the loser effect appears to be that winner effects create a situation in which pairs of individuals primarily interact by fighting, which makes assigning position in the hierarchy fairly straightforward. Conversely, loser effects quickly produce individuals that are not going to be aggressive because of their low estimate of their own fighting ability after a few losses. Hence, with loser effects, most

TABLE 14.4. Winning, losing, and hierarchy formation. When winner effects alone are at play (green box), a clear linear hierarchy exists, with the position of each individual clearly delineated. When loser effects alone are at play (orange box), only the alpha individual is clear. Loser effects tend to outweigh winner effects (red box). *(Based on Dugatkin, 1997b, p. 585)*

Loser Effect		Winner Effect 0				Winner Effect 0.1				Winner Effect 0.2			
		A	B	C	D	A	B	C	D	A	B	C	D
0	A					—	146	136	140	—	146	172	160
	B					8	—	8	172	6	—	168	158
	C					21	170	—	168	7	9	—	167
	D					6	18	7	—	2	2	3	—
0.1	A	—	0	0	0	—	161	195	156	—	177	173	157
	B	171	—	171	163	0	—	2	0	0	—	0	0
	C	0	0	—	0	0	0	—	0	0	0	—	0
	D	0	0	0	—	10	159	153	—	0	0	0	—
0.2	A	—	156	168	174	—	156	178	149	—	1	1	0
	B	0	—	3	3	0	—	0	0	0	—	0	0
	C	0	0	—	0	0	0	—	0	157	173	—	148
	D	0	0	0	—	0	0	1	—	1	3	0	—

interactions will end in neither individual opting to fight, making it difficult to assign relative ranks to most individuals in a hierarchy.

It is important to note that the winner and loser effects models mentioned above do not examine the *evolution* of winner and loser effects, but rather the ramifications of these effects, should they exist. Thus, these models assume winner and loser effects exist and look at the consequences of such an assumption (see Mesterton-Gibbons, 1999, for a model of the evolution of winner and loser effects).

BYSTANDER EFFECTS

Bystander effects occur when the observer of an aggressive interaction between two other individuals changes its assessment of the fighting abilities of those it has observed. Bystanders then *learn* beforehand something about the opponents they may face one day in the future (Coultier et al., 1996; Johnsson and Akerman, 1998; Oliveira et al., 1998). Verbal models and computer simulations have found that bystander effects can have important consequences on

FIGURE 14.13. Chickens pecking. Chicken groups have been a model system for examining aggression, including hierarchy formation and winner, loser, and bystander effects. *(Photo credit: E.R. Degginger/Color-Pic, Inc.)*

hierarchy formation (Chase, 1974, 1982, 1985; Dugatkin, 2001a). Experimental work has found bystander effects in chickens (Chase, 1985; Figure 14.13) and in fish (for example, green swordtails; Earley and Dugatkin, 2002). On the flip side of the coin, not only do bystanders change their estimation of the fighting ability of those they observe, but individuals involved in aggressive interactions change their behavior *if they are watched*. This latter phenomenon is referred to as the audience effect (Evans and Marler, 1994; McGregor and Peake, 2000; Doutrelant et al., 2001).

Endocrinology, Neurotransmitters, and Aggression

With the possible exception of mating, more work has been done on the role of hormones and aggression than on any other behavior of interest to ethologists. The focus of work in this field runs the gamut from interest in the evolution of hormones to how we can learn to control aggression by a deeper understanding of its proximate basis. So much work has been undertaken on the endocrinology of aggression, however, that we can touch on only the tip of the iceberg here (Moore, 2000). As an introduction to this topic, we will examine the role of corticosterone and testosterone in shaping aggressive behavior in animals and humans. We then will turn to the role of neurotransmitters in aggressive behavior.

CORTICOSTERONE AND AGGRESSION

Let's return for a moment to the copperhead snakes we first encountered in our discussion of winners and losers earlier in this chapter. Recall that copperhead losers pay a big price in terms of lost mating opportunities. On top of the purely behavioral work, Schuett also studied endocrinology and aggression in these snakes and tackled the question of whether there were any hormonal correlates to the loser effect (Schuett et al., 1996). Subordinate males in many vertebrate species show elevated levels of plasma corticosterone as a result of stress (Leshner, 1981; Greenberg and Crews, 1983; Hannes, 1985; Greenberg and Wingfield, 1987; Moore and Lindzey, 1992; Sapolsky, 1993), and consequently Schuett examined the possible role that this hormone played in explaining some proximate factors underlying the loser effect in copperhead snakes.

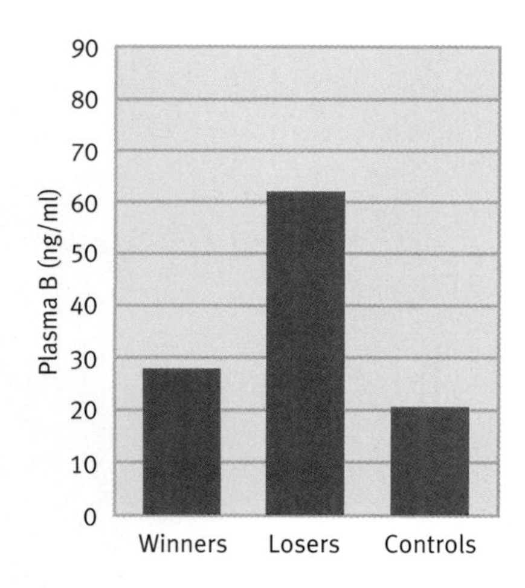

FIGURE 14.14. Hormones, winning, and losing. Losers show increased levels of plasma corticosterone compared to controls. No such change was found in winners. *(Based on Schuett et al., 1996)*

Schuett and his colleagues (1996) allowed pairs of males to fight when a female was present. The snakes were kept together until one male was judged dominant to the other, at which point the individuals were separated, and a blood sample was collected for hormonal analysis. In addition to this treatment, two controls were run: in the first a lone male in his home cage was used, while in the second a single male and a female were placed in the arena. The researchers took radioimmunoassays of plasma corticosterone on winners, losers, and both classes of control males, and they found that plasma corticosterone was significantly greater in losers than in winners or controls (Figure 14.14).

Increased levels of corticosterone produced dramatic effects in terms of both fighting and courting behavior in male copperheads. Not only do males who lose fights, and consequently have raised corticosterone levels, act subordinate and rarely, if ever, challenge other males, they also almost never court any females who are in the vicinity of where they fought. Somewhat surprisingly, levels of circulating corticosterone in the copperhead snakes involved in the study were not affected by how long fights lasted—males who lost in quick fights did not differ from males who lost more prolonged aggressive interactions.

Recent work in a number of other species suggests that the stress-related hormones such as cortisol often inhibit learning and/or memory (de Quervain et al., 1998; de Kloet et al., 1999; Sapolsky, 2000; Lemaire et al., 2000). Whether winners and losers differ in their ability to learn as a function of hormonal changes that occur as a result of aggressive interactions is a fascinating, but unexplored question.

TESTOSTERONE AND AGGRESSION

While hundreds, perhaps thousands, of studies on the hormone testosterone generally find that increased levels of this hormone are associated with increased aggression in male animals, this dramatically oversimplifies the effects of testosterone on violence (Moore, 2000) particularly so in humans. To see this, let us briefly examine the question of whether testosterone levels in human males are associated with violence, and if so, how.

Mazur and Booth (1998) hypothesize that high levels of endogenous testosterone (T) are associated with behavioral dominance in men. Dominance behavior is used broadly here to mean everything from actual violence to more subtle behaviors such as rebellion and breaking the law. Mazur and Booth argue that testosterone and dominance interact in a complex, but fascinating manner. Individuals with relatively high baseline levels of testosterone are in fact more aggressive and dominant. T also changes, however, in anticipation of conflict, in essence "priming" an individual for what is to come. Furthermore, a feedback loop exists: T rises in winners of aggressive interactions and is lowered in those who lose such interactions.

Mazur and Booth set up a study to compare their feedback model of dominance and testosterone with an alternative model in which T is assumed to be relatively fixed through time (Mazur and Booth, 1998). This latter model still assumes that high levels of T lead to dominance, but the feedback mechanism of Mazur and Booth's model is now lacking. To compare the two models, Mazur and Booth used data on Air Force veterans, amassing information a total of four times over a decade. Results indicate that T fluctuates just enough that the feedback model does a better job of predicting aggression and related acts of dominance in military men.

While Mazur and Booth's work indicates that testosterone plays an important part in our understanding of some aspects of human dominance and aggression, other studies suggest a much more limited role for this hormone's role in aggression. For example, Campbell and colleagues examined the relationship between testosterone and aggression in 119 men (Campbell et al., 1997). Their results suggest no significant relationship between testosterone and any measure of aggression.

More generally, Archer and his colleagues ran a meta-analysis on the relationship between testosterone and aggression among young men (Archer et al., 1998). Recall from Chapter 7 that a meta-analysis is used to examine the results of many studies simultaneously, and it essentially takes the data from published work

Interview with Dr. JOHN MAYNARD SMITH

As a theoretician, what drew you to the study of animal aggression?
When I was an undergraduate student of zoology, *circa* 1950, I read Konrad Lorenz's *The Study of Instinct,* in which he describes how animal contests are often settled by ritualized signaling rather than by lethal fighting. This is puzzling. At the time the usual explanation was that, if escalated fighting was common, animals would get injured or killed, and that would be bad for the population; for example, Julian Huxley wrote that it would "militate against the survival of the species." Even as an undergraduate, I knew that had to be wrong: if it paid an individual to fight all out, then natural selection would lead it to do so, whatever the effect on the species.

So I knew there was a problem, but I did not think further about it at the time. I returned to it some fifteen years later, stimulated partly by an unpublished paper by George Price.

How did you and George Price come up with the idea of using game theory to address animal aggression?
In 1965 or thereabouts I was asked by *Nature* to referee a manuscript by George Price (an author I had never heard of) pointing to the puzzle of ritualized contest behavior, and suggesting that the reason why an individual does not escalate is that, if it did, its opponent might retaliate, and it might be seriously injured. Although it was not obvious to me that Price's explanation would work, his paper seemed to me important, because it pointed to an unexplained aspect of animal behavior. Unfortunately, his paper was far too long for *Nature*. I therefore suggested that he be asked to submit a

shortened version, and told him that he should submit the existing draft to a different journal—perhaps the *Journal of Theoretical Biology,* which I thought would publish it in full.

I then decided I should think about Price's suggestion in more detail. I knew that there was a subject called the theory of games, developed to analyze human conflicts. So I got a textbook on game theory out of the library, and read the first few chapters. At that time the only thing I got out of the book was the idea of a payoff matrix. I wrote down a matrix for three "strategies"—hawk (always escalate), dove (always display, and run away if your opponent escalates), and retaliator (display, but retaliate if your opponent escalates). I then asked myself what would happen if a population of individuals were to play this game, with the payoffs representing their fitnesses. The obvious answer is that the population would approach what is now called an Evolutionarily Stable Strategy, or ESS—that is, a strategy such that, if almost everyone in the population does it, it would not pay an individual to do anything different. It turned out that, given certain conditions, retaliator is an ESS.

When I came to write this up, it was clear that I must quote Price's paper. But it had never been published. With *Nature*'s help I contacted him—he was living in London, with no job—and we published a joint paper describing his idea about retaliation, and the ESS method of analyzing such problems.

There is a final curious twist to the story. What I should have known was that my idea of an ESS is very similar to a "Nash Equilibrium," proposed by John Nash some years earlier, and familiar to people working in classical game theory. I had not read enough of the textbook. I finally learned about the Nash Equilibrium from Peter Hammerstein, at the time a graduate student jointly supervised by myself and Reinhardt Selten, who later won a Nobel Prize in economics. In effect, the ESS describes how a population will evolve under natural selection, and the Nash Equilibrium how individuals should behave if they are rational. I think the ESS works better in practice, because animal behavior is indeed the product of natural selection, but human behavior is not usually rational.

I think there should be a moral to this story, but I'm not sure what it is.

The hawk-dove game produces some paradoxical outcomes. Why is that particularly appealing?

It is always exciting when a mathematical model leads to an unexpected conclusion: one may be on the point of understanding something that was previously mysterious. One surprising outcome of the hawk-dove game was the conclusion that the "bourgeois" strategy (if owner, play hawk; if intruder, play dove) is an ESS. Given the concept of a payoff matrix, and of an ESS, the conclusion is at once obvious. It also explains a lot of observed behavior. But I remember thinking that it is absurd to suppose that animals have a concept of property. After some further thought, I realized that no such concept is needed; it is sufficient that the intensity with which an animal will fight to defend a resource should increase the longer it has been in possession of that resource.

T IS ALWAYS EXCITING WHEN A MATHEMATICAL MODEL LEADS TO AN UNEXPECTED CONCLUSION: ONE MAY BE ON THE POINT OF UNDERSTANDING SOMETHING THAT WAS PREVIOUSLY MYSTERIOUS.

One great advantage of a mathematical model over a verbal one is that it is easier to see what assumptions have been made. So, faced with a pattern of behavior that needs explanation, make a simple mathematical model. Does it predict the behavior? If not, which of the assumptions you have made needs altering? If it does, which of the assumptions were really necessary—for example, does the bourgeois strategy depend on an animal having the concept "I am the owner," or can that assumption be replaced by a more plausible one?

Should natural selection ever favor a lethal fighting strategy?

Yes. In the hawk-dove game, if $V > C$ then it pays always to escalate. For example, if survival depends on obtaining some resource (e.g., a winter territory), and if that resource can be obtained only by fighting, then it pays to fight, even if fighting is potentially lethal. In fact, lethal fights are not all that uncommon. For example, male fig wasps fight for control of the females emerging from a fig, and have mandibles able to cut a rival in half; male narwhals injure and sometimes kill their opponents with their tusks, which are up to 3m long; male gladiator frogs have blades on their thumbs, and fights can be lethal.

DR. JOHN MAYNARD SMITH is professor emeritus at the University of Sussex in England. A foreign member of the National Academy of Sciences, he is regarded as one of the most important evolutionary biologists of the twentieth century. His work (in collaboration with George Price) introduced the idea of an evolutionarily stable strategy.

and tests whether, summed over all the studies being examined, a statistically relevant pattern can be discerned. The Archer team's meta-analysis of eighteen hormone studies found no differences between testosterone levels in violence-prone versus nonviolence-prone populations.

NEUROTRANSMITTERS AND AGGRESSION

In reviewing the literature on the relationship between serotonin, aggression, and social status, Edwards and Kravitz (1997) note that, confusing as some of the work in this area is, most researchers are in agreement that serotonin plays an important role in animal aggression. The role, however, appears to differ across animal systems. In mammals, low serotonin levels are often linked with high levels of aggression (Sheard, 1983; Raleigh et al., 1991; Coccaro, 1992), but lower social status. Yet, the situation can be even more complex, as the effect of serotonin (and its precursors) may depend on whether animals were raised in social or asocial environments (Raleigh and McGuire, 1991) and on what specific type of aggressive behavior is being studied (Balaban et al., 1996). For example, Evan Balaban and his colleagues (1996) ran a meta-analysis on thirty-nine studies of aggression and serotonin to test the assertion that low levels of serotonin (in this case, the serotonin metabolite 5-hydroxyindoleacetic acid, 5-HIAA) are associated with impulsive and violent behavior in humans. The meta-analysis compared very violent psychiatric patients to two control groups, one of nonviolent psychiatric patients and one of healthy volunteers, and found no association between 5-HIAA and *impulsive violent behavior*. But there were indeed lower levels of 5-HIAA in all psychiatric patients (violent and nonviolent) when compared with controls.

In fish, as opposed to mammals, enhanced serotonergic function is seen in more subordinate individuals, leading to reduced fighting behavior (Winberg et al., 1992, 1997; Winberg and Nilsson, 1993; Winberg and Lepage, 1998). For example, Winberg and his colleagues (1992) studied brain serotonergic activity and hierarchy formation in Arctic char (*Salvelinus alpinus*). In phase 1 of their work, the researchers constructed four groups, each consisting of four fish. After a stable hierarchy had formed in each group, the groups were disbanded. In phase 2, Winberg and his colleagues took all four alpha males and moved them into one group, all four beta males into a second group, all four gamma males into a third group, and all four bottom-ranking males into a fourth group. When correlating serotonin and serotonin-related chemicals with social status, fish that were subordinate in phase 2 showed higher serotonergic activity, which Winberg and his team hypothesized inhibited the neural circuitry associated with aggression.

A different picture emerges when we look at the relationship between serotonin and aggression and social status in crustaceans (Kravitz, 1988; Huber et al., 1997; Yeh et al. 1997). In crustaceans, increased serotonergic function leads to enhanced aggression and high social status. Take the case of lobsters. When lobsters are paired up in fights, they generally escalate aggression through a series of ritualized combats (Huber and Kravitz, 1995). Once an individual loses a fight, it avoids aggressive interactions for days. But losers can be made more aggressive if they are given injections of serotonin (Huber et al., 1997; Figure 14.15). Furthermore, if fluoxetine (Prozac), an inhibitor of serotonin, is injected at the same time as serotonin, this effect disappears, suggesting a strong role for serotonin in lobster aggression.

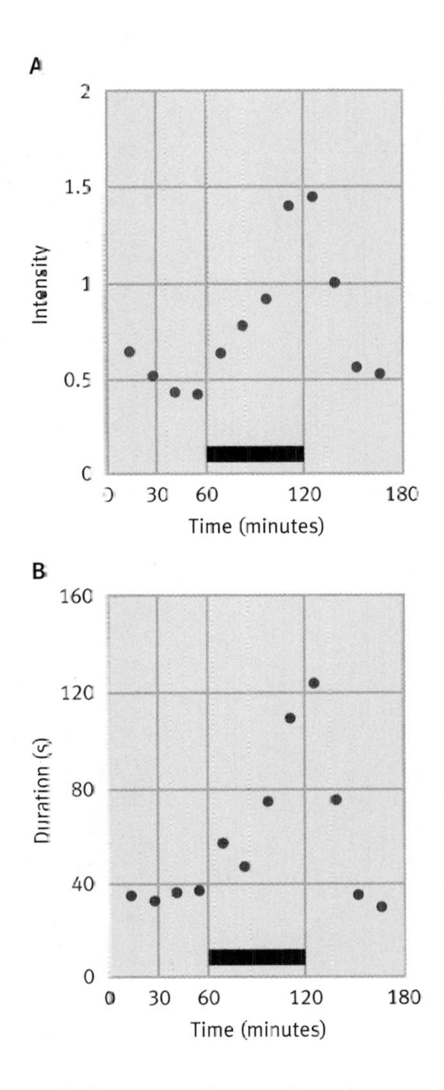

FIGURE 14.15. Serotonin and aggression. A small individual was made subordinate by being matched against an individual that was 30 percent larger than it was. When serotonin was continuously infused into subordinates (red bar), their aggressive levels surged. Subordinates returned to their pre-infusion levels within 30 minutes after serotonin infusions were turned off. **(A)** The intensity of the aggression over time. **(B)** The duration of the aggression. *(Based on Huber et al., 1997)*

SUMMARY

1. Ethologists and evolutionary game theorists have built a series of models that examine the evolution of fighting behavior. The three most well-developed game theory models of aggression are: the hawk-dove game, the war of attrition game, and the sequential assessment game. In all three of these game theory models, a cost to fighting is assumed.
2. There are many complicated varieties of the hawk-dove game. One version adds two new strategies: bourgeois and antibourgeois. A bourgeois strategy instructs an individual to play hawk if it is a territory holder, and dove if it does not own a territory. Antibourgeois codes for the opposite—play dove if you are a territory holder, and hawk if you are not.
3. The war of attrition model was created to examine aggression when the choice available to individuals is more continuous—for example, "fight for x seconds, then stop." Rather than getting a set time for fighting, the war of attrition predicts an ESS distribution of contest lengths.
4. The underlying foundation of the sequential assessment model rests on the idea that assessing an opponent's fighting abilities is critical to behavior in an aggressive contest. Assessing one's opponent is analogous to statistical sampling. A single sample (assessment) introduces significant random error; the more sam-

pling (assessment) one does, the lower the error rate and hence the more confident one can be in whatever is being estimated. At the evolutionarily stable solution to the sequential assessment game, individuals should begin with the least dangerous type of aggressive behavior (A) and "sample" each other with respect to A for some period of time. After that, the next most dangerous behavior (B) should be the dominant action among the protagonists. Depending on the behavioral repertoire of the animals in question, more and more dangerous behaviors are tacked on to the sequence.

5. Winner and loser effects are usually defined as an increased probability of winning at time T, based on victories at times T-1, T-2, and so on, and an increased probability of losing at time T, based on losing at times T-1, T-2, and so on.
6. Bystander effects occur when the observer of an aggressive interaction between two other individuals changes its assessment of the fighting abilities of those it has observed. Not only do bystanders change their estimation of the fighting ability of those they observe, but individuals involved in aggressive interactions change their behavior if they are watched. This latter phenomenon is referred to as the audience effect.
7. Most researchers are in agreement that serotonin plays an important role in animal aggression. Nonetheless, the role appears to differ across animal systems. In mammals, low serotonin levels are often linked with high levels of aggression, but lower social status. The situation can be even more complex, as the effect of serotonin (and its precursors) may depend on whether animals were raised in a social or an asocial environment and on what specific type of aggressive behavior is being studied. In fish, as opposed to mammals, enhanced serotonergic function is seen in more subordinate individuals, leading to reduced fighting behavior. In crustaceans, increased serotonergic function leads to enhanced aggression and high social status.

DISCUSSION QUESTIONS

1. In 1990, Enquist and Leimar published a paper, "The evolution of fatal fighting" in *Animal Behaviour* (vol. 39, pp. 1–9). When do you think fighting to the death might evolve?
2. The classic hawk-dove game we examined in this chapter assumes that losers pay a cost (C) that is not paid by winners. In

the matrix below, we are assuming that both hawks in a fight pay a cost.

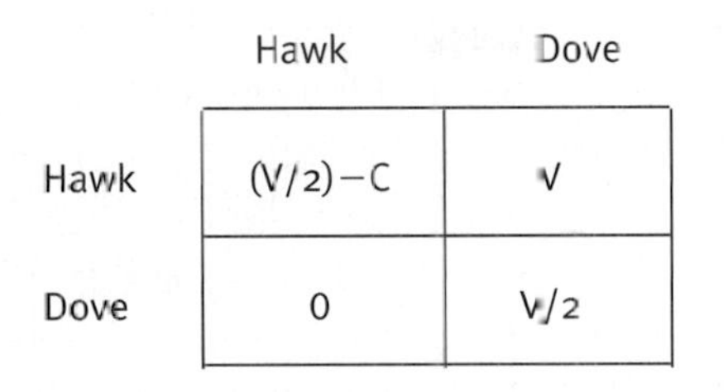

	Hawk	Dove
Hawk	(V/2) − C	V
Dove	0	V/2

For the case of both V > C and V < C, calculate the ESS for this new game.

3. A number of studies have suggested that loser effects are both more common and more dramatic than winner effects. Construct a hypothesis as to why this might be. How could you test your hypothesis?
4. If stress-related hormones such as cortisol often inhibit learning and/or memory, how might that compound the difficulties subordinate fish face in trying to raise their rank in hierarchies?

SUGGESTED READING

Archer, J. (1988). *The behavioural biology of aggression.* Cambridge: Cambridge University Press. A short book that provides an overview of biological and psychological approaches to animal aggression.

Dugatkin, L. A. (2002). Winning streak. *New Scientist, 173,* 32–35. A popular science article on winner, loser, and bystander effects.

Edwards, D., & Kravitz, E. (1997). Serotonin, social status and aggression. *Current Opinion in Neurobiology, 7,* 812–819. An overview of the effect of one neurotransmittor (serotonin) on aggression across species.

Maynard Smith, J., & Price, G. (1973). The logic of animal conflict. *Nature, 246,* 15–18. The classic paper introducing ethologists to mathematical models (game theory) of aggression.

Riechert, S. (1998). Game theory and animal contests. In L. A. Dugatkin and H. K. Reeve (Eds.), *Game theory and animal behavior* (pp. 64–93). New York: Oxford University Press. A review chapter on game theory and aggression.

15

Defining Play

Types and Functions of Play

- Object Play
- Locomotor Play
- Social Play
- A General Theory for the Function of Play

Some Proximate Aspects of Play

- Hormones, Energy, and Play in Young Belding's Ground Squirrels
- The Neurobiology of Play in Young Rats

A Phylogenetic Approach to Play

Interview with Dr. Bernd Heinrich

Play

Play behavior in nonhumans is something of a paradox. On the one hand, everyone who has a cat or dog will attest to the large role that play occupies in their pet's life. Pet play stories are legendary and numerous (Figure 15.1). On the other hand, compared to other topics that we are covering (for example, foraging, cooperation, mate choice), until about twenty years ago, ethologists and behavioral ecologists did relatively little in the way of *controlled* studies of play. One possible explanation for this paradox is that pet owners' stories are typically anecdotes at best, and huge exaggerations at worst. With ethology attempting to gain ground as a serious scientific endeavor, there was probably a tendency to focus on other subjects—subjects not as easily made light of (Figure 15.2).

A second possible reason that play behavior was (and is) relatively understudied is that ethologists tend to study behaviors that appear to have some evolutionary function. When it comes to play, function is often hard to ascertain, and hence there is a natural reluctance to shy away from experiments in this area. A third, related reason for the relative dearth of controlled studies of play is tied to theory. Robert Fagen argues that a lack of concrete theory for the function of play explains why play research lags behind other areas of interest in ethology (Fagen, 1981).

The situation with respect to the study of play has undoubtedly improved (somewhat) since E. O. Wilson wrote that "no behavioral concept has proved more ill-defined, elusive, controversial and even unfashionable than play" (E. O. Wilson, 1975). There is now a sizable literature on many aspects of animal play, and we shall try to cover the salient aspects of this growing area of interest (Bekoff, 1974; Bruner et al., 1976; Symons, 1978;

FIGURE 15.1. Play stories. Play stories are legendary among pet owners. Here a dog looks "sad" when it is not included in the game of baseball that is being played by its owner and the other boys.

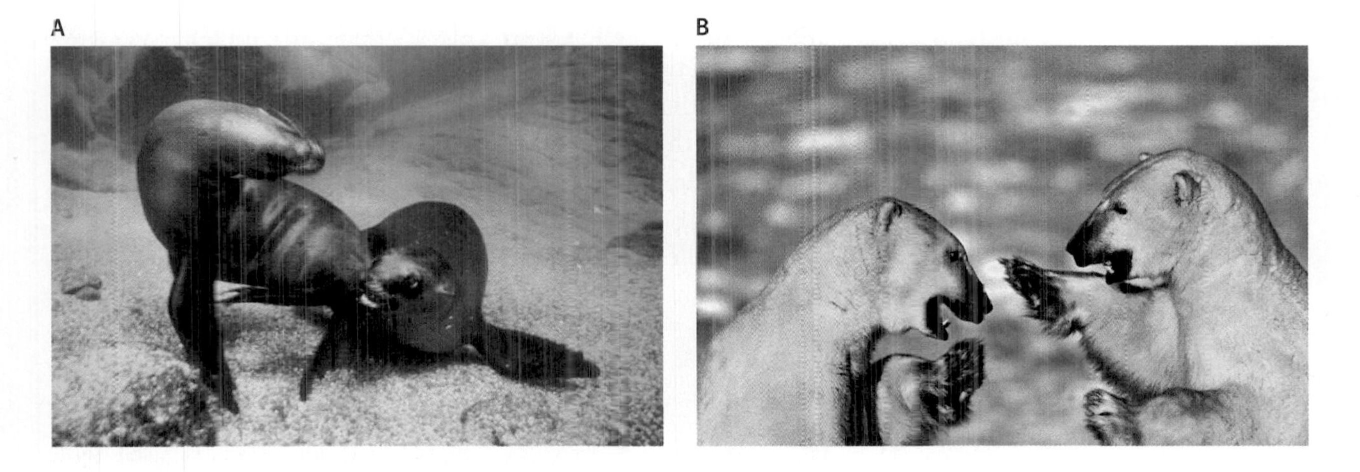

FIGURE 15.2. Play. Play behavior in **(A)** sea lions (playing tag), and **(B)** polar bears (play boxing). *(Photo credits: Marty Snyderman/Visuals Unlimited; Jeff Vanuga/Corbis)*

Fagen, 1981; Bekoff and Byers, 1998; Power, 2000). Here we shall examine

- How play is defined
- The different types of play behavior (object, locomotor, and social play) and their purported functions
- Some proximate aspects of play behavior

Defining Play

Despite the increased theoretical rigor that now underlies much of ethological thinking, there is one area in which this discipline appears "softer" than other scientific disciplines, and that is in the area of precise definitions. By their very nature, behaviors can often be more difficult to define precisely than inanimate objects. Throughout this book, whenever possible, we have relied on definitions that already exist in the ethological literature and that oftentimes match the everyday definition of the subject we are tackling. Since everyone knows what "foraging" means, for example, we shouldn't run into much in the way of definitional problems when we discuss how animals obtain food. The same holds true for subjects like "parental care" and "antipredator activities" and for a whole host of other behaviors as well. Definitional issues, however, do surface when it comes to other topics we cover, particularly the subject of this chapter—**play**.

Some of the reluctance of ethologists to study play centers on the notion that play is some amorphous, fuzzy behavior that defies definition. If that is true, then examining the function, type, physiology, and mathematics of play would appear to be difficult at best.

In the animal behavioral literature there are numerous definitions of play (Fagen, 1981; Bekoff, 1984; Martin and Caro, 1985; Bekoff and Allen, 1998; Burghardt, 1998). The most widely cited of these definitions is that of Marc Bekoff and John Byers who argue that:

> Play is all motor activity performed postnatally that appears to be purposeless, in which motor patterns from other contexts may often be used in modified forms and altered temporal sequencing. If the activity is directed toward another living being it is called social play. (Bekoff and Byers, 1981, pp. 300–301)

This definition clearly centers on the structure of play rather than its function and is problematic even in the eyes of those who suggested it to begin with.

Bekoff and Allen (1998) assert that one problem with the above definition is that behaviors such as repetitive pacing (pacing back and forth for long periods of time in what appears to be a purposeless manner that is occasionally seen in zoo animals), which no one would consider play, still fall under its domain. In addition, it is not only extremely difficult to determine when a behavior "appears purposeless," but behaviors that are apparently purposeless may be so for three very different reasons (Heinrich and Smokler, 1998): (1) observers may simply fail to decipher what the immediate benefit of the play behavior in question truly is, (2) the purpose and potential benefit may not be accrued until long after play has occurred, and (3) the benefits may be multiple and confounding. The situation is actually more complex, because, as Heinrich and Smokler note, Bekoff and Byers's definition does not claim that play is purposeless, but only that it *appears* to be so.

Other definitions besides that of Bekoff and Byers have as many, if not more, problems. So where does that leave us with respect to a definition of play? It seems that those who study play have adopted a definition similar to the United States Supreme Court's definition of pornography: they can't say exactly what it is, but they know it when they see it. This is not quite as bad a state of affairs as it may seem. For example, in an edited volume on play behavior in animals, many of the authors adopt this view of play. They argue that, since experimental work on play lags behind controlled work in other areas of animal behavior, it is best to take a "wait and see" approach, wherein observers study play in many species. The hope is that such work will eventually uncover certain commonalities in play behavior, and that such commonalities will then be used to construct a definition. Whether this proves true or not remains to be seen, but it does appear to be the state of the field of play research today, and as such we will have to adopt the "I know it when I see it" definition of play until something more concrete comes along.

Types and Functions of Play

Ethologists studying play behavior in animals generally speak of three different types of play: object, locomotor, and social play. We shall examine each of these, and in so doing, attempt not only to describe and provide examples of what constitutes each type of play, but also to determine if current work allows us to examine the function of each of the three types of play.

OBJECT PLAY

As the name implies, object play refers to play that centers on the use of inanimate objects such as sticks, rocks, leaves, feathers, fruit, and human-provided objects, and the pushing, throwing, tearing, or manipulating of such objects (S. Hall, 1998). Object play has been documented in animals in a wide array of taxa, and is particularly well-studied in captive populations where "toys" are given to animals to provide them with new items in an otherwise relatively constant environment. Object play has been distinguished from object exploration (Figure 15.3), with play often following exploration

FIGURE 15.3. Play or exploration? Here a cheetah cub comes upon a novel object: a can. Exploring the can appears designed to address the "what is it?" question, while play appears designed to tackle the "what does this do?" question.

(Hutt, 1966; Wood-Gush and Vestergaard, 1991). Animals undertaking object exploration appear to be examining what an object is, while animals engaged in object play appear to be addressing the question: "What can I do with this object?" (Hutt, 1970; Power, 2000). Since the nature and possible function of object play differ across age groups and are far better understood in juveniles, let us look at juvenile play in ravens and cheetahs.

OBJECT PLAY AMONG JUVENILE RAVENS. While the hypotheses surrounding object play in juveniles may differ in detail, they tend to center around two areas. First, from a functional perspective, object play in young animals tends to be associated with some aspect of "practice" that will benefit the animal in either the short or long term (for example, young predatory animals may use object play to practice hunting; Beck, 1980; Fagen, 1981; P. K. Smith, 1982; Martin and Caro, 1985). Clearly, such practice-based benefits are tied tightly to learning. Second, from a time-budget perspective, young animals often have more "free time" to engage in play (Burghardt, 1988), thus allowing for the necessary time inherent in the functional component of juvenile object play.

In his delightful and thought-provoking book *Mind of the Raven,* Bernd Heinrich presents a strong case that natural selection has operated in such a way as to provide ravens (also called wolf birds) with an uncanny intelligence (Heinrich, 1999). Perhaps as a result of this, play occupies a raven's time more than one might predict from a general survey of the avian behavioral literature (Figure 15.4). For example, young ravens undertake object play with virtually every new kind of object they encounter: leaves, twigs, pebbles, bottle caps, seashells, glass fragments, and inedible berries (Heinrich and Smokler, 1998). Heinrich, who has studied these birds for thousands of hours, writes of "young birds' seemingly obsessive drive to contact and manipulate literally all kinds of objects" (Heinrich and Smokler, 1998, p. 32).

FIGURE 15.4. Hanging around. Various "hanging games" Heinrich observed in ravens. *(Based on Heinrich, 1999, p. 289)*

Object play in young ravens sets the stage for what individuals fear, or don't fear, when they mature (Heinrich, 1988a; Heinrich et al., 1996). Adult ravens continue to manipulate objects as they mature. Yet, compared with how they react to those items that they have played with when younger, they treat items that they have never encountered before—including potential food sources—with great trepidation. To demonstrate the potential fitness benefit to object play, Heinrich (1995) examined object play in an interesting experiment involving both observation and manipulation.

Heinrich examined object play in four young juvenile ravens reared by experimenters in a controlled, but relatively natural, forest environment in Maine (Heinrich, 1995). Young ravens were observed in thirty-minute sessions for more than thirty days. During the first ten observation periods, Heinrich noted all the naturally occurring objects the birds encountered. Nine hundred and eighty naturally occurring items that fell into ninety-five different categories were encountered by the young birds during their first ten trials, and such encounters often involved some combination of exploration and play.

After the first ten trials, Heinrich then added forty-four "novel" items to the ravens' environment (Table 15.1) and observed the

TABLE 15.1. Raven play. Young ravens play with almost any new item they encounter. *(Based on Heinrich, 1995)*

ORGANIC INCONSPICUOUS	ORGANIC CONSPICUOUS	INORGANIC CONSPICUOUS
Tenebrio larvae*	Small green apples	Seashell fragments
Squash stems	*Raja* egg cases	Bottle caps
Fucus seaweed bladders	*Ovalipes* crab shells	Red glass chips
Promethea (empty) cocooons	Pistachio (closed) nuts	Beach pebbles
Caddisfly *(Ephemeroptera)**	*Libinia* crab shells	Cigarette butts
Tenebrio adults*	*Busycon* egg cases	White glass chips
Leather chips	Mushrooms	Buttons
Spongomorpha seaweed	*Emerita* crabs*	Glass vials
Wood burls	*Ovalipes* crab claws	
Leaf rolls	*Papilio* butterfly wings	
Onion flower buds	Sparrow eggs*	
Dry beans	Wild strawberries *(Fragaria)**	
Brown birch cones	Red checkerberries *(Gaultheria)*	
Silver maple samaras	Carrot slices	
Red maple samaras	Green *(Ribes)* gooseberries	
Red spruce cones	Wild raspberries *(Rubus)**	
Green birch cones	Blueberries *(Vaccinium)**	
	Red bunchberries *(Cornus)*	
	Soft maize kernels*	

**Edible items.*

manner in which juveniles interacted with these new objects. What he found was that exploration and play were directed at novel items. Novelty, per se, rather than other characteristics such as shininess, palatability, or conspicuousness, explained which items they chose. Yet, while they did not choose items based on their palatability, they quickly treated inedible items as background material (that is, they did not handle these items much after their initial encounters with them), and edible novel items as their preferred foods. This suggests that juvenile play and exploration in ravens has the important benefit of enabling the juvenile ravens to identify new food sources. Ravens are scavengers (as well as predators), and in a world where many objects may be food, play and curiosity in ravens seem to be selected as a means to decipher what counts as food and what doesn't (Heinrich, 1999).

OBJECT PLAY IN YOUNG CHEETAHS. In order to better understand both the costs and benefits of play in young animals, Tim Caro observed cheetah cubs for a total of 2,600 hours over three years in the Serengeti National Park in Tanzania (Caro, 1995b; Figure 15.5). During this time, Caro saw much in the way of object, social, and locomotor play (Table 15.2). In particular, he was interested in the role of each type of play in shaping prey capture in cheetahs and what the costs associated with each type of play might be.

Given that cheetahs spend a considerable amount of time play-

FIGURE 15.5. Cheetah play. Young cheetahs play often, even occasionally while they sit in trees. *(Photo credit: Tim Caro)*

TABLE 15.2. Types of cheetah play. Young cheetahs benefit from a wide variety of play activities in nature. *(From Caro, 1995b, p. 335)*

TYPE OF PLAY	BEHAVIOR PATTERNS*	RECIPIENT
Locomotor play	Bounding gait Rushing around	No recipient
Contact social play	Patting Biting Kicking Grasping	Any family member
Object play	Patting Biting Kicking Carrying	Object
Noncontact social play	Stalking† Crouching† Chasing Fleeing Rearing up	Any family member

**Definitions: bounding gait: slow run with stiff legs causing a rocking motion; rushing around: short sprint often including turns; pat: slap or touch with forepaw; bite: close jaws on animal or object; kick: strike with hindfeet; grasp: hold with forepaws or forelegs; carry: move with object in mouth; stalk: slow approach with body held low; crouch: stationary posture with body low and belly often on ground; chase: run after another animal; flee: run away from another animal; rear up: forelegs off the ground; sniff: place nose close to object.*

†Crouches and stalks that continued from one cub's focal minute to the next were scored just once.

ing as cubs (Figure 15.6), Caro examined the possible costs associated with juvenile play. Estimating the cost of any behavior, particularly a behavior as complex and difficult to measure as play, is never an easy task, but a number of studies, involving various species, indicate that under certain conditions play may be very costly (Lawick-Goodall, 1968; Douglas-Hamilton and Douglas-

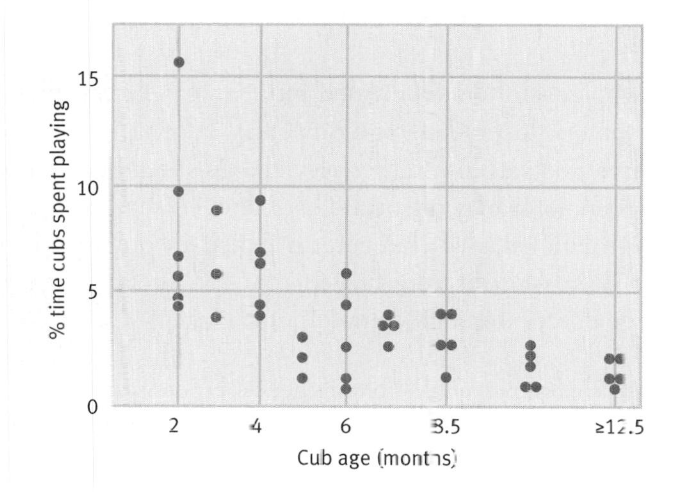

FIGURE 15.6. Juvenile cheetahs spend more time at play than adults. In cheetahs, play progressively disappears with age. *(From Caro, 1995b)*

Hamilton, 1975; Berger, 1980; Arnold and Trillmich, 1985; Harcourt, 1991). As an extreme case, Harcourt (1991) found that while play accounted for just over 6 percent of young fur seals' time, twenty-two of the twenty-six pups that were killed over the course of Harcourt's observations were engaged in some sort of play.

To estimate the potential costs of play in young cheetahs, Caro measured both the distance that a cub rushed around or chased during play, and the distance cubs moved from their mothers during play. The former measure was an attempt to measure energetic costs, while the latter was an estimate of predation or general injury costs associated with play (the further the cub was from its mother, the greater was its risk of being harmed either during play or by a predator such as a lion or a spotted hyena).

Caro presents a strong case for cheetah play being a very low cost activity. To begin with, while it turned out to be nearly impossible to measure the direct energy costs of play, cubs were *never* seriously injured during play. Furthermore, while young involved in play were slightly farther from their mothers than cubs not playing, in all cases the cubs' mothers were so close that their cubs were not under any serious predation threat. Costs of play in juvenile cheetahs may still exist, but Caro's study was about as comprehensive as one can get under field conditions, and if such costs do exist, they are in all likelihood minor.

On the benefits side of the cost/benefit ledger, play (including object play) seemed to have significant functional consequences. In particular, litters of cubs that displayed higher levels of object play demonstrated increased rates of patting, grasping, and biting live prey that their mother had just released. Such predatory-like behavior on the part of the cubs should, in all likelihood, make them more successful hunters when they mature.

LOCOMOTOR PLAY

Locomotor play is often subdivided into two categories: play chasing and solitary locomotor-rotational play (Power, 2000). While some debate exists as to whether play chases should be considered a kind of locomotor play or social play, here we shall only consider locomotor-rotational play (henceforth called locomotor play) and leave our discussion of play chasing for the social play section. Here is one description of locomotor play:

> The single most frequent and phylogenetically widespread locomotor act of play must surely be a leap upward . . . Hops, springs, bounces

> and bucks are variations on the basic vertical leap . . . Animals may somersault, roll, flip forward or backward, spin, whirl, pirouette, make handstands, chase their tails, rear and kick up their heels . . . Often a vertical leap is decorated with body-twists, rear-kicks or head shakes. These acrobatics can be spectacular. (Fagen, 1981, pp. 287–291)

Two major hypotheses have been put forth to explain the function of locomotor play in animals and humans. The first, and most obvious, is that locomotor play both provides general exercise and training for specific motor skills needed later in life (Byers, 1984, 1998). A second potential benefit is that it provides animals with a better understanding of "the lay of the land" (where things are in relation to one another; Power, 2000), and this may provide immediate benefits (Symons, 1978; Stamps, 1995). We shall focus here on exercise-related benefits.

Locomotor play has been studied primarily in rodents, primates, and ungulates, and it includes such actions as leaps, jumps, twists, shakes, whirls, and somersaults (Figure 15.7). Byers and Walker (1995) have argued that while play incorporates a wide array of behaviors, the specific benefits of play may nonetheless be much more limited than originally speculated. Since Brownlee (1954) first raised the issue of locomotor play, the focus of research in this area has been on the benefits accrued via exercise, physical training, and practice. Yet, these benefits are hard to determine. In an attempt to quantify the possible fitness consequences of locomotor play, Byers and Walker (1995) reviewed a list of nineteen potential anatomical and physiological benefits associated with exercise and physical training (Table 15.3). They began by searching for benefits that were available to individuals as juveniles, but not as adults, and that were long-lasting in their effects. Only two of the nineteen possible benefits met their criteria.

One likely benefit accrued by locomotor play focuses on cerebellar synaptogenesis (the creation and distribution of synapses in the cerebellum). The cerebellum plays a critical role in interlimb

FIGURE 15.7. Pronghorn play. Pronghorns undertake various forms of locomotor play, including **(A)** "high speed running," **(B)** "fast turns," and **(C)** "stots" (jumping with all four legs simultaneously off the ground). *(Photo credits: John Byers)*

TABLE 15.3. Physiological effects of elevated motor activity. In examining the potential benefits of locomotor play, Byers and Walker listed nineteen benefits that might be associated with elevated motor activity. *(Based on Byers and Walker, 1995)*

SPECIFIC EFFECT	PRESUMED BENEFIT TO FITNESS	EFFECT AVAILABLE TO JUVENILES?	EFFECT PERMANENT?	EFFECT AVAILABLE TO ADULTS?
Increase in VO_2 max*	Greater endurance	Yes	No	Yes
Decrease in heart rate during exercise	Greater endurance	Yes	No	Yes
Decrease in blood lactate level during exercise	Greater endurance	Yes	No	Yes
Increased heart weight : body weight ratio	Greater endurance	Yes	No	Yes
Increased myoglobin	Greater endurance	Yes	No	Yes
Greater numbers and size of skeletal muscle mitochondria	Greater endurance	Yes	No	Yes
Increased muscle glycogen and triglyceride stores	Greater endurance	Yes	No	Yes
Greater capacity to oxidize fat	Greater endurance	Yes	No	Yes
Greater slow-twitch fiber area	Greater endurance	Yes	No	Yes
Greater total blood volume	Greater endurance	Yes	No	Yes
Greater muscle capillary density	Greater endurance	Yes	No	Yes
Greater maximal ventilation rate	Greater endurance	Yes	No	Yes
Increased maximal muscle blood flow	Greater endurance	Yes	No	Yes
Bone remodeling	Increased strength	Yes	No	Yes
Fast-twitch fiber hypertrophy	Increased strength	Yes	No	Yes
Increased recruitment of motor units	Increased strength	Yes	No	Yes
Modification of cortical areas involved in movement	Increased motor skill/ energetic economy of movement	Yes	No	Yes
Modification of muscle fiber type differentiation	Increased motor skill/ energetic economy of movement	Yes	Probably	Unlikely
Modification of cerebellar synapse distribution	Increased motor skill/ energetic economy of movement	Yes	Yes	Diminished

$*VO_2max$ measures maximum oxygen reuptake.

coordination, smooth movement, postural changes, eye-limb coordination, and many other aspects of movement in mammals (Thatch et al., 1992; Byers and Walker, 1995). During development, more cerebellar synapses are created than are used in later life, and the number of such synapses appears to be a function of experience (Greenough and Juraska, 1979; Pysh and Weiss, 1979; Purves and Lichtman, 1980; M. Brown et al., 1991; Jacobson, 1991).

Byers and Walker hypothesize that "play is timed to occur during the sensitive period of terminal cerebral development to modify

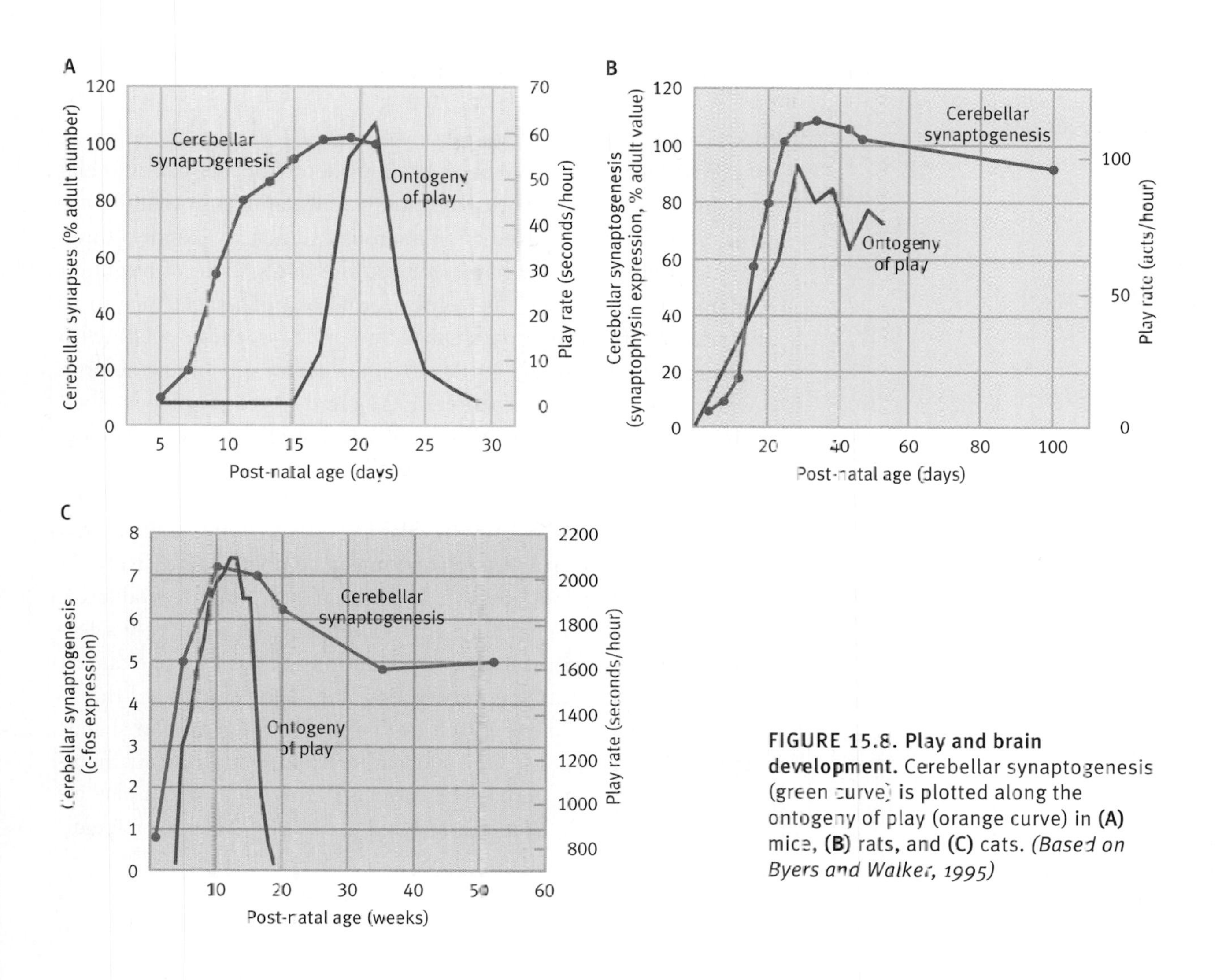

FIGURE 15.8. Play and brain development. Cerebellar synaptogenesis (green curve) is plotted along the ontogeny of play (orange curve) in **(A)** mice, **(B)** rats, and **(C)** cats. *(Based on Byers and Walker, 1995)*

synapse formation and/or elimination" (Byers and Walker, 1995, pp. 32–33). With respect to motor play in juvenile mice, the fit between cerebellar synapse distribution and the ontogeny (development) of play is quite good (Figure 15.8A). Mice start playing at about fifteen days of age and peak in their locomotor play activities at nineteen to twenty-five days, and this corresponds nicely to cerebellar synaptogenesis. The same general pattern is found when examining locomotor and social play in rats (Figure 15.8B) and social play in cats (Figure 15.8C). A correlation between cerebellar synapse distribution and the ontogeny of play does not, however, necessarily mean that changes in the latter *caused* changes in the former. To demonstrate causation, experiments are needed. That said, more and more evidence is mounting in support of Byers and Walker's hypothesis. A second major developmental change—the differentiation of muscle fiber into "fast" or "slow" fibers—also maps nicely onto the ontogeny of play (Close, 1972; Edgarton, 1978; Roy et al., 1988).

SOCIAL PLAY

Social play is perhaps the most well-studied variety of the three play types we shall cover here. As is evident from its name, social play incorporates play with other individuals, and its potential benefits have been espoused in numerous studies. Generally, three functions of social play have received the lion's share of attention (Thompson, 1996): (1) like other forms of play, social play may provide much-needed physical skills, such as those relating to fighting and hunting, (2) social play may lead to the forging of long-lasting social bonds (Carpenter, 1934), and (3) social play may aid in the development of cognitive skills. The studies we review below on social play touch on each of these functions.

SOCIAL PLAY AND BIGHORN SHEEP. In bighorn sheep *(Ovis canadensis),* as in many ungulates, males compete aggressively for females (Geist, 1966, 1974b). In this species, male reproductive variance is much higher than that of females, because only a few males win many aggressive contests and hence mate, while almost all females have mating opportunities. Joel Berger reasoned that if males and females differ in the degree to which aggression shapes their adult interactions, then such differences may manifest themselves much earlier, perhaps as early as during lamb social play interactions. As such, Berger examined social play in lambs (Berger, 1980).

Berger studied juvenile social play in two populations of bighorn sheep: one in the sloping grassy hills of British Columbia and one in the Colorado Desert. In line with his prediction, Berger found that male lambs engaged in much more contact play than did female lambs. This contact play may help males prepare for the battles that will dominate their struggle for reproductive success in later life.

Three other interesting results emerged from the bighorn sheep study. First, social "contact" play (butting, pushing, and so on) was often preceded by "rotational" movements that appeared to serve as a signal that contact was part of a play sequence (see below for more on play signals). Second, it turns out that play behavior does not automatically appear at a certain age in sheep; rather, it is manifested after young individuals have associated with each other for a certain amount of time. Young that stayed by the side of their mothers began contact play at later ages than did lambs that associated with other lambs from an early age (Figure 15.9). Lastly, Berger also found differences in the frequency of play across his two populations. In both populations, lambs were involved in play at very early ages, but play tapered off much more quickly in the desert population than in the grassy slope popula-

FIGURE 15.9. Bighorn play. Juvenile bighorn sheep who **(A)** stayed close to their mothers for extended periods of time play less often than **(B)** other juveniles.

tion. One obvious explanation—that young sheep in the desert were more energetically stressed—does not appear to be supported by the data, as lambs in the desert actually receive more milk from their mothers than do lambs in the grassy hill population. Rather, it appears that desert lambs often bump into cactus while playing, and the consequent pain they suffer acts as a negative reinforcement for play (Berger, 1980).

SOCIAL PLAY AND COGNITION IN PRIMATES AND CARNIVORES. Researchers focusing on play in primates and carnivores have speculated on various cognitively based benefits that accrue as a result of social play. Mendozagranados and Sommer (1995) studied play in immature chimps in the Arnhem Zoo in the Netherlands. In addition to their findings on partner preferences during play, they found that male-male social play was especially common. One possible benefit of numerous male-male social play sessions is to provide young males with the cognitive skills necessary for the coalition formation that plays such an important part in their adult life (de Waal, 1992). Another possible cognitively related benefit of social play revolves around the idea of "self-assessment." Here, animals use social play as a means to monitor their developmental progress as compared to others. For example, in infant sable antelope *(Hippotragus niger)*, individuals prefer same-aged play partners. While this preference could be due to numerous factors, Thompson (1996) hypothesizes that it is primarily a function of infants attempting to choose play partners that provide them with a reasonable comparison from which to gauge their own development (Figure 15.10).

For a more detailed examination of the role of cognition in primate and carnivore social play, let us turn to two studies: one conceptual/theoretical and one empirical. Marc Bekoff's work on play concentrates primarily on carnivores who have a social system not unlike that hypothesized in our own human ancestors—namely, a social

FIGURE 15.10. Antelope play. Young sable antelope like this pair often engage in play, particularly with same-aged partners. *(Photo credits: Kaci Thompson)*

system encompassing a division of labor, food sharing, parental care, and both within- and between-sex dominance hierarchies.

Bekoff (2000) tackles head-on a fundamental question about play: How do animals, especially young animals, know that they are engaged in play? And even more to the point, how do they communicate this information to each other? Since many of the behavior patterns seen during play are also part and parcel of other behavioral venues (aggression, hunting, mating), how do animals know they are playing and not involved in the real activity? Bekoff puts forth three possible solutions to this important, but often overlooked, question.

One manner in which animals may distinguish play from related activities is that, while play may share some similarities to other behaviors, the order and frequency of behavioral components in play are often quite different from those of the "real" activity (Hill and Bekoff, 1977). That is, when play behavior is compared with the adult functional behavior that it resembles, behavioral patterns during play are often exaggerated and misplaced. If young animals are able to distinguish these exaggerations and misorderings of behavioral patterns, for example, by observing adults not involved in play, a relatively simple explanation exists for how animals know they are playing.

A second, somewhat related, means by which animals may be able to distinguish play from other activities is by the placement of "play markers" (see Chapter 12; Bekoff, 1977, 1995; Berger, 1980). A third means by which young animals may be able to distinguish play from related behaviors is by **role reversal,** or **self-handicapping,** on the part of any older playmates they may have. In role reversal and handicapping, older individuals either allow subordinate younger animals to take on the dominant role during play or per-

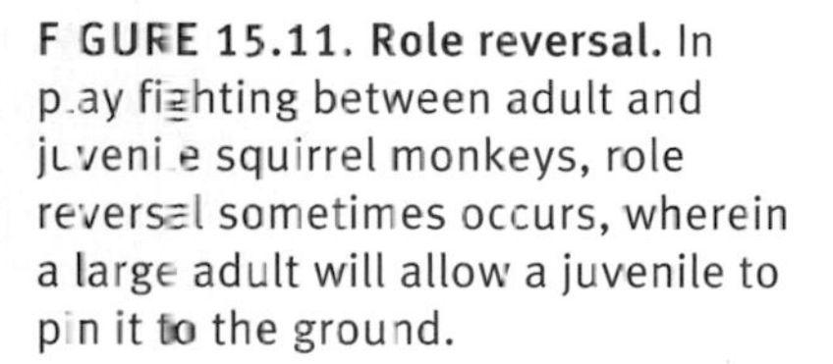
FIGURE 15.11. Role reversal. In play fighting between adult and juvenile squirrel monkeys, role reversal sometimes occurs, wherein a large adult will allow a juvenile to pin it to the ground.

form some act (for example, an aggressive act) at a level clearly below what they are capable of (Figure 15.11). Either of these provides younger playmates with the opportunity to recognize that they are involved in a play encounter.

The obvious follow-up question to "how do animals know they are engaged in play?" is "even when you know it's playtime, why play fairly?" One likely answer is that "cheating" at play probably does net an individual many real benefits. But cheating may lead others to ostracize an individual at social playtime. If the benefits of social play are great enough (Figure 15.12), such ostracism could have a significant negative impact on a cheater.

FIGURE 15.12. The benefits of play. Bekoff has hypothesized that young animals learn to cooperate during play (left) and that this might have benefits later in life, when cooperation is necessary—for example, when adult lions are hunting a wildebeest (right).

PLAY FIGHTING AND COGNITIVE TRAINING IN SQUIRREL MONKEYS. Over the last thirty years, play fighting has been studied in squirrel monkeys (*Saimiri sciureus*) in both the lab and the field (Dumond, 1968; Baldwin, 1969, 1971; Biben, 1986). Play begins as early as five weeks old, when infants start interacting with each other while still on their respective mothers' backs. Young of each sex prefer playing with same-sex partners, and in fact, they play very differently with opposite-sex individuals (Biben, 1986). One fascinating aspect of play fighting in young squirrel monkeys is the important position that role reversal plays, particularly in male-male play bouts.

As in young human children, young male squirrel monkeys prefer to play with others lower in dominance rank (Biben, 1986; Humphreys and Smith, 1987; Boulton, 1991). This finding presents something of a paradox, at least at first glance. After all, if every male prefers to engage in play fighting with subordinates, and males can opt not to play (as is the case for squirrel monkeys), the subordinate in any potential play pair should always search out another partner, spiraling down to the case of no one pairing up to play with him (Altmann, 1962). Clearly, this doesn't occur, as males engage in a great deal of play, suggesting that subordinates often accept a dominant play partner. But why? The answer appears to be that individuals that are clearly dominant *outside* the context of play often allow subordinates to take on the dominant role when it comes to play—that is, they engage in role reversal, thus providing normally subordinate individuals with an incentive for playing. Somewhat surprisingly, role reversal breaks down in between-sex play. Males, who are typically dominant to females, do not engage in role reversal when playing with females.

Why are dominant males so quick to engage in role reversal when play fighting with same-sex partners? Part of the answer probably lies in the fact that role reversal during play does not appear to influence the dominant/subordinate relationship *outside* of play. So, the cost of role reversal is probably minimal. Yet, why bother with role reversal in the first place? The answer, as we alluded to earlier, is that without role reversal, few play partners would be available, and if play has certain inherent benefits, such benefits would be hard to come by without role reversal.

What might such benefits be for young male squirrel monkeys? One obvious benefit may be that play fighting trains males for true aggressive behavior later in life. Yet, the direct evidence for such benefits later in life are absent in squirrel monkeys—it is not as if those who play more win more fights later in life or those who win play fights win real fights. Instead, Biben (1998) offers three cognitively based possible benefits to play fighting in squirrel monkeys:

1. Behavioral flexibility—since play involves little in the way of real costs to squirrel monkeys, it may be an ideal means for individuals to learn how to be behaviorally flexible. This may be particularly relevant with respect to role reversal in male squirrel monkey play fights.
2. Gauging the intention of others—real fighting in adult squirrel monkey life can be potentially dangerous. Play fighting might provide males with training in gauging the intentions of others in adult life.
3. Experience in both the subordinate and dominant roles—males that end up as dominant in adult life must "work their way up" a dominance hierarchy. As they do, they will lose as well as win many encounters. Play fighting may teach males how to act both as a subordinate and as a dominant—roles they will encounter throughout life (again, this is particularly relevant to the role reversal aspect of play fighting in male squirrel monkeys).

Which, if any, of these possible benefits drives play in the squirrel monkey system remains to be tested.

A GENERAL THEORY FOR THE FUNCTION OF PLAY

Spinka and his colleagues have put forth a fascinating new hypothesis for the function of play in mammals. They argue that play allows animals to develop the physical and psychological skills to handle unexpected events in which they experience a loss of control. Specifically, they "propose that play functions to increase the versatility of movements used to recover from sudden shocks such as loss of balance and falling over, and to enhance the ability of animals to cope emotionally with unexpected stressful situations" (Spinka et al., 2001, p. 141). So, for example, the loss of control and balance associated with being chased by predators or losing an aggressive interaction may be dealt with more effectively if play allows animals to prepare for such events.

Spinka and his collaborators list twenty-four predictions that emerge from their hypotheses. Below we shall touch on a number of these hypotheses and the evidence available to judge them:

- At the most general level, they predict that the amount of play experienced will affect an animal's ability to handle unexpected events. While this prediction has not been directly tested, some correlational work in both humans and nonhumans supports it. In rats, for example, individuals deprived of social play often react more negatively to unexpected stimuli than those not deprived of play (Potegal and Einon, 1989). In humans, measures of rough-and-tumble play are sometimes correlated with scores on social problem-solving tests (Pellegrini, 1995; Saunders et al., 1999).

- Self-handicapping, where dominant animals allow subordinates to defeat them during play fights, should be ubiquitous in species that play. Self-handicapping is thought to be an excellent means for preparing for the unexpected, as individuals put themselves in a position very different from that in which they normally find themselves. Evidence from many species supports the omnipresence of self-handicapping in species that exhibit play.
- Play should have measurable effects on an animal's somatosensory, motor, and emotion centers. Spinka and his colleagues argue that during play the brain must deal with sensory inputs that are different from other behaviors and "must be solved by kinematic improvisation and emotional flexibility." In support of this prediction, rats that have been deprived of social play have long-term changes in μ- and κ-opioid receptors, and they have permanently altered levels of dopamine and other neurotransmitters (van den Berg et al., 1999), all of which "are important for coordinating an organism's response to stress" (Spinka et al., 2001).
- Locomotor play should be most common in species that live in the most variable environments. If locomotor play allows one to experience loss of locomotor control, this affect might be most beneficial in environments that change most rapidly. More generally speaking, individuals who engage in play should be more prepared for the unexpected, which is more likely to occur where there is environmental change. There is currently very little evidence available to test this particular prediction generated by Spinka and his collaborators.
- Spinka and his colleagues argue that any differences in play behavior that exist between the sexes should increase over developmental time. The underlying logic here is that if between-sex differences in encountering the unexpected exist and if play behavior prepares one for the unexpected, then between-sex differences in play should be magnified by time, as the between-sex differences in encountering the unexpected (fights, predators) increase with time. This prediction is generally supported by data from a number of species that engage in play.

Some Proximate Aspects of Play

While we have seen the critical role of learning in shaping play behavior, most, but not all, of what we have discussed has been evolutionary in nature. That is, we have been primarily focusing on the ultimate causation of play. We will now examine proximate causes of play.

HORMONES, ENERGY, AND PLAY IN YOUNG BELDING'S GROUND SQUIRRELS

As in many species of rodents, young male and female Belding's ground squirrels (*Spermophilus beldingi*) differ in both the duration and type of play in which they engage (Holekamp et al., 1984; Jamieson and Armitage, 1987; Meder, 1990; Pederson et al., 1990). Since male-female differences in mammalian play are often related to sex hormones (Goy and Phoenix, 1972; Abbott, 1984; Thor and Holloway, 1986; Orgeur, 1995), an endocrinological perspective on this sexual dimorphism immediately suggests that such differences may be a function of gonadal hormones in ground squirrels (Holekamp et al., 1984; Nunes et al., 1999a, 1999b).

In natural populations of Belding's ground squirrels, males initiate much more sexual play than do females, but no differences in play behavior are evident across the sexes with respect to play fighting (Figure 15.13). In order to better understand the proximate basis for squirrel play, Nunes and his colleagues (1999b) treated one group of newborn females with testosterone (in an oil vehicle), and a second control group with an oil vehicle alone. In both treatments total play behavior was highest near weaning and gradually decreased during post-weaning time. As such, testosterone had no effect per se on the temporal aspects of play fighting in females. Nonetheless, females in the testosterone treatment displayed a significantly increased frequency of sexual play behavior. In fact, testosterone-treated females displayed sexual behavior at near the level displayed by same-aged males.

FIGURE 15.13. Squirrel play. Belding's ground squirrels undertake social play. Social play often resembles adult copulation and fighting. *(Photo credit: Scott Nunes)*

In addition to examining proximate underpinnings to play via hormonal manipulation, the researchers also manipulated another proximate agent that may affect play—available nutrients. In a "provisioned treatment," pregnant females, and subsequently their newborns, had their diet supplemented with a high fat/high energy nutrient (peanut butter), while controls received no such supple-

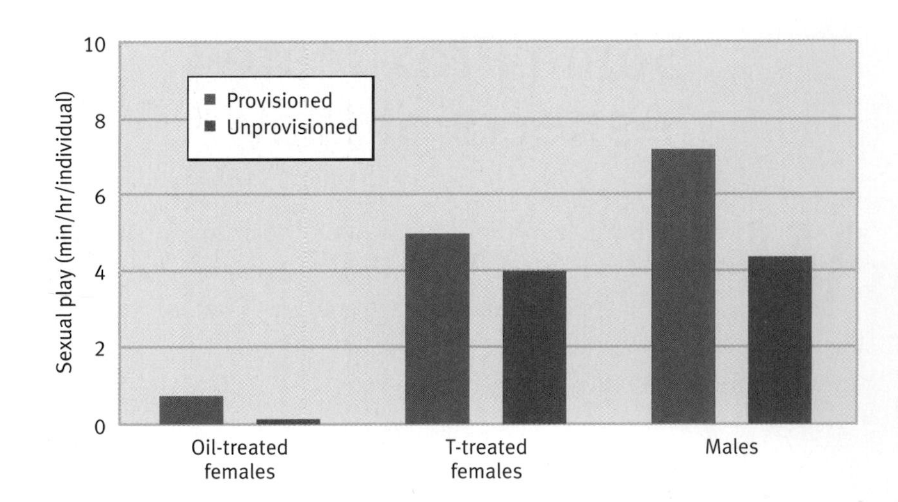

FIGURE 15.14. Testosterone, play, and food. In juvenile Belding's ground squirrels, males typically display much higher levels of sexual play than do females. Females in testosterone treatment (T-treated females) increase sexual play. For both young males and females, provisioning (supplementing the diet with a high fat/high energy nutrient) increased rates of sexual play. *(Based on Nunes et al., 1999b)*

ment. The provisioning treatment had a very significant impact on juvenile play behavior. In all provisioned groups, some of which received testosterone and some of which didn't, rates of play increased, while no change in play was uncovered in the unprovisioned groups (Figured 15.14).

Results such as those of Nunes and his collaborators shed light not only on the proximate causation but on ultimate issues in play as well. For example, if play is generally seen only when food resource levels are high (a proximate response), then it may be that play typically evolves only in species where resource levels are generally high, for it is only there that the benefit-to-cost ratio may be great enough for play to evolve (an ultimate issue).

THE NEUROBIOLOGY OF PLAY IN YOUNG RATS

One thing that most researchers of play agree on is that play often involves some of the most complex behavioral patterns—including complex motor patterns—undertaken by young animals. Given this, and the fact that learning complex behavioral patterns may be one of the functions of play, an understanding of the neurobiology and neurochemistry of play certainly seems in order. An investigation of such proximate factors underlying play, as we shall soon see, may indeed provide us with some clues about play's evolutionary origins.

Broadly speaking, neuroethologists use one of two techniques to study play behavior. In the first, neurotransmitters are targeted to examine their role in play (inhibition, stimulation, and so on), and in the second, the actual neural pathways involved with a particular form of play are targeted, either by making surgical lesions or by some pharmacological means that "lights up" a neural path-

FIGURE 15.15. Rat city. This experimental apparatus served as a "play city" for Siviy's work on rats, play, and neurotransmitters. *(Photo credit: Steven Siviy)*

way. Both of these techniques have proven useful in studying the biochemistry of play fighting in rats (Siviy, 1998).

With respect to examining neurotransmitters and their role in play, researchers systemically administer a neuroactive compound that either blocks or enhances a particular neurotransmitter. If they do this with enough neurotransmitters, a broad picture of the neurochemistry of play emerges. For example, dopamine, norepinephrine, and serotonin seem to be involved in rat play fighting.

A number of lines of evidence suggest that dopamine plays an important role in play fighting in rats. Dopamine inhibitors, for example, typically reduce play (Beatty et al., 1984; Holloway and Thor, 1985; Niesink and Van Ree, 1989). Rather than looking at dopamine in terms of increasing or decreasing play activities, however, a number of researchers have argued that dopamine's most important function may be to invigorate or "prime" an animal to prepare for play (Blackburn et al., 1992; Salamone, 1994; Siviy, 1998).

Rats can be trained to anticipate play (Humphreys and Einon, 1981; Normansell and Panksepp, 1990), and hence it is possible to directly examine whether anticipation of play is concurrent with any changes in dopamine levels. Stephen Siviy (1998) constructed a rat experimental apparatus that consisted of two chambers connected by a tube (Figure 15.15). After counting the number of times a rat crossed the tube for five minutes rats were divided into one of two treatment groups. Those in the play group were allowed to play with another rat for five minutes. Those in the control group had five minutes more in the apparatus, but no other individual was present and hence no social play occurred. Rats in the play treatment crossed back and forth in the tube before their partner was placed in the experimental apparatus much more than did rats in the control condition. One interpretation of this is that rats in the play treatment anticipated the opportunity for play and searched it out, accounting for the increased number of crossings. In addition, half the rats were given a dopamine inhibitor drug. These rats reduced their tunnel crossings significantly, but their play behavior, once a partner was present, remained level, providing support for the notion that dopamine acts to increase the anticipation of play (Figure 15.16).

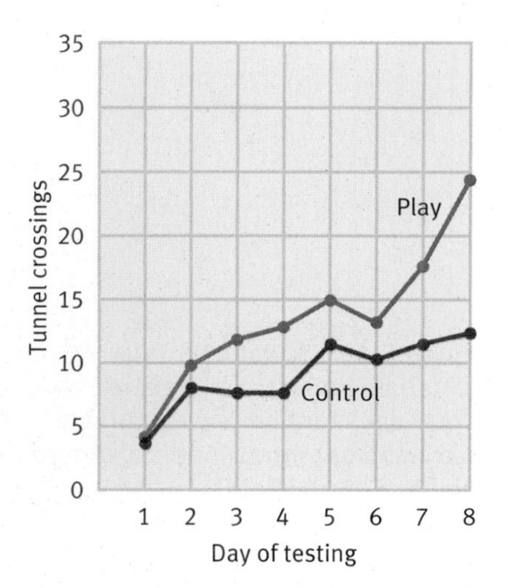

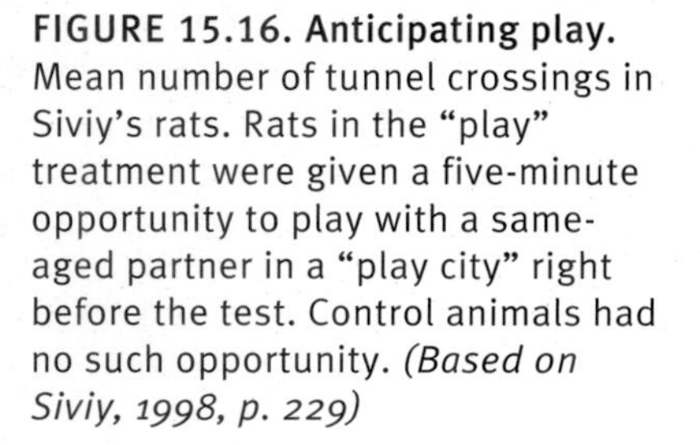
FIGURE 15.16. Anticipating play. Mean number of tunnel crossings in Siviy's rats. Rats in the "play" treatment were given a five-minute opportunity to play with a same-aged partner in a "play city" right before the test. Control animals had no such opportunity. *(Based on Siviy, 1998, p. 229)*

A second general question addressed in the neurobiological literature on play is: What neural pathway(s) underlie play? One technique used in searching for neural pathways is lesioning various areas of the brain and testing for the effect of such a lesion on play. While this technique has had some limited success (Panksepp et al., 1984; Pellis et al., 1992b, 1993; Siviy and Panksepp, 1985, 1987), Siviy suggests that modern biochemical tools may provide an even greater window into the neurobiology of play. For example, early work had shown that lesions to the parafascicular area (PFA) of a rat's brain reduces play fighting (Siviy and Panksepp, 1985, 1987). Siviy followed up this work using biochemical and immunohistological tools to better understand whether a neural pathway underlying play could be identified in the PFA. He hypothesized that if the PFA was critical to play, then cells in this portion of the rat brain should be very active during play. This activity can be measured by quantifying the amount of a protein product (associated with the c-fos gene) that is found in the PFA.

Siviy found that rats that had just been involved in play (and then were sacrificed for brain extraction) indeed had much higher neural activity in the PFA than did control rats. Somewhat surprisingly, however, this increased neural activity was not limited to the PFA, but rather it was found in other areas of the brain (in the cortex and hypothalamus) as well. As such, play seemed to require the use of much of a rat's brain, not just its PFA.

The neurobiological approach to play outlined above has some interesting implications for understanding the evolution of play. Siviy raises two hypotheses for the functions of play that emerge from the neurobiological work that he and many others have undertaken. First, play may serve as a mechanism for coping with stress throughout development. This hypothesis is generated from the finding that the same monoamines (dopamine, norepinephrine, and serotonin) that are involved in play behavior are also critical in animal stress responses (Goldstein et al., 1996). Second, the neural circuit work described earlier supports the hypothesis that play may facilitate learning and creativity (Piaget, 1951). That is, while the lack of a precise neural circuit for play may frustrate some researchers, since learning and creativity appear to be driven by general brain activity levels, this finding suggests a large potential benefit associated with play. In fact, it turns out that transcription of the c-fos (and related genes) appears to be an important component of the molecular machinery associated with learning (Morgan and Curran, 1991; Kaczmarek, 1993; Siviy, 1998).

Creativity may also be a benefit associated with play. Research into play in young school-age children suggests that academic tasks are performed better after a recess period (Pellegrini and Smith, 1998). Exactly why this is so is still unclear, but it may be that play, in some indirect manner, facilitates learning in humans as well as in other animals.

A Phylogenetic Approach to Play

As is the case whenever possible, ethologists wish to understand the role that phylogeny plays in the expression of a given behavior in animals. In essence, such analyses ask whether the variation in a behavior that we see across species is related to variation in ancestral species. To even undertake a phylogenetic analysis of behavior, the behavior in question must have been the subject of investigation in many closely related species. Social play behavior in rodents meets that criteria, as it has been investigated in detail in at least thirteen species of muriod rodents.

The critical piece of information on which a phylogenetic analysis of rodents revolves is that there is a wide range in the complexity of play seen in this group. The question then becomes: Do species with relatively complex play share a common ancestor(s), while those with simpler play repertoires share a different common ancestor(s)? Is the variation we see today a result of a few ancestral forms of rodents differing in their play complexity and passing such differences down to the species that evolved from such ancestors? If so, then to speak of play complexity as being a result of natural selection acting in *each* of the thirteen rodent species independently would most likely be an error.

Pellis and Iwaniuk (1999) tested whether phylogeny might account for some of the variation in the play fighting seen so often in rodents. Play fighting is made up not only of aggressive acts, but also of precopulatory behavior on the part of interactants. To begin their analysis, they assigned a play fighting "complexity" score to each of the thirteen muriod rodent species they examined. These

Interview with Dr. BERND HEINRICH

Were you ever worried that people wouldn't take the idea of play in ravens seriously? Was that part of the reason you documented the phenomenon so meticulously?
I suppose most people do not associate play with birds. For the most part, they think of play as a human activity, or an activity of one of the higher mammals. But I can't say I was worried about what anyone might think. I was interested in ravens, and play seemed like a very important aspect of their lives that gave me a window to see them and understand them better.

Isn't play a waste of an animal's time?
Play is traditionally defined by the fact that it does not have an immediate visible payback. The emphasis here is on immediate and visible. For example, young ravens play with (peck, manipulate) all sorts of objects that are new to them. Anyone watching only that would conclude this activity is a waste of time, unless they realized that the birds thereby get to sample and hence learn about almost all available potential food items in whatever environment they might live. By approaching predators cautiously, and jumping back and advancing again, they learn almost precisely what they can risk in a later feeding situation when scrounging food from the same predator. So play is no waste of time at all. In proximate terms it is fun. In ultimate terms it is investing in the future, especially an uncertain one requiring flexibility of response that can't be predetermined.

What does an investigator have to see before he can feel comfortable calling an activity play?

I don't feel totally comfortable calling any activity play, any more than I feel totally comfortable characterizing any activity as demonstrating intelligence. There are so many gradations and angles. I think that the more precisely we try to define "it," the less we may see, because play like intelligence is an emergent property that is likely revealed as something "extra" above its components. It's a little bit like the analogy of putting an elephant under the microscope to try to see it more precisely. Play includes and grades into all sorts of activities with no discrete boundaries. Take just one activity—bathing. The ravens that I study bathe obsessively in the first few months of their lives after they leave the nest. They jump into water whenever they get the chance, vocalize with many comfort sounds, and thrash till they are soaked, come out, preen and dry off, and do "it" again and again. This is long before, and greatly in excess of, any potentially useful function such as cleaning themselves; for example, dirtying the feathers does not affect the bathing frequency. As the birds get older they bathe more infrequently (but show no apparent behaviorally better or more practiced response), but even if they did they probably still do it because it is an enjoyable experience for them. However, watching them bathe then, one is not likely to mistake it for the "pure" play of the youngsters any more. Play is supposedly practice for future benefits, but even in the young a few days out of the nest the activity already seems "perfect." Nevertheless, I doubt that anyone watching the young would not agree that their activity in a washbasin is not the purest of fun. I have no hesitation calling it playful. Is it play? I'm not sure. Nor am I sure that it matters.

What's the most unexpected thing you've learned from your work on play in ravens?

I think the most unexpected thing I learned was that the raven's attraction to shiny things, such as coins or rings, etc., has less to do with conspicuousness but more with variety. These items are rare to them, and hence when we see the birds play with them it is because they have not yet learned that they are useless.

Do you think play exists outside birds, mammals, and the occasional reptile?

I could envision play only being important for long-lived animals who may experience a diversity of situations. If they are short-lived and/or face a constantly uniform environment, then specific programming would be better. Thus, I do not expect play in insects.

How important is it to study play behavior in the field versus in the laboratory?

If it doesn't occur in the laboratory, then that does not mean it does not exist in the field. So I'd start with the field. However, some details are best observed under controlled conditions. That is, both field and lab have unique deficiencies, and both also have unique possibilities. It depends on what one wants to find out, but generally one complements the other.

Is it possible to experimentally examine whether animals enjoy playing?

No. First, what do we mean by "enjoy"? When we run a marathon, we could equally well say it is an extremely painful experience, or a very satisfying one. We could stick electrodes in the pain and pleasure center and add up discharges and define play by a number. Even if we could for one species, we could not transfer that equation to another.

> SO PLAY IS NO WASTE OF TIME AT ALL. IN PROXIMATE TERMS IT IS FUN. IN ULTIMATE TERMS IT IS INVESTING IN THE FUTURE, ESPECIALLY AN UNCERTAIN ONE REQUIRING FLEXIBILITY OF RESPONSE THAT CAN'T BE PREDETERMINED.

I personally think we can't avoid some subjectivity, and I doubt that our understanding of animals is greatly reduced because of it.

Do you think our constantly improving understanding of the brain will send play researchers down new paths?

Unfortunately I'm not well-informed on the frontiers of brain research. Of course, we are all aware that through medical research we've now identified brain regions in humans (and possibly other mammals) that are associated with sensation, emotions, sensory and motor processing, etc. It is conceivable that by NMI or some other method we can detect the location of neural activity associated with unambiguous play, and then see if we can also detect the same activity in the same area during some other activity that we also perceive as play. But, even if we might find a correlation, we would still be no closer to knowing what it *is*.

DR. BERND HEINRICH is a professor at the University of Vermont. His long-term work on raven behavior, as discussed in his book Mind of the Raven *(HarperCollins), is a wonderful example of how to test hypotheses on everything from foraging to cognition in the wild.*

scores were a composite based on the three broad components of play fighting, which themselves were composites of seven actual measurements used.

The higher the value on any of Pellis and Iwaniuk's measures, the greater the play fighting complexity score. Values for this complexity measure varied from 0 in *Psammomys obesus* to 0.94 in *Rattus norvegicus*. Using the phylogenetic analysis program called MacClade (Maddison and Maddison, 1992), Pellis and Iwaniuk tested whether similarities in complexity scores could be a function of phylogeny, in the sense that complexity scores mapped well onto the phylogenetic relationship between the species they studied. None of the aspects of play fighting that were examined by the researchers could be explained by phylogenetic relationships, in that closely related species of rodents were no more likely to share similar play attributes than were species that were much more distantly related.

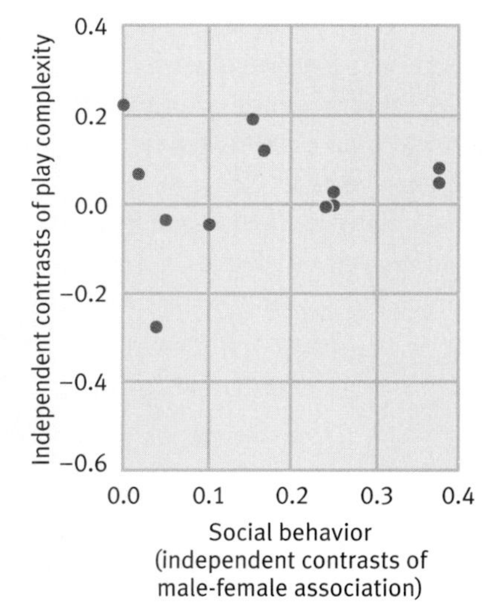

FIGURE 15.17. Play association. In a multi-species comparison of many rodent species, researchers found no relationship between levels of sociality (male-female association) and play. *(Based on Pellis and Iwaniuk, 1999)*

Pellis and Iwaniuk proceeded to examine whether a second possible factor—sociality—might explain the variation in play fighting across the species they studied. Prior work in a variety of taxa had suggested that species differences with respect to sociality may map onto differences in play (Cheney, 1978; Miller and Nadler, 1981; Zucker et al., 1986). Consequently, Pellis and Iwaniuk devised a "sociality" index for the muriod rodents they were studying. While sociality can be defined in many ways, Pellis and Iwaniuk opted for a measure that focuses on male-female interaction. The score system they devised varied from 0, when males and females associated with each other only during mating periods, to 1, when males and females associated with each other year-round. The question of interest then was whether high sociality scores are associated with greater play complexity values. As with the phylogenetic comparisons, the researchers did not find that variation in sociality correlated with variation in play complexity in muriod rodents (Figure 15.17).

Based on their analyses, Pellis and Iwaniuk put forth the following scenario for the evolution of play fighting in muriod rodents. Rather than depicting the ancestral state of play in muriods as simple, they argue that the ancestral state of play was moderately complex in muriod rodents. From this state of moderate complexity, species independently evolved more complex and some less complex play fighting repertoires. This hypothesis remains to be tested, but if it is correct it will reshape the way play evolution is conceptualized. Now, not only must we explain *what* selection forces acted to make play more complex, but *how* natural selection can favor less complex play as well. In other words, if correct, Pellis and Iwaniuk's hypothesis forces us to test whether both complex and simple play are adaptive in their own right.

SUMMARY

1. Play behavior has proven to be very hard to clearly define. The most widely accepted definition is that "play is all motor activity performed postnatally that appears to be purposeless, in which motor patterns from other contexts may often be used in modified forms and altered temporal sequencing. If the activity is directed toward another living being it is called social play." Yet, this definition is not without its problems, including determining when a behavior is purposeless.
2. Ethologists studying play behavior generally speak of three different types of play: object, locomotor, and social play.
3. From a functional perspective, object play in young animals tends to be associated with some aspect of "practice" that will benefit the animal either in the short or long term.
4. One hypothesis for the function of locomotor play in animals and humans is that it both provides general exercise and trains specific motor skills needed later in life.
5. With respect to social play, three functional hypotheses center on: sharpening physical skills, such as those relating to fighting and hunting; promoting long-lasting social bonds; and aiding in the development of cognitive skills.
6. Neuroethologists often use two techniques to study play behavior. In the first, certain neurotransmitters are targeted and either inhibited or stimulated to examine their role in play. In the second, the actual neural pathways involved with a particular form of play are targeted and studied making surgical lesions or using some pharmacological means to "light up" a neural pathway.
7. Spinka and his colleagues have hypothesized that play allows animals to develop the physical and psychological skills to handle unexpected events in which they experience a loss of control. From this, the researchers generated twenty-four specific predictions regarding animal and human play.
8. Play has been studied extensively across many species of rodents. With such data available, a phylogenetic analysis of play is possible, and ethologists have asked whether species with relatively complex play share a common ancestor(s), while those with simpler play repertoires share a different common ancestor(s).

DISCUSSION QUESTIONS

1. Based on what you have learned in this chapter, try to construct a definition of play. After you have done so, answer the following questions: Does your definition cover all cases of play? Does it cover behaviors that you don't consider to be play?
2. In the interview, Bernd Heinrich argues that we can never know whether play is enjoyable to nonhumans. Can you construct an argument that it is, at least in principle, possible to know the answer to this question? If so, what would your argument be? If not, can you give other reasons besides those covered by Heinrich that explain why you believe we can't know whether animals enjoy play?
3. Gordon Burghardt has described object play in a turtle in the Washington Zoo. The turtle, named Pigface, played with new objects thrown in its pool. Before objects were introduced into its environment, Pigface was in the habit of clawing on his own limbs and neck, causing infection and fungal growth. Once the play objects were introduced, however, this self-destructive behavior dissipated to a great extent. Why do you think that might be? How could this sort of study help in the design of animal habitats in zoos?
4. Think about play in young children. Does reading a book for pleasure count as play? Does watching a movie or television or playing a video game count as play? If they are considered play, how might these activities fit into Spinka's hypotheses about play?
5. Some researchers have suggested that play facilitates "creativity." After constructing your own definition of creativity, how would you test this hypothesis? Can you construct tests that measure both behavior and neurobiological/endocrinological correlates of play?

SUGGESTED READING

Bekoff, M., & Byers, J. (Eds.). (1998). *Animal play: Evolutionary, comparative and ecological perspectives.* Cambridge, Eng.: Cambridge University Press. The most comprehensive overview of the ecology and evolution of play, mostly in animals.

Heinrich, B. (1999). *Mind of the raven.* New York: HarperCollins. A beautifully written account of the complexities of raven life, including a nice section on raven play.

Pellis, S., & Iwaniuk, A. (1999). The roles of phylogeny and sociality in the evolution of social play in murid rodents. *Animal Behaviour, 58*, 361–373. One of the few attempts to examine the role of phylogeny on the distribution of play behavior.

Power, T. G. (2000). *Play and exploration in children and animals.* Hillsdale, NJ: Lawrence Erlbaum Associates. A nice attempt at integrating work on animal and human play.

Spinka, M., et al. (2001). Mammalian play: Training for the unexpected. *Quarterly Review of Biology, 76,* 141–168. A recent attempt to integrate all aspects of play under one umbrella.

16

Senescence in the Wild?

Theoretical and Empirical Perspectives on Senescence

- The Antagonistic Pleiotropy Model of Senescence
- Disposable Soma Theory and Longevity
- Longevity and Extending Life Spans

Hormones, Heat-Shock Proteins, and Aging

- Glucocorticoids and Aging
- Heat-Shock Proteins and Aging

Disease and Animal Behavior

- Avoidance of Disease-filled Habitats
- Avoidance of Diseased Individuals
- Self-Medication
- Why Some Like It Hot

Interview with Dr. Richard Wrangham

Aging and Disease

Compare two dogs from the same population who were born at the same time, and what you will almost always find is that, while they are obviously getting older at the same rate, they are not responding to the aging process in the same way. One dog will often be more active and will have suffered fewer of the problems associated with aging than the other. They are both the same age, but aging affects the two individuals differently.

While play behavior is often a luxury of youth, aging and disease are the scourge of later years. In this chapter, we shall first cover aging, and then disease, one of the major causes of death in animals. We begin by examining aging, senescence, and their relation to animal behavior. **Senescence** is defined as an age-specific decline in survival (Promislow, 1991); it is said to occur when the probability of dying increases with increasing age (Figure 16.1). In principle, it is possible for an organism to get older and older, but never senesce. For example, it might be that in a population of animals, the probability that an organism makes it from year x to year $x + 1$ is always 80 percent. That is, whether we are calculating the chance of a three-year-old organism making it to age four or a twenty-year-old

FIGURE 16.1. Aging and senescence. Bill is shown here at four different ages: 20, 21, 50, and 51. If the probability of Bill's surviving from age 20 to 21 (which in this case is shown as 0.95) is greater than the corresponding probability of surviving from age 50 to 51 (shown here as 0.75), then senescence is occurring.

making it to twenty-one, the odds are four in five. In that case, clearly individuals are *aging*, but they are not senescing.

Through the course of this chapter we shall touch on the following age- and disease-related topics:

- Is there senescence in animals in the wild?
- What theories have been put forth to explain senescence and aging?
- What does the empirical literature reveal about the various theories of senescence and aging?
- How do animals respond to disease?

Senescence in the Wild?

At first glance, the question "Is there senescence in animals in the wild?" might seem like a silly one. As we have seen in earlier chapters, animal life in the wild is often fraught with danger and obstacles. As such, natural historians and ethologists long believed that senescence was rare in nature, since the prevailing wisdom was that animals rarely live long enough to begin senescing. Despite the fact that some observational work suggested that this view may in fact be wrong (Nesse, 1988), it was not until Daniel Promislow undertook a rigid and systematic approach to senescence in natural populations that ethologists realized just how wrong (Promislow, 1991).

Promislow analyzed detailed demographic data on forty-nine different species of mammals in fifty-six populations to examine whether individuals in natural populations senesce or not. Using sophisticated mathematical models that correct for errors made by prior investigators, he found that in forty-six of the fifty-six populations making up his study, senescence had taken place, and in twenty-six of these cases, not only did age-specific mortality decline, but it did so at a statistically significant rate (Figure 16.2). In addition to his basic finding that senescence often occurs in natural populations, Promislow also found that senescence:

- Often did not begin until sexual maturity.
- Accelerated in species in which individuals produced many litters.
- Decelerated in species in which individuals came into sexual maturity at a later age.
- Decelerated in large-brained species.

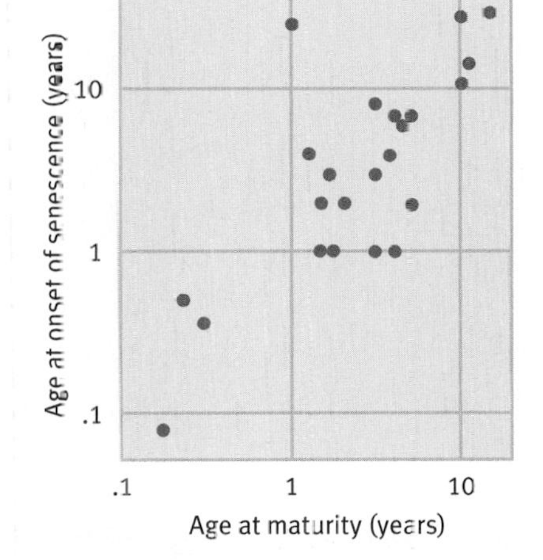

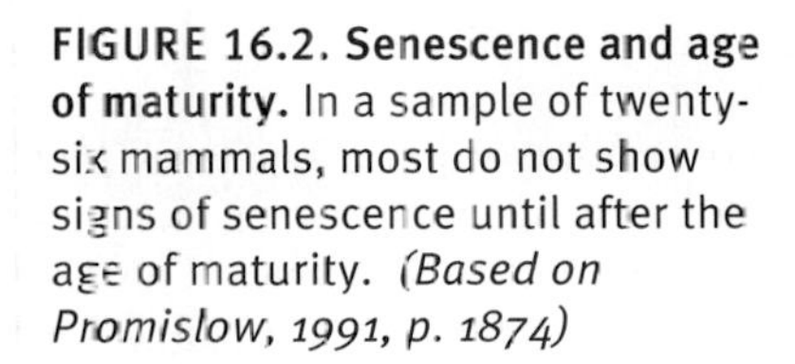
FIGURE 16.2. Senescence and age of maturity. In a sample of twenty-six mammals, most do not show signs of senescence until after the age of maturity. *(Based on Promislow, 1991, p. 1874)*

Theoretical and Empirical Perspectives on Senescence

Perhaps because of its fundamental scientific importance in conjunction with its implications for the health sciences, senescence has been the subject of much speculation. In this section, we shall examine one theoretical model for the evolution of senescence, and then we shall take a look at a few studies testing such a model (Hamilton, 1966; Charlesworth, 1980; Rose, 1991; Gosden, 1996; Austad, 1997).

THE ANTAGONISTIC PLEIOTROPY MODEL OF SENESCENCE

Perhaps the most well-known theory of senescence is that of George Williams (1957). Williams's "antagonistic pleiotropy" idea centers on a population-genetic phenomenon known as **pleiotropy**. Pleiotropic genes are those genes that have more than one effect on an organism, and **antagonistic pleiotropy** refers to cases in which these effects work in opposite directions with respect to fitness. Pleiotropic effects can be manifested in at least two very different ways. First, a gene can affect more than one trait at any given time. For example, gene 1 might affect both hunting behavior and mating behavior in adult organisms. Second, pleiotropic effects can be manifested sequentially. In this case, gene 1 can affect trait 1 at time 1 and trait 2 at time 2 (Figure 16.3). Williams's pleiotropic model of senescence focuses on the latter effects (Fisher, 1930; Haldane, 1941; Medawar, 1952).

Williams's idea begins by noting that natural selection should act more strongly on traits that manifest themselves early in life, rather than later. The reason is that younger individuals, by defini-

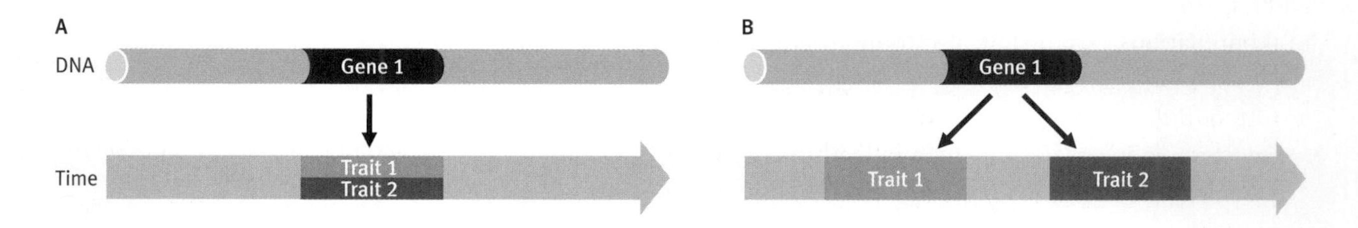

FIGURE 16.3. Pleiotropy. Pleiotropy can act in two different ways. Here we have gene 1 acting on two different traits during the ontogeny of an organism, either **(A)** at the same time, or **(B)** at two different times.

tion, have more of their reproductive life ahead of them, and since successful reproduction is the currency of genetic evolution, selection will act more strongly on traits that appear in young stages of an individual's life.

To see how this might work, let's imagine that a gene has two separate effects on an individual, but that both of these effects occur at time 1, and that time 1 represents youth. For example, suppose our gene makes individuals bigger and also reduces their visual acuity. Suppose further that being bigger provides individuals with a 10 percent increase in mating success, but reduced visual acuity reduces mating success by 10 percent. In this scenario, selection should be neutral with respect to the gene of interest. Now, imagine that a mutation occurs such that our gene still makes younger individuals bigger, but now the negative effects of reduced visual acuity are not manifested until the organism is older. All of a sudden, selection now favors our gene, as its positive effects occur early in life, when most reproduction is still to come, while its negative effects are not manifested until an age when much less lies ahead in terms of reproduction (Figure 16.4). Switch the scenario—make the negative effect occur in early life and the positive effect occur later life—and selection should eliminate our hypothetical pleiotropic gene.

Effect of gene 1 early in life	Effect of gene 1 late in life
+	+
+	−
−	+
−	−

FIGURE 16.4. The effect of timing. If gene 1 has an effect (+ or −) both early and late in life, four possible scenarios exist. The boxed combinations should be favored by natural selection.

If selection acts the way we model it above, then Williams's antagonistic pleiotropy model helps explain senescence. In this context, senescence is conceptualized as the accumulation of negative effects of pleiotropic genes in aging animals (for an alternative view, see Bernstein and Bernstein, 1991).

Michael Rose (1991) notes that there are at least two ways to try to test the antagonistic pleiotropy model of senescence. One can either search for genes that appear to have positive effects early in life and negative effects later on or one can examine selection experiments in which selection of one trait appears to have an antagonistic effect on a second trait (we shall touch on this approach in the section on "disposable soma theory" discussed below). The former approach is reflected in work on the *abnormal abdomen* gene in fruit flies (*Drosophila mercatorum*; Templeton et al., 1985; Figure 16.5). An abnormal abdomen is characterized by the juvenile cuticle remaining into adulthood in flies. For a number of reasons—for example, changes in egg-to-adult development time, early ovarian development, and early age at first oviposition—this gene has just the effects envisioned in Williams's antagonistic pleiotropy model. Not only does the *abnormal abdomen* gene greatly increase early reproduction, but it also speeds up senescence and decreases longevity.

FIGURE 16.5. Fruit flies and the *abnormal abdomen* gene. The *abnormal abdomen* gene, expressed here in *Drosophila mercatorum*, increases early reproduction, but it also accelerates senescence and decreases longevity. *(Photo credit: Alan Templeton)*

Albin (1993) suggests that the antagonistic pleiotropy model is the best model available to understand the evolution of senescence and longevity in humans. While he has not found pleiotropic genes per se, Albin discusses suggestive evidence that four genetic diseases in humans—Huntington's disease, idiopathic hemochromatosis, myotonic dystrophy, and Alzheimer's disease—may be caused by genes that have such pleiotropic effects. For example, Albin argues that Huntington's disease is caused by a single gene, and this disease clearly shortens an individual's life span. Yet, at the same time, the relative fecundity of Huntington's patients is about 9 percent greater than the fecundity of controls. Although it is not clear that the same gene that causes increased age-specific mortality later in life causes increased fertility earlier in those who develop Huntington's disease, further study of this possibility may prove fruitful.

DISPOSABLE SOMA THEORY AND LONGEVITY

One aspect of aging that has long fascinated ethologists, evolutionary biologists, and the general public alike is longevity. We are intensely interested in how long we live, and more generally, in how long creatures in general live. Part of this interest stems from a never-ending desire to figure out how to increase our life span, but whatever the reason, this interest has translated into a fair share of work on longevity in both humans and nonhumans. It is important for us to keep in mind that longevity and senescence, though usually strongly related to one another, can interact in complex ways. Consider males and females in a given species. Suppose that male age-specific mortality went from 0.01 to 0.02 to 0.04 over three years, while over the same time course, female age-specific mortality went from 0.2 to 0.21 to 0.22. Here males are senescing faster, as their age-specific mortality is doubling each year while that of females is increasing at a slower rate. At the same time, however, males are also living lives that are, on average, longer than those of females, as their absolute age-specific mortality is lower than that of females.

One theory that tackles the question of longevity has been dubbed the **disposable soma theory** (Kirkwood, 1977; Kirkwood and Holliday, 1979; Kirkwood and Rose, 1991). This theory, in a slightly different form, can also be used to understand senescence. The disposable soma theory suggests that, under some circumstances, it may pay for organisms to divert energy and resources from normal maintenance and repair associated with traits that have little connection to reproduction (that is; somatic traits) and spend energy on traits related to reproduction. Under this theory, decreased longevity is simply the consequence of such a diversion of resources—ignore maintenance and repair, and the body will start to fall apart sooner.

The disposable soma theory differs from the antagonistic pleiotropy theory in that the disposable soma theory does not assume that a single gene has multiple effects, just that allocation decisions are made with respect to somatic repair and maintenance. It is important to realize, however, that just because the disposable soma theory does not assume pleiotropic effects, it does not mean that the disposable soma theory can't accommodate pleiotropic effects. Rather, when the reallocation of resources is due to pleiotropic effects, the disposable soma theory becomes a subset of the antagonistic pleiotropy theory. For example, selection experiments on fruit flies have found that flies that are selected based on increased longevity often show decreased early fecundity (Rose, 1991; Promislow, 1995; Austad, 1997). From the perspective of the disposable soma model, individuals who reproduce less when they are young have more resources for repair and maintenance, and hence they live longer and senesce more slowly. While this is in line with the predictions of the disposable soma theory, these results also strongly suggest a pleiotropic gene(s) that influences both senescence and fecundity. So, in this case, the disposable soma "reallocation" may be pleiotropically controlled. Yet, as we mentioned earlier, this need not be the case, in that the reallocation prediction underlying the disposable soma idea is not wed to any particular causal agent (for example, pleiotropic genes).

The disposable soma theory predicts that longevity should be generally lower in: (1) species where large investments are made in traits relating to reproduction, and (2) the sex that invests more in reproduction. Let us examine some of the evidence for each of these claims.

DISPOSABLE SOMA THEORY AND HUMAN LONGEVITY. The basic idea underlying the disposable soma theory is that individuals only have so much in the way of resources. According to evolutionary theory, there are only two ways to spend one's resources : (1) on reproduction and reproduction-related activities—that is, on traits related to making new copies of genes for the next generation, or (2) on growth and normal body maintenance—that is, on traits related to sustaining what already exists. Given this finite pie, disposable soma theory predicts that, on average, the more an individual reproduces, the shorter its life span. If you have only so much to spend, any resources you use on maintenance are not available for reproduction.

To test the disposable soma theory, Lycett and his colleagues (2000) used demographic information from a population of 16,500 families that inhabited the Krummhorn region of Germany between 1720 and 1870 (Figure 16.6). They focused on two fundamental predictions of the disposable soma theory: namely, that married women should have shorter life spans than their unmarried counterparts, and that there should arise a negative relationship between fecundity (number of offspring produced) and life span. The first of these tests—married versus unmarried—is a coarse test of the theory in that married women should have more children and hence live shorter lives, while the second is a specific test of the disposable soma theory. Furthermore, in order to gauge the impact of economic class on the predictions of the disposable soma theory, Lycett and his colleagues divided up their population into three economic strata based on land holdings: wealthy farmers, intermediate small landowners, and the poor landless.

Somewhat surprisingly, at first no difference in average age at death was found between married and unmarried women in any economic class, nor was there any discernible pattern between fecundity and life span. What's more, it wasn't even the case that wealthy women with fewer children lived any longer than poor women with many children.

Nonetheless, Lycett and his group weren't about to abandon the disposable soma theory without further investigation. They reasoned that other factors may have confounded their analysis and masked evidence for the disposable soma theory. They then proposed that women who married late might live longer and have fewer children. Thus, duration of marriage per se might be important in the sense that it might have masked the effect they predicted.

FIGURE 16.6. Evolution and human aging. The photograph shows the family of Wiardus Alberts Ohling (1804–1868), a wealthy farmer from the Krummhorn region of Germany. *(Photo credit: Eckart Voland)*

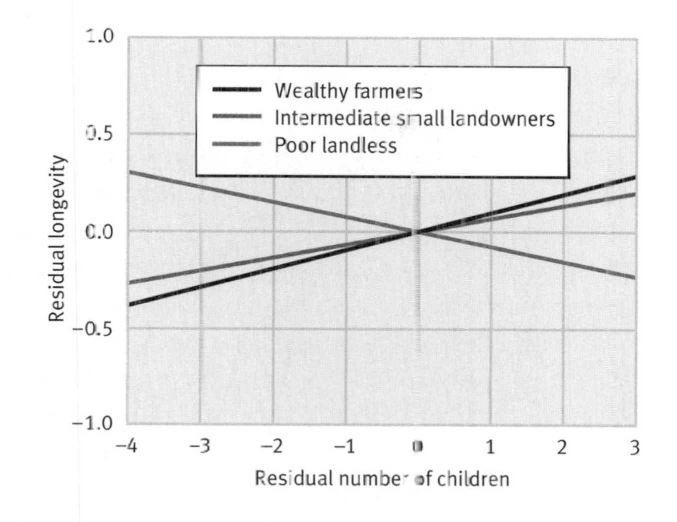

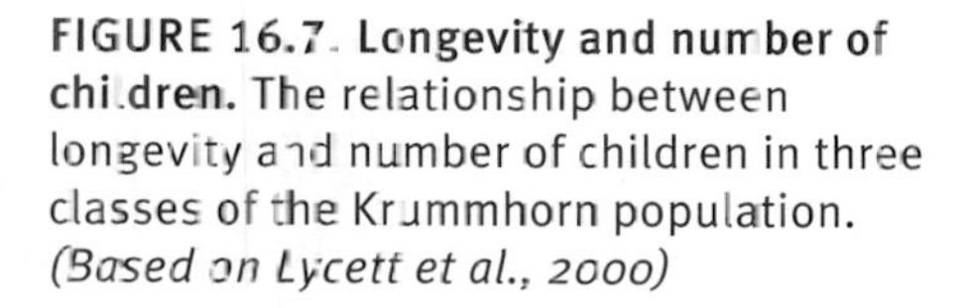

FIGURE 16.7. Longevity and number of children. The relationship between longevity and number of children in three classes of the Krummhorn population. *(Based on Lycett et al., 2000)*

When duration of marriage is removed from the analysis (using statistical techniques), a very interesting pattern emerges. For the wealthy farmer class, women who had lots of children lived *longer*; for the middle class, there was no relationship between reproduction and life span; and for those on the lowest rung of the economic ladder, more children did indeed translate into *earlier* death (Figure 16.7). Lycett and his colleagues interpreted this as partial support for the disposable soma theory, in that one might expect the negative relationship between childbearing and longevity to be most pronounced when economic deprivation was greatest. To further support this interpretation, they found that there was a higher cost of reproduction to their study population in the early 1700s, when economic conditions were really severe, than later on, when living conditions improved overall.

TSETSE FLIES AND LONGEVITY. Clutton-Brock and Langley (1997) tackled reproduction and longevity in a series of laboratory experiments using tsetse flies (*Glossina morsitans morsitans*; Figure 16.8). They began by examining whether just being mated had any impact on a female tsetse fly's life span. In this experiment, females were either mated with a male or remained unmated, and all females were raised in cages separately from males. That is to say, females who mated with males were separated from their mate, as well as from all other males, twenty-four hours after mating. Somewhat surprisingly, unmated and mated females lived equally long (Figure 16.9). In their second, and more critical experiment, Clutton-Brock and Langley raised males and females together in circular cages. Some cages were very male biased (twenty-five males and five females); other cages were highly biased toward females (twenty-five females

FIGURE 16.8. Tsetse longevity. The life span of this fly depends, in part, on the sex ratio of the group it finds itself in. *(Photo credit: Ann and Rob Simpson)*

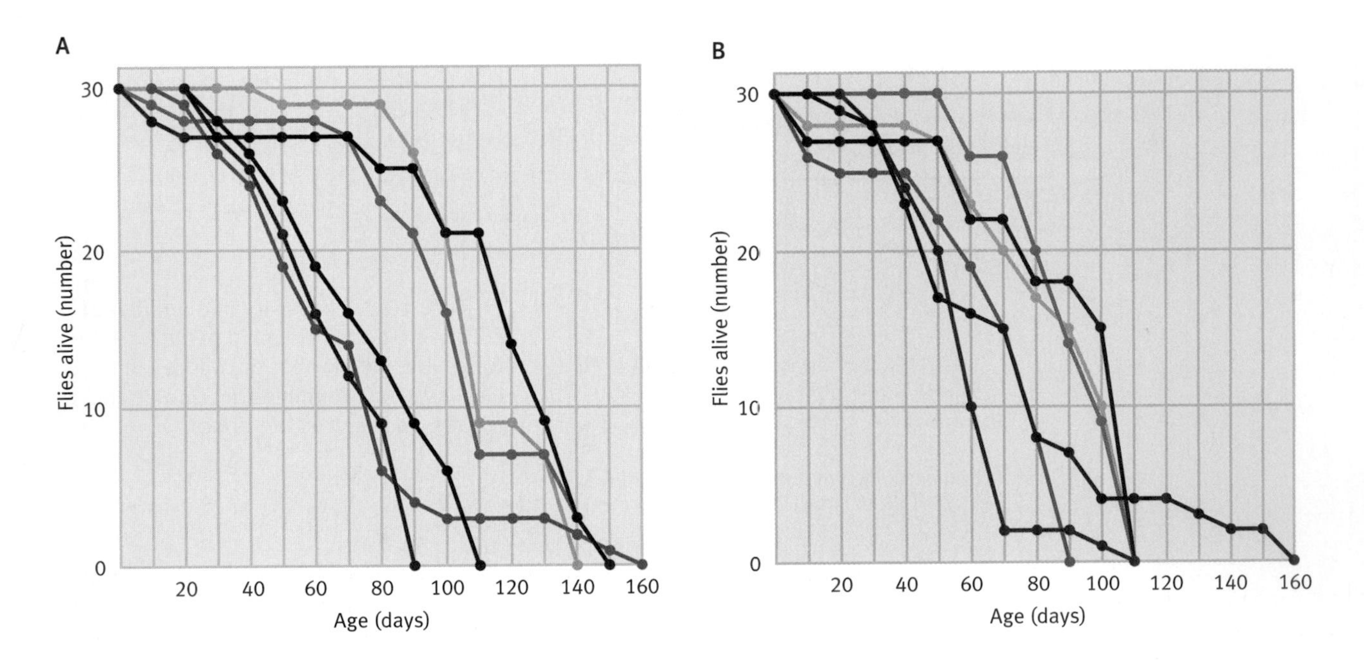

FIGURE 16.9. Mated versus unmated females. Over several replications of the experiment (as represented by different lines on the graphs), no significant differences in longevity were found between **(A)** mated and **(B)** unmated tsetse flies when males were absent from the experimental cages. *(Based on Clutton-Brock and Langley, 1997)*

and five males); other cages spanned the gamut between the extremes (Figure 16.10).

The results were striking. In cages where females were in the vast minority, female longevity was significantly reduced. Females were constantly courted by so many males so much of the time that it actually shortened their life spans significantly (Figure 16.11A). On the other hand, when males were in the minority, it was their life

FIGURE 16.10. Longevity and sex ratios. The experimental protocol in Clutton-Brock and Langley's study was to manipulate sex ratios to examine the effect of courtship on longevity in tsetse flies.

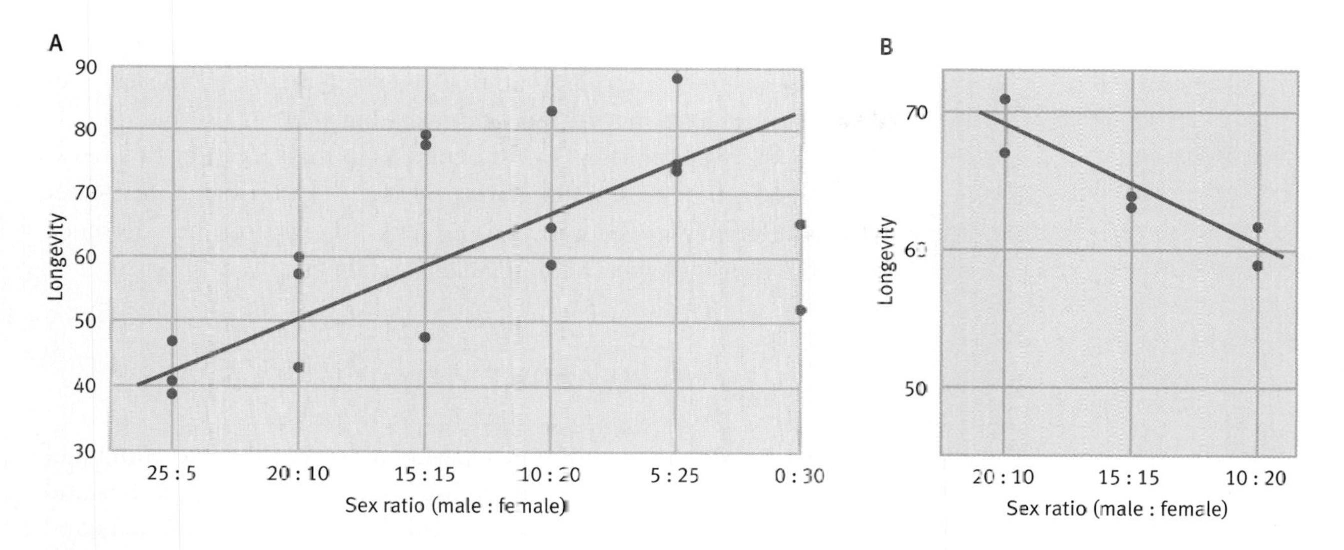

FIGURE 16.11. The effect of sex ratios on longevity in females and males. (A) Female longevity is strongly related to sex ratio. The more biased the sex ratio is toward females, the longer the females live. **(B)** Male longevity is also affected by sex ratio. When sex ratios are skewed toward males, their longevity increases. *(Based on Clutton-Brock and Langley, 1997)*

spans, and not those of females, that were reduced (Figure 16.11B). In those cages, females were not harassed and could afford to be choosier about whom they mated with. The few males present, then, needed to court females vigorously to have any chance of obtaining a mate—so vigorously that it decreased their average life spans from about seventy to sixty days. Merely shifting the sex ratio of the flies changed the costs of courtship, and who paid those costs, dramatically (Partridge and Farquhar, 1981; Partridge et al., 1986, 1987; Partridge and Fowler, 1990; Cordts and Partridge, 1996).

COMPARATIVE STUDIES AND THE DISPOSABLE SOMA THEORY. As we saw in Chapter 6, sexual selection is often strongest in the male sex, and this can often lead to male-male combat as well as to very exaggerated male epigametic traits—that is, male secondary traits related to reproduction. Given this relationship between reproduction and male traits, Promislow (1992) used the comparative approach to test some of the basic predictions of disposable soma theory. More specifically, Promislow examined the relationship between sexual size dimorphism and mortality in males and females.

Using data on thirty-five natural populations of mammals, Promislow used the difference in size between males and females as a measure for the strength of sexual selection. The idea underlying this assumption is that when competition between males is strongest, selection based on male size should be greatest, thus producing a big discrepancy between male and female size. This should translate into a greater cost of reproduction in males (since more resources are needed to reach the large size), and hence a shorter average life span than that seen in females. Promislow's results were striking—when contrasts between male and female body size were plotted against contrasts in male and female life expectancy, the result was a strong positive relationship. The greater the difference in size between the sexes, the greater the difference in life expectancy between females and males.

LONGEVITY AND EXTENDING LIFE SPANS

FIGURE 16.12. Fruit fly longevity. In *Drosophila subobscura*, if reproduction is pushed back six to eight weeks, female longevity increases. *(Photo credit: George W. Gilchrist)*

From before the time of Ponce de Leon's search for the fountain of youth, humans have had an almost insatiable desire to understand how to prolong the human life span. In an attempt to understand how such a goal might be reached, there have been many artificial selection experiments on animals designed to see how life spans can be lengthened (or shortened).

Since many evolutionary models of aging assume that age-specific mortality does not begin to decline until the time of first reproduction (Rose, 1991), many investigators have examined what would happen if the age of first reproduction was artificially delayed. If in each generation only those females that reproduced after age X were selected, then the age of first reproduction should increase, and as a consequence, so might longevity (Edney and Gill, 1968). Wattiaux (1968a, 1968b) tested this idea in two species of fruit flies, *Drosophila subobscura* (Figure 16.12) and *Drosophila pseudoobscura*. In the former, a six- to eight-week postponement of first reproduction was enforced, while in the latter, the delay was for four weeks. Over the course of many generations, when compared to controls, females in the "late-reproducing line" had significantly increased longevity. In addition, males from the late-reproducing female line also had significantly increased longevity. This sort of experiment has been reproduced many times in fruit flies, with results that are similar to those of Wattiaux (Rose and Charlesworth, 1981; Luckinbill et al., 1984; Rose, 1984, 1999).

Hormones, Heat-Shock Proteins, and Aging

Not only do size and age at first reproduction affect longevity, so too do stress and the presence of stress hormones and stress proteins.

GLUCOCORTICOIDS AND AGING

Glucocorticoids—or so-called "stress" hormones—are a mixed blessing. On the one hand, these hormones are critically important in that they often underlie many of the adaptive responses animals use when placed in stressful situations (for example, being attacked by a predator, aggressive interactions, environmental shock; Nelson, 2000). But excessive, prolonged exposure to glucocorticoids can have serious adverse effects on the animal's health, leading to such diseases as diabetes and hypertension (Sapolsky, 1999). In addition, high levels of glucocorticoids are often associated with suppression of the immune system (Munck et al., 1984; Sapolsky, 1996). Robert Sapolsky has argued that many of these symptoms are characteristic of the damage associated with aging, and that, as a consequence, a better understanding of the adverse effects of glucocorticoids will lead to a more thorough understanding of some of the proximate causes of aging (Sapolsky et al., 1986; Sapolsky, 1999).

The neurological impact of glucocorticoids on the hippocampus often mimics normal hippocampal deterioration associated with aging (Sapolsky et al., 1986; Sapolsky, 1999). Studies of the effect of glucocorticoids on hippocampal-dependent learning normally take one of three forms. "Replication" studies address whether administering glucocorticoids to younger animals at levels normally found in older animals will cause degeneration of the hippocampus in the younger animals. In contrast, "prevention" studies examine whether creating the glucocorticoid milieu typically associated with that of young individuals can prevent the normal degeneration of hippocampal function in older individuals. Lastly, "correlation" studies examine whether prior exposure to glucocorticoids predicts degeneration of the hippocampus in older animals. In his review of glucocorticoids, stress, aging, and neurobiology, Sapolsky uncovered a number of general patterns emerging from studies on animals (particularly rats) and humans. In particular, Sapolsky reviewed work that demonstrates that exposure to high levels of glucocorticoids may do the following:

- Disrupt hippocampal-dependent learning, often in a manner similar to the normal decay of hippocampal-dependent learning associated with aging (McEwen and Sapolsky, 1995; Talmi et al., 1996).
- Inhibit nerve growth (Gould, 1994; Sapolsky, 1996; Reagen and McEwen, 1997).
- Increase the probability that neuron repair will fail (McEwen and Sapolsky 1995; Sapolsky, 1996).
- Promote neuron death in a manner similar to the way that neuron loss occurs with normal degeneration of the hippocampus (McEwen, 1992).

Clearly, future work on the relationship between glucocorticoids, stress, and hippocampal function will shed more light on proximate issues surrounding aging (Finch and Rose, 1995).

HEAT-SHOCK PROTEINS AND AGING

A class of interesting proximate players underlying senescence are the heat-shock proteins (hsp) (Feder, 1999; Feder and Hoffmann, 1999). These proteins are inducible by heat, and they act as "molecular chaperones" that minimize damage to other proteins facing a particular stress (for example, heat). When environmental factors such as heat stress cause certain proteins to assume a structure that differs from their normal structure, this can have devastating effects, such as the buildup of proteins that are now toxic to cells. Molecular chaperones like hsp bind to such deformed and dangerous proteins and, via a number of different pathways, defuse the danger. Heat-shock proteins are universal, they play many roles in response to stress, and they are very similar in structure across an extremely diverse set of organisms. Given their both ubiquitous and important role in the life of the organism, the role of hsp in senescence has received significant attention recently (Lithgow, 1996; Lithgow and Kirkwood, 1996).

In mammals, a number of studies have found a correlation between senescence and heat-shock proteins. Older rats, for example, are unable to transcribe heat-shock proteins (hsp70) at the same rate as younger rats (Blake et al., 1991; Udelsman et al., 1993; Gutsmann-Conrad et al., 1998). These correlational sorts of studies, however, make any arrows of causality impossible to draw. That is, it is not possible to determine whether the process of senescence caused hsp70 to be less effective, whether less effective hsp70 promoted senescence, or whether both processes were acting completely independently of one another. To try to determine which alternative explains what is occurring, Tatar and his colleagues (1997) inserted twelve extra copies of hsp70 genes into one strain of fruit flies, while they had a second control line retain their normal complement of these genes and placed dummy (not functioning) inserts in their genome. What Tatar and his colleagues found was that when heat shock was applied to both groups at four days of age, the group of flies with extra copies of the hsp70 gene had reduced mortality rates over time. Exposure to the heat appears then to activate hsp70 genes in a manner that somehow decreases mortality.

Disease and Animal Behavior

One factor that plays an important role in both aging and senescence is disease. Animals tend to be more susceptible to disease as they age, and at least some of the age-specific increase in mortality we discussed earlier is a result of disease. In this section, we shall examine the behavioral means available to animals in their constant struggle against disease. In particular, we shall examine whether animals:

- Avoid areas that contain disease-causing agents.
- Avoid sick conspecifics.
- Self-medicate.

AVOIDANCE OF DISEASE-FILLED HABITATS

In the late 1990s, millions of cows in Britain were incinerated as a result of an outbreak of "mad cow" disease (also known as bovine spongiform encephalopathy). This disease devastated herds of cattle across Britain, terrified farmers all over the world, and caused billions of dollars worth of economic devastation. Given that virulent diseases can wipe out entire populations of animals, we might expect that natural selection would act strongly on any behavioral trait that helped to minimize an animal's exposure to disease. The two most likely ways this might occur are through (1) the avoidance of areas that contain pathogens, and (2) the avoidance of individuals who are already ill. We shall touch on the former here, and return to the latter mechanism in the next section.

One means by which individuals may reduce the risk of infection from parasitic diseases is by producing offspring in areas that have low parasite levels (Kiesecker and Skelly, 2000). Amphibians are particularly good for testing whether such disease behaviors are in play as they host many parasitic pathogens (Worthylake and Hovingh, 1989; Blaustein et al., 1994; Kiesecker and Blaustein, 1995, 1997), and at the same time they can clearly distinguish between oviposition sites based on a wide variety of characteristics (Howard, 1978; Seale, 1982; Resetarits and Wilbur, 1989; Hopey and Petranks, 1994; Laurila and Aho, 1997). Since choice of oviposition site in gray treefrogs (*Hyla versicolor*) is known to affect both larval performance and mortality (Kiesecker and Blaustein,

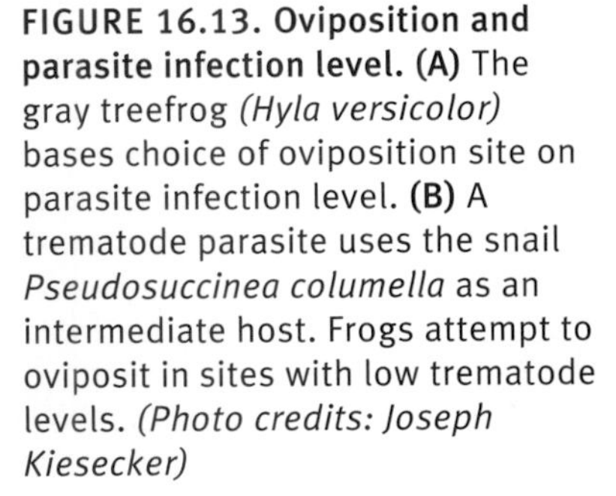

FIGURE 16.13. Oviposition and parasite infection level. (A) The gray treefrog *(Hyla versicolor)* bases choice of oviposition site on parasite infection level. **(B)** A trematode parasite uses the snail *Pseudosuccinea columella* as an intermediate host. Frogs attempt to oviposit in sites with low trematode levels. *(Photo credits: Joseph Kiesecker)*

1997, 1999), Kiesecker and Skelly (2000) examined whether these frogs (Figure 16.13A) base their oviposition decisions on parasite infection level. A co-inhabitant of the ponds containing *H. versicolor* is the snail *Pseudosuccinea columella*, which is an intermediate host for a trematode parasite (Figure 16.13B). This snail produces free-swimming trematodes that infect *H. versicolor* larvae and cause a drop in the treefrog's growth and survival rates.

Kiesecker and Skelly addressed two related questions in their study of ethology and disease avoidance in frogs: (1) Do ovipositing frogs distinguish between sites with and without *P. columella*, and (2) do frogs respond to the density of these snails? To tackle these questions, they employed twenty-five artificial ponds and five treatments: no snails (control), five infected snails, five uninfected snails, ten infected snails, and ten uninfected snails (Figure 16.14). The researchers then assessed oviposition behavior of natu-

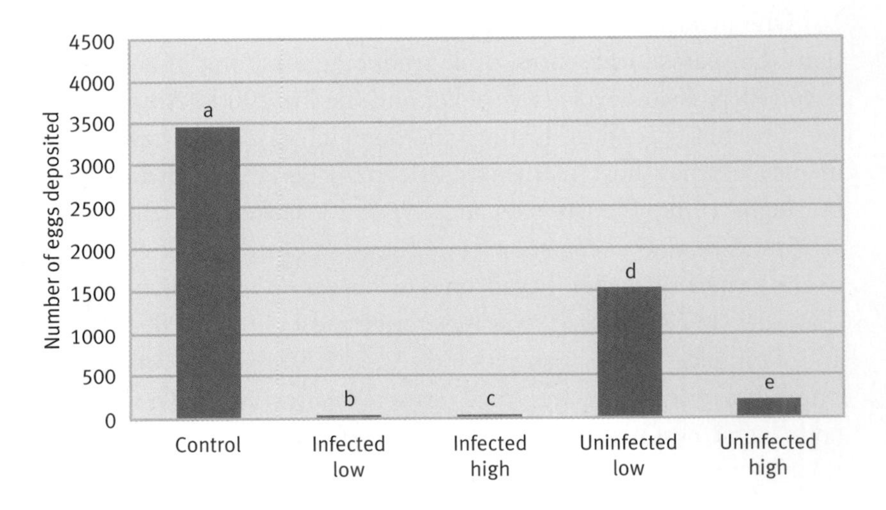

FIGURE 16.14. Parasites and oviposition sites. *Hyla versicolor* laid more eggs at control sites (with no snails) (a) than at sites with experimentally low (b) or high (c) levels of parasites. Frogs also preferred sites with low densities of snails (d) rather than high densities of snails (e), even if the snails were uninfected. *(Based on Kiesecker and Skelly, 2000, p. 2941)*

rally occurring gray treefrogs. The results were unambiguous. Despite controls making up only 20 percent of the ponds, 66.1 percent of all eggs deposited by *H. versicolor* were laid in such ponds, demonstrating clearly that gray treefrogs were favoring pools with no snails. In addition, the frogs also responded to the density treatments, with pools containing uninfected snails receiving 33.5 percent of the eggs laid, and pools containing infected snails receiving only 0.4 percent of the eggs laid.

FIGURE 16.15. Bullfrog quarantine. Healthy bullfrogs prefer to associate with other individuals uninfected with intestinal pathogens. *(Photo credit: Ken Lucas/Visuals Unlimited)*

AVOIDANCE OF DISEASED INDIVIDUALS

In Chapter 6, we learned that females are hypothesized to choose males with low parasite loads, and some evidence supports this hypothesis. Outside the context of mating, however, the broader question of whether individuals avoid diseased conspecifics, and how such avoidance is manifested, has been much less widely investigated (Brown and Brown, 1992; Dugatkin et al., 1994). Nonetheless, such studies are needed to resolve what is really occurring, as the basic evolutionary hypothesis that individuals should avoid diseased conspecifics is in direct contrast with the epidemiological assumption that no such avoidance tactics exist in animals.

Kiesecker and his colleagues (1999) studied avoidance of diseased conspecifics in the bullfrog, *Rana catesbeiana* (Figure 16.15). In *Rana,* as well as in a number of other amphibian genera, individuals are often infected with the intestinal pathogen, *Candida humicola*, which can spread quickly and decrease growth rate and increase mortality by a variety of different means (Beebee and Wong, 1992; Beebee, 1995; Griffiths, 1995). Kiesecker and his colleagues tested whether *R. catesbeaina* individuals recognized and avoided infected groupmates, and if so, how they knew that others were infected.

Tadpoles in close proximity to infected individuals are likely to become infected. The researchers sought to determine if uninfected individuals therefore kept their distance from infected tadpoles. In order to determine whether tadpoles behaviorally "quarantine" infected individuals, Kiesecker and his collaborators undertook the sort of "choice" experiment that is so common in experiments on mating. A "stimulus" tadpole was placed in each end of a long rectangular plastic arena behind a mesh partition. One of these stimulus tadpoles was infected with *C. humicola* and the other was not. A "focal" tadpole was placed in the center of this arena, and then the amount of time that it spent near each of the other tadpoles was recorded. The researchers found that uninfected focal individuals showed a strong preference to spend time near the uninfected stimulus tadpole (Figure 16.16). In an interesting twist, infected focals did not display a preference for either infected or uninfected individuals. Presumably,

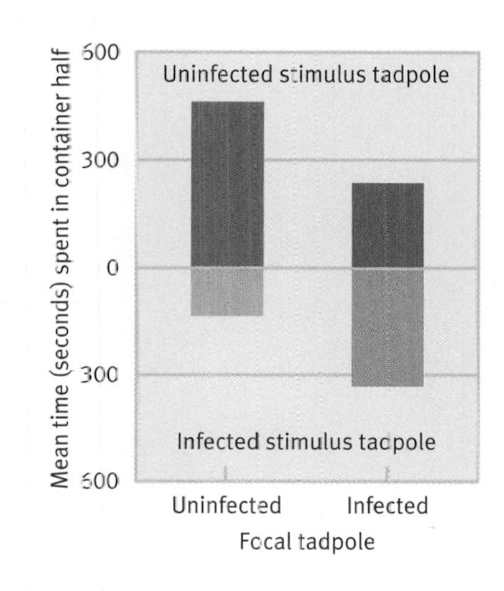

FIGURE 16.16. Infectious association. Infected and uninfected focal animals were given a choice to associate with another infected or uninfected tadpole. Uninfected focal tadpoles preferred to associate with uninfected conspecifics, while infected focals showed no preference. *(Based on Kiesecker et al., 1999)*

once frogs were infected, the choice presented by Kiesecker and his colleagues had no fitness consequences, and hence no preference was uncovered.

Kiesecker and his collaborators followed up their initial experiment to better understand how focal individuals were determining whether conspecifics were ill or not. The experimental protocol was identical to their initial experiment, with a few important exceptions. First, based on the results from their initial experiment, they opted to test only uninfected focals. Second, in addition to the choice test they initially ran—a test that allowed for both visual and chemical cues—they now ran a treatment that allowed for only visual or only chemical cues. In addition to replicating the results of their first experiment (that is, avoidance of infected stimulus individuals when both types of cues were available), Kiesecker and his colleagues found that tadpoles were assessing illness by chemical, not visual cues.

SELF-MEDICATION

Intriguing evidence is starting to mount that animals may modify their diet in a way that has the effect of self-medicating (Janzen, 1978; Rodriguez and Wrangham, 1993; Huffman, 1997; Lozano, 1998). Lozano divides self-medicating into two broad categories: (1) preventative—that is, actions taken to prevent sickness, and (2) therapeutic—that is, actions taken to alleviate sickness once an animal is ill. Consider the following behaviors that *may* represent self-medication in nonhumans:

- The use of potentially antibacterial plant substances in bird nests (Wimberger, 1984; Clark and Mason, 1985, 1988).
- The consumption of soil, dirt, and rock by gorillas, chimpanzees, and Japanese macaques as an antidiarrheal mechanism (Mahaney, 1993; Mahaney et al., 1993, 1996).
- "Anting" behavior in birds, wherein birds rub crushed ants on their plumage, in an apparent attempt to soothe irritated skin and reduce the number of parasites (Potter, 1970; Clunie, 1976; Ehrlich et al., 1986). This has also been observed in mammals such as grey squirrels and capuchin monkeys (Bagg, 1952; Hauser, 1964; Longino, 1984).
- "Fur rubbing" in primates. Here, primates rub various fruits, leaves, and vines against their fur for reasons similar to anting in birds (Baker, 1996; Gompper and Holyman, 1993).

With respect to therapeutic self-medication, it has long been thought that chimpanzees undertake certain forms of leaf swallowing as a means to suppress parasite infection. To test this idea,

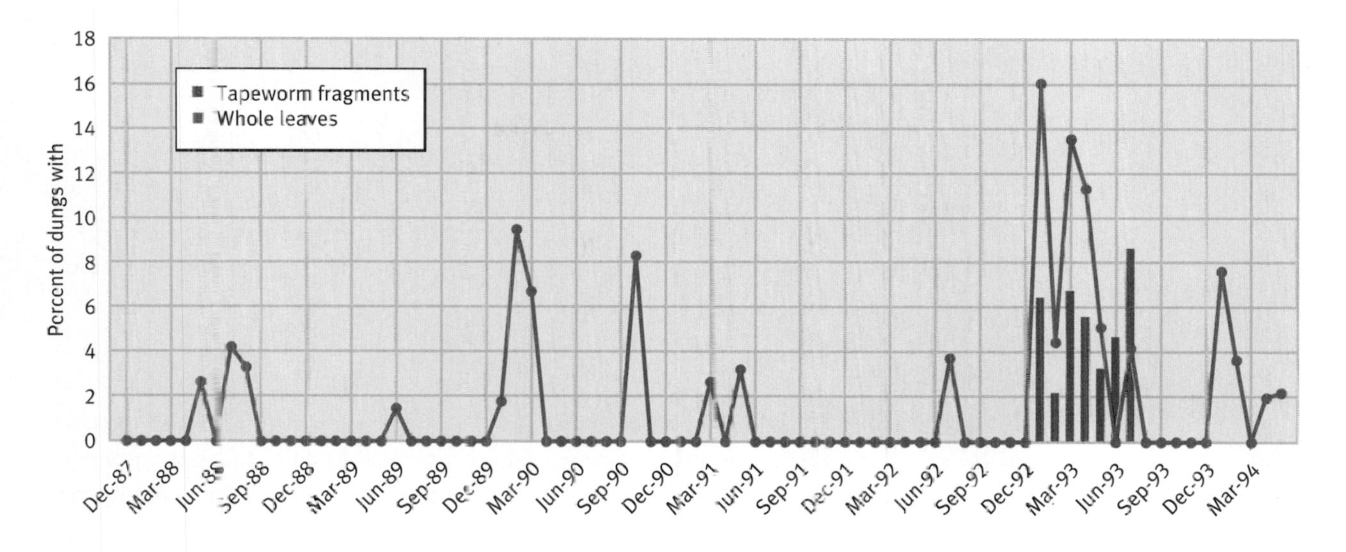

FIGURE 16.17. Tapeworms and chimp self-medication. Chimps increased consumption of whole leaves in concordance with an increase in tapeworm fragments found in their dung. *(Based on Wrangham, 1995)*

Richard Wrangham (1995) examined the relationship between leaf swallowing and the presence or absence of tapeworms in the guts of chimpanzees. What he found was that during a seven-month period when tapeworm infestation was high (no tapeworms were found outside this time window), chimps showed extremely high levels of leaf swallowing compared to when tapeworms were absent (Figure 16.17) While it is not clear how effective a medicating procedure this is, chimps nonetheless appear to modify their diet in ways commensurate with self-medication. Similar sorts of work on self-medication have been undertaken in the Gombe and Mahale chimp populations of Africa (Wrangham and Goodall, 1989; Huffman, 1993; Huffman and Wrangham, 1994).

Clearly, the study of self-medicating in nonhumans is still in its infancy, based largely on anecdote, and not without major conceptual traps (Sapolsky, 1994). Nonetheless, a number of fascinating hypotheses have already been put forth and call out for testing. For example, Huffman and Lozano raise the provocative claim that since the same plants are often used for medicinal reasons in humans and other primates, ethological work may shed light on the nature of self-medication in early hominids (Huffman, 1997; Lozano, 1998).

WHY SOME LIKE IT HOT

Humans are quite concerned about how their diet affects their health, including the effect of diet on disease and disease prevention. From an evolutionary perspective, to study the effect of diet

Interview with Dr. RICHARD WRANGHAM

How did you ever think to test whether chimps ate leaves to fight tapeworm infections? What other sorts of things were you working on when this idea hit you?
The possibility of a relationship between leaf-swallowing and tapeworm infestation emerged while my research team was working to habituate two "new" communities of wild chimpanzees in Kibale National Park in Uganda. Since the chimpanzees themselves were still too shy to allow observation, we studied their feeding habitats partly through fecal analysis. Infections by the tapeworm *Bertiella* are very obvious in chimpanzees, because they often lead to a thick white worm-like object being egested in the feces. This worm-like thing is a terminal segment of the tapeworm called a proglottid. *Bertiella* proglottids are remarkable because they are mobile, scattering fertilized eggs as they loop along on the forest floor. So when we found the feces quickly, the proglottid was still present and very obvious—it's up to 2–3 cm long. The evidence of leaf-swallowing was equally clear. There's no mistaking it for leaf-eating, because instead of finding barely recognizable fragments of chewed leaf, we found entire leaves that had passed through the chimpanzee gut without receiving a single tooth mark. They were always from the same few species of plant, none of which we saw being eaten until a few years later, when the chimpanzees were sufficiently well habituated to allow us to watch them closely at dawn. I soon noticed that the presence of proglottids was associated with whole leaves in the dung. Having previously suspected that the function of leaf-swallowing by chimpanzees in Gombe National Park in Tanzania was to expel worms, I was primed to look for this kind of correlation, of course.

How did it feel watching chimps apparently self-medicate themselves? Were you shocked at what you were seeing?
It was exciting, naturally enough, because nothing like it had been reported before. Why would chimpanzees put all this care into selecting leaves of a particular and rather uncommon species, and then swallow them slowly one by one despite their rough texture and considerable size? Admittedly, the behavior wasn't completely surprising. In the 1960s, Jane Goodall had found whole leaves of *Aspilia* in chimpanzee feces. But because no one at that time saw chimpanzees swallowing these leaves, none of the observers had paid any attention to the problem of why the leaves hadn't been chewed, or what they meant to the chimpanzees. It took seeing the swallowing itself to focus my attention.

The puzzle was exacerbated by the peculiar time of day that chimpanzees swallowed their leaves. Normally when chimpanzees leave their night-nests, the first thing that they do is to walk directly to a productive fruit-tree, where they feed for an hour or more. But on leaf-swallowing days, they would postpone their morning meal. They would walk in the gray dawn to a patch of *Aspilia* and spend 20 minutes or more selecting and picking leaves that were very rough and obviously difficult to swallow. Sometimes they grimaced as they did it, so it seemed to take an effort to swallow the leaves. I found these leaves completely tasteless, unlike everything else they ate. There had to be a good reason for this postponement of feeding and ingestion of an apparently rather unpleasant object, but no ordinary explanations fit. So whatever the explanation was going to be—selection of a rare nutrient, psychoactive drug, or self-medication—it would obviously be fascinating.

How do the chimps know to eat more leaves when they are infected with tapeworms? Do they learn this from watching others?

There's no experimental evidence about how tapeworm infections make chimpanzees feel (and for ethical reasons, I trust there never will be). So we can only guess how chimpanzees learn to swallow leaves when infected. Different populations of chimpanzees swallow different species of leaves, in a pattern that suggests observational learning of the particular tradition of choice. Certainly infants see their mothers swallowing leaves from an early age, and they have a strong tendency to try unusual behaviors for themselves. They are also very interested in their mother's bodies. For instance, they pay strong attention to their mother's fresh wounds. Perhaps the most reasonable possibility is that infants copy their mother's behavior, and eventually learn for themselves that if they had not been feeling well, a few swallowed leaves can make them feel better.

Aside from the case of leaf-eating, what is the best case of self-medication you know of in nonhumans?

Once again, the most intriguing evidence comes from chimpanzees. In the Mahale Mountains in Tanzania, Michael Huffman has studied why chimpanzees occasionally chew the juicy tips of young stems of a particular shrub, *Vernonia amygdalina*. Like the leaf-swallowing case, this "bitter-pith-chewing" is odd because the rate of ingestion is slow compared to ordinary feeding. In addition, the *Vernonia* pith is intensely bitter, much more so than any other chimpanzee foods. I once saw a chimpanzee "gag" after chewing *Vernonia* pith. Finally, *Vernonia* pith is selected very rarely. On the basis of a few detailed observations, Huffman has hypothesized that chimpanzees chew *Vernonia* pith only when they have gut infections of pathogenic bacteria, responsible for diarrhea and lethargy. His cases are so striking that my guess is he's right.

If animals do indeed self-medicate, is it possible we can learn something valuable about fighting illness in humans by watching animals?

One of the most exciting possibilities when we first investigated leaf-swallowing by chimpanzees was that they might have been selecting leaves that contained drugs that humans could also use. Unfortunately for that idea, there's now good evidence that leaf-swallowing works through physical, not chemical, actions. Chemical extracts of the leaves have no special toxins, and don't harm parasitic worms. But the leaves are all peculiarly rough-surfaced, and probably remove worms partly by scouring the wall of the small intestine. Other cases of self-medication, however, point clearly to the potential importance of specific chemicals. For example, *Vernonia amygdalina* contains a cocktail of powerful toxins that appear to follow the delicate medical principle of a drug being a poison that's a little worse for the pathogen than for the host. Huffman's team is investigating whether any of the components of *Vernonia amygdalina* pith might have medicinal value for people.

In a similar way, various birds line their nests with leaves. Toxins in the leaves appear to reduce pathogen loads for the chicks. Examples like this could provide ideas for human application.

Do chimps show any sign of using *preventative* medicine—i.e., do they appear to act in ways to reduce the chance of getting ill in the first place?

There's no indication yet of preventive medicine exactly. But like many other frugivores (including monkeys and parrots), chimpanzees regularly eat small amounts of termite-mound clays and other soils. People used to think of this as a way to get rare minerals, but it turns out that the only mineral consistently common in the soils eaten by chimpanzees is iron, which is not in short supply. On the other hand, all of the soil types eaten by chimpanzees have the kaolin-like property of being effective anti-toxins. In Kibale, chimpanzees tend to eat a walnut-sized chunk of soil late in the day, mostly after 3 P.M., around the time when they tend to start eating more leaves and piths. They take the soil off the root mass of a fallen tree, or from sinkholes where they can select below the humus layer. The cationic properties of these soils enable them to absorb toxins such as alkaloids and terpenes. It's still not clear what regulates this form of soil ingestion, but since it happens almost daily, one possibility is that the chimpanzees are systematically dosing themselves against the possibility of a high toxin load while eating vegetable matter in their evening meals.

DR. RICHARD WRANGHAM is a professor at Harvard University. His long-term studies on primates in their natural habitat has reshaped the way ethologists view primate (including human) behavior. Some of this work is described in his immensely readable book Demonic Males *(Mariner Books), written with Dale Peterson.*

on disease requires us to imagine how such an effect would have been manifested before modern-day contraptions such as refrigeration came into use. Imagining ourselves as hunter/gatherers during the time before refrigerators, how might our decisions about what to eat have prevented disease? One obvious way would be to avoid eating rotting meats—clearly repulsion to smells associated with decaying food would be strongly selected. But are there equally useful, if perhaps more subtle, ways that diet could be used to prevent disease in humans? Jennifer Billing and Paul Sherman in their paper "Antimicrobial Functions of Spices: Why Some Like It Hot" argue that the answer to this question is a definitive "yes" (Billing and Sherman, 1998).

Billing and Sherman are primarily interested in the evolution of spice use, and they focus on the following question: Over the course of thousands of generations, why should natural selection have favored adding spices to the human diet? To tackle this question, they examined forty-three spices in meat-based cuisines from thirty countries. After working their way through 4,578 recipes and scouring the literature for studies on the antimicrobial functions of the spices they studied, Billing and Sherman came up with some rather amazing findings. Let's start by taking a look at what does *not* explain the pattern of spice use across the world. The most obvious prediction regarding spice is that people simply eat the spices that are native to their region. Yet, people from areas of greater spice availability do not eat more spices. Billing and Sherman also ruled out several other alternative hypotheses. It turns out that we can't explain patterns of spice use by the nutritional value of spices, the odors they omit (potentially masking nasty smells), or the ability of spices to make people perspire and thus feel more comfortable in hot climates.

What Billing and Sherman did find was that all forty-three spices they examined had some antibacterial properties (they acted to thwart some potential bacterial disease). Billing and Sherman found that as mean annual temperature increases and food spoilage and bacterial diseases become more likely, the proportion of recipes containing spices increases, as does the number of spices per recipe (Figure 16.18). Likewise, with increasing temperature, the total number of spices used in recipes increases, and reliance on the spices with the strongest antibacterial properties increases. They found that the relationship between annual temperature and bacterial inhibition was particularly strong for garlic and onion, the two spices that were shown to have the strongest antibacterial effects.

Billing and Sherman go on to show that the exception to the rule often proves the rule. In addition to antimicrobial properties,

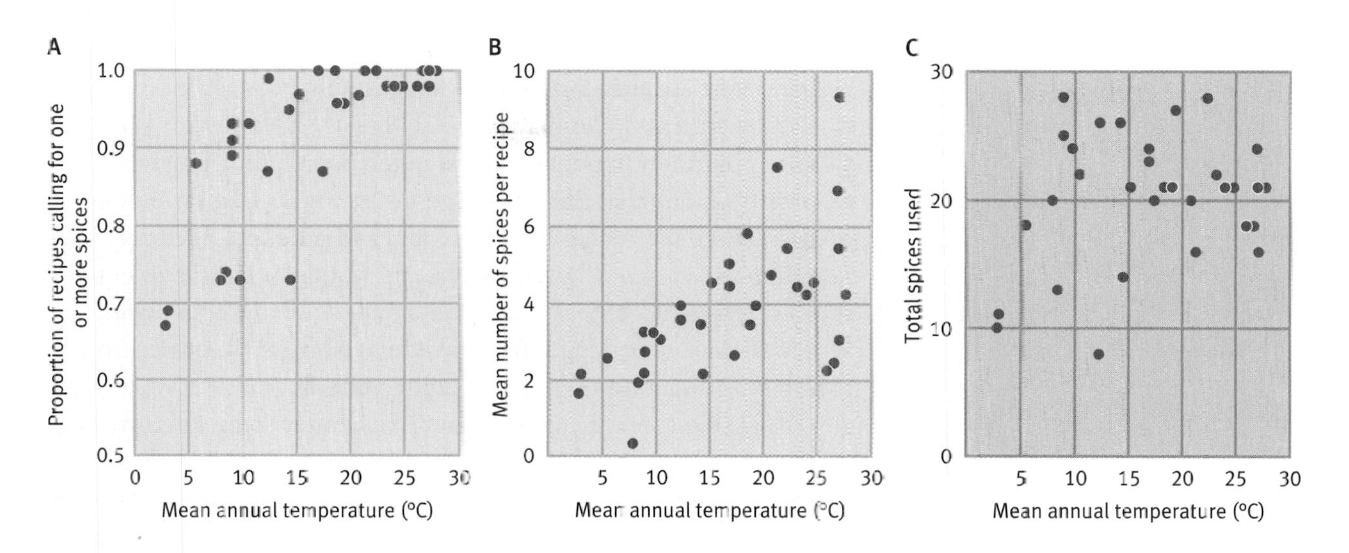

FIGURE 16.18. Spicy temperatures. A strong relationship exists between temperature and **(A)** the proportion of meat recipes calling for one or more spices, **(B)** the mean number of spices per recipe, and **(C)** the total number of spices used in the thirty-four countries that were examined. *(Based on Billing and Sherman, 1998, p. 15)*

spices also have the negative effect of slightly increasing mutation rates in those who consume them. For the vast majority of people, the benefits of destroying potentially deadly diseases outweigh the slight increase in mutation rate, but for some groups this might not be the case. Following up some work done by Dr. Margie Profet, Billing and Sherman suggest that the costs of spice intake for two groups particularly susceptible to the dangers of increased mutation rate—pregnant women and young children—might explain why such individuals often have strong aversions to spices (Profet, 1992; Sherman and Flaxman, 2002).

SUMMARY

1. With respect to aging, it is important to separate how long an animal lives from the process of senescence. Senescence is as an age-specific decline in survival. Under this definition, when the probability of dying increases with increasing age, we have senescence.
2. Natural historians and ethologists long believed that senescence was rare in nature, since they believed that animals rarely lived long enough to begin senescing. Despite the fact that some observational work suggested that this view may in fact be wrong, very few rigid and systematic tests on senescence in natural

populations have been undertaken. Promislow's work involving forty-nine species of mammals suggests that senescence does in fact occur in wild animals.

3. One prominent model for senescence is the "antagonistic pleiotropy" model. Pleiotropic genes are those that have more than one effect on an organism, and antagonistic pleiotropy refers to the case when these effects work in opposite directions with respect to fitness. In the antagonistic pleiotropy model, senescence is conceptualized as the accumulation of negative effects of pleiotropic genes in aging animals.
4. Disposable soma theory suggests that, under some circumstances, it may pay for organisms to divert energy and resources from normal maintenance and repair associated with traits that have little connection to reproduction (that is, somatic traits) and "spend" energy on traits related to reproduction. Under this theory, decreased longevity is simply the consequence of such a diversion of resources—ignore maintenance and repair, and the body starts to fall apart sooner.
5. Much work has been done examining glucocorticoids and heat-shock proteins as proximate players in the process of aging and senescence.
6. With respect to disease and animal behavior, two means by which individuals may reduce the risk of infection from parasitic diseases are: (a) preferentially producing offspring in areas that have low parasite levels, and (b) avoiding parasitized conspecifics.
7. Intriguing, but so far anecdotal, evidence is starting to mount that animals may modify their diet in a way that has the effect of self-medicating. This may be preventive or therapeutic self-medication.
8. Potential cases involving self-medication in nonhumans include: (a) antibacterial plant substances in bird nests, (b) eating soil, dirt, and rock as an antidiarrheal mechanism, (c) "anting" behavior in birds, and (d) "fur rubbing" in primates.

DISCUSSION QUESTIONS

1. What is the difference between aging and senescence? Given the definition in the chapter, is it possible that animals might not only fail to senesce, but they might actually have survival probabilities that are the exact opposite of those associated with

senescence? What would a chart of survival probabilities look like? Do you think this is likely in nature? Why or why not?

2. What might be an adaptationist's most basic argument against the antagonistic pleiotropy hypothesis? Hint: This hypothesis assumes that pleiotropic effects cannot be broken down.
3. If glucocorticoids are linked to disease, senescence, and aging, why hasn't natural selection significantly reduced levels of these hormones?
4. If animals, particularly primates, use self-medication, what impact might this have on the relationship between conservation biology and the medical sciences?
5. Imagine you are working with some species of fish in which individuals form associations with other group members when foraging, avoiding predators, and so on. How might you go about constructing laboratory experiments to examine how important avoiding diseased conspecifics is when compared to other factors in choosing associates?

SUGGESTED READING

Austad, S. (1997). *Why we age: What science is discovering about the body's journey through life*. New York: John Wiley. A popular book on aging in animals and humans.

Gosden, R. (1996). *Cheating time: Science, sex, and aging*. New York: W. H. Freeman & Co. Another popular book on aging.

Huffman, M. (1997). Current evidence for self-medication in primates: A multidisciplinary perspective. *Yearbook of Physical Anthropology, 40*, 171–200. A nice review on self-medication in our closest relatives.

Lozano, G. (1998). Parasitic stress and self-medication in wild animals. *Advances in the Study of Behavior, 27*, 291–317. A review article that ties aging to the idea of stress in wild populations of animals.

Rose, M. (1991). *The evolutionary biology of aging*. New York: Oxford University Press. A more technical book on evolution and aging. Some background in advanced evolutionary theory should be helpful to the reader.

17

Boldness and Shyness

- Bold and Inhibited Pumpkinseeds
- Guppies, Boldness, and Predator Inspection

Some Case Studies

- Hyena Personalities
- Octopus Personalities
- Ruff Satellites
- Learning and Personality in Great Tits
- Chimpanzee Personalities and Cultural Transmission

Coping Styles

Some Practical Applications of Animal Personality Research

- Predators and Domesticated Prey
- Guide Dog Personalities

Interview with Dr. Jerome Kagan

Animal Personalities

Watch almost any group of animals long enough, and you will start noticing individual differences among group members. For example, some primatologists, such as Jane Goodall, will tell you that chimpanzees have personalities in the same manner that you and I do. Once you get to know a chimp's personality, you can even predict, in a general fashion, how it will act when placed in a new behavioral scenario. Take a chimp with an "anxious" personality and place it in a novel environment, and it will act differently than a chimp with a less anxious personality. But the same general argument can be made with respect to many animals, most of whom are not nearly as closely related to humans as chimps are. For example, in the guppies that are discussed numerous times throughout this book, we can find what might arguably be called "bold" and "inhibited" personalities. Test a series of male guppies over and over again in the absence and presence of danger, and we end up with very distinct behavioral types. Some fish are willing to take risks and inspect this danger, and others aren't. We'll return to this subject shortly.

One common undercurrent to the work described in this book is that, although there are certainly instances in which a single behavior is the best solution to the problem an animal faces, oftentimes multiple behavioral solutions to problems animals face co-exist. Animals can be cooperators and cheaters, fighters (hawks) and nonfighters (doves), and so on—a combination of different behaviors can often exist concurrently in a population. Here we shall extend this concept even further and consider the possibility that consistent behavioral differences among individuals can amount to "personalities." In so doing, we will conceptualize **personality differences** as consistent long-term phenotypic behavioral differences among individuals (Schleidt, 1976; Slater, 1981; Caro and Bateson, 1986; Clark and Ehlinger, 1987; Pervin and John, 1997). This admittedly fuzzy definition is in fact close to what most psychologists mean when speaking of personality in humans. In fact, there is a growing psychological literature that relies on animal models to understand personality in the broadest sense (Zuckerman, 1996; Gosling, 2001), as well as with respect to curing personality "disorders" (Henry and Stephens, 1977; Wenegrat, 1984; Stevens and Price, 1996; McGuire and Troisi, 1998; Koolhaas et al., 1999).

Our discussion of animal personalities can be shored up theoretically by casting such individual differences in the language of evolutionary game theory, and as such, we will be referring back to some examples of game theory that we have touched on in earlier chapters. To solidify the connection between game theory and personality, recall that game theory models often produce a solution that contains more than one behavioral strategy. If individuals

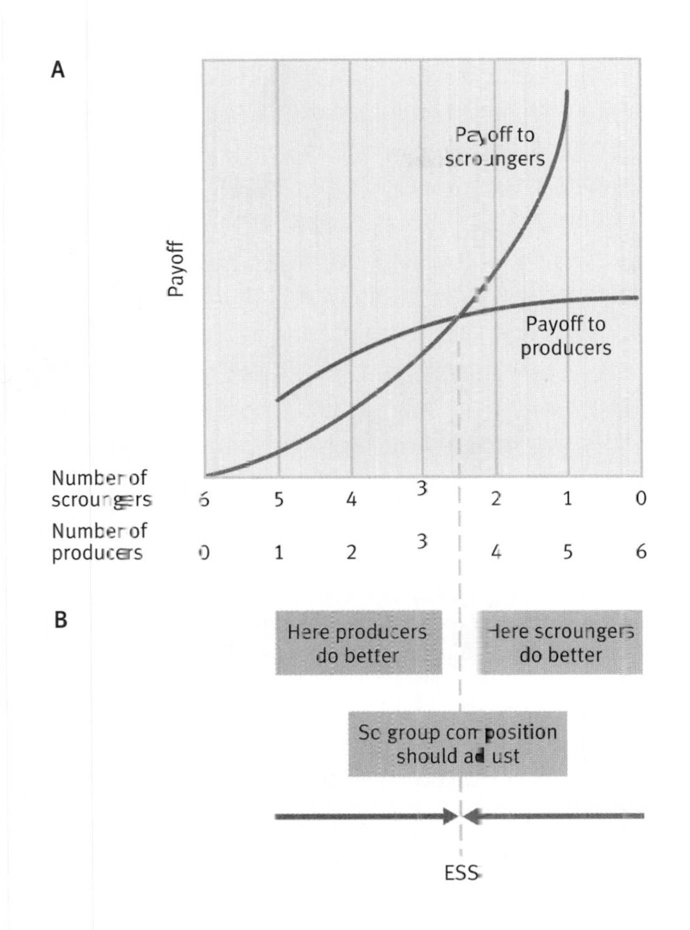

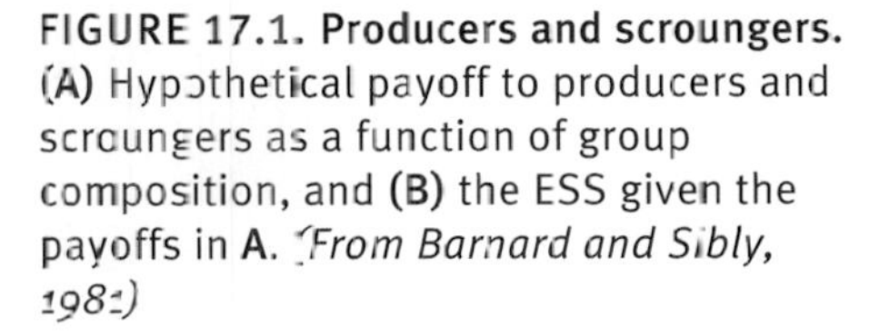
FIGURE 17.1. Producers and scroungers. (A) Hypothetical payoff to producers and scroungers as a function of group composition, and **(B)** the ESS given the payoffs in **A.** *(From Barnard and Sibly, 1981)*

adopt such strategies for long periods of time, these strategies can be thought of as personality variables. To see how this might work, let's return to the producers and scroungers we first encountered in Chapter 10. When it comes to foraging in groups, animals often adopt one of two very different strategies. Producers search for food, and hence they accrue the costs associated with uncovering new food patches. Scroungers, on the other hand, watch producers, and learn where new food patches are by parasitizing the work of producers.

Barnard and Sibly (1981) constructed a game theory model of the producer/scrounger scenario (Figure 17.1) that we shall envision as a model of personality types (producers and scroungers). The solution to this game is some combination of both producers and scroungers, with the equilibrium frequency of each strategy depending on the exact costs and benefits associated with producing and scrounging. As with any game theory model, solutions that contain more than one behavioral strategy can be thought of in at least two ways. For example, imagine that the equilibrium frequency of producers is 75 percent, and hence that of scroungers is 25 percent. This might translate into each individual in a group playing producer 75 percent of the time, and scrounger the remain-

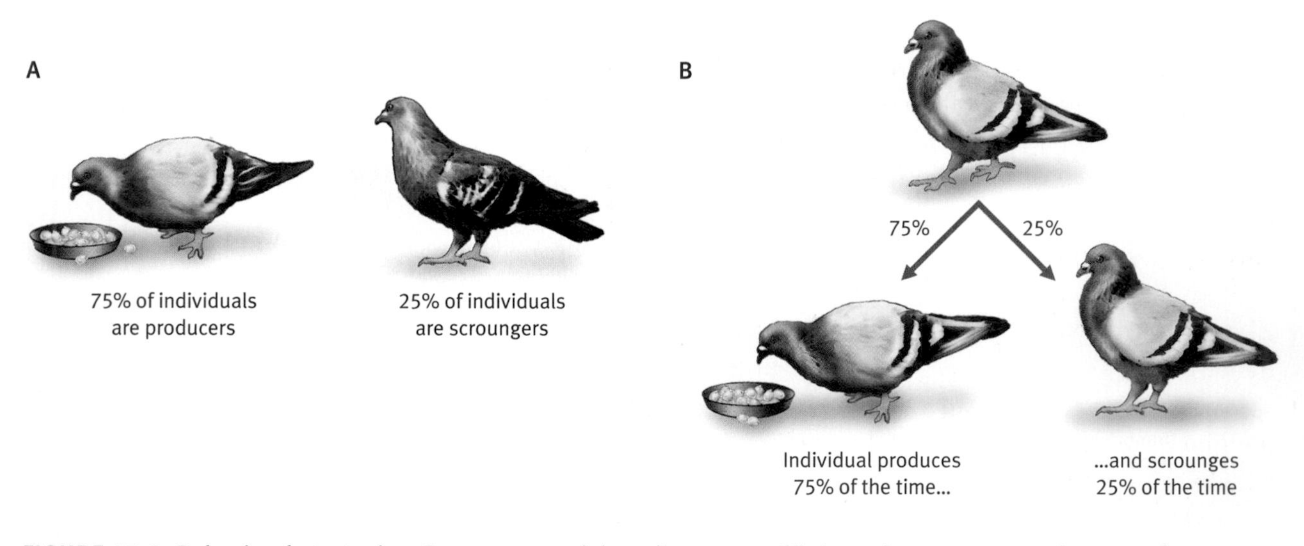

FIGURE 17.2. Behavioral strategies. Suppose a model predicts an equilibrium of 75 percent producers and 25 percent scroungers. This can occur by either **(A)** having 75 percent of individuals play producer and 25 percent scrounger, or **(B)** having the probability that an individual will act as a producer set at 75 percent.

ing 25 percent. Alternatively, we might see 75 percent of the individuals in a group consistently adopting the producer strategy, and 25 percent of the individuals opting to be scroungers (Figure 17.2). When the latter is the manner in which the equilibrium frequency is in fact expressed, which is often the case (Giraldeau and Caraco, 2000), the producer/scrounger game can be thought of in terms of predicting personality traits.

Food producers and scroungers are, of course, only one way in which personality traits manifest themselves in animals and humans. In this chapter, we shall examine:

- Bold and "inhibited" personality types.
- A series of case studies of personality across a wide array of species and ecological conditions.
- Differences in "coping" styles.
- Practical implications of animal personalities.

Boldness and Shyness

Psychologists have long argued that where a person falls on the "shy/bold" continuum is one of the most stable personality variables yet studied. If you are shy when you are young, chances are very good that you will be shy when you get older

(Kagan et al., 1987, 1988; Kagan and Reznick, 1989; Kagan, 1994). While definitions of shyness and boldness vary considerably, normally **boldness** refers to the tendency to take risks in both familiar and unfamiliar situations, while **shyness** refers to the reluctance to take such risks, or even to engage in unfamiliar activity. Many aspects of shyness also map onto another psychological term—behavioral inhibition—while boldness is similar to "sensation seeking" (Zuckerman, 1979, 1994). Because of the semantic luggage associated with these terms, we shall adopt the relatively neutral terms "bold" and "inhibited" throughout this section.

Psychologists often, but not always, speak of boldness and inhibition as general personality traits—that is, individuals who are bold in one context are likely to be bold in other contexts, and likewise for shy individuals. Evolutionarily based animal behaviorists argue, however, that there is no reason that natural selection should necessarily select for boldness or shyness as general personality traits. Rather, individuals may be bold in some contexts and shy in others, depending on the costs and benefits associated with the behavioral contexts being examined (D. S. Wilson et al., 1993, 1994; Coleman and Wilson, 1998). This is the view that we shall adopt here. At first, adopting such a view might lead us to believe that shyness and boldness are not truly personality traits, as they are not generalizable across different contexts. There is, however, no reason that the context-specific term "bold forager," for example, should not be considered a personality trait, as long as individuals are *consistent* about being bold when foraging.

It is a promising sign that one emerging area of common interest between behavioral ecologists and those in the mental health community is the *evolution* of behaviorally inhibited versus bold personality types (Marks, 1987; Kagan et al., 1988; Kagan, 1994; Marks and Nesse, 1994; D. S. Wilson et al., 1994; Zuckerman, 1994; Nesse and Williams, 1995; Stevens and Price, 1996; McGuire and Troisi, 1998). While a great deal of work has been undertaken in understanding both inhibition and boldness from psychological, psychiatric, and physiological perspectives (Kogan and Wallach, 1964; Zuckerman, 1979, 1991, 1994, 1996; Koopsman et al.,1995; Moller et al., 1998; Gerra et al., 1999), until recently, little in the way of controlled experimental work has examined the *costs and benefits* of inhibition and boldness. From a behavioral ecology perspective, this omission is striking, as a thorough understanding of the evolution of inhibition and boldness can only be attained through a knowledge of the costs and benefits associated with these traits. As such, we shall examine two systems in which the costs and benefits of boldness and inhibition have been studied in depth

FIGURE 17.3. Bold and shy fish. Pumpkinseed sunfish have been studied extensively in order to understand the evolution of boldness and shyness. *(Photo credit: Dwight R. Kuhn)*

BOLD AND INHIBITED PUMPKINSEEDS

David Wilson, Christine Coleman, and their colleagues' work on the pumpkinseed sunfish (*Lepomis gibbosus*), which can be found in ponds and lakes across the United States, is one of the most comprehensive studies of boldness and inhibition undertaken to date (Figure 17.3). Wilson and his team created an ingenious trapping method that allowed them to segregate shy and bold fish from natural populations (Wilson et al., 1993). They trapped fish from a given pond using two techniques (Figure 17.4). The first technique was to place traps in the water. In effect, these traps were meant to mimic the "novel object" test that psychologists use to classify humans along the shy/bold continuum. These traps were designed so that a pumpkinseed fish would have to actively swim into them to be collected. Once a fish was inside, however, the construction of the trap was such that it was very difficult for the fish to get out. Presumably, this technique would primarily capture bold sunfish, who would be the only ones willing to enter the trap in the first place.

The second technique involved dragging a large net called a seine through the pond. Seining occurred immediately after all traps were collected and was undertaken in such a manner as to capture every sunfish in the vicinity of the traps. Seining, then, should capture a combination of both inhibited and bold fish,

FIGURE 17.4. Experimental set-up to study bold and shy fish. To examine boldness and shyness in pumpkinseed sunfish, David Sloan Wilson and his colleagues used two different experimental techniques. In one, a large seine was dragged through a pond; in the other, underwater traps were used to capture fish.

while the trapping method should have captured, on average, bolder fish than the seining method. In addition, they ran an ingenious control in which trapped fish were placed in a seine and run through the pond for thirty seconds (about how long a normal sweep of the pond takes). This was done to test whether the seining process itself was more traumatic than the trapping process, and hence might account for any differences found between seined and trapped fish.

Wilson and his colleagues examined the diets of trapped and seined fish, their parasite loads, and their growth rates. In addition, they tagged and released fish back into the pond, and then made detailed observations of these marked fish. In one experiment in which both trapped and seined fish were tagged and released back into their pond, behavioral observations indicated that trapped fish, as might be expected from bolder fish, were less likely to flee from human observers. Behavioral observations of pumpkinseeds in their natural habitat suggest that the researchers indeed collected different proportions of bold and inhibited fish using the two techniques.

Trapped fish were more likely to forage away from other fish, and their diet contained three times as many copepods (small crustaceans that are the usual food of pumpkinseeds) as did the diet of seined fish (Ehlinger and Wilson, 1988; D. S. Wilson et al., 1996). In addition, trapped and seined fish differed in terms of the parasites they carried, suggesting different habitat use. The diet and parasite data in conjunction with the behavioral observations suggest that trapped and seined fish, though caught from the same population, behave very differently in nature. Somewhat surprisingly, trapped and seined fish were not segregating (that is, forming groups of bold and groups of shy fish) when swimming in natural aggregations in ponds. Trapped and seined fish had the same average distribution of trapped and seined neighbors and, as such, it was not as if bold fish were swimming only with bold fish and shy fish only with other shy fish.

Once seined and trapped fish were brought into the laboratory, Wilson and his colleagues (1993) ran a barrage of behavioral and physiological tests on them over the course of approximately three months. Their results can be summarized as follows:

- There were no differences between trapped fish and trapped fish that were subsequently run through the pond in the seine (differences between groups were not due to the effect of the capturing technique itself).
- There were no differences between trapped and seined fish with respect to age or sex (close to a 1:1 sex ratio was found in both groups).

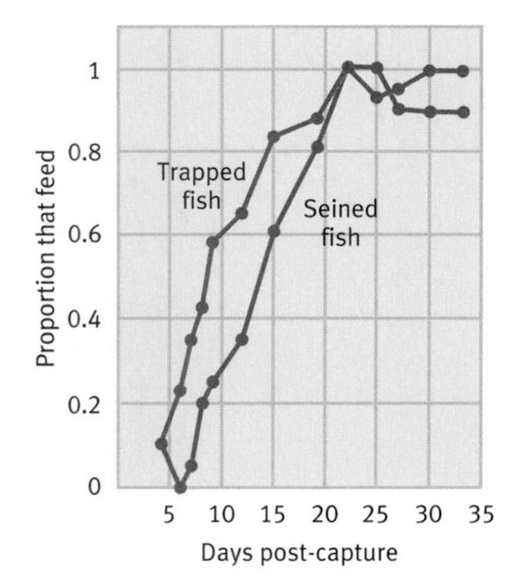

FIGURE 17.5. Feeding and boldness. Two techniques were used by Wilson and his colleagues to capture pumpkinseed sunfish: trapping, which caught the boldest fish, and seining, which caught a mixture of bold and shy fish. Once brought into the lab, trapped fish acclimated to feeding more quickly than seined fish (this effect disappears with time). *(Based on D. S. Wilson et al., 1993)*

- Trapped fish acclimated to feeding in the lab more quickly than did seined fish (Figure 17.5), as might be expected if in fact trapped fish were bolder.
- After twenty-five days of acclimation, Wilson and his colleagues ran a series of behavioral and physiological post-acclimation tests. These tests included "response to handling," novel object tests, aggressive contests between seined and trapped fish, and physiological tests of stress. In no case did the trapped and seined fish differ from each other in any of these experiments (see Francis, 1990, for more on aggressive personalities in fish).

Exactly how all the variables involved interact to produce the frequency of shy and bold individuals seen in nature is very complex and hard to pin down. Despite this, however, Wilson and his collaborators suggest that a true bold/inhibited continuum does exist in pumpkinseeds, both in nature and in the lab, and that where individuals fall along this continuum has important consequences for the fish, such as differences in diet, safety, parasite load, and so forth. That being said, the differences between trapped and seined fish disappeared after thirty days of social and ecological isolation, during which time fish were housed in individual aquaria in the laboratory. Wilson and his colleagues argue that while there are numerous possible explanations for the disappearance of differences across samples over time in the laboratory, it may be that ecologically and socially relevant cues solidify the differences between bold and inhibited fish. They believe that without such cues, differences in boldness and inhibition either disappear or become too small to notice.

In a follow-up study, Coleman and Wilson (1998) collected and tagged trapped and seined fish in ponds and exposed them to three stimuli in their ponds: a red-tipped stick extended toward the fish, a novel food source, and a new predator. For each stimulus, consistent individual differences were found, with some fish bolder than others. Nonetheless, these individual-level differences were not the same for each individual across stimuli. That is, boldness and inhibition were found to be context-specific, as evolutionary models predict they often will be.

GUPPIES, BOLDNESS, AND PREDATOR INSPECTION

Predator inspection behavior in fish (Chapters 9 and 11) is ideal for measuring the costs and benefits of inhibition and boldness. Pitcher and his colleagues (1986) coined the term predator inspection to describe the case in which one to a few individuals move away from their school and approach a putative predator to gain information about this potential danger. Predator inspection has now been documented in more than half a dozen species of fish (see Dugatkin and Godin, 1992a, 1992b; Pitcher, 1992; Dugatkin, 1997a, for reviews of predator inspection behavior in fish). Potential benefits of inspection behavior include: (1) determining whether the potential danger is in fact a predator, (2) announcing to an ambushing predator that it has been spotted, (3) assessing the motivational state of the predator (for example, is it hunting?), and (4) obtaining information on the distance between the school of fish and the predator (Magurran and Pitcher, 1987).

Individual guppies differ in their tendency to inspect a predator, but within-individual variability with respect to inspection is relatively low; both inspectors and noninspectors are consistent in their behavior when predators are present (Dugatkin and Alfieri, 1991b; Dugatkin and Wilson, 2000; Figure 17.6). As such, it seems reasonable to treat bold and inhibited inspection behavior as personality traits and to move on to the question of what are the underlying costs and benefits associated with boldness and behavioral inhibition and inspection in guppies (see Fraser et al., 2001, for an analysis of the costs and benefits to "movers" and "stayers" in the fish *Rivulus hartii*).

In terms of the cost of boldness during inspection, Pitcher and his colleagues used such provocative phrases as "Dicing with Death" and "Who Dares Wins" in the titles of their papers on predator inspection, but is it the case that predator inspection behavior is in fact dangerous? Do bold inspectors truly put themselves at risk, compared

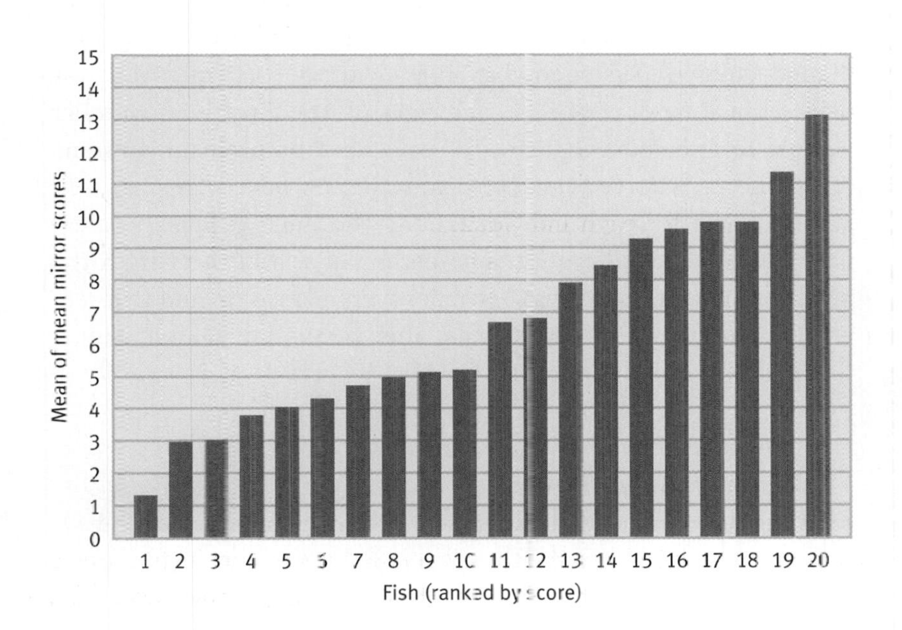

FIGURE 17.6. Variation in risk taking. Guppies are ranked by their risk-taking scores during predator inspection. Significant differences in boldness exist between individuals, but fish were fairly consistent in their risk-taking tendencies over time. *(Based on Dugatkin and Alfieri, 1991)*

with their (inhibited) noninspecting shoal mates? Isn't it possible that bold individuals simply "warn" potential predators that they have been seen and that any attack would be fruitless, thus making bold fish as least as safe, if not safer, than inhibited fish? In terms of inhibition and boldness, do inhibited or bold individuals survive with higher probabilities?

In order to address the costs of potential predation pressure on bolder individuals, Dugatkin (1992c) used ten pools, each of which was 1 meter in diameter and contained a predator, a prey refuge, and a group of six male guppies. Before being placed in a pool, each male was tested to determine its boldness score, which was an aggregate measure of how close a fish was willing to come to a potential predator. Each group of six males was composed of two bold individuals, two "moderate inspectors," and two inhibited individuals. After thirty-six and sixty hours, the pools were checked to see how many fish remained, and it was found that predation had been greatest on the boldest individuals and weakest on the inhibited fish (see Milinski et al., 1997, for a similar finding in sticklebacks, and Godin and Davis, 1995a, 1995b, and Milinski and Bolthauser, 1995, for a debate on mortality and inspection in guppies and sticklebacks). Thus, there is a clear and evolutionarily relevant cost that is being paid by bold individuals. The cost is even more dramatic when it is noted that any information obtained by bold individuals is passed on, in one way or another, to all fish in their group (Magurran and Higham, 1988).

From an adaptationist perspective, one would hypothesize that there must be some compensating benefits(s) obtained by bold individuals, or natural selection would have culled this trait from the population. Godin and Dugatkin (1996) uncovered one such benefit in a study of inspection and mate choice. The experimental protocol was such that in each trial a pair of males was placed in a large tank. On one side of this tank was another aquaria that served as housing for a pike cichlid predator. In some trials, the predator was present; in some, its tank was empty. In addition, in the large tank containing the male guppies, a clear partition was placed on the side of the aquaria farthest from the predator tank. In some trials, females were placed behind this see-through partition; in some trials, this section remained empty. Godin and Dugatkin found:

- More colorful male guppies were more likely to inspect predators (Figure 17.7A).
- Differences in inspection behavior between colorful and drab males were manifested only when females were observing (that is, no significant differences in inspection were found when males could not see females observing them; Figure 17.7B).

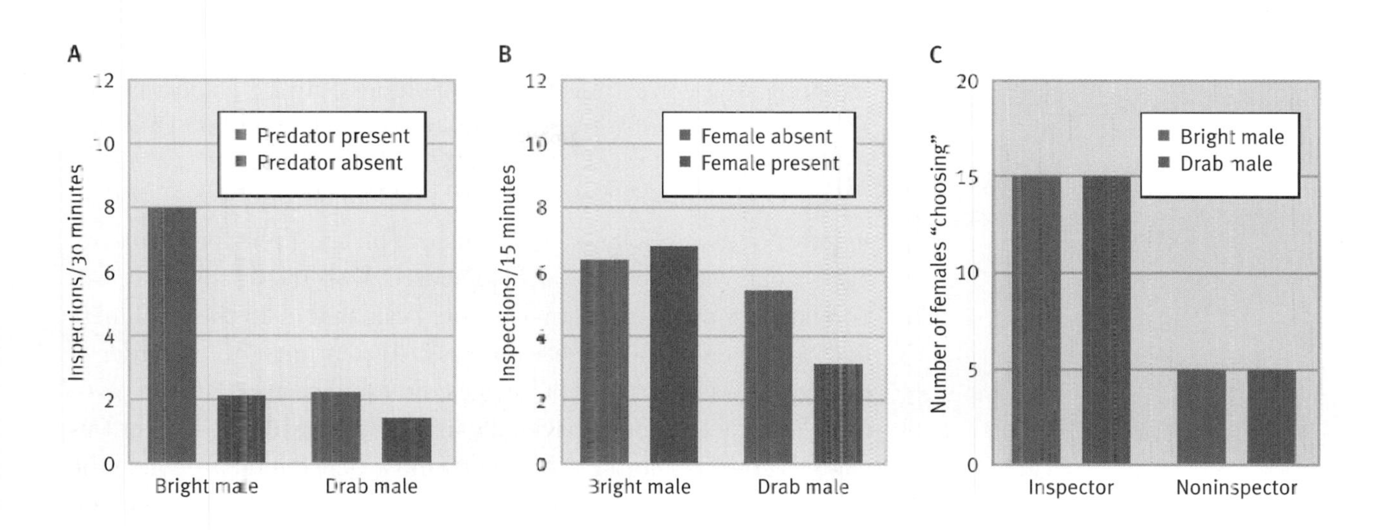

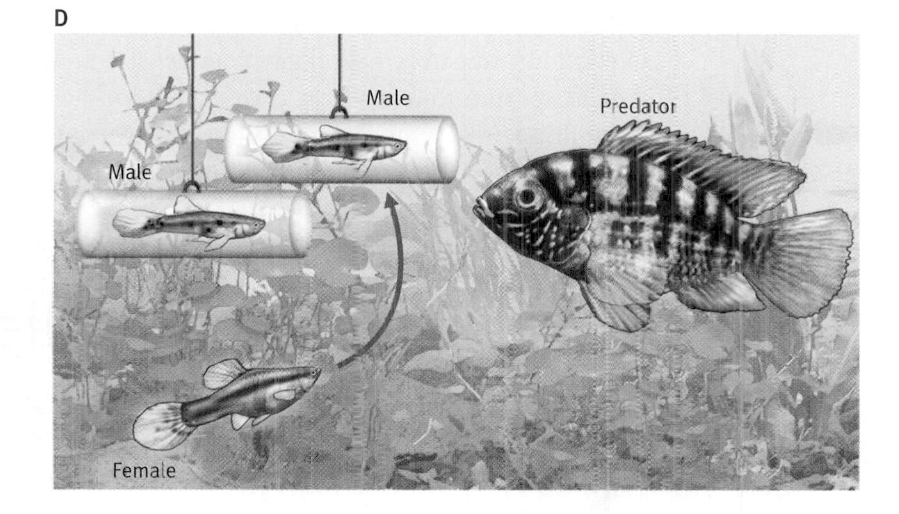

FIGURE 17.7. Color, boldness, and attractiveness. (A) When observed by females, brightly colored males inspected a predator more often than did drab males. The difference disappears if the predator is removed from his tank. **(B)** The difference between colorful and drab males in terms of boldness exists only when female guppies are observing males. **(C)** When color and boldness are experimentally decoupled in male guppies, females prefer bolder individuals, regardless of color. *(Godin and Dugatkin, 1996)* **(D)** To experimentally decouple boldness and color patterns in guppies, a motorized pulley system was built. Using this system, either colorful or drab males could be made bold by placing them in a small tube that was moved back and forth. *(Based on Dugatkin and Godin, 1998)*

- Females preferred colorful, bold males as potential mates.
- When color and boldness were experimentally uncoupled, experimental work clearly showed that it is a male's boldness, and not his color, that makes him attractive to females (Figures 17.7C and 17.7D).

These studies together demonstrate that one clear benefit to being a bold (versus an inhibited) male guppy is that it makes one much more attractive as a potential mate. The question then arises as to whether there are any other benefits associated with being a bolder inspector. While there is no such direct evidence, there is an interesting correlation between boldness during inspection behavior and associative learning abilities in male guppies (Dugatkin et al., 2002).

In Dugatkin and Alfieri (2003), male guppies were tested both in terms of their boldness during inspection and their ability to associate a cue with a particular food source. Two treatments were undertaken. In treatment I, males were evaluated as to their boldness first, and then given the learning task. In treatment II, the order of testing was reversed. Results suggest that bolder male guppies learn more quickly than do behaviorally inhibited individuals. But for this advantage to appear, bold behaviors must be expressed prior to the learning opportunities. It appears that predator inspection itself either "primes" bold fish to learn better, or it has a negative effect on learning in behaviorally inhibited inspectors. Whether behaviorally inhibited inspectors have higher levels of stress-related hormones after exposure to a predator than do bold individuals and whether this difference in hormone levels translates into lowered rates of learning are interesting topics for future exploration.

Some Case Studies

With the exception of work on humans (Kagan, 1994), compared to many of the subjects discussed throughout this book, the number of personality studies ethologists have undertaken is relatively small. As such, the detailed cost/benefit analyses that we'd ideally like to have are rare. That being said, there are enough case studies across a wide array of species to provide us with some understanding of animal personality traits (Gosling, 2001).

HYENA PERSONALITIES

Spotted hyenas (*Crocuta crocuta*) are native to African savannas (Figure 17.8) and they live in stable clans that often engage in cooperative hunting. Females hold the alpha status in groups, dominance is inherited maternally, and the overall dynamics of social life in hyenas are complex (Frank, 1986a, 1986b; Zabel et al., 1992; Holekamp et al., 1996; Holekamp and Smale, 1998; Gosling, 1998; Engh et al., 2000). This complexity sets the stage for Samuel Gosling's (1998) study of personality in the spotted hyena.

FIGURE 17.8. Hyena personalities. Gosling studied forty-four personality traits in spotted hyenas. A detailed analysis revealed that hyena personalities are most easily understood in terms of assertiveness, excitability, human-directed agreeableness, sociability, and curiosity. *(Photo credit: E. R. Degginger/Color-Pic, Inc.)*

Gosling studied thirty-four hyenas in an attempt to determine what constitutes personality in these animals, and whether or not personality can be boiled down to a few variables, or is, by its very nature, too difficult to decompose into separate traits. As far as animal behavior and behavioral ecology go, Gosling's work on hyenas started in a somewhat unusual way. Early on, three hyena experts were provided with a list of forty-two traits that had been used in other work on animal personality, and they were asked which traits they believed would apply to hyenas. Then these experts were provided with a list of forty traits used in human personality work, and they were asked which applied to hyenas. From these lists, as well as items the experts could add on, a list of sixty traits was amassed. Subsequent to this, two experts were asked to go through this list of sixty traits and remove any redundancies (for example, "fearful" and "apprehensive").

The forty-four traits that remained after the final culling process were then defined specifically with respect to detailed hyena behavior, including six traits that scored how hyenas interacted with humans. At least with respect to the four observers who gathered data on hyenas, this list of forty-four was a concrete one, as the average reliability across observers was 0.71 (out of a score of 1, where 1 indicates that all observers scored every behavior exactly the same way). With data on forty-four traits in thirty-four hyenas in hand, Gosling employed a statistical technique called principle component analysis to determine if certain traits grouped together, and whether such groups seemed to make any sense in terms of personality attributes. Results of this analysis suggest that hyena personalities are most easily understood in terms of five aggregate traits—assertiveness (which incorporated fifteen of the original forty-four traits), excitability (twelve of the original traits), human-directed agreeableness (seven of the original traits), sociability (four of the original traits), and curiosity (six of the original traits) (Table 17.1).

TABLE 17.1. Hyena personality traits. A list of the forty-four variables Gosling used in his study of hyena personality. Trait labels with high absolute values clustered together to form five aggregate personality indices. The clusters can be identified by the groups of boldface numbers in each column. *(Based on Gosling, 1998, p. 110)*

TRAIT LABELS	ASSERTIVENESS	EXCITABILITY	HUMAN-DIRECTED AGREEABLENESS	SOCIABILITY	CURIOSITY
Assertive	**0.95**	−0.08	−0.14	−0.01	−0.01
Argumentative	**0.93**	0.13	−0.10	−0.02	−0.17
Aggressive	**0.90**	0.11	−0.22	0.03	−0.15
Bold	**0.88**	−0.32	0.08	0.11	0.05
Confident	**0.88**	−0.25	0.00	0.01	0.13
Persistent	**0.84**	0.07	−0.03	0.08	0.38
Fearful	**−0.80**	0.34	−0.31	−0.14	−0.08
Jealous	**0.79**	0.41	0.15	0.09	−0.09
Strong	**0.77**	−0.08	0.05	0.21	0.34
Irritable	**0.74**	0.19	−0.39	−0.27	−0.27
Greedy	**0.69**	0.27	0.20	0.00	0.19
Careful	**−0.66**	−0.25	−0.17	0.29	−0.30
Scapegoating	**0.65**	0.25	−0.27	−0.08	−0.23
Opportunistic	**0.62**	−0.06	0.31	0.18	0.52
Friendly	**−0.57**	−0.32	0.41	0.38	0.27
Vigilant	0.05	**0.88**	0.09	0.15	0.07
Excitable	0.34	**0.87**	−0.09	0.03	0.08
High strung	−0.23	**0.84**	−0.27	−0.23	0.08
Slow	0.06	**−0.82**	0.24	−0.01	−0.17
Calm	−0.01	**−0.82**	0.06	0.36	0.20
Active	0.03	**0.81**	−0.21	0.10	0.28
Lazy	0.00	**−0.75**	0.25	−0.18	−0.30
Vocal	0.16	**0.73**	0.46	0.09	0.13
Nervous	−0.44	**0.71**	−0.34	−0.15	−0.06
Moody	0.27	**0.67**	−0.41	−0.17	−0.25
Nurturant	0.01	**−0.52**	0.11	0.23	0.50
Eccentric	0.15	**0.50**	0.45	0.24	0.17
Testing	0.22	0.21	**−0.83**	−0.01	0.11
Social	0.04	0.01	**0.81**	0.47	0.04
Tame	−0.12	−0.37	**0.78**	0.22	−0.10
Warm	−0.08	−0.11	**0.75**	0.50	0.05
Obedient	0.26	0.08	**0.75**	−0.12	0.40
Deceitful	0.18	0.36	**−0.69**	−0.18	0.00
Flexible	0.28	−0.20	**0.68**	−0.02	0.44
Warm	−0.08	0.09	0.16	**0.86**	0.15
Affiliative	0.15	0.16	0.19	**0.86**	0.03
Sociable	0.18	−0.10	−0.04	**0.85**	0.20
Cold	0.14	0.10	−0.45	**−0.79**	−0.06
Exploratory	0.09	0.23	0.17	0.00	**0.79**
Impulsive	0.13	0.31	−0.29	0.06	**0.73**
Curious	0.07	0.22	0.44	0.02	**0.61**
Imaginative	0.00	−0.13	0.29	0.34	**0.52**
Playful	−0.13	0.12	−0.08	0.14	**0.51**
Intelligent	0.22	−0.19	0.35	0.32	**0.37**

Gosling proceeded to examine whether any of these aggregate personality traits correlated with an individual's sex, age, or rank in the hierarchy. Age correlated with none of these aggregate traits. Nor did dominance rank correlate with any of these aggregate traits. Yet, sex did matter, with females being much more assertive than males, as might be expected in a social system with matriarchal dominance hierarchies.

In order to get a broader picture of what the hyena work meant for personality research in animals, Gosling compared the five aggregate traits he uncovered in hyenas with major traits found in studies of rhesus monkeys (Stevenson-Hinde and Zunz, 1978; Stevenson-Hinde et al., 1980; Bolig et al., 1992) and gorillas (Gold and Maple, 1994). With the exception of "human-directed agreeableness," which was not measured in the primate studies, there was generally good agreement between major personality traits across hyenas and nonhuman primates, suggesting some cross-species commonalities with respect to personality. Naturally, the specific details of how various personality traits manifest themselves are species-specific (and perhaps even population-specific), but Gosling is well aware of this, and adroitly notes:

> One would expect nervousness to be manifested differently in chimpanzees and octopuses. Despite these challenges, researchers should make every effort to ensure that their item pool is as comparable as possible with other research. . . . Ultimately, a standard taxonomy of terms will be available from which animal personality researchers can choose. (Gosling, 1998, p. 114)

OCTOPUS PERSONALITIES

The vast majority of work on animal personality work has been undertaken in vertebrates. This may reflect an inherent bias in the way we think about the complex nature of personality, but whatever the reason for the focus away from invertebrates, in so doing we are ignoring a large portion of the animal world. Jennifer Mather has put forth the argument that because learning plays a fundamental role in their development and because they display an array of complex behaviors, octopuses are ideal subjects for a sortie into personality in invertebrates (Mather and Anderson, 1993, 1999; Mather, 1995).

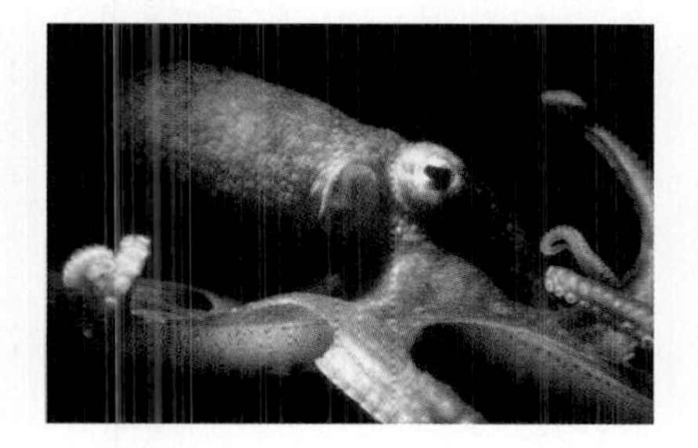

FIGURE 17.9. Octopus personalities. Studies on personality in the red octopus *(Octopus rubescens)* represent some of the only work on invertebrates done in this field. *(Photo credit: Ken Howard/Sea Images)*

To obtain a sense of what, if anything, personality might encompass in octopuses, Mather and Anderson (1993) tested the response of forty one-year-old *Octopus rubescens* (Figure 17.9) to three different treatments: alert, threat, and feed. In the "alert" treatment, an observer opened the lid to an octopus tank and brought his head

TABLE 17.2. Octopus personality traits. Three "factors" (or dimensions) incorporate the personality variables used in Mather and Anderson's work on personality in *Octopus rubescens*. *(Based on Mather and Anderson, 1993)*

DIMENSION	PREDICTOR BEHAVIOR
Factor 1: Activity (active/inactive)	In den[a]
	At rest[a]
	Grasp[b]
Factor 2: Reactivity (anxious/calm)	Squirt[b]
	Shrink[b]
	Swim[b]
	Crawl[b]
Factor 3: Avoidance (avoiding/bold)	In den[b]
	In den[c]
	Color change[a]
	Ink[b]
	Alert[c]

[a]*Alerting test.* [b]*Threat test.* [c]*Feeding test.*

down to where the octopus could see it. The "threat" treatment involved using a brush to touch, and presumably frighten, an octopus. The "feed" treatment involved recording an octopus's response when a food item (a shore crab, *Hemigrapsus nudus* or *H. oregonensis*) was put into its tank. The nineteen octopus behaviors displayed across these three treatments were subject to a principle component analysis, and three "components" emerged: active/inactive, anxious/calm, and bold/inhibited (Table 17.2). These three personality traits accounted for 45 percent of all the behavioral variance uncovered, and hence they explained a good chunk of the behavior observed by Mather and Anderson. These personality traits were long-lasting (Mather, 1991), and they were remarkably similar to Buss and Plomin's (1986) characterization of human infant personality, as well as Stevenson-Hinde and his colleagues' (1980) description of rhesus monkey personality, providing further support for some cross-species similarities in terms of personality variables.

RUFF SATELLITES

Let us return for a moment to the subject of alternative male mating strategies first discussed in Chapters 6 and 13. Here we shall reexamine this phenomenon, but in the light of personality, rather than of territoriality per se. To do so, we turn to one of the better-studied cases of alternative mating tactics. David Lank and his colleagues have been studying lekking male strategies in ruffs (*Philomachus pugnax*), a lekking sandpiper bird (Figure 17.10). Two male mating morphs exist in this species. "Independent" males form leks and actively guard an area within such a lek from other independent males. Satellite males, on the other hand, temporarily share an independent male's lek spot and form "an uneasy alliance" with an independent male (Lank et al., 1999). Such alliances may be selected among males, as females appear to prefer lek areas that contain both morphs (Hugie and Lank, 1997; Widemo, 1998). Independents and satellites differ not only in mating strategy, but in coloration and body mass, with satellites being generally smaller (Bachman and Widemo, 1999) and in possession of lighter plumage (Hogan-Warburg, 1966).

Independent males are generally much more aggressive than satellites, and since this polymorphism in behavior, as well as the difference in general mating tactics, is stable throughout a ruff's life, it seems fair to consider independent and satellite strategies in the light of animal personalities. This is a particularly interesting personality case, as in the ruff we have associated morphological changes that go along with different personality traits, similar to the bluegill sunfish case discussed in Chapter 6. In light of this, Lank and his colleagues have addressed two questions: (1) Is the behavioral polymorphism in mating strategy the result of a genetic polymorphism?, and (2) Is there any way to produce similar phenotypic differences in females, who normally express neither the independent nor satellite personality traits?

FIGURE 17.10. Two types of ruff. Two male ruffs of different morphs co-occupying a single display court on a lek. The dark male is an independent territory holder, while the white male is a nonterritorial satellite male. The independent resident male defends the court against other independent males, but satellite males have access to all courts. The independent resident male benefits from having a satellite because females prefer to visit and mate on co-occupied courts. Nonetheless, independent residents also incur reproductive costs, because visiting females may mate with the satellite instead, or mate with both males. As seen here, the independent male often tries to dominate the satellite by standing with its bill, which is a weapon, over the satellite's head. *(Photo credit: David Lank)*

To address the issue of an underlying genetic polymorphism, Lank and his colleagues (1995) reared chicks collected from eggs at forty-three ruff nests in Finland. Behavioral observations were used to determine the mother of the chicks, and DNA fingerprinting was employed to ascertain likely fathers. Once these chicks matured into adults, they were bred under controlled conditions to determine whether male mating strategy was inherited. Independent males sired twenty-seven sons, and of these, twenty-five developed into independent males themselves. For satellite males, fourteen of their thirty sons matured into satellites—not as high a figure as for independents, but still fairly high for a complex behavioral trait. Males then appear to inherit their mating strategy, in part, from their fathers. To quantify this, the researchers tested the observed number of satellites and independents against various different genetic models and found that a very simple inheritance system was likely at work. It appears that mating strategy (and hence one aspect of personality) in males is inherited via a single locus with two alleles, with the allele for the satellite being dominant. For at least this one personality trait, then, the genetics are surprisingly straightforward.

Lank and his colleagues' (1995) genetic analysis found that male mating strategy was *not* inherited via the sex chromosome, which might seem counterintuitive. Since only males display this polymorphism, one might assume that the trait would be linked to the sex chromosome. Yet, this need not be the case if females carry the gene(s) for the trait of interest but simply don't express the phenotype associated with the gene(s). Genetic analysis suggested that this was true for ruffs, so Lank and his colleagues (1999) ran a fascinating test to ascertain for certain whether this was the case.

If the satellite and independent strategies are not sex-linked, then in principle it should be possible to express these personality traits in females, in whom they are normally absent. Bird systems are particularly good models for these sorts of experiments, as prior work has demonstrated that testosterone implants in female birds can induce phenotypes that are usually only expressed in males (Emlen and Lorenz, 1942; Witschi, 1961). In addition, in ruffs, males that are castrated fail to molt into breeding plumage, further suggesting a critical role for testosterone in the expression of traits normally only expressed in males (van Oordt and Junge, 1936). To determine whether independent and satellite personality types could be induced in female ruffs, Lank and his colleagues (1999) took females with known pedigrees, gave them testosterone implants and examined which phenotypes were expressed in the females. The use of individuals with known pedigrees was particularly helpful, as without such knowledge the researchers could have ascertained only whether male phenotypes could be ex-

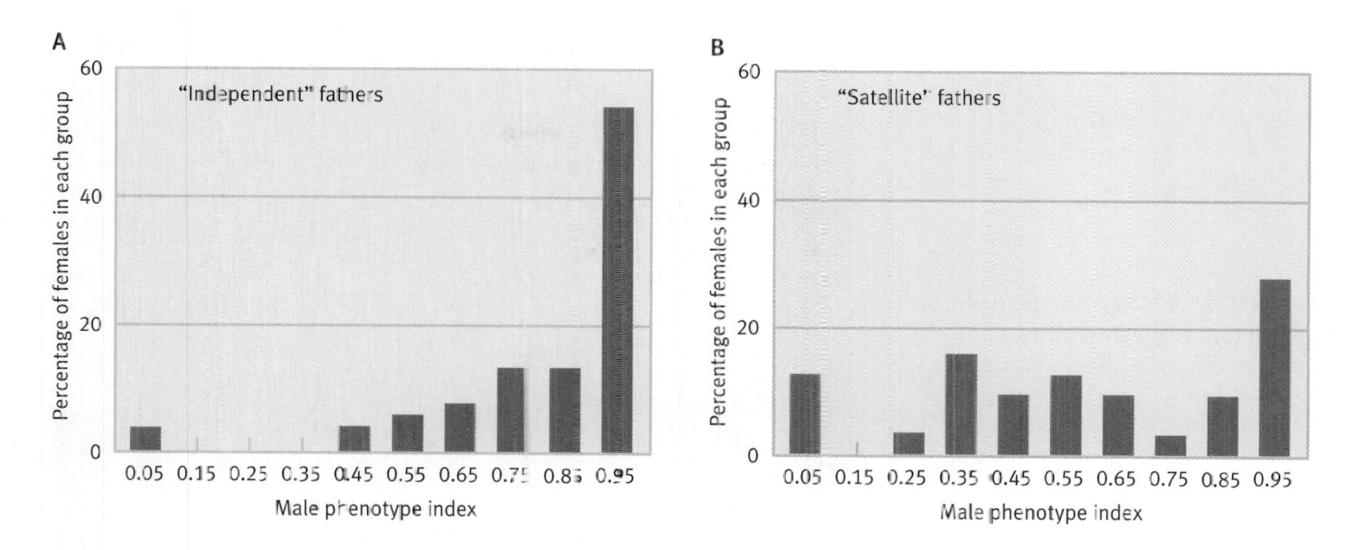

FIGURE 17.11. Hormones, heritability, and ruff personalities. In ruffs, two male phenotypes, independent and satellite, are well established. No such divergence in females is typically found. When females have testosterone experimentally implanted, **(A)** those from "independent" fathers show much higher "male phenotype" indices than did **(B)** those from "satellite" fathers. *(Based on Lank et al., 1999, p. 2327)*

pressed in females, but not whether a *particular* female expressed the phenotype that one would expect based on the males in her family. With pedigree information, such a test is easily done.

Two days after testosterone implants, female ruffs (known as reeves) were displaying typical male behavior, and a week later they were actually forming leks! After three to five weeks, reeves had grown neck ruffs and male-like display feathers, leaving no doubt of the effects of testosterone on both female behavior and morphology. Results from the pedigree analysis in reeves were remarkably similar to those found in the genetic study mentioned above. Females that had independent fathers nearly always displayed the independent personality type. The distribution of personality types from reeves that had satellite fathers was much flatter, with some females displaying satellite behavior, and some displaying much more independent traits (Figure 17.11). These results, in conjunction with the finding that the females tested in the pedigree study often displayed the independent or satellite behavior displayed by their brothers, strongly suggest that, at least in some cases, the inheritance of differences in personality can be understood using one of the simplest known modes of genetic inheritance (autosomal, single locus control).

LEARNING AND PERSONALITY IN GREAT TITS

In a study of exploratory behavior and reaction to novel objects in great tits (*Parus major*), Verbeek and his collaborators (1994) found

FIGURE 17.12. Personality types in great tits. In great tit birds, two personality types, fast and slow, appear to exist. Fast individuals approach novel objects quickly and spend little time with them. Slow individuals approach novel objects more reluctantly, but learn more about such items. *(Photo credit: Jack Ballard/Visuals Unlimited)*

two very different personality types in juvenile birds (Figure 17.12). "Fast" individuals quickly approach novel objects and explore new environments in a rapid manner, spending relatively short periods in any particular area. Furthermore, fast individuals are aggressive (Verbeek et al., 1996, 1998), and once they develop a food-searching pattern, they stick with it and are averse to changing such patterns. On the other end of the spectrum, "slow" birds are more reluctant to approach novel objects, are reluctant to form any foraging routine, and never start fights (Figure 17.13). At the same time, however, slow birds, while taking longer to explore an environment, also spend more time learning about each aspect of a new environment.

FIGURE 17.13. Fast and slow birds. An overview of Verbeek's study on "fast" and "slow" great tit birds.

Great tits are relatively consistent in where they fall on the fast/slow spectrum; indeed, recent work suggests that these personality types are heritable (Drent et al., 2002). In order to better understand whether other aspects of tit behavior correlate with the fast/slow personality types, Marchetti and Drent (2000) examined birds in a scenario in which they could learn about changes in their foraging environment from a "tutor" bird. The theoretical backdrop for the protocol they devised was that slow birds might be similar to the "producers" of the producer-scrounger game, while fast birds might be akin to the scroungers.

The protocol for this experiment was complex (as was the analysis), but essentially the question posed was whether slow and fast birds respond to a tutor's information in different ways. Marchetti and Drent found that while slow birds change feeders often when alone, they do not do so on the information provided by tutors—that is, they are *not* using tutors to help them uncover new food sources. Conversely, fast birds, though reluctant to change feeders once they have established a routine on their own, are quick to change when paired with tutors who provide them with information about new food sources. Overall, then, the slow tits match the notion of a producer fairly well, and the fast birds behave as would be expected from scroungers.

CHIMPANZEE PERSONALITIES AND CULTURAL TRANSMISSION

In the context of human personality studies, it would not strike anyone as surprising to hear that the culture an individual lives in affects his personality. In certain limited conditions, this appears to be true for other primates as well. By far the best-documented case of cultural transmission of information affecting animal personality involves seven long-term studies of chimpanzees in Africa, studies that ranged from seven to thirty years in length (Whiten et al., 1999; Whiten and Boesch, 2001). Senior researchers in these projects constructed a list of sixty-five behaviors that qualified as "cultural variants" that were probably, but not definitively, spread by imitation. These variants ranged from using leaves as sponges to picking out bone marrow to displaying at the start of a rainstorm. Somewhat remarkably, of these sixty-five behaviors, thirty-nine were present at some sites, but completely absent at others. In addition, this list only includes behaviors that were in principle possible at all sites. For example, since "algae fishing" is not possible when algae are absent, as at some chimpanzee sites, algae fishing is not included among these thirty-nine variants.

If you were to visit these seven sites in Africa and watch the chimpanzees for years and years, you would see relatively stable innovative behavior patterns being spread within each group, but across groups you would observe very different suites of behaviors being used. Given that we are talking about a long list of behaviors, and since individual behavior is fairly consistent within sites, it seems fair to conclude that the personalities of chimps across these seven African sites are very different, and that much of the difference is the result of the cultural transmission of behavior within groups.

It is admittedly frustrating that it is hard to say more about chimpanzee personality per se than what was said above (for further discussion, see Weiss et al., 2000). That is, given the large database available on chimps in the wild, why can't we say more about personality variables that appear to be under cultural control? While there are some anecdotes on specific personality traits in specific individuals, the controlled experimental studies of personality per se that we see in other parts of this chapter have simply not been undertaken on a large scale for chimps. There are no doubt many reasons for this, not the least of which are the ethical considerations of personality work in chimps, the tendency to want to minimize experimental work in natural populations, the long generation time of chimpanzees, and perhaps the fear that personality work on chimpanzees would be viewed as an attempt to anthropomorphize animal behavior.

Coping Styles

How an animal copes with everyday, as well more extreme, stressors can have a profound impact on its health (Koolhaas et al., 1999). In their review of the literature on stress and coping, Koolhaas and his colleagues found two coping styles—proactive and reactive—across a wide variety of animals, where a **coping style** "can be defined as a coherent set of behavioral and physiological stress responses which is consistent over time and which is characteristic to a certain group of animals" (Koolhaas et al., 1999, p. 925; Table 17.3). Henry and Stephens (1977) have charac-

TABLE 17.3. Coping styles. A list of species in which individual differences in coping style have been studied. Some studies examined behavioral variables; others examined physiological variables. A single plus indicates a single-parameter study; two pluses indicate a multi-parameter study. *(Based on Koolhaas et al., 1999, p. 926)*

SPECIES	BEHAVIORAL PARAMETERS	PHYSIOLOGICAL PARAMETERS
Mouse (*Mus musculus domesticus*)	+ +	+ +
Rat (*Rattus novegicus*)	+ +	+ +
Pig (*Sus scrofa*)	+ +	+ +
Tree shrew (*Tupaja belangeri*)	+ +	+ +
Cattle (*Bos taurus*)	+	+ +
Great tit (*Parus major*)	+ +	
Chicken (*Gallus domesticus*)	+	+ +
Beech marten (*Martes foina*)	+	
Stickleback (*Gasterosteus aculeatus*)	+ +	
Rainbow trout (*Oncorhynchus mykiss*)	+	+
Rhesus monkey (*Macaca mulatta*)	+ +	+ +
Human (*Homo sapiens*)	+ +	+ +
Octopus (*Octopus rubescens*)	+ +	

TABLE 17.4. Proactive and reactive coping styles. A summary of the behavioral differences between proactive and reactive male rats and mice. *(Based on Koolhaas et al., 1999)*

BEHAVIORAL CHARACTERISTICS	PROACTIVE	REACTIVE
Attack latency	Low	High
Active avoidance	High	Low
Defensive burying	High	Low
Nest-building	High	Low
Routine formation	High	Low
Cue dependency	Low	High
Conditioned immobility	Low	High
Flexibility	Low	High

terized these two coping styles as the "active response" (Cannon, 1915) and the "conservation-withdrawal" response (Engel and Schmale, 1972). The proactive (active response) coping style entails territorial control and aggression, while the reactive (conservation-withdrawal) coping style is characterized by immobility and low levels of aggression (Koolhaas et al., 1999; Table 17.4).

Benus and his colleagues (1991) tested the more general notion that the aggression associated with territoriality was somehow related to a more general reaction to environmental stress. They found that individual mice who were aggressive to conspecifics handled new challenges in a more proactive manner than did individuals who were less aggressive. The same sort of results were uncovered in experiments on rats. Sluyter and his colleagues (1995), for example, studied different strains of rats that had been selected for high or low levels of aggression. Subsequent to being tested in an intruder experiment (in which another male was placed in an individual's home cage), male rats were tested in a "defensive burying" experiment. In such an experiment, a small electric prod was placed in a male's cage, and if the male investigated the prod and touched it, he received a mild shock. Once shocked, a rat had essentially two ways to avoid future shock: he could either bury the prod under his bedding (an active response) or significantly curtail his movements. The researchers found that individuals from the aggressive strain of mice were much more likely to bury the prod than were mice from the less aggressive strain, suggesting that two coping styles map nicely onto personality types.

Proactive and reactive personalities have been found not only in laboratory and natural populations of animals, but also in domesticated animals such as pigs (Spoolder et al., 1996) and cows

(Hopster, 1998). While data on the costs and benefits of these two coping styles are scant, Koolhaas suggests they may both be adaptive. For example, data on disease in animals suggests that proactive and reactive animals differ in their susceptibility to diseases such as hypertension, atheroscleroris (Ely, 1981; Manuck et al., 1983; Sgoifo et al., 1997; Hessing et al., 1994), gastric ulcers (Weiss, 1972; Driscoll and Kugler, 1984), and immunosuppresive capabilities (Hessing et al., 1995). From a more proximate perspective, a better understanding of the endocrinological basis for proactive and reactive personalities is coming to light. For example, returning to the "prod burying" experiment discussed above (Sluyter et al., 1995), the proactive rodents (those who bury the prod) generally have high plasma noradrenaline and low plasma adrenaline and corticosterone, while reactive (withdrawn) rats typically have low plasma noradrenaline and high plasma corticosterone (again, similar results have been found in farm animals as well; Hessing et al., 1994).

Some Practical Applications of Animal Personality Research

The study of animal personality has potential practical implications for conservation biology, human-animal interactions, farming, and many other areas. Here we shall examine the practical implications of animal personality work for preventing the death of livestock and aiding people with disabilities.

PREDATORS AND DOMESTICATED PREY

Farmers and breeders have long been interested in the processes of natural and artificial selection with respect to both the anatomy and behavior of domesticated livestock. An understanding of these processes can provide a road map for breeding programs designed to produce some optimal product (fruit, vegetable, meat, and so forth). But an ethological understanding of personality traits may also prove handy for farmers and ranchers in another respect—that of minimizing the number of their animals killed by large predators that have been deemed "problem individuals" (Linnell et al., 1999, 2000).

As a result of species recovery and conservation plans, a number of large carnivores have been successfully reintroduced into the wild, or have had their natural population numbers increase dramatically over the last few decades. On the negative side of this conservation success has been the rekindling of an old rivalry between herders and large carnivores that feed on their animals (Blanco et al., 1992; Quigley and Cragshaw, 1992; Oli et al., 1994; Cozza et al., 1996; Kaczensky, 1996). Reducing this conflict may be facilitated by an understanding of the ethology of the carnivores that are causing the problem (Linnell et al., 1996; Sagor et al., 1997).

Because there is widespread opposition to large-scale killing of carnivore populations, one way that the carnivore versus livestock problem has been tackled is to focus on "problem individuals"—carnivore individuals that repeatedly attack and kill livestock. The data on the detailed hunting behavior of large carnivores preying on domestic prey are difficult to obtain, but studies of hunting behavior in cougars (Ross et al., 1997), leopards (Mitzutani, 1993), lions (Stander, 1990), and bears (Claar et al., 1986) suggest that certain individuals are more likely to prey upon domesticated animals. If such individuals do in fact exist, and if their behavior is both consistent and markedly different from that of their groupmates, then a case can be made for personality differences in carnivores, at least with respect to their hunting tactics.

Exactly *how to use* the information on personality differences and problem individuals, should it prove to be a general phenomenon, is still a matter of debate (Linnell et al., 1999, 2000), but *that this information should be used* is not. That fact alone is a testament to the practical power of ethological work on animal personality.

GUIDE DOG PERSONALITIES

In order to become a guide dog for the blind, individual dogs must pass a very rigorous test. It turns out that most dogs can, in fact, learn the difficult tasks needed to complete their training. Fear, however, often interferes with a dog's performance and is the single most prominent reason dogs fail to qualify to aid the blind (Baillie, 1972; Scott and Bielfelt, 1976; Goddard and Beilharz, 1982, 1983). This is not only an inherently interesting finding, but it suggests that a personality trait—fear or fearlessness (Boissy, 1995)—underlies which dogs become guide dogs and which do not.

In order to understand the genetics of fear more completely and in order to suggest a good breeding program for guide dogs, Goddard and Beilharz (1985) mated four breeds of dogs: Labrador retrievers, Australian kelpies, boxers, and German shepherds. Pups

Interview with Dr. JEROME KAGAN

Why study personality?
The study of personality has theoretical and practical significance because the variation among humans in stable mood and behavior has consequences for the individual's adaptation, location of choice, and the risk for psychopathology. Unfortunately, current conclusions about basic personality types are based primarily on questionnaire evidence rather than behavioral observations combined with biological measurements. It is likely therefore that the currently popular categories are too general and will be replaced with a larger number of personality types, and these will take into account the contexts in which the person acts.

Your long-term study of shyness in children has uncovered many new, important findings. How did you first get started with this study at Harvard? What is the single most unexpected result you obtained?
Our study of inhibited and uninhibited children had its origins in three different sets of observations. In the book *Birth to Maturity*, which summarized in 1962 the longitudinal sample at the Fells Research Institute, the only profile stable from early childhood to adulthood resembled the children we call inhibited. The second observation came from our study of the effects of day care in a sample of Boston children that contained different ethnic groups. The temperamentally inhibited infants preserved their profile, which seemed to be minimally affected by whether they were attending a day-care center or being raised at home. The third result came from Cynthia

Garcia-Coll's thesis, over twenty years ago, which demonstrated that children selected to be inhibited or uninhibited in the second year preserved their respective behavioral styles. The most unexpected result of our work over the last twenty-five years was the observation that four-month-old infants who showed vigorous motor activity and crying to varied classes of stimuli were biased to become inhibited children in the second year.

How has an evolutionary framework changed the way you study human personality?
The importance of an evolutionary framework derives from the fact that closely related species or strains differ in their typical behavior to appropriate incentives. In addition, there is early evidence that these differences are derived from inherited differences in brain neurochemistry.

Darwin wrote *The Expression of Emotions in Man and Animals* more than 125 years ago. Why did it take so long for the serious study of animal personality to develop?
Investigators of animal behavior were not interested in variation because they saw each species as an essential prototype and were more interested in species differences than in differences within a species. Further, there had been no interest in predicting differential adaptation in animals reared in a laboratory because the context was identical for all animals.

What is the single most important thing psychologists studying human personality have gleaned from work on animal personality?
The most important fact gleaned from work on animals is the demon-

stration of different biological features correlated with different behavioral preferences.

What is the single most important thing psychologists studying animal personality have gleaned from work on human personality?

I believe that those investigators studying animals have been motivated to study the approach-avoidance categories because of the work on human personality types.

Could you talk a little about the role of color in the study of animal personality?

I have little to say about the role of color in animal behavior. The structures that produce differential coloration of skin or fur are ectodermal in origin and therefore might be expected to correlate with behavior. One of the most relevant studies comes from Belayev's study of silver fox on a Siberian farm. After twenty generations of breeding tame fox, the offspring were likely to be tame, and showed changes in the distribution of melanin in the fur.

How do you define "temperament?" How does this differ from other related concepts?

Temperament is defined as an inherited biological profile that is linked initially to preferences for emotions and behavior. The initial bias is acted on by others to produce personality types. Thus, by two years of age, it is difficult to see the original temperamental bias, for it has become integrated with experience to produce a personality profile.

> THE MOST IMPORTANT FACT GLEANED FROM WORK ON ANIMALS IS THE DEMONSTRATION OF DIFFERENT BIOLOGICAL FEATURES CORRELATED WITH DIFFERENT BEHAVIORAL PREFERENCES.

How do you think the various genome-sequencing projects will affect our understanding of personality in humans and nonhumans?

The Genome Project will make an important contribution to our understanding of personality because it will eventually provide evidence regarding the inherited neurochemical profiles that create a large number of human temperaments.

What's the next frontier in the study of animal personality?

The next frontier in the study of animals will be the detection of a large number of specific functional relations among the neurobiology of the animal, associated behavioral profiles, and the contexts in which the animal acts.

DR. JEROME KAGAN is a professor at Harvard University. He is currently conducting the longest-running study ever on personality development in humans. Dr. Kagan's work, as represented in his book Galen's Prophecy: Temperament in Nature *(Basic Books), integrates evolutionary approaches into what has historically been an area reserved for cognitive and developmental psychologists.*

from all crosses were allowed to remain with their mothers until they were six weeks old, at which point they were weaned and moved to the Guide Dog Association's kennel. At twelve weeks of age, pups were given to "puppy walkers," who began training them to become guide dogs.

Goddard and Beilharz then measured fear and fearlessness using thirty-eight different behaviors that were subsequently transformed into twelve components via principle component analysis. The first three of these components map onto "general fearfulness," "fearfulness of objects," and "inhibited response to fear." Individual dog scores were consistent on these three components when the dogs were tested at six and twelve months of age.

Based on the observations of the puppy walkers and the individuals who tested the dogs on the guide dog exam, only general fearfulness was a good predictor of which animals would pass their final guide dog exam. This is particularly important because only "general fearfulness" has a strong underlying genetic component, whereas environmental factors explain much of the variance in the other two components of fear in these four species. The genetic crosses also show that Labrador retrievers were most easily trained to be guide dogs (German shepherds were the most difficult to train). Goddard and Beilharz (1985) used the genetics of fear, along with the ethologically based information described above, to suggest a breeding program for guide dogs.

SUMMARY

1. A combination of different behaviors can often exist concurrently in a population, and consistent behavioral differences among individuals can often be considered "personalities." Personality differences can be conceptualized as consistent long-term phenotypic behavioral differences among individuals.
2. Boldness/shyness is one of the most stable personality variables. But evolutionarily based animal behaviorists argue that there is no reason that natural selection should *necessarily* select for boldness or shyness as a general personality trait. Rather, individuals may be bold in some contexts and shy in others, depending on the costs and benefits associated with the behavioral contexts being examined.
3. With the exception of work on humans, compared to many of the subjects discussed throughout this book, the number of personality studies ethologists have undertaken is relatively small.

As such, the detailed cost/benefit analyses that we'd ideally like to have are rare. That being said, there are enough case studies across a wide array of species to provide us with some understanding of animal personality traits.

4. Studies such as those on the behavior and genetics of alternative strategies in lekking birds allow us to rethink mating strategies and heritability in the framework of personality research.
5. Cultural transmission of information affecting animal personality has been uncovered in seven long-term studies of chimpanzees in Africa. Stable innovative patterns of behavior are spread within groups of chimpanzees through imitation. But behaviors and personalities across different populations are very different.
6. How an animal copes with everyday, as well more extreme, stressors can have a profound impact on its health. Two general coping styles emerge from studies across a wide variety of animals, where a coping style can be defined as "a coherent set of behavioral and physiological stress responses which is consistent over time and which is characteristic to a certain group of animals." These two styles are referred to as the proactive or "active response" and the reactive or "conservation-withdrawal" response. The proactive coping style entails territorial control and aggression, while the reactive coping style is characterized by immobility and low levels of aggression.
7. The study of animal personality has potential practical implications for conservation biology, human-animal interactions, farming, and many other areas, including preventing the death of livestock and aiding people with disabilities.

DISCUSSION QUESTIONS

1. Go to the monkey exhibit at your local zoo and pick four individual monkeys to observe for at least four hours. Record all the information you can on each individual ("grooms," "eats," "attacks," "retreats," "sleeps," "plays," and so on), and if possible, note the proximity of the animals you are studying to others in the group. From your observations, can you suggest a list of behaviors that you might focus on during a longer, more controlled, study of personality in the population you are observing?
2. How would you construct an experiment to examine whether boldness and/or behavioral inhibition are heritable traits?

3. Besides the ones mentioned in this chapter, what general costs and benefits might you associate with being bold or inhibited? Next pick a particular species you are familiar with and create a list of the potential costs and benefits of boldness and shyness in that species.
4. Pick up a few recent issues of the journal *Animal Behaviour* and scan the titles for anything on "alternative strategies." Once you have found one or two such articles, read them in detail. Is there any mention of personality in these papers? If not, how might you reanalyze the data to see whether the alternative strategies could be construed in light of the work on animal personalities? What other sorts of data could you collect to better understand whether the alternative strategies represent personality types?
5. Besides preventing the death of livestock and selecting guide dogs, can you think of any other practical applications of personality work in animals? How might you construct some experiments to better understand whether the applications you suggest are feasible?

SUGGESTED READING

Buss, D. (1999). *Evolutionary psychology*. Boston: Allen & Bacon. A textbook on the evolution of traits traditionally studied by psychologists.

Gosling, S. D. (2001). From mice to men: What can we learn about personality from animal research? *Psychological Bulletin, 127*(1): 45–86. A recent review article on personality and animal research.

Kagan, J. (1994). *Galen's prophecy: Temperament in human nature*. New York: Basic Books. Reviews the longest-term study of shyness and inhibition in children and the history of personality and temperament studies.

Whiten, A., & Boesch, C. (2001). The culture of chimpanzees. *Scientific American*, January 2001, 60–67. An interesting review that suggests how culture can potentially explain some personality traits seen within and across chimp populations.

Wilson, D. S., et al. (1994). Shyness and boldness in humans and other animals. *Trends in Ecology and Evolution*; *9*, 442–446. A short, but thought-provoking article.

Glossary

adaptation A trait that results in the highest fitness among a specified set of behaviors in a particular environment. Adaptations are typically the result of the process of natural selection.

alliance A long-term coalition. *See also* coalition.

allogrooming The grooming of another individual.

antagonistic pleiotropy The case in which pleiotropic effects work in opposite directions with respect to fitness. *See also* pleiotropy.

appetitive stimulus Any stimulus that is considered positive, pleasant, or rewarding.

artificial selection Natural selection in which man is the selective agent. *See also* natural selection.

aversive stimulus Any stimulus that is associated with some unpleasant event.

axons Nerve cell fibers that transmit information.

boldness The tendency to take risks in both familiar and unfamiliar situations.

broad-sense heritability *See* heritability.

brood parasitism A form of behavioral parasitism that involves raising the young of others, often from another species.

byproduct mutualism A type of cooperation in which an individual pays an immediate cost or penalty for not acting cooperatively, such that the immediate net benefit of cooperating outweighs that of cheating.

classical conditioning *See* Pavlovian conditioning.

coalition Cooperative action taken by at least two individuals or groups against another individual or group. *See also* cooperation.

communication The transfer of information from a signaler to a receiver.

conceptual approach An approach that usually entails importing ideas generated in different disciplines and combining them in a new, cohesive way to address a topic in ethology. Generally speaking, natural history and experimentation play a role in concept generation, but a broad-based concept itself is not usually directly tied to any specific observation or experiment.

conditioned stimulus (CS) A stimulus that initially fails to elicit a particular response, but comes to do so when it is associated with a second (unconditioned) stimulus. *See also* unconditioned stimulus.

contagion The case in which the performance of an instinctive pattern of behavior by one individual will act as a releaser for the same behavior in other individuals.

cooperation An outcome that despite possible costs to the individual, provides some benefits to group members. "To cooperate" means to behave in a way to make cooperation possible.

coping style A set of behavioral and related stress responses that are consistent over time.

copying Behavior that occurs when an observer repeats the actions of a demonstrator. The copier is often rewarded for whatever behavior it has copied.

cultural transmission The transfer of information by teaching or social learning. *See also* social learning; teaching.

culture A system of information transfer that affects an individual's phenotype, in the sense that part of the phenotype is acquired from others by teaching or social learning.

dear enemy effect The dissipation of aggression between neighbors after territorial boundaries have been established.

dendrites Nerve cell fibers that receive electrical information from other cells.

developmental stability A measure of how well an organism handles changing environments as it matures.

disposable soma theory The theory that it may be beneficial for organisms to divert energy and resources from normal maintenance and repair associated with traits that have little connection to reproduction (i.e., somatic traits) and invest such energy in traits related to reproduction.

dyadic interactions Interactions involving only two individuals.

empirical approach An approach that entails gathering data in one form or another. Empirical work in ethology can take many forms, but most often is either observational or experimental.

endocrine system A system of ductless glands that secrete chemical messengers.

ethology The scientific study of animal behavior.

eusociality The most advanced form of sociality. A eusocial society is defined by three characteristics: cooperative brood care, a division of labor, and overlapping generations.

evolutionarily stable strategy (ESS) A strategy that if used by almost everyone in a population will not decrease in frequency when new, mutant strategies arise. "Mutant" refers to a new strategy introduced into a population at a very low frequency.

extinction curves A graph that depicts how long an animal will remember some paired association once the pairing itself has stopped.

extrapair copulations Copulations that occur outside the context of a pair bond.

fitness Lifetime reproductive success. Usually fitness is measured in relative terms.

fluctuating asymmetry Random deviations from perfect symmetry of the left and right side of the body.

genetic recombination In sexually reproducing organisms, genetic recombination occurs when pairs of chromosomes line up during cell division and sections of one chromosome "cross over" and swap positions with sections of their pairmate. This swapping can create new genetic variation.

genetic variation Variation caused by genetic differences.

goal-directed learning *See* instrumental conditioning.

good genes model A model of sexual selection in which females choose to mate with males that possess genes that are best suited to their particular environment. Females who choose the males with such genes receive "indirect" benefits, in the sense that their offspring receive some of the good genes that led their mother to choose a male as a mate in the first place.

group selection A hierarchical model in which natural selection operates at two levels: within-group selection and between-group selection.

habituation Becoming less sensitive to stimuli over time.

handicap principle A theory for the evolution of "honest indicator" traits. This theory suggests that honest indicator traits should be generally quite "costly" to produce, since the costlier the trait, the more difficult it is to fake.

hawk-dove game The most well-known model for the evolution of aggression.

heritability Broad-sense heritability measures the total proportion of variance in a trait that is due to genetic variance. Narrow-sense heritability is more technically complex, but measures the proportion of genetic variance in a trait that is accessible to natural selection.

home range A delineated, undefended area in which an animal spends most of its time.

horizontal cultural transmission Cultural transmission in which information is passed across individuals of the same age or peer group. *See also* cultural transmission.

ideal free distribution model A mathematical model developed to predict how animals will distribute themselves among habitats with varying levels of resource availability.

imitation The acquisition of a topographically novel response through observation.

inclusive fitness A method for calculating fitness that takes into account not only the effect a gene has on its bearer but also the effect it has on its blood relatives (descendant and nondescendant kin).

individual learning A relatively permanent change in behavior as a result of experience. Individual learning differs from social learning in that it does not involve learning from others.

instrumental conditioning Learning that occurs when a response made by an animal is somehow reinforced. An animal must undertake some action or response in order for the conditioning process to produce learning.

intersexual selection A form of sexual selection in which individuals of one sex choose which individuals of the other sex to take as mates. *See also* sexual selection.

intrasexual selection A form of sexual selection whereby members of one sex compete with each other for access to the other sex. *See also* sexual selection.

kamikaze sperm hypothesis The hypothesis that natural selection might favor sperm that are designed to kill other males' sperm rather than to fertilize eggs.

law of effect A principle that maintains that if a response in the presence of a stimulus is followed by a satisfying event, the association between the stimulus and the response will be strengthened.

local enhancement A phenomenon in which an individual is drawn to a particular area because it observed another individual in that location.

loser effect The increased probability of losing a fight at time T, based on losing at time T-1, T-2, and so on, respectively.

marginal value theorem A mathematical model developed to predict how long a forager will spend in a patch of food before moving on to new patches.

mate-choice copying The act of copying the mate choice of others in a population.

migration The movement of organisms over long distances (often in a seasonal manner).

monogamous mating system A mating system in which a male and female mate only with one another during a given breeding season.

mushroom bodies A cluster of small neurons located at the front of the brain of some invertebrates.

mutation Any change in genetic structure.

mutualism Any interaction that benefits all parties involved. Interspecies cooperation is often referred to as mutualism.

narrow-sense heritability *See* heritability.

natural selection The process at the heart of Darwin's theory of evolution. Natural selection occurs when variants of a trait that best suit an organism to its environment and that are heritable increase in frequency over evolutionary time. This process requires variation, fitness differences, and heritability.

nomads Individuals who lack a home range or a territory and who rarely frequent the same area over time. Occasionally, reference will be made to a nomadic species.

nuptial gifts Prey presented by males to females during courtship.

oblique cultural transmission Cultural transmission in which information is passed across generations, but not via parent to offspring transmission.

operant learning *See* instrumental conditioning.

optimal foraging theory (OFT) A family of mathematical models developed to predict animal foraging behavior. Models based on this theory examine questions concerning what to eat, where to eat, how long to eat, and so forth.

optimal skew theory A family of models that predicts the distribution of breeding within a group, as well as the degree of cooperation or conflict over reproductive activities.

parental investment The amount of energy parents invest in raising their offspring.

parent-offspring conflict A zone of conflict between how much a given offspring wants in terms of parental resources and how much a parent is willing to give.

parent-offspring regression A technique for measuring narrow-sense heritability that involves measuring a trait in parents *and* offspring.

Pavlovian conditioning The experimental pairing of a conditioned and unconditioned stimulus. *See also* conditioned stimulus; unconditioned stimulus.

personality differences Consistent long-term phenotypic behavioral differences among individuals.

phenotype The observable characteristics of an organism.

phenotype variance Variance that is due to genetic factors, environmental factors, or an interaction of the two.

phenotypic plasticity The ability of an organism to produce different phenotypes depending on environmental conditions. *See also* phenotype.

play While there are numerous definitions of play, the most commonly accepted one is that of Bekoff and Byers (1981): "Play is all motor activity performed postnatally that appears to be purposeless, in which motor patterns from other contexts may often be used in modified forms and altered temporal sequencing."

play markers Behavioral indicators that denote that some action that is about to be undertaken should be considered playful.

pleiotropy The case when a gene has more than one effect on an organism.

plural breeders Multiple pairs of breeders that nest together and help raise each other's offspring.

polyadic interactions Interactions involving more than two individuals.

polyandry A mating system wherein females mate with more than one male per breeding season.

polygenic Caused by the action of more than one gene.

polygynandry A mating system wherein several males form pair bonds with several females simultaneously.

polygyny A mating system wherein males mate with more than one female per breeding season.

polygyny threshold model A model developed to predict the conditions under which a mating system will move from monogamy to polygyny (or vice versa).

predator inspection An antipredator behavior in which one to a few individuals break away from a group and slowly approach a potential predator to obtain various sorts of information.

preexisting bias *See* sensory exploitation.

prisoner's dilemma A game theory payoff matrix that is used to study the evolution of cooperation.

producers Individuals that find and procure food.

promiscuity A type of mating system in which both polyandry and polygyny are occurring. In one form of promiscuity, both males and females mate with many partners and no pair bonds are formed. In the second type of promiscuous breeding system, labeled polygynandry, several males form pair bonds with several females simultaneously.

proximate Referring to factors that are immediate, rather than evolutionary, in nature.

resource holding power A measure of an animal's fighting ability.

risk-sensitive optimal foraging models A family of models that examine how variance in food supplies affects foraging behavior.

role reversal Behavior wherein older individuals allow subordinate, younger animals to take on the dominant role during play or in which older individuals perform some act (e.g., an aggressive act toward a subordinate individual) that is at a level clearly below that of which the older individual is capable. Also known as self-handicapping.

runaway sexual selection A model of sexual selection in which the genes for mate choice in the female and the genes for preferred traits in males become genetically linked.

scroungers Individuals that obtain a portion of their diet by parasitizing the food that producers have uncovered.

self-handicapping. *See* role reversal.

senescence An age-specific decline in the probability of survival.

sensitization Becoming more sensitive to stimuli over time.

sensory drive *See* sensory exploitation.

sensory exploitation A theory of sexual selection that posits that females may initially prefer male traits that elicit the greatest amount of stimulation from their sensory systems. Also known as sensory exploitation, sensory drive, or preexisting bias.

sexual selection A form of natural selection that depends "not on the struggle for existence in relation to other organic beings or to external conditions, but on a struggle between the individuals of one sex, generally the males, for the possession of the other sex." (Darwin, 1871).

shyness The reluctance to take risks, or to engage in unfamiliar activity.

singular breeders A single pair of breeders in a group; they sire virtually all of the group's offspring and have other members of the group helping them to raise the young.

social facilitation When the mere presence of a model, regardless of what it does, is thought to facilitate learning on the part of an observer.

social grooming *See* allogrooming.

social learning Learning via the observation of others. Observation is used in a broad sense, and can include obtaining information via olfaction and other senses besides sight.

sperm competition A form of sexual selection that occurs directly between sperm after insemination.

syrinx The vocal organ of birds.

teaching A behavior that occurs when one individual serves as an instructor and at least one other individual acts as a pupil who learns from the teacher. Teachers have a much more active and complicated role than just being a model that others mimic (as in social learning).

territory A delineated, defended area.

territory budding The formation of new territories by aggression on the part of the territory holder and his family, which produces an expanded area that includes space for newly developing breeders.

theoretical approach An approach that entails the generation of a model, usually a mathematical model, of some behavior.

tit for tat A behavioral strategy that instructs a player to initially cooperate with a new partner, and subsequently to do whatever that partner does.

ultimate questions Questions relating to the evolution of a trait.

unconditioned stimulus (US) A stimulus that elicits a vigorous response in the absence of training.

vertical cultural transmission Cultural transmission in which information is passed directly from parent(s) to offspring. *See also* cultural transmission.

waggle dance A dance performed by forager bees. The waggle dance provides information on the spatial location of food located at some distance from the hive.

winner effect The increased probability of winning a fight at time T, based on victories at times T-1, T-2, and so on.

References

Abbott, D. (1984). Differentiation of sexual behavior in female marmoset monkeys: Effects of neonatal testosterone or a male co-twin. *Programs in Brain Research, 61,* 349–358.

Abbott, J. C., Dunbrack, R. L., & Orr, C. D. (1985). The interaction of size and experience in dominance relationships of juvenile steelhead trout. *Behaviour, 92,* 243–251.

Abdel-malek, S. A. (1963). Diurnal rhythm of feeding of the three-spine stickleback of Kandalaksha Bay of the White Sea. *Vopr Ikhtiol (in Russian), 3,* 326–335.

Able, K. P. (1999). *Gatherings of angels: Migrating birds and their ecology.* Ithaca, NY: Comstock Books.

Abrahams, M. V., & Pratt, T. C. (2000). Hormonal manipulations of growth rate and its influence on predator avoidance: Foraging trade-offs. *Canadian Journal of Zoology-Revue Canadienne De Zoologie, 78*(1), 121–127.

Adams, E., & Mesterton-Gibbons, M. (1995). The cost of threat displays and the stability of deceptive communication. *Journal of Theoretical Biology, 175,* 405–421.

Ågren, G. (1984). Pair formation in the Mongolian gerbil. *Animal Behaviour, 32,* 528–535.

Alatalo, R., Lundberg, A., & Stahlbrandt, K. (1984). Female mate choice in the pied flycatcher *Ficedula hypoleuca. Behavioral Ecology and Sociobiology, 14,* 253–261.

Albin, R. (1988). The pleiotropic theory of senescence: Supportive evidence from human genetic disease. *Ethology and Sociobiology, 9,* 371–382.

Albin, R. (1993). Antagonistic pleiotropy, mutation accumulation and human genetic disease. *Genetica, 91,* 279–286.

Alcock, J. (1998). *Animal behavior* (6th ed.). Sunderland, MA: Sinauer Associates.

Alcock, J. (2001). *The triumph of sociobiology.* New York: Oxford University Press.

Alcock, J., & Sherman, P. (1994). The utility of the proximate-ultimate dichotomy in ethology. *Ethology, 96,* 58–62.

Alerstam, T. (1990). *Bird migration.* Cambridge: Cambridge University Press.

Alexander, R. D. (1974). The evolution of social behavior. *Annual Review of Ecology and Systematics, 5,* 325–383.

Alexander, R. D. (1979). *Darwinism and human affairs.* Seattle: University of Washington Press.

Alexander, R. D. (1991). Social learning and kin recognition. *Ethology and Sociobiology, 12,* 387–399.

Alexander, R., & Borgia, G. (1978). Group selection, altruism and the levels of organization of life. *Annual Review of Ecology and Systematics, 9,* 449–475.

Alexander, R. D., & Sherman, P. W. (1977). Local mate competition and parental investment in insects. *Science, 196,* 494–499.

Alexander, R. D., & Tinkle, D. W. (Eds.). (1981). *Natural selection and social behavior.* New York: Chiron Press.

Alfieri, M. (2000). *The interaction effects of experience, learning, and social associations on survival in the guppy, Poecilia reticulate.* Ph.D. thesis, University of Louisville, Louisville, KY.

Allan, S., & Suthers, R. (1994). Lateralization and motor stereotypy of song production in the brown-headed cowbird. *Journal of Neurobiology, 25,* 1154–1166.

Allee, W. (1951). *Cooperation among animals.* New York: Henry Schuman.

Altmann, J. (1973). Observational study of behavior: Sampling methods. *Behaviour, 49,* 227–267.

Altmann, S. A. (1956). Avian mobbing behavior and predator recognition. *Condor, 58,* 241–253.

Altmann, S. (1962). Social behavior of anthropoid primates: Analysis of recent concepts. In E. Bliss (Ed.), *Roots of behavior* (pp. 277–285). New York: Harper and Brothers.

Altmann, S., Wagner, S., & Lenington, S. (1977). Two models for the evolution of polygyny. *Behavioral Ecology and Sociobiology, 2,* 397–410.

Andersson, M. (1994). *Sexual selection.* Princeton: Princeton University Press.

Aquilino, W. (1991). Family structure and home leaving: A further specification of the relationship. *Journal of Marriage and the Family, 53,* 999–1010.

Archer, J. (1988). *The behavioural biology of aggression.* Cambridge: Cambridge University Press.

Archer, J., Birring, S., & Wu, F. C. W. (1998). The association between testosterone and aggression among young men: Empirical findings and a meta-analysis. *Aggressive Behavior, 24,* 411–420.

Arnold, W., & Trillmich, F. (1985). Time budget in Galapagos fur seals: The influence of mother's presence and absence on pup activity and play. *Behaviour, 92,* 302–321.

Atkinson, D., & Begon, M. (1987). Reproductive variation and adult size in two co-occurring grasshopper species. *Ecological Entomology, 12,* 119–127.

Austad, S. (1997). *Why we age: What science is discovering about the body's journey through life.* New York: John Wiley.

Avital, E., & Jablonka, E. (2000). *Animal traditions: Behavioural inheritance in evolution.* Cambridge: Cambridge University Press.

Axelrod, R. (1984). *The evolution of cooperation.* New York: Basic Books.

Axelrod, R., & Hamilton, W. D. (1981). The evolution of cooperation. *Science, 211,* 1390–1396.

Bachman, G., & Widemo, F. (1999). Relationships between body composition, body size and alternative reproductive tactics in a lekking sandpiper, the ruff (*Philomachus pugnax*). *Functional Ecology, 13,* 411–416.

Baerends, G. P., Wanders, J., & Vodegel, R. (1986). The relationship between marking patterns and motivational state in pre-spawning behavior of the cichlid fish *Chromidotilapia guentheri* (Sauvage). *Netherlands Journal of Zoology, 36,* 88–116.

Baerends-van Roon, J., & Baerends, G. (1979). *The morphogenesis of the behavior of the domestic cat, with special emphasis on the development of prey-catching.* Amsterdam: New Holland.

Bagg, A. (1952). Anting not exclusively an avian trait. *Journal of Mammalogy, 33,* 243.

Bailey, W. (1991). *Acoustic behaviour of insects: An evolutionary perspective.* London: Chapman and Hall.

Baillie, J. (1972). The behavioural requirements necessary for guide dogs for the blind in the United Kingdom. *British Veterinary Journal, 128,* 477.

Baker, M. (1996). Fur rubbing: Use of medicinal plants by capuchin monkeys (*Cebus capucinus*). *American Journal of Primatology, 38,* 362–370.

Baker, R. R. (1978). *The evolutionary ecology of animal migration.* New York: Holmes & Meier Publishers.

Baker, R. R., & Bellis, M. A. (1988). Kamikaze sperm in mammals. *Animal Behaviour, 36,* 936–939.

Baker, R. R., & Bellis, M. A. (1993a). Sperm competition: Ejaculate adjustment by males and the function of masturbation. *Animal Behaviour, 46,* 861–885.

Baker, R. R., & Bellis, M. A. (1993b). Sperm competition: Ejaculate manipulation by females and a function for the female orgasm. *Animal Behaviour, 46,* 887–909.

Bakker, T. C. M. (1993). Positive genetic correlation between female preference and preferred male ornament in sticklebacks. *Nature, 363,* 255–257.

Balaban, E., Alper, J. S., & Kasamon, Y. L. (1996). Mean genes and the biology of aggression: A critical review of recent animal and human research. *Journal of Neurogenetics, 11,* 1–43.

Balda, R., & Kamil, A. (1989). A comparative study of cache recovery by three corvid species. *Animal Behaviour, 38,* 486–495.

Balda, R., Pepperberg, I., & Kamil, A. (Eds.). (1998). *Animal cognition in nature.* San Diego: Academic Press.

Baldwin, J. (1969). The ontogeny of social behavior of squirrel monkeys (*Saimiri sciureus*) in a seminatural environment. *Folia Primologica, 11,* 35–79.

Baldwin, J. (1971). The social organization of a semifree-ranging troop of squirrel monkeys (*Saimiri sciureus*). *Folia Primologica, 14,* 23–50.

Ball, N., Amlaner, C., Shaffery, J., & Opp, M. (1988). Asynchronous eye-closure and unihemispheric quiet sleep of birds. In W. Koella, F. Obal, H. Schulz, & P. Visser (Eds.), *Sleep '86* (pp. 151–153). New York: Gustav Fischer Verlag.

Bandura, A. (1977). *Social learning theory.* Englewood Cliffs, NJ: Prentice Hall.

Bandura, A. (1986). *Social foundations of thought and action.* Englewood Cliffs, NJ: Prentice Hall.

Bandura, A., Ross, D., & Ross, S. (1961). Transmission of aggression through imitation of aggressive models. *Journal of Abnormal and Social Psychology, 63,* 572–582.

Banks, E. (1985). Warder Clyde Allee and the Chicago school of animal behavior. *Journal of the History of the Behavioral Sciences, 21,* 345–353.

Barkley, R., Neill, W., & Gooding, R. (1978). Skipjack tuna, *Katsuwonus pelamis,* habitat based in temperature and oxygen requirements. *U.S. Fish and Wildlife Service Fishery Bulletin, 76,* 653–662.

Barlow, G. W. (1963). Ethology of the Asian teleost *Badis badis.* II. Motivation and signal value of colour patterns. *Physiological Zoology, 11,* 97–105.

Barnard, C. J. (Ed.). (1984). *Producers and scroungers.* London: Croom Helm/Chapman and Hall.

Barnard, C. J., & Brown, C. A. (1984). A payoff asymmetry in resident-resident disputes between shrews. *Animal Behaviour, 32,* 302–304.

Barnard, C. J., & Brown, C. A. (1985). Risk-sensitive foraging in common shrews (*Sorex araneus*). *Behavioral Ecology and Sociobiology, 16,* 161–164.

Barnard, C. J., & Sibly, R. M. (1981). Producers and scroungers: A general model and its application to captive flocks of house sparrows. *Animal Behaviour, 29,* 543–550.

Basil, J., Kamil, A., Balda, R., & Fite, K. (1996). Differences in hippocampal volume among food storing corvids. *Brain, Behavior and Evolution, 7,* 156–164.

Basolo, A. (1990). Female preference predates the evolution of the sword in swordfish. *Science, 250,* 808–811.

Basolo, A. (1995). A further examination of a pre-existing bias favouring a sword in the genus *Xiphoporus. Animal Behaviour, 50,* 365–375.

Bass, A. H. (1992). Dimorphic male brains and alternative reproductive tactics in a vocalizing fish. *Trends in Neuroscience, 15,* 139–145.

Bass, A. H. (1996). Shaping brain sexuality. *American Scientist, 84,* 352–363.

Bass, A. H. (1998). Behavioral and evolutionary neurobiology: A pluralistic approach. *American Zoologist, 38,* 97–107.

Bass, A. H., & Baker, R. (1990). Sexual dimorphisms in the vocal control system of a teleost fish: Morphological and physiological identified cells. *Journal of Neurobiology, 21,* 1155–1168.

Bass, A. H., & Baker, R. (1991). Adaptive modification of homologous vocal control traits in teleost fish. *Brain, Behavior and Evolution, 38,* 240–254.

Bass, A. H., Denizot, J., & Marchaterre, M. (1986). Ultrastructural features and hormone-dependent sex differences of mormyrid electric organs. *Journal of Comparative Neurology, 254,* 511–528.

Bass, A. H., & Hopkins, C. (1983). Hormonal control of sex differentiation: Changes in the electric organ discharge. *Science, 220,* 971–974.

Bass, A. H., & Hopkins, C. (1985). Hormonal control of sex differences in the electric organ discharge of mormyrid fishes. *Journal of Comparative Physiology, 156,* 587–604.

Bass, A. H., & Marchaterre, M. (1989). Sound-generating (sonic) motor system in a teleost fish (*Porichthys notatus*): Sexual polymorphism in the ultrastructure of myofibrils. *Journal of Comparative Neurology, 286,* 141–153.

Bass, A. H., & Volman, S. (1987). From behavior to membranes: Testosterone-induced changes in action potential duration in electric organs. *Proceedings of the National Academy of Sciences, U.S.A., 84,* 9295–9298.

Bateson, P. (1978). Early experience and sexual preferences. In J. B. Hutchinson (Ed.), *Biological determinants of sexual behavior* (pp. 29–53). New York: Wiley.

Bateson, P. (1990). Choice, preference and selection. In M. Bekoff & D. Jamieson (Eds.), *Interpretation and explanation in the study of animal behaviour* (Vol. I. Interpretation, intentionality and communication, pp. 149–156). Boulder: Westview Press.

Baum, D., & Larson, A. (1991). Adaptation reviewed: A phylogenetic methodology for studying character macroevolution. *Systematic Zoology, 40,* 1–18.

Beatty, W., Costello, K., & Berry, S. (1984). Suppression of play fighting by amphetamines: Effects of catecholamine antagonists, agonists and synthesis inhibitors. *Pharmacology Biochemistry and Behavior, 20,* 747–755.

Beauchamp, G. (1997). Determinants of interspecific brood amalgamation in waterfowl. *Auk, 114,* 11–21.

Beck, B. (1980). *Animal tool behavior: The use and manufacture of tools by animals.* New York: Garland STPM Press.

Bednarz, J. C. (1988). Cooperative hunting in Harris' hawks (*Parabueto unicintus*). *Science, 239,* 1525–1527.

Beebee, T. (1995). Tadpole growth: Is there an interference effect in nature. *Herpetological Journal, 5,* 204–205.

Beebee, T., & Wong, A. (1992). Prototheca-mediated interference competition between anuran larvae operates by resource diversion. *Physiological Zoology, 65,* 815–831.

Beecher, M. D., Beecher, I., & Hahn, S. (1981a). Parent-offspring recognition in bank swallows: II. Development and acoustic basis. *Animal Behaviour, 29,* 95–101.

Beecher, M. D., Beecher, I., & Lumpkin, S. (1981b). Parent-offspring recognition in bank swallows: I. Natural history. *Animal Behaviour, 29,* 86–94.

Beecher, M. D., Campbell, S. E., & Nordby, J. C. (1998). The cognitive ecology of song communication and song learning in the song sparrow. In R. Dukas (Ed.), *Cognitive ecology* (pp. 175–199). Chicago: University of Chicago Press.

Beecher, M. D., Medvin, B., Stoddard, P., & Loesch, P. (1986). Acoustic adaptations for parent-offspring recognition in swallows. *Experimental Biology, 45,* 179–193.

Bekoff, M. (Ed.). (1974). Social play in mammals. *American Zoologist, 14* (special issue on social play).

Bekoff, M. (1977). Social communication in canids: Evidence for the evolution of a stereotyped mammalian display. *Science, 197,* 1097–1099.

Bekoff, M. (1984). Social play behavior. *Bioscience, 34,* 228–233.

Bekoff, M. (1995). Play signals as punctuation: The structure of play in social canids. *Behaviour, 132,* 419–429.

Bekoff, M. (2000). Social play behavior: Cooperation, fairness, trust, and the evolution of morality. *Journal of Consciousness Studies, 8,* 81–90.

Bekoff, M., & Allen, C. (1992). Intentional icons: Toward an evolutionary cognitive ethology. *Ethology, 91,* 1–16.

Bekoff, M., & Allen, C. (1998). Intentional communication and social play: How and why animals negotiate and agree to play. In M. Bekoff & J. Byers (Eds.), *Animal play: Evolutionary, comparative and ecological perspectives* (pp. 97–114). Cambridge: Cambridge University Press.

Bekoff, M., & Byers, J. (1981). A critical reanalysis of the ontogeny and phylogeny of mammalian social and locomotor play: An ethological hornet's nest. In K. Immelman, G. Barlow, L. Petrinivich, & M. Main (Eds.), *Behavioral development* (pp. 296–337). Cambridge: Cambridge University Press.

Bekoff, M., & Byers, J. (Eds.). (1998). *Animal play: Evolutionary, comparative and ecological perspectives.* Cambridge: Cambridge University Press

Bell, G. (1997). *Selection: The mechanism of evolution.* New York: Chapman and Hall.

Bell, M., & Foster, S. (Eds.). (1994). *The evolutionary biology of the threespine stickleback.* New York: Oxford University Press.

Bellis, M. A., Baker, R. R., & Gage, M. J. G. (1990). Variation in rat ejaculates consistent with the kamikaze-sperm hypothesis. *Journal of Mammalogy, 71,* 479–480.

Bellrose, F., & Holm, D. (1994). *Ecology and management of the wood duck.* Mechanicsburg, PA: Stackpole Books.

Belovsky, G. E. (1978). Diet optimization in a generalist herbivore: The moose. *Theoretical Population Biology, 14,* 105–134.

Belovsky, G. E. (1984). Herbivore optimal foraging: A comparative test of 3 models. *American Naturalist, 124,* 97–115.

Bennett, A., Cuthill, I., Partridge, J., & Lunau, K. (1997). Ultraviolet plumage colors predict mate preferences in starlings. *Proceedings of the National Academy of Sciences, U.S.A., 94,* 8618–8621.

Bennett, A., Cuthill, I. C., Partridge, J. C., & Maier, E. (1996). Ultraviolet vision and mate choice in zebra finches. *Nature, 380,* 433–435.

Bent, A. C. (1938). *Life histories of North American birds.* Washington, DC: U.S. Government Printing Office.

Bentley, P. J. (1998). *Comparative vertebrate endocrinology.* Cambridge: Cambridge University Press.

Benus, R. F., Bohus, B., Koolhaas, J. M., & Oortmerssen, G. (1991). Heritable variation in aggression as a reflection of individual coping scores. *Experientia, 47,* 1008–1019.

Bercovitch, F. (1988). Coalitions, cooperation and reproductive tactics among adult male baboons. *Animal Behaviour, 36,* 1198–1209.

Berger, J. (1979). Weaning conflict in desert and mountain bighorn sheep: An ecological interpretation. *Zeitschrift fur Tierpsychologie, 50,* 187–200.

Berger, J. (1980). The ecology, structure and function of social play in bighorn sheep (*Ovis canadensis*). *Journal of Zoology (London), 192,* 531–542.

Berglund, A., Rosenqvist, G., & Svennson, I. (1986). Reversed sex roles and parental energy investment in zygotes of two pipefish species. *Marine Ecology–Progress Reports, 29,* 209–215.

Berglund, A., Rosenqvist, G., & Svennson, I. (1988). Multiple matings and paternal care in the pipefish *Syngnathus typhle. Oikos, 51,* 184–188.

Berglund, A., Rosenqvist, G., & Svennson, I. (1989). Reproductive success of females limited by males in two pipefish species. *American Naturalist, 133,* 506–516.

Bergman, A., & Feldman, M. (1995). On the evolution of learning: Representation of a stochastic environment. *Theoretical Population Biology, 48,* 251–276.

Bernstein, C., & Bernstein, H. (1991). *Aging, sex and DNA repair.* New York: Academic Press.

Berthold, P. (1990). Wegzugbeginn und Einsetzen der zugunruhe bei 19 Vogelpopulationen – eine vergleichende untersuchung. In R. van den Elzen, K. Schuchmann, & K. Schmidt-Koenig (Eds.), *Proceedings of the International Centennial Meeting of the DO-G, Current topics in avian biology* (pp. 217–222). Bonn: DO-G.

Bertram, B. (1978). Living in groups: Predators and prey. In J. R. Krebs & N. Davies (Eds.), *Behavioural ecology: An evolutionary approach* (pp. 64–96). London: Blackwell Scientific Publications.

Betzig, L. (1986). *Despotism and differential reproduction: A Darwinian view of history.* New York: Aldine.

Betzig, L., Borgerhoff-Mulder, M., & Turke, P. (Eds.). (1988). *Human reproductive behaviour: A Darwinian perspective.* Cambridge: Cambridge University Press.

Biben, M. (1986). Individual- and sex-related strategies of wrestling play in captive squirrel monkeys. *Ethology, 71,* 229–241.

Biben, M. (1998). Squirrel monkey playfighting: Making the case for a cognitive function for play. In M. Bekoff & J. Byers (Eds.), *Animal play: Evolutionary, comparative and ecological perspectives* (pp. 161–182). Cambridge: Cambridge University Press.

Bildstein, K. L. (1983). Why white-tailed deer flag their tails. *American Naturalist, 121,* 709–715.

Billen, J., & Morgan, E. D. (1997). Pheromonal communication in social insects: Sources and secretions. In R. Vander Meer, M. Breed, K. Espelie, & M. Winston (Eds.), *Pheromone communication in social insects* (pp. 3–33). Boulder, CO: Westview Press.

Billing, J., & Sherman, P. W. (1998). Antimicrobial functions of spices: Why some like it hot. *Quarterly Review of Biology, 73,* 3–49.

Birch, L. (1980). Effects of peer models' food choices and eating behaviors in preschoolers' food preferences. *Child Development, 51,* 14–18.

Birkhead, T., & Møller, A. (1992). *Sperm competition in birds.* London: Academic Press.

Birkhead, T., & Møller, A. (1998). *Sperm competition and sexual selection.* London: Academic Press.

Bishop, D. T., Cannings, C., & Maynard Smith, J. (1978). The war of attrition with random rewards. *Journal of Theoretical Biology, 74,* 377–388.

Bitterman, M. (1975). The comparative analysis of learning. *Science, 188,* 699–709.

Bjorklund, M., & Westman, B. (1986). Adaptive advantages of monogamy in the great tit (*Parus-Major*): An experimental test of the polygyny threshold-model. *Animal Behaviour, 34,* 1436–1440.

Bjorksten, T., David, P., Pomiankowski, A., & Fowler, K. (2000). Fluctuating asymmetry of sexual and nonsexual traits in stalk-eyed flies: A poor indicator of developmental stress and genetic quality. *Journal of Evolutionary Biology, 13,* 89–97.

Black, J., Issacs, K., Anderson, B., Alcantara, A., & Greenough, W. (1990). Learning causes synaptogenesis, whereas motor activity causes angiogenesis in cerebellar cortex of adult rats. *Proceedings of the National Academy of Sciences, U.S.A., 87,* 5568–5572.

Blackburn, J., Pfaus, J., & Phillips, A. (1992). Dopamine functions in appetitive and defensive behaviors. *Progress in Neurobiology, 39,* 247–249.

Blake, M., Udelsman, R., Feulner, G., Norton, D., & Holbrook, N. (1991). Stress induced shock protein 70 expression in adrenal cortex: An adrenocorticotropic hormone-sensitive, age-dependent response. *Proceedings of the National Academy of Sciences, U.S.A., 88,* 9873–9877.

Blanco, J., Reig, S., & Cuesta, L. (1992). Distribution, status and conservation problems of the wolf (*Canis lupis*) in Spain. *Biological Conservation, 60,* 73–80.

Blaustein, A. (1983). Kin recognition mechanisms: Phenotype matching or recognition alleles. *American Naturalist, 121,* 749–754.

Blaustein, A., Hokit, D., O'Hara, R., & Holt, R. (1994). Pathogenic fungus contributes to amphibian losses in the Pacific Northwest. *Biological Conservation, 67,* 251–254.

Boccia, M. L. (1987). The physiology of grooming: A direct test of the tension reduction mechanism. *American Journal of Primatology, 12,* 330.

Boehm, C. (1992). Segmentary warfare and management of conflict: A comparison of East African chimpanzees and patrilineal-patrilocal humans. In A. H. Harcourt & F. B. M. de Waal (Eds.), *Coalitions and alliances in humans and other animals* (pp. 137–173). Oxford: Oxford University Press.

Boesch, C. (1994a). Chimpanzees-red colobus: A predator-prey system. *Animal Behaviour, 47,* 1135–1148.

Boesch, C. (1994b). Cooperative hunting in wild chimpanzees. *Animal Behaviour, 48,* 653–667.

Boesch, C. (1994c). Hunting strategies of Gombe and Tai chimpanzees. In R. Wrangham, W. C. McGrew, F. de Waal, & P. Heltne (Eds.), *Chimpanzee cultures* (pp. 77–91). Cambridge, MA: Harvard University Press.

Boesch, C., & Boesch, H. (1989). Hunting behavior of wild chimpanzees in the Tai National Park. *American Journal of Physical Anthropology, 78,* 547–573.

Boissy, A. (1995). Fear and fearfulness in animals. *Quarterly Review of Biology, 70,* 165–191.

Bolig, R., Price, C., O'Neill, P., & Suomi, S. (1992). Subjective assessment of reactivity level and personality traits of rhesus monkeys. *International Journal of Primatology, 13,* 287–306.

Bonner, J. T. (1980). *The evolution of culture in animals.* Princeton: Princeton University Press.

Borgerhoff-Mulder, M. (1988). Kipsigis bridewealth payments. In L. Betzig, M. Borger-Mulder, & P. Turke (Eds.), *Human reproductive behaviour: A Darwinian perspective* (pp. 65–82). Cambridge: Cambridge University Press.

Borgerhoff-Mulder, M. (1990). Kipsigis women's preferences for wealthy men: Evidence for female choice in mammals. *Behavioral Ecology and Sociobiology, 27,* 255–264.

Bottger, K. (1974) Zur biologie von *Sphaerodema grassei ghesquieri.* Studien an zentralafrikanischen belostomatiden (Heteroptera: Insecta) I. *Archive fur Hydrobiologie, 74,* 100–122.

Boucher, D. (Ed.). (1985). *The biology of mutualism: Ecology and evolution.* Oxford: Oxford University Press.

Boulton, M. J. (1991). Partner preferences in middle school children's playful fighting and chasing. *Ethology and Sociobiology, 12,* 177–193.

Boyd, R., & Richerson, P. J. (1985). *Culture and the evolutionary process.* Chicago: University of Chicago Press.

Boyse, E., Beauchamp, G., Yamazaki, K., & Bard, J. (1991). Genetic components of kin recognition in mammals. In P. G. Hepper (Ed.), *Kin recognition* (pp. 162–219). Cambridge: Cambridge University Press.

Bradbury, J. W., & Andersson, M. B. (Eds.). (1987). *Sexual selection: Testing the alternatives.* New York: Wiley-Interscience.

Bradbury, J. W., & Vehrencamp, S. (1977). Social organization and foraging in emballonurid bats. III. Mating systems. *Behavioral Ecology and Sociobiology, 2,* 1–17.

Bradbury, J. W., & Vehrencamp, S. (1998). *Animal communication.* New York: Oxford University Press.

Brandon, J., & Coss, R. (1982). Rapid dendritic spine stem shortening during one-trial learning: The honeybee's first orientation flight. *Journal of Neuroscience, 12,* 133–142.

Brandon, R. (1990). *Adaptation and environment.* Princeton: Princeton University Press.

Brantley, R., & Bass, A. H. (1994). Alternative male spawning tactics and acoustic signalling in the plainfin midshipman fish, *Porichthys notatus. Ethology, 96,* 213–232.

Bray, J. (1988). Children's development in during early remarriage. In E. Hetherington & J. Arasteh (Eds.), *Impact of divorce, single parenting and stepparenting on children* (pp. 279–298). Hillsdale, NJ: Lawrence Erlbaum Associates.

Brick, O. (1999). A test of the sequential assessment game: The effect of increased cost of sampling. *Behavioral Ecology, 10,* 726–732.

Briskie, J., Montgomery, R., & Birkhead, T. (1997). The evolution of sperm size in birds. *Evolution, 51,* 937–945.

Brodie, E., Moore, A., & Janzen, F. (1995). Visualizing and quantifying natural selection. *Trends in Ecology & Evolution, 10,* 313–318.

Bronstein, J. L. (1994). Our current understanding of mutualism. *Quarterly Review of Biology, 69,* 31–51.

Brooks, R. (1996). Copying and the repeatability of mate choice. *Behavioral Ecology and Sociobiology, 39,* 323–329.

Brown, C. R. (1986). Cliff swallow colonies as information centers. *Science, 234,* 83–85.

Brown, C. R. (1988). Social foraging in cliff swallows: Local enhancement, risk sensitivity and the avoidance of predators. *Animal Behaviour, 36,* 780–792.

Brown, C. R., & Brown, M. B. (1992). Ectoparasitism as a cause of natal dispersal in cliff swallows. *Ecology, 73,* 1718–1723.

Brown, C. R., & Brown, M. B. (2000). Heritable basis for choice of group size in a colonial bird. *Proceedings of the National Academy of Sciences, U.S.A., 97,* 14825–14830.

Brown, C. R., Brown, M. B., & Shaffer, M. L. (1991a). Food-sharing signals among socially foraging cliff swallows. *Animal Behaviour, 42,* 551–564.

Brown, G. R. (2001). Sex-biased investments in non-human primates: Can Trivers and Willard's theory be tested? *Animal Behaviour, 61,* 683–694.

Brown, J. L. (1969a). The buffer effect and productivity in tit populations. *American Naturalist, 103,* 347–354.

Brown, J. L. (1969b). Territorial behavior and population regulation in birds. *Wilson Bulletin, 81,* 293–329.

Brown, J. L. (1970). Cooperative breeding and altruistic behavior in the Mexican jay, *Apelocoma ultramarina. Animal Behaviour, 18,* 366–378.

Brown, J. L. (1974). Alternate routes to sociality in jays—with a theory for the evolution of altruism and communal breeding. *American Zoologist, 14,* 63–80.

Brown, J. L. (1975). *The evolution of behavior.* New York: W. W. Norton & Co.

Brown, J. L. (1982). Optimal group size in territorial animals. *Journal of Theoretical Biology, 95,* 793–810.

Brown, J. L. (1983). Cooperation: A biologist's dilemma. In J. S. Rosenblatt (Ed.), *Advances in the study of behavior* (Vol. 13; pp. 1–37). New York: Academic Press.

Brown, J. L. (1987). *Helping and communal breeding in birds.* Princeton: Princeton University Press.

Brown, J. L. (1994). Historical patterns in the study of avian social behavior. *Condor, 96,* 232–243.

Brown, J. L. (1998). The new heterozygosity theory of mate choice and the MHC. *Genetica, 104,* 215–221.

Brown, J. L., Brown, E., Brown, S. D., & Dow, D. D. (1982). Helpers: Effects of experimental removal on reproductive success. *Science, 215,* 421–422.

Brown, J. L., & Eklund, A. (1994). Kin recognition and the major histocompatibility complex: An integrative review. *American Naturalist, 143,* 435–461.

Brown, J. S. (1998). Game theory and habitat selection. In L. A. Dugatkin & H. K. Reeve (Eds.), *Game theory and animal behavior* (4th ed., pp. 188–220). New York: Oxford University Press.

Brown, J. S., & Rosenzweig, M. L. (1985). Habitat selection in slowly regenerating environments. *Journal of Theoretical Biology, 123,* 151–171.

Brown, M., Hopkins, W., & Keynes, R. (1991b). *Essentials of neural development.* Cambridge: Cambridge University Press.

Brownlee, A. (1954). Play in domestic cattle in Britain: An analysis of its nature. *British Veterinary Journal, 110,* 48–68.

Bruner, J., Jolly, A., & Sylva, K. (Eds.). (1976). *Play: Its role in development and evolution.* New York: Basic Books.

Brunton, D. H. (1990). The effects of nesting stage, sex and type of predator on parental defense by killdeer (*Charadrius vociferous*): Testing models of avian parental defense. *Behavioral Ecology and Sociobiology, 26,* 181–190.

Brush, A. (1990). Metabolism of carotenoid pigments in birds. *Federation of American Societies for Experimental Biology Journal, 4,* 2969–2977.

Brush, A., & Power, D. (1976). House finch pigmentation, carotenoid pigmentation and the effect of diet. *Auk, 93,* 725–739.

Budiansky, S. (1998). *If a lion could talk: Animal intelligence and the evolution of consciousness.* New York: Free Press.

Bull, J. J. (1980). Sex determination in reptiles. *Quarterly Review of Biology, 55,* 3–21.

Bull, J. J. (1983). *Evolution of sex determining mechanisms.* Menlo Park, CA: Benjamin Cummings.

Burgess, J. W. (1976). Social spiders. *Scientific American,* March 1976, 100–106.

Burghardt, G. (1988). Precocity, play and the ectotherm-endotherm transition. *Behaviour, 36,* 246–257.

Burghardt, G. (1998). Play. In G. Greenberg & M. Haraway (Eds.), *Comparative psychology: A handbook* (Vol. 13; pp. 725–735). New York: Garland Publishing.

Burley, N. (1985). Leg band color and mortality patterns in captive breeding populations of zebra finches. *Auk, 30,* 647–651.

Burley, N. (1986a). Sex ratio manipulation in color banded populations of zebra finches. *Evolution, 40,* 1191–1206.

Burley, N. (1986b). Sexual selection for aesthetic traits in species with biparental care. *American Naturalist, 127,* 415–445.

Burley, N. (1988). Wild zebra finches have band-colour preferences. *Animal Behaviour, 36,* 1235–1237.

Burley, N., Kranntzberg, G., & Radman, P. (1982). Influence of colour-banding on the conspecific preferences of zebra finches. *Animal Behaviour, 30,* 444–445.

Buss, A., & Plomin, R. (1986). The EAS approach to temperament. In R. Plomin & J. Dunn (Eds.), *The study of temperament: Changes, continuities and challenges* (pp. 67–79). Hillsdale, NJ: Lawrence Erlbaum Associates.

Busse, C. (1978). Do chimps hunt cooperatively? *American Naturalist, 112,* 767–770.

Butcher, G., & Rohwer, S. (1989). The evolution of conspicuous and distinctive coloration for communication in birds. *Current Ornithology, 6,* 51–107.

Byers, J. (1984). Play in ungulates. In P. K. Smith (Ed.), *Play in animals and humans* (pp. 43–65). Oxford: Blackwell Scientific Publications.

Byers, J. (1998). Biological effects of locomotor play: Getting into shape or something more specific. In M. Bekoff & J. Byers (Eds.), *Animal play: Evolutionary, comparative and ecological perspectives* (pp. 205–220). Cambridge: Cambridge University Press.

Byers, J., & Walker, C. (1995). Refining the motor training hypothesis for the evolution of play. *American Naturalist, 146,* 25–40.

Bygott, J. D. (1979). Agonistic behavior, dominance and social structure in wild chimpanzees of the Gombe National Park. In D. A. Hamburg & E. R. McCown (Eds.), *The great apes.* Menlo Park, CA: Benjamin Cummings.

Bygott, J. D., Bertram, B., & Hanby, J. P. (1979). Male lions in large coalitions gain reproductive advantages. *Nature, 282,* 839–841.

Byrne, R. W. (1993). Do larger brains mean greater intelligence? *Behavioral and Brain Sciences, 16,* 696–697.

Cade, T. J. (1982). *The falcons of the world.* Ithaca, NY: Comstock/Cornell University Press.

Caldwell, R. (1986). Withholding information on sexual condition as a competitive mechanism. In L. Drickamer (Ed.), *Behavioral ecology and population biology* (pp. 83–88). Toulouse: Privat.

Caldwell, R. (1992). Recognition, signalling and reduced aggression between former mates in a stomatopod. *Animal Behaviour, 44,* 1 –19.

Campbell, A., Muncer, S., & Odber, J. (1997). Aggression and testosterone: Testing a bio-social model. *Aggressive Behavior, 23,* 229–238.

Cannon, W. (1915). *Bodily changes in pain, hunger, fear and rage.* New York: Appleton.

Capaldi, E., Robinson, G., & Fahrbach, S. (1999). Neuroethology of spatial learning: The birds and the bees. *Annual Review of Psychology, 50,* 651–682.

Caraco, T. (1979). Time budgeting and group size: A test of theory. *Ecology, 60,* 618–627.

Caraco, T. (1980). On foraging time allocation in a stochastic environment. *Ecology, 61,* 119–128.

Caraco, T. (1981). Risk-sensitivity and foraging groups. *Ecology, 62,* 527–531.

Caraco, T. (1983). White-crowned sparrows: Foraging preferences in a risky environment. *Behavioral Ecology and Sociobiology, 12,* 63–69.

Caraco, T., Martindale, S., & Whitham, M. (1980). An empirical demonstration of risk sensitive foraging preferences. *Animal Behaviour, 28,* 820–830.

Carlier, P., & Lefebvre, L. (1996). Differences in individual learning between group-foraging and territorial Zenaida doves. *Behaviour, 133,* 1197–1207.

Carlson, B. A., Hopkins, C. D., & Thomas, P. (2000). Androgen correlates of socially induced changes in the electric organ discharge waveform of a mormyrid fish. *Hormones and Behavior, 38,* 177–186.

Caro, T. M. (1980). Effects of the mother, object play and adult experience on predation in cats. *Behavioural and Neural Biology, 29,* 29–51.

Caro, T. M. (1994a). *Cheetahs of the Serengeti Plains.* Chicago: University of Chicago Press.

Caro, T. M. (1994b). Ungulate antipredator behavior: Preliminary and comparative data from African Bovids. *Behaviour, 128,* 189–228.

Caro, T. M. (1995a). Pursuit-deterrence revisited. *Trends in Ecology and Evolution, 10,* 500–503.

Caro, T. M. (1995b). Short-terms costs and correlates of play in cheetahs. *Animal Behaviour, 49,* 333–345.

Caro, T. M., & Bateson, P. (1986). Organization and ontogeny of alternative tactics. *Animal Behaviour, 34,* 1483–1499.

Caro, T. M., & Hauser, M. D. (1992). Is there teaching in nonhuman animals? *Quarterly Review of Biology, 67,* 151–174.

Caro, T. M., Lombardo, L., Goldizen, A. W., & Kelly, M. (1995). Tail-flagging and other antipredator signals in white deer: New data and synthesis. *Behavioral Ecology, 6,* 442–450.

Carpenter, C. R. (1934). A field study of the behavior and social relations of howling monkeys. *Comparative Psychology Monographs, 10,* 1–168.

Carpenter, C. R. (1942). Sexual behavior of free-ranging rhesus monkeys, *M. mulatta. Journal of Comparative Psychology, 33,* 113–162.

Catchpole, C., & Slater, P. (Eds.). (1995). *Bird song: Biological themes and variation.* New York: Cambridge University Press.

Cavalli-Sforza, L. L., & Feldman, M. W. (1981). *Cultural transmission and evolution: A quantitative approach.* Princeton: Princeton University Press.

Chapais, B. (1992). The role of alliances in social inheritance of rank among female primates. In A. H. Harcourt, & F. B. M. de Waal (Eds.), *Coalitions and alliances in humans and other animals* (pp. 29–59). Oxford: Oxford University Press.

Charlesworth, B. (1980). *Evolution in age-structured populations.* Cambridge: University of Cambridge Press.

Charnov, E. L. (1973). *Optimal foraging: Some theoretical explorations.* Seattle: University of Washington Press.

Charnov, E. L. (1976). Optimal foraging, the marginal value theorem. *Theoretical Population Biology, 9,* 129–136.

Charnov, E. L. (1977). An elementary treatment of the genetical theory of kin selection. *Journal of Theoretical Biology, 66,* 541–550.

Charnov, E. L. (1978). Evolution of eusocial behavior: Offspring choice or parental manipulation? *Journal of Theoretical Biology, 75,* 451–465.

Charnov, E. L. (1982). *The theory of sex allocation.* Princeton: Princeton University Press.

Charnov, E. L., & Krebs, J. R. (1975). The evolution of alarm calls: Altruism or manipulation? *American Naturalist, 109,* 107–112.

Chase, I. D. (1974). Models of hierarchy formation in animal societies. *Behavioral Science, 19,* 374–382.

Chase, I. D. (1982). Dynamics of hierarchy formation: The sequential development of dominance relationships. *Behaviour, 80,* 218–240.

Chase, I. D. (1985). The sequential analysis of aggressive acts during hierarchy formation: An application of the "jigsaw puzzle" approach. *Animal Behaviour, 33,* 86–100.

Cheney, D. L. (1977). The acquisition of rank and the development of reciprocal alliances among free-ranging immature baboons. *Behavioral Ecology and Sociobiology, 2,* 303–318.

Cheney, D. L. (1978). The play partners of immature baboons. *Animal Behaviour, 26,* 1038–1050.

Cheney, D. L., & Seyfarth, R. (1981). Selective forces affecting the predator calls of vervet monkeys. *Animal Behaviour, 28,* 362–367.

Cheney, D. L., & Seyfarth, R. (1985). Vervet alarm calls: Manipulation through shared information. *Behaviour, 94,* 150–166.

Cheney, D. L., & Seyfarth, R. M. (1990). *How monkeys see the world.* Chicago: University of Chicago Press.

Chisholm, J. S., & Burbank, V. K. (1991). Monogamy and polygyny in Southeast Arnhem-Land: Male coercion and female choice. *Ethology and Sociobiology, 12,* 291–313.

Chivers, D. P., Wisenden, B. D., & Smith, R. J. F. (1996). Damselfly larvae learn to recognize predators from chemical cues in the predator's diet. *Animal Behaviour, 52,* 315–320.

Christie, J. (1983). Female choice in the resource defence mating system of the sand fiddler crab, *Uca pugilator. Behavioral Ecology and Sociobiology, 12,* 169–180.

Church, S., Bennett, A., Cuthill, I., & Partridge, J. (1998). Ultraviolet cues affect the foraging behaviour of blue tits. *Proceedings of the Royal Society of London, 265,* 1509–1514.

Ciofi, C., & Swingland, I. R. (1997). Environmental sex determination in reptiles. *Applied Animal Behaviour, 51,* 251–265.

Claar, J., Klaver, R., & Servheen, C. (1986). Grizzly bear management on the Flathead Indian Reservation, Montana. *International Conference of Bear Research and Management, 6,* 203–208.

Clark, A. B., & Ehlinger, T. J. (1987). Pattern and adaptation in individual behavioral differences. In P. P. G. Bateson & P. H. Klopfer (Eds.), *Perspectives in ethology* (Vol. 7; pp. 1–45). New York: Plenum Press.

Clark, L., & Mason, J. (1985). Use of nest material as insecticidal and antipathogenic agents by the European starling. *Oecologia, 67,* 169–176.

Clark, L., & Mason, J. (1988). Effect of biologically-active plants used as nest material and the derived benefit to starling nestlings. *Oecologia, 77,* 174–180.

Clark, M., Crews, D., & Galef, B. (1991). Sex steroid levels of pregnant and fetal Mongolian gerbils. *Physiology and Behavior, 49,* 239–243.

Clark, M., & Galef, B. (1999). A testosterone mediated trade-off between parental care and sexual effort in male Mongolian gerbils, *Meriones unguiculatus. Journal of Comparative Psychology, 113,* 388–395.

Clark, M., & Galef, B. (2000). Why some male Mongolian gerbils may help at the nest: Testosterone, asexuality and alloparenting. *Animal Behaviour,* 59, 801–806.

Clark, M., vom Saal, F., & Galef, B. (1992). Fetal intrauterine position correlates with endogenous testosterone levels of adult Mongolian gerbils. *Physiology and Behavior, 51,* 957–960.

Clark, M., Vonk, J., & Galef, B. (1998). Intrauterine position, parenting and nest-sites attachment in male Mongolian gerbils. *Developmental Psychobiology, 32,* 177–181.

Clarke, J. A. (1983). Moonlight's influence on predator/prey interactions between short-eared owls (*Asio flammeus*) and deer mice (*Peromyscus maniculatus*). *Behavioral Ecology and Sociobiology, 13,* 205–209.

Clayton, D. H. (1990). Mate choice in experimentally parasitized rock doves: Lousy males lose. *American Zoologist, 30,* 251–262.

Clayton, N. S., & J., K. (1995). Memory in food-storing birds: From behaviour to brain. *Current Opinions in Neurobiology,* 5, 149–154.

Clement, T., Feltus, J., Kaiser, D., & Zentall, T. (2000). "Work ethic" in pigeons: Reward value is directly related to the effort or time required to obtain the reward. *Psychonomic Bulletin and Review, 7,* 100–106.

Clements, K. C., & Stephens, D. W. (1995). Testing models of non-cooperation: Mutualism and the prisoner's dilemma. *Animal Behaviour, 50,* 527–535.

Close, R. (1972). Dynamic properties of fast and slow skeletal muscles of the rat during development. *Journal of Physiology, 173,* 74–95.

Clunie, F. (1976). Jungle mynah "anting" with a millipede. *Notornis, 23,* 77.

Clutton-Brock, T. H. (Ed.). (1988). *Reproductive success.* Chicago: University of Chicago Press.

Clutton-Brock, T. H. (Ed.). (1991). *The evolution of parental care.* Princeton: Princeton University Press.

Clutton-Brock, T. H., & Albon, S. (1979). The roaring of red deer and the evolution of honest advertising. *Behaviour, 49,* 145–169.

Clutton-Brock, T. H., Albon, S., Gibson, R. M., & Guinness, F. (1979). The logical stag: Adaptive aspects of fighting in the red deer. *Animal Behaviour, 27,* 211–225.

Clutton-Brock, T. H., Albon, S., & Guinness, F. (1984). Maternal dominance, breeding success and birth sex ratios in red deer. *Nature, 308,* 358–360.

Clutton-Brock, T. H., & Godfrey, C. (1991). Parental investment. In J. Krebs & N. Davies (Eds.), *Behavioural ecology* (3rd ed., pp. 234–262). Oxford: Blackwell Scientific Publications.

Clutton-Brock, T. H., & Harvey, P. (1980). Primates, brain and ecology. *Journal of Zoology, London, 190,* 309–323.

Clutton-Brock, T. H., & Iason, G. (1986). Sex ratio variation in mammals. *Quarterly Review of Biology, 61,* 339–374.

Clutton-Brock, T. H., & Langley, P. (1997). Persistent courtship reduces male and female longevity in captive tsetse flies *Glossina morsitans morsitans* Westwood (Diptera: Glossinidae). *Behavioral Ecology, 8,* 392–395.

Clutton-Brock, T. H., & McComb, K. (1993). Experimental tests of copying and mate choice in fallow deer (*Dama dama*). *Behavioral Ecology, 4,* 191–193.

Coccaro, E. (1992). Impulsive aggression and central serotonergic system function in humans: An example of a dimensional brain-behavior relationship. *International Clinical Psychopharmacology, 7,* 3–12.

Coleman, K., & Wilson, D. S. (1998). Shyness and boldness in pumpkinseed sunfish: Individual differences are context-specific. *Animal Behaviour, 56,* 927–936.

Connor, R. C. (1992). Dolphin alliances and coalitions. In A. H. Harcourt & F. B. M. de Waal (Eds.), *Coalitions and alliances in humans and other animals* (pp. 415–443). Oxford: Oxford University.

Connor, R. C. (1995). The benefits of mutualism: A conceptual framework. *Biological Review, 70,* 427–457.

Connor, R. C., Heithaus, M., & Barre, L. (1999). Superalliance of bottlenose dolphins. *Nature, 397,* 571–572.

Connor, R. C., Smolker, R. A., & Richards, A. F. (1992). Two levels of alliance formation among male bottleneck dolphins. *Proceedings of the National Academy of Sciences, U.S.A., 89,* 987–990.

Conover, M. R., Reese, J. G., & Brown, A. D. (2000). Costs and benefits of subadult plumage in mute swans: Testing hypotheses for the evolution of delayed plumage maturation. *American Naturalist, 156,* 193–200.

Cook, M., Mineka, S., Wolkenstein, B., & Laitsch, K. (1985). Observational conditioning of snake fear in unrelated rhesus monkeys. *Journal of Abnormal Psychology, 94,* 591–610.

Cooke, F., Finney, G. H., & Rockwell, R. (1976). Assortative mating in lesser snow geese (*Anser carulescens*). *Behavior Genetics, 6,* 127–139.

Cooke, F., & McNally, C. (1975). Mate selection and color preferences in lesser snow geese. *Behaviour, 53,* 151–170.

Cooke, F., Mirsky, P., & Seiger, M. (1972). Color preferences in lesser snow geese and their possible role in mate selection. *Canadian Journal of Zoology, 50,* 529–536.

Cordts, R., & Partridge, L. (1996). Courtship reduces longevity of male *Drosophila melanogaster*. *Animal Behaviour, 52,* 269–278.

Coss, R. G. (1991). Context and animal behavior III. The relationship between early development and evolutionary persistence of ground squirrel antisnake behavior. *Ecological Psychology, 3,* 277–315.

Coss, R. G., Brandon, J., & Globus, A. (1980). Changes in the morphology of dendritic spines on honeybee calycal interneurons associated with cumulative nursing and foraging experiences. *Brain Research, 192,* 49–59.

Coss, R. G., & Owings, D. H. (1985). Restraints on ground squirrel antipredator behavior: Adjustments over multiple time scales. In T. D. Johnston & A. T. Pietrewicz (Eds.), *Issues in the ecological study of learning* (pp. 167–200). Hillsdale, NJ: Lawrence Erlbaum Associates.

Coultier, S., Beaugrand, J. P., & Lague, P. C. (1995). The role of individual differences and patterns of resolution in the formation of dominance orders in domestic hen triads. *Behavioural Processes, 38,* 227–239.

Cowie, R. (1977). Optimal foraging in great tits. *Nature, 268,* 137–139.

Cox, C., & Le Boeuf, B. J. (1977). Female incitation of male competition: A mechanism in sexual selection. *American Naturalist, 111,* 317–335.

Cozza, K., Fico, R., Batistini, L., & Rogers, E. (1996). The damage-conservation interface illustrated by predation on domestic livestock in Italy. *Biological Conservation, 78,* 329–336.

Crawley, M. (1983). *Herbivory.* Oxford: Blackwell Scientific Publications.

Creel, S. R. (1990). How to measure inclusive fitness. *Proceedings of the Royal Society of London, 241,* 229–231.

Creel, S. R. (1996). Variation in reproductive suppression among dwarf mongooses: Interplay between mechanisms and evolution. In N. G. Solomon & J. A. French (Eds.), *Cooperative breeding in mammals.* Cambridge: Cambridge University Press.

Creel, S. R. (2001). Cooperative hunting and sociality in African wild dogs, *Lycaon pictus.* In L. Dugatkin (Ed.), *Model systems in behavioral ecology.* Princeton: Princeton University Press.

Creel, S. R., Creel, N., Wildt, D. E., & Monfort, S. L. (1992). Behavioral and endocrine mechanisms of reproductive suppression in Sergenti dwarf mongooses. *Animal Behaviour, 43,* 231–245.

Creel, S. R., Monfort, S. L., Wildt, D. E., & Waser, P. M. (1991). Spontaneous lactation is an adaptive result of pseudopregnancy. *Nature, 351,* 660–662.

Creel, S. R., & Waser, P. (1991). Failure of reproductive suppression in dwarf mongooses: Accident or adaptation? *Behavioral Ecology, 2,* 7–15.

Crews, D., & Moore, M. (1986). Evolution of mechanisms controlling mating behavior. *Science, 231,* 121–125.

Crook, J. H., & Crook, S. (1988). Tibetan polyandry: Problems of adaptation and fitness. In L. Betzig, M. Borgerhoff-Mulder, & P. Turke (Eds.), *Human reproductive behaviour: A Darwinian perspective* (pp. 97–114). Cambridge: Cambridge University Press.

Crowl, T., & Covich, A. (1990). Predator-induced life-history traits in a freshwater snail. *Science, 247,* 949–951.

Crowley, P. H., Gillett, S., & Lawton, J. H. (1988). Contests between larval damselflies: Empirical steps toward a better ESS model. *Animal Behaviour, 36,* 1496–1510.

Crozier, R. (1987). Genetic aspects of kin recognition: Concepts, models and synthesis. In D. C. Fletcher & C. D. Michener (Eds.), *Kin recognition in animals* (pp. 55–74). Chichester: Wiley.

Curio, E. (1978). The adaptive significance of avian mobbing. I. Teleonomic hypotheses and predictions. *Zeitschrift fur Tierpsychologie, 48,* 175–183.

Curio, E., Ernest, U., & Vieth, W. (1978a). The adaptive significance of avian mobbing. II. Cultural transmission and enemy recognition in blackbirds: Effectiveness and some constraints. *Zeitschrift fur Tierpsychologie, 48,* 185–202.

Curio, E., Ernest, U., & Vieth, W. (1978b). Cultural transmission of enemy recognition: One function of mobbing. *Science, 202,* 899–901

Curio, E., Klump, G., & Regelmann, K. (1983). An antipredator response in the great tit *(Parus major)*: Is it tuned to predator risk? *Oecologia, 60,* 83–88.

Curio, E., & Regelmann, K. (1985). The behavioral dynamics of great tits (*Parus major*) approaching a predator. *Zeitschrift fur Tierpsychologie, 69,* 3–18.

Curio, E., & Regelmann, K. (1986). Predator harassment implies a real deadly risk: A reply to Hennessy. *Ethology, 72,* 75–78.

Currie, C. R., Mueller, U. G., & Malloch, D. (1999a). The agricultural pathology of ant fungus gardens. *Proceedings of the National Academy of Sciences, U.S.A., 96,* 7998–8002.

Currie, C. R., Scott, J. A., Summerbell, R. C., & Malloch, D. (1999b). Fungus-growing ants use antibiotic-producing bacteria to control garden parasites. *Nature, 398,* 701–704.

Curtsinger, J. W. (1986). Stay times in *Scatophaga* and the theory of evolutionarily stable strategies. *American Naturalist, 128,* 130–136.

Daly, M., & Wilson, M. (1988). *Homicide.* New York: Aldine de Gruyter.

Daly, M., & Wilson, M. (1994). Some differential attributes of lethal assaults on small children by stepfathers versus genetic fathers. *Ethology and Sociobiology, 15,* 207–217.

Daly, M., & Wilson, M. (1996). Violence against stepchildren. *Current Directions in Psychological Science, 5,* 77–81.

Daly, M., & Wilson, M. (1999). *The truth about Cinderella: A Darwinian view of parental love.* New Haven: Yale University Press.

Darwin, C. (1845). *Journal of researches into the natural history and geology of the countries visited during the voyage of H.M.S. 'Beagle' round the world, under the command of Captain FitzRoy, R. N.* (2nd ed.). London: J. Murray.

Darwin, C. (1859). *On the origin of species.* London: J. Murray.

Darwin, C. (1868). *The variation of animals and plants under domestication* (Vol. 1–2). London: J. Murray.

Darwin, C. (1871). *The descent of man and selection in relation to sex.* London: J. Murray.

Darwin, C. (1872). *The expression of emotions in man and animals.* London: J. Murray.

Dasser, V. (1987). Slides of group members as representations of the real animals. *Ethology, 76,* 65–73.

Dasser, V. (1988). A social concept in Java monkeys. *Animal Behaviour, 36,* 225–230.

David, P., Hingle, A., Fowler, K., & Pomiankowski, A. (1999). Measurement bias and fluctuating asymmetry estimates. *Animal Behaviour, 57,* 251–253.

Davies, N. B. (1976). Food, flocking and territorial behavior in the pied wagtail *Motacilla alba* in winter. *Journal of Animal Ecology, 45,* 235–254.

Davies, N. B. (1978). Territorial defence in the speckled wood butterfly (*Pararge aegeria*): The resident always wins. *Animal Behaviour, 26,* 138–147.

Davies, N. B. (1986). Reproductive success of dunnocks in a variable mating system.II. Factors influencing provisioning rate, nestling weight and fledgling success. *Journal of Animal Ecology, 55,* 123–138.

Davies, N. B. (1989). Sexual conflict and the polygyny threshold. *Animal Behaviour, 38,* 226–234.

Davies, N. B. (1991). Mating systems. In J. R. Krebs & N. B. Davies (Eds.), *Behavioural ecology* (pp. 263–299). Oxford: Blackwell Scientific Publications.

Davies, N. B. (1992). *Dunnock behavior and social evolution.* Oxford: Oxford University Press.

Davies, N. B., & Halliday, T. (1978). Deep croaks and fighting assessment in toads, *Bufo bufo. Nature, 391,* 56–58.

Davies, N. B., & Houston, A. I. (1981). Owners and satellites: The economics of territory defence in the pied wagtail *Motacilla alba. Journal of Animal Ecology, 50,* 157–180.

Davies, N. B., & Houston, A. I. (1984). Territory economics. In J. R. Krebs & N. B. Davies (Eds.), *Behavioral ecology: An evolutionary approach.* Sunderland, MA: Sinauer Associates.

Davies, N. B., & Lundberg, A. (1984). Food distribution and a variable mating system in the dunnock. *Journal of Animal Ecology, 53,* 895–912.

Davis, D. (1942). The phylogeny of social nesting in the Crotophaginae. *Quarterly Review of Biology, 17,* 115–134.

Davis, J., & Daly, M. (1997). Evolutionary theory and the human family. *Quarterly Review of Biology, 72,* 407–435.

Davis, R. (1993). Mushroom bodies and Drosophila learning. *Neuron, 11,* 1–14.

Dawkins, R. (1976). *The selfish gene.* New York: Oxford University Press.

Dawkins, R. (1979). Twelve misunderstandings of kin selection. *Zeitschrift fur Tierpsychologie, 51,* 184–200.

Dawkins, R. (1982). *The extended phenotype.* Oxford: Oxford University Press.

Dawkins, R. (1987). *The blind watchmaker.* New York: W. W. Norton & Co.

Dawkins, R., & Krebs, J. R. (1978). Animal signals: Information for manipulation? In J. R. Krebs & N. B. Davies (Eds.), *Behavioural ecology* (pp. 282–315). Sunderland, MA: Sinauer Associates.

DeCola, J., & Faneslow, M. (1995). Differential inflation with short and long term CS-US intervals: Evidence of a nonassociative process in long-delay taste avoidance. *Animal Learning and Behavior, 23,* 154–163.

Dejong, G. (1994). The fitness of fitness concepts and the description of natural selection. *Quarterly Review of Biology, 69,* 3–29.

de Kloet, E. R., Oitzl, M. S., & Joels, M. (1999). Stress and cognition: Are corticosteroids good or bad guys? *Trends in Neuroscience, 22,* 422–426.

De Laet, J. (1985). Dominance and anti-predator behavior of great tits *Parus major*: A field study. *Ibis, 127,* 372–377.

Delehanty, D., Fleischer, R., Colwell, M., & Oring, L. (1998). Sex-role reversal and the absence of extra-pair fertilization in Wilson's phalaropes. *Animal Behaviour, 55,* 995–1002.

Delph, L., & Havens, K. (1998). Pollen competition in flowering plants. In T. Birkhead & A. Møller (Eds.), *Sperm competition and sexual selection* (pp. 149–173). London: Academic Press.

DeMartini, E. (1990). Annual variations in fecundity, egg size and condition of the plainfin midshipman (*Porichthys notatus*). *Copeia, 3,* 850–855.

Dennett, D. C. (1978). Intentional systems. In D. C. Dennett (Ed.), *Brainstorms: Philosophical essays on mind and philosophy* (pp. 3–22). Cambridge, MA: MIT Press.

Dennett, D. C. (1987). *The intentional stance.* Cambridge, MA: Bradford/MIT Press.

de Quervain, D. J. F., Roozendaal, B., & McGaugh, J. L. (1998). Stress and glucocorticoids impair retrieval of long-term spatial memory. *Nature, 394,* 787–790.

DeSalle, R., & Schierwater, B. (Eds.). (1998). *Molecular approaches to ecology and evolution.* Basel: Birkhauser Verlag.

de Waal, F. (1989a). *Peacemaking among primates.* Cambridge, MA: Harvard University Press.

de Waal, F. B. M. (1989b). Food sharing and reciprocal obligations among chimpanzees. *Journal of Human Evolution, 18,* 433–459.

de Waal, F. B. M. (1992). Coalitions as part of reciprocal relations in the Arnhem chimpanzee colony. In A. H. Harcourt & F. B. M. de Waal (Eds.), *Coalitions and alliances in humans and other animals* (pp. 233–257). Oxford: Oxford University Press.

Dewsbury, D. (1992). On problems studied in ethology, comparative psychology and animal behavior. *Ethology, 92,* 89–107.

Dewsbury, D. (1994). On the utility of the proximate-ultimate distinction in the study of animal behavior. *Ethology, 96,* 63–68.

Dhondt, A. A. (1987). Polygynous blue tits and monogamous great tits: Does the polygyny-threshold model hold. *American Naturalist 129,* 213–220.

Dickemann, M. (1979). The ecology of mating systems in hypergynous dowry societies. *Social Science Information, 18,* 163–195.

Dill, L. (1983). Adaptive flexibility in the foraging behavior of fishes. *Canadian Journal of Fisheries and Aquatic Sciences, 40,* 398–408.

Dill, L., & Fraser A. (1984). Risk of predation and the feeding behaviour of juvenile Coho salmon (*Oncorhynchus kisutch*). *Behavioral Ecology and Sociobiology, 16,* 65–72.

DiMarco, F. P., & Hanlon, R. T. (1997). Agonistic behavior in the squid *Loligo plei* (*Loliginidae, Teuthoidea*): Fighting tactics and the effects of size and resource value. *Ethology, 103,* 89–108.

Dingle, H. (1996). *Migration: The biology of life on the move.* New York: Oxford University Press.

Dingle, H., & Caldwell, R. (1972). Reproductive and maternal behavior of the mantis shrimp *Gonodactylus bredini Manning* (*Crustacea: Stomatopoda*). *Biological Bulletin, 142,* 417–426.

Dittman, A., & Quinn, T. P. (1996). Homing in Pacific salmon: Mechanisms and ecological basis. *Journal of Experimental Biology, 199,* 83–91.

Dodson, J. J. (1988). The nature and role of learning in the orientation and migratory behavior of fishes. *Environmental Biology of Fishes, 23,* 161–182.

Dominey, W. J. (1983). Mobbing in colonially nesting fish, especially the bluegill, *Lepomis macrochirus. Copeia, 4,* 1086–1088.

Domjan, M. (1987). Photoperiodic and endocrine control of social proximity in male Japanese quail (*Coturnix coturnix japonica*). *Behavioral Neuroscience, 15,* 147–153.

Domjan, M. (1992). Adult learning and mate choice: Possibilities and experimental evidence. *American Zoologist, 32,* 48–61.

Domjan, M. (1998). *The principles of learning and behavior.* Pacific Grove, CA: Brooks/Cole Publishing Co.

Domjan, M., Blesbois, E., & Williams, J. (1998). The adaptive significance of sexual conditioning: Pavlovian control of sperm release. *Psychological Science, 9,* 411–415.

Domjan, M., & Hall, S. (1986). Determinants of social proximity in Japanese quail (*Coturnix coturnix japonica*): Male behavior. *Journal of Comparative Psychology, 100,* 59–67.

Domjan, M., & Hollis, K. L. (1988). Reproductive behavior: A potential model system for adaptive specializations in learning. In R. C. Bolles & M. D. Beecher (Eds.) *Evolution and learning* (pp. 213–237). Hillsdale, NJ: Lawrence Erlbaum Associates.

Domjan, M., & Holloway, K. (1997). Sexual learning. In G. Greenberg & M. Harraway (Eds.), *The encyclopedia of comparative psychology* (pp. 602–612). New York: Garland.

Domjan, M., Lyons, R., North, N., & Bruell, J. (1986). Sexual Pavlovian conditioned approach behavior in male Japanese quail (*Coturnix coturnix japonica*). *Journal of Comparative Psychology, 100,* 413–421.

Donaldson, T. J. (1984). Mobbing behavior by *Stegastes albifasciatus* (*Pomacentridae*), a territorial mosiac of damselfish. *Japanese Journal of Ichthyology, 31,* 345–348.

Douglas-Hamilton, I., & Douglas-Hamilton, O. (1975). *Among the elephants.* Glasgow: Collins.

Doutrelant, C., McGregor, P., & Oliveira, R. (2001). The effect of an audience on intra-sexual communication in male Siamese fighting fish. *Betta splendens. Behavioral Ecology, 12,* 283–286.

Downhower, J. F., & Armitage, K. (1971). The yellow-bellied marmot and the evolution of polygamy. *American Naturalist, 105,* 355–370.

Drent, P., van Oers, K. & van Noordwijk, A. J. (2002). Realized heritability of personalities in the great tit (*Parus major*). *Proceedings of the Royal Society of London, 270,* 45–51.

Driscoll, P., & Kugler, P. (1984). Genetic and histological aspects of stomach lesions induced by systematic injection of phenylbutazone in the rat. *Experienta, 40,* 967–969.

Drummond, H., & Canales, C. (1998). Dominance between booby nestlings involves winner and loser effects. *Animal Behaviour, 55,* 1669–1676.

Drummond, H., & Osorno, J. L. (1992). Training siblings to be submissive losers: Dominance between booby nestlings. *Animal Behaviour, 44,* 881–893.

Duellman, W. E., & Trueb, L. (1994). *Biology of amphibians.* Baltimore: Johns Hopkins University Press.

Dugatkin, L. A. (1991). Dynamics of the TIT FOR TAT strategy during predator inspection in guppies. *Behavioral Ecology and Sociobiology, 29,* 127–132.

Dugatkin, L. A. (1992a). The evolution of the con artist. *Ethology and Sociobiology, 13,* 3–18.

Dugatkin, L. A. (1992b). Sexual selection and imitation: Females copy the mate choice of others. *American Naturalist, 139,* 1384–1389.

Dugatkin, L. A. (1992c). Tendency to inspect predators predicts mortality risk in the guppy, *Poecilia reticulata. Behavioral Ecology, 3,* 124–128.

Dugatkin, L. A. (1995). Formalizing Allee's idea on dominance hierarchies: An intra-demic selection model. *American Naturalist, 146,* 154–160.

Dugatkin, L. A. (1996a). Copying and mate choice. In C. Heyes & B. G. Galef (Eds.), *Social learning in animals: The roots of culture* (pp. 85–105). New York: Academic Press.

Dugatkin, L. A. (1996b). The interface between culturally-based preferences and genetic preferences: Female mate choice in *Poecilia reticulata. Proceedings of the National Academy of Sciences, U.S.A., 93,* 2770–2773.

Dugatkin, L. A. (1997a). *Cooperation among animals: An evolutionary perspective.* New York: Oxford University Press.

Dugatkin, L. A. (1997b). Winner effects, loser effects and the structure of dominance hierarchies. *Behavioral Ecology, 8,* 583–587.

Dugatkin, L. A. (1998a). Breaking up fights between others: A model of intervention behaviour. *Proceedings of the Royal Society of London, 265,* 433–437.

Dugatkin, L. A. (1998b). A model of coalition formation in animals. *Proceedings of the Royal Society of London, 265,* 2121–2125.

Dugatkin, L. A. (2000). *The imitation factor: Evolution beyond the gene.* New York: Free Press.

Dugatkin, L. A. (2001a). Bystander effects and the structure of dominance hierarchies. *Behavioral Ecology, 12,* 348–352.

Dugatkin, L. A. (Ed.). (2001b). *Model systems in behavioral ecology: Integrating conceptual, theoretical and empirical perspectives.* Princeton: Princeton University Press.

Dugatkin, L. A. (2001c). Preface to *Model systems in behavioral ecology.* In L. A. Dugatkin (Ed.), *Model systems in behavioral ecology: Integrating conceptual, theoretical and empirical perspectives.* Princeton: Princeton University Press.

Dugatkin, L. A., & Alfieri, M. (1991a). Guppies and the tit for tat strategy: Preference based on past interaction. *Behavioral Ecology and Sociobiology, 28,* 243–246.

Dugatkin, L. A., & Alfieri, M. (1991b). TIT FOR TAT in guppies: The relative nature of cooperation and defection during predator inspection. *Evolutionary Ecology, 5,* 300–309.

Dugatkin, L. A., & Alfieri, M. (1992). Interpopulational differences in the use of the tit for tat strategy during predator inspection in the guppy. *Evolutionary Ecology, 6,* 519–526.

Dugatkin, L. A., & Alfieri, M. (2003). Boldness, behavioral inhibition and learning: An evolutionary approach. *Ethology Ecology & Evolution,* in press.

Dugatkin, L. A., FitzGerald, G. J., & Lavoie, J. (1994). Juvenile three-spined sticklebacks avoid parasitized conspecifics. *Environmental Biology of Fishes, 39,* 215–218.

Dugatkin, L. A., & Godin, J.-G. J. (1992a). Predator inspection, shoaling and foraging under predation hazard in the Trinidadian guppy, *Poecilia reticulata. Environmental Biology of Fishes, 34,* 265–276.

Dugatkin, L. A., & Godin, J.-G. J. (1992b). Prey approaching predators: A cost-benefit perspective. *Annales Zoologici Fennici, 29,* 233–252.

Dugatkin, L. A., & Godin, J. G. (1998). How females choose their mates. *Scientific American,* April 1998, 46–51.

Dugatkin, L. A., Lucas, S., & Godin, J.-G. J. (2002). Serial effects of mate-choice copying in the guppy (*Poecilia reticulata*). *Ethology Ecology & Evolution, 14,* 45–52.

Dugatkin, L. A., & Reeve, H. K. (1994). Behavioral ecology and "levels of selection": Dissolving the group selection controversy. *Advances in the Study of Behaviour, 23,* 101–133.

Dugatkin, L. A., & Wilson, D. S. (1991). ROVER: A strategy for exploiting cooperators in a patchy environment. *American Naturalist, 138,* 687–701.

Dugatkin, L. A., & Wilson, D. S. (2000). Assortative interactions and the evolution of cooperation in guppies. *Evolutionary Ecology, 2,* 761–767.

Dukas, R. (Ed.). (1998a). *Cognitive ecology.* Chicago: University of Chicago Press.

Dukas, R. (1998b). Constraints on information processing and their effects on behavior. In R. Dukas (Ed.), *Cognitive ecology* (pp. 89–127). Chicago: University of Chicago Press.

Dukas, R. (1998c). Evolutionary ecology of learning. In R. Dukas (Ed.), *Cognitive ecology* (pp. 129–174). Chicago: University of Chicago Press.

Dukas, R. (1999). Ecological relevance of associative learning in fruit fly larvae. *Behavioral Ecology and Sociobiology, 45,* 195–200.

Dukas, R., & Bernays, E. A. (2000). Learning improves growth rate in grasshoppers. *Proceedings of the National Academy of Sciences, U.S.A., 97,* 2637–2640.

Dumond, F. (1968). The squirrel monkey in a seminatural environment. In L. Rosenblum & R. Cooper (Eds.), *The squirrel monkey* (pp. 87–145). New York: Academic Press.

Dunbar, R. (1991). Functional significance of social grooming in primates. *Folia Primatologica, 57,* 121–131.

Dunbar, R. (1992). Neocortex size as a constraint on group size in primates. *Journal of Human Evolution, 20,* 469–493.

Dunbar, R., Knight, C., & Power, C. (Eds.). (1999). *The evolution of culture: An interdisciplanary view.* New Brunswick, NJ: Rutgers University Press.

Duncker, K. (1938). Experimental modification of children's food preferences through social suggestion. *Journal of Abnormal and Social Psychology, 33,* 489–507.

Dyer, F. (1998). Cognitive ecology of navigation. In R. Dukas (Ed.), *Cognitive ecology* (pp. 201–260). Chicago: University of Chicago Press.

Eadie, J., Sherman, P., & Semel, B. (1998). Conspecific brood parasitism, population dynamics and the conservation of cavity-nesting birds. In T. Caro (Ed.), *Behavioral ecology and conservation biology* (pp. 306–340). New York: Oxford University Press.

Earley, R., & Dugatkin, L. A. (2002). Eavesdropping on visual cues in green swordtails (*Xiphophorus helleri*): A case for networking. *Proceedings of the Royal Society of London, 269,* 943–952.

Eason, P. K., Cobbs, G. A., & Trinca, K. G. (1999). The use of landmarks to define territorial boundaries. *Animal Behaviour, 58,* 85–91.

East, M. (1981). Alarm calling and parental investment in the robin *Erithacus rubecula. Ibis, 123,* 223–230.

Eberhard, W. (1996). *Female control: Sexual selection by cryptic female choice.* Princeton: Princeton University Press.

Edgerton, V. (1978). Mammalian muscle fibers and their adaptability. *American Zoologist, 18,* 113–125.

Edney, E., & Gill, R. (1968). Evolution of senescence and specific longevity. *Nature, 220,* 281–282.

Edwards, D., & Kravitz, E. (1997). Serotonin, social status and aggression. *Current Opinion in Neurobiology, 7,* 812–819.

Edwards, S. V., & Naeem, S. (1993). The phylogenetic component of cooperative breeding in perching birds. *American Naturalist, 141,* 754–789.

Eggebeen, D., & Hogan, D. (1990). Giving between generations in Americans. *Human Nature, 1,* 211–232.

Eggers, D. M. (1976). Theoretical effects of schooling by planktivorous fish predators on rate of prey consumption. *Journal of the Fisheries Research Board of Canada, 33,* 1965–1971.

Ehlinger, T. J., & Wilson, D. S. (1988). Complex foraging polymorphism in bluegill sunfish. *Proceedings of the National Academy of Sciences, U.S.A., 85,* 1878–1882.

Ehrenpreis, A. (1650/1978). *An epistle on brotherly community as the highest command of law.* Farmington, PA: Plough Publishing.

Ehrlich, P., Dobkin, D., & Wheye, D. (1986). The adaptive significance of anting. *Auk, 103,* 835.

Ehrman, L., & Parsons, P. (1981). *Behavior genetics and evolution.* New York: McGraw-Hill.

Eibesfeldt, I., & Eible-Eibesfeldt, E. (1967). Das parasitenabwehren der minima-arbeiterinnen der blattschneider-ameise (*Atta cephalotes*). *Zeitschrift fur Tierpsychologie, 24,* 278–281.

Eisenberg, J. F. (1966). The social organization of mammals. In W. Kukenthal, *Handbook of zoology* (Vol. 10; pp. 1–92). Berlin: Walter de Gruyter & Co.

Eisenberg, J., & Wilson, D. (1978). Relative brain size and feeding strategies in the Chiropotera. *Evolution, 32,* 740–751.

Ekman, J., & Askenmo, C. (1984). Social rank and habitat use in willow tit groups. *Animal Behaviour, 32,* 508–514.

Elgar, M. (1986). House sparrows establish foraging flocks by giving chirrup calls if the resource is divisible. *Animal Behaviour, 34,* 169–174.

Elgar, M., & Crespi, B. (Eds.). (1992). *Cannibalism: Ecology and evolution among diverse taxa.* Oxford: Oxford University Press.

Elgar, M., & Pierce, N. (1988). Mating success and fecundity in an ant-tended lycaenid butterfly. In T. Clutton-Brock (Ed.), *Reproductive success* (pp. 59–75). Chicago: University of Chicago Press.

Ely, D. (1981). Hypertension, social rank and aortic arterosclerosis in CBA/J mice. *Physiology & Behavior, 26,* 655–661.

Emlen, J. M. (1966). The role of time and energy in food preference. *American Naturalist, 100,* 611–617.

Emlen, J. M. (1968). Optimal choice in animals. *American Naturalist, 102,* 385–389.

Emlen, J. M. (1973). *Ecology: An evolutionary approach.* Reading, MA: Addison-Wesley Pub. Co.

Emlen, J. T., & Lorenz, F. W. (1942). Pairing responses of free-living valley quail to sex hormone pellet tablets. *Auk, 59.*

Emlen, S. T. (1967). Migratory orientation in the Indigo bunting, *Passerina cyanea.* II. Mechanisms of celestial orientation. *Auk, 84,* 463–469.

Emlen, S. T. (1970). Celestial rotation: Its importance in the development of migratory orientation. *Science, 170,* 1198–1201.

Emlen, S. T. (1975). The stellar-orientation system of a migratory bird. *Scientific American, 233,* 102–111.

Emlen, S. T. (1982a). The evolution of helping: An ecological constraints model. *American Naturalist, 119,* 29–39.

Emlen, S. T. (1982b). The evolution of helping: II. The role of behavioral conflict. *American Naturalist, 119,* 40–53.

Emlen, S. T. (1991). Evolution of cooperative breeding in birds and mammals. In J. Krebs & N. Davies (Eds.), *Behavioural ecology* (3rd ed., pp. 301–335). Oxford: Blackwell Scientific Publications.

Emlen, S. T. (1995a). Can avian biology be useful to the social sciences? *Journal of Avian Biology, 26,* 273–276.

Emlen, S. T. (1995b). An evolutionary theory of the family. *Proceedings of the National Academy of Sciences, U.S.A., 92,* 8092–8099.

Emlen, S. T., & Oring, L. W. (1977). Ecology, sexual selection and evolution of mating systems. *Science, 197,* 215–223.

Emlen, S. T., & Wrege, P. H. (1992). Parent-offspring conflict and the recruitment of helpers among bee-eaters. *Nature, 356,* 331–333.

Emlen, S., Wrege, P., & Webster, M. (1998). Cuckoldry as a cost of polyandry in the sex-role-reversed wattled jacana, *Jacana jacana. Proceedings of the Royal Society of London, 265,* 2359–2364.

Endler, J. (1980). Natural selection on color patterns in *Poecilia reticulata. Evolution, 34,* 76–91.

Endler, J. (1986). *Natural selection in the wild.* Princeton: Princeton University Press.

Endler, J. (1995). Multiple trait co-evolution and environmental gradients in guppies. *Trends in Ecology & Evolution, 10,* 22–29.

Endler, J. (1999). Adaptive genetic variation in the wild: Concluding remarks. In T. Mousseau, B. Sinervo, & J. Endler (Eds.), *Adaptive genetic variation in the wild* (pp. 251–260). New York: Oxford University Press.

Endler, J. A., & Houde, A. E. (1995). Geographic variation in female preferences for male traits in *Poecilia reticulata. Evolution, 49,* 456–468.

Endler, J., & McLellan, T. (1988). The process of evolution: Toward a newer synthesis. *Annual Review of Ecology and Systematics, 19,* 395–421.

Engel, G., & Schmale, A. (1972). Conservation-withdrawal: A primary regulatory process for organic homeostasis. In R. Porter & J. Knight (Eds.), *Physiology, emotions and psychosomatic illness* (pp. 57–75). Amsterdam: Elsevier.

Engh, A. L., Esch, K., Smale, L., & Holekamp, K. E. (2000). Mechanisms of maternal rank "inheritance" in the spotted hyena, *Crocuta crocuta. Animal Behaviour, 60,* 323–332.

Enquist, M., & Leimar, O. (1983). Evolution of fighting behavior: Decision rules and assessment of relative strength. *Journal of Theoretical Biology, 102,* 387–410.

Enquist, M., & Leimar, O. (1987). Evolution of fighting behavior: The effect of variation in resource value. *Journal of Theoretical Biology, 127,* 185–207.

Enquist, M., & Leimar, O. (1990). The evolution of fatal fighting. *Animal Behaviour, 39,* 1–9.

Enquist, M., Leimar, O., Ljungberg, T., Mallner, Y., & Segardahl, N. (1990). A test of the sequential assessment game: Fighting in the cichlid fish, *Nannacara anomala. Animal Behaviour, 40,* 1–15.

Enquist, M., Ljungberg, T., & Zandor, A. (1987). Visual assessment of fighting ability in the cichlid fish *Nannacara anomala. Animal Behaviour, 35,* 1262–1263.

Ens, B. J., & Goss-Custard, J. (1984). Interference among oyster catchers (*Haematopus ostralegus*) feeding on mussels (*Mytilus edulis*) on the Exe Estuary. *Journal of Animal Ecology, 53,* 217–231.

Essapian, F. S. (1963). Observations on abnormalities of parturition in captive bottle-nosed dolphins. *Journal of Mammalogy, 44,* 405–414.

Estes, R. D., & Goddard, J. (1967). Prey selection and hunting behavior of the African dog. *Journal of Wildlife Management, 31,* 52–70.

Estoup, A., Scholl, A., Pouvreau, A., & Solignac, M. (1995). Monoandry and polyandry in bumble bees (*Hymenoptera–Bombinae*) as evidenced by highly variable microsatellites. *Molecular Ecology, 4,* 89–93.

Etheredge, J., Perez, S., Taylor, O., & Jander, R. (1999). Monarch butterflies (*Danaus plexippus*) use a magnetic compass for navigation. *Proceedings of the National Academy of Sciences, U.S.A., 96,* 13845–13846.

Evans, C. S., Evans, L., & Marler, P. (1993). On the meaning of alarm calls: Functional reference in an avian vocal system. *Animal Behaviour, 46,* 23–38.

Evans, C. S., & Marler, P. (1994). Food calling and audience effects in male chickens, *Gallus gallus:* Their relationships to food availability, courtship and social facilitation. *Animal Behaviour, 47,* 1159–1170.

Even, M., Dahr, M., & vom Saal, F. (1992). Transport of steroids between fetuses via amniotic fluid in relation to the intrauterine position in rats. *Journal of Reproduction and Fertility, 96,* 709–712.

Ewald, P. (1985). Influences of asymmetries in resource quality and age on aggression and dominance in black-chinned hummingbirds. *Animal Behaviour, 33,* 705–719.

Ewald, P. (1994). *Evolution of infectious diseases.* New York: Oxford University Press.

Ewer, R. (1963). The behaviour of the meerkat (*Suricata suricatta*). *Zeitschrift fur Tierpsychologie, 20,* 570–607.

Ewer, R. (1969). The "instinct to teach." *Nature, 222,* 698.

Ezeh, A. C. (1997). Polygyny and reproductive behavior in sub-Saharan Africa: A contextual analysis. *Demography, 34,* 355–368.

Fagen, R. M. (1981). *Animal play behavior.* Oxford: Oxford University Press.

Fagnou, D., & Tuchek, J. (1995). The biochemistry of learning and memory. *Molecular and Cellular Biochemistry, 149/150,* 279–286.

Fahrbach, S. E., Giray, T., & Robinson, G. E. (1995). Volume changes in the mushroom bodies of adult honey-bee queens. *Neurobiology of Learning and Memory, 63,* 181–191.

Fairbanks, L. (1980). Relationships among adult females in captive vervet monkeys: Testing a model of rank-related attractiveness. *Animal Behaviour, 28,* 853–859.

Falls, J. B. (1982). Individual recognition by song in birds. In D. Kroodsma & E. Miller (Eds.), *Acoustic communication in birds* (pp. 237–278). New York: Academic Press.

Farris, S. M., Robinson, G. E., & Fahrbach, S. E. (2001). Experience- and age-related outgrowth of intrinsic neurons in the mushroom bodies of the adult worker honeybee. *Journal of Neuroscience, 21,* 6395–6404.

Faulkes, C. G., Abbott, D. H., & Jarvis, J. (1990). Social suppression of ovarian cyclicity in captive and wild colonies of the naked mole rat, *Heterocephalus glaber*. *Journal of Reproduction and Fertility*, 88, 559–568.

Feder, M. (1999). Organismal, ecological and evolutionary aspects of heat-shock proteins and the stress response: Established conclusion and unresolved issues. *American Zoologist*, 39, 857–864.

Feder, M., & Hoffmann, G. (1999). Heat-shock proteins, molecular chaperones and the stress response: Evolutionary and ecological physiology and the stress response: Established conclusion and unresolved issues. *Annual Review of Physiology*, 61, 243–282.

Feener, D. H., & Moss, K. A. (1990). Defense against parasites by hitchhikers in leaf-cutting ants: A quantitative assessment. *Behavioral Ecology and Sociobiology*, 26, 17–29.

Fentress, J. C., & Ryon, J. (1986). A multidimensional approach to agonistic behavior in wolves. In H. Frank (Ed.), *Man and wolf: Advances, issues and problems in captive wolf research* (pp. 253–273). Dordrecht: Junk Publishers.

Ficken, M. S., Witkin, S. R., & Weise, C. M. (1980). Associations among members of a black-capped chickadee flock. *Behavioral Ecology and Sociobiology*, 8, 245–249.

Fiedler, K., & Maschwitz, U. (1988). Functional analysis of the myrmecophilous relationships between ants (*Hymenoptera: formicidae*) and lycaenids (*Lepidoptera: Lycaenidae*). *Oecologia*, 75, 204–206.

Finch, C., & Rose, M. (1995). Hormones and the physiological architecture of life history evolution. *Quarterly Review of Biology*, 70, 1–52.

Fisher, D. C. (1985). Evolutionary morphology: Beyond the analogous, the anecdotal and the ad hoc. *Paleobiology*, 11, 120–138.

Fisher, J. (1954). Evolution and bird sociality. In J. Huxley, A. C. Hardy, & E. B. Ford (Eds.), *Evolution as a process* (pp. 71–83). London: Allen and Unwin.

Fisher, J., & Hinde, R. (1949). The opening of milk bottles by birds. *British Birds*, 42, 347–357.

Fisher, R. A. (1915). The evolution of sexual preference. *Eugenics Review*, 7, 184–192.

Fisher, R. A. (1930). *The genetical theory of natural selection*. New York: Dover.

Fisher, R. A. (1958). *The genetical theory of natural selection* (2nd ed.). New York: Dover.

FitzGerald, G., & van Havre, N. (1987). The adaptive significance of cannibalism in sticklebacks. *Behavioral Ecology and Sociobiology*, 20, 125–128.

Fitzgibbon, C. D. (1994). The costs and benefits of predator inspection behaviour in Thomson's gazelles. *Behavioral Ecology and Sociobiology*, 34, 139–148.

Fitzpatrick, J. W., & Woolfenden, G. E. (1984). The helpful shall inherit the scrub. *Natural History*, 93, 55–63.

Fitzpatrick, J. W., & Woolfenden, G. E. (1988). Components of lifetime reproductive success in the Florida scrub jay. In T. H. Clutton-Brock (Ed.), *Reproductive success* (pp. 305–320). Chicago: University of Chicago Press.

Fleishman, L., Lowe, E., & Leal, M. (1993). Ultraviolet vision in lizards. *Nature*, 365, 397.

Fletcher, D., & Michener, C. (Eds.). (1987). *Kin recognition in animals*. New York: Wiley.

Flinn, M., & Low, B. (1986). Resource competition, social competition and mating patterns in human societies. In D. Rubenstein & R. Wrangham (Eds.), *Ecological aspects of social evolution* (pp. 217–243). Princeton: Princeton University Press.

Flint, J., Corley, R., DeFries, J., Fulker, D., Gray, J., Miller, S., & Collins, A. C. (1995). Chromosomal mapping of three loci determining quantitative variation of emotionality in the mouse. *Science*, 269, 1432–1435.

Foltz, D. (1981). Genetic evidence for long-term monogamy in a small rodent, *Peromyscus polionotus*. *American Naturalist*, 117, 665–675.

Ford, J. K. (1991). Vocal traditions among resident killer whales (*Orcinus orca*) in coastal waters of British Columbia. *Canadian Journal of Zoology*, 69, 1454–1483.

Ford, R., & McLaughlin, F. (1983). Variation of male fidelity in monogamous birds. In R. Johnstone (Ed.), *Current ornithology* (Vol. 1; pp. 329–356). New York: Plenum Press.

Forslund, P. (2000). Male-male competition and large size mating advantage in European earwigs, *Forficula auricularia*. *Animal Behaviour*, 59, 753–762.

Foster, K. R., & Ratnieks, F. L. W. (2000). Social insects: Facultative worker policing in a wasp. *Nature*, 407, 692–693.

Foster, K. R., & Ratnieks, F. L. W. (2001). Convergent evolution of worker policing by egg eating in the honeybee and common wasp. *Proceedings of the Royal Society of London*, 268, 169–174.

Francis, R. (1990). Temperament in a fish: A longitudinal study of the development of individual differences of aggression and social rank in the midas cichlid. *Ethology*, 86, 311–325.

Frank, L. G. (1986a). Social organization of the spotted hyena (*Crocuta crocuta*): 1. Demography. *Animal Behaviour*, 34, 1500–1509.

Frank, L. G. (1986b). Social organization of the spotted hyena (*Crocuta crocuta*): 2. Dominance and reproduction. *Animal Behaviour*, 34, 1510–1527.

Fraser, D. F., Gilliam, J. F., Daley, M., Le, A., & Skalski, G. (2001). Explaining leptokurtic movement distributions: Intrapopulation variance in boldness and exploration. *American Naturalist*, 158, 124–135

Freeberg, T. (1998). The cultural transmission of courtship patterns in cowbirds, *Molothrus ater*. *Animal Behaviour*, 56, 1063–1073.

Freedman, H. I., Addicott, J. F., & Rai, B. (1987). Obligate mutualism with a predator: Stability and persistence of three-species models. *Theoretical Population Biology*, 32, 157–175.

Fretwell, S. (1972). *Populations in a seasonal environment*. Princeton: Princeton University Press.

Fretwell, S., & Lucas, H. (1970). On territorial behavior and other factors influencing habitat distribution in birds. I. Theoretical development. *Acta Biotheoretica, 19,* 16–36.

Fritz, R., & Simms, E. (Eds.). (1992). *Plant resistance to herbivores and pathogens: Ecology, evolution, and genetics.* Chicago: University of Chicago Press.

Fry, C. (1977). The evolutionary significance of cooperative breeding in birds. In B. Stonehouse & C. Perrins (Eds.), *Evolutionary ecology.* London: MacMillan.

Fujii, R., & Oshima, N. (1986). Control of chromaphore movements in teleost fishes. *Zoological Science, 3,* 13–47.

Gadagkar, R. (1997). *Survival strategies: Cooperation and conflict in animal societies.* Cambridge, MA: Harvard University Press.

Gaffrey, G. (1957). Zur fortpflanzungbsbiologie bei hunden. *Zool. Gart. Jena, 23,* 251.

Galef, B. G. (1988). Imitation in animals: History, definition, and interpretation of data from the psychological laboratory. In T. R. Zentall & B. G. Galef (Eds.), *Social learning: Psychological and biological perspectives* (pp. 3–28). Hillsdale, NJ: Lawrence Erlbaum Associates.

Galef, B. G., & Giraldeau, L.-A. (2001). Social influences on foraging in vertebrates: Causal mechanisms and adaptive functions. *Animal Behaviour, 61,* 3–15.

Galef, B. G., Manzig, L., & Field, R. (1986). Imitation learning in budgerigars: Dawson and Foster Revisited. *Behavioral Processes, 13,* 191–202.

Galef, B. G., & White, D. J. (1998). Mate-choice copying in Japanese quail, *Coturnix coturnix japonica. Animal Behaviour, 55,* 545–552.

Galef, B. G., & Wigmore, S. (1983). Transfer of information concerning distant foods: A laboratory investigation of the "information-centre" hypothesis. *Animal Behaviour, 31,* 748–758.

Gallagher, J. (1976). Sexual imprinting: Effects of various regimens of social experience on the mate preference in Japanese quail (*Coturnix coturnix japonica*). *Behaviour, 57,* 91–115.

Gamboa, G. J., Reeve, H. K., & Pfennig, D. W. (1986). The evolution and ontogeny of nestmate recognition in social wasps. *Annual Review of Entomology, 31,* 431–454.

Gangestad, S. W., & Thornhill, R. (1998). Menstrual cycle variation in women's preferences for the scent of symmetrical men. *Proceedings of the Royal Society of London, 265,* 927–933.

Garcia, J., Ervin, F., & Koelling, R. (1966). Learning with prolonged delay of reinforcement. *Psychonomic Science, 5,* 121–122.

Garcia, J., & Koelling, R. (1966). Relation of cue to consequence in avoidance learning. *Psychonomic Science, 4,* 123–124.

Garcia, J., McGowan, B., & Green, K. (1972). Biological constraints on conditioning. In A. Black & W. Prokasy (Eds.), *Classic conditioning II* (pp. 3–27). New York: Appleton.

Gardini, F., Talarico, E., & Brambilla, F. (1999). Neurotransmitters, neuroendocrine correlates of sensation-seeking temperament in normal humans. *Neuropsycholobiology, 39,* 207–213.

Garson, P. J., Pleszczynska, W. K., & Holm, C. H. (1981). The polygyny threshold model: A reassessment. *Canadian Journal of Zoology-Revue Canadienne De Zoologie, 59,* 902–910.

Gauthreaux, S. (Ed.). (1980). *Animal migration, orientation, and navigation.* New York: Academic Press.

Gauthreaux, S. (Ed.). (1997). *Animal migration, orientation, and navigation* (2nd ed.). New York: Academic Press.

Gazzaniga, M., Ivry, R., & Mangun, G. (1998). *Cognitive neuroscience.* New York: W. W. Norton & Co.

Geist, V. (1966). The evolution of horn-like organs. *Behaviour, 27,* 175–224.

Geist, V. (1974a). The development of social behavior through play in the Stellar sea lion. *American Zoologist, 14,* 205–220.

Geist, V. (1974b). On fighting strategies in animal combat. *Nature, 250,* 354.

Gelowitz, C., Mathis, A., & Smith, R. J. F. (1993). Chemosensory recognition of northern pike (*Esox lucius*) by brook stickleback (*Culaea inconstans*): Population differences and the influence of predator diet. *Behaviour, 127,* 106–118.

George, C. (1960). *Behavioral interactions in the pickerel and the mosquitofish.* Unpublished Ph.D., Harvard University.

Gerra, G., Avanzini, P., Zaimovic, A., Sartor, I. R., Bocchi, C., Timpano, M., Zambelli, U., Delsignore, R., Gardini, F., Talarico, E., & Brambilla, F. (1999). Neurotransmitters, neuroendocrine correlates of sensation-seeking temperament in normal humans. *Neuropsycholobiology, 39,* 207–213.

Getty, T. (1987). Dear enemy and the prisoner's dilemma: Why should territorial neighbors form defensive coalitions. *American Zoologist, 27,* 327–336.

Gibson, R. M., Bradbury, J. W., & Vehrencamp, S. L. (1991). Mate choice in lekking sage grouse: The roles of vocal display, female site fidelity and copying. *Behavioral Ecology, 2,* 165–180.

Gibson, R. M., & Hoglund, J. (1992). Copying and sexual selection. *Trends in Ecology and Evolution, 7,* 229–232.

Giles, N., & Huntingford, F. A. (1984). Predation risk and interpopulational variation in anti-predator behavior in the three-spined stickleback. *Animal Behaviour, 32,* 264–272.

Giraldeau, L. A., & Caraco, T. (2000). *Social foraging theory.* Princeton: Princeton University Press.

Giraldeau, L. A., & Lefebvre, L. (1986). Exchangeable producer and scrounger roles in a captive flock of feral pigeons: A case for the skill pool effect. *Animal Behaviour, 34,* 797–803.

Giraldeau, L. A., & Lefebvre, L. (1987). Scrounging prevents cultural transmission of food-finding in pigeons. *Animal Behaviour, 35,* 387–394.

Girondot, M., Fouillet, H., & Pieau, C. (1998). Feminizing turtle embryos as a conservation tool. *Conservation Biology, 12,* 353–365.

Godard, R. (1991). Long-term memory of neighbors in a migratory songbird. *Nature, 350,* 228–229.

Godard, R. (1993). Tit-for-tat among neighboring hooded warblers. *Behavioral Ecology and Sociobiology, 33,* 45–50.

Goddard, M. E., & Beilharz, R. (1982). Genetic and environmental factors affecting the suitability of dogs as guide dogs for the blind. *Theoretical and Applied Genetics, 62,* 97–102.

Goddard, M. E., & Beilharz, R. (1983). Genetics of traits which determine the suitability of dogs as guide dogs for the blind. *Applied Animal Ethology, 9,* 299–315.

Goddard, M. E. & Beilharz, R. (1984). A factor analysis of fearfulness in potential guide dogs. *Applied Animal Behaviour Science, 12,* 253–265.

Goddard, M. E., & Beilharz, R. (1985). A multivariate analysis of the genetics of fearfulness in potential guide dogs. *Behavior Genetics, 15,* 69–89.

Godin, J. G. (Ed.). (1997). *Behavioural ecology of fishes.* Oxford: Oxford University Press.

Godin, J. G., & Davis, S. A. (1995a). Who dares, benefits: Predator approach behaviour in the guppy (*Poecilia reticulata*) deters predator pursuit. *Proceedings of the Royal Society of London, 259,* 193–200.

Godin, J. G., & Davis, S. A. (1995b). Boldness and predator deterrence: A reply to Milinski and Boltshauser. *Proceedings of the Royal Society of London, 262,* 107–112.

Godin, J. G., & Dugatkin, L. A. (1996). Female mating preference for bold males in the guppy, *Poecilia reticulata. Proceedings of the National Academy of Sciences, U.S.A., 93,* 10262–10267.

Gold, K. C., & Maple, T. (1994). Personality assessment in the gorilla and its utility as a management tool. *Zoo Biology, 13,* 509–522.

Goldstein, L., Rasmussen, A., Bunney, B., & Roth, R. (1996). Role of the amygdala in the coordination of behavioral, neuroendocrine and prefrontal cortical monoamine responses to psychological stress in the rat. *Journal of Neuroscience, 16,* 4787–4798.

Gomendio, M., Harcourt, A., & Roldan, E. (1998). Sperm competition in mammals. In T. Birkhead & A. Moller (Eds.), *Sperm competition and sexual selection* (pp. 667–756). London: Academic Press.

Gompper, M., & Holyman, A. (1993). Grooming with *Trattinnickia* resin: Possible pharmaceutical plant use by coatis in Panama. *Journal of Tropical Ecology, 9,* 533–540.

Goodall, J. (1986). *The chimpanzee of Gombe: Patterns of behavior.* Cambridge, MA: Belknap Press.

Goodwin, D. (1986). *Crows of the world* (2nd ed.). London: British Museum.

Goodwin, T. (1950). Carotenoids and reproduction. *Biological Reviews of the Cambridge Philosophical Society, 25,* 391–413.

Gosden, R. (1996). *Cheating time: Science, sex and aging.* New York: W. H. Freeman & Co.

Gosling, S. D. (1998). Personality dimensions in spotted hyenas (*Crocuta crocuta*). *Journal of Comparative Psychology, 112,* 107–118.

Gosling, S. D. (2001). From mice to men: What can we learn about personality from animal research? *Psychological Bulletin, 127,* 45–86.

Gould, E. (1994). The effects of adrenal steroids and excitatory input in neuronal birth and survival. *Annals of the New York Academy of Sciences, 743,* 73–93.

Gould, J., & Keeton, W. (1996). *Biological sciences* (6th ed.). New York: W. W. Norton & Co.

Gould, J. L., & Marler, P. (1984). Ethology and the natural history of learning. In P. Marler & H. Terrace (Eds.), *The biology of learning* (pp. 47–74). Berlin: Springer-Verlag.

Gould, S. J. (1977). *Ontogeny and phylogeny.* Cambridge, MA: Harvard University Press.

Gould, S. J., & Lewontin, R. (1979). The spandrels of San Marcos and the Panglossian paradigm: A critique of the adaptationist programme. *Proceedings of the Royal Society of London, 205,* 581–598.

Gouzoules, S. (1984). Primate mating systems, kin associations and cooperative behavior: Evidence for kin recognition. *Yearbook of Physical Anthropology, 27,* 99–134.

Gouzoules, S., & Gouzoules, H. (1987). Kinship. In B. Smuts, D. L. Cheney, R. Seyfarth, R. Wrangham, & T. Struhsaker (Eds.), *Primate societies* (pp. 299–305). Chicago: University of Chicago Press.

Goy, R., & Phoenix, C. (1972). The effects of testosterone propionate administered before birth on the development of behavior in genetic female rhesus monkeys. In C. Sawyer & R. Gorski (Eds.), *Steroid hormones and brain function* (pp. 193–201). Berkeley, CA: University of California Press.

Grafen, A. (1984). Natural selection, kin selection and group selection. In J. Krebs & N. Davies (Eds.), *Behavioural ecology: An evolutionary approach* (2nd ed., pp. 62–84). Oxford: Blackwell Scientific Publications.

Grafen, A. (1985). A geometric view of relatedness. *Oxford Surveys in Evolutionary Biology, 2,* 28–89.

Grafen, A. (1988). On the uses of data on lifetime reproductive success. In T. Clutton-Brock (Ed.), *Reproductive success* (pp. 454–471). Chicago: University of Chicago Press.

Grafen, A. (1990a). Biological signals as handicaps. *Journal of Theoretical Biology, 144,* 517–546.

Grafen, A. (1990b). Sexual selection unhandicapped by the Fisher Process. *Journal of Theoretical Biology, 144,* 473–476.

Graham, J., & Desjardins, C. (1980). Classical conditioning: Induction of luteinizing and testosterone secretion in anticipation of sexual activity. *Science, 210,* 1039–1041.

Grant, B. R., & Grant, P. R. (1996). Cultural inheritance of song and its role in the evolution of Darwin's finches. *Evolution, 50,* 2471–2487.

Grant, J. W. A., & Green, L. D. (1995). Mate copying versus preference for actively courting males by female Japanese medaka (*Oryzias latipes*). *Behavioral Ecology, 7,* 165–167.

Grant, P. R. (1986). *Ecology and evolution of Darwin's finches.* Princeton: Princeton University Press.

Grant, P. R., & Grant, B. R. (1994). Phenotypic and genetic effects of hybridization in Darwin's finches. *Evolution, 48,* 297–316.

Grant, P. R., & Grant, B. R. (1995). Predicting microevolutionary responses to directional selection on heritable variation. *Evolution, 49,* 241–251.

Green, R. F. (1980). Bayesian birds: A simple example of Oaten's stochastic model of optimal foraging. *Theoretical Population Biology, 18,* 244–256.

Greenberg, N., & Crews, D. (1983). Physiological ethology of aggression in amphibians and reptiles. In B. Svare (Ed.), *Hormones and aggressive behavior* (pp. 469–506). New York: Plenum Press.

Greenberg, N., & Wingfield, J. (1987). Stress and reproduction: Reciprocal relationships. In D. Norris & R. Jones (Eds.), *Hormones and reproduction in fishes, amphibians and reptiles* (pp. 461–503). New York: Plenum Press.

Greenewalt, C. (1968). *Bird song: Acoustics and physiology.* Washington, DC: Smithsonian Press.

Greenough, W., & Chang, F. (1988). Plasticity of synapse structure and pattern in the cerebral cortex. In A. Peters & E. Jones (Eds.), *Cerebral cortex* (pp. 391–440). New York: Plenum Press.

Greenough, W., & Juraska, J. (1979). Experience-induced changes in brain fine structure: Their behavioral implications. In M. Hahn, C. Jensen, & B. Dudek (Eds.), *Development and evolution of brain size: Behavioral implications* (pp. 295–320). New York: Academic Press.

Griffin, D. R. (1964). *Bird migration.* Garden City, NY: Natural History Press.

Griffiths, R. (1995). Determining competition mechanisms in tadpole assemblages. *Herpetological Journal, 5,* 208–210.

Griffiths, S. W., & Magurran, A. E. (1997). Familiarity in schooling fish: How long does it take to acquire? *Animal Behaviour, 53,* 945–949.

Griffiths, S. X., & Magurran, A. E. (1999). Schooling decisions in guppies *(Poecilia reticulata)* are based on familiarity rather than kin recognition by phenotype matching. *Behavioral Ecology and Sociobiology, 45,* 437–445.

Grobecker, D., & Pietsch, T. (1978). Crows use automobiles as nutcrackers. *Auk, 95,* 760.

Gronenberg, W., Heeren, S., & Holldobler, B. (1996). Age-dependent and task-related morphological changes in the brain and the mushroom bodies of the ant *Camponotus floridanus. Journal of Experimental Biology, 199,* 2011–2019.

Gross, M. R. (1982). Sneakers, satellites, and parentals: Polymorphic mating strategies in North American sunfishes. *Zeitschrift fur Tierpsychologie, 60,* 1–26.

Gross, M. (1985). Disruptive selection for alternative life histories in salmon, *Nature, 313,* 47–48.

Gross, M., & Charnov, R. (1980). Alternative male life histories in bluegill sunfish. *Proceedings of the National Academy of Sciences, U.S.A., 77,* 6937–6940.

Guthrie, D. M., & Muntz, W. (1993). Role of vision in fish behavior. In T. Pitcher (Ed.), *Behaviour of teleost fishes.* New York: Chapman and Hall.

Guthrie, R. D. (1971). A new theory of mammalian rump patch evolution. *Behaviour, 38,* 132–145.

Gutierrez, G., & Domjan, M. (1996). Learning and male-male sexual competition in Japanese quail (*Coturnix japonica*). *Journal of Comparative Psychology, 110,* 170–175.

Gutsmann-Conrad, A., Heydari, A., You, S., & Richardson, A. (1998). The expression of heat shock protein 70 decreases with cellular senescence in vitro and in cells derived from young and old human subjects. *Experimental Cell Research, 241,* 404–413.

Haas, C., & D., J. (1993). Social play among juvenile bighorn sheep: Structure, development and relationship to adult behavior. *Ethology, 93,* 105–116.

Hack, M. A. (1997). Assessment strategies in the contests of male crickets, *Acheta domesticus (L). Animal Behaviour, 53,* 733–747.

Hagenguth, H., & Rembold, H. (1978). Identification of juvenile hormone 3 as the only JH homolog in all developmental stages of the honeybee. *Zeitschrift fur Naturforschun C, 33,* 847–850.

Haig, D. (1993). Genetic conflicts in human pregnancy. *Quarterly Review of Biology, 68,* 495–532.

Hailman, J. (1982). Ontogeny: Toward a general theoretical framework for ethology. *Perspectives in Ethology, 15,* 1–159.

Haldane, J. B. S. (1941). *New paths in genetics.* London: Allen and Unwin.

Hall, K. R., & Devore, I. (1965). Baboon social behavior. In I. Devore (Ed.), *Primate behavior: Field studies of monkeys and apes* (pp. 53–110). New York: Holt, Rinehart and Winston.

Hall, S. (1998). Object play by adult animals. In M. Bekoff & J. Byers (Eds.), *Animal play: Evolutionary, comparative and ecological perspectives* (pp. 45–60). Cambridge: Cambridge University Press.

Hames, R. (1996). Costs and benefits of monogamy and polygyny for Yanomamo women. *Ethology and Sociobiology, 17,* 181–199.

Hamilton, W. D. (1963). The evolution of altruistic behavior. *American Naturalist, 97,* 354–356.

Hamilton, W. D. (1964). The genetical evolution of social behaviour. I and II. *Journal of Theoretical Biology, 7,* 1–52.

Hamilton, W. D. (1966). The moulding of senescence by natural selection. *Journal of Theoretical Biology, 12,* 12–45.

Hamilton, W. D. (1967). Extraordinary sex ratios. *Science, 156,* 477–487.

Hamilton, W. D., & Zuk, M. (1982). Heritable true fitness and bright birds: A role for parasites. *Science, 218,* 384–387.

Hammerstein, P., & Parker, G. A. (1982). The asymmetric war of attrition. *Journal of Theoretical Biology, 96,* 647–682.

Hannes, R. P. (1985). The influence of standard opponent tests on blood androgens and corticoid levels of high and low ranking swordtail males before and after social isolation. *Aggressive Behavior, 11,* 9–15.

Harcourt, A. H., & de Waal, F. B. M. (Eds.). (1992). *Coalitions and alliances in humans and other animals.* Oxford: Oxford University Press.

Harcourt, A. H., Harvey, P., Larson, S., & Short, R. (1981). Testis weight, body size and breeding system in primates. *Nature, 293,* 55–57.

Harcourt, R. (1991). Survivorship costs of play in the South American fur seal. *Animal Behaviour, 42,* 509–511.

Hardy, J. (1961) Studies of behavior and phylogeny of certain New World jays (*Garrulinae*). *University of Kansas Science Bulletin, 42,* 13–149.

Harlow, H. (1959). Learning set and error factor theory. In S. Koch (Ed.), *Psychology: A study of science* (Vol. 2; pp. 492–537). New York: McGraw-Hill.

Harper, D. C. (1982). Competitive foraging in mallards: "Ideal free" ducks. *Animal Behaviour, 30,* 575–584.

Hartl, D., & Clark, A. (1989). *Principles of population genetics.* Sunderland, MA: Sinauer Associates.

Hartung, J. (1982). Polygyny and inheritance of wealth. *Current Anthropology, 23,* 1–12.

Harvell, C. D. (1991). Coloniality and inducible polymorphism. *American Naturalist, 138,* 1–14.

Harvell, C. D. (1994). The evolution of polymorphism in colonial invertebrates and social insects. *Quarterly Review of Biology, 69,* 155–185.

Harvell, C. D. (1998). Genetic variation and polymorphism in the inducible spines of a marine bryozoan. *Evolution, 52,* 80–86.

Harvey, P., & Harcourt, S. (1984). Sperm competition, testis size and breeding systems in primates. In R. L. Smith (Ed.), *Sperm competition and the evolution of animal mating systems* (pp. 589–600). New York: Academic Press.

Harvey, P. H., & Pagel, M. D. (1991). *The comparative method in evolutionary biology.* Oxford: Oxford University Press.

Hauser, D. (1964). Anting by gray squirrels. *Journal of Mammalogy, 45,* 136–138.

Hauser, M. (1997). *The evolution of communication.* Cambridge, MA: MIT Press.

Hauser, M. (2000). *Wild minds.* New York: Henry Holt.

Hauser, M., Cheney, D., & Seyfarth, R. (1986). Group extinction and fusion in free-ranging vervet monkeys. *American Journal of Primatology, 11,* 63–77.

Hauser, M., & Konishi, M. (Eds.). (1999).*The design of animal communication.* Cambridge, MA: MIT Press.

Hauser, M. D., & Nelson, D. A. (1991). Intentional signaling in animal communication. *Trends in Ecology & Evolution, 6,* 186–189.

Hausfater, G., & Hrdy, S. B. (Eds.). (1984). *Infanticide.* New York: Aldine.

Hay, M., & Fuller, P. (1981). Seed escape from hetromyids rodents: The importance of microhabitat and seed preference. *Ecology, 62,* 1395–1399.

Healey, S. (1992). Optimal memory: Toward an evolutionary ecology of animal cognition. *Trends in Ecology & Evolution, 8,* 399–400.

Healey, S., & Krebs, J. R. (1992). Food storing and the hippocampus in corvids: Amount and volume are correlated. *Proceedings of the Royal Society of London, 248,* 241–245.

Heape, W. (1931). *Emigration, migration and nomadism.* Cambridge: Cambridge University Press.

Hegner, R. (1985). Dominance and anti-predator behaviour in blue tits (*Parus caeruleus*). *Animal Behaviour, 33,* 762–768.

Heinrich, B. (1988a). Why do ravens fear their food? *Condor, 90,* 950–952.

Heinrich, B. (1988b). Winter foraging at carcasses by three sympatric corvids, with emphasis on recruitment by the raven, *Corvus corax. Behavioral Ecology and Sociobiology, 23,* 141–156.

Heinrich, B. (1989). *Ravens in winter.* New York: Simon and Schuster.

Heinrich, B. (1995). Neophilia and exploration in juvenile common ravens, *Corvus corax. Animal Behaviour, 50,* 695–704.

Heinrich, B. (1999). *Mind of the raven.* New York: HarperCollins.

Heinrich, B., & Marzluff, J. M. (1991). Do common ravens yell because they want to attract others? *Behavioral Ecology and Sociobiology, 28,* 13–21.

Heinrich, B., Marzluff, J., & Adams, W. (1996). Fear and food recognition in naive common ravens. *Auk, 112,* 499–503.

Heinrich, B., & Smolker, R. (1998). Play in common ravens (*Corvus corax*). In M. Bekoff & J. Byers (Eds.), *Animal play: Evolutionary, comparative and ecological perspectives* (pp. 27–44). Cambridge: Cambridge University Press.

Helfman, G., & Schultz, E. (1984). Social transmission of behavioral traditions in a coral reef fish. *Animal Behaviour, 32,* 379–384.

Hemelrijk, C. K., & Ek, A. (1991). Reciprocity and interchange of grooming and "support" in captive chimpanzees. *Animal Behaviour, 41,* 923–935.

Hennessy, D. F. (1986). On the deadly risk of predator harassment. *Ethology, 72,* 72–74

Hennessy, D. F., & Owings, D. H. (1978). Snake species discrimination and the role of olfactory cues in the snake-directed behavior of the California ground squirrel. *Behaviour,* 115–124.

Henry, J., & Stephens, P. (1977). *Stress, health and the social environment: A sociobiological approach to medicine.* Berlin: Springer-Verlag.

Hepper, P. G. (Ed.). (1991). *Kin recognition.* Cambridge: Cambridge University Press.

Hepper, P., & Cleland, J. (1998). Developmental aspects of kin recognition. *Genetica, 104,* 199–205.

Hessing, M., Coenen, G., Vaiman, M., & Renard, C. (1995). Individual differences in cell-mediated and humoral immunity in pigs. *Veterinary Immunology and Immunopathology, 45,* 97–113.

Hessing, M., Hagelso, A., Schouten, W., Wiepkema, P., & Van Beek, J. (1994). Individual, behavioral and physiological strategies in pigs. *Physiology & Behavior, 55,* 39–46.

Hetherington, E., Stanley-Hagan, M., & Anderson, E. (1989). Marital transitions: A child's perspective. *American Psychologist, 44,* 303–312.

Heyes, C. M. (1994). Social learning in animals: Categories and mechanisms. *Biological Reviews of the Cambridge Philosophical Society, 69,* 207–231.

Heyes, C. M., & Galef, B. G. (Eds.). (1996). *Social learning in animals: The roots of culture.* London: Academic Press.

Hill, C., & Pierce, N. (1989). The effect of adult diet on the biology of butterflies, 1: The common imperial blue, *Jalmenus evagoras. Oecologia, 81,* 249–257.

Hill, G. (1990). Female house finches prefer colourful males; sexual selection for a condition-dependent trait. *Animal Behaviour, 40,* 563–572.

Hill, G. (1991). Plumage coloration is a sexually selected indicator of male quality. *Nature, 350,* 337–339.

Hill, G. (1992). The proximate basis of variation in carotenoid pigmentation in male house finches. *Auk, 109,* 1–12.

Hill, G. (1993a). Geographic variation in the carotenoid plumage pigmentation of male house finches (*Carpodacus mexicanus*). *Biological Journal of the Linnean Society, 49,* 63–86.

Hill, G. (1993b). Male mate choice and the evolution of female plumage coloration in the house finch. *Evolution, 47,* 1515–1525.

Hill, G. (1993c). The proximate basis of inter- and intrapopulation variation in female plumage coloration in the house finch. *Canadian Journal of Zoology, 71,* 619–627.

Hill, H., & Bekoff, M. (1977). The variability of some motor components of social play and agonistic behavior in infant Eastern coyotes, *Canis latrans. Animal Behaviour, 25,* 907–909.

Hinde, R., & Fisher, J. (1951). Further observations on the opening of milk bottles by birds. *British Birds, 44,* 393–396.

Hinton, G., & Nowlan, S. (1987). How learning can guide evolution. *Complex Systems, 1,* 495–502.

Hirth, D. H., & McCullough, D. R. (1977). Evolution of alarm signals in ungulates with special reference to white-tailed deer. *American Naturalist, 111,* 31–42.

Hodge, M. A., & Uetz, G. (1995). A comparison of agonistic behaviour of colonial web-building spiders from desert and tropical habitats. *Animal Behaviour, 50,* 963–972.

Hoffman-Goetz, L., & Pedersen, B. (1994). Exercise and the immune system: A model of the stress response? *Immunology Today, 15,* 382–387.

Hoffmann, A. (1999). Laboratory and field heritabilities: Some lessons from Drosophila. In T. Mousseau, B. Sinervo, & J. Endler (Eds.), *Adaptive genetic variation in the wild* (pp. 200–218). New York: Oxford University Press.

Hogan, J. (1994). The concept of cause in the study of behavior. In J. Hogan & J. Bolhuis (Eds.), *Causal mechanisms of behavioural development* (pp. 3–15). Cambridge: Cambridge University Press.

Hogan-Warburg, A. J. (1966). Social behavior of the ruff, *Philomachus pugnax (L). Ardea, 54,* 109–229.

Hoglund, J., & Alatalo, R. (1995). *Leks.* Princeton: Princeton University Press.

Hoglund, J., Alatalo, R., Gibson, R., & Lundberg, A. (1995). Mate-choice copying in black grouse. *Animal Behaviour, 49,* 1627–1633.

Hoglund, J., Alatalo, R. V., & Lundberg, A. (1990). Copying the mate choice of others? Observations on female black grouse. *Behaviour, 114,* 221–236.

Hoglund, J., Widemo, F., Sutherland, W. J., & Nordenfors, H. (1998). Ruffs, *Philomachus pugnax,* and distribution models: Can leks be regarded as patches? *Oikos, 82,* 370–376.

Hogstad, O. (1995). Alarm calling by willow tits, *Paras montanus,* as mate investment. *Animal Behaviour, 49,* 221–225.

Holden, P., & Sharrock, J. (1988). *The Royal Society for the protection of birds book of the British Isles.* London: MacMillan.

Holekamp, K. E., & Smale, L. (1998). Behavioral development in the spotted hyena. *Bioscience, 48,* 997–1005.

Holekamp, K., Smale, L., Simpson, H., & Holekamp, N. (1984). Hormonal influences on natal dispersal in free-living Belding's ground squirrels (*Spermophilus beldingi*). *Hormones and Behavior, 18,* 465–483.

Holekamp, K. E., Smale, L., & Szykman, M. (1996). Rank and reproduction in the female spotted hyena. *Journal of Reproduction and Fertility, 108,* 229–237.

Holldobler, B., & Roces, F. (2001). The behavioral ecology of stridulatory communication in leaf-cutting ants. In L. Dugatkin (Ed.), *Model systems in behavioral ecology.* Princeton: Princeton University Press.

Holldobler, B., & Wilson, E. O. (1977). The number of queens: An important trait in ant evolution. *Naturwissenschaften, 64,* 8–15.

Holldobler, B., & Wilson, E. O. (1990). *The ants.* Cambridge, MA: Harvard University Press.

Hollis, K. L. (1984). The biological function of Pavlovian conditioning: The best defense is a good offense. *Journal of Experimental Psychology—Animal Behavior Processes, 10,* 413–425.

Hollis, K. L., Cadieux, E. L., & Colbert, M. M. (1989). The biological function of Pavlovian conditioning: A mechanism for mating success in blue gourami (*Trichogaster trichopterus*). *Journal of Comparative Psychology, 103*, 115–121.

Hollis, K. L., Dumas, M., Singh, P., & Fackelman, P. (1995). Pavlovian conditioning of aggressive behavior in blue gourami fish (*Trichogaster trichopterus*): Winners become winners and losers stay losers. *Journal of Comparative Psychology, 109*, 123–133.

Hollis, K. L., Pharr, V., Dumas, M., Britton, G., & Field, J. (1997). Classic conditioning provides paternity advantage for territorial male blue gouramis (*Trichogaster trichopterus*). *Journal of Comparative Psychology, 111*, 219–225.

Holloway, W., & Thor, D. (1985). Interactive effects of caffeine, 2-cholroadenosine and haloperidol on activity, social investigation and play fighting of juvenile rats. *Pharmacology Biochemistry and Behavior, 22*, 421–426.

Holmes, W., & Sherman, P. W. (1983). Kin recognition in animals. *American Scientist, 71*, 46–55.

Hoogland, J. L. (1983). Nepotism and alarm calls in the black-tailed prairie dog (*Cynomys ludovicianus*). *Animal Behaviour, 31*, 472–479.

Hoogland, J. L. (1995). *The black-tailed prairie dog*. Chicago: University of Chicago Press.

Hoogland, J., & Sherman, P. W. (1976). Advantages and disadvantages of bank swallow coloniality. *Ecological Monographs, 46*, 33–58.

Hopey, M., & Petranks, J. (1994). Restriction of wood frogs to fish free habitats: How important is adult choice? *Copeia, 1994*, 1023–1025.

Hopkins, C. D. (1981). On the diversity of electric signals in a community of mormyrid electric fishes in west Africa. *American Zoologist, 21*, 211–222.

Hopkins, C. D. (1986). Behavior of mormyridae. In T. Bullock & W. Heilgenberg (Eds.), *Electroreception* (pp. 527–576). New York: Wiley.

Hopkins, C. D. (1999a). Signal evolution in electric communication. In M. Hauser & M. Konishi (Eds.), *The design of animal communication* (pp. 461–491). Cambridge, MA: MIT Press.

Hopkins, C. D. (1999b). Design features for electric communication. *Journal of Experimental Biology, 202*, 1217–1228.

Hopkins, C. D., & Bass, A. (1981). Temporal coding of recognition in an electric fish. *Science, 212*, 85–87.

Hopster, H. (1998). *Coping strategies in dairy cows*. Wageningen: Agricultural University.

Houde, A. E. (1988). Genetic difference in female choice in two guppy populations. *Animal Behaviour, 36*, 510–516.

Houde, A. E. (1997). *Sex, color and mate choice in guppies*. Princeton: Princeton University Press.

Houde, A. E., & Endler, J. A. (1990). Correlated evolution of female mating preference and male color pattern in the guppy, *Poecilia reticulata*. *Science, 248*, 1405–1408.

Howard, R. D. (1978). The influence of male defended oviposition sites on early embryo mortality in bullfrogs. *Ecology, 59*, 789–798.

Howe, N., & Harris, L. (1978). Transfer of the sea anemone pheromone, anthropleurine, by the nudibranch *Aeolidia papillosa*. *Journal of Chemical Ecology, 4*, 551–561.

Hrdy, S. B. (1979). Infanticide among animals. *Ethology and Sociobiology, 1*, 13–40.

Hrdy, S. B. (1999). *Mother nature: A history of mothers, infants and natural selection*. New York: Pantheon Books.

Hsu, Y., & Wolf, L. (1999). The winner and loser effect: Integrating multiple experiences. *Animal Behaviour, 57*, 903–910.

Huang, Z., Robinson, G., Tobe, S., Yagi, K., Strambi, C., Strambi, A., & Stay, B. (1991). Hormonal regulation of behavioral development in the honey bee is based on changes in the rate of juvenile hormone biosynthesis. *Journal of Insect Physiology, 37*, 733–741.

Huber, R., & Kravitz, E. (1995). A quantitative analysis of agonistic behavior in juvenile American lobsters (*Homarus americanus*). *Brain, Behavior and Evolution, 46*, 72–83.

Huber, R., Smith, K., Delago, A., Isakson, K., & Kravitz, E. (1997). Serotonin and aggressive motivation in crustaceans: Altering the decision to retreat. *Proceedings of the National Academy of Sciences, U.S.A., 94*, 5939–5942.

Huffman, M. (1993). An investigation of the use of medicinals by chimpanzees current status and future prospects. *Primate Research, 9*, 179–187.

Huffman, M. (1996). Acquisition of innovative cultural behaviors in nonhuman primates: A case study of stone handling, a socially transmitted behavior in Japanese macaques. In C. M. Heyes & B. G. Galef (Eds.), *Social learning in animals: The roots of culture* (pp. 267–289). London: Academic Press.

Huffman, M. (1997). Current evidence for self-medication in primates: A multidisciplinary perspective. *Yearbook of Physical Anthropology, 40*, 171–200.

Huffman, M., & Wrangham, R. (1994). Diversity of medicinal plant use by chimpanzees in the wild. In R. Wrangham, W. C. McGrew, F. de Waal, & P. Heltne (Eds.), *Chimpanzee culture* (pp. 129–148). Cambridge, MA: Harvard University Press.

Hughes, L., Chang, B., Wagner, D., & Pierce, N. (2000). Effects of mating history on ejaculate size, fecundity and copulation duration in the ant-tended lycaenid butterfly, *Jalmenus evagoras*. *Behavioral Ecology and Sociobiology, 47*, 119–128.

Hugie, D. M., & Lank, D. B. (1997). The resident's dilemma: A female choice model for the evolution of alternative mating strategies in lekking male ruffs (*Philomachus pugnax*). *Behavioral Ecology*, 8, 218–225.

Hull, C. L. (1943). *Principles of behavior*. New York: Appleton.

Hulscher-Emeis, T. (1992). The variable patterns of *Tilapia zillii* (Cichlidae): Integrating ethology, chromatophore regulation and the physiology of stress. *Netherlands Journal of Zoology, 42,* 525–560.

Humphreys, A., & Einon, D. (1981). Play as a reinforcer for maze learning in juvenile rats. *Animal Behaviour, 29,* 259–270.

Humphreys, A., & Smith, P. (1987). Rough and tumble play, friendship and dominance in school children: Evidence for continuity and change with age. *Child Development, 58,* 201–212.

Hunt, S., Cuthill, I., Swaddle, J., & Bennett, A. (1997). Ultraviolet vision and band-colour preferences in female zebra finches, *Taeniopygia guttata. Animal Behaviour, 54,* 1383–1392.

Huntingford, F. A. (1976). The relationship between anti-predator behaviour and aggression among conspecifics in the three-spined stickleback, *Gasterosteus aculeatus. Animal Behaviour, 24,* 245–260.

Huntingford, F. A. (1984). Some ethical issues raised by studies of predation and aggression. *Animal Behaviour, 32,* 210–215.

Huntingford, F., & Turner, A. (1987). *Animal conflict.* London: Chapman and Hall.

Huntingford, F. A., & Wright, P. J. (1992). Inherited population differences in avoidance conditioning in three-spined sticklebacks, *Gasterosteus aculeatus. Behaviour, 122,* 264–273.

Hurd, P. (1997). Cooperative signalling between opponents in fish fights. *Animal Behaviour, 54,* 1309–1315.

Husberger, E., & Snowdon, C. (Eds.). (1997). *Social influences on vocal development.* Cambridge: Cambridge University Press.

Hutchins, D., & Barash, D. (1976). Grooming in primates: Implications for its utilitarian function. *Primates, 17,* 145–150.

Hutt, C. (1966). Exploration and play in children. *Zoological Society of London (Symposium), 18,* 61–81.

Hutt, C. (1970). Specific and diverse exploration. *Advances in Child Development and Behavior, 5,* 119–180.

Huxley, J. (1932). *Problems of relative growth.* London: MacVeagh.

Huxley, J. (1938). Darwin's theory of sexual selection and the data subsumed by it, in light of recent research. *American Naturalist, 72,* 416–433.

Huxley, J. (1942). *Evolution: The modern synthesis.* New York: Harper.

Huxley, T. H. (1888). The struggle for existence: A programme. *Nineteenth Century, 23,* 161–180.

Ibara, R., Penny, T., Ebeling, A., van Dykhuizen, G., & Caillet, G. (1983). The mating call of plainfin midshipman fish, *Porichthys notatus.* In D. Noakes (Ed.), *Predators and prey in fish* (pp. 205–212). The Hague: W. Junk Pubs.

Immelman, K. (1972). Sexual and other long-term aspects of imprinting in birds and other species. In L. D. R. Hinde & E. Shaw (Eds.), *Advances in the study of behavior* (Vol. 4; pp. 147–174). New York: Academic Press.

Ims, R. A. (1987). Responses in social organization and behaviour manipulation of the food resource in the vole *Clethrionomys rufocanus. Journal of Animal Ecology, 56,* 585–596.

Ims, R. A. (1988). Spatial clumping of sexually receptive females induces space sharing among male voles. *Nature, 335,* 541–543.

Irons, W. (1983). Human female reproductive strategies. In S. Wasser (Ed.), *Social behavior of female vertebrates* (pp. 169–213). New York: Academic Press.

Jacobs, G. H. (1992). Ultraviolet vision in vertebrates. *American Zoologist, 32,* 544–554.

Jacobson, M. (1991). *Developmental neurobiology.* New York: Plenum Press.

Jamieson, S. H., & Armitage, K. (1987). Sex difference in the play behavior of yearling yellow-bellied marmots. *Ethology, 74,* 237–253.

Janik, V., & Slater, P. (1997). Vocal learning in mammals. *Advances in the Study of Behavior, 26,* 59–99.

Janzen, D. (1978). Complications in interpreting the chemical defenses of trees against tropical arboreal plant-eating vertebrates. In G. Montgomerie (Ed.), *The ecology of arboreal foliovores* (pp. 73–84). Washington, DC: Smithsonian Press.

Janzen, F., & Paukstis, G. (1991). Environmental sex determination in reptiles: Ecology, evolution and experimental design. *Quarterly Review of Biology, 66,* 149–179.

Jarvis, J. (1981). Eusociality in a mammal: Cooperative breeding in the naked mole-rat. *Science, 212,* 571–573.

Jeffreys, A., Wilson, V., & Thein, S. (1985). Hypervariable "minisatellite" regions in human DNA. *Nature, 314,* 67–73.

Jenni, D., & Collier, G. (1972). Polyandry in the American jacana (*Jacana spinosa*). *Auk, 89,* 743–765.

Jennions, M., & Backwell, P. (1996). Residency and size affect fight duration and outcome in the fiddler crab *Uca annulipes. Biological Journal of the Linnean Society, 57,* 293–306.

Jensen, P., & Yngvesson, J. (1998). Aggression between unacquainted pigs: Sequential assessment and effects of familiarity and weight. *Applied Animal Behavior Science, 58,* 49–61.

Jerison, H. (1973). *Evolution of the brain and intelligence.* New York: Academic Press.

John, J. (1994). The avian spleen: A neglected organ. *Quarterly Review of Biology, 69,* 327–351.

Johnsson, J., & Akerman, A. (1998). Watch and learn: Preview of the fighting ability of opponents alters contest behaviour in rainbow trout. *Animal Behaviour, 56,* 771–776.

Johnston, T., & Pietrewicz, A. (Eds.). (1985). *Issues in the ecological study of learning.* Hillsdale, NJ: Lawrence Erlbaum Associates.

Johnstone, R. A. (1995). Sexual selection, honest advertisement and the handicap principle: Reviewing the evidence. *Biological Reviews of the Cambridge Philosophical Society, 70,* 1–65.

Johnstone, R. A. (1998a). Game theory and communication. In L. A. Dugatkin & H. K. Reeve (Eds.), *Game theory and animal behavior* (4th ed., pp. 94–117). New York: Oxford University Press.

Johnstone, R. A. (1998b). Conspiratorial whispers and conspicuous displays: Games of signal detection. *Evolution, 52,* 1554–1563.

Johnstone, R. A., & Dugatkin, L. A. (2000). Coalition formation in animals and the nature of winner and loser effects. *Proceedings of the Royal Society of London, 267,* 17–21.

Johnstone, R. A., Woodroffe, R., Cant, M., & Wright, J. (2000). Reproductive skew in multimember groups. *American Naturalist, 155,* 315–331.

Johnstone, T. D. (1982). The selective costs and benefits of learning. *Advances in the Study of Behavior, 12,* 65–106.

Jolicoeur, P., Pirlot, P., Baron, G., & Stephan, H. (1984). Brain structure and correlation patterns in insectivora, chiroptera and primates. *Systematic Zoology, 33,* 14–29.

Jones, A., Rosenqvist, G., Berglund, A., Arnold, S., & Avise, J. (2000). The Bateman gradient and the cause of sexual selection in a sex-role-reversed pipefish. *Proceedings of the Royal Society of London, 267,* 677–680.

Jungreis, S. A. (1987). Biomagnetism: An orientation mechanism in migrating insects. *Florida Entomologist, 70,* 277–283.

Kacelnik, A., & Abreu, E. (1998). Risky choice and Weber's Law. *Journal of Theoretical Biology, 194,* 289–298.

Kacelnik, A., Krebs, J., & Bernstein, C. (1992). The ideal free distributions and predator-prey populations. *Trends in Ecology and Evolution, 7,* 50–55.

Kaczensky, P. (1996). *Livestock-carnivore conflicts in Europe.* Munich: Munich Wildlife Society.

Kaczmarek, L. (1993). Molecular biology of vertebrate learning: Is c-fos a new beginning? *Journal of Neuroscience Research, 34,* 377–381.

Kagan, J. (1994). *Galen's prophecy: Temperament in human nature.* New York: Basic Books.

Kagan, J., & Reznick, S. (1989). Inhibited and uninhibited types of children. *Child Development, 60,* 838–845.

Kagan, J., Reznick, S., & Snidman, N. (1987). Temperamental variation in response to the unfamiliar. In N. A. Krasnegor, E. M. Blass, M. A. Hofer, & W. P. Smotherman (Eds.), *Perinatal development: A psychobiological perspective* (pp. 397–419). New York: Academic Press.

Kagan, J., Reznick, J. S., & Snidman, N. (1988). Biological bases of childhood shyness. *Science, 240*(April), 167–171.

Kamil, A., Krebs, J., & Pulliam, H. R. (Eds.). (1987). *Foraging behavior.* New York: Plenum Press.

Kamil, A., & Yoerg, S. (1982). Learning and foraging behavior. *Perspectives in Ethology, 5,* 325–364.

Kamin, L. J. (1968). "Attention-like" processes in classic conditioning. In M. R. Jones (Ed.), *Miami Symposium on the Prediction of Behavior: Aversive stimulation* (pp. 9–31). Miami: Miami University Press.

Kamin, L. J. (1969). Predictability, surprise, attention and conditioning. In B. Campbell & R. Church (Eds.), *Punishment and aversive behavior* (pp. 279–296). New York: Appleton-Century-Crofts.

Kandel, E., & Abel, T. (1995). Neuropeptides, adenyl cyclase and memory storage. *Science, 268,* 825–826.

Kaplun, D., & Reich, R. (1976). The murdered child and his killers. *American Journal of Psychiatry, 133,* 809–813.

Kawai, M. (1965). Newly acquired precultural behavior of the natural troop of Japanese monkeys on Koshima Islet. *Primates, 6,* 1–30.

Kawamura, S. (1959). The process of sub-culture propagation among Japanese macaques. *Primates,* 2, 43–60.

Kawamura, S., Blow, N., & Yokoyama, S. (1999). Genetic analyses of visual pigments of the pigeon (*Columbia livia*). *Genetics, 153,* 1839–1850.

Kawanabe, H., Cohen, J., & Iwasaki, K. (Eds.). (1993). *Mutualism and community organization: Behavioural, theoretical, and food-web approaches.* New York: Oxford University Press.

Kear, J. (1972). Reproduction and family life. In P. Scott (Ed.), *The swans* (pp. 80–124). Boston: Houghton Mifflin.

Keefe, M. (1992). Chemically mediated avoidance behaviour in wild brook trout, *Salvelinus fontinalis*: The response to familiar and unfamiliar predaceous fishes and the influences of diet. *Canadian Journal of Zoology, 70,* 288–292.

Keenleyside, M. (1955). Some aspects of the schooling behavior of fish. *Behaviour, 8,* 83–248.

Keenleyside, M., & Yamamoto, F. (1962). Territorial behaviour of juvenile Atlantic salmon (*Salmo salar L.*). *Behaviour, 19,* 139–169.

Keller, L., & Reeve, H. K. (1994). Partitioning of reproduction in animal societies. *Trends in Ecology and Evolution, 9,* 98–102.

Kennedy, M., & Gray, R. (1993). Can ecological theory predict the distribution of foraging animals? A critical analysis of experiments on the ideal free distribution. *Oikos, 68,* 158–166.

Keverne, E. B., Martel, F. L., & Nevison, C. M. (1996). Primate brain evolution: Genetic and functional considerations. *Proceedings of the Royal Society of London, 263,* 689–696.

Keverne, E. B., Martenz, N. D., & Tuite, B. (1989). Beta-endorphin concentration in cerebrospinal fluid of monkeys are influenced by grooming relationships. *Psychoneuroendocrinology, 14,* 155–161.

Kiesecker, J., & Blaustein, A. (1995). Synergism between UV-B radiation and a pathogen magnifies amphibian embryo mortality in nature. *Proceedings of the National Academy of Sciences, U.S.A., 92,* 11049–11052.

Kiesecker, J., & Blaustein, A. (1997). Influences of egg-laying behavior on pathogenic infection of amphibian eggs. *Conservation Biology, 11,* 214–220.

Kiesecker, J., & Blaustein, A. (1999). Pathogen reverses competition between larval anurans. *Ecology, 80,* 2442–2448.

Kiesecker, J., & Skelly, D. K. (2000). Choice of oviposition site by gray treefrogs: The role of potential parasitic infection. *Ecology, 81,* 2939–2943.

Kiesecker, J., Skelly, D. K., Beard, K., & Preisser, E. (1999). Behavioral reduction of infectious risk. *Proceedings of the National Academy of Sciences, U.S.A., 96,* 9165–9168.

Kingdon, J. (1977). *East African mammals: Volume 3: Carnivores.* New York: Academic Press.

Kirkpatrick, M. (1982). Sexual selection and the evolution of female choice. *Evolution, 36,* 1–12.

Kirkpatrick, M., & Dugatkin, L. A. (1994). Sexual selection and the evolutionary effects of mate copying. *Behavioral Ecology and Sociobiology, 34,* 443–449.

Kirkpatrick, M., & Ryan, M. (1991). The evolution of mating preferences and the paradox of the lek. *Nature, 350,* 33–38.

Kirkwood, T. (1977). Evolution and aging. *Nature, 270,* 301–304.

Kirkwood, T., & Holliday, R. (1979). The evolution of aging and longevity. *Proceedings of the Royal Society of London, 205,* 531–546.

Kirkwood, T., & Rose, M. (1991). Evolution of senescence: Late survival sacrificed for reproduction. *Philosophical Transactions of the Royal Society of London, 332,* 15–24.

Kitchen, W. D. (1972). *The social behavior and ecology of the pronghorn.* Unpublished Ph.D., University of Michigan.

Kleiman, D. (1977). Monogamy in mammals. *Quarterly Review of Biology, 52,* 39–69.

Kodric-Brown, A., & Brown, J. H. (1984). Truth in advertising: The kinds of traits favored by sexual selection. *American Naturalist, 124,* 309–323.

Koenig, W. D., & Pitelka, F. (1981). Ecological factors and kin selection in the evolution of cooperative breeding in birds. In R. Alexander & D. Tinkle (Eds.), *Natural selection and social behavior* (pp. 261–280). New York: Chiron Press.

Koenig, W., Pitelka, F. A., Carmen, W. J., Mumme, R. L., & Stanback, M. T. (1992). The evolution of delayed dispersal in cooperative breeders. *Quarterly Review of Biology, 67,* 111–150.

Koga, T., Murai, M., & Yong, H. S. (1999). Male-male competition and intersexual interactions in underground mating of the fiddler crab *Uca paradussumieri. Behaviour, 136,* 651–667.

Kogan, N., & Wallach, M. (1964). *Risk taking.* New York: Holt, Rinehart and Winston.

Komdeur, J. (1992). Importance of habitat saturation and territory quality for the evolution of cooperative breeding in the Seychelles warbler. *Nature, 358,* 493–495.

Komers, P. E. (1989). Dominance relationships between juvenile and adult black-billed magpies. *Animal Behaviour, 37,* 256–265.

Konopka, R., & Benzer, S. (1971). Clock mutants of *Drosophila melanogaster. Proceedings of the National Academy of Sciences, U.S.A., 68,* 2112–2116.

Koolhaas, J. M., Korte, S. M., De Boer, S. F., Van Der Vegt, B. J., Van Reenen, C. G., Hopster, H., De Jong, I. C., Ruis, M. A. W., & Blokhuis, H. J. (1999). Coping styles in animals: Current status in behavior and stress-physiology. *Neuroscience and Biobehavioral Reviews, 23,* 925–935.

Koops, M. A., & Grant, J. W. A. (1993). Weight symmetry and sequential assessment in convict cichlid contests. *Canadian Journal of Zoology, 71,* 475–479.

Koopsman, J., Boomsma, D., Heath, A., & Vandoornen, L. (1995). A multivariate genetic analysis of sensation seeking. *Behavior Genetics, 25,* 349–356.

Korona, R. (1989). Ideal free distribution of unequal competitors can be determined by the form of competition. *Journal of Theoretical Biology, 138,* 347–352.

Kraus, W. F. (1989). Surface-wave communication during courtship in the giant water bug, *Abedus indentatus* (Heteroptera, Belostomatidae). *Journal of the Kansas Entomological Society, 62,* 316–328.

Kravitz, E. (1988). Hormonal control of behavior: Amines and the biasing of behavioral output in lobsters. *Science, 241,* 1775–1781.

Krebs, J. (1978). Optimal foraging: Decision rules for predators. In J. Krebs & N. Davies (Eds.), *Behavioral ecology: An evolutionary approach* (2nd ed., pp. 23–63). Sunderland, MA: Sinauer Associates.

Krebs, J. R. (1980). Optimal foraging, predation risk and territory defence. *Ardea, 68,* 83–90.

Krebs, J. R. (1982). Territorial defence in the great tit (*Parus major*): Do residents always win? *Behavioral Ecology and Sociobiology, 11,* 185–194.

Krebs, J. R., Clayton, N., Healy, S., Cristol, D., Patel, S., & Jolliffe, A. (1996). The ecology of avian brain: Food-storing memory and the hippocampus. *Ibis, 138,* 343–346.

Krebs, J. R., & Davies, N. (Eds.). (1979). *Behavioral ecology: An evolutionary approach.* Sunderland, MA: Sinauer Associates.

Krebs, J. R., & Davies, N. (Eds.). (1984). *Behavioral ecology: An evolutionary approach* (2nd ed.). Sunderland, MA: Sinauer Associates.

Krebs, J. R., & Davies, N. (1987). *An introduction to behavioural ecology* (2nd ed.). Oxford: Blackwell Scientific Publications.

Krebs, J. R., & Davies, N. (Eds.). (1991). *Behavioral ecology: An evolutionary approach* (3rd ed.). Sunderland, MA: Sinauer Associates.

Krebs, J. R., & Davies, N. (1993). *An introduction to behavioural ecology.* Oxford: Blackwell Scientific Publications.

Krebs, J. R., & Davies, N. (Eds.). (1997). *Behavioral ecology: An evolutionary approach* (4th ed.). Sunderland, MA: Sinauer Associates.

Krebs, J. R., & Dawkins, R. (1984). Animal signals: Mind-reading and manipulation? In J. R. Krebs & N. B. Davies (Eds.), *Behavioural ecology* (2nd ed., pp. 380–401). Sunderland, MA: Sinauer Associates.

Krebs, J. R., & Inman, A. (1994). Learning and foraging: Individuals, groups, and populations. In L. Real (Ed.), *Behavioral mechanisms in evolutionary ecology* (pp. 46–65). Chicago: University of Chicago Press.

Krebs, J. R., Kacelnik, A., & Taylor, P. (1978). Test of optimal sampling in foraging great tits. *Nature, 275*, 27–30.

Krebs, J. R., & McCleery. (1984). Optimization in behavioural ecology. In J. Krebs & N. Davies (Eds.) *Behavioral ecology: An evolutionary approach* (2nd ed., pp. 91–121). Sunderland, MA: Sinauer Associates.

Krebs, J. R., Sherry, D. F., Healey, S. D., Perry, V. H., & Vaccarino, A. L. (1989). Hippocampal specialization of food-storing birds. *Proceedings of the National Academy of Sciences, U.S.A., 86*, 1388–1392.

Krebs, R., & Loeschcke, V. (1994). Costs and benefits of activation of the heat shock response in *Drosophila melanogaster. Functional Ecology, 8*, 730–737.

Krimbas, C. B. (1984). On adaptation, neo-Darwinian tautology and population fitness. *Evolutionary Biology, 17*, 1–57.

Kroeber, A. L., & Kluckhohn, C. (1952). Culture, a critical review of the concepts and definitions. *Papers of the Peabody Museum, 47*, 1–223.

Kroodsma, D. (1999). Making ecological sense of song development by songbirds. In M. Hauser & M. Konishi (Eds.), *The design of animal communication* (pp. 319–342). Cambridge, MA: MIT Press.

Kropotkin, P. (1902). *Mutual aid* (3rd ed.). London: William Heinemann.

Krushinskaya, N. (1966). Some complex forms of feeding behaviour of nutcracker, *Nucfraga caryocatactes*, after removal of old cortex. *Zhurnal Evoliutsionnoi Biokhimii I Fiziologii, 11*, 563–568.

Krushinskaya, N. (1970). On the problem of memory. *Priroda, 9*, 75–78.

Kruuk, H. (1972). *The spotted hyena: A study of predation and social behavior.* Chicago: University of Chicago Press.

Kuhn, T. (1962). *The structure of scientific revolutions.* Chicago: University of Chicago Press.

Kummer, H. (1978). On the value of social relationships to nonhuman primates: A heuristic scheme. *Social Science Information, 12*, 687–705.

Kunz, T. H., & Allgaier, A. L. (1994). Allomaternal care: Helper-assisted birth in the Rodrigues fruit bat, *Pteropus rodricensis (Chiroptera Pteropodidae). Journal of Zoology, London, 232*, 691–700.

Kynard, B. (1979). Breeding behaviour of a lacustrine population of threespine sticklebacks. *Behaviour, 67*, 178–207.

Kyriacou, C., Oldroyd, M., Wood, J., Sharp, M., & Hill, M. (1990). Clock mutations alter developmental timing in Drosophila. *Heredity, 64*, 395–401.

Lacey, E. A., & Sherman, P. W. (1991). Social organization of naked mole-rats: Evidence for divisions of labor. In P. W. Sherman, J. Jarvis, & R. D. Alexander (Eds.), *The biology of the naked mole-rat* (pp. 275–336). Princeton: Princeton University Press.

Lafleur, D. L., Lozano, G. A., & Sclafini, M. (1997). Female mate-choice copying in guppies, *Poecilia reticulata*: A reevaluation. *Animal Behaviour, 54*, 579–586.

LaGory, K. E. (1987). The influence of habitat and group characteristics on the alarm and flight response of white-tailed deer. *Animal Behaviour, 35*, 20–25.

Laland, K. N. (1994a). On the evolutionary consequences of sexual imprinting. *Evolution, 48*, 477–489.

Laland, K. N. (1994b). Sexual selection with a culturally transmitted mating preference. *Theoretical Population Biology, 45*, 1–15.

Laland, K. N., & Williams, K. (1998). Social transmission of maladaptive information in the guppy. *Behavioral Ecology, 9*, 493–499.

LaMunyon, C., & Ward, S. (1999). Evolution of sperm size in nematodes: Sperm competition favors larger sperm. *Proceedings of the Royal Society of London, 266*, 263–267.

Landau, H. G. (1951a). On dominance relations and the structure of animal societies II. Some effects of possible social causes. *Bulletin of Mathematical Biophysics, 13*, 245–262.

Landau, H. G. (1951b). On dominance relations and the structure of animal societies: I. Effects of inherent characteristics. *Bulletin of Mathematical Biophysics, 13*, 1–19.

Lank, D. B., Coupe, M., & Wynne-Edwards, K. E. (1999). Testosterone-induced male traits in female puffs *(Philomachus pugnax)*. Autosomal inheritance and gender differentiation. *Proceedings of the Royal Society of London, 266*, 2323–2330.

Lank, D. E., Smith, C. M., Hanotte, O., Burke, T., & Cooke, F. (1995). Genetic polymorphism for alternative mating-behavior in lekking male ruff *Philomachus pugnax. Nature, 378*, 59–62.

Laurila, A., & Aho, T. (1997). Do female common frogs choose their breeding habitats to avoid predation in tadpoles. *Oikos, 78*, 585–591.

Law, R., & Koptur, S. (1986). On the evolution of non-specific mutualism. *Biological Journal of the Linnean Society, 27*, 251–267.

Lawick-Goodall, J. (1968). The behaviour of free-living chimpanzees in the Gombe Stream Reserve. *Animal Behavior Monographs, 1*, 161–311.

Lawrence, D., & Fetsinger, L. (1962). *Deterrents and reinforcement: The psychology of insufficient rewards.* Stanford: Stanford University Press.

Le Boeuf, B. J. (1973). Social behaviour in some marine and terrestrial carnivores. In E. Reese & F. Lighter (Eds.), *Contrasts in behavior* (pp. 251–279). New York: Wiley.

Le Boeuf, B. J., & Reiter, J. (1988). Lifetime reproductive success on Northern elephant seals. In T. H. Clutton-Brock (Ed.). (1988). *Reproductive success.* Chicago: University of Chicago Press.

Lee, A., & Bass, A. (1994). *Acoustic behavior in the plainfin midshipman fish.* Paper presented at the Animal Behavior Society 31st meeting.

Lee, J., & Bernays, E. (1990). Food tastes and toxic effects: Associative learning by the polyphagous grasshopper *Schisto-*

cerca americana (Drury) (*Orthoptera: Acrididae*). *Animal Behaviour, 39,* 163–173.

Lefebvre, L., Gaxiola, A., Dawson, S., Timmermans, S., Rosza, L., & Kabai, P. (1998). Feeding innovations and forebrain size in Australasian birds. *Behaviour, 135,* 1077–1097.

Lefebvre, L., Whittle, P., Lascaris, E., & Finkelstein, A. (1997). Feeding innovations and forebrain size in birds. *Animal Behaviour, 53,* 549–560.

Leffler, J. (1993). *An introduction to free radicals.* New York: Wiley.

Leigh, E. (1971). *Adaptation and diversity.* San Francisco: Freeman.

Leimar, O., Austad, S., & Enquist, M. (1991). A test of the sequential assessment game: Fighting in the bowl and doily spider *Frontinella pyramitela. Evolution, 45,* 862–874.

Leimar, O., & Enquist, M. (1984). Effects of asymmetries in owner-intruder conflicts. *Journal of Theoretical Biology, 111,* 475–491.

Lemaire, V., Koehl, M., Le Moal, M., & Abrous, D. (2000). Prenatal stress produces learning deficits associated with an inhibition of neurogenesis in the hippocampus. *Proceedings of the National Academy of Sciences, U.S.A., 97,* 11032–11037.

Lenington, S. (1980). Female choice and polygyny in redwinged blackbirds. *Animal Behaviour, 28,* 347–361.

Leopold, A. S. (1977). *The California quail.* Los Angeles: University of California Press.

Leshner, A. (1981). The role of hormones in the control of submissiveness. In P. Brain & D. Benton (Eds.), *Multidisciplinary approaches to aggression research* (pp. 309–322). New York: Elsevier.

Levins, R. (1968). *Evolution in changing environments.* Princeton: Princeton University Press.

Levitan, D. (2000). Sperm velocity and longevity trade off each other and influence fertilization in the sea urchin *Lytechinus variegatus. Proceedings of the Royal Society of London, 267,* 531–534.

Lewis, W. L., & Takasu, K. (1990). Use of learned odors by a parasitic wasp in accordance with host and food needs. *Nature, 348,* 635–636.

Lewontin, R. C. (1961). Evolution and the theory of games. *Journal of Theoretical Biology, 1,* 382–403.

Licht, T. (1989). Discriminating between hungry and satiated predators: The response of guppies (*Poecilia reticulata*) from high and low predation sites. *Ethology, 82,* 238–243.

Liers, E. (1951). Notes on the river otter (*Lontra canadensis*). *Journal of Mammalogy, 32,* 1–9.

Liley, N., & Seghers, B. (1975). Factors affecting the morphology and behavior of the guppy (*Poecilia reticulata*) in Trinidad. In G. Baerands & A. Manning (Eds.), *Functions and evolution in behavior* (pp. 92–118). Oxford: Oxford University Press.

Lill, A., & Wood-Gush, D. G. M. (1965). Potential ethological isolating mechanism of assortative mating in the domestic fowl. *Behaviour, 25,* 16–44.

Lima, S. L. (1985). Foraging-efficiency-predation-risk tradeoff in the grey squirrel. *Animal Behaviour, 33,* 155–165.

Lima, S. L. (1988a). Initiation and termination of daily feeding in dark-eyed juncos: Influences of predation risk and energy reserves. *Oikos, 53,* 3–11.

Lima, S. L. (1988b). Vigilance during the initiation of daily feeding in dark-eyed juncos. *Oikos, 53,* 12–16.

Lima, S. L., & Dill, L. (1990). Behavioral decisions made under the risk of predation: A review and prospectus. *Canadian Journal of Zoology, 68,* 619–640.

Lima, S. L., & Valone, T. J. (1986). Influence of predation risk on diet selection: A simple example in the grey squirrel. *Animal Behaviour, 34,* 536–544.

Linnell, J. D. C., Odden, J., Smith, M. E., Aanes, R., & Swenson, J. E. (1999). Large carnivores that kill livestock: Do "problem individuals" really exist? *Wildlife Society Bulletin, 27,* 698–705.

Linnell, J. D. C., Smith, M. E., Odden, J., Kaczensky, P., & Swenson, J. E. (1996). Strategies for the reduction of carnivore-livestocks: A review. *Norwegian Institute for Nature Research Oppdragsmelding, 443,* 1–118.

Linnell, J. D. C., Swenson, J. E., & Andersen, R. (2000). Conservation of biodiversity in Scandinavian boreal forests: Large carnivores as flagships, umbrellas, indicators, or keystones? *Biodiversity and Conservation, 9,* 857–868.

Linsdale, J. (1946). *The California ground squirrel.* Berkeley: University of California Press.

Lithgow, G. (1996). Molecular genetics of *Caenorhabditis elegans.* In E. Schneider & J. Rowe (Eds.), *Handbook of the biology of aging.* San Diego: Academic Press.

Lithgow, G., & Kirkwood, T. (1996). Mechanisms and evolution of aging. *Science, 273,* 80.

Littledyke, M., & Cherrett, J. (1976). Direct ingestion of plant sap from leaves cut by the leaf-cutting ants *Atta cephalotes* and *Acromyrmex octospinas. Bulletin of Entomological Research, 66,* 205–217.

Longino, J. (1984). True anting by a capuchin (*Cebus capucinus*). *Primates, 25,* 243–245.

Lorenz, K. (1935). Der Kumpan in der Umwelt des Vogels. *Journal of Ornithology, 83,* 137–213, 289–413.

Lorenz, K. (1966). *On aggression.* New York: Harcourt, Brace and World.

Losey, G. S., Stanton, F. G., Jr., Telecky, T. M., Tyler, W. A. I., & Class, Z. G. S. (1986). Copying others, an evolutionarily stable strategy for mate choice: A model. *American Naturalist, 128,* 653–664.

Loughry, W. J. (1987). The dynamics of snake harassment by black-tailed prairie dogs. *Behaviour, 103,* 27–48.

Low, B. (2000). *Why sex matters: A Darwinian look at human behavior.* Princeton: Princeton University Press.

Lozano, G. (1998). Parasitic stress and self-medication in wild animals. *Advances in the Study of Behavior, 27,* 291–317.

Lucas, N. S., Hume, E. M., & Henderson, H. (1937). On the breeding of the common marmoset (*Hapale jacchus*) in captivity when irradiated with ultra-violet rays. II. A ten years family history. *Proceedings of the Zoological Society of London (A), 107,* 205–211.

Luckinbill, L., Arking, R., Clare, M., Cirocco, W., & Buck, S. (1984). Selection for delayed senescence in *Drosophila melanogaster. Evolution, 38,* 996–1003.

Lycett, J., Dunbar, R., & Voland, E. (2000). Longevity and the costs of reproduction in a historical human population. *Proceedings of the Royal Society of London, 267,* 31–35.

Lyon, B., & Eadie, J. (1991). Mode of development and interspecific avian brood parasitism. *Behavioral Ecology, 2,* 309–318.

MacArthur, R., & Pianka, E. (1966). On optimal use of a patchy environment. *American Naturalist, 100,* 603–609.

Mace, G., Harvey, P., & Clutton-Brock, T. (1981). Brain size and ecology in small mammals. *Journal of Zoology, 193,* 333–354.

Macedonia, J. M. (1990). What is communicated in the antipredator calls of lemurs: Evidence from playback experiments with ringtailed and ruffed lemurs. *Ethology, 86,* 177–190.

MacFadden, B., & Jones, D. S. (1985). Magnetic butterflies: A case study of the monarch. In J. Kirschvink, D. S. Jones, & B. MacFadden (Eds.), *Magnetite biomineralization and magnetoreception in organisms* (pp. 407–425). New York: Plenum Press.

Macphail, E. (1987). The comparative psychology of intelligence. *Behavioral and Brain Sciences, 10,* 645–656.

Maddison, P., & Maddison, W. (1992). MacClade (Version 3.05). Sunderland, MA: Sinauer.

Maestripieri, D. (1993). Vigilance costs of allogrooming in macaque mothers. *American Naturalist, 141,* 744–753.

Magurran, A. E. (1990). The inheritance and development of minnow anti-predator behaviour. *Animal Behaviour, 39,* 834–842.

Magurran, A. E., & Higham, A. (1988). Information transfer across fish shoals under predator threat. *Ethology, 78,* 153–158.

Magurran, A. E., & Pitcher, T. J. (1987). Provenance, shoal size and the sociobiology of predator-evasion in minnow shoals. *Proceedings of the Royal Society of London, 229,* 439–465.

Magurran, A. E., & Seghers, B. H. (1990). Population differences in predator recognition and attack cone avoidance in the guppy *Poecilia reticulata. Animal Behaviour, 40,* 443–452.

Magurran, A. E., & Seghers, B. H. (1991). Variation in schooling and aggression amongst guppy populations in Trinidad. *Behaviour, 118,* 214–234.

Magurran, A. E., & Seghers, B. H. (1993).Evolution of adaptive variation in antipredator behaviour. *Marine Behaviour and Physiology, 23,* 29–44.

Magurran, A. E., Seghers, B. H., Carvalho, G. R., & Shaw, P. W. (1992). Behavioral consequences of an artificial introduction of guppies (*Poecilia reticulata*) in N. Trinidad: Evidence for the evolution of anti-predator behavior in the wild. *Proceedings of the Royal Society of London, 248,* 117–122.

Magurran, A. E., Seghers, B. H., Carvalho, G., & Shaw, P. (1993). Evolution of adaptive variation in antipredator behaviour. *Marine Behaviour and Physiology, 23,* 29–44.

Magurran, A. E., Seghers, B. H., Shaw, P., & Carvalho, G. (1994). Schooling preferences for familiar fish in the guppy, *Poecilia reticulata. Journal of Fish Biology, 45,* 401–406.

Magurran, A. E., Seghers, B. H., Shaw, P. W., & Carvalho, G. R. (1995). The behavioral diversity and evolution of guppy, *Poecilia reticulata,* populations in Trinidad. *Advances in the Study of Behavior, 24,* 155–202.

Mahaney, W. C. (1993). Scanning electron microscopy of earth mined and eaten by mountain gorilla in the Virunga Mountains, Rwanda. *Primates, 34,* 311–319.

Mahaney, W. C., Hancock, R., Aufreiter, S., & Huffman, M. (1996). Geochemistry and clay mineralogy of termite mound soil and the role of geophagy in chimpanzees of the Mahale Mountains, Tanzania. *Primates, 37,* 121–134.

Mahaney, W. C., Hancock, R., & Inoue, M. (1993). Geochemistry and clay mineralogy of soils eaten by Japanese macaques. *Primates, 34,* 85–91.

Malm, K., & Jensen, P. (1997). Weaning and parent-offspring conflict in the domestic dog. *Ethology, 103,* 653–664.

Manuck, S., Kaplan, J., & Clarkson, T. (1983). Behaviorally induced heart rate reactivity and atherosclerosis in synomolgous monkeys. *Psychosomatic Medicine, 45,* 95–108.

Marchetti, C., & Drent, P. (2000). Individual differences in the use of social information in foraging by captive great tits. *Animal Behaviour, 60,* 131–140.

Markl, H. (1968). Die verstandigung durch stridulationssignale bei blattschneiderameisesn. II Erzeugung und eigenschaften der signale. *Zeitschrift fur Vergleichende Physiologie, 60,* 103–150.

Marks, I. (1987). *Fear, phobias and rituals.* New York: Oxford University Press.

Marks, I., & Nesse, R. (1994). Fear and fitness: An evolutionary analysis of anxiety disorders. *Ethology and Sociobiology, 15,* 247–261.

Marler, P. (1968). Visual systems. In T. Sebeok (Ed.), *Animal communications* (pp. 103–126). Bloomington, IN: Indiana University Press.

Marler, P. (1999). On innateness: Are sparrow songs "learned" or "innate." In M. Hauser & M. Konishi (Eds.), *The design of animal communication* (pp. 293–318). Cambridge, MA: MIT Press

Marler, P., & Terrace, H. (Eds.). (1984). *The biology of learning.* New York: Springer-Verlag.

Marshall, A., Wrangham, R., & Arcadi, A. (1999). Does learning affect the structure of vocalizations in chimpanzees? *Animal Behaviour, 58,* 825–830.

Marshall, J. C. (1970). The biology of communication in man and animals. In J. Lyons (Ed.), *New horizons in linguistics.* Harmondsworth: Penguin.

Martin, P., & Caro, T. (1985). On the functions of play and its role in behavioral development. *Advances in the Study of Behavior, 15,* 59–103.

Masseti, M. (2000). Did the study of ethology begin in Crete 4000 years ago? *Ethology, Ecology and Evolution, 12,* 89–96.

Mather, J. A. (1991). Foraging, feeding and prey remains in middens of juvenile *Octopus vulgaris* (mollusca, cephalopoda). *Journal of Zoology, 224,* 27–39.

Mather, J. A. (1995). Cognition in cephalopods. *Advances in the Study of Behavior, 24,* 317–353.

Mather, J., & Anderson, R. C. (1993). Personalities of octopuses (*Octopus rubescens*). *Journal of Comparative Psychology, 107,* 336–340.

Mather, J. A., & Anderson, R. C. (1999). Exploration, play, and habituation in octopuses (*Octopus dofleini*). *Journal of Comparative Psychology, 113,* 333–338.

Mathis, A., & Smith, R. J. F. (1993a). Chemical labeling of northern pike (*Esox lucius*) by the alarm pheromone of fathead minnows (*Pimephales promelas*). *Journal of Chemical Ecology, 19,* 1967–1979.

Mathis, A., & Smith, R. J. F. (1993b). Fathead minnows, *Pimephales promelas,* learn to recognize northern pike, *Esox lucius,* as predators on the basis of chemical stimuli from minnows in the pike's diet. *Animal Behaviour, 46,* 645–656.

Matsuoka, S. (1980). Pseudo warning call in tit mice. *Tori, 29,* 87–90.

Maynard Smith, J. (1974a). *Models in ecology.* Cambridge: University of Cambridge Press.

Maynard Smith, J. (1974b). The theory of games and the evolution of animal conflicts. *Journal of Theoretical Biology, 47,* 209–221.

Maynard Smith, J. (1982). *Evolution and the theory of games.* Cambridge: Cambridge University Press.

Maynard Smith, J., & Parker, G. A. (1976). The logic of asymmetric contests. *Animal Behaviour, 24,* 159–175.

Maynard Smith, J., & Price, G. (1973). The logic of animal conflict. *Nature, 246,* 15–18.

Mayr, E. (1961). Cause and effect in biology. *Science, 134,* 1501–1506.

Mayr, E. (1982). *The growth of biological thought.* Cambridge, MA: Harvard University Press.

Mayr, E. (1983). How to carry out the adaptationist program. *American Naturalist, 121,* 324–334.

Mazur, A., & Booth, A. (1998). Testosterone and dominance in men. *Behavioral and Brain Sciences, 21,* 353–362.

McComb, K., & Clutton-Brock, T. H. (1994). Is mate choice copying or aggregation responsible for skewed distributions of females on leks? *Proceedings of the Royal Society of London, 255,* 13–19.

McCullough, D. R. (1969). The tule elk: Its history, behavior and ecology. *University of California Publications in Zoology, 88,* 1–209.

McEwen, B. (1992). Reexamination of the glucocorticoid hypothesis of stress and aging. *Brain Research, 93,* 365–387.

McEwen, B., & Sapolsky, R. M. (1995). Stress and cognitive function. *Current Opinions in Neurobiology, 5,* 205–210.

McGregor, P. K., & Peake, T. (2000). Communication networks: Social environments for receiving and signalling behaviour. *Acta Ethologica, 2,* 71–81.

McGuire, M., & Troisi, A. (1998). *Darwinian psychiatry.* New York: Oxford University Press.

McKibben, J., Bodnar, D., & Bass, A. (1995). *Everybody's humming but is anybody listening: Acoustic communications in a marine teleost fish.* Paper presented at the Fourth International Congress of Neuroethology.

McKinney, F., Cheng, K., & Bruggers, D. (1984). Sperm competition in apparently monogamous birds. In R. Smith (Ed.), *Sperm competition and the evolution of animal mating systems* (pp. 523–545). New York: Academic Press.

McMann, S. (1993). Contextual signaling and the structure of dyadic encounters in *Anolis carolinensis. Animal Behaviour, 46,* 657–668.

McNab, B. K. (1973). Energetics and distribution of vampires. *Journal of Mammalogy, 54,* 131–144.

McNamara, J. M., & Houston, A. I. (1990). State-dependent ideal free distributions. *Evolutionary Ecology, 4,* 298–311.

McNamara, J. M., & Houston, A. I. (1992). Risk sensitive foraging: A review of the theory. *Bulletin of Mathematical Biology, 54,* 355–378.

Medawar, P. (1952). *An unsolved problem in biology.* London: H. K. Lewis.

Meder, A. (1990). Sex differences in the behavior of immature captive lowland gorillas. *Primates, 31,* 51–63.

Meltzoff, A. N. (1999). Origins of theory of mind, cognition, and communication. *Journal of Communication Disorders, 32,* 251–269.

Meltzoff, A. N., & Moore, M. K. (1977). Imitation of facial and manual gestures by human neonates. *Science, 198,* 75–78.

Mendozagranados, D., & Sommer, V. (1995). Play in chimpanzees of the Arnhem Zoo: Self-serving compromises. *Primates, 36,* 57–68.

Mesterton-Gibbons, M. (1992). Ecotypic variation in an asymmetric Hawk-Dove game: When is bourgeois an ESS? *Evolutionary Ecology 6,* 198–222.

Mesterton-Gibbons, M. (1999). On the evolution of pure winner and loser effects: A game-theoretic model. *Bulletin of Mathematical Biology, 61,* 1151–1186.

Mesterton-Gibbons, M., & Adams, E. (1998). Animal contests as evolutionary games. *American Scientist, 86,* 334–341.

Mesterton-Gibbons, M., & Dugatkin, L. A. (1992). Cooperation among unrelated individuals: Evolutionary factors. *Quarterly Review of Biology, 67,* 267–281.

Meyer, A., Morrissey, J. M., & Schartl, M. (1994). Recurrent origin of a sexually selected trait in Xiphophorus fishes inferred from a molecular phylogeny. *Nature, 368,* 539–542.

Milinski, M. (1977). Do all members of a swarm suffer the same predation? *Zeitschrift fur Tierpsychologie, 45,* 373–378.

Milinski, M. (1979a). Can an experienced predator overcome the confusion of swarming prey more easily? *Animal Behaviour, 27,* 1122–1126.

Milinski, M. (1979b). An evolutionarily stable feeding strategy in sticklebacks. *Zeitschrift fur Tierpsychologie, 51,* 36–40.

Milinski, M. (1987). TIT FOR TAT and the evolution of cooperation in sticklebacks. *Nature, 325,* 433–435.

Milinski, M., & Bakker, T. (1990). Female sticklebacks use male coloration in mate choice and hence avoid parasitized sticklebacks. *Nature, 344,* 330–333.

Milinski, M., & Boltshauser, P. (1995). Boldness and predator deterrence: A critique to Godin and Davis. *Proceedings of the Royal Society of London, 262,* 103–105.

Milinski, M., Kulling, D., & Kettler, R. (1990). Tit for Tat: Sticklebacks "trusting" a cooperating partner. *Behavioral Ecology, 1,* 7–12.

Milinski, M., Luthi, J., Eggler, R., & Parker, G. (1997). Cooperation under predation risk: Experiments on costs and benefits. *Proceedings of the Royal Society of London, 264,* 831–837.

Milinski, M., & Wedekind, C. (2001). Evidence for MHC-correlated perfume preferences in humans. *Behavioral Ecology, 12,* 140–149.

Miller, J., Crowder, L., & Moser, M. (1985). Migration and utilization of estuarine nurseries by juvenile fishes: An evolutionary perspective. In M. Rankin (Ed.), *Migration: Mechanisms and adaptive significance.* Austin, TX: Marine Science Center.

Miller, L., & Nadler, R. (1981). Mother-infant development in captive chimpanzees and orangutans. *International Journal of Primatology, 2,* 247–261.

Mineka, S., & Cook, M. (1983). Social learning and the acquisition of snake fear in monkeys. In T. Zentall & B. G. Galef (Eds.), *Social learning. Psychological and biological perspectives* (pp. 51–73). Hillsdale, NJ: Lawrence Erlbaum Associates.

Mineka, S., Davidson, M., Cook, M., & Keir, R. (1984). Observational conditioning of snake fear in rhesus monkeys. *Journal of Abnormal Psychology 93,* 355–372.

Mitani, J., Hasegawa, T., Gros-Louis, J., Marler, P., & Byrne, R. (1992). Dialects in wild chimpanzees? *American Journal of Primatology, 27,* 233–243.

Mitchell, W. A., & Valone, T. J. (1990). Commentary: The optimization approach—studying adaptations by their function. *Quarterly Review of Biology, 65,* 43–52.

Mitman, G. (1988). From the population to society: The cooperative metaphors of W. C. Allee and A. E. Emerson. *Journal of the History of Biology 21,* 173–194.

Mitman, G. (1992). *The state of nature: Ecology, community and American social thought, 1900–1950.* Chicago: University of Chicago Press.

Mittlebach, G. (1984). Group size and feeding rate in bluegills. *Copeia, 1984.* 998–1000.

Mitzutani, F. (1993). Home range of leopards and their impact on livestocks on Kenyan ranches. *Zoological Society of London (Symposium), 65,* 425–439.

Mock, D. (1980). White-dark polymorphism in herons. In D. L. Drawe (Ed.), *Proceedings of the First Welder Wildlife Symposium* (pp. 145–161). Sinton, TX: Welder Wildlife Foundation.

Mock, D., & Parker, G. (1997). *The evolution of sibling rivalry.* New York: Oxford University Press.

Molina-Borja, M., Padron-Fumero, M., & Alfonso-Martin, T. (1998). Morphological and behavioural traits affecting the intensity and outcome of male contests in *Gallotia galloti galloti* (family *Lacertidae*). *Ethology, 104,* 314–322.

Moller, A., Hell, D., & Krober, H. L. (1998). Sensation seeking: A review of the theory and its applications. *Fortschritte der Neurologie Psychiatrie, 66,* 487–495.

Møller, A. P. (1988). False alarm calls as a means of resource usurpation in the great tit *Parus major. Ethology, 79,* 25–30.

Møller, A. P. (1990a). Deceptive use of alarm calls by male swallows, *Hirundo rustica:* A new paternity guard. *Behavioral Ecology, 1,* 1–6.

Møller, A. P. (1990b). Effects of an haematophagous mite on the barn swallow (*Hirundo rustica*). *Ecology, 71,* 2345–2357.

Møller, A. P. (1994). *Sexual selection and the barn swallow.* Oxford: Oxford University Press.

Møller, A. P., & Erritzøe, J. (1996). Parasite virulence and host immune defense: Host immune response is related to nest re-use in birds. *Evolution, 50,* 2066–2072.

Møller, A. P., & Erritzøe, J. (1998). Host immune defence and migration in birds. *Evolutionary Ecology, 12,* 945–953.

Møller, A. P., & Swaddle, J. (1998). *Asymmetry, developmental stability, and evolution.* Oxford: Oxford University Press.

Møller, A. P., & Thornhill, R. (1998). Bilateral symmetry and sexual selection: A meta-analysis. *American Naturalist, 151,* 174–192.

Montgomerie, R. (2000). Sweet FA: The trouble with fluctuating asymmetry. *Behavioral and Brain Sciences, 23,* 616+.

Montgomery, R., Lyon, B., & Holder, K. (2001). Dirty ptarmigan: Behavioral modification of conspicuous male plumage. *Behavioral Ecology, 12,* 429–438.

Moore, H. D. M., Martin, M., & Birkhead, T. R. (1999). No evidence for killer sperm or other selective interactions between human spermatozoa in ejaculates of different males in vitro. *Proceedings of the Royal Society of London, 266,* 2343–2350.

Moore, M., & Lindzey, J. (1992). The physiological basis of sexual behavior in male reptiles. In C. Gans & D. Crews (Eds.), *Biology of reptilia* (Vol. 18, pp. 70–113). Chicago: University of Chicago Press.

Moore, R. (2000). *An introduction to behavioral endocrinology* (2nd ed.). Sunderland, MA: Sinauer Associates.

Moore, R., Newton, B., & Sih, A. (1996). Delayed hatching as a response of streamside salamander eggs to chemical cues from predatory sunfish. *Oikos, 77,* 331–335.

Morgan, E. (1984). Chemical words and phrases in the language of pheromones for foraging and recruitment. In T. D. Lewis (Ed.), *Communication* (pp. 169–194). London: Academic Press.

Morgan, J., & Curran, T. (1991). Stimulus transmission using coupling in the nervous system: Involvement of the inducible proto-oncogenes fos and jun. *Annual Review of Neuroscience, 14,* 421–451.

Morris, D. W. (1994). Habitat matching: Alternatives and implications to populations and communities. *Evolutionary Ecology, 8,* 387–406.

Morse, D. H. (1970). Ecological aspects of some mixed species foraging flocks of birds. *Ecological Monographs, 4,* 119–168.

Morton, D., & Chiel, H. (1994). Neural architecture for adaptive behavior. *Trends in Neuroscience, 10,* 413–420.

Morton, E. S. (1977). On the occurrence and significance of motivation-structural rules in some bird and animal sounds. *American Naturalist, 111,* 855–869.

Mouget, F., & Bretagnolle, V. (2000). Predation as a cost of sexual communication in nocturnal seabirds: An experimental approach using acoustic signals. *Animal Behaviour, 60,* 647–656.

Mousseau, T. A., & Roff, D. A. (1987). Natural selection and the heritability of fitness components. *Heredity, 59,* 181–197.

Mousseau, T., Sinervo, B., & Endler, J. (Eds.). (1999). *Adaptive genetic variation in the wild.* New York: Oxford University Press.

Moynihan, M. (1998). *The social regulation of competition and aggression in animals.* Washington, DC: Smithsonian Institution Press.

Mueller, U. (1991). Haplodiploidy and the evolution of facultative sex ratios in a primitively eusocial bee. *Science, 254,* 442–444.

Mulder, R., Dunn, P., Cockburn, A., Lazenby-Cohen, K., & Howell, M. (1994). Helpers liberate female fairy-wrens from constraints on extra-pair mate choice. *Proceedings of the Royal Society of London, 255,* 223–229.

Munck, A., Guyre, P., & Holbrook, N. (1984). Physiological functions of glucocorticoids in stress and the relation to pharmacological actions. *Endocrine Reviews, 5,* 25–41.

Mundinger, P. C. (1999). Genetics of canary song learning: Innate mechanisms and other neurobiological considerations. In M. Hauser & M. Konishi (Eds.), *The design of animal communication* (pp. 369–390). Cambridge, MA: MIT Press.

Munn, C. A. (1986). Birds that "cry wolf." *Nature, 319,* 143–145.

Munro, R., Smith, L. T., & Jupa, J. (1968). The genetic basis of color differences observed in mute swan. *Auk, 85,* 504–505.

Nash, S., & Domjan, M. (1991). Learning to discriminate the sex of conspecifics in male Japanese quail (*Coturnix coturnix japonica*): Tests of "biological constraints." *Journal of Experimental Psychology—Animal Behavior Processes, 17,* 342–353.

Nelissen, M. H. J. (1985). Structure of the dominance hierarchy and dominance determining "group factors" in *Melanochromis auratus. Behaviour, 94,* 85–107.

Nelissen, M. H. J. (1991). Communication. In M. Keenleyside (Ed.), *Cichlid fishes: Behaviour, ecology and evolution* (pp. 225–241). London: Chapman and Hall.

Nelson, R. (2000). *An introduction to behavioral endocrinology.* Sunderland, MA: Sinauer Associates.

Nesse, R. M. (1988). Life table tests of evolutionary theories of senescence. *Experimental Gerontology, 23,* 445–453.

Nesse, R., & Williams, G. C. (1995). *Why we get sick: The new science of Darwinian medicine.* New York: Vintage Books.

Niesink, R., & Van Ree, J. (1989). Involvement of opioid and dopaminergic systems in isolation-induced pinning and social grooming of young rats. *Neuropharmacology, 28,* 411–418.

Nishida, T. (1979). The social structure of chimpanzees of the Mahale Mountains. In D. A. Hamburg & E. R. McCown (Eds.), *The great apes.* Menlo Park, CA: Benjamin/Cummings.

Nishida, T., Hasegawa, T., Hayaki, H., Takahata, Y., & Uehara, S. (1992). Meat-sharing as a coalition strategy by an alpha male chimpanzee? In T. Nishida (Ed.), *The chimpanzees of the Mahale Mountains.* Tokyo: University of Tokyo Press.

Nishida, T., Hiraiwa-Hasegawa, M., Hasegawa, T., & Takahata, Y. (1985). Group extinction and female transfer in wild chimpanzees in the Mahale National Park, Tanzania. *Zeitschrift fur Tierpsychologie, 67,* 284–301.

Nishida, T., Uehara, S., & Nyondo, R. (1983). Predatory behaviour among wild chimpanzees of the Mahale Mountains. *Primates, 20,* 1–20.

Noe, R. (1986). Lasting alliances among adult male Savannah baboons. In J. Else & P. Lee (Eds.), *Primate ontogeny, cognition and social behaviour* (pp. 381–392). Cambridge: Cambridge University Press.

Normansell, L., & Panksepp, J. (1990). Effects of morphine and naloxone on play-rewarded social discrimination in juvenile rats. *Developmental Psychobiology, 23,* 75–83.

Norris, D. (1996). *Vertebrate endocrinology.* New York: Academic Press.

Nottebohm, F. (1999). The anatomy and timing of vocal learning in birds. In M. Hauser & M. Konishi (Eds.), *The de-*

sign of animal communication (pp. 63–110). Cambridge, MA: MIT Press.

Nottebohm, F., & Nottebohm, M. E. (1976). Left hypoglossal dominance in control of canary and white-crowned sparrow song. *Journal of Comparative Physiology, 108,* 171–192, 1368–1370.

Nowak, M., & Krakauer, D. (1999). The evolution of language. *Proceedings of the National Academy of Sciences, U.S.A., 96,* 8028–8033.

Nowak, M., Plotkin, J., & Jansen, V. (2000). The evolution of syntactic communication. *Nature, 404,* 495–498.

Nunes, S., Duniec, T., Schweppe, S., & Holekamp, K. (1999a). Energetic and endocrine mediation of natal dispersal behavior in Belding's ground squirrels. *Hormones and Behavior, 35,* 113–124.

Nunes, S., Muecke, E. M., Anthony, J., & Batterbee, A. S. (1999b). Endocrine and energetic mediation of play behavior in free-living Belding's ground squirrels. *Hormones and Behavior, 36,* 153–165.

Nursall, J. R. (1973). Some behavioral interactions of spottail shiners (*Notropis hudsonius*), yellow perch (*Perca flavescens*) and northern pike (*Esox lucius*). *Journal of the Fisheries Research Board of Canada, 30,* 1161–1178.

O'Brien, S. J., Wildt, D. E., Goldman, D., Merril, C. R., & Bush, M. (1983). The cheetah is depauperate in genetic variation. *Science, 221,* 459–462.

O'Connor, K. I., Metcalfe, N. B., & Taylor, A. C. (1999). Does darkening signal submission in territorial contests between juvenile Atlantic salmon, *Salmo salar*? *Animal Behaviour, 58,* 1269–1276.

O'Donald, P. (1980). *Genetic models of sexual selection.* Cambridge: University of Cambridge Press.

Oli, M., Taylor, K., & Rogers, M. (1994). Snow leopard (*Panthera uncia*) predation of livestock: An assessment of local perceptions in the Annapurna conservation area, Nepal. *Biological Conservation, 68,* 63–68.

Oliveira, R. F., McGregor, P. K., & Latruffe, C. (1998). Know thine enemy: Fighting fish gather information from observing conspecific interactions. *Proceedings of the Royal Society of London, 265,* 1045–1049.

Ollson, O., Wiktander, U., Holmgren, N., & Nilsson, S. (1999). Gaining ecological information about Bayesian foragers through their behaviour II. A field test with woodpeckers. *Oikos, 87,* 264–276.

Orgeur, P. (1995). Sexual play behavior in lambs androgenized in utero. *Physiology and Behavior, 57,* 185–187.

Orians, G. (1962). Natural selection and ecological theory. *American Naturalist, 96,* 257–263.

Orians, G. (1969). On the evolution of mating systems in birds and mammals. *American Naturalist, 103,* 589–603.

Oring, L. W., Fleisher, R., Reed, J., & Marsden, K. (1992). Cuckoldry through stored sperm in the sequentially polyandrous spotted sandpiper. *Nature, 359,* 631–633.

Ortega, C. (1998). *Cowbirds and other brood parasites.* Tucson, AZ: University of Arizona Press.

Oster, G. F., Eshel, I., & Cohen, D. (1977). Worker-queen conflict and the evolution of social insects. *Theoretical Population Biology, 12,* 49–85.

Oster, G. F., & Wilson, E. O. (1978). *Caste and ecology in the social insects.* Princeton: Princeton University Press.

Owens, O., Dixon, A., Burke, T., & Thompson, D. (1995). Strategic paternity assurance in the sex-role reversed Eurasian dotterel (*Charadrius morinellus*): Behavior and genetic evidence. *Behavioral Ecology, 6,* 14–21.

Owings, D. H., & Coss, R. G. (1977). Snake mobbing by California ground squirrels: Adaptive variation and ontogeny. *Behaviour, 62,* 50–69.

Owings, D. H., & Leger, D. W. (1980). Chatter vocalization of California ground squirrels: Predator- and social-role specificity. *Zeitschrift fur Tierpsychologie, 54,* 163–184.

Packer, C. (1977). Reciprocal altruism in *Papio anubis. Nature, 265,* 441–443.

Packer, C., Gilbert, D., Pusey, A., & O'Brien, S. (1991). A molecular genetic analysis of kinship and cooperation in African lions. *Nature, 351,* 562–565.

Packer, C., Herbst, L., Pusey, A. E., Bygott, J. D., Hanby, J. P., Cairns, S. J., & Borgerhoff-Mulder, M. (1988). Reproductive success of lions. In T. H. Clutton-Brock (Ed.), *Reproductive success* (pp. 363–383). Chicago: University of Chicago Press.

Packer, C., & Pusey, A. E. (1982). Cooperation and competition within coalitions of male lions: Kin selection or game theory? *Nature, 296,* 740–742.

Packer, C., & Rutton, L. (1988). The evolution of cooperative hunting. *American Naturalist, 132,* 159–194.

Palameta, B., & Lefebvre, L. (1985). The social transmission of a food-finding technique in pigeons: What is learned? *Animal Behaviour, 33,* 892–896.

Palmer, A. R. (1996). Waltzing with asymmetry. *Bioscience, 46,* 518–532.

Palmer, A. R. (1999). Detecting publication bias in meta-analyses: A case study of fluctuating asymmetry and sexual selection. *American Naturalist, 154,* 220–223.

Panksepp, J., Siviy, S., & Normansell, L. (1984). The psychobiology of play: Theoretical and methodological considerations. *Neuroscience and Biobehavioral Reviews, 8,* 465–492.

Papaj, D., & Lewis, A. (Eds.). (1993). *Insect learning: Ecological and evolutionary perspectives.* New York: Chapman and Hall.

Papaj, D., & Prokopy, R. J. (1989). Ecological and evolutionary aspects of learning in phytophagus insects. *Annual Review of Entomology, 34,* 315–350.

Parker, G. A. (1970a). Sperm competition and its evolutionary effect on copula duration in the fly *Scatophagua stercoraria. Journal of Insect Physiology, 16,* 1301–1328.

Parker, G. A. (1970b). Sperm competition and its evolutionary consequences in insects. *Biological Reviews of the Cambridge Philosophical Society, 45,* 525–567.

Parker, G. A. (1974a). Assessment strategy and the evolution of fighting behaviour. *Journal of Theoretical Biology, 47,* 223–243.

Parker, G. A. (1974b). Courtship persistence and female-guarding as male investment strategies. *Behaviour, 48,* 157–184.

Parker, G. A. (1974c). The reproductive behavior and nature of sexual selection in *Scatophaga stercoraria.* IX. Spatial distribution of fertilization rates and evolution of male search strategy within the reproductive area. *Evolution, 28,* 93–108.

Parker, G. A. (2001). Golden flies, sunlit meadows: A tribute to the yellow dungfly. In L. A. Dugatkin (Ed.), *Model systems in behavioral ecology: Integrating conceptual, theoretical and empirical perspectives* (pp. 3–26). Princeton: Princeton University Press.

Parker, G. A., & Maynard Smith, J. (1987). The distribution of stay times in *Scatophaga*: A reply to Curtsinger. *American Naturalist, 129,* 621–628.

Parker, G. A., & Rubinstein, D. (1981). Role assessment, reserve strategy and the acquisition of information in asymmetric animal contests. *Animal Behaviour, 29,* 221–240.

Parker, G. A., & Stuart, R. A. (1976). Animal behavior as a strategy optimizer: Evolution of resource assessment strategies and optimal emigration thresholds. *American Naturalist, 110,* 1055–1076.

Parker, G. A., & Sutherland, W. J. (1986). Ideal free distribution when individuals differ in competitive ability: Phenotype-limited ideal free models. *Animal Behaviour, 34,* 1222–1242.

Parker, G. A., & Thompson, E. (1980). Dung fly struggles: A test of the war of attrition. *Behavioral Ecology and Sociobiology, 7,* 37–44.

Parker, P. G., Waite, T. A., Heinrich, B., & Marzluff, J. M. (1994). Do common ravens share ephemeral food resources with kin? DNA fingerprinting evidence. *Animal Behaviour, 48,* 1085–1093.

Parmigian, S., & Vom Saal, F. (Eds.). (1995). *Infanticide and parental care.* New York: Gordon & Breach Science Publishers.

Partridge, L., & Farquhar, M. (1981). Sexual activity reduces lifespan of male fruit flies. *Nature, 294,* 580–582.

Partridge, L., & Fowler, K. (1990). Non-mating costs of exposure to males in female *Drosophila melanogaster. Journal of Insect Physiology, 36,* 419–425.

Partridge, L., Fowler, K., Trevitt, S., & Sharp, W. (1986). An examination of the effects of males on the survival and egg-production rates of female *Drosophila melanogaster. Journal of Insect Physiology, 32,* 925–929.

Partridge, L., Green, A., & Fowler, K. (1987). Effects of egg production and of exposure to males on female survival in *Drosophila melanogaster. Journal of Insect Physiology, 33,* 745–749.

Pavlov, I. P. (1927). *Conditioned reflexes.* New York: Oxford University Press.

Payne, R. (1977). The ecology of brood parasitism in birds. *Annual Review of Ecology and Systematics, 8,* 1–28.

Payne, R., Tyack, P., & Payne, L. (1983). Progressive changes in the songs of humpback whales (*Megaptera novaeangliae*): A detailed analysis of two seasons in Hawaii. In R. Payne (Ed.), *Communication and behavior of whales* (pp. 9–57). Boulder, CO: Westview Press.

Pederson, J., Glickma, S., Frank, L., & Beach, F. (1990). Sex differences in the play behavior of immature spotted hyeanas, *Crocuta crocuta. Hormones and Behavior, 24,* 403–420.

Pellegrini, A. D. (1995). Boys rough and tumble play and social competence: Contemporaneous and longitudinal relations. In A. D. Pellegrini (Ed.), *The future of play theory: A multidisciplinary inquiry into the contributions of Brian Sutton-Smith* (pp. 107–126). Albany, NY: SUNY Albany Press.

Pellegrini, A., & Smith, P. K. (1998). Physical activity play: The nature and function of a neglected aspect of play. *Child Development, 69,* 577–598.

Pellis, S., Casteneda, E., McKenna, M., Tran-Nguten, L., & Whishaw, I. B. (1993). The role of the striatum in organizing sequences of play fighting in neonatally dopamine depleted rats. *Neuroscience Letters, 158,* 13–15.

Pellis, S., & Iwaniuk, A. (1999). The roles of phylogeny and sociality in the evolution of social play in muriod rodents. *Animal Behaviour, 58,* 361–373.

Pellis, S., Pellis, V., & Kolb, B. (1992a). Neonatal testosterone augmentation increases juvenile play fighting but does not influence the adult dominance relationships of male rats. *Aggressive Behavior, 18,* 437–442.

Pellis, S., Pellis, V., & Whishaw, I. B. (1992b). The role of the cortex in play fighting by rats: Developmental and evolutionary implications. *Brain, Behavior and Evolution, 39,* 270–284.

Penn, D., & Potts, W. (1998). How do major histocompatibility complex genes influence odor and mating preferences? *Advances in Immunology, 69,* 411–436.

Penn, D., & Potts, W. (1999). The evolution of mating preferences and the major histocompatibility complex genes. *American Naturalist, 153,* 145–164.

Pepper, J. W., Braude, S. H., Lacey, E. A., & Sherman, P. W. (1991). Vocalizations of the naked mole-rat. In P. W. Sherman, J. Jarvis, & R. D. Alexander (Eds.), *The biology of the naked mole-rat* (pp. 243–274). Princeton: Princeton University Press.

Perez, S. M., Taylor, O. R., & Jander, R. (1997). A sun compass in monarch butterflies. *Nature,* 387(6628), 29–29.

Perez, S. M., Taylor, O. R., & Jander, R. (1999). The effect of a strong magnetic field on monarch butterfly (*Danaus plexippus*) migratory behavior. *Naturwissenschaften, 86,* 140–143.

Perrins, C. (1968). The purpose of the high-intensity alarm call in small passerines. *Ibis, 110,* 200–201.

Perusse, D. (1994). Mate choice in modern societies. Testing evolutionary hypotheses with behavioral data. *Human Nature—An Interdisciplinary Biosocial Perspective, 5,* 235–278.

Pervin, L., & John, O. P. (1997). *Personality: Theory and research.* New York: Wiley.

Peters, J. M., Queller, D. C., Imperatriz-Fonseca, V. L., Roubik, D. W., & Strassmann, J. E. (1999). Mate number, kin selection and social conflicts in stingless bees and honeybees. *Proceedings of the Royal Society of London, 266,* 379–384.

Petit, L. J. (1991). Experimentally induced polygyny in a monogamous bird species: Prothonotary warblers and the polygyny threshold. *Behavioral Ecology and Sociobiology, 29,* 177–187.

Pfennig, D. W. (1995). Absence of joint nesting advantage in desert seed harvester ants: Evidence from a field experiment. *Animal Behaviour, 49,* 567–575.

Pfennig, D. W., Reeve, H. K., & Sherman, P. W. (1993). Kin recognition and cannibalism in spadefoot toads. *Animal Behaviour, 46,* 87–94.

Pfennig, D. W., & Sherman, P. W. (1995). Kin recognition. *Scientific American,* June 1995, 98–103.

Pfungst, O. (1911). *Clever Hans: The horse of Mr. Von Osten.* New York: Holt, Rinehart and Winston.

Piaget, J. (1951). *Play, dreams and imitation in childhood.* New York: W. W. Norton & Co..

Pierce, N., Kitching, R., Buckley, R., Talor, M., & Benbow, K. (1987). The costs and benefits of cooperation between the Australian lycaenid butterfly, *Jalmenus evagoras,* and its attendant ants. *Behavioral Ecology and Sociobiology, 21,* 237–248.

Pietrewicz, A., & Richards, J. (1985). Learning to forage: An ecological perspective. In T. Johnston & A. Pietrewicz (Eds.), *Issues in the ecological study of learning* (pp. 99–118). Hillsdale, NJ: Lawrence Erlbaum Associates.

Pinker, S. (1991). Rules of language. *Science, 253,* 530–535.

Pinker, S. (1994). *The language instinct.* New York: Morrow.

Pitcher, T. J. (1986). Functions of shoaling behaviour. In T. Pitcher (Ed.), *The behavior of teleost fishes* (pp. 294–338). Baltimore, MD: John Hopkins University Press.

Pitcher, T. J. (1992). Who dares wins: The function and evolution of predator inspection behaviour in shoaling fish. *Netherlands Journal of Zoology, 42,* 371–391.

Pitcher, T. J., Green, D. A., & Magurran, A. E. (1986). Dicing with death: Predator inspection behavior in minnow shoals. *Journal of Fish Biology, 28,* 439–448.

Pitcher, T. J., & Wyche, C. (1983). Predator avoidance behaviour in sand-eel schools: Why do schools seldom split. In D. Noakes, B. Lindquist, G. Helfman, & J. Ward (Eds.), *Predators and prey in fishes* (pp. 193–204). The Hague: Junk.

Pitnick, S., Markow, T., & Spicer, G. (1995). Delayed male maturity is a cost of producing large male sperm in *Drosophila. Proceedings of the National Academy of Sciences, U.S.A., 92,* 10614–10618.

Pleszczynska, W. K., & Hansell, R. (1980). Polygeny and decision theory: Testing of a model in lark buntings, *Calamospiza melanocorys. American Naturalist, 116,* 821–830.

Plomin, R., DeFries, J. C., McClearn, G. E., & McGuffin, P. (2000). *Behavioral genetics.* New York: Worth Publishers.

Polhemus, J. (1990). Surface wave communication in water striders. Field observations of unreported taxa (*Heteroptera: Gerridae, Veliidae*). *Journal of the New York Entomological Society, 98,* 383–384.

Polhemus, J., & Karunaratne, P. (1993). A review of the genus *Rhagadotarsus,* with descriptions of three new species (*Heteroptera: Gerridae*). *Raffles Bulletin of Zoology, 41,* 95–112.

Poppleton, F. (1957). The birth of an elephant. *Oryx, 4,* 180–181.

Poran, N. S., & Coss, R. G. (1990). Development of antisnake defenses in California ground squirrels (*Spermophilus beecheyi*): I. Behavioral and immunological relationships. *Behaviour, 112,* 222–245.

Portmann, A. (1947). Etudes sur la cerebralisation des oiseaux: Les indices intra-cerebraux. *Alauda, 15,* 1–15.

Potegal, M., & Einon, D. (1989). Aggressive behaviors in adult rats deprived of playfighting experience as juveniles. *Developmental Psychobiology, 22,* 159–172.

Potter, E. (1970). Anting in wild birds: Its frequency and probable purpose. *Auk, 87,* 692–713.

Potts, G. W. (1970). The schooling ethology of *Lutianus monostigma* in the shallow reef environment of Aldabra. *Journal of Zoology, 161,* 223–235.

Poundstone, W. (1992). *Prisoner's dilemma: Jon Von Neuman, game theory and the puzzle of the bomb.* New York: Doubleday.

Power, T. G. (2000). *Play and exploration in children and animals.* Hillsdale, NJ: Lawrence Erlbaum Associates.

Pravosudov, V., & Clayton, N. (2001). Effects of demanding foraging conditions on cache retrieval accuracy in food-caching mountain chickadees (*Poecile gambeli*). *Proceedings of the National Academy of Sciences, U.S.A., 268,* 363–368.

Price, T., Schluter, D., & Heckman, N. E. (1993). Sexual selection when the female benefits directly. *Biological Journal of the Linnean Society, 48,* 187–211.

Price, T., & Schulter, D. (1991). On the low heritabilility of life-history traits. *Evolution, 45,* 853–861.

Profet, M. (1992). Pregnancy sickness as adaptation: A deterrent to maternal ingestion of teratogens. In J. Barkow, L. Cosmides, & J. Tooby (Eds.), *The adapted mind* (pp. 324–365). New York: Oxford University Press.

Profet, M. (1993). Menstruation as a defense against pathogens transported by sperm. *Quarterly Review of Biology, 68,* 335–386.

Promislow, D. E. (1991). Senescence in natural populations of mammals: A comparative study. *Evolution, 45,* 1869–1887.

Promislow, D. E. (1992). Costs of sexual selection in natural populations of mammals. *Proceedings of the Royal Society of London, 247,* 203–207.

Promislow, D. E. (1995). New perspectives on comparative tests of antagonistic pleiotropy using *Drosophila. Evolution, 49,* 394–397.

Pruett-Jones, S. G. (1992). Independent versus non-independent mate choice: Do females copy each other? *American Naturalist, 140,* 1000–1009.

Pruett-Jones, S., & Lewis, M. (1990). Sex ratio and habitat limitation promote delayed dispersal in superb fairy wrens. *Nature, 348,* 541–542.

Pulido, F., Berthold, P., Mohr, G., & Querner, U. (2001). Heritability of the timing of autumn migration in a natural bird population. *Proceedings of the Royal Society of London, 268,* 953–959.

Pulliam, R. (1973). On the advantages of flocking. *Journal of Theoretical Biology, 38,* 419–422.

Pulliam, R. (1974). On the theory of optimal diets. *American Naturalist, 108,* 59–75.

Purves, D., & Lichtman, J. (1980). Elimination of synapses in the developing nervous system. *Science, 210,* 153–157.

Pysh, J., & Weiss, G. (1979). Exercise during development induces an increase in Purkinje cell dendritic tree size. *Science, 206,* 230–231.

Queller, D. C. (1992). Quantitative genetics, inclusive fitness and group selection. *American Naturalist, 139,* 540–558.

Quigley, H., & Cragshaw, P. (1992). A conservation plan for the jaguar (*Panthera onca*) in the Pantanal region of Brazil. *Biological Conservation, 61,* 149–157.

Quinn, J. S., Woolfenden, G. E., Fitzpatrick, J. W., & White, B. N. (1999). Multi-locus DNA fingerprinting supports genetic monogamy in Florida scrub-jays. *Behavioral Ecology and Sociobiology, 45,* 1–10.

Quinn, T. (1984). Homing and straying in Pacific salmon. In E. McCleav, J. G. Arnold, J. Dodson, & W. Neill (Eds.), *Mechanisms of migration in fishes* (pp. 357–362). New York: Plenum Press.

Quinn, W. G., Harris, W. A., & Benzer, S. (1974). Conditioned behavior in *Drosophila melanogaster. Proceedings of the National Academy of Sciences, U.S.A., 71,* 708–712.

Quinney, T. E. (1983). Tree swallows cross a polygyny threshold. *Auk, 100,* 750–754.

Radakov, D. V. (1973). *Schooling in the ecology of fish.* New York: Wiley.

Raleigh, M., & McGuire, M. (1991). Bidirectional relationships between tryptophan and social behavior in vervet monkeys. In R. Schwarcz, S. Young, & R. Brown (Eds.), *Kynurenine and serotonin pathways* (pp. 289–298). New York: Plenum Press.

Raleigh, M., McGuire, M., Brammer, G., Pollack, D., & Yuwiler, A. (1991). Serotonergic mechanisms promote dominance in adult vervet monkeys. *Brain Research, 559,* 181–190.

Rankin, M. (1985). *Migration: Mechanism and adaptive significance.* Austin, TX: Marine Science Center.

Rasa, O. A. (1986). Coordinated vigilance in dwarf mongoose family groups: The "Watchman's song" hypothesis and the costs of guarding. *Ethology, 71,* 1986.

Ratnieks, F. L. (1995). Evidence for queen-produced egg-marking pheromone and its use in worker policing in the honey bee. *Journal of Apicultural Research, 34,* 31–37.

Ratnieks, F. L., & Visscher, P. K. (1988). Reproductive harmony via mutual policing by workers in eusocial hymenoptera. *American Naturalist, 132,* 217–236.

Ratnieks, F. L., & Visscher, P. K. (1989). Worker policing in the honeybee. *Nature, 342,* 796–797.

Rattenborg, N., Lima, S., & Amlaner, C. (1999a). Facultative control of avian unihemispheric sleep under the risk of predation. *Behavioral Brain Research, 105,* 163–172.

Rattenborg, N., Lima, S., & Amlaner, C. (1999b). Half-awake to the risk of predation. *Nature, 397,* 397–398.

Raubenheimer, D., & Tucker, D. (1997). Associative learning by locusts: Pairing of visual cues with consumption of protein and carbohydrate. *Animal Behaviour, 54,* 1449–1159.

Read, A., & Weary, D. (1992). The evolution of bird song: Comparative analysis. *Philosophical Transactions of the Royal Society of London, Series B, 338,* 165–187.

Reader, S. M., & Laland, K. N. (2002). Social intelligence, innovation and enhanced brain size in primates. *Proceedings of the National Academy of Sciences, U.S.A., 99,* 4436–4441.

Reagen, L., & McEwen, B. (1997). Controversies surrounding glucocorticoid-mediated cell death in the hippocampus. *Journal of Chemical Neuroanatomy, 13,* 149–163.

Real, L. (1980). Fitness, uncertainty and the role of diversification in evolution and behavior. *American Naturalist, 115,* 623–638.

Real, L., & Caraco, T. (1986). Risk and foraging in stochastic environments. *Annual Review of Ecology and Systematics, 17,* 371–390.

Reeve, H. K. (1989). The evolution of conspecific acceptance thresholds. *American Naturalist, 133,* 407–435.

Reeve, H. K. (1992). Queen activation of lazy workers in colonies of the eusocial naked mole rat. *Nature, 358,* 147–149.

Reeve, H. K. (2003). *Transactional models of social evolution.* Princeton: Princeton University Press, in press.

Reeve, H. K., & Ratnieks, F. L. (1993). Queen-queen conflicts in polygynous societies: Mutual tolerance and reproductive skew. In L. Keller (Ed.), *Queen number and sociality in insects* (pp. 45–86). Oxford: Oxford University Press.

Reeve, H. K., & Sherman, P. W. (1991). Intracolonial aggression and nepotism by the breeding female naked mole rat. In P. W. Sherman, J. Jarvis, & R. D. Alexander (Eds.), *The biology of the naked mole rat* (pp. 337–357). Princeton: Princeton University Press.

Reeve, H. K., & Sherman, P. W. (1993). Adaptation and the goals of evolutionary research. *Quarterly Review of Biology, 68,* 1–32.

Reeve, H. K., Westneat, D. F., Noon, W. A., Sherman, P. W., & Aquadro, C. F. (1990). DNA "fingerprinting" reveals high levels of inbreeding in colonies of the eusocial naked mole rat. *Proceedings of the National Academy of Sciences, U.S.A., 87,* 2496–2500.

Reeve, H. K., Westneat, D. F., & Queller, D. C. (1992). Estimating average within group relatedness from DNA fingerprints. *Molecular Ecology, 2,* 223–232.

Ren, R., Su, Y., Yan, K., Qi, H., & Bao, W. (1990a). Social relationships among golden monkeys in breeding cages [in Chinese]. *Acta Psychologica Sinica, 22,* 277–282.

Ren, R., Yan, K., Su, Y., Qi, H., & Bao, W. (1990b). Social communication by posture and facial expression in golden monkeys (*Rhinopithecus roxellanae*) [in Chinese]. *Acta Psychologica Sinica, 21,* 159–167.

Ren, R., Yan, K., Su, Y., Qi, H., Liang, B., Bao, W., & Waal, F. B. M. D. (1991). The reconciliation behavior of golden monkeys (*Rhinopithecus roxellanae roxellanae*) in small breeding groups. *Primates,* 32, 321–327.

Rendell, E., & Whitehead, H. (2001). Culture in whales and dolphins. *Behavioral and Brain Sciences, 24,* 309–382.

Resetarits, W., & Wilbur, H. (1989). Choice of oviposition site by *Rana chrysoscelis*: Role of predators and competitors. *Ecology, 70,* 220–228.

Reusch, T. B. H., Haberli, M. A., Aeschlimann, F. B., & Milinski, M. (2001). Female sticklebacks count alleles in a strategy of sexual selection explaining MHC polymorphism. *Nature, 414,* 300–302.

Reyer, H.-U. (1986). Avian helpers at the nest: Are they psychologically castrated? *Ethology, 71,* 216–228.

Reznick, D. (1996). Life history evolution in guppies: A model system for the empirical study of adaptation. *Netherlands Journal of Zoology, 46,* 172–190.

Reznick, D., Brygae, H., & Endler, J. (1990). Experimentally induced life-history evolution in a natural population. *Nature, 346,* 357–359.

Reznick, D., Shaw, F., Rodd, H., & Shaw, R. (1997). Evaluation of the rate of evolution in natural populations of guppies (*Poecilia reticulata*). *Science, 275,* 1934–1937.

Rice, D. W. (1989). Sperm whale. *Physeter macrocephalus* Linnaeus 1758. In S. Ridgeway & R. Harrison (Eds.) *Handbook of marine animals* (Vol. 4; pp. 177–233). London: Academic Press.

Richard, K., Dillon, M., Whitehead, H., & Wright, J. (1996). Patterns of kinship in groups of free-living sperm whales (*Physeter macrocephalus*) revealed by multiple molecular genetic techniques. *Proceedings of the National Academy of Sciences, U.S.A., 93,* 8792–8795.

Richerson, P., & Boyd, R. (1989). The role of evolved predispositions in cultural evolution, or human sociobiology meets Pascal's wager. *Ethology and Sociobiology, 10,* 195–219.

Riddell, W., & Corl, K. (1977). Comparative investigation of the relationship between cerebral indices and learning abilities. *Behavioral and Brain Sciences, 14,* 305–308.

Riddiford, L. (1994). Cellular and molecular actions of juvenile hormone. I. General considerations and premetamorphic actions. *Advances in Insect Physiology, 24,* 213–274.

Ridley, M. (1996). *Evolution* (2nd ed.). Oxford: Blackwell Scientific Publications.

Riechert, S. (1998). Game theory and animal contests. In L. A. Dugatkin & H. K. Reeve (Eds.), *Game theory and animal behavior* (pp. 64–93). New York: Oxford University Press.

Rijnsdorp, A. D., Broekman, P. L. V., & Visser, E. G. (2000a). Competitive interactions among beam trawlers exploiting local patches of flatfish in the North Sea. *Ices Journal of Marine Science, 57,* 894–902.

Rijnsdorp, A. D., Dol, W., Hoyer, M., & Pastoors, M. A. (2000b). Effects of fishing power and competitive interactions among vessels on the effort allocation on the trip level of the Dutch beam trawl fleet. *Ices Journal of Marine Science,* 57, 927–937.

Rissing, S., & Pollock, G. (1986). Social interaction among pleometric queens of *Veromessor pergandei* during colony foundation. *Animal Behaviour, 34,* 226–234.

Rissing, S., & Pollock, G. (1987). Queen aggression, pleometric advantage and brood raiding in the ant *Veromessor pergandei*. *Animal Behaviour,* 35, 975–982.

Rissing, S., & Pollock, G. (1988). Pleometrosis and polygyny in ants. In R. Jeanne (Ed.), *Interindividual behavioral variability in social insects* (pp. 170–222). New York: Westview Press.

Rissing, S., & Pollock, G. (1991). An experimental analysis of pleometric advantage in *Messor pergandei*. *Insectes Sociaux, 63,* 205–211.

Rissing, S., Pollock, G., Higgins, M., Hagen, R., & Smith, D. (1989). Foraging specialization without relatedness or dominance among co-founding ant queens. *Nature,* 338, 420–422.

Robinson, G. E. (1985). Effects of a juvenile hormone analog on honey bee foraging behavior and alarm pheromone. *Journal of Insect Physiology, 31,* 277–282.

Robinson, G. E. (1987). Modulation of alarm pheromone perception in the honey bee: Evidence for division of labor based on hormonally regulated response thresholds. *Journal of Comparative Physiology A, 160,* 613–619.

Robinson, G. E. (1992). Regulation of division of labor in insect societies. *Annual Review of Ecology and Systematics,* 37, 637–665.

Robinson, G. E., Fahrbach, S., & Winston, M. (1999). Insect societies and the molecular biology of social behavior. *BioEssays, 19,* 1099–1108.

Robinson, G. E., Page, R. E., Strambi, C., & Strambi, A. (1989). Hormonal and genetic control of behavioral integration in honeybee colonies. *Science, 246,* 109–112.

Robinson, S., & Smotherman, W. (1991). Fetal learning: Implications for the development of kin. In P. G. Hepper

(Ed.), *Kin recognition* (pp. 308–334). Cambridge: Cambridge University Press.

Roces, F., & Holldobler, B. (1995). Vibrational communication between hitchhikers and foragers in leaf-cutting ants (*Atta cephalotes*). *Behavioral Ecology and Sociobiology, 37,* 297–302.

Roces, F., & Holldobler, B. (1996). Use of stridulation in foraging leaf-cutting ants: Mechanical support during cutting or short-range recruitment signal? *Behavioral Ecology and Sociobiology, 39,* 293–299.

Roces, F., Tautz, J., & Holldobler, B. (1993). Stridulation in leaf-cutting ants: Short range recruitment through plant-borne substances. *Naturwissenschaften, 80,* 521–524.

Rodd, F. H., Hughes, K. A., Grether, G. F., & Baril, C. T. (2002). A possible non-sexual origin of mate preference: Are male guppies mimicking fruit? *Proceedings of the Royal Society of London, 269,* 475–481.

Rodriguez, E., & Wrangham, R. (1993). Zoopharmacognosy: The use of medicinal plants by animals. *Recent Advances in Phytochemistry, 27,* 89–105.

Roell, A. (1978). Social behaviour of the jackdaw, *Corvus monedula,* in relation to its niche. *Behaviour, 64,* 1–124.

Roguet, C., Dumont, B., & Prache, S. (1998). Selection and use of feeding sites and feeding stations by herbivores: A review. *Annales De Zootechnie, 47,* 225–244.

Rohwer, F. C., & Freeman, S. (1989). The distribution of conspecific nest parasitism in birds. *Canadian Journal of Zoology, 67,* 239–253.

Rohwer, S. (1978). Parental cannibalism of offspring and egg raiding as a courtship strategy. *American Naturalist, 112,* 429–440.

Rohwer, S. (1982). The evolution of reliable and unreliable badges of fighting ability. *American Zoologist, 22,* 531–546.

Romanes, G. J. (1884). *Mental evolution in animals.* New York: AMS Press.

Romanes, G. J. (1889). *Mental evolution in man.* New York: Appleton.

Romanes, G. J. (1892). *Darwin and after Darwin: An exposition of the Darwinian theory and a discussion of post-Darwin ideas* (7th ed.). Chicago: Open Court Publishing Co.

Romanes, G. J. (1898). *Animal intelligence* (7th ed.). London: Kegan Paul, Trench, Trubner and Co.

Rood, J. P. (1980). Mating relationships and breeding suppression in the dwarf mongoose. *Animal Behaviour, 28,* 143–150.

Rood, J. P. (1986). Ecology and social evolution in the mongooses. In D. I. Rubenstein & R. W. Wrangham (Eds.), *Ecological aspects of social evolution* (pp. 131–152). Princeton: Princeton University Press.

Rood, J. P. (1990). Group size, survival, reproduction and routes to breeding in dwarf mongooses. *Animal Behaviour, 39,* 566–572.

Roper, K., Kaiser, D., & Zentall, T. (1995). True directed forgetting in pigeons may occur only when alternative working memory is required on forget: Cue trials. *Animal Learning and Behavior, 23,* 280–285.

Roper, T. J. (1986). Badges of status in avian societies. *New Scientist, 109,* 38–40.

Rose, M. R. (1984). Laboratory evolution of postponed senescence in *Drosophila melanogaster. Evolution, 38,* 1004–1010.

Rose, M. R. (1991). *The evolutionary biology of aging.* New York: Oxford University Press.

Rose, M. R. (1999). Genetics of aging in *Drosophila. Experimental Gerontology, 34,* 577–585.

Rose, M. R., & Charlesworth, B. (1981). Genetics of life history in *Drosophila melanogaster*: Sib analysis of adult females. *Genetics, 97,* 173–186.

Rosenzweig, M. L. (1981). A theory of habitat selection. *Ecology, 62,* 327–335.

Rosenzweig, M. L. (1985). Some theoretical aspects of habitat selection. In M. Cody (Ed.), *Habitat selection in birds* (pp. 517–540). New York: Academic Press.

Rosenzweig, M. L. (1990). Do animals choose habitats? In M. Bekoff & D. Jamieson (Eds.), *Interpretation and explanation in the study of animal behaviour* (Vol. I. Interpretation, intentionality and communication, pp. 157–179). Boulder, CO: Westview Press.

Rosenzweig, M. L. (1991). Habitat selection and population interactions. *American Naturalist, 137,* S5-S28.

Roskaft, E., Wara, A., & Viken, A. (1992). Reproductive success in relation to resource-access and parental age in a small Norwegian farming parish during the period 1700–1900. *Ethology and Sociobiology, 13,* 443–461.

Ross, P., Jalkotzy, G., & Festa-Bianchet, M. (1997). Cougar predation on bighorn sheep in Southwestern Alberta during winter. *Canadian Journal of Zoology, 74,* 771–775.

Rothstein, S., & Pirotti, R. (1987). Distinctions among reciprocal altruism, kin selection and cooperation and a model for the initial evolution of beneficent behavior. *Ethology and Sociobiology, 9,* 189–209.

Rowe, L., Repasky, R. R., & Palmer, A. R. (1997). Size-dependent asymmetry: Fluctuating asymmetry versus antisymmetry and its relevance to condition-dependent signaling. *Evolution, 51,* 1401–1408.

Rowland, W. J., & Sevenster, P. (1985). Sign stimuli in three-spined sticklebacks (*Gasterosteus aculeatus*): A reexamination of some classic experiments. *Behaviour, 93,* 241–257.

Roy, R., Graham, S., & Peterson, J. (1988). Fiber type composition of the plantaflexors of giraffes (*Giraffa camelopardalis*) at different postnatal stages of development. *Comparative Biochemistry and Physiology, A Physiology, 91,* 347–352.

Rozin, P. (1976). The selection of foods by rats, humans, and other animals. In J. Rosenblatt, R. A. Hinde, C. Beer, & E. Shaw (Eds.), *Advances in the study of behavior* (Vol. 6; pp. 21–76). New York: Academic Press.

Rozin, P. (1988). The adaptive-evolutionary point of view in experimental psychology. In R. Atkinson, R. Herrnstein, G.

Lindzey, & R. Luce (Eds.) *Steven's handbook of experimental psychology* (pp. 503–546). New York: Wiley.

Rozin, P., & Schiller, D. (1980). The nature and acquisition of a preference for chili peppers by humans. *Motivation and Emotion, 4,* 77–101.

Russell, J. K. (1983). Altruism in coati bands: Nepotism or reciprocity? In S. Wasser (Ed.), *Social behavior of female vertebrates* (pp. 263–290). New York: Academic Press.

Ryan, M. J. (1990). Sexual selection, sensory systems and sensory exploitation. *Oxford Surveys in Evolutionary Biology, 7,* 157–195.

Ryan, M. J., Fox, J., Wilczynski, W., & Rand, S. (1990). Sexual selection for sensory exploitation in the frog, *Physalaemus pustulosus. Nature, 343,* 66–67.

Ryan, M. J., & Keddy-Hector, A. (1992). Directional patterns of female mate choice and the role of sensory biases. *American Naturalist, 139,* S4–S35.

Ryan, M. J., & Rand, S. (1990). The sensory basis of sexual selection for complex calls in the tungara frog, *Physalaemus pustulosus. Evolution, 44,* 305–314.

Ryti, R. T., & Case, T. J. (1984). Spatial arrangement and diet overlap between colonies of desert ants. *Oecologia, 62,* 401–404.

Sagør, J., Swenson, J., & Røskaft, E. (1997). Compatibility of brown bear (*Ursus arctos*) and free ranging sheep in Norway. *Biological Conservation, 81,* 91–95.

Salamone, J. (1994). The involvement of nucleus accumbens dopamine in appetitive and aversive motivation. *Behavioral Brain Research, 61,* 117–133.

Sapolsky, R. M. (1993). Neuroendocrinology of the stress-response. In J. Becker, S. Breedlove, & D. Crews (Eds.), *Behavioral endocrinology* (pp. 287–324). Cambridge, MA: MIT Press.

Sapolsky, R. M. (1994). Fallible instinct: A dose of skepticism about medicinal "knowledge" of animals. *Science, 34,* 13–15.

Sapolsky, R. M. (1996). Stress, glucocorticoids and damage to the nervous system: The current state of confusion. *Stress, 1,* 1–15.

Sapolsky, R. M. (1999). Glucocorticoids, stress, and their adverse neurological effects: Relevance to aging. *Experimental Gerontology, 34,* 721–732.

Sapolsky, R. M. (2000). Glucocorticoids and hippocampal atrophy in neuropsychiatric disorders. *Archives of General Psychiatry, 57,* 925–935.

Sapolsky, R. M., Krey, L., & McEwen, B. (1986). The endocrinology of stress and aging: The glucocorticoid cascade hypothesis. *Endocrine Reviews, 7,* 284–306.

Sasvari, L. (1985). Keypeck conditioning with reinforcement in two different locations in thrush, tit and sparrow species. *Behavioural Processes, 11,* 245–252.

Sauer, E. G. (1957). Die sternenorientierung nactlich ziehender grasmucken. *Zeitschrift fur Tierpsychologie, 14,* 29–70.

Saunders, I., Sayer, M., & Goodale, A. (1999). The relationship between playfulness and coping in preschool children: A pilot study. *American Journal of Occupational Therapy, 53,* 221–226.

Sawaguchi, T., & Kudo, H. (1990). Neocortical development and social structure in primates. *Primates, 31,* 283–289.

Schaller, G. (1967). *The deer and the tiger.* Chicago: University of Chicago Press.

Schenkel, R. (1966). Play, exploration and territoriality in the wild lion. *Symposia of the Zoological Society of London, 18,* 11–22.

Schino, G., Scuccio, S., Maestrippi, D., & Turillazzi, P. G. (1988). Allogrooming as a tension reduction mechanism: A behavioral approach. *American Journal of Primatology, 16,* 43–50.

Schjelderup-Ebbe, T. (1922). Beitrage zur sozialpsychologie des haushuhns. *Zeitschrift fur Tierpsychologie, 88,* 225–252.

Schleidt, W. (1976). On individuality: The constituents of distinctiveness. In P. Bateson & P. Klopfer (Eds.), *Perspectives in ethology* (Vol. 2; pp. 299–310). New York: Plenum Press.

Schlupp, I., Marler, C., & Ryan, M. (1994). Males benefit by mating with heterospecific females. *Science, 263,* 373–374.

Schneider, K. (1984). Dominance, predation and optimal foraging in white-throated sparrow flocks. *Ecology, 65,* 1820–1827.

Schoener, T. (1971). Theory of feeding strategies. *Annual Review of Ecology and Systematics, 11,* 369–404.

Schuett, G. (1997). Body size and agonistic experience affect dominance and mating success in male copperheads. *Animal Behaviour, 54,* 213–224.

Schuett, G. W., & Gillingham, J. C. (1989). Male-male agonistic behaviour of the copperhead, *Agkistrodon contortrix. Amphibia-Reptilia, 10,* 243–266.

Schuett, G., Harlow, H., Rose, J., van Kirk, E., & Murdoch, W. (1996). Levels of plasma corticosterone and testosterone in male copperheads (*Agkistrodon contortrix*) following staged fights. *Hormones and Behavior, 30,* 60–68.

Scott, J. P. (1958). *Aggression.* Chicago: University of Chicago Press.

Scott, J. P., & Bielfelt, S. (1976). Analysis of the puppy training program, *Guide dogs for the blind: Their selection, development and training.* Amsterdam: Elsevier.

Scott, S. (1987). *Field guide to the birds of North America.* Washington, DC: National Geographic Society.

Seale, D. (1982). Physical factors influencing oviposition by the woodfrog, *Rana sylvatica,* in Pennsylvania. *Copeia, 1982,* 627–635.

Searcy, W. A. (1986). Sexual selection and the evolution of song. *Annual Review of Ecology and Systematics, 17,* 507–533.

Searcy, W. A. (1992). Song repertoire and mate choice in birds. *American Zoologist, 32,* 71–80.

Searcy, W. A., & Brenowitz, E. (1988). Sexual differences in species recognition of avian song. *Nature, 332,* 152–154.

Searcy, W. A., & Yasukawa, K. (1996). The reproductive success of secondary females relative to that of monogamous and primary females in red-winged blackbirds. *Journal of Avian Biology, 27,* 225–230.

Sebeok, T., & Rosenthal, R. (Eds.). (1981). *The Clever Hans phenomenon: Communication with horses, whales, apes, and people.* New York: New York Academy of Science.

Seeley, T. (1985). *Honeybee ecology: A study of adaptation in social life.* Princeton: Princeton University Press.

Seeley, T. (1995). *The wisdom of the hive.* Cambridge, MA: Harvard University Press.

Seeley, T. (1997). Honey bee colonies are group-level adaptive units. *American Naturalist, 150 (Supplement),* S22–S41.

Seger, J. (1981). Kinship and covariance. *Journal of Theoretical Biology, 91,* 191–213.

Seger, J. (1989). All for one, one for all, that is our device. *Nature, 338,* 374–375.

Seghers, B. H. (1973). *An analysis of geographic variation in the antipredator adaptations of the guppy, Poecilia reticulata.* Unpublished Ph.D., University of British Columbia.

Selander, R., Smith, M., Yang, S., Johnson, W., & Gentry, J. (1971). Biochemical polymorphism and systematics in the genus *Peromyscus*: I. Variation of the oldfield mouse (*Peromyscus polionotus*). *Studies in Genetics, 6,* 49–90.

Seligman, M., & Hager, J. (1972). *Biological boundaries of learning.* New York: Appleton.

Semel, B., & Sherman, P. (1986). Dynamics of nest parasitism in wood ducks. *Auk, 103,* 136–139.

Semel, B., Sherman, P., & Byers, S. (1988). Effects of brood parasitism and nest box placement on wood duck breeding ecology. *Condor, 90,* 920–930.

Semel, B., Sherman, P., & Byers, S. (1990). Nest boxes and brood parasitism in wood ducks: A management dilemma. In L. Fredickson, G. Burger, S. Havera, D. Graber, R. Kirkby, & T. Taylor (Eds.), *Proceedings of the 1988 North American Wood Duck Symposium* (pp. 163–170). St. Louis, MO.

Servedio, M. R., & Kirkpatrick, M. (1996). The evolution of mate choice copying by indirect selection. *American Naturalist, 148,* 848–867.

Seyfarth, R. M. (1980). The distribution of grooming and related behaviours among adult female vervet monkeys. *Animal Behaviour, 28,* 798–813.

Seyfarth, R. M., & Cheney, D. L. (1997). Some general features of vocal development in nonhuman primates. In E. Husberger & C. Snowdon (Eds.), *Social influences on vocal development.* Cambridge: Cambridge University Press.

Seyfarth, R. M., Cheney, D. L., & Marler, P. (1980). Vervet monkey alarm calls: Semantic communication in a free-ranging primate. *Animal Behaviour, 28,* 1070–1094.

Sgoifo, A., De Boer, S., Westenbrock, C., Maes, F., Beldhuis, H., Suzuki, T., & Koolhaus, J. (1997). Incidence of arrhythmias and heart rate variability in wild-type rats exposed to stress. *American Journal of Physiology, 273,* H1754–1760.

Shapiro, L. (1980). Species identification in birds: A review and synthesis. In M. Roy (Ed.), *Species identity and attachment: A phylogenetic evaluation.* New York: Garland Press.

Shaw, P. W., Carvalho, G. R., Seghers, B. H., & Magurran, A. E. (1992). Genetic consequences of an artificial introduction of guppies (*Poecilia reticulata*) in N. Trinidad. *Proceedings of the Royal Society of London, 248,* 111–116.

Sheard, M. (1983). Aggressive behavior: Effects of neural modulation by serotonin. In E. Simmell, M. Hahn, & J. Walters (Eds.), *Aggressive behavior* (pp. 167–181). Hillsdale, NJ: Lawrence Erlbaum Associates.

Sherman, P. W. (1977). Nepotism and the evolution of alarm calls. *Science, 197,* 1246–1253.

Sherman, P. W. (1980). The meaning of nepotism. *American Naturalist, 116,* 604–606.

Sherman, P. W. (1981). Kinship, demography, and Belding's ground squirrel nepotism. *Behavioral Ecology and Sociobiology, 8,* 251–259.

Sherman, P. W. (1985). Alarm calls of Belding's ground squirrels to aerial predators: Nepotism or self-preservation? *Behavioral Ecology and Sociobiology, 17,* 313–323.

Sherman, P. (1988). The levels of analysis. *Animal Behaviour, 36,* 616–619.

Sherman, P. W. (2001). Wood ducks: A model system for investigating conspecific parasitism in cavity-nesting birds. In L. A. Dugatkin (Ed.), *Model systems in behavioral ecology: Integrating conceptual, theoretical and empirical perspectives* (pp. 311–337). Princeton: Princeton University Press.

Sherman, P., & Flaxman, M. (2002). Nausea and vomiting of pregnancy in an evolutionary perspective. *American Journal of Obstetrics and Gynecology, 186,* S109–S197.

Sherman, P. W., & Holmes, W. (1985). Kin recognition: Issues and evidence. *Fortschritte der Zoologie, 31,* 437–460.

Sherman, P. W., Jarvis, J., & Alexander, R. (Eds.). (1991). *The biology of the naked mole-rat.* Princeton: Princeton University Press.

Sherman, P. W., & Reeve, H. K. (1997). Forward and backward: Alternative approaches to studying human social evolution. In L. Betzig (Ed.), *Human nature: A critical reader* (pp. 147–158). New York: Oxford University Press.

Sherry, D. F., & Vaccarino, A. (1989). Hippocampus and memory for food caches in black-capped chickadees. *Behavioral Neuroscience, 103,* 308–318.

Sherry, D. F., Vaccarino, A., Buckenham, K., & Herz, R. S. (1989). The hippocampal complex of food storing birds. *Brain, Behavior and Evolution, 34,* 308–317.

Shettleworth, S. J. (1984). Learning and behavioural ecology. In J. R. Krebs & N. B. Davies (Eds.), *Behavioural ecology: An evolutionary approach* (Vol. 2; pp. 170–194). Oxford: Blackwell Scientific Publications.

Shettleworth, S. J. (1998). *Cognition, evolution and behavior.* New York: Oxford University Press.

Shettleworth, S. J. (2000). Modularity and the evolution of cognition. In C. Heyes & L. Huber (Eds.), *The evolution of cognition* (pp. 43–60). Cambridge, MA: MIT Press.

Shettleworth, S. J. (2002). Spatial behavior, food storing, and the modular mind. In M. Bekoff, C. Allen, & G. Burghardt (Eds.), *The cognitive animal*. Cambridge, MA: MIT Press.

Shuster, S. M., & Caldwell, R. (1989). Male defense of the breeding cavity and factors affecting the persistence of breeding pairs in the stomatopod, *Gonodactylus bredini (Crustacea: Hoplocarida). Ethology, 82,* 192–207.

Sigg, D., Thompson, C. M., & Mercer, A. R. (1997). Activity-dependent changes to the brain and behavior of the honeybee, *Apis mellifera (L.). Journal of Neuroscience, 17,* 7148–7156.

Sih, A., & Christensen, B. (2001). Optimal diet theory: When does it work, and when it does it fail? *Animal Behaviour, 61,* 379–390.

Sih, A., & Moore, R. D. (1993). Delayed hatching of salamander eggs in response to enhanced larval predation risk. *American Naturalist, 142,* 947–960.

Silk, J. B. (1982). Altruism among female *Macaca radiata*: Explanations and analysis of patterns of grooming and coalition formation. *Behaviour, 79,* 162–188.

Silk, J. B. (1992a). The patterning of intervention among male bonnet macaques: Reciprocity, revenge and loyalty. *Current Anthropology, 33,* 319–324.

Silk, J. B. (1992b). Patterns of intervention in agonistic contests among male bonnet macaques. In A. Harcourt & F. B. M. de Waal (Eds.), *Coalitions and alliances in humans and other animals* (pp. 214–232). Oxford: Oxford University Press.

Silk, J. B., Smaules, A., & Rodman, P. (1981). The influence of kinship, rank, and sex on affiliation and aggression between adult female and immature bonnett macaques (*Macaca radiata*). *Behaviour, 78,* 111–137.

Simmons, L. W. (2001). *Sperm competition and its evolutionary consequences.* Princeton: Princeton University Press.

Simmons, L. W., Tomkins, J. L., Kotiaho, J. S., & Hunt, J. (1999). Fluctuating paradigm. *Proceedings of the Royal Society of London, 266,* 593–595.

Singer, W. (1994). Putative functions of temporal correlations in neocorticol processing. In C. Koch & J. Davis (Eds.), *Large-scale neuronal theories of the brain* (pp. 201–237). Cambridge, MA: MIT Press.

Siviy, S. (1998). Neurobiological substrates of play behavior: Glimpses into the structure and function of mammalian playfulness. In M. Bekoff & J. Byers (Eds.), *Animal play: Evolutionary, comparative and ecological perspectives* (pp. 221–242). Cambridge: Cambridge University Press.

Siviy, S., & Panksepp, J. (1985). Dorsomedial diencephalic involvement in the juvenile play of rats. *Behavioral Neuroscience, 99,* 1103–1113.

Siviy, S., & Panksepp, J. (1987). Juvenile play in the rat: Thalamic and brain stem involvement. *Physiology and Behavior, 41,* 103–114.

Skinner, B. F. (1938). *The behavior of organisms.* New York: Appleton-Century-Crofts.

Skinner, B. F. (1959). A case history in scientific method. In S. Koch (Ed.), *Psychology: A study of science* (Vol. 2; pp. 359–379). New York: McGraw-Hill.

Skutch, A. F. (1935). Helpers at the nest. *Auk, 52,* 257–273.

Skutch, A. F. (1987). *Helpers at bird's nests.* Iowa City: University of Iowa Press.

Slagsvold, T., Lijfield, J., Stenmark, G., & Breiehagen, T. (1988). On the costs of searching for a male in female pied flycatchers *Ficedula hypoleuca. Animal Behaviour, 36,* 433–442.

Slansky, F., Jr., & Scriber, J. M. (1985). Food consumption and utilization. In G. A. Kerkut & L. I. Gilbert (Eds.), *Comprehensive insect physiology, biochemistry and pharmacology* (Vol. 4; pp. 87–163). Oxford: Pergamon Press.

Slater, P. J. B. (1981). Individual differences in animal behavior. In P. Bateson & P. Klopfer (Eds.), *Perspectives in ethology* (Vol. 4; pp. 35–49). New York: Plenum Press.

Sluyter, F., Bult, A., Lynch, C., Oortmerssen, G., & Koolhaus, J. (1995). A comparison between house mouse lines selected for attack latency or nest building: Evidence for genetic basis for alternative behavioral strategies. *Behavior Genetics, 25,* 247–252.

Smith, A. P., & Alcock, J. (1980). A comparative study of the mating systems of Australian eumenid wasps (*Hymenoptera*). *Zeitschrift fur Tierpsychologie, 53,* 41–60.

Smith, C. (1975). Quantitative inheritance. In G. Fraser & O. Mayo (Eds.), *The textbook of human genetics* (pp. 382–441). Oxford: Blackwell Scientific Publications.

Smith, I. C., Huntingford, F. A., Atkinson, R. J. A., & Taylor, A. C. (1994). Strategic decisions during agonistic behaviour in the velvet swimming crab, *Necora puber. Animal Behaviour, 47,* 885–894.

Smith, J., Benkman, C., & Coffey, K. (1999). The use and mis-use of public information. *Behavioral Ecology, 10,* 54–62.

Smith, J. N. M. (1977a). Feeding rates, search paths and surveillance for predators in great tailed grackle flocks. *Canadian Journal of Zoology, 55,* 891–898.

Smith, P. K. (1982). Does play matter? Functional and evolutionary aspects of animal and human play. *Behavioral and Brain Sciences, 5,* 139–184.

Smith, P. S. (1991). Ontogeny and adaptiveness of tail-flagging behavior in white-tailed deer. *American Naturalist, 138,* 190–200.

Smith, R. J. F. (1986). Evolution of alarm signals: Role of benefits of retaining members or territorial neighbors. *American Naturalist, 128,* 604–610.

Smith, R. L. (1979). Paternity assurance and altered roles in the mating behavior of a giant water bug, *Abedus herberti (Heteroptera: Belostomatidae). Animal Behaviour, 27,* 716–725.

Smith, R. L. (Ed.). (1984). *Sperm competition and the evolution of animal mating systems.* New York: Academic Press.

Smith, S. M. (1975). Innate recognition of coral snake pattern by a possible avian predator. *Science, 187,* 759–760.

Smith, S. M. (1977b). Coral-snake pattern recognition and stimulus generalization by naive great kiskadees (*Aves: Tyrannidae*). *Nature, 265,* 535–536.

Smith, W. J. (1968). Message meaning analysis. In T. Sebeok (Ed.), *Animal communications* (pp. 44–60). Bloomington: Indiana University Press.

Smith, W. J. (1977). *The behavior of communicating: An ethological approach.* Cambridge, MA: Harvard University Press.

Smuts, B. (1985). *Sex and friendship in baboons.* New York: Aldine.

Smythe, N. (1977). The function of mammalian alarm advertising: Social signals or pursuit invitation. *American Naturalist, 111,* 191–194.

Snodgrass, R. E. (1935). *Principles of insect morphology.* New York: McGraw-Hill.

Snyder, R. J. (1984). Seasonal variation in the diet of the three spine stickleback. *California Fish and Game, 70,* 167–172.

Sober, E. (1984). *The nature of selection: Evolutionary theory in philosophical focus.* Cambridge, MA: Bradford/MIT Press.

Sober, E. (1987). What is adaptationism? In J. Dupre (Ed.), *The latest on the best: Essays on evolution and optimality* (pp. 105–118). Cambridge, MA: MIT Press.

Sober, E., & Wilson, D. S. (1998). *Unto others.* Cambridge, MA: Harvard University Press.

Sol, D., & Lefebvre, L. (2000). Behavioural flexibility predicts invasion success in birds introduced to New Zealand. *Oikos, 90,* 599–605.

Solomon, N. G., & French, J. A. (Eds.). (1996). *Cooperative breeding in mammals.* Cambridge: Cambridge University Press.

Sordahl, T. A. (1990). The risks of avian mobbing and distraction behavior: An anecdotal review. *Wilson Bulletin, 102,* 349–352.

Sparks, J. (1967). Allogrooming in primates: A review. In D. Morris (Ed.), *Primate ethology* (pp. 148–175). London: Weidenfeld and Nicolson.

Spinka, M., Newberry, R., & Bekoff, M. (2001). Mammalian play: Training for the unexpected. *Quarterly Review of Biology, 76,* 141–168.

Spinks, A. C., O'Riain, M. J., & Polakow, D. A. (1998). Intercolonial encounters and xenophobia in the common mole rat, *Cryptomys hottentotus hottentotus* (Bathyergidae): The effects of aridity, sex and reproductive status. *Behavioral Ecology, 9,* 354–359.

Spoolder, H., Burbidge, J., Lawrence, A., Simmins, P., & Edwards, S. (1996). Individual behavioral differences in pigs: Intra-and inter-test consistency. *Applied Animal Behaviour Science, 49,* 185–198.

Stacey, P. B., & Koenig, W. (Eds.). (1990). *Cooperative breeding in birds: Long-term studies of ecology and behavior.* Cambridge: Cambridge University Press.

Stacey, P. B. (1979). Habitat saturation and communal breeding in the acorn woodpecker. *Animal Behaviour, 27,* 1153–1166.

Stacey, P. B. (1982). Female promiscuity and male reproductive success in social birds and mammals. *American Naturalist, 120,* 51–64.

Stacey, P. B., & Ligon, J. D. (1987). Territory quality and dispersal options in the acorn woodpecker, and a challenge to the habitat-saturation model of cooperative breeding. *American Naturalist, 130,* 654–676.

Stammbach, E. (1988). Group responses to specially skilled individuals in a *Macaca fascicularis* group. *Behaviour, 107,* 687–705.

Stamps, J. A. (1987a). Conspecifics as cues to territory quality: A preference of juvenile lizards (*Anolis aeneus*) for previously used territories. *American Naturalist, 129,* 629–642.

Stamps, J. A. (1987b). The effect of familiarity with a neighborhood on territory acquisition. *Behavioral Ecology and Sociobiology, 21,* 273–277.

Stamps, J. A. (1995). Motor learning and the adaptive value of familiar space. *American Naturalist, 146,* 41–58.

Stamps, J. A. (2001). Learning from lizards. In L. A. Dugatkin (Ed.), *Model systems in behavioral ecology* (pp. 149–168). Princeton: Princeton University Press.

Stamps, J. A., & Krishnan, V. V. (1999). A learning-based model of territorial assessment. *Quarterly Review of Biology, 74,* 291–318.

Stander, P. (1990). A suggested management strategy for stock raiding lions in Namibia. *South African Journal of Wildlife Research, 20,* 37–43.

Stanford, C. B. (1998). *Chimpanzee and red colobus: The ecology of predator and prey.* Cambridge, MA: Harvard University Press.

Stearns, S. (1989). The evolutionary significance of phenotypic plasticity. *BioScience, 39,* 436–444.

Stein, R. C. (1968). Modulation in bird sound. *Auk, 94,* 229–243.

Stephens, D. (1991). Change, regularity and value in the evolution of learning. *Behavioral Ecology, 2,* 77–89.

Stephens, D. W. (1993). Learning and behavioral ecology: Incomplete information and environmental predictability. In D. Papaj & A. Lewis (Eds.), *Insect learning: Ecological and evolutionary perspectives* (pp. 195–218). New York: Chapman and Hall.

Stephens, D. W., Anderson, J. P., & Benson, K. E. (1997). On the spurious occurrence of tit-for-tat in pairs of predator-approaching fish. *Animal Behaviour, 53,* 113–131.

Stephens, D., & Krebs, J. (1986). *Foraging theory.* Princeton: Princeton University Press.

Stevens, A., & Price, J. (1996). *Evolutionary psychiatry: A new beginning.* New York: Routledge.

Stevenson-Hinde, J., Stillwell-Barnes, R., & Zunz, M. (1980). Subjective assessment of rhesus monkeys over four successive years. *Primates, 21,* 66–82.

Stevenson-Hinde, J., & Zunz, M. (1978). Subjective assessment of individual rhesus monkeys. *Primates, 19,* 473–482.

Stockley, P., Gage, M., Parker, G., & Møller, A. (1997). Sperm competition in fishes: The evolution of testis size and ejaculate characteristics. *American Naturalist, 149,* 933–954.

Strager, H. (1995). Pod-specific call repertoires and compound calls of killer whales (*Orcinus orca* Linnaeus 1758) in the waters of Northern Norway. *Canadian Journal of Zoology, 73,* 1037–1047.

Strassmann, B. (1996). The evolution of endometrial cycles and menstruation. *Quarterly Review of Biology, 71,* 181–220.

Strassmann, B. (1997). Polygyny as a risk factor for child mortality among the Dogon. *Current Anthropology, 38,* 688–695.

Struhsaker, T. T. (1967a). Auditory communication among vervet monkeys (*Cercopithecus aethiops*). In S. A. Altmann (Ed.), *Social communication among primates* (pp. 281–324). Chicago: University of Chicago Press.

Struhsaker, T. T. (1967b). Behavior of vervet monkeys (*Cercopithecus aethiops*). *University of California Publications in Zoology, 82,* 1–74.

Sullivan, J., Jassim, O., Fahrbach, S., & Robinson, G. (2000). Juvenile hormone paces behavioral development in the adult worker honeybee. *Hormones and Behavior, 37,* 1–14.

Sullivan, K. (1984). The advantages of social foraging in downy woodpeckers. *Animal Behaviour, 32,* 16–32.

Sullivan, K. (1985). Selective alarm calling by downy woodpeckers in mixed species flocks. *Auk, 102,* 184–187.

Summers-Smith, J. D. (1963). *The house sparrow* (Vol. 34). London: Collins.

Sutherland, W. J. (1983). Aggregation and the ideal free distribution. *Journal of Animal Ecology, 52,* 821–828.

Sutherland, W. J., & Parker, G. (1985). Distribution of unequal competitors. In R. Sibly & R. Smith (Eds.), *Behavioural ecology* (pp. 255–274). Oxford: Blackwell Scientific Publications.

Suthers, R. A. (1997). Peripheral control and lateralization of birdsong. *Journal of Neurobiology, 33,* 632–652.

Suthers, R. A. (1999). The motor basis of vocal performance in songbirds. In M. Hauser & M. Konishi (Eds.), *The design of animal communication* (pp. 37–62). Cambridge, MA: MIT Press.

Suthers, R. A., & Goller, F. (1996). Respiratory and syringeal dynamics of song production in northern cardinals. In M. Burrows, T. Matheson, P. Newland, & H. Schuppe (Eds.), *Nervous systems and behavior* (Proceedings of the 4th International Congress of Neuroethology, p. 335). Stuttgart: Georg Thieme Verlag.

Suthers, R. A., Goller, F., & Hartley, R. S. (1994). Motor dynamics of song production by mimic thrushes. *Journal of Neurobiology, 25,* 917–936.

Suthers, R. A., Goller, F., & Hartley, R. S. (1996). Motor stereotypy and diversity of songs of mimic thrushes. *Journal of Neurobiology, 30,* 231–245.

Suthers, R. A., & Hartley, R. S. (1990). Effect of unilateral denervation on the acoustic output from each side of the syrinx in the singing mimic thrushes. *Society for Neuroscience (Abst), 16,* 1249.

Swaddle, J. P., & Cuthill, I. C. (1995). Asymmetry and human facial attractiveness: Symmetry may not always be beautiful. *Proceedings of the Royal Society of London, 261,* 111–116.

Symons, D. (1978). *Play and aggression: A study of rhesus monkeys.* New York: Columbia University Press.

Symons, D. (1990). Adaptiveness and adaptation. *Ethology and Sociobiology, 11,* 427–444.

Taborsky, M. (1994). Sneakers, satellites, and helpers: Parasitic and cooperative behavior in fish reproduction. *Advances in the Study of Behavior, 23,* 1–97.

Talbot, C., Nicod, A., Cherny, S., Fulker, D., Collins, A. C., & Flint, J. (1999). High-resolution mapping of quantitative trait loci in outbred mice. *Nature Genetics, 21,* 305–308.

Talmi, M. C., E., Bengelloun, W., & Soumireu-Mourat, B. (1996). Chronic RU486 treatment reduces age-related alterations of mouse hippocampal function. *Neurobiology of Aging, 17,* 9–17.

Tang-Halpin, Z. (1991). Kin recognition cues of vertebrates. In P. G. Hepper (Ed.), *Kin recognition* (pp. 220–258). Cambridge: Cambridge University Press.

Tarboton, W. (1992). Aspects of the breeding biology of the African jacana. *Ostrich, 63,* 141–157.

Tatar, M. (1999). Transgenes and in the analysis of lifespan and fitness. *American Naturalist, 154,* 567–581.

Tatar, M., Khazaeli, A., & Curtsinger, J. (1997). Chaperoning extended life. *Nature, 390,* 30.

Taub, D. (1980). Female choice and mating strategies among wild Barbary macaques. In D. Lindburg (Ed.), *The macaques: Studies in ecology, behavior and evolution.* New York: Van Nostrand Reinhold.

Taub, D. (1984). Male caretaking behavior among wild Barbary macaques. In D. Taub (Ed.), *Primate paternalism.* New York: Van Nostrand Reinhold.

Taylor, M. (1986). Receipt of support from family of black Americans: Demographic and familial differences. *Journal of Marriage and the Family, 48,* 647–677.

Telecki, G. (1973). *The predatory behavior of chimpanzees.* Lewisburg: Bucknell University Press.

Templeton, A., Crease, T., & Shah, F. (1985). The molecular through ecological genetics of abnormal abdomen in *Drosophila mercatorum.* I. Basic genetics. *Genetics, 111,* 805–818.

Templeton, J. J., & Giraldeau, L.-A. (1996). Vicarious sampling: The use of personal and public information by starlings foraging in a simple patchy environment. *Behavioral Ecology and Sociobiology, 38,* 105–114.

Temrin, H., Buchmayer, S., & Enquist, M. (2000). Step-parents and infanticide: New data contradict evolutionary predictions. *Proceedings of the Royal Society of London, 267,* 943–945.

ten Cate, C., & Vos, D. R. (1999). Sexual imprinting and evolutionary processes in birds: A reassessment. *Advances in the Study of Behavior, 28,* 1–31.

Terry, R. L. (1970). Primate grooming as a tension reduction mechanism. *Journal of Psychology, 76,* 129–136.

Thatch, W., Goodkin, H., & Keating, J. (1992). The cerebellum and adaptive coordination of movement. *Annual Review of Neuroscience, 15,* 403–442.

Thiessen, D., & Yahr, P. (1977). *The gerbil in behavioral investigations.* Austin: University of Texas Press.

Thompson, J. N. (1982). *Interaction and coevolution.* New York: John Wiley.

Thompson, K. V. (1996). Play-partner preferences and the function of social play in infant sable antelope, *Hippotragus niger. Animal Behaviour, 52,* 1143–1152.

Thor, D., & Holloway, W. (1986). Social play soliciting by male and female juvenile rats: Effects of neonatal androgenization and sex of cagemates. *Behavioral Neuroscience, 2,* 275–279.

Thorndike, E. L. (1898). *Animal intelligence: Experimental studies.* New York: MacMillan.

Thorndike, E. L. (1911). Animal intelligence: An experimental study of the association processes in animals. *Psychological Monographs, 2,* whole issue.

Thornhill, R. (1976). Sexual selection and nuptial feeding behavior in *Bittacus apicalis. American Naturalist, 110,* 529–548.

Thornhill, R. (1979a). Adaptive female-mimicking in a scorpionfly. *Science, 295,* 412–414.

Thornhill, R. (1979b). Male and female sexual selection and the evolution of mating systems in insects. In M. Blum & N. Blum (Eds.), *Sexual selection and reproductive competition in insects.* New York: Academic Press.

Thornhill, R. (1980a). Mate choice in *Hylobittacus apicalis* and its relation to some models of female choice. *Evolution, 34,* 519–538.

Thornhill, R. (1980b). Sexual selection in the black-tipped hangingfly. *Scientific American, 242,* 162–172.

Thornhill, R., & Alcock, J. (1983). *The evolution of insect mating systems.* Cambridge, MA: Harvard University Press.

Thorpe, W. H. (1956). *Learning and instinct in animals.* London: Methuen.

Thorpe, W. H. (1961). *Bird-song: The biology of vocal communication and expression in birds.* Cambridge: Cambridge University Press.

Thorpe, W. H. (1963). *Learning and instinct in animals* (2nd ed.). London: Methuen.

Tinbergen, N. (1964). The evolution of signalling devices. In W. Etkin (Ed.), *Social behavior and organization among vertebrates* (pp. 206–230). Chicago: University of Chicago Press.

Todt, D., & Naguib, M. (2000). Vocal interactions in birds: The use of song as a model in communication, *Advances in the Study of Behavior, 29,* 247–296.

Toivanen, P., & Toivanen, A. (Eds.). (1987). *Avian immunology.* Boca Raton, FL: CRC Press.

Tollrian, R., & Harvell, C. (Eds.). (1998). *The ecology and evolution of inducible defenses.* Princeton: Princeton University Press.

Toma, D., Bloch, G., Moore, D., & Robinson, G. (2000). Changes in *period* mRNA levels in the brain and division of labor in honey bee colonies. *Proceedings of the National Academy of Sciences, U.S.A., 97,* 6914–6919.

Tooby, J., & Cosmides, L. (1989a). Evolutionary psychology and the generation of culture, Part I. *Ethology and Sociobiology, 10,* 29–49.

Tooby, J., & Cosmides, L. (1989b). Evolutionary psychology and the generation of culture, Part II. *Ethology and Sociobiology, 10,* 51–97.

Tortora, G. J., & Grabowski, R. (1996). *Principles of anatomy and physiology.* New York: Addison Wesley Longman.

Travasso, M., & Pierce, N. (2000). Acoustics, context and function of vibrational signalling in a lycaenid butterfly-ant mutualism. *Animal Behaviour, 60,* 13–26.

Trehene, J., & Foster, W. (1981). Group transmission of predator avoidance in a marine insect: The Trafalger Effect. *Animal Behaviour, 29,* 911–917.

Trivers, R. (1971). The evolution of reciprocal altruism. *Quarterly Review of Biology, 46,* 189–226.

Trivers, R. (1974). Parent-offspring conflict. *American Zoologist, 14,* 249–265.

Trivers, R. (1985). *Social evolution.* Menlo Park, CA: Benjamin Cummings.

Trivers, R., & Hare, H. (1976). Haplo-diploidy and the evolution of the social insects. *Science, 191,* 249–263.

Trivers, R., & Willard, D. (1973). Natural selection of parental ability to vary the sex ratio of offspring. *Science, 179,* 90–92.

Turlings, T., Wackers, F., Vet, L., Lewis, W., & Tumlinson, H. (1993). Learning of host-finding cues by hymenopterous parasitoids. In D. Papaj & A. Lewis (Eds.), *Insect learning* (pp. 51–78). London: Chapman and Hall.

Udelsman, R., Blake, M., Stagg, C., Li, D., & Holbrook, N. (1993). Vascular heat chock protein expression in response to shock. *Journal of Clinical Investigation, 91,* 465–473.

Uehara, S., Nishida, T., Hamai, M., Hasegawa, T., Hayaki, H., Huffman, M., Kawanaka, K., Kobayashi, S., Mitani, J., Takahata, Y., Takasaki, H., & Tsukahara, T. (1992). Characteristics of predation by the chimpanzees in the Mahale National Park, Tanzania. In T. Nishida, W. C. McGrew, P. Marler, M. Pickford, & F. de Waal (Eds.), *Topics in primatology* (Vol. 1; pp. 143–158). Tokyo: University of Tokyo Press.

Valone, T. J. (1989). Group foraging, public information and patch estimation. *Oikos, 56,* 357–363.

Valone, T. J. (1991). Bayesian and prescient assessment: Foraging with pre-harvest information. *Animal Behaviour, 41,* 569–577.

Valone, T. J., & Benkman, C. (1999). Public information as a mechanism favouring social aggregations: A brief review of empirical evidence. In N. Adams & R. Slotow (Eds.), *Proceedings of the 22nd International Ornithological Congress* (pp. 1238–1336). Johannesburg: Birdlife.

Valone, T. J., & Brown, J. S. (1989). Measuring patch assessment abilities of desert granivores. *Ecology, 70,* 1800–1810.

Valone, T. J., & Giraldeau, L.-A. (1993). Patch estimation by group foragers: What information is used? *Animal Behaviour, 45,* 721–728.

Valone, T. J., & Lima, S. (1987). Carrying food items to cover for consumption: The behavior of ten bird species under the risk of predation. *Oecologia, 71,* 286–294.

Valone, T. J., & Templeton, J. J. (2000). Public information for the assessment of quality: A widespread social phenomenon. *Philosophical Transactions of the Royal Society of London, 357,* 1549–1557.

van den Berg, C., van Ree, J., Spruijt, B., & Kitchen, I. (1999). Effects of juvenile isolation and morphine treatment on social interactions and opioid receptors in adult rats: Behavioural and autoradiographic studies. *European Journal of Neuroscience, 11,* 3023–3032.

Vandermeer, J. (1984). The evolution of mutualism. In B. Shorrocks (Ed.), *Evolutionary ecology* (pp. 221–232). Oxford: Blackwell Scientific Publications.

Vander Meer, R., Breed, M., Espelie, K., & Winston, M. (Eds.). (1997). *Pheromone communication in social insects.* Boulder, CO: Westview Press.

Vander Wall, S. B. (1990). *Food hoarding in animals.* Chicago: University of Chicago Press.

Vanderwolf, C., & Cain, D. (1994). The behavioral neurobiology of learning and memory: A conceptual reorientation. *Brain Research Review, 19,* 264–297.

van Oordt, G., & Junge, G. (1936). Die hormonal Wirkung der Gonaden auf Sommer and Prachtklied. III. Der Einfluss der kastration auf mannliche Kampflaufer (*Philomachus pugnax*). *Wilhem Roux Archiv fur Entwicklungsmechanik der Organismen, 134,* 112–121.

Veblen, T. (1899). *The theory of the leisure class.* New York: MacMillan.

Vehrencamp, S. (1983). A model for the evolution of despotic versus egalitarian societies. *Animal Behaviour, 31,* 667–682.

Verbeek, M., Boone, A., & Drent, P. (1996). Exploration, agonistic behaviour and dominance in juvenile male great tits. *Behaviour, 133,* 945–963.

Verbeek, M., de Goede, P., Drent, P., & Wiepkema, P. (1998). Individual behavioural characteristics and dominance in aviary groups of great tits. *Behaviour, 136,* 23–48.

Verbeek, M., Drent, P., & Wiepkema, P. (1994). Consistent individual differences in early exploration behaviour of male great tits. *Animal Behaviour, 48,* 1113–1121.

Vermej, G. (1987). *Evolution and escalation.* Princeton: Princeton University Press.

Verner, J., & Willson, M. (1966). The influence of habitats on mating systems of North American passerine birds. *Ecology, 47,* 143–147.

Verrell, P. (1986). Wrestling in the red spotted newt: Resource value and contest asymmetry determine contest duration and outcome. *Animal Behaviour, 34,* 398–402.

Via, S., & Lande, R. (1985). Genotype-environment interaction and the evolution of phenotypic plasticity. *Evolution, 39,* 505–522.

Viitala, J., Korpimaki, E., Palokangas, P., & Koivula, M. (1995). Attraction of kestrels to vole scent marks visible in ultraviolet light. *Nature, 373,* 425–427.

Villarreal, R., & Domjan, M. (1998). Pavlovian conditioning of social-affiliative behavior in the Mongolian gerbil (*Meriones unguiculatus*). *Journal of Comparative Psychology, 112,* 26–35.

Visscher, P. K., & Seeley, T. (1982). Foraging strategies of honeybee colonies in a temperate forest. *Ecology, 63,* 1790–1801.

vom Saal, F. (1989). Sexual differentiation in litter-bearing mammals: Influence of sex of adjacent fetuses in utero. *Journal of Animal Science, 67,* 1824–1840.

von Frisch, K. (1967). *The dance language and orientation of bees.* Cambridge, MA: Harvard University Press.

Vonesh, J. (2000). Dipteran predation on the eggs of four *Hyperolius* frog species in western Uganda. *Copeia, 2000,* 560–566.

Wade, M. J., & Pruett-Jones, S. G. (1990). Female copying increases the variance in male-mating success. *Proceedings of the National Academy of Sciences, U.S.A., 87,* 5749–5753.

Wahlstrom, L. K., & Kjellander, F. (1995). Ideal free distribution and natal dispersal in female roe deer. *Oecologia, 103,* 302–308.

Waldman, B. (1987). Mechanisms of kin recognition. *Journal of Theoretical Biology, 128,* 159–185.

Walker, P. (Ed.). (1989). *Cambridge dictionary of biology.* Cambridge: Cambridge University Press.

Wallace, B. (1984). On adaptation, neo-Darwinian tautology and population fitness: A reply. *Evolutionary Biology, 17,* 59–71.

Walter, M. (1973). Effects of parental colouration on the mate preference of offspring in the zebra finch, *Taeniopygia guttata castanotis. Behaviour, 46,* 154–173.

Walters, J. R., Doerr, P. D., & Carter, J. H. I. (1992). Delayed dispersal and reproduction as a life-history tactic in cooperative breeders: Fitness calculations from red-cockaded woodpeckers. *American Naturalist, 139,* 623–643.

Walther, F. R. (1969). Flight behaviour and avoidance of predators in Thomson's gazelles (*Gazella thomsoni* Guenther 1884). *Behaviour, 34,* 184–221.

Ward, P., & Zahavi, A. (1973). The importance of certain assemblages of birds as "information centres" for finding food. *Ibis, 115,* 517–534.

Waring, H. (1963). *Color change mechanism in cold-blooded vertebrates.* New York: Academic Press.

Warkentin, K. (1995). Adaptive plasticity in hatching age: A response to predation risk trade-offs. *Proceedings of the National Academy of Sciences, 92,* 3507–3510.

Warkentin, K. (1999). The development of behavioral defenses: A mechanistic analysis of vulnerability in red-eyed tree frog hatchlings. *Behavioral Ecology, 10,* 251–262.

Warkentin, K. (2000). Wasp predation and wasp-induced hatching of red-eyed treefrog eggs. *Animal Behaviour, 60,* 503–510.

Warner, R. R. (1988). Traditionality of mating-site preferences in a coral-reef fish. *Nature, 335,* 719–721.

Warriner, C., Lemmon, W., & Ray, T. (1957). Early experience as a variable in mate selection. *Animal Behaviour, 11,* 221–224.

Waser, P. (1981). Sociality or territorial defense? The influence of resource renewal. *Behavioral Ecology and Sociobiology, 8,* 231–237.

Waser, P., & Waser, M. (1985). *Ichneumia alhicauda* and the evolution of viverrid gregariousness. *Zeitschrift fur Tierpsychologie, 68,* 137–151.

Watkins, W., & Schevill, W. (1977). Sperm whale codas. *Journal of the Acoustical Society of America, 62,* 1486–1490.

Wattiaux, J. (1968a). Cumulative parental effects in *Drosophila subobscura. Evolution, 22,* 406–421.

Wattiaux, J. (1968b). Parental age effects in *Drosophila pseudoobscura. Experimental Gerontology, 3,* 55–61.

Webb, P. W., & Skadsen, J. M. (1980). Strike tactics of *Esox. Canadian Journal of Zoology, 58,* 1462–1469.

Webster, M., & Westneat, D. (1998). The use of molecular markers to study kinship in birds: Techniques and questions. In R. DeSalle & B. Schierwater (Eds.), *Molecular approaches to ecology and evolution* (pp. 7–36). Basel: Birkhauser.

Wedekind, C., & Furi, S. (1997). Body odour preferences in men and women: Do they aim for specific MHC combinations or simply heterozygosity? *Proceedings of the Royal Society of London, 264,* 1471–1479.

Weigensberg, I., & Roff, D. (1996). Natural heritabilities: Can they be reliably estimated in the laboratory. *Evolution, 50,* 2149–2157.

Weilgart, L., & Whitehead, H. (1997). Group-specific dialects and geographical variation in coda repertoire in South Pacific sperm whales. *Behavavioral Ecology and Sociobiology, 40,* 277–285.

Weiner, J. (1995). *The beak of the finch: A story of evolution in our time.* New York: Vintage Books.

Weisbard, C., & Goy, R. W. (1976). Effect of parturition and group composition on competitive drinking order in Stumptail macaques (*Macaca arctiodes*). *Folia Primatologia, 25,* 95–121.

Weiss, A., King, J., & Figueredo, A. (2000). The heritability of personality factors in chimpanzees (*Pan troglodytes*). *Behavior Genetics, 30,* 213–221.

Weiss, J. (1972). Influence of psychological variables on stress-induced pathology. In R. Porter & J. Knight (Eds.), *Physiology, emotion and psychosomatic illness.* Amsterdam: Elsevier.

Wenegrat, B. (1984). *Sociobiology and mental disorders: A new view.* Menlo Park, CA: Addison/Wesley.

Werner, E. (1986). Amphibian metamorphosis: Growth rate, predation risk and optimal size at transformation. *American Naturalist, 128,* 319–341.

Werner, E. (1988). Size, scaling, and the evolution of complex life cycles. In B. Ebenman & L. Persson (Eds.), *Size-structured populations* (pp. 60–81). Berlin: Springer-Verlag.

Werner, E., & Gilliam, J. (1984). The ontogenetic niche and species interactions in size structured populations. *Annual Review of Ecology of Ecology and Systematics, 15,* 393–425.

Werner, E. E., & Hall, D. J. (1974). Optimal foraging and the size selection of prey by the bluegill sunfish *Lepomis macrochirus. Ecology, 55,* 1042–1055.

West, H., Clancy, A., & Michael, P. (1992). Enhanced responses of nucleaus accumbens neurons in male rats to novel odors associated with sexually receptive males. *Brain Research, 585,* 49–55.

West-Eberhard, M. J. (1975). The evolution of social behavior by kin selection. *Quarterly Review of Biology, 50,* 1–35.

West-Eberhard, M. J. (1979). Sexual selection, social competition and evolution. *Proceedings of the American Philosophical Society, 123,* 222–234.

West-Eberhard, M. J. (1983). Sexual selection, social competition and speciation. *Quarterly Review of Biology, 58,* 155–183.

West-Eberhard, M. J. (1989). Phenotypic plasticity and the origins of diversity. *Annual Review of Ecology and Systematics, 20,* 249–278.

Westneat, D. F. (1987a). Extra-pair copulations in a predominantly monogamous bird: Genetic evidence. *Animal Behaviour, 35,* 877–886.

Westneat, D. F. (1987b). Extra-pair copulations in a predominantly monogamous bird: Observations of behavior. *Animal Behaviour, 35,* 865–876.

Westneat, D. F. (2000). A retrospective and prospective look at the role of genetics in mating systems. In M. Fiest-Bianchet & M. Appollonio (Eds.), *Vertebrate mating systems.* River Edge, NJ: World Scientific Publishing Co.

Westneat, D. F., Fredrick, P., & Wiley, R. (1987). The use of genetic markers to estimate the frequency of successful alternative mating tactics. *Behavioral Ecology and Systematics, 21,* 35–45.

Westneat, D. F., Sherman, P. W., & Morton, M. L. (1990). The ecology and evolution of extra-pair copulations in birds. In D. Power (Ed.), *Current ornithology* (Vol. 7; pp. 331–369). New York: Plenum Press.

Westneat, D. F., Walters, A., McCarthy, T. Hatch, M., & Hein, W. (2000). Alternative mechanisms of nonindependent mate choice. *Animal Behaviour, 59,* 467–476.

Wheeler, J., & Rissing, S. (1975). Natural history of *Veromessor pergandei* I. The nest. *Pan-Pacific Entomologist, 51,* 205–216.

White, D. J., & Galef, B. G. (1999). Social effects on mate choices of male Japanese quail, *Coturnix coturnix japonica. Animal Behaviour, 57,* 1005–1012.

White, D. J., & Galef, B. G. (2000). "Culture" in quail: Social influences on mate choices of female *Coturnix japonica. Animal Behaviour, 59,* 975–979.

White, L. (1994a). Coresidence and leaving home: Young adults and their parents. *Annual Review of Sociology, 20,* 81–102.

White, L. (1994b). Stepfamilies over the life course: Social support. In A. Booth & J. Dunn (Eds.), *Stepfamilies: Who benefits? Who does not?* (pp. 109–137). Hillsdale, NJ: Lawrence Earlbaum and Associates.

White, L., & Reidmann, A. (1992). Ties among adult siblings. *Social Forces, 71,* 85–102.

Whitehead, H. (1998). Cultural selection and genetic diversity in matrilineal whales. *Science, 282,* 1708–1711.

Whitehead, H., Waters, S., & Lyrholm, T. (1991). Social organization of female sperm whales and their offspring: Constant companions and casual acquaintances. *Behavioral Ecology and Sociobiology, 29,* 385–389.

Whitehead, H., & Weilgart, L. (1991). Patterns of visually observable behaviour and vocalizations in groups of female sperm whales. *Behaviour, 118,* 275–296.

Whiten, A., & Boesch, C. (2001). The culture of chimpanzees. *Scientific American,* January 2001, 60–67.

Whiten, A., Goodall, J., McGrew, W., Nishida, T., Reynolds, V., Sugiyama, Y., Tutin, C., Wrangham, R., & Boesch, C. (1999). Cultures in chimpanzees. *Nature, 399,* 682–685.

Whitlock, M. (1998). The repeatability of fluctuating asymmetry: A revision and extension. *Proceedings of the Royal Society of London, 265,* 1429–1431.

Whitlock, M. C., & Fowler, K. (1997). The instability of studies of instability: Comment. *Journal of Evolutionary Biology, 10,* 63–67.

Whitten, P. L. (1987). Infants and adult males. In B. Smuts, D. L. Cheney, R. Seyfarth, W. R., & T. Struhsaker (Eds.), *Primate societies* (pp. 343–357). Chicago: University of Chicago Press.

Whoriskey, F., & FitzGerald, G. (1985). Sex, cannibalism and sticklebacks. *Behavioral Ecology and Sociobiology, 18,* 15–18.

Whyte, M. (1980). Cross-cultural codes dealing with the relative status of women. In H. Barry & A. Schlegel (Eds.), *Cross-cultural samples and codes.* Pittsburgh: University of Pittsburgh Press.

Widemo, F. (1998). Alternative reproductive strategies in the ruff (*Philomachus pugnax*): A mixed ESS. *Animal Behaviour, 56,* 329–336.

Wilcox, R. S. (1972). Communication by surface waves: Mating behavior of a water strider (*Gerridae*). *Journal of Comparative Physiology, 80,* 255–266.

Wilcox, R. S. (1995). Ripple communication in aquatic and semiaquatic insects. *Ecoscience, 2,* 109–115.

Wilcoxon, H., Dragoin, W., & Kral, P. (1971). Illness-induced aversions in rats and quail: Relative salience of visual and gustatory cues. *Science, 171,* 826–828.

Wilkinson, G. (1984). Reciprocal food sharing in vampire bats. *Nature, 308,* 181–184.

Wilkinson, G. (1985). The social organization of the common vampire bat. I. Patterns and causes of association. *Behavioral Ecology and Sociobiology, 17,* 111–121.

Wilkinson, G. S. (1990). Food sharing in vampire bats. *Scientific American,* February 1990, 76–82.

Wilkinson, G. S. (1993). Artificial sexual selection alters allometry in the stalk-eyed fly *Cyrtodiopsis dalmanni* (*Diptera: Diopsidae*). *Genetical Research, 62,* 213–222.

Wilkinson, G. S., Kahler, H., & Baker, R. (1998a). Evolution of female mating preferences in stalk-eyed flies. *Behavioral Ecology, 9,* 525–533.

Wilkinson, G. S., Presgraves, D., & Crymes, L. (1998b). Male eye span in stalk-eyed flies indicates genetic quality by meiotic drive suppression. *Nature, 391,* 276–278.

Wilkinson, G. S., & Reillo, P. (1994). Female choice response to artificial selection on an exaggerated male trait in a stalk-eyed fly. *Proceedings of the Royal Society of London, 255,* 1–6.

Williams, G. C. (1957). Pleiotropy, natural selection, and the evolution of senescence. *Evolution, 11,* 398–411.

Williams, G. C. (1966). *Adaptation and natural selection.* Princeton: Princeton University Press.

Williams, G. C. (1992). *Natural selection: Domains, levels and challenges.* Oxford: Oxford University Press.

Wills, C. (1991). Maintenance of multiallelic polymorphism at the MHC region. *Immunological Reviews, 124,* 165–220.

Wilson, D. J., & Lefcort, H. (1993). The effect of predator diet on the alarm response of red-legged frog, *Rana aurora,* tadpoles. *Animal Behaviour, 46,* 1017–1019.

Wilson, D. S. (1980). *The natural selection of populations and communities.* Menlo Park, CA: Benjamin Cummings.

Wilson, D. S. (1990). Weak altruism, strong group selection. *Oikos, 59,* 135–140.

Wilson, D. S., Clark, A. B., Coleman, K., & Dearstyne, T. (1994). Shyness and boldness in humans and other animals. *Trends in Ecology & Evolution, 9,* 442–446.

Wilson, D. S., Coleman, K., Clark, A., & Biederman, L. (1993). Shy-bold continuum in pumpkinseed sunfish (*Lep-*

omis gibbosus): An ecological study of a psychological trait. *Journal of Comparative Psychology, 107,* 250–260.

Wilson, D. S., Muzzall, P. M., & Ehlinger, T. J. (1996). Parasites, morphology, and habitat use in a bluegill sunfish (*Lepomis macrochirus*) population. *Copeia, 2,* 348–354.

Wilson, D. S., & Sober, E. (1994). Re-introducing group selection to the human behavioral sciences. *Behavioral and Brain Sciences, 17,* 585–654.

Wilson, E. O. (1971). *The insect societies.* Cambridge, MA: Harvard University Press.

Wilson, E. O. (1975). *Sociobiology: The new synthesis.* Cambridge, MA: Harvard University Press.

Wilson, E. O. (1980a). Caste and division of labor in leaf-cutter ants (*Hymenoptera: Fromicidae: Atta*). I. The ergonomic optimization of leaf cutting. *Behavioral Ecology and Sociobiology, 7,* 157–165.

Wilson, E. O. (1980b). Caste and division of labor in leaf-cutter ants (*Hymenoptera: Fromicidae: Atta*). I. The overall pattern in *A. sexdens. Behavioral Ecology and Sociobiology, 7,* 143–156.

Wiltschko, R., & Wiltschko, W. (1995). *Magnetic orientation in animals.* Berlin: Springer-Verlag.

Wimberger, P. (1984). The use of green plant material in birds nests to avoid ectoparasites. *Auk, 101,* 615–618.

Winberg, S., & Lepage, O. (1998). Elevation of brain 5-HT activity, POMC expression, and plasma cortisol in socially subordinate rainbow trout. *American Journal of Physiology, 274,* 645–654.

Winberg, S., & Nilsson, G. (1993). Time course of changes in brain serotonergic activity and brain trytophan levels in dominant and subordinate juvenile Artic charr. *Journal of Experimental Biology, 179,* 181–195.

Winberg, S., Nilsson, G., & Olsen, K. H. (1992). Changes in brain serotonergic activity during hierarchy behavior in Artic charr (*Salvelinus alpinus L.*) are socially mediated. *Journal of Comparative Physiology, 170,* 93–99.

Winberg, S., Winberg, Y., & Fernald, R. (1997). Effect of social rank on brain monoaminergic activity in a cichlid fish. *Brain Behavior and Evolution, 49,* 230–236.

Withers, G., Fahrbach, S., & Robinson, G. (1993). Selective neuroanatomical plasticity and division of labour in the honeybee. *Nature, 354,* 238–240.

Withers, G. S., Fahrbach, S. E., & Robinson, G. E. (1995). Effects of experience and juvenile-hormone on the organization of the mushroom bodies of honey-bees. *Journal of Neurobiology, 26,* 130–144.

Witkin, S. R., & Ficken, M. S. (1979). Chickadee alarm calls: Does mate investment pay dividends. *Animal Behaviour, 27,* 1275–1276.

Witschi, E. (1961). Sex and secondary sexual characteristics. In M. A. (Ed.), *Biology and comparative physiology of birds* (pp. 115–168). New York: Academic Press.

Witte, K., & Ryan, M. (1998). Male body length influences mate-choice copying in the sailfin molly, *Poecilia latipinna. Behavioral Ecology, 9,* 534–539.

Wood-Gush, D. G. M., & Vestergaard, K. (1991). The seeking of novelty and its relation to play. *Animal Behaviour, 42,* 599–606.

Woodland, D. J., Jaafar, Z., & Knight, M.-L. (1980). The "pursuit deterrent" function of alarm calls. *American Naturalist, 115,* 748–753.

Woolfenden, G. E., & Fitzpatrick, J. W. (1978). Inheritance of territory in group-breeding birds. *Bioscience, 28,* 104–108.

Woolfenden, G. E., & Fitzpatrick, G. E. (1990). Florida scrub jays: A synopsis after 18 years. In P. B. Stacey & W. D. Koenig (Eds.), *Cooperative breeding in birds* (pp. 241–266). Cambridge: Cambridge University Press.

Wooton, R. J. (1976). *The biology of sticklebacks.* New York: Academic Press.

Worthylake, K., & Hovingh, P. (1989). Mass mortality of salamanders (*Ambystoma tigrinum*) by bacteria (*Acinetobacter*) in an oligotrophic seepage mountain lake. *Great Basin Naturalist, 49,* 364–372.

Wrangham, R. (1995). Relationship of chimpanzee leaf-swallowing to a tapeworm infection. *American Journal of Primatology, 37,* 297–303.

Wrangham, R., & Goodall, J. (1989). Chimpanzee use of medicinal leaves. In P. Heltne & L. Marquardt (Eds.), *Understanding chimpanzees* (pp. 22–37). Cambridge, MA: Harvard University Press.

Wrangham, R., & Peterson, D. (1996). *Demonic male: Apes and the origins of human violence.* New York: Houghton Mifflin.

Wright, S. (1969). *Evolution and genetics of populations: The theory of gene frequencies* (Vol. 2). Chicago: University of Chicago Press.

Wyatt, G., & Davey, K. (1996). Cellular and molecular action of juvenile hormones: II. Roles of juvenile hormone in adult insects. *Advances in Insect Physiology, 26,* 2–155.

Wyles, J., Kunkel, J., & Wilson, A. (1983). Birds, behaviour and anatomical evolution. *Proceedings of the National Academy of Sciences, U.S.A., 80,* 4394–4397.

Yamamoto, I. (1987). Male parental care in the raccoon dog (*Nyctereutes procyonoides*) during the early rearing stages. In Y. Eto, J. L. Brown, & J. Kikkawa (Eds.), *Animal societies* (Vol. 4; pp. 189–196). Tokyo: Japan Societies Scientific Press.

Yanega, D. (1988). Social plasticity and early diapausing females in a primitively social bee. *Proceedings of the National Academy of Sciences, U.S.A., 85,* 4374–4377.

Yanega, D. (1989). Caste determination and differential diapause within the first brood of *Halictus rubicundus* in New York. *Behavioral Ecology and Sociobiology, 24,* 97–107.

Ydenberg, R. C., & Dill, L. M. (1986). The economics of fleeing from predators. *Advances in the Study of Behavior, 16,* 229–249.

Ydenberg, R. C., Giraldeau, L. A., & Falls, J. B. (1988). Neighbours, strangers and the asymmetric war of attrition. *Animal Behaviour, 36,* 433–437.

Yeh, S., Musolf, B., & Edward, D. (1997). Neuronal adaptations to changes in social dominance status in crayfish. *Journal of Neuroscience, 17,* 697–708.

Yoerg, S. I. (1991). Ecological frames of mind: The role of cognition in behavioral ecology. *Quarterly Review of Biology, 66,* 288–301.

Yokoyama, S., Radlwimmer, B., & Blow, N. (2000). Ultraviolet pigments in birds evolved from violet pigments by a single amino acid change. *Proceedings of the National Academy of Sciences, U.S.A., 97,* 7366–7371.

Yom-Tov, Y. (1980). Intraspecific nest parasitism in birds. *Biological Reviews of the Cambridge Philosophical Society, 55,* 93–108.

Young, H. (1987). Herring gull preying on rabbits. *British Birds, 80,* 363.

Zabel, C., Glickman, S., Frank, L., Woodmansee, K., & Keppel, G. (1992). Coalition formation in a colony of prepubertal spotted hyenas. In A. H. Harcourt & F. B. M. de Waal (Eds.), *Coalitions and alliances in humans and other animals* (pp. 112–135). Oxford: Oxford University Press.

Zahavi, A. (1975). Mate selection: A selection for a handicap. *Journal of Theoretical Biology, 53,* 205–214.

Zahavi, A. (1977). The cost of honesty (further remarks on the handicap principle). *Journal of Theoretical Biology, 67,* 603–605.

Zahavi, A., & Zahavi, A. (1997). *The handicap principle.* New York: Oxford University Press.

Zajonc, R. B. (1965). Social facilitation. *Science, 149,* 269–274.

Zamble, E., Hadad, G., Mitchell, J., & Cutmore, T. (1985). Pavlovian conditioning of sexual arousal: First- and second-order effects. *Journal of Experimental Psychology-Animal Behavior Behavior Processes,* 598–610.

Zamble, E., Mitchell, J., & Findlay, H. (1986). Pavlovian conditioning of sexual arousal: Parametric and background manipulations. *Journal of Experimental Psychology-Animal Behavior Processes, 12,* 403–411.

Zentall, T. R., & Galef, B. G. (1988). *Social learning: Psychological and biological perspectives.* Hillsdale, NJ: Lawrence Erlbaum Associates.

Zill, N. (1988). Behavior, achievement and health problems among children in stepfamilies: Findings from a national survey on child health. In E. Hetherington & J. Arasteh (Eds.), *Impact of divorce, single parenting and stepparenting on children* (pp. 325–368). Hillsdale, NJ: Lawrence Erlbaum Associates.

Zucker, E., Dennon, M., Puleo, S., & Maple, T. (1986). Play profiles of captive adult orangutans: A developmental perspective. *Developmental Psychobiology, 19,* 315–326.

Zuckerman, M. (1979). *Sensation seeking.* Hillsdale, NJ: Lawrence Erlbaum Associates.

Zuckerman, M. (1991). *Psychobiology of personality.* Cambridge: Cambridge University Press.

Zuckerman, M. (1994). *Behavioral expressions and biosocial bases of sensation seeking.* Cambridge: Cambridge University Press.

Zuckerman, M. (1996). The psychobiological model for impulsive unsocialized sensation seeking: A comparative approach. *Neuropsychobiology, 34,* 125–129.

Zuckerman, S. (1932). *The social life of monkeys and apes.* New York: Harcourt Brace.

Credits

PHOTOS

p. ii Two cheetahs/© Michele Burgess/Index Stock Imagery/PictureQuest.

CHAPTER 1 opener Detail of cave painting/© Pierre Colombel/Corbis. 1.1© E. R. Degginger/Color-Pic, Inc. 1.2 Archaeological Museum of Herakleion. 1.4 © Pierre Colombel/Corbis. 1.6A, B: Ann & Rob Simpson. 1.7 Graham Hickman. 1.9 Manfred Milinski. 1.11 L. Bernays; courtesy of Reuven Dukas. 1.12A Reuven Dukas. 1.15 Paul Hobson, © Copyright 2002 Nature Picture Library. All rights reserved. 1.16 Jeff Galef. 1.20 Alissa Crandall/Corbis.

CHAPTER 2 opener Detail from Four dogs in a row/Getty Images. 2.1A, B Ann & Rob Simpson. 2.2A Getty Images. 2.2B © Yann Arthus-Bertrand/Corbis. 2.6 Dwight R. Kuhn. 2.7 Tim Caro. 2.12 Jan Rees. 2.13A, B Charles Brown. 2.15 © A. J. Copley/Visuals Unlimited. 2.16A, B Lee Dugatkin. 2.17 Lee Dugatkin. 2.19A, B James Gilliam. 2.22A Brad Semel; courtesy of Paul Sherman. 2.26 Stanley Braude. 2.28A, B Tim Caro. 2.29 Craig Packer.

CHAPTER 3 opener Detail of zebra finch/Ann & Rob Simpson. 3.3 Geoff Hill. 3.4 Geoff Hill. 3.7 G & H Denzau/naturepl.com. 3.9A, B, C, D Serge Pellis. 3.12A, B Andrew Bass. 3.14 Birgit Ehmer. 3.17 Ann & Rob Simpson. 3.18 Ann & Rob Simpson. 3.20 Ann & Rob Simpson.

CHAPTER 4 opener A male *Gonodactylus smithii* in a threat position/Roy Caldwell. 4.1 Ann & Rob Simpson. 4.3 © Ken Lucas/Visuals Unlimited. 4.8 Sovfoto. 4.13 Eric S. Murphy. 4.15 Roy Caldwell. 4.16 © James Mountjoy; courtesy of L. Lefebvre. 4.19 Tom Walker/Photri. 4.20A K. Hollis. 4.20B Jeff Galef. 4.21 © E. R. Degginger/Color-Pic, Inc.

CHAPTER 5 opener Stone play in Japanese macaque/Michael Huffman. 5.1 Umeyo Mori. 5.2A, B Frans de Waal. 5.3 Michael Huffman. 5.6 Andrew Meltzoff. 5.7 Andrew Meltzoff. 5.9A Damon Fourie/Stone; supplied by Getty Images. 5.9B Jacques Copeau/FPG; supplied by Getty Images. 5.11 Albert Bandura. 5.15 Sarah Colter. 5.16A, B Tim Caro. 5.18 © Gary Carter/Visuals Unlimited. 5.19A © Brandon Cole. 5.19B © François Gohier. 5.19C © Philip Rosenberg.

CHAPTER 6 opener Male horses competing for females/© Rich Pomerantz/www.richpomerantz.com. 6.1A © Mark F. Wallner. 6.1B © Rich Pomerantz/www.richpomerantz.com. 6.5 Bill Beatty. 6.8 © Maslowski/Visuals Unlimited. 6.10 © Corbis. 6.12A © Michael & Patricia Fogden. 6.12B Johann Schumacher. 6.18 © Mark Moffett; courtesy Jerry Wilkinson. 6.21 © Art Morris/Visuals Unlimited. 6.26 niallbenvie.com. 6.33 © François Gohier.6.35 Ryan L. Earley. 6.36 Michael Ryan.

CHAPTER 7 opener Detail of polygamist with his wives/© Nik Wheeler/Corbis. 7.1 © Nik Wheeler/Corbis. 7.3 Compliments of the Peromyscus Genetic Stock Center and Clint Cook. 7.6 Ann & Rob Simpson. 7.8A Nick Davies. 7.8B W. B. Carr. 7.13 Ann & Rob Simpson. 7.15A, B, C Monique Borgerhoff-Mulder. 7.16A, B David Westneat. 7.20A, B, C, D Geoff Parker.

CHAPTER 8 opener White-fronted bee-eaters/N. J. Demong. 8.1A George D. Lepp; courtesy of Paul W. Sherman. 8.1B Paul W. Sherman. 8.2 John Hooglund and Wind Cave National Park. 8.9 Roger Brown/Auscape International. 8.11 © F. K. Schleicher/Vireo. 8.13 Lynn M. Stone/Auscape International. 8.15A, B N. J. Demong. 8.17A, B Francis Ratnieks. 8.19A, B Kevin Foster. 8.23 Yves Lanceau/Auscape International. 8.25A © Ariel Skelley/Corbis. 8.25B © Science VU/CDC/Visuals Unlimited. 8.29 © Gary Carter/Visuals Unlimited. 8.31 David Pfennig. 8.34 © Eric and David Hosking/Corbis.

CHAPTER 9 opener Dwarf mongooses/SailfishExisNet. 9.1 Michael Alfieri. 9.3 © Chris Crowley/Visuals Unlimited. 9.4 © Jack Milchanowski/Visuals Unlimited. 9.5 Dwight R. Kuhn. 9.12 Jerry Wilkinson. 9.16 Jeff Stevens. 9.21 Annie Griffiths Belt/Corbis. 9.23A, B Scott Creel. 9.26A Richard Connor. 9.26B Frans de Waal. 9.28 Naomi Pierce.

CHAPTER 10 opener Richardson ground squirrel foraging/Ron Erwin. 10.1A, B Ron Erwin. 10.2 Christian Ziegler. 10.3A © Glenn Oliver/Visuals Unlimited. 10.3B © William Grentell/Visuals Unlimited. 10.11 © Joe McDonald/Visuals Unlimited. 10.15 © Gary Meszares/Visuals Unlimited. 10.18 © Gary Meszares/Visuals Unlimited. 10.21A Scott Creel. 10.21B © Paul Souders/Worldfoto. 10.22 Christopher Boesch. 10.23 © Bob Newman/Visuals Unlimited. 10.27 © Nancy Alexander/Visuals Unlimited. 10.28 Jennifer Templeton. 10.29 © George D. Lepp/Corbis.

CHAPTER 11 opener Thomson's gazelle/© Gerald and Buff Corsi/Visuals Unlimited. **11.1A, B, C** Richard Coss. **11.7A, B** Steven Lima. **11.8A, B** Karen Warkentin. **11.11** © Adam Jones/Visuals Unlimited. **11.13** © Gary Carter/ Visuals Unlimited. **11.14A** © Gerald and Buff Corsi/Visuals Unlimited. **11.14B** © Joe McDonald/Visuals Unlimited. **11.16** © Johnny Jensen/Visuals Unlimited.

CHAPTER 12 opener Yellow-bellied marmots communicating tactilely/Elinor Osborn Photography. **12.2A** Elinor Osborn Photography. **12.2B, C** Michael Quinton. **12.5** © Leroy Simon/Visuals Unlimited. **12.7** Bert Holldobler. **12.9** Bert Holldobler. **12.10A, B** Marc Bekoff. **12.11A, B, C** Stim Wilcox. **12.12** © E. R. Degginger/Color-Pic, Inc. **12.16** Bernd Heinrich. **12.19** Frans de Waal. **12.21** Frans de Waal. **12.22** © E. R. Degginger/Color-Pic, Inc. **12.24** Hal Whitehead.

CHAPTER 13 opener Indigo bunting/© Maslowski/ Visuals Unlimited. **13.3A** © Marc Epstein/Visuals Unlimited. **13.5A, B** Judy Stamps. **13.7** © Rob Simpson/Visuals Unlimited. **13.9** © Charles McRae/Visuals Unlimited. **13.11** © Charlie Heidecker/Visuals Unlimited. **13.12A** © William J. Weber/Visuals Unlimited. **13.12B** © Joe McDonald/ Visuals Unlimited. **13.13** © Maslowski/Visuals Unlimited. **13.15** © Fritz Polking/Visuals Unlimited. **13.16** © Science/Visuals Unlimited.

CHAPTER 14 opener Grizzly bears fighting/© Kennan Ward/Corbis. **14.2A** Tom Vezo. **14.2B** Jeff Mondragon/ mondragonphoto.com. **14.8** Magnus Enquist. **14.10** Hugh Drummond. **14.12A** Gordon Schuett. **14.13** © E. R. Degginger/Color-Pic, Inc.

CHAPTER 15 opener Panda playing/© AFP/Corbis. **15.2A** © Marty Snyderman/ Visuals Unlimited. **15.2B** © Jeff Vanuga/Corbis. **15.5** Tim Caro. **15.7A, B, C** John Byers. **15.10A, B** Kaci Thompson. **15.13** Scott Nunes. **15.15** Steven Siviy.

CHAPTER 16 opener Giant Galápagos tortoise/Fred J. Eckert/Eckert Images. **16.5** Alan Templeton. **16.6** Eckart Voland. **16.8** Ann & Rob Simpson. **16.12** George W. Gilchrist. **16.13A, B** Joseph Kiesecker. **16.15** © Ken Lucas/Visuals Unlimited.

CHAPTER 17 opener Dog with cocked head/© Royalty-Free/Corbis. **17.3** Dwight R. Kuhn. **17.8** © E. R. Degginger/Color-Pic, Inc. **17.9** Ken Howard/Sea Images. **17.10** David Lank. **17.12** © Jack Ballard/Visuals Unlimited.

ILLUSTRATIONS

FIGURE 1.3 From M. Masseti, Did the study of ethology begin in Crete 400 years ago? *Ethology, Ecology and Evolution, 12,* pp. 89–96. Drawing by Marco Masseti. Reproduced by permission of *Ethology, Ecology and Evolution*.

FIGURE 9.2 T. H. Kunz & A. L. Allagaier, Allomaternal care: Helper assisted birth in the Rodrigues fruit bat. *Journal of Zoology, London, 232*: 691–700. Copyright © 1994. Illustrations by Fiona Reid. Reprinted with the permission of the author, Thomas Kunz.

FIGURE 11.5 From T. J. Pitcher & C. J. Wyche, Predator avoidance behaviour in sand-eel schools: Why schools seldom split. In D. I. G. Noakes, B. G. Linquist, G. S. Helfman, & J. A. Ward (Eds.), *Predators and prey in fish* (pp. 193–204). Copyright © Junk, The Hague, Netherlands. Reproduced by permission of the author, Tony Pitcher.

FIGURE 13.14 From Emlen & Emlen, A new technique for measuring orientation of migratory songbirds. Copyright © 1969, 1975. Reprinted with the permission of the author, Stephen Emlen.

FIGURE 14.4 From M. Mesterton-Gibbons and E. Adams, Animal contests as evolutionary games, in *American Scientist* (July-August 1998, Vol. 86:334–341). Illustration credit: Linda Huff/American Scientist. Copyright © 1998 American Scientist. Reprinted by permission from American Scientist, magazine of Sigma Xi, The Scientific Research Society.

FIGURE 14.12B From G. W. Schuett & J. C. Gillingham, Male-male agonistic behaviour of the copperhead, *Agkistrodon contortrix. Amphibia-Reptilia, 10*: 243–66. Copyright © 1989 *Amphibia-Reptilia.*. Reprinted with the permission of the authors, Gordon Schuett, Arizona State University West, and James Gillingham, Central Michigan University.

Index

Page numbers in *italics* refer to photos and illustrations; those in **bold** refer to graphs and tables.

Abbott, D, 529
Abbott, J. C., 430
Abdel-malek, S. A., 300
Abel, T., 93
Able, K. P., 132, 465, 466
Abreu, E., 357
Ambystoma tigrinum (tiger salamander), *287*
acetylcholine, 89
Acinonyx jubatus, see cheetah
acorn woodpecker (*Melanerpes formicivorus*), 266–67, *266*, **267**
Acromyrmex octospinosus (leaf-cutter ant), *341*
Acromyrmex versicolor (ant), 318–19, *318*, **319**
Actinomycetous (bacterium), *341*
active (proactive) coping style, 589–90, 595
Adams, E., 418, 483, 485
adaptation, 57–61, 70
  definitions of, 57–58
  fitness and, 57–58
Adaptation and Natural Selection (Williams), 50
adrenal glands, 82
adrenaline, 590
adrenocorticotropic hormone (ACTH), 81, 82
African hunting dog (*Lycaon pictus*), 298
Agalychnis callidryas (red-eyed treefrog), 387–89, *387*, **388**
aggression, 477–507, *479–81*, **481**, **483–84**, *485–86*, **487**, *490*, *492*, **493–94**, *495*, **498**, *499*, **500**, **505**
  audience effect and, 499
  bystander effects and, 498–99, 506
  communication and, 427–34, *428*, **429**, *430*, **431**, 443
  cooperative breeding and, 325–26
  game theory models of, 479–91, 505
  hawk-dove game model of, 480, 482–87, **483**, **484**, *485*, *486*, 503, 505
  hierarchy formation and, 492–93, 496–98, **498**
  hormones and, 499–504, **500**
  learning and, 132, 136–40, **137**, **140**, 141, 498–99
  neurotransmitters and, 499, 504–5, **505**, 506
  play fighting and, 85–87
  reconciliation and, 432–34, *433*
  resource holding power (RHP) and, 497
  resource value and, 480–83, *480*, **481**
  sequential assessment model of, 480, 489–91, 505–6
  testosterone and, 80–81, **81**, 85–87
  war of attrition model of, 480, 487–88, **488**, 505
  winner-loser effects and, 137–40, **140**, 491–98, *491*, **493**, **494**, *495*, *498*, *499*, 506
aging, 542–52
  defined, 542–43
  disposable soma theory of, 546–49, 551–52, 564
  heat-shock proteins and, 554, 564
  hormones and, 552–54, 564
  see also longevity; senescence
Agkistrodon contortrix (copperhead snake), 495–96, *495*
Agren, G., 134
Aho, T., 555
Aix sponsa (wood duck), 58–61, *59*
Akerman, A., 498
alarm-calling hypothesis, 254–56, *254*, **255**, *256*
alarm calls, 393
  as antipredator behavior, 389–92, *389–90*, *391*, 408
  deceptive, 434–35
  kinship and, 254–56, *254*, **255**, *256*
  predation and, *254*, 255–56, 434–36
  of vervet monkey, 389–91, 389, *390*
Alatalo, R., 200, 234
Albin, R., 545–46
Albon, S., 206, 207–8, **208**
Alces alces (moose), 352, *352*, *353*
Alcock, J., 50, 51, 75, 76, 125, 183, 185, 221, 223, 225
aldosterone, 82
Alerstam, T., 465
Alexander, R. D., 10, 51, 66–67, 278, 286, 315
Alfieri, Michael, 308, 402, 403, *403*, 575, 578
Allan, S., 438
Allee, W. C., 300–301
allele, 36, 45, 166
Allen, C., 389, 512

Allgaier, A. L., 297–98
alliances, 326
  first-order, 330
  second-order, 330
allogrooming (social grooming), 298–99, *299*
Alpine chough (*Pyrrhocorax graculus*), 341–42, **342**
Altmann, J., 324
Altmann, S. A., 231, 392, 406, 526
altruism, 10–11, 136, 287, 289, 300, 332–33
  handicap principle and, 440–41
  see also cooperation
Alzheimer's disease, 546
American cockroach, 4, *4*
American jacana (*Jacana spinosa*), 205
amino acid, 38, 96–97
Anas platyrhynchos (mallard), 94–95, **94**, *95*, 453–54, **454**, *454*
Anderson, R. C., 581, 582, **582**
Andersson, M., 180, 181, 186, 193, 212–13
androgens, 80, 426–27
Anolis aeneus (lizard), 455–57, *456*, **457**
antagonistic pleiotropy model, 545–47, **545**, *545*, 564
"anting" behavior, 558, 564
antipredator behavior, 51–53, *53*, 379–409, *380–81*, **381–82**, *383*, *385–87*, **388**, *389–91*, *393*, **394**, *396*, **397–99**, *400*, **401**, **403**, *403*, **406**
  alarm signals and, 389–92, *389–90*, *391*, 408
  direct fitness and, 402–6
  group size and, 381–83, **382**, *383*, 391
  hatching time as, 387–89, *387*, **388**
  interpopulation differences and, 395–401, *396*, **397**
  learning and, 402–6, **403**, *403*, **406**, 408
  mobbing and, 392, 404, 406–7
  predator inspection and, *see* predator inspection
  predator-prey "arms race" and, 400–401, *400*, **401**, 404, 408
  social learning and, 406–7, 408
  tail flagging as, 391–92, *391*
  tradeoffs in, 384–89, *385*, 404, 407
ants, 26, 443
  Acromyrmex versicolor, 318–19, *318*, **319**
  brood raiding in, 316–19, **317**, *318*, **319**
  foraging in, 340–41, *341*, 421–23, *422–23*
  fungi and, 340–41, *341*
  group selection in, 316–19, **317**, *318*, **319**
  Iridomyrmex anceps, 331, *331*, **334–35**
  leaf-cutter, *341*, 421–23, *422–23*
  minim workers, caste of, 423, *423*
  and mutualism with butterflies, 331, 334–35
  seed harvester, 316–19, **317**
Ants, The (Hölldobler and Wilson), 27
Aphelocoma coerulescens (Florida scrub jay), 267, *267*, 462–64, *462*
Apis mellifera, see honeybee
appetitive stimulus, 116–17
Aquilino, W., 274
Arabian babblers, 440
Archer, J., 137, 274, 478, 501, 504
Arctic char (*Salvelinus alpinus*), 504–5
Aristotle, 5
Armitage, K., 229, 529
Arnold, W., 518
artificial selection, 32–34, *33–34*, 69, 552, 590
Askenmo, C., 385
Atkinson, D., 15
Atlantic leatherback sea turtle, *97*
Atlantic salmon (*Salmo salar*), 428–30, *428*, **429**
audience effect, 499
Austad, S., 544, 547
Australian wasp (*Epsilon spp*), 223
autoshaping, 138
aversive stimulus, 117
Avital, E., 146
Axelrod, R., 11, 301, 303, 305
axons, 88–89, *89*

babblers, 440
baboon:
  chacma, 160
  olive, 328–30
  yellow, 160
Bachman, G., 583
Backwell, P., 490
Baerends, G. P., 160, 428
Baerends-van Roon, J., 160
Bagg, A., 558
Baker, M., 558
Baker, R., 92
Baker, R. R., 243–44
Bakker, T. C. M., 189, 193
Balaban, E., 504
Balda, R., 127, 341, 357
Baldwin, J., 526
Ball, N., 94
Bandura, A., 152–53, *154*
Banks, E., 300
bank swallow (*Riparia riparia*), 258, 291
Barash, D., 298
Barbary macaque (*Macaca sylvanus*), 226
bare-necked umbrellabird, *188*
Barkley, R., 133, 470
Barlow, G. W., 428
Barnard, C. J., 356, 362, 366, 482
barn swallows, 185, **185**, *186*, 436, **436**, *436*
base mutations, 38
Basil, J., 94
Basolo, A., 211–14, *214*
Bass, A. H., 90, 92, 426
Bateson, P., 195, 451, 568
Baum, D., 57
Beatty, W., 531
Beauchamp, G., 58
Beck, B., 514
Bednarz, J., 299
Beebee, T., 557
Beecher, M., 132, 291
bees, *4*, 26, *125*, *349*
  bumblebee, 271
  see also honeybee

Begon, M., 15
behavioral genetics, 44–49
  behavioral variation and, 46–49, *47*, **48**
  gene location and, 45–46
  Mendel's laws and, 44, 45
  polygenetic traits and, 45–46
Beilharz, R., 591, 594
Bekoff, M., 389, 512, 523–24, 525
Belding's ground squirrel (*Spermophilus beldingi*), 254–56, *254*, **255**, *256*, 258, 434, 529–30, *529*
Bell, G., 37
Bell, M., 396
Bellis, M. A., 242, 243–44
Bellrose, F., 60
Belovsky, G., 352
Benkman, C., 369
Bennett, A., 96
Bent, A. C., 299
Bentley, P. J., 83
Benus, R. F., 589
Benzer, S., 371
Bercovitch, F., 329, 330
Berger, J., 518, 522, 523
Berglund, A., 206
Bergman, A., 130
Bernays, E. A., 13
Bernstein, C., 545
Bernstein, H., 545
Berthold, P., 469–70
Bertiella (tapeworm), 559–61, **559**
Bertram, B., 361
Betzig, L., 234
Biben, M., 526–27
bighorn sheep (*Ovis canadensis*), 522–23, *523*
Bigson, R., 198, 201
Bildstein, K. L., 391
Billen, J., 421
Billing, J., 562–63
Birch, L., 368–69
birds, 94, 171, 172, 244, 249, 336, 392
  aggression in, *480*, 492–93, *492*, **493**
  "anting" behavior in, 558, 564
  antipredator behavior in, 406–7
  brain size of, 357–58
  communication in, 416–17, 430, **430**, *430*, **431**, 434, 435–39, **436**, *436*, **438**
  cooperative breeding in, 261, 264–65, 269–70, **269**, *269*, 289, 321–22, **322**, 462–64
  cooperative hunting in, 299, *299*
  cultural transmission in, 162, *162*, 164–65, 406–7
  deceitful communication in, 435, **436**, *436*, 443
  dynasty-building in, 266–67
  extrapair copulations (EPCs) in, 237–38, *237*, **238**, 250
  food storage by, 341–42, **342**, *342*
  foraging in, **348**, 350–51, **351**, 355–58, *355*, **356**, **358**, 364–68, *364*, **365**, *366*, **367**, 370, *370*, 374
  helpers-at-the-nest and, 261, **261**, 264, *264–65*, 462–64
  hierarchy formation in, 492–93
  imitation in, 157, *157*
  kin recognition in, *290*
  learning in, 359–60
  mate choice in, 231–33, *231–32*, **232–33**
  memory in, 341–42, **342**, *342*
  migration of, 449, 465, 466–70, *466*, **470**
  offspring rule and, 261
  personality in, 582–87, *583*, **585**, *586*, 595
  play in, 514–16, *514*, **515**, 534–35
  plumage coloration in, 76–80, *76*, *78*, **79**, 188–89, *188*
  polyandry in, 224
  PTM in, 231–33, *231–32*, **232–33**
  sensory bias in, 215, **215**
  sibling rivalry in, 283–85, *283–84*, **285**
  social learning in, 364–68, *364*, **365**, *366*, **367**
  song learning and mate choice in, 203–5, *205*
  syrinx of, 437–39
  territorial defense in, 460–64, *460*, *462*
  territorial dispersion in, 264, *264–65*
  ultraviolet vision in, 96–97, *96*, 105
  weaning in, 277–78, *278*
  winner-loser effects and, *480*, 492–93, *492*, **493**
  see also specific species
Birkhead, T., 240, 244
Bishop, D. T., 487
Bitterman, M., 122, 127
Bjorklund, M., 234
Bjorksten, T., 192
Black, J., 94
black bear, *340*
Blackburn, J., 531
blackcap (*Sylvia atricapilla*), 470, **470**
black-capped chickadee (*Parus atricapillus*), 434
black grouse, 200–201, *200*, **201**
black rat, *15*
black-tailed prairie dog, 255
Blake, M., 554
Blanco, J., 591
Blaustein, A., 290, 555
blocking, 118–19, 141
blue-footed booby (*Sula nebouxii*), 492–93, *492*, **493**
bluegill sunfish, 209–10, *210*, 347, **348**
  foraging by, 361–62, **361**, *361–62*
  sneaker and satellite males in, 459–60, 474, 583
blue gourami (*Trichogaster trichopterus*), 135, *135*, 137–40, **137**
  learning and mate choice in, 197–98, **198**
blue jay (*Cyanocitta cristata*), 311–14, *311*, **312–13**
blue petrel (*Halobaena caerulea*), 384
blue tit (*Parus caeruleus*), 159, *159*
"Bobo" doll experiment, 153–54, *154*
Boccia, M. L., 298
Boesch, C., 363–64, 433, 587
Boesch, H., 363–64
boldness, 570–78
  defined, 571
Bolig, R., 581

Bolthauser, P., 576
Bonner, J. T., 146
boomerang factor, 310
Booth, A., 501
Borgerhoff-Mulder, M., 230, 234, 235
Borgia, G., 315
Bottger, K., 425, 426
bottle-nosed dolphin (*Tursiops truncatus*), 65, 298
  coalition formation in, 327, 328, 330
Boucher, D., 331
Boulton, M. J., 526
Boyd, R., 130, 146, 150–51, 198, 200
Boyse, E., 286
Bradbury, J. W., 180, 228, 412
brain:
  cultural transmission and size of, 169–73, **172**
  hippocampal region of, 341–42, **342**
  learning and size of, 357–58
  mushroom bodies of, 92–94, *92*
  PFA area of, 532
  play and synapse development in, 519–21, **521**
  play fighting and, 530–33
  unihemispheric sleep and, 94–95, **94**, *95*
Brandon, J., 93
Brandon, R., 57
Brantley, R., 90
Bray, J., 273
Brenowitz, E., 437
Bretagnolle, V., 384
Brick, O., 490
Brienomyrus brachyistius (electric fish), 426–27, **427**, 443
Briskie, J., 244
broad-sense heritability, 41–43, **43**
Brodie, E., 37
Bronstein, J. L., 331
brood parasitism, 58–61, *59*
brood raiding, 316–18
Brooks, R., 203
Brown, C. A., 47–48, *47*, **48**, 155, 356, 419, 482
Brown, G. R., 99
Brown, J. H., 186
Brown, J. L., 20, 51, 82, 83, 189, 230, 261, 262, 266, 301, 309, 321, 322, 323, 463
Brown, J. S., 369, 451
Brown, M., 47–48, *47*, **48**, 520
brown-headed cowbird (*Molothrus ater*), 203–5, *205*, 438
Brownlee, A., 519
brown skua (*Catharacta antarctica lonnbergi*), 384
brown thrush (*Toxostoma rufum*), 438
Brunton, D. H., 392
Brush, A., 76, 77
bryozoan (*Membranipora membranacea*), 112–13, *112*, 402
budding, territorial, 267, *267*, *462*, 463
budgerigars, 157
Budiansky, S., 412
Bufo bufo (toad), 206, 417, **417**
Bull, J., 97–98
bullfrog (*Rana catesbeiana*), 557–58, 557
bumblebee, 271
Burbank, V. K., 234
Burgess, J. W., 484
Burghardt, G., 512, 514
Burley, N., 215
bursa of Fabricus, 466
Buss, A., 582
Busse, C., 363
Butcher, G., 77
Buteo jamaicensis (red-tailed hawk), 385
butterfly-ant mutualism, 331–35, *331*, **334–35**
Byers, J., 512, 519, 520–21
Bygott, J. D., 68
byproduct mutualism, 309–14, **310**, **312–13**, **315**, 336, 418, 462
  in blue jays, 311–14, *311*, **312–13**, **315**
  boomerang factor and, 310
  food calls and, 314, **315**
  reciprocity and, 310–13, **312**
Byrne, R. W., 357, 358
bystander effects, 498–99, 506

cactus finch (*Geospiza scadens*), 164–65
Cade, T. J., 299
Cain, D., 104
Calamospiza melanocorys (lark bunting), 231–32, *231*
Caldwell, R., 125–26
Callithrix jacchus (marmoset), 298
Campbell, A., 501
Canadian General Social Survey (GSS) of 1990, 262, 265, 268
Canales, C., 493
canary (*Serinus canaria*), 438
Canis lupus (wolf), 65, 327
Cannon, W., 589
canthaxanthin, 77
capuchin monkey, 558
Caraco, T., 155, 353, 355–56, 362, 570
Carlier, P., 127–28
Caro, T., 158–61, 327, 362, 391, 392, 512, 514, 516–18, 568
carotenoid pigments, 76–77
Carpodacus mexicanus (house finch), 76–80, *76*, *78*, **79**
Carpenter, C. R., 298, 522
Case, T. J., 316
Catchpole, C., 437
Catharacta antarctica lonnbergi (brown skua), 384
causation, 75
  correlation and, *87*
  hormones and, 80–87, **81**, *83*, *86*, *87*
  proximate, 75–76
Centrocercus urophasianus (sage grouse), 201, **201**
Cervus elaphus (red deer), 100–103, *100*, **101**, 206–8, *207*, **208**, 213
c-fos gene, 532–33
chacma baboon, 160
Chang, F., 94
Chapais, B., 65
Charlesworth, B., 544, 552
Charnov, E., 99, 261, 279, 343, 347–48, 374, 389, 391
Chase, I. D., 499
cheetah (*Acinonyx jubatus*), *21*, 25, *39*, 65, *65*, *159*, 160, 323, *343*
  object play in, *513*, 516–18, *516*, **517**
  and predator inspection by gazelles, 393–95, *393*, **394**

Cheney, D. L., 136, 299, 389, 390–91, 392, 431, 435
Cherrett, J., 421
chickens, 196, 499, *499*
Chiel, H., 125
child abuse, 273–74, **273**
chimpanzees, *49, 149,* 160, 170, 412
  aggression in, 430–34, **432,** *432, 433*
  between-group raiding in, 455
  cooperative hunting by, 363–64, *363*
  cultural transmission in, 587–88, 595
  learning in, *153,* 430–32, **432,** *432*
  pant hoots of, 430–32, **432,** *432, 433*
  personality in, 49, 568, 587–88, 595
  reconciliation in, 432–34
  self-medication in, 558–61, **559**
  soil ingestion in, 561
  see also primates
Chisholm, J. S., 234
Chivers, D., 110
Christie, J., 230
chromosomes, 39, 46, 584
  sex determination and, 97
Church, S., 96
cicada, 66–67
Ciofi, C., 96
circadian rhythm, 371
Cissa erythrorhyncha (red-billed blue magpie), 342, **342**
Claar, J., 591
Clark, A. B., 39, 41, 568
Clark, M., 83–85
Clarke, J. A., 385
classical conditioning, *see* Pavlovian conditioning
Clayton, N., 15
Cleland, J., 286
Clement, T., 359–60
Clements, K. C., 311, *311*
Clethrionomys rufocanus (grey-sided vole), 229
Clever Hans, 413–14, *413*
cliff swallow (*Petrochelidon pyrrhonota*), 47–48, *47,* **48,** 155
  food calling in, 418, 419–20, **419**
Close, R., 521
Clunie, F., 558
Clutton-Brock, T. H., 40, 99, 100, *100,* 102–3, 203, 206–8, **208,** 276, 357, 358, 549–50
coalition formation, 65-69, 298, 326–31
  alliances and, 326, 330
  defined, 326, 336
  in mammals, 330
  in primates, 327, 328–29, *328–29*
  reciprocal, 329
coati (*Nasua narica*), 65, 327
Coccaro, E., 504
cognitive ecology, 138–39
coho salmon (*Oncorhynchus kisutch*), 133, 470–71, **471,** *471*
Coleman, K., 571, 572, 574
collared flycatcher, 213
Collier, G., 205
Columbia livia, see pigeon
common crow, 357
common mole rat (*Cryptomys hottentotus*), 9–10, *9,* **10**
common raven (*Corvus corax*), 430, **430,** *430,* **431,** 534–35
  object play in, 514–16, *514,* **515**
common wasp (*Vespula vulgaris*), 271
communication, 411–15, *413–14,* **415, 417, 419,** *420, 421,* **421,** *422–25,* **425,** *426,* **427,** *428,* **429,** *430,* **430–32,** *432–34, 436,* **436, 438,** *439,* **442**
  aggression and, 427–34, *428,* **429,** *430,* **431,** 443
  alarm calls and, 434–36, *434,* **436,** *436*
  animal-human, 412–14
  "arms race" in, 416
  birdsong as, 436–39, **438**
  and case of "Clever Hans," 413–14
  by chemical signaling, 422
  classical approach to, 416
  deceptive, 434–35, 443
  defined, 412, 440
  foraging and, 418–23, 443
  handicap principle and, 416–17, 440–41, 443
  honesty in, 415–18, 443
  learning and, 430–32, **432,** *432*
  mate investment and, 434
  mating and, 424–27
  modes of, 414, *414,* 443
  play and, 423–24, *424,* 443
  predation and, 434–36, *434,* **436,** *436*
  by ripples, 425, *425,* 426, 443
  by vibrations, 422–23, 443
  whale song as, 439–43, *439,* **442**
conditioned response (CR), *117,* 118
conditioned stimulus (CS), 116, 118, 141
Connor, R., 65, 309, 327–28, 330, 331
Conover, M. R., 45
conservation biology, 27
conspecific cueing hypothesis, 456
conspecific threshold model, 286–87
contagion (response facilitation effect), 156–57
Cook, M., 163
Cooke, F., 195
cooperation, 10–11, 295–337, *296, 298–300,* **301–8,** *308–9,* **310,** *311,* **312–13, 315, 317,** *318,* **319,** *320,* **322,** *324,* **325, 327,** *328–29, 331,* 334–35
  alliances and, 326, 330
  byproduct mutualism and, 301, **301,** 309–14, **310,** *311,* **312–13, 315,** 336
  coalitions and, 298, 326–31, *328–29,* 336
  communication and, 441
  cooperative breeding and, 321–26, **322,** *324,* **325,** 336
  definition of, 296–97, 335–36
  food sharing and, 308–9, *308–9,* **310**
  foraging and, 362–64, *363*
  game theory and, 301–2
  group hunting and, 299, *299*
  group selection and, 301, **301,** 315–21, **317,** *318,* **319,** *320,* 336

cooperation (*continued*)
group size and, 362–64, *363*
and helping in birthing process, 297–98, *298*
hormones and, 323–26, *324*
interspecific mutualism and, 331–35, *331,* **334–35,** 336
learning and, 314
nest raiding and, 300, *300*
phylogenetic component of, 321–22, **322,** 336
play and, *525*
reciprocity and, 301–9, **302–8,** 336
reproductive suppression and, 323–26, *324,* **325**
social grooming and, 298–99, *299*
cooperative breeding, *324,* **325, 327,** 336
aggression and, 325–26
hormones and, 323
in birds, 82–85, 261, 264–65, 269–70, 289, 321–22, 462–64
among dwarf mongooses, 323–26
in social insects, 269–71
social rank and, 325–26, **325**
cooperative polyandry, 224
coping style, 588–90, **588, 589,** 595
copperhead snake (*Agkistrodon contortrix*), 495–96, *495*
Cordts, R., 551
Corl, K., 357, 358
corpora pedunculata (mushroom bodies), 92–94, *92*
corpus allatum, 373
corticosterone, 83, 499–500, 590
corticotropin-releasing hormone (CRH), 82
cortisol, 82, 83, 500
Corvidae, 341–42, *342*
Corvus:
C. corax, see common raven
C. corone (European crow), 342
C. frugilegus (rook), 342, **342**
C. monedula (jackdaw), 341–42, *342,* **342**
Coss, R., 3, 9, 380, 392, 396
Coturnix japonica (quail), 203
Coultier, S., 496
Covich, A., 110
Cowie, R., 350, **351**
Cox, C., 208, **209**
Cozza, K., 591
Cragshaw, P., 591
Creel, S., 261, 274–75, **275,** 323–26, 362
Crenicichla alta (pike cichlid), *53, 57,* 296, *296*
Crespi, B., 287
Crews, D., 99, 500
Crocuta crocuta, see spotted hyena
Crook, J. H., 275–76
Crook, S., 275–76
cross-fostering experiment, 48, 204–5, *205*
crow, 341, **342**
Crowl, T., 110
Crowley, P. H., 487
Crozier, R., 286
crustaceans, role of serotonin in, 505, 506
cryptic choice, 242, 249
Cryptomys hottentotus (common mole rat), 9–10, *9,* **10**
cuckoldry, 209–10, *210*
cultural transmission, 15–18, **16,** *16, 17,* 28, 145–75, *146–47,* **148,** *149–50,* **152,** *152, 153–57, 159, 161–62, 165,* **172**
aggression and, 153–54, *154*
animal culture and, 150–52
of birdsong, 203–5, *205*
brain size and, 169–73, **172**
"chain" of, 407
contagion and, 156–57, 173
copying and, 157–58, 168–69, 173, 198–99, **199,** 200–203, **201**
definition of, 6
of foraging skills, 364–68, *366,* **367**
generational effects of, 16–17
genes and, 164–69, *166,* 173
horizontal, 161, *161,* 162–63, 173
in humans, *150, 152,* 153–54, *154,* 162
imitation and, *150,* 157, 173
learning and, 151
local enhancement and, 154–55, *155*
mate choice and, 198–205, **199, 201,** *202,* **203,** 216
modes of, 161–64
oblique, 161, *161,* 163–65, 173
opportunity for selection models and, 199–200
personality and, 587–88, 595
population genetic models of, 200–203, *202*
social facilitation and, 155–56, *156*
social learning and, 364–69, **366–67,** *366, 368–69*
speed of, 151–52
teaching and, 158–61, *159,* 169, 173
types of, 152–61
vertical, 161, *161,* 162, 164–65, 173
culture, defined, 150
Culture and the Evolutionary Process (Boyd and Richerson), 150
Curio, E., 392, 406–7, 434
Curious Naturalists (Tinbergen), 212
Curran, T., 533
Currie, C., 340–41
Curtsinger, J. W., 488
Cuthill, I. C., 192
Cyanocitta cristata (blue jay), 311–14, *311,* **312–13**
Cygnus olor (mute swan), 45, *45*

Dachsbracken dogs, 278
Daly, M., 256–57, 262–68, **268,** 272–75
Dama dama (fallow deer), 203
damselfly (*Enallagma spp*), 110–11, *110,* **111,** 402
Daphnia magna (water flea), **348,** 452, *453*
Darwin, Charles Robert, 6–7, 26, 27, 32, 33, 39, 50, 66, 149, 164, 178, 179–80, 205, 213, 215, 249, 300, 332, 427, 478, 592

Darwinism and Human Affairs (Alexander), 66, *67*
Darwin's finches, 7–9, *7–8*, 173
cultural transmission in, 164–65
Dasser, V., 136
Davey, K., 373
David, P., 192
Davies, N. B., 51, 127, 206, 221, 224, 226–27, 228, 230, 231, 245–47, 248–49, 417, 460–61, 486, *486*
Davis, D., 321
Davis, J., 262–68, **268**, 272, 275
Davis, R., 93
Davis, S. A., 576
Dawkins, R., 50–51, 415–16
dear enemy effect, 457–58, *458*, 459, 474
DeCola, J., 123
deer, 180, *180*
De Jong, I. C., 40
de Kloet, E. R., 500
De Laet, J., 385
Delehanty, D., 224
Delph, L., 240
DeMartini, E., 90
dendrites, 88, *89*
Dennett, D. C., 389
de Quervain, D. J. F., 500
Descent of Man, The (Darwin), 35, 178
Desjardins, C., 134
Desmodus rotundus (vampire bat), 308–9, *308–9*, **310**
developmental stability, 190–91
Devore, I., 163
de Waal, F., 65, 299, 326, 327, 328, 432, 523
Dewsbury, D., 75
Dhondt, A. A., 234
diabetes, 281, 553
Dickemann, M. 234
Dill, L. M., 385, 393
DiMarco, F. P., 490
Dingle, H., 125, 132
direct benefits model, 182–85
nuptial gifts and, 183–85
direct fitness, 19
antipredator behavior and, 402–6
disease, 542, 555–63, **556–57**, *556–57*, **559**, **563**, 564
avoidance of habitat and, 555–57, 564
avoidance of individuals and, 557–58, **557**, *557*, 564
diet and, 559–63, **563**, 564
self-medication and, 558–61, **559**, 564
disposable soma theory, 546–49, 551–52
Dittman, A., 133, 471
Djungarian hamster (*Phodopus campbelli*), 86
DNA, 38, 46, 51, 97, 442
mitochondrial, 165–67, 173
transposition of, 38
DNA fingerprinting, 63–64, 68, 237, 248, 584
Dobzhansky, T., 67
Dodson, J. J., 132, 133, 470
dogs, 33–34, *33–34*
guide, 591–94
play and communication in, 423–24, *424*
weaning in, 278, *278*
wild, 35, 362–63, *363*, 393, 394
Dolichovespula saxonica (vespine wasp), 271, 272
dolphin (*Tursiops truncatus*), 65, 298, 327, 328
dominance, 100–101
Dominey, W. J., 392
Domjan, M., 116, 120, 134–35, 194, 195, 196–97
Donaldson, T. J., 392
dopamine, 531, 532
Doppler vibrometry, 422
Douglas-Hamilton, I., 517–18
Douglas-Hamilton, O., 517–18
Doutrelant, C., 499
Downhower, J. F., 229
downy woodpecker (*Picoides pubescens*), 434, *434*
dragonflies, 225
Drent, P., 586
Driscoll, P., 590
Drummond, H., 492–93
Dugatkin, L. A., 18, 54, 136, 146, 152, 198, 199, 200, 202, 297, 301, 305–6, 308, 310, 316, 328, 351, 392, 478, 496, 497, 499, 575, 576, 578
Dukas, R., 13, 113
Dumetella carolinensis (gray catbird), 438
Dumond, F., 526
Dunbar, R., 298, 357
Duncker, K., 368, 369
dungfly, 225, 240–41, **241**, *241*, 487–88, **488**
dunnock (*Prunella modularis*), 226–27, *226–27*, 245–47, **247**, *290*
dwarf mongoose (*Helogale parvula*), 274–75, **275**, 323–26, *324*, **325**, **327**
dyadic interactions, 326, 328
Dyer, F., 132
dynasty-building hypothesis, 266–67

Eadie, J., 58
eagles, 389
Eason, P. K., 132
East, M., 434
Eberhard, W., 242
ecological constraint theory, 262
ectoparasites, 185, *186*, 187
Edgerton, V., 521
Edney, E., 552
Edward, D., 90
Edwards, D., 504
Edwards, S. V., 321–22
effect, law of, 121
egg-dumping, 58–61, 59
Eggebeen, D., 267
egrets, 283–85, *284*
Ehlinger, T. J., 39, 568, 573
Ehrenpreis, A., 319
Ehrlich, 558
Ehrman, L., 44
Eibesfeldt, I., 423
Eible-Eibesfeldt, E., 423
Eichenbaum, H., 139
Einon, D., 528, 531
Eisenberg, J., 357, 358
Ek, A., 330
Eklund, A., 189
Ekman, J., 385
elective group size, 381–82, **382**
electric catfish (*Malapterurus electricus*), 426–27, *426*, **427**
electric fish (*Brienomyrus brachyistius*), 426–27, **427**, 443

electric organ discharges (EODs), 426–27, **427**
electrophoresis, 237–38
elephant seal (*Mirounga angustirostris*), 208–9, *208,* **209,** *480*
Elephas maximus (Indian elephant), 298
Elgar, M., 287, 314, 334
Ely, D., 590
Emerson, A. E., 300–301
Emlen, J. M., 212, 343
Emlen, J. T., 374, 584
Emlen, S. T., 83, 221, 224, 228, 261–66, **268,** 272, 274, 288–89, 323, 463, 466–68
Enallagma spp (damselfly), 110–11, *110,* **111,** 402
Endler, J., 37, 39, 41, 51, 52, 54, 57
endocrine system, 80–82, 105
  nervous system compared with, **88**
endoparasites, 185, 187–88, 197–98
Engh, A. L., 578
Enquist, M., 428, 480, 489, 490, 491
Ens, B. J., 362
epinephrine, 82
Epsilon spp (Australian wasp), 223
Erichsen, J., 374
Erritzøe, J., 465–66
Esox lucius (pike), 110–15, 397, 398
Essapian, F. S., 298
Estes, R. D., 391
Estoup, A., 271
estrogen, 81, 83
estrogen conjugate (EC), 326
Etheredge, J., 468–69
European crow (*Corvus corone*), 342
European earwig (*Forficula auricularia*), 208
European jay (*Garrulus glandarius*), 342, **342**
European magpie (*Pica pica*), 342
eusociality, 62
Evans, C. S., 389, 499
Evans, L., 389
Even, M., 83
evolutionarily stable strategies (ESS), 303, **303–4,** 483, 485, 502, 503, 505
evolutionary psychology, 27
Evolution in Mind (Plotkin), 138
"Evolution of Reciprocal Altruism, The" (Trivers), 301
Ewald, P., 227, 482
Ewer, R., 160, 169
Expression of Emotions in Man and Animals, The (Darwin), 427, 592
extinction curves, 124
extrapair copulations (EPCs), 236–38, **237,** *237,* 250, 436, *436*
Ezeh, A. C., 234

Fagen, R., 510, 512, 514
Fagnou, D., 104
Fairbanks, L., 298
fallow deer (*Dama dama*), 203
Falls, J. B., 457
family dynamics, 261–76
  cooperative breeding and, 269–72, **269,** *269–70,* **271,** *272*
  dynasty-building hypothesis and, 266–67, *266*
  evolutionary theory of family and, 261–62, *261,* **263**
  family stability and, 264–66, *264–65*
  infanticide and, 272–73, 281–82, **282**
  optimal skew model of, 274–75, **275,** *275*
  in stepfamilies, 272–74, **273**
  territorial budding and, 267, *267, 426,* 463
  wealth and, 267–68, **268**
  see also kinship
Faneslow, M., 123
Farquhar, M., 551
Farris, S. M., 93
fat sand jirds (*Psammomys obesus*), 86
Faulkes, C. G., 324
Feder, M., 554
Feener, D. H., 423
Feldman, M., 130
female defense polygyny, 223–24
Fentress, J. C., 65, 327
Ficken, M. S., 434
fiddler crab (*Uca paradussumieri*), 206
Fiedler, K., 334
"fight or flight" response, 82
fig wasp, 503
Finch, C., 554
finches, 162, *162,* 164–65
  and natural selection, 8–9
first-order alliances, 330
fish, 171, 244, 336, 392
  aggression in, 490–91, *490,* 493–94 antipredator behavior in, 381–83, **382,** *383,* 396–400, *396,* **397–99,** 402–6, **403,** *403,* **406**
  boldness and inhibition in, 568, 572–74, *572,* **574**
  bystander effect and, 499
  communication in, 426–27, *426,* **427,** 428–30, *428,* **429**
  cuckoldry in, 209–10, *210*
  cultural transmission in, 162–63
  foraging in, 361–62, *361–62,* **362**
  hierarchy formation in, 496–98
  interpopulation differences in, 396–400, *396,* **397–99**
  learning and mate choice in, 197–98, **198**
  learning in, 470–71
  mate-choice copying in, 202–3, *202,* **203**
  migration of, 132–33, *133,* 470–71
  nest raiding in, 300, *300*
  neurotransmitter function in, 504–5, **505,** 506
  personality in, 568, 572–74, *572,* **574**
  sensory bias in, 211–14, *214*
  sequential assessment in, 490–91, *490*
  subordination in, 428–30, *428,* **429**
  TFT strategy of, 305–8, **305–8**
  winner-loser effects in, 493–95, **494**
  see also specific species
Fisher, D. C., 57, 544
Fisher, J., 132, 159, 182

Fisher, R. A., 40, 66, 98, 186, 193, 198, 212, 44
fitness, 49, 51, 213
adaptation and, 57–58
cultural selection and, 167
defined, 40
direct and indirect, *19*
inclusive, *see* inclusive fitness hypothesis
natural selection and, 37, 40–41
as relative concept, *40*
FitzGerald, G., 300
Fitzgibbon, C., 393–94
Fitzpatrick, J., 462
flat-claw defense, 206
Flaxman, M., 553
Fleishman, L., 96
Fletcher, D., 136, 286
Flinn, M., 234
Flint, J., 46
Florida scrub jay (*Aphelocoma coerulescens*), 267, *267*
territoriality in, 462–64, *462*
fluctuating symmetry, 190–92, *191*, **192**
fluoxetine (Prozac), 505
follicle stimulating hormone, 83
Foltz, D., 222
foraging, 20, 37, 166, 339–77, *340–42*, *344–45*, **346–48**, *349*, **350–51**, *352–55*, **356**, **358–59**, **360**, *360*, *361–63*, **365**, *366*, **367**, *368*, *370*, *371*, **372**, 373
brain size and, 341–42, **342**, 357–59, **358–59**
communication and, 418–23
conservation and, 375
cooperation and, 362–64, *363*
copying and, 158
encounter rate in, 344, 345
energy value in, 344
group life and, 360–70
group size and, 361–64, **361**, *361–62*
handling time in, 344
hormonal aspects of, 373, 373
hunger state and, 354–55
IFD model of, 452–54
information-center hypothesis and, 15–16
innovation in 357–59, **358–59**
learning and, 127–28, 141, 356–60, **358–59**, 376
mathematical optimality theory of, 21, *21*
molecular aspects of, 371–73, *371*, **372**
optimal theory of, *see* optimal foraging theory
period gene and, 371–72, *371*
predation and, 22–24, *23*, 385–87, *386*
producers and, 366–67, *366*
public information and, 369–70, *370*, *376*
risk-sensitive model of, 353–56, *354*, *355*, **356**, 376
"rule of thumb" in, 375
scroungers and, 366–67, *366*, **367**
social facilitation and, 155–56, *156*
social learning and 364–69, *364*, **365**, *366*, *368–69*, 376
spatial memory and, 341–42, *342*
specific nutrient restraints in, 351–52, *352*, *353*
for trace elements, 351–52, *352*, 376
Ford, J. K., 439
Ford, R., 235
Forficula auricularia (European earwig), 208
Foster, K. R., 271
Foster, S., 396
Foster, W., 383
Fowler, K., 192, 551
Frank, L. G., 578
Fraser, A., 335
Fraser, D. F., 575
Freeberg, T. 204–5, *205*
Freedman, H. I., 426
Freeman, S., 58
free-operant procedure, 121
French, J. A , 82
Fretwell, S., 451
friarbird (*Philemon corniculatus*), 406–7
frog, 214–15, *214*
see also specific frogs
fruit fly, 37, 40–41, 371, 545–46, *545*, 547, 552, *552*, 554
Fry, C., 321
Fujii, R., 428
Fuller, P., 385
fungi, ants and, 340–41, *341*
Furi, S., 189
"fur-rubbing" behavior, 558, 564

Gadagkar, R., 136–37, 478
Gaffrey, G., 298
Galápagos Islands, 7–9, 164–65, 439
Galef, B. G., 15, 16, 83–85, 146, 152, 154, 156–57, 170–71, 203, 360, 361
Gamboa, G. J., 478
gametes, 178–79, **179**
game theory, 11, 301–2, **306**, 333, 336
aggression and, 479–91, 505
personality and, 568–70, **569**
see also prisoner's dilemma game
Gangestad, S., 192
Garcia, J., 122–24, 138
Garcia-Coll, C., 592
Garrulus glandarius (European jay), 342, **342**
Garson, P. J., 230
Gasterosteus aculeatus (three-spined stickleback), *11*, 54, **54**, 128–30, **129**, 141, *188*, 189, 300, *300*, 308, 396, 404, 452, *453*, 576
Gauthreaux, S., 132, 465
Gazzaniga, M., 88
Geist, V., 428, 522
Gelowitz, C., 110
general adaptation syndrome (GAS), 82
genes, 36, 41, 49, 67, 70, 146, 283
abnormal abdomen, 545–46, *545*
calculating relatedness and, 259–61
c-fos, 532–33
cultural transmission and, 164–69, *166*, 171, 173
Hamilton's Rule and, 260–61
"hitchhiking," 165–66
in-utero conflict and, 280–81
kin recognition and, 286
linkage disequilibrium and, 166
mate choice and, *see* mate choice
period (*per*), 371

genes (*continued*)
pleiotropic, 544–46, 547
polygenetic traits and, 45–46
as proximate explanation, 101–4, 106
selfish, 50–51, 168–69
"Genetic Evolution of Social Behavior I, The" (Hamilton), 260
"Genetic Evolution of Social Behavior II, The" (Hamilton), 260
genetic monogamy, 236
genetic techniques, 61–69, **63**
Genome Project, 593
genotype, 45
Geospiza:
G. fortis (medium ground finch), 7–9, *8*, 164–65
G. magnirostris, 7–9, *8*
G. scandens (cactus finch), 164–65
Gerra, G., 571
Getty, T., 132, 457–58
giant water bugs, 425, 426
Giles, N., 396
Gill, R., 552
Gilliam, J., 387
Giraldeau, L.-A., 155, 361, 364, *366*, 367–68, **367**, 369, 370, 570
Girondot, M., 99
gladiator frog, 503
Glossina morsitans morsitans (tsetse fly), 549–51, *549*, *550*, **551**
glucocorticoids, 81
aging and, 553–54, 564
goal-directed learning, *see* instrumental conditioning
Godard, R., 132, 458
Goddard, J., 391
Goddard, M. E., 591, 594
Godfrey, C., 99
Godin, J.-G., 54, 128, 306, 392, 575, 576
Gold, K. C., 582
golden monkeys, 433–34
Goldstein, L., 532
Goller, F., 438
Gombe Preserve, Tanzania, 363–64, 559, 560
Gomendio, M., 244
Gompper, M., 558
Gonodactylus bredini (stomatopod), 125–26, *126*
Goodall, J., 363, 364, 431, 559, 560, 568
good genes model, 182, 186–92
fluctuating asymmetry and, 190–92, *191*, **192**
Hamilton-Zuk hypothesis and, 188–89
MHC and, 189–90, **190**, *190*
parasites and, 187–89, *188*, **189**
Goodwin, D., 341
Goodwin, T., 77
gopher snake, 380, *380*
gorillas, 160, 170, 558
Gosden, R., 544
Gosling, S., 568, 578–79, *579*, 581
Goss-Custard, J. D., 362, 374, 375
Gould, E., 553
Gould, J., 39, 88, 90
Gould, S. J., 41, 58, 75
Gouzoules, H., 136
Gouzoules, S., 136
Goy, R., 529
Goy, R. W., 299
Grafen, Alan, 40, 261, 417, 441
Graham, J., 134
Grant, J. W. A., 203, 490
Grant, P. R., 7, 8, 162, 164-65
Grant, B. R., 7, 8, 162, 164–65
grasshopper (*Schistocerca americana*), 13–15, *13–15*, **14**
Gray, R., 451
gray catbird (*Dumetella carolinensis*), 438
gray-crowned babbler, **261**
gray squirrel (*Sciurus carolinensis*), 385, 558
gray treefrog (*Hyla versicolor*), 555–57, **556**, 556
great tit (*Parus major*), 345–46, *345*, **348**, 350, 374
learning and personality in, 585–87, *586*
Green, L. D., 203
Green, R. F., 369
Greenberg, N., 500
Greenewalt, C., 437
Greenough, W., 94, 520
green swordtail (*Xiphophorus helleri*), 211–14, *214*, 499
grey-sided vole (*Clethrionomys rufocanus*), 229
Griffin, D. R., 465
Griffiths, R., 557
Grobecker, D., 357
Gronenberg, W., 94
Gross, M., 209, 210, 460
ground squirrel, 380–81, *380–81*, **381**, 396
group selection, 315–21, 336
between-group, 315–16, 320, 336
brood raiding and, 316–17
in humans, 319–21, **320**
in insects, 316–19, **317**
within-group, 315–16, 320, 336
growth hormone releasing hormone (GHRH), 82
guppy (*Poecilia reticulata*), 51–57, *52–53*, **54–55**, 58, 128, 168, 171, 396
cultural transmission in, 162–63
learning in, 402–6, **403**, *403*, **406**
mate-choice copying in, 202–3, *202*, **203**
personality in, 568, 574–78, **575**, **577**, *577*
predator inspection in, 305–8, 574–78, **575**, **577**, *577*
risk taking and cooperation in, 296, *296*
TFT strategy in, 305–8
Guthrie, D. M., 428
Gutierrez, G., 135, 195
Gutsmann-Conrad, A., 554

habitat choice, 448–51, *449*, 451–54
abiotic factors in, 450, 474
biotic factors in, 450, 474
foraging success and, 452–54, **454**, *454*
home range in, 448–49
IFD model of, 451–54, 474
resource matching rule and, 452–53
habituation, 114–15, *114–15*, 141
Hack, M. A., 490
Hagenguth, H., 373
Hager, J., 123
Haig, D., 279–81

Hailman, J., 75
Haldane, J. B. S., 258–59, 544
Hall, D. J., 347
Hall, K. R., 163
Hall, S., 196
Halliday, T., 206, 417
Halobaena caerulea (blue petrel), 384
Hames, R., 234
Hamilton, W. D., 10, 11, 19–20, *19*, 50, 62, 66, 186, 187, 248, 257, 258, 291, 301–2, 303, 332, 482, 544
Hamilton's Rule, 260–61, 269
Hamilton-Zuk hypothesis, 187–89
Hammerstein, P., 487, 503
handicap principle, 186–87, 212, 416–17, 440–41, 443
Hanlon, R. T., 490
Hannes, R. P., 500
Hansell, R., 231–32, 233
haplodiploidy, 269–70
Harcourt, A. H., 65, 239, 326, 327, 328
Hardy, J., 321
Hare, H., 278, 279
Harlow, H., 122
Harper, D., 453–54, **454**
Harris, L., 110
Harris's hawk (*Parabuteo unicinctus*), 299, 299
Hartl, D., 41
Hartley, R. S., 438
Hartung, J., 234
Harvell, C., 112–13
Harvey, P., 75, 239, 357, 358
Haskins, C. P., 56
Hatch, M., 299
Hauser, D., 558
Hauser, M., 158–61, 389, 412, 437
Hausfater, G., 272, 273
Havens, K., 240
hawk-dove game model, 480, 482–87, **483**, **484**, *485*, *486*
  antibourgeois strategy in, 483, 484–85, 505
  bourgeois strategy in, 483–87, *486*, 505
  resource value in, 482–83
Hay, M., 385
Healey, S., 125, 341–42, 458
Heape, W., 465
heat-shock proteins, 554, 564
Hegner, R., 385
Heinrich, B., 430, 512, 514–16, 534–35
Helfman, G., 163
Helogale parvula (dwarf mongoose), 274–75, **275**, 323–26, *324*, **325**, **327**
helping-at-the-nest, 82–85, *83–84*, 261, **261**, 264, *264–65*, 321–22, 462–64
Hemelrijk, C. K., 330
Hennessy, D. F., 392, 393
Henry, J., 568, 588–89
Hepper, P. G., 136, 283
heritability analyses, 41
herring gull, 357
Hessing, M., 590
Heterocephalus glaber (naked mole rat), 62–64, *62*, **63–64**, 67, 332
Hetherington, E., 273
Heyes, C., 113–14, 146, 152, 154, 157
Higham, A., 306, **307**, 576
Hill, C., 334
Hill, G., 76–80, *78*, 79
Hill, H., 524
Hinde, R., 159, 182
Hinton, G., 130
hippocampus, 92–93, 94, 341–42, **342**, 553, 554
Hippotragus niger (sable antelope), 523, *524*
Hirth, D. H., 391
Hodge, M. A., 484–85
Hoffman-Goetz, L., 465
Hoffmann, A., 44
Hoffmann, G., 554
Hogan, D., 267
Hogan, J., 75
Hogan-Warburg, A. J., 583
Hoglund, J., 198, 200–201
Hogstad, O., 434, 452
Holden, P., 357
Holekamp, N., 529, 578
Hölldobler, B., 27, 316, 421–23
Holliday, R., 546
Hollis, K., 134–35, 137, 140, 194, 197
Holloway, K., 194
Holloway, W., 529, 531
Holm, D., 60
Holmes, W., 290
Holyman, A., 558
home range, 448–49
homicide, 256–57, **257**
honeybee (*Apis mellifera*), **63**
  foraging by, 371–73, *371*, **372**, *373*
  "policing" by, 270–71, *270*, **271**
  spatial learning in, 92–93
  waggle dance of, 418, 420–21, *420–21*, **421**
hooded warbler (*Wilsonia citrina*), 458, *458*, *459*
Hoogland, J., 255, 291
Hopey, M., 555
Hopkins, Carl, 426
Hopster, H., 590
horizontal cultural transmission, 161, *161*, 162–63
hormones, 105
  aggression and, 499–504, **500**
  aging and, 552–54, 564
  behavior and, 80–82
  cooperative breeding and, 323
  helping-at-the-nest and, 82–85, **83–84**
  personality and, 584–85, **585**
  placentally produced, 281
  play and, 529–30, **530**
  proximate causation and, 80–87, **81**, *83*, *86*, *87*
  "tropic," 81
  see also specific hormones
horses, *180*
Houde, A. E., 52, 54, 55, 396
house finch (*Carpodacus mexicanus*), 76–80, *76*, *78*, **79**
house sparrow (*Passer domesticus*), 314, **315**
Houston, A., 353, 451, 460–61
Hovingh, P., 555
Howard, R. D., 555
Howe, N., 110
Hrdy, S., 263, 272, 273
Hsu, Y, 494–95, **494**, 496
Huang, Z., 373
Huber, R., 505
Huffman, M., 147–48, *147*, 558, 559, 561
Hughes, L., 334
Hugie, D. M., 583

Hull, C. L., 360
Hulscher-Emeis, T., 428
human chorionic gonadotropin, 281
human growth hormone, 81
human placental lactogen, 281
humans:
  aggression in, 501
  alliances and, 330
  cultural transmission in, *150, 152,* 153–54, *154,* 162
  family stability in, 264–66
  group selection in, 319–21, *320*
  homicide in, 256–57, **257,** 258
  infanticide in, 281–82, **282**
  in-utero conflicts in, 279–81, *280*
  polygyny in, 233–35, *234,* **235**
  PTM and mate choice in, 233–35, *234,* **235**
  social learning in, 368–69, *368–69*
  social play in, 528
  sperm number in, **245**
  stepfamily violence in, 272–74, **273**
  wealth and kinship in, 267–68, **268**
humpback whale, 443
Humphreys, A., 526, 531
Hunt, S., 96
Huntingford, F. A., 115, 128–30, 137, 396, 478
Huntington's disease, 45, 546
Husberger, E., 431
Hutchins, D., 298
Hutt, C., 514
Hutterites, 319–21, *320*
Huxley, J., 32, 75, 180, 502
Huxley, T. H., 478
Hyla versicolor (gray treefrog), 555–57, **556,** 556
Hylobittacus apicalis (scorpionfly), *37,* 183–85, *183,* **184**
hypersecretion, 80, 105
hypertension, 553, 590
hypothalamus, 82, 532

Iason, G., 99
Ibara, R., 90
ideal free distribution (IFD) model, 451–54
identity by descent, 259, 291
idiopathic hemochromatosis, 546
imitation (observational learning), *150,* 157
Immelman, K., 195
imperial blue butterfly (*Jalmenus evagoras*), 331, *331,* **334–35**
imprinting, 195–96
Ims, R. A., 229
inclusive fitness (kinship) theory, 19–20, 257–61, *260,* 262, 264, 276, 291, 321
  calculating, 259–60
  relatedness and, 259–61
  see also family dynamics
Indian elephant (*Elephas maximus*), 298
indigo bunting (*Passerina cyanea*), 236–38, **237,** *237*
  stellar navigation in, 466–68, *466*
indirect fitness, *19*
individual learning, 111–13
infanticide, 272–73
  in humans, 281–82, **282**
information-center hypothesis, 15–16
Inman, A., 357
insects, 249, 300, 336
  aggression in, **478,** *479,* 484–88, *485, 486,* **488**
  antipredator behavior in, 387–89, *387,* **388**
  communication in, 420–23, *420–23,* **421,** 424–26, **425,** *425*
  cooperative breeding in, 269–72, *270,* **271,** *272*
  eusocial, **63**
  female defense polygyny in, 223–24
  foraging in, 340–41, *341,* 371–73, *371,* **372,** *373,* 421–23
  group selection in, 316–19, **317,** *318,* **319**
  haplodiploidy in, 269–70
  learning in, 13–15, **13–15,** 92–94, *92*
  migration of, 449
  ovipositor length in, 37–38, *37–38, 40–41, 42,* 43
  polyandry in, 224–25, *225*
  sex ratio in, 278–79, *279*
  sperm competition in, 240–41, **241,** *241*
  sperm morphology of, 244, *244*
  see also specific species
Insect Societies, The (Wilson), 26
instincts, 26
instrumental (operant) conditioning, 119–21, *121*
intersexual selection, 178–81, *180,* 215–16
  see also mate choice
interspecific mutualism, 331–35, *331,* **334–35,** 336
intrasexual selection, 178–81, *180,* 215–16
intrauterine position (IUP), 83–84, *84*
in-utero conflict, 279–81, *280*
Iridomyrmex anceps, 331, *331,* **334–35**
Irons, W., 234
Iwaniuk, A., 533, 536–37

Jablonka, E., 146
Jacana:
  J. jacana (wattled jacana), 224, *224*
  J. spinosa (American jacana), 205
jackdaw (*Corvus monedula*), 341–42, **342,** *342*
Jackson's widowbird, 212
Jacobs, G. H., 96
Jacobson, M., 520
Jalmenus evagoras (imperial blue butterfly), 331, *331,* **334–35**
Jamieson, S. H., 529
Janik, V., 431
Janzen, D., 558
Janzen, F., 97, 99
Japanese macaque monkey (*Macaca fuscata*), 146–49, *146–47,* **148,** 151, 389, 390–91, 558
Japanese quail, 135, *135*
  learning and mate choice in, 196–97, **196–97**
Java monkey (*Macaca fascicularis*), 136, *136*
jays, 322, 341
  blue, 311–14, *311,* **312–13**
  Mexican, 20, 261
Jenni, D., 205
Jennions, M., 490
Jensen, P., 278, 490

Jerison, H., 400–401
John, J., 466
John, O. P., 568
Johnsson, J., 493
Johnston, T., 127
Johnstone, R. A., 323, 328, 416, 417, 418
Johnstone, T. D., 357
Jolicoeur, P., 172, 357, 358
Jones, A., 206
Junge, G., 584
Juraska, J., 520
juvenile hormone (JH), 373, *373*

Kacelnik, A., 357, 451, **452**
Kaczensky, P., 591
Kagan, J., 571, 578, 592–93
Kalotermes flavicollis (wood termite), 63
kamikaze sperm hypothesis, 243–44
Kamil, A., 20–21, 341, 357
Kamin, L. J., 119
Kandel, E., 93
Kaplun, D., 234
Karunaratne, P., 426
Kawamura, S., 146
Kawanabe, H., 331
Kear, J., 45
Keddy-Hector, A., 211
Keefe, M., 110
Keenleyside, M., 54, 428
Keeton, W., 39, 88, 90
Keller, L., 323
Kennedy, M., 451
ketotestorone, 427
Keverne, E. B., 172, 298
Kiesecker, J., 555–58
killer whale, 165, *165*, 173, 439
killifish (*Rivulus hartii*), 52–53, *53*, 575
Kingdon, J., 324
king pigeons, 196
kin recognition, 286–91, **287**, *287*, **290**, *290*, 292
  conspecific threshold model of, 286–87
  matching models of, 286–90, **287**, *287*, **290**, *290*
  rule-of-thumb models of, 290–91, 292
  spatial cues and, 291
kinship, 253–93, *254–55*, **255**, 256, **257**, 258, *260*, **261–63**, *264–67*, **267–69**, *269–70*, **271**, *272*, **273**, **275**, *275*, *278–80*, **282**, *283–84*, **287**, *287*, **290**, *290*
  alarm-calling hypothesis and, 254–56, *254*, **255**, 256
  animal behavior and, 255–58
  coalition formation and, 65–69, *65–68*, **69**
  definition of, 259
  eusocial behavior and, 62–64, **63**
  family dynamics and, *see* family dynamics
  Hamilton's Rule and, 260–61
  inclusive fitness theory and, 257–61, *260*
  kin recognition and, 286–91, **287**, *287*, **290**, *290*, 292
  kin (nepotistic) selection, 27, 66, 440
  parent-offspring conflict and, *see* parent-offspring conflict
  relatedness and, 259–61
  sibling rivalry and, 276, 282–85, *283–84*, **285**
  teaching and, 160–61
  as term, 289–90
  weaning and, 277–78, *278*
  see also family dynamics
Kipsigi tribe, 234–35, *234*, **235**
Kirkpatrick, M., 182, 193, 200, 211
Kirkwood, T., 546, 554
Kitchen, W. D., 391
Kleiman, D., 222
Kluckhohn, C., 150
Kodric-Brown, A., 196
Koelling, R., 123
Koenig, W. D., 262, 321
Koga, T., 206
Kogan, N., 571
Komdeur, J., 264
Komers, P. E., 484
Konishi, M., 158, 437
Konopka, R., 371
Koolhaas, J. M., 568, 588–90
Koops, M. A., 490
Koopsman, J., 571
Koptur, S., 331
Korona, R., 451
Krakauer, D., 430–31
Kraus, W. F., 425, 426
Kravitz, E., 90, 504, 505
Krebs, J., 20–21, 22, 51, 94, 127, 155, 228, 230, 231, 341–42, 343, 352, 357, 374–75, 385, 389, 391, 415–16, 457
Krimbas, C. B., 57
Kroeber, A. L., 150
Kroodsma, D., 437
Kropotkin, P., 300–301, 478
Krummhorn region, Germany, 548, *548*
Krushinskaya, N., 341
Kruuk, H., 392
Kudo, H., 172
Kugler, P., 590
Kuhn, T., 257
Kummer, H., 299
Kunz, T. H., 297–98
Kynard, B., 300
Kyriacou, C., 371

Lacey, E. A., 62
Lack, D., 248, 374
Lafleur, D. L., 203
Lagopus mutus (rock ptarmigan), 416–17
Laland, K. N., 162–63, 172, **172**, 200
LaMunyon, C., 244
Landau, H. G., 137, 496, 497
Lande, R., 112
Langley, P., 549, *550*
Lank, D., 583–84
lark bunting (*Calamospiza melanocorys*), 231–32, *231*
Larson, A., 57
Laurila, A., 555
Law, R., 331
Lawick-Goodall, J., 363, 517
law of effect, 121
leaf-cutter ant (*Acromyrmex octospinosus*), *341*
leaf-cutter ant (*Atta cephalotes*), 422–23
  foraging in, 421–23, *422–23*
learning, 6, 12–15, *12–15*, **14**, 28, 109–43, *110*, **111**, **113**, *114–20*, *125*, *127*, **128–29**, **131**, *133*, *135*, *136*, **137**, **140**, 286
  adaptationist school of, 122–24

learning (*continued*)
aggression and, 132, 136–40, **137, 140,** 141, 498–99
antipredator behavior and, 402–6, **403,** *403,* **406,** 408
in birds, 359–60
of birdsong, 203–5
blocking and, 118–19, 141
contagion and, 157
culture and, 151
defined, 112, 140–41
environmental stability and, 122, 130–32, **131**
evolution of, 130–32, **131**
foraging and, 127–28, 356–60, **358–59,** 376
group living and, 127–28, **128**
habitat selection and, 132–33, 141
habituation and, 114–15, *114–15,* 141
hippocampal-dependent, 553
imitation and, 157
in insects, 13–15, **13–15,** 92–94, *92*
instrumental (operant) conditioning and, 119–21, *121,* 141
kinship relations and, 132, 135–36, 141
law of effect and, 121
mate choice and, 194–98, **196–97**
mate selection and, 132, 133–35
memory and, 124–25
natural selection and, 12–13, 28, 122, 124–26, 129–30
overshadowing and, 118–19, 141
parental investment and, 135, *135*
Pavlovian (classical) conditioning and, 115–19, *116–18,* 121, 141
personality and, 585–87, *586*
phenotypic plasticity and, 112–13, **113,** 141
play and, 532–33
population comparisons and, 127–30, *127,* **128–29**
predation and, 110–11, **111,** 128–30, **129**
in primates, 430–32, **432,** *432*
as proximate factor, 104
reinforcement and, 123–24
relational theory of, 360
second-order conditioning and, 118, *118,* 141
sensitization and, 114–15, *114–15,* 141
single-stimulus experience and, 114–15, *114–15,* 141
social, *see* social learning
spatial, 92–93
species-specific, 122-24
territoriality and, 472–73
within-species studies of, 122–26
of work ethic, 359–60
Le Boeuf, B. J., 208, **209,** 229
Lee, A., 90
Lee, J., 13
Lefebvre, L., 127–28, 172, 357–59, **358,** 364–68, *366,* **367**
Leffler, J., 465
Leger, D. W., 380
Leigh, E., 57
Leimar, O., 428, 478, 489, 490
Lemaire, V., 500
Lenington, S., 231, 234
Leopold, A. S., 434
Lepage, O., 504
Lepomis gibbosus (pumpkinseed sunfish), 572–74, *572,* **574**
Leptodeira septentrionalis (snake), **388**
Leshner, A., 500
lesser snow goose, 195, *195*
Levins, R., 112
Levitan, D., 242–43
Lewis, A., 13
Lewis, M., 264
Lewis, W. L., 13
Lewontin, R. C., 14, 58, 482
Licht, T., 128
Lichtman, J., 520
Liers, E., 160
life-dinner principle, 405
Ligon, J. D., 266–67
Liley, N., 54
Lill, A., 196
Lima, S., 385, 386
Lindzey, J., 500
linear programming, 352, 353
linkage disequilibrium, 166
Linnell, J. D. C., 590, 591
Linsdale, J., 381
lion (*Panthera leo*), 50, 65–66, 68, *68,* **69,** 160, 321, 393, 394, 525, 591
Lithgow, G., 554
little blue heron, **285**
Littledyke, M., 421
lizard (*Anolis aeneus*), 455–57, *456,* **457**
lobsters, 505
local enhancement, concept of, 154–55, *155*
locomotor play, 518–21, *519,* **520**
cerebral synapse development and, 519–21, **521**
environment and, 528
longevity, 546–50
extended life span and, 552
fecundity and, 548–49
sex ratio and, 550–51, **550, 551**
Longino, J., 558
long-tailed widowbird, 212, 213
Lorenz, Konrad, 22, 195, 374, 478, 502, 584
Losey, G., 200
Loughry, W. J., 392
Low, B., 234
Loxia curvirostra (red crossbill), 370
Lozano, G., 558, 559
Lucas, H., 451
Lucas, N. S., 298
Luckinbill, L., 552
Lundberg, A., 247
lutenizing hormone, 83
Lycaon pictus (African hunting dog), 298
Lycett, J., 547–49
Lyon, B., 58

Macaca:
M. fascicularis (Java monkey), 136, *136*
M. fuscata (Japanese macaque monkey), 146–49, *146–47,* **148,** 151, 389, 390–91, 558
M. sylvanus (Barbary macaque), 226
MacArthur, R. H., 343, 374
MacClade analysis program, 536
McComb, K., 203
McCullough, D. R., 391
Mace, G., 357, 358
Macedonia, J. M., 389

McEwen, B., 553
McGregor, P. K., 499
McGuire, M., 504, 568, 571
McKibben, J., 90
McKinney, F., 236
McLaughlin, F., 236
McMann, S., 490
McNally, C., 195
McNamara, J. M., 353, 451
Macphail, E., 127
Maddison, P., 536
Maddison, W., 536
Maestripieri, D., 299
magpie, 341, *342*
Magurran, A., 54, 55, 56, 128, 203, 306, **307**, 396, 398–99, 575, 576
Mahale Mountains, Tanzania, 363, 455, 559, 560
Mahaney, W. C., 558
Major Histocompatability Complex (MHC), 189–90, **190**, *190*
Malapterurus electricus (electric catfish), 426–27, *426*, **427**
male-male competition, 205–10, *207*, **208**, **209**, *210*
  by interference, 208–9
  via cuckoldry, 209–10, *210*
mallard (*Anas platyrhynchos*), 94–95, *94*, *95*, 453–54, **454**, *454*
Malm, K., 278
Malurus cyaneus (superb fairy wren), 264, *264–65*
mammals, 244, 336, 392, 473, *480*
  alarm signals in, 389–91, *389–90*, 393
  antipredator behavior in, 385–86, *386*
  and cooperation in birthing process, 297–98, *298*
  food storage by, 341–42, **342**, *342*
  foraging in, 351–52, *352*, 353
  "fur-rubbing" behavior in, 558, 564
  locomotor play in, 518–19, *519*
  memory in, 341–42, **342**, *342*
  migration of, 449, *465*
  object play in, *513*, 516–18, *516*, **517**
  play fighting in, 530–31
  self-medication in, 558–59, **589**
  senescence in, 543, 554, 564
  sexual selection in, 551–52
  social grooming in, 298–99, *298*
  social play in, 512–14, 523, *524*, 528, 529–30
  weaning in, 277–78, 278
  see also specific species
Manuck, S., 590
Maple, T., 581
Marchaterre, M., 92
Marchetti, C., 586
marginal value theory, 348–51, **350**, **351**, 376
Markl, H., 422
Marks, I., 571
Marler, C., 416
Marler, P., 127, 195, 389, 437, 499
marmoset (*Callithrix jacchus*), 298
Marshall, A., 431–32
Marshall, J. C., 389
Martin, P., 512, 514
Marzluff, J., 430
Maschwitz, U., 334
Masseti, M., 4
mate choice, 78
  birdsong learning and, 203–5, *205*
  communication and, 424–27, **425**, *425*, *426*, **427**
  copying and, 198–203, **199**, 200, 202–3
  cultural transmission and, 198–205, **199**, **201**, *202*, *205*, 216
  direct benefits model of, 182–85, **182**, 216
  genetics and, 181–94, **182**, **184**, **185**, *186*, *188*, **190**, 216
  good genes model of, 182, 186–92, *186*, 216
  learning and, 194–98, *196–97*
  neuroethology and, 211–15, *214*
  opportunity for selection models and, 199–200
  population genetic models of, 200
  runaway selection model of, 182, 193–94, 200, 216
  strategic models of, 200
Mather, J., 581–82, *582*
Mathis, A., 110
mating systems, 219–51, *220*, **221**, *222*, **223**, *224–27*, **228**, **230**, *231–32*, **232–33**, *234*, **235**, *237*, **237**, *239*, **240–43**, *241*, *244*, **245**, **247**
  battle of the sexes and, 246–47
  extrapair copulations (EPCs) and, 236–38, **237**, *237*, 250
  forms of, 220–28, **221**
  pair bonds and, 220–21
  PTM and, 229–35, **230**, *231–32*, **232–33**, *234*, **235**, 250
  resource dispersion and, 228–29, 250
  sexual selection and, 249
  sperm competition in, *see* sperm competition
  variable, 245–47, **247**, 249
  see also specific systems
Matsuoka, S., 389
Maynard Smith, J., 10, 248, 303, 343, 482, 487, 488, 502–3
Mayr, E., 41, 75, 125
Mazur, A., 501
meadow grasshopper, 38
medaka (*Oryzias latipes*), 203
Medawar, P., 544
Meder, A., 529
medium ground finch (*Geospiza fortis*), 7–9, *8*, 164–65
meerkats, 102, 160
Melanerpes formicivorus (acorn woodpecker), 266–67, *266*, **267**
Melanochromis auratus (cichlid fish), 328
Meltzoff, A., *150*
Membranipora membranacea (bryozoan), 112–13, *112*, 402
memory:
  learning and, 122, 124–26, 138
  optimal, 124–25
Mendel, Gregor, 146
Mendel's laws, 44, 45
Mendozagranados, D., 523
menstruation, 227
Meriones unguiculatus (Mongolian gerbil), 83–85, *83–84*, 134, 135

Merops bullockoides (white-fronted bee-eater), 269, **269,** *269,* 289, 463
Mesocricetus auratus (Syrian golden hamster), *86*
messenger RNA (mRNA), 371–73
Messor pergandei (seed harvester ant), 316–19, **317**
Mesterton-Gibbons, M., 297, 310, 418, 483, 485, 496, 498
methoprene, 373
Mexican jay, 20, 261
Mexican spider (*Oecibus civitas*), 484–85
Meyer, A., 214
Michener, C., 136, 286
migration, 39, 449–50, 451, 464–71, *465*
  as defense against parasites, 465–66, 474
  of fish, 132–33, *133*
  heritability in, 469–70
  irruptive, 464–65
  learning and, 470–71, *470,* **471**
  magnetic orientation in, 468–69, **469**
  stellar navigation in, 466–68, *466–67*
  "zugunruhe" (restlessness) in, 470
Milinski, M., 189, 306, 308, 383, 392, 404–5, 452–53, *453,* 576
Miller, J., 132, 470
Mind of the Raven (Heinrich), 514
Mineka, S., 163–64
minnow (*Phoxinus phoxinus*), 54, **54, 307,** 396
  interpopulation differences in, 396–400, *396,* **397–99**
Mirounga angustirostris (elephant seal), 208–9, *208,* **209,** *480*
Mitani, J., 431
Mitchell, W. A., 57
Mitman, G., 300
mitochondrial DNA (mtDNA), 165–67, 173
Mittlebach, G., 361–62
Mitzutani, F., 591
mobbing behavior, 392, 404, 406–7
Mock, D., 283, *283,* 361, 478, 492
"modular mind," 139
Molina-Borja, M., 490
Møller, A., 185, *186,* 190, 191–92, 240, 244, 389, 436, *436,* 465–66, 571
Molothrus ater (brown-headed cowbird), 203–5, *205,* 438
monarch butterfly, 464, 468–69, *468*
Mongolian gerbil (*Meriones unguiculatus*), 83–85, *83–84,* 134, 135
monoamines, 532
monogamy, 221–23, *221, 226–27,* 245–46, 248–49, 250
  defined, 221
  genetic, 236
  in oldfield mouse, 222–23, *222*
  serial, 220, 221
  social, 236
Montgomerie, R., 192, 417
Moore, H. D. M., 244
Moore, M., 99, 500, 501
Moore, R. D., 387, 499
Moore, T., 66–67
moose (*Alces alces*), 352, *352, 353*
Morgan, E. D., 421, 422
Morgan, J., 533
Morris, D. W., 451
Morse, D. H., 361
Morton, D., 125
Morton, E. S., 434
mosquito fish, 308
Moss, K. A., 423
Motacilla alba (pied wagtail), 460–61, *460–61,* **461**
motmots, 406
Mouget, F., 384
mountain chickadee (*Poecile gambeli*), 15
Mousseau, T. A., 37, 38, 40, 41, 44, 51
Moynihan, M., 478
Mueller, U., 279
Mulder, R., 238
Munck, A., 553
Mundinger, P. C., 437
Munn, C. A., 389
Munro, R., 45
Muntz, W., 428
mushroom bodies (corpora pedunculata), 92–94, *92*
mutation, 38
mute swan (*Cygnus olor*), 45, *45*
Mutual Aid among Animals (Kropotkin), 478
mutualism, 331–35, *331,* **334–35,** 336

Naeem, S., 321–22
Naguib, M., 437
naked mole rat (*Heterocephalus glaber*), 62–64, *62,* **63–64,** 67, 332
Nannacara anomala (cichlid fish), 490–91, *490*
narrow-sense heritability, 41, 43–44, **44**
  parent-offspring regression and, 46–47
Nash, J., 502
Nash, S., 196–97, 333
Nash Equilibrium, 502–3
Nasua narica (coati), 65, 327
natural selection, 6–7, 16, 32–71
  adaptation and, 57–61
  altruistic behavior and, *see* altruism
  animal behavior and, 49–57
  antipredator behavior and, 51–57, *52–53,* **54–55,** 57
  artificial selection and, 32–34, *33–34,* 69
  behavioral genetics and, 44–49
  coalition formation and, 65–69, *65, 68,* **69**
  in Darwin's finches, 7–9
  defined, 6
  fitness and, 37, 40–41, 49–50, 51, 69
  gamete size and, 178–79, **179**
  genetic recombination and, 39
  genetic techniques and, 61–69, **63,** 70
  genetic variation and, 35–36, 38–39, 50
  group selection and, 315
  group size and, 55–56, **55**
  impact of, 149
  learning and, 12–13, 28, 122, 124–26, 129–30
  limited resources and, 37
  memory and, 125–26
  Mendel's Laws and, 69–70
  mode of inheritance and, 37, 41–44, *42*

- molecular genetics and, 67
- mutation and, 38
- operation of, 37–44
- process of, 35–44
- resources and, 69
- selective advantage of traits and, 35–36, *35*
- selfish genes and, 50–51
- speed of, 152, **152**
- variation and, 37, 38–39, 69
- xenophobia and, 9–10
- *see also* evolution

Nelissen, M. H. J., 328, 428
Nelson, D. A., 389
Nelson, R., 553
neocortex, 172
nerve cell, *89*, *90*
nervous impulse, 88–90, *89*
nervous system, 87–92, **88**, *89*, **91**, *91*, *92*, 105
Nesse, R. M., 543, 571
nest parasitism, 58–61, *59*
nest raiding, 300
neuroethology, 211–15, *214*
neurons, 88–89, 105
neurotransmitters, 89–90, 499, 504–5, **505**, 530–33, *531*, *532*, 537
Niesink, R., 531
Nilsson, G., 504
Nishida, T., 363, 455
Noe, R., 329
nonshared environments, 49
noradrenaline, 590
norepinephrine, 82, 531, 532
Normansell, L., 531
Norris, D., 83
northern cardinal, *188*, 438
Norway rat (*Rattus norvegicus*), 15–16, *16*, 150–51, 536
Nottebohm, F., 437, 438
Nottebohm, M. E., 438
Nowak, M., 430–31
Nowlan, S., 130
Nunes, S., 529, 530
nuptial gifts, 183–85, **184**, 225
Nursall, J. R., 332
Nyctereutes procyonoides (raccoon dog), 298

object play, 513–18, *513*, *514*, **515**, *516*, **517**
oblique cultural transmission, 161, *161*, 163–64
O'Brien, S. J., 39
observational learning (imitation), 157
O'Connor, K. I., 429
Octopus rubescens (red octopus), 581–82, *581*, **582**
O'Donald, P., 193
Oecibus civitas (Mexican spider), 484–85
offspring rule, 261
Ohling, W. A., 548
oldfield mouse (*Peromyscus polionotus*), 222–23, *222*
Oli, M., 591
olive baboon (*Papio cyanocephalus anubis*), 329–30
Oliveira, R. F., 498
Ollson, O., 369
"On Aims and Methods of Ethology" (Tinbergen), 139
Oncorhynchus kisutch (coho salmon), 133, 470–71, **471**, *471*
On the Origin of Species (Darwin), 6–7, 32, 35, 50, 66
open field behavior, 45
operant (instrumental) conditioning, 119–21, *121*
optimal diet model, **346–47**
optimal foraging theory, 342–51, *345*, 374–76, 386
- marginal value theory and, 343–51, **350–51**
- what to eat question in, 343–47, **346–47**
- where to eat question in, 347–51, *348*, *349*, **350–51**

optimal skew model, 274–75, **275**, *275*, 463
Orgeur, P., 529
Orians, G., 75, 229, 451
Oring, L. W., 221, 224, 228
Ortega, C., 291
Oryzias latipes (medaka), 203
Oshima, N., 428
Osorno, J. L., 493
Osten, W. von, 413
Oster, G. F., 279, 353
otter, 160
overshadowing, 118–19, *119*, 141
Ovis canadensis (bighorn sheep), 522–23, 523
Owens, O., 224
Owings, Donald, 380, 392, 396
owner's dilemma model, 462

Packer, Craig, 65, 68, 327, 328–30, 362
Pagel, M. D., 75
pair bonds, 220–21
Palameta, B., 364–65, 366
Palmer, A. R., 192
Panksepp, J., 531, 532
Panthera leo (lion), 50, 65–66, 68, *68*, **69**, 160, 321, 393, 394, 525, 591
"pant hoots," 430–32, **432**, *432*
Papaj, D., 13
paper wasp (*Polistes fuscatus*), **63**, 332
Papio cyanocephalus anubis (olive baboon), 328–30
Parabuteo unicinctus (Harris's hawk), 299, *299*
parafascicular area (PFA), 532
parallel walks, 207–8, *207*
Pararge aegeria (speckled wood butterfly), 485–87, *486*
parasites, 185–86
- avoidance of, 555–57, **556**, *556*
- good genes model and, 187–89, *188*, **189**
- migration as defense against, 465–66

parental investment, 99–101, 103
- by step-parents, 272–74, **273**

parent-offspring conflict, 276–82, *278–80*, **281**, 291
- infanticide and, 281–82, **282**
- in-utero, 279–81, *280*
- sex ratio and, 278–79, *279*
- weaning and, 277–78, *278*

parent-offspring regression, 46–47
Parker, G. A., 137, 240–41, 248, 249, 274, 283, *283*, 348, 374, 451, 452, 454, 478, 487–88, 492
Parker, P. G., 430
Parmigian, S., 272
Parsons, P., 44
Partridge, L., 551

Parus:
P. atricapillus (black-capped chickadee), 434
P. caeruleus (blue tit), 159, *159*
P. major (great tit), 345–46, *345*, **348**, 350, 374, 585–87, *586*
Passer domesticus (house sparrow), 314, **315**
Passerina cyanea, see indigo bunting
patrilines, 224–25
Paukstis, G., 97, 99
Pavlov, Ivan, 116, *117*, 122
Pavlovian (classical) conditioning, 115–19, *116–18*, 121, 141, **195**
blocking and, 118–19, *120*
learnability and, 118
overshadowing and, 118–19
second-order conditioning and, 118, *118*
Payne, R., 58
peacock, *187*, 440
Peake, T., 499
Pedersen, B., 465
Pederson, J., 529
Pellegrini, A. D., 528, 533
Pellis, S., 86–87, 532, 533, 536–37
Pellis, V., 532
Penn, D., 189
Pepper, J. W., 62
peptide hormones, 83
Perez, S. M., 468
period (*per*) gene, 371–72, *371*
Peromyscus polionutus (oldfield mouse), 222–23, *222*
Perrins, C., 389
personality, 567–96, **569**, *570*, *572*, **574–75**, *577*, **577**, *579*, **580**, *581*, **582**, *583*, **585**, *586*, **588–89**
boldness and shyness in, 570–78, *572*, **574**, 594
coping styles and, 588–90, **588**, **591**, 595
cultural transmission and, 587–88, 595
defined, 568
in domesticated animals, 589–90
game theory model of, 568–70, **569**
of guide dogs, 591–94
hormones and, 584–85, **585**, 590
learning and, 585–87, *586*
practical application of research in, 590–94, 595
Perusse, D., 234
Pervin, L., 568
Peters, J. M., 271
Peterson, D., 478
Petit, L. J., 232, 233
Petranks, J., 555
Petrochelidon pyrrhonota (cliff swallow), 47–48, *47*, **48**
Pfennig, D. W., 286, 287, 318
Pfungst, O., 413
phenotypes, 44, 45
culture and, 150
defined, 112
testosterone and, 584–85
phenotypic plasticity, 112–13, **113**, 141
phenotypic variance, 42
Philemon corniculatus (friarbird), 406–7
Philomachus pugnax (ruff), 582–85, *583*, **585**
Phodopus campbelli (Djungarian hamster), *86*
Phoenix, C., 529
Phoxinus phoxinus, see minnow
Physalaemus (frog):
P. coloradorum, 214–15, *214*
P. pustulosus, 214–15, *214*
Physeter macrocephalus (sperm whale), 165, *165*, 173, 439–43, *439*, **442**
Piaget, J., 532
Pianka, E. R., 313, 374
Pica pica (European magpie), 342
Picoides pubescens (downy woodpecker), 434, *434*
pied flycatcher, 234
pied wagtail (*Motacilla alba*), 460–61, *460–61*, **461**
Pierce, N., 331–35
Pietrewicz, A., 357
Pietsch, T., 357, 575
pigeon (*Columba livia*), 32, *32*, 97
learning in, 359–60
social learning in, 364–68, *364*, **365**, *366*, **367**
pigs, 589–90
pike (*Esox lucius*), 110–11, 397, 398
pike cichlid (*Crenicichla alta*), 53, *57*, 296, *296*
pilot whale, 165, *165*, 173
Pinker, S., 430
pipefish (*Sygnathus typhle*), 205–6
Pirotti, R., 309
Pitcher, T. J., 54, 296, 306–7, 381, 382, *382*, 383, 392, 396, 574
Pitelka, F., 262
pituitary gland, 81, 82
plainfin midshipman (*Porichthys notatus*), 90–92, **91**, *91–92*
platyfish (*Xiphophorus maculatus*), 211, 214
play, 503–39, *510–11*, *513–14*, **515**, *516*, **517**, *519*, **520–21**, *423–25*, *529*, **530**, *531*, **532**, **536**
between-sex differences and, 528–29
cognitive training and, 526–27
creativity and, 532–33
defining, 511–13, 534, 537
functions of, 527–28
hormones and, 529–30, **530**
learning and, 532–33
locomotor, 518–21, *519*, **520**, **521**, 528, 537
neurotransmitters and, 530–33, *531*, *532*, 537
object, 513–18, *513*, *514*, **515**, *516*, **517**, 537
phylogeny and, 533–37
play markers, 423–24, *424*, 443
proximate aspects of, 529–33
role reversal in, 524, *525*
self-handicapping in, 524, 528
social, 522–27, *525*, 537
sociality and, 536, **536**
testosterone and, 529–30, **530**
types of, 513–27, 537
play fighting, 85–87, *86*, *87*, 526–27, 530–33, **530**
pleiotropy, 544–46, **544**
Pleszczynska, W. K., 231–32, 233
Plomin, R., 44, 582
Plotkin, H., 138
plumage coloration, 76–80, *76*, *78*
plural breeders, 323
Poecile gambeli (mountain chickadee), 15

Polhemus, J., 426
Polistes fuscatus (paper wasp), **63,** 332
pollen competition, 240
Pollock, G., 316, 317
polyadic interactions, 326
polyandry, *221,* 223, 225–27, 245–46, 247, 248, 249, 250
  concurrent, 275–76
  cooperative, 224
  resource defense, 224
  see also promiscuous mating systems
polygamy, *226–27*
polygynandry, *221,* 226, *226–27,* 245–46, 247, 250
polygyny, 220, *221,* 223–25, 238, 245–46, 248–49, 250
  female defense, 223–24
  in humans, 233–35, *234,* **235**
  induced, 232–33, **232–33**
  resource dispersion in, 228–29, 250
  threshold model of, *see* polygyny threshold model (PTM)
  variance and, 223, **223**
  see also promiscuous mating systems
polygyny threshold model (PTM), 229–35, **230**
  in birds, 231–33, *231–32,* **232–33**
  in humans, 233–35, *234,* **235**
poodles, 33–34, *33–34*
Poppleton, F., 298
Poran, N., 380
Porichthys notatus (plainfin midshipman), 90–92, 91, *91–92*
Portmann, A., 357
potato washing, 146–47, *146–47,* 149, 151
Potegal, M., 528
Potter, E., 558
Potts, G. W., 382
Potts, W., 189
Poundstone, W., 302
Power, D., 76, 77
Power, T. G., 514, 518, 519
Pravosudov, V., 15
predation:
  alarm calls and, *254,* 255–56, 434–36
  behavior and, 51–53, *53,* **54–55**
  brain size and, 401, **401**
  domesticated prey and, 590–91
  effects of, **54**
  foraging and, 385–87, *386*
  group size and, 55–56, **55,** 381–83, **382,** 383
  immunology and, 380–81, *380–81,* **381**
  inspection in, *see* predator inspection
  learning and, 110–11, 111
  phenotypic plasticity and, 112–13, **113**
  shoaling (schooling) and, 54–56
  sleep in mallard ducks and, 94–95, **94,** 95
  see also antipredator behavior
predator inspection, *11,* 54–56, 53, 296, *296,* 305–8, **305–8,** 392–95, *393,* **394**
  boldness and inhibition in, 574–78, **575, 577,** *577*
  costs and benefits of, 393–95, *393,* **394,** 575–77, **577,** *577*
  interpopulation differences and, 396–400, *396,* **397–99**
preexisting bias, 211–15
"prey selection" model, 22
prey-stealing, 184–85
Price, G., 482, 502
Price, J., 568, 571
Price, T., 44, 182
primates, 289, 333
  alarm signals in, 389–91, *389, 390,* 393, 408
  brain size of, 172, **172**
  coalition formation in, 327, 328–29, *328–29*
  communication in, 430–34, **430,** *433,* 443
  cooperative hunting in, 363–64, 363
  cultural transmission in, 146–49, *146–47,* **148,** 153–64, 170, 171, 587–88
  dimorphism in, **240**
  "fur rubbing" in, 558, 564
  kinship recognition in, 136
  personality in, 568, 587–88
  promiscuous mating systems in, 226
  reconciliation in, 430–34, **430,** *433,* 443
  self-medication in, 558–61, **559**
  social grooming in, 298–99, 299
  social play and cognition in, 523–24
  soil ingestion in, 561
  sperm competition in, 239–40, **240**
  step-parenting in, 272–73
  testes size of, 239–40, **240**
  tool use in, 172
  see also specific species
principle component analysis, 579, 582
prisoner's dilemma game, 11, 24, 301–2, **302,** 306–7, **306,** 336
  dear enemy model and, 457–58
  ESS analysis of, **304**
  iterated version of, 303
  territoriality and, 457–58
  TFT strategy and, 303–5, **304**
proactive (active) coping style, 589–90, 595
producers, at foraging, 365–67, 569–70, **569**
Profet, Margaret, 227, **228,** 563
progesterone, 83
promiscuous mating systems, 226–27, *226–27,* **228,** 236–44, 250
  extrapair copulations (EPCs) and, 236–38, **237,** *237*
  reproductive health and, 227
  sperm competition and, *see* sperm competition
Promislow, D., 542–43, 547, 551, 552, 564
pronghorn antelope, *519*
prothonotary warbler (*Protonotaria citrea*), 232–33, **232–33,** *232*
proximate causes:
  biochemical factors in, 95–97, *96,* 105
  Clutton-Brock on, 102–3
  defined, 76
  genes and, 101–4, 106
  hormones and, 80–87, **81,** 83, *86, 87*
  nervous system and, 87–92, **88,** *89,* **91,** *91, 92*

proximate causes (*continued*)
ultimate causes and, 75–82, **76,** *76, 78,* **79**
Prozac (fluoxetine), 505
Pruett-Jones, S., 198–99, **199,** 264, *264*
Prunella modularis (dunnock), 226–27, *226–27,* 245–47, **247,** *290*
Psammomys obesus (fat sand jirds), *86,* 536
Pteropus rodricensis (Rodrigues fruit bat), 297–98, *298*
Pulido, F., 470
Pulliam, H. R., 20–21
Pulliam, R., 343
pumpkinseed sunfish (*Lepomis gibbosus*), 572–74, *572,* **574**
Purves, D., 520
P., Anne, 65, 68, 327
Pyrrhocorax graculus (Alpine chough), 341–42, **342**
Pysh, J., 520

quail (*Coturnix japonica*), 203
quantitative trait loci (QTL), 46
Queller, D. C., 261
Quigley, H., 591
Quinn, T., 132, 470, 471
Quinn, T. P., 133
Quinn, W. G., 13
Quinney, T. E., 234

raccoon dog (*Nyctereutes procyonoides*), 298
Radakov, D. V., 382
Raleigh, M., 504
Rana catesbeiana (bullfrog), 557–58, *557*
Rand, S., 214
Rankin, M., 465
Ratnieks, F. L., 270–71, 274
rats, 15–16, *15, 16,* 154, 171, 553, 554, 589
learning in, 114–24, *114–18, 121,* 139, 150–51, 170, 360–61
play fighting in, 85–87, *86*
play in, 528, 530–32, *531,* **532**
Rattenborg, N., 95
Rattus norvegicus (Norway rat), 15–16, **16,** 150–51, 536
Raubenheimer, D., 13
reactive (conservation-withdrawal) coping style, 588–89, 595
Read, A., 437
Reader, S. M., 172, **172**
Reagen, L., 553
Real, L., 353
reciprocity, 301–9, 336, 440
byproduct mutualism and, 310–13, **312**
ESS strategy and, 303–4, **303–4**
food sharing and, 308–9, *308–9,* **310**
game theory and, 301–2
"index opportunity" for, 309
prisoner's dilemma game and, 301–3, **302,** 306–7, **306**
TFT strategy and, 303–8, **304, 305–8**
recombination, genetic, 39
red-billed blue magpie (*Cissa erythrorhyncha*), 342, **342**
red crossbill (*Loxia curvirostra*), 370
red deer (*Cervus elaphus*), 99–103, *100,* **101,** 213
male-male competition in, 206–8, *207,* **208**
red-eyed treefrog (*Agalychnis callidryas*), 387–89, *387,* **388**
red-tailed hawk (*Buteo jamaicensis*), 385
redwinged blackbird, 22–24, *23,* 234, *449*
Reeve, H. K., 40, 57, 58, 60–64, 75, 274, 286, 316, 323, 332–33
Regelmann, K., 392, 434
Reich, R., 274
Reidmann, A., 267
Reillo, P., 193, 194, **194**
Reiter, J., 206
Rembold, H., 373
Ren, R., 433–34
Rendell, E., 165, 439
reproductive skew theory, 262
reptiles:
aggression in, 495–96, *495*
sex determination in, 97–98, *98*
Resetarits, W., 555
"residual reproductive value," 281–82
resource defense polyandry, 224
resource holding power (RHP), 497
resource matching rule, 452–53
Reusch, T. B. H., 190
Reyer, H.-U., 324
Reznick, D., 52–54, 56
Reznick, S., 571
Rhagadotarsus:
R. anomalus (water strider), 425, *425,* 426, 443
R. kraepelin, 425
R. remigis, 425
rhesus monkeys, 160, *433,* 581, 582
cultural transmission in, 163–64
Richards, J., 357
Richards, K., 439
Richardson's ground squirrel, *340*
Richerson, P., 130, 146, 150–51, 198, 200
Riddell, W., 357, 358
Riddiford, L., 373
Ridley, M., 37, 400
Riechert, S., 274, 480, 483, 487
Rijnsdorp, A. D., 452
Riparia riparia (bank swallow), *258,* 291
ripple communication, 424–26, **425,** *425*
Rissing, S., 316–19
ritualization, 27
Rivulus:
R. hartii (killifish), 52–53, *53,* 575
R. marmoratus, 493–95
robin, 74–75, *74*
Robinson, G., 373
Robinson, G. E., 371, 373
Robinson, S., 286
Roces, F., 422, 423
rock dove, **189**
rock ptarmigan (*Lagopus mutus*), 416–17
rodents, 94
Rodrigues fruit bat (*Pteropus rodricensis*), 297–98, *298*
Rodriguez, E., 558
Roell, A., 430
Roff, D., 44
Roff, D. A., 44
Rohwer, F. C., 58
Rohwer, S., 77, 300, 428
role reversal, in social play, 524, *525*
Romanes, G., 149

Rood, J. P., 324, 326
rook (*Corvus frugilegus*), 342, **342**
Roper, K., 126
Roper, T. J., 423
Rose, M., 544–47, 552, 554
Rosenthal, R., 413
Rosenzweig, M. L., 451
Roskaft, E., 234
Ross, P., 591
Rothstein, S., 209
Rowe, L., 192
Rowland, W. J., 115
Roy, R., 521
Rozin, P., 127, 357, 369
Rubinstein, D., 274
ruff (*Philomachus pugnax*), 582–85, *583*, **585**
ruffed grouse, *414*
runaway sexual selection, 182, 193–94, *193*, **194**, 200, 216
Russell, J. K., 65, 327
Rutton, L., 362
Ryan, M. J., 182, 203, 211, 214
Ryon, J., 65, 327
Ryti, R. T., 316

sable antelope (*Hippotragus niger*), 523, *524*
sage grouse (*Centrocercus urophasianus*), 201, **201**
Sagør, J., 591
sailfin mollie (*Poecilia latipinna*), 203
Saimiri sciureus (squirrel monkey), 160, 525, *526–27*
Salamone, J., 531
Salmo salar (Atlantic salmon), 428–30, *428*, **429**
Salvelinus alpinus (Arctic char), 504–5
Sapolsky, R., 500, 553
Sauer, E. G., 456
Saunders, I., 528
Sawaguchi, T., 172
Scaphiopus bombifrons (spadefoot toad), 287–90, **287**, *287*, **290**
Schaller, G., 160
Schenkel, R., 160
Schiller, D., 369
Schino, G., 298
Schistocerca americana (grasshopper), 13–15, *13–15*, **14**
Schjelderup-Ebbe, T., 496
Schleidt, W., 568
Schlupp, I., 203
Schneider, K., 385
Schoener, T., 343
schooling, *see* shoaling
Schuett, G., 495–96, 500
Schulter, D., 44
Schultz, E., 163
Sciurus carolinensis (gray squirrel), 385, 558
scorpionfly (*Hylobittacus apicalis*), 37, 183–85, *183*, **184**
Scott, J. P., 478
Scott, S., 357
Scriber, J. M., 15
scroungers, in foraging, 366–67, 569–70, **569**
Seale, D., 555
searching image, concept of, 374
Searcy, W. A., 234, 437
sea urchins, 242–43, **242–43**
Sebeok, T., 413
secondary (epigametic) sexual characteristics, 181
second-order alliances, 330
second-order conditioning, 118, *118*, 141
seed harvester ant (*Messor pergandei*), 316–19, **317**
Seeley, T., 225, 420, 421
Seger, J., 261, 319
Seghers, B. H., 52, 54, 203
segregation, principle of, 45
Selander, R., 222
self-handicapping, in social play, 524
selfish gene, 316
group selection and, 316
Selfish Gene, The (Dawkins), 50
Seligman, M., 123
Selten, R., 503
Semel, B., 58, 60–61, **61**
senescence:
antagonistic pleiotropy model of, 544–46, **544**, **545**, 564
definition of, 542–43, 563
disposable soma theory of, 546–49, 551–52, 564
longevity and, *see* longevity
in the wild, 543, **543**
see also aging
sensation seeking, *see* boldness
sensitization, 114–15, *114–15*, 141
sensory bias, 211–15, *214*, 216
sequential assessment model, 480, 489–91, *490*
sequential polygamy, 223
Serengeti National Park, Tanzania, 39, 393, 516
Serinus canaria (canary), 438
serotonin, 504–5, **505**, 506, 531, 532
serum-to-venom binding levels, 380–81, **381**
Servedio, M. R., 200
Sevenster, P., 115
sex determination, 97–99
sex ratios, 97–99, *98*, 278–79, *279*
longevity and, 550–51, **550**, **551**
social rank and, 100–101, **101**
temperature and, 98–99, *99*
theory of, 99
sexual imprinting, 195–96
sexual selection, 103, 177–217, **179**, *180*, **182–83**, *183*, **184–85**, *186–88*, **189–90**, *190–91*, **192**, *193*, **194–99**, **201**, *202*, **203**, *205*, *207*, **208–9**, *210*, *214*, **215**
birdsong learning and, 203–5, *205*
cultural transmission and, 198–205
dimorphism and, 551–52
direct benefits model of, 182–85, **184**, **185**, 216
genetics and, 181–94, **182**, **184**, **185**, *186*, *188*, **190**, 216
good genes model of, 182, 186–92, *186*, *188*, **190**, 216
intersexual, 178–81, *180*; *see also* mate choice
intrasexual, 178–81, *180*
learning and, 194–98, **196–97**
male-male competition and, 205–10, *207*, **208**, **209**
mate-choice copying and, 198–203, **199**, **201**
mating systems and, 249
neuroethology and, 211–15, *214*
runaway selection model of, 182, 193–94, 200, 216
sensory bias and, 211–15, *214*

Sexual Selection (Andersson), 212
Seyfarth, R., 136, 163, 389, 390–91, 392, 431, 435
Sgoifo, A., 590
Shapiro, L., 195
Sharrock, J., 357
Shaw, P. W., 56
Sheard, M., 504
Sherman, P., 40, 57, 58, 60–61, **61**, 62, 75, 125, 127, 254, 256, 291, 356, 562–63
Sherry, D. F., 341
Shettleworth, S., 112, 122, 125, 127, 138–39, 356
shoaling (schooling) behavior, 54, 55, 56, 58
 as antipredator behavior, 381–83
 confusion effect in, 383
 flash expansion in, 383
 fountain effect in, 382
 predator inspection and, 383
 Trafalgar effect in, 383
shore crab (*Hemigrapsus nudus*), 582
Shuster, S. M., 125–26
shyness (inhibition), 570–78
 defined, 571
sibling rivalry, 282–85, *283–84*, **285**, 291
Sibly, R. M., 362
Sigg, D., 93
Sigmund, K., 67
sign stimuli, 27
Sih, A., 37
silent mutation, 38
Silk, J. B., 298-99, 328
silver fox, 593
Simmons, L. W., 192, 240
simultaneous polygamy, 223
Singer, W., 125
singular breeders, 323
Siviy, S., 531–33, **532**
Skadsen, J. M., 383
Skelly, D. K., 555–56
Skinner, B. F., 121, 122, 360
Skinner box, 121, *121*, 311
Skutch, A. F., 82, 261
Slagsvold, T., 230
Slansky, F., 15
Slater, P., 431, 437
Slater, P. J. B., 568
sleep, predation and, 94–95, **94**, *95*
slow-wave sleep, 95
Sluyter, F., 589
Smale, L., 578
Smith, I. C., 490
Smith, J., 370
Smith, J. N. M., 362, 406
Smith, P., 526
Smith, P. K., 514, 533
Smith, P. S., 391
Smith, R. J. F., 110, 391
Smith, R. L., 425, 426
Smith, S. M., 389, 406
Smith, W. J., 416
Smokler, R., 512, 514
Smotherman, W., 286
Smuts, B., 329
Smythe, N., 391
Snodgrass, R. E., 38
Snowdon, C., 431
Snyder, R. J., 300
Sober, E., 57, 315, 320
social facilitation, 155–56, *156*
social grooming (allogrooming), 298–99, *299*
social learning, 6, 15, **16**, 18, 111–12, 152–58, 173, 286
 antipredator behavior and, 406–7, 408
 foraging and, 364–69, *364*, **365**, *366*, **367**, *368–69*, 376
 in humans, 368–69, *368–69*
 in preschoolers, 368–69, *368–69*
 in primates, 430–32, **432**, *432*
 in rats, 16
 in whales, 442
 see also cultural transmission
social monogamy, 236
social play, 522–27
 cognition and, 523–25
 role reversal in, 524, *525*
 self-assessment in, 523
 self-handicapping in, 524
Sociobiology: The New Synthesis (Wilson), 26–27, 50
sodium, 352, *352*
Sol, D., 359
Solomon, N. G., 82
Sommer, V., 523
Sordahl, T. A., 406
spadefoot toad (*Scaphiopus bombifrons*), 287–90, **287**, *287*, **290**
Sparks, J., 298
spatial learning, 92–93
speckled wood butterfly (*Pararge aegeria*), 485–87, *486*
sperm competition, 238–44, 249, 250
 cryptic choice in, 242, 249
 in dungflies, 240–41, **241**, *241*
 evolution of genitalia and, 239–40
 kamikaze sperm hypothesis and, 243–44
 last male preference in, 241
 in primates, 239–40, **240**
 in sea urchins, 242–43, **242–43**
 sperm morphology and, 244, *244*
 sperm velocity and, 242–43, **242–43**
Spermophilus beldingi (Belding's ground squirrel), 254–56, *254*, **255**, *256*, 258, 434, 529–30, *529*
sperm whale (*Physeter macrocephalus*), 165, *165*, 173
 songs of, 439–43, *439*, **442**
spices, ingestion of, 559–63, **563**
spider monkey, 160
Spinka, M., 527–28, 537
Spinks, A. C., 9
Spoolder, H., 589
spotted hyena (*Crocuta crocuta*), 65, 327, 393, 394
 personalities of, 578–81, *579*, **580**
squirrel monkey (*Saimiri sciureus*), 160, 525, 526–27
squirrels, 341, 385–86, *386*
Stacey, P., 266–67, 274, 321, 323
stag beetles, 180
stalk-eyed flies, 193–94, *193*, **194**
Stammbach, E., 299
Stamps, J., 455–57, 472–73, 519
Stander, P., 591
starling (*Sturnus vulgaris*), 370, *370*, *385*
Stearns, S., 112
Stein, R. C., 437
Stentor (single-celled organism), 138
Stephens, D. W., 20, 22, 130–31, **131**, 141, 155, 311, *311*, 343, 352, 357

Stephens, P., 568, 588–89
Stevens, A., 568, 571
Stevenson-Hinde, J., 581, 582
Stockley, P., 244
stomatopod (*Gonodactylus bredini*), 125–26, *126*
stone play, 147–49, *147*, **148**
Strager, H., 439
Strassmann, B. I., 227, 234
stress, 82, 500, 532, 554, 589
Struhsaker, T. T., 163, 389
Stuart, R. A., 348
Study of Instinct, The (Lorenz), 502
Sturnus vulgaris (starling), 370, *370*, 385
Sula nebouxii (blue-footed booby), 492–93, *492*, **493**
Sullivan, J., 373, *373*
Sullivan, K., 362, 434
Summers-Smith, J. D., 314
superalliances, 330
superb fairy wren (*Malurus cyaneus*), 264, *264–65*
surreptitious promiscuity, 236–44
  extrapair copulations, 236–38
  sperm competition, 238–44
Sutherland, W. J., 375, 451, 454
Suthers, R. A., 437, 438
Swaddle, J., 190
Swaddle, J. P., 192
Swingland, I. R., 96
Sygnathus typhle (pipefish), 205–6
Sylvia atricapilla (blackcap), 470, **470**
Symons, D., 57, 519
synaptic terminals, 88, 89
Syrian golden hamster (*Mesocricetus auratus*), *86*
syrinx, avian, 437–39

Taborsky, M., 82
Taeniopygia guttata (zebra finch), 96–97, *96*, 196, 215, **215**
tail-flagging, 391–92, *391*
Tai National Forest, Ivory Coast, 160, 363–64, *363*
Takasu, K., 13
Talbot, C., 46
Talmi, M. C., 553
Tang-Halpin, Z., 290
Tatar, M., 554
Taub, D., 226
Tautz, J., 423
Taylor, M., 267
teaching, 158–61, *159*
  definition of, 158–59
  parent/offspring relationship and, 160–61
Telecki, G., 363
temperament, definition of, 593
Templeton, A., 545
Templeton, J., 370
Temrin, H., 274
ten Cate, C., 195
Terkel, J., 170
termites, 26
Terrace, H., 127, 195
territory and territoriality, 127, 448–49, 450, 454–64
  antibourgeois strategy and, 484–87, *485*, *486*
  "budding" of, 267, *267*, *462*, 463
  dear enemy effect and, 457–58, *458–59*
  defense of, 460–61, *460–61*, **461**
  family conflict and, 463–64
  family dynasties and, 451
  learning and, 450–51, 455–58, *456*, **457**, 458, 472–73
  owner's dilemma model and, 462
  prior residency advantage in, 473
  prisoner's dilemma and, 457–58
  raiding behavior and, 455
  site tenacity in, 473
  sneaker and satellite behavior and, 458, 459–61, *460–61*, **461**
Terry, R. L., 298
testosterone, 80–82
  aggression and, 80–81, **81**, 499, 501–4
  helping-at-the-nest and, 83–85, *84*
  intrauterine position and, 83–84
  parental care and, 84–85
  phenotypes and, 584–85
  play fighting and, 85–87, *86*, *87*, 529–30, **530**
Thatch, W., 520
theory, scientific, 22
Thiessen, D., 134
Thompson, E., 387, 487–88, 523
Thompson, J. N., 331
Thompson, K. V., 522
Thomson's gazelle, 393–95, *393*, **394**
Thor, D., 529, 531
Thorndike, Edward, 121–22
Thornhill, R., 183, 184–85, 191–92, 221, 223, 225
Thorpe, W. H., 154, 156, 437
three-spined stickleback (*Gasterosteus aculeatus*), *11*, 54, **54**, 128–30, **129**, 141, *188*, 189, 300, *300*, 308, 396, 404, 452, *453*, 576
threshold, of nerve cell, 89
thyroid stimulating hormone, 81
thyrotropin-releasing hormone (TRH), 82
thyroxin, 471
tiger salamander (*Ambystoma tigrinum*), *287*
Tinbergen, Niko, 22, 139, 171, 212, 374, 404, 416
Tinkle, D. W., 51
tit for tat (TFT) strategy, 303–5, **304, 305–8**, 336, 458
toad (*Bufo bufo*), 206, 417, **417**
Todt, D., 437
Toivanen, A., 466
Toivanen, P., 466
Tollrian, R., 112–13
Toma, D., 371–72, **372**
Toxostoma rufum (brown thrush), 438
transactional skew theory, 332
transposition, mutation and, 38
Travasso, M., 334–35
treecreepers, 322
Trehene, J., 383
Tribulus cistoides (caltrop), 8
Trichogaster trichopterus (blue gourami), 135, *135*, 137–40, **137**, 197–98, **198**
Trillmich, F., 518
Trinidad, 51, 52
Trivers, R., 10–11, 99, 134, 179, 248, 277, 279, 301, 391, 482
Troisi, A., 548, 571
tropic hormones, 81
truncation selection experiment, 43–44
tsetse fly (*Glossina morsitans morsitans*), 549–51, *549*, *550*, **551**
Tuchek, J., 104

Tucker, D., 13
Turke, P., 234
Turlings, T., 13
Turner, A., 137, 478
Tursiops truncatus (bottle-nosed dolphin), 65, 298, 327, *328,* 330
turtles, 97–98, *98*
twins, 49, 64

Uca paradussumieri (fiddler crab), 206
Udelsman, R., 554
Uehara, S., 363
Uetz, G., 484–85
ultimate causes, 33, 75–82, **76,** *76, 78,* **79,** 105
ultraviolet vision, 96–97, *96*
unconditioned stimulus (US), 116, 118, 141
unihemispheric sleep, 94–95, **94,** *95*

Vaccarino, A., 341
Valone, T. J., 57, 369, 385, 386
vampire bat (*Desmodus rotundus*), 308–9, *308–9,* **310**
Vandermeer, J., 331
Vander Meer, R., 421
Vander Wall, S. B., 341
Vanderwolf, C., 104
van Havre, N., 300
van Oordt, G., 584
Van Ree, J., 531
variation, 32, 44
behavioral, 46–49, *47,* **48**
broad-sense heritability and, 41–43, **43**
in Darwin's finches, 7–9, *7–8*
environmental, 48–49
genetic, 35–36, 38–39
lack of, 39
migration and, 39
narrow-sense heritability and, 41, 43–44, **44**
natural selection and, 38–39
of traits, 35–36
vasopressin, 83
Veblen, T., 418
Vehrencamp, S., 228, 274, 323, 412
Verbeek, M., 585–86
Vermej, G., 400
Verner, J., 229
Verrell, P., 482
vertical cultural transmission, 161, *161,* 162
vervet monkeys, *19,* 160, 435
alarm calls in, 389–91, *389, 390*
deceptive alarm calls in, 435
vespione wasp (*Dolichovespula saxonica*), 271, 272
Vespula vulgaris (common wasp), 271
Vestergaard, K., 514
Via, S., 112
Viitala, J., 96
Villarreal, R., 134, 135
Visscher, P. K., 270–71, 420
Volman, S., 426
vom Saal, F., 83, 272
Vonesh, J., 387
von Frisch, K., 22, 420
Vonk, J., 84
Vos, D. R., 195
vultures, *480*

Wade, M. J., 199, **199**
waggle dance, 418, 420–21, *420–21,* **421**
Waldman, B., 286, 290
Walker, C., 519, 520–21
Walker, P., 112
Wallace, B., 57
Wallach, M., 571
Walter, M., 196
Walters, J. R., 264
Walther, F. R., 392
Ward, P., 15, 419
Ward, S., 244
Waring, H., 428
Warkentin, K., 387–88
Warner, R. R., 163
war of attrition game model, 480, 487–88, **488**
Warriner, C., 196
Waser, M., 324
Waser, P., **275,** 324
wasps, 4, 26, 225, 332–33, 478, *479*
antipredator behavior in, 387–89, *387,* **388**
Australian, 223
common, 271
fig, 503
paper, **63,** 332
vespine, 271, 272
water flea (*Daphnia magna*), **348,** 452, *453*
water strider (*Rhagadotarsus anomalus*), 425, *425,* 426, 443
Wattiaux, J., 552
wattled jacana (*Jacana jacana*), 224, *224*
weaning, 277–78, *278,* 291
Weary, D., 437
Webb, P. W., 383
Webster, M., 238
Wedekind, C., 189–90
Weigensberg, I., 44
Weilgart, L., 437, 439, 442
Weiner, J., 7
Weisbard, C., 299
Weiss, A., 49, 587
Weiss, G., 520
Weiss, J., 590
Wenegrat, B., 568
Werner, E., 347, 387
West, H., 134
West-Eberhard, M. J., 112, 211, 261, 309
Westman, B., 234
Westneat, D., 198, 236–38
whales, 165–67, *165,* 173
songs of, 439–43, *439,* **442**
Wheeler, J., 316
White, D. J., 203
White, L., 267, 273
white-fronted bee-eater (*Merops bullockoides*), 269, **269,** *269,* 289, 463
Whitehead, H., 165–67, 173, 437, 439, 442
Whiten, A., 433, 587
white-tailed deer, 391–92, *391*
Whitlock, M. C., 192
Whitten, P. L., 226
Whoriskey, F., 300
Whyte, M., 334
Widemo, F., 583
Wigmore, S., 16
Wilbur, H., 555
Wilcox, R. S., 424, 425, 426
Wilcoxon, H., 123
wild dogs, *35,* 393, 394
cooperative hunting in, 362–63, *363*

wildebeest (gnu), 464, *465*, 525
Wilkinson, G., 37, 308
Wilkinson, J., 193, *193*, **194**, 308–9, **310**
Willard, D., 99
Williams, G. C., 10, 40, 50, 66, 212, 315, 544–45, 571
Williams, Kerry, 162–63
Wills, C., 189
Willson, M., 229
Wilson, D. S., 301, 308, 315, 316, 320, 351, 357, 358, 375, 571–74, **574**
Wilson, E. O., 10, 26–27, 50, 51, 169, 316, 353, 421, 422, 510
Wilson, M., 256–57, 273, 274
Wilsonia citrina (hooded warbler), 458, *458*, 459
Wiltschko, R., 468
Wiltschko, W., 468
Winberg, S., 504, 505
Wingfield, J., 500
winner and loser effects, 137–40, **140**, 491–98, *492*, **493**, **494**, *495*, 506
  hierarchy formation and, 492–93, 496–98, **498**
  mathematical models of, 496–98, **498**
Withers, G., 93
Witkin, S. R., 434
Witschi, E., 584
Witte, K., 203
Wittgenstein, L., 412
Wolf, L., 494–95, **494**, 496
wolf (*Canis lupus*), 65, 327
Wong, A., 557
wood duck (*Aix sponsa*), 58–61, *59*
Wood-Gush, D. G. M., 196, 514
Woodland, D. J., 391
wood termite (*Kalotermes flavicollis*), 63
Woolfenden, G., 462
Wootton, R. J., 128, 300
Worthylake, K., 555
Wrangham, R., 478, 558, 559, 560–61
Wrege, P. H., 463
wrens, 322
Wright, P. J., 128–30
Wright, S., 75
Wyatt, G., 373
Wyche, C., 382
Wyles, J., 357, 358

X chromosome, 45
xenophobia, 9
Xiphophorus:
  X. helleri (green swordtail), 211–14, *214*, 499
  X. maculatus (platyfish), 211–14

Yahr, P., 134
Yamamoto, F., 428
Yamamoto, I., 298
Yanega, D., 279
Yasukawa, K., 234
Ydenberg, R. C., 393, 457
Yeh, S., 505
yellow baboon, 160
yellow-bellied marmot, *414*
yellow-eyed junco, 355–56, **356**
Yngvesson, J., 490
Yoerg, S. I., 357
Yokoyama, S., 96
Yom-Tov, Y., 58
Young, H., 357

Zabel, C., 65, 327, 578
Zahavi, A., 15, 186, 206, 212, 416–17, 418, 419, 440–41, 443
Zajonc, R. B., 155
Zamble, E., 134, 195
zebra finch (*Taeniopygia guttata*), 96–97, *96*, 196
  sensory bias in, 215, **215**
Zenaida dove (*Zenaida aurita*), 127–28, *127*, **128**
Zentall, T. R., 146, 152
Zill, N., 274
Zuckerman, M., 568, 571
Zuckerman, S., 298
zugunruhe (migratory restlessness), 470
Zuk, M., 186, 187
Zunz, M., 581